Erosion and Sedimentation Manual

U.S. Department of the Interior
Bureau of Reclamation
Technical Service Center
Sedimentation and River Hydraulics Group
Denver, Colorado

November 2006

Acknowledgements

This manual was prepared by hydraulic engineers of the Bureau of Reclamation, U.S. Department of the Interior, under the leadership and direction of Chih Ted Yang, Ph.D., P.E. Dr. Yang (retired) was Reclamation's Manager of the Sedimentation and River Hydraulics Group, Technical Service Center in Denver, Colorado. Dr. Yang organized the overall outline of chapters in this manual and provided numerous reviews of the entire document.

The authors appreciate the financial support provided by the Bureau of Reclamation's Technical Service Center and its clients for the development, testing, and application of the technology summarized in this manual. And a special thank you to Teri Manross for an excellent job as technical editor of this manual.

Table of Contents

Page

Chapter 1 – Introduction 1-1
1.1 Background and Needs 1-1
1.2 Objectives 1-1
1.3 Manual Organization 1-2
1.4 Summary 1-5
1.5 References 1-5

Chapter 2 – Erosion and Reservoir Sedimentation 2-1
2.1 Introduction 2-1
2.2 Empirical Approach for Erosion Estimation 2-1
2.2.1 Universal Soil Loss Equation 2-2
2.2.2 Revised Universal Soil Loss Equation 2-8
2.2.3 Modified Universal Soil Loss Equation 2-16
2.2.4 Direct Measurement of Sediment Yield and Extension of Measured Data 2-17
2.2.5 Sediment Yield as a Function of Drainage Area 2-18
2.2.6 Sediment Yield Classification Procedure 2-19
2.3 Physically Based Approach for Erosion Estimates 2-19
2.4 Computer Model Simulation of Surface Erosion Process 2-29
2.4.1 Total Maximum Daily Load of Sediment 2-34
2.4.2 Generalized Sediment Transport Model for Alluvial River River Simulation (GSTARS) 2-36
2.4.3 Rainfall-Runoff Relationship 2-38
2.4.4 GSTAR-W Model 2-39
2.4.5 Erosion Index Map 2-41
2.5 Example Case Studies 2-41
2.5.1 Drainage Area Descriptions 2-41
2.5.2 Example Computations of Sediment Yield 2-42
2.5.3 Example Based on the RUSLE 2-42
2.5.4 Example Based on Drainage Area 2-44
2.5.5 Example Based on the Sediment Yield Classification Procedure 2-44
2.5.6 Example Based on Unit Stream Power 2-46
2.5.6.1 Flood Hydrology 2-46
2.5.6.2 Application of the Sheet Erosion Equation 2-48
2.5.6.3 Results 2-49

Table of Contents (continued)

Page

2.5.7 Comparison of Different Approaches 2-54

2.6 Reservoir Sedimentation 2-57

2.6.1 Reservoir Sediment Trap Efficiency 2-57

2.6.2 Density of Deposited Sediment 2-60

2.63 Sediment Distribution Within a Reservoir 2-64

2.6.4 Delta Deposits 2-73

2.6.5 Minimum Unit Stream Power and Minimum Stream Power Method 2-77

2.7 Summary 2-85

2.8 References 2-86

Chapter 3 – Noncohesive Sediment Transport 3-1

3.1 Introduction 3-1

3.2 Incipient Motion 3-1

3.2.1 Shear Stress Approach 3-2

3.2.2 Velocity Approach 3-7

3.3 Sediment Transport Functions 3-12

3.3.1 Regime Approach 3-12

3.3.2 Regression Approach 3-14

3.3.3 Probabilistic Approach 3-16

3.3.4 Deterministic Approach 3-17

3.3.5 Stream Power Approach 3-23

3.3.5.1 Bagnold's Approach 3-23

3.3.5.2 Engelund and Hansen's Approach 3-25

3.3.5.3 Ackers and White's Approach 3-25

3.3.6 Unit Stream Power Approach 3-28

3.3.7 Power Balance Approach 3-32

3.3.8 Gravitational Power Approach 3-34

3.4 Other Commonly Used Sediment Transport Functions 3-36

3.4.1 Schoklitsch Bedload Formula 3-36

3.4.2 Kalinske Bedload Formula 3-37

3.4.3 Meyer-Peter and Müller Formula 3-39

3.4.4 Rottner Bedload Formula 3-40

3.4.5 Einstein Bedload Formula 3-41

3.4.6 Laursen Bed-Material Load Formula 3-41

3.4.7 Colby Bed Material Load Formula 3-42

Table of Contents (continued)

Page

3.4.8 Einstein Bed-Material Load Formula 3-44
3.4.9 Toffaleti Formula 3-44
3.5 Fall Velocity 3-45
3.6 Resistance to Flow 3-47
3.6.1 Einstein's Method 3-49
3.6.2 Engelund and Hansen's Method 3-54
3.6.3 Yang's Method 3-58
3.7 Nonequilibrium Sediment Transport 3-63
3.8 Comparison and Selection of Sediment Transport Formulas 3-63
3.8.1 Direct Comparisons with Measurements 3-64
3.8.2 Comparison by Size Fraction 3-73
3.8.3 Computer Model Simulation Comparison 3-77
3.8.4 Selection of Sediment Transport Formulas 3-83
3.8.4.1 Dimensionless Parameters 3-85
3.8.4.2 Data Analysis 3-86
3.8.4.3 Procedures for Selecting Sediment Transport Formulas 3-102
3.9 Summary 3-104
3.10 References 3-104

Chapter 4 – Cohesive Sediment Transport 4-1
4.1 Introduction 4-1
4.2 Cohesive Sediment Processes 4-1
4.2.1 Aggregation 4-1
4.2.2 Deposition 4-5
4.2.3 Consolidation 4-7
4.2.4 Toxicant Adsorption and Desorption 4-9
4.2.5 Erosion 4-10
4.2.5.1 Physical Factors Affecting Erodibility 4-11
4.2.5.2 Electrochemical Factors Affecting Erodibility 4-12
4.2.5.3 Biological Factors Affecting Erodibility 4-12
4.2.6 Experimental Methods to Determine Erosion Parameters 4-14
4.2.6.1 Rotating Cylinder 4-16
4.2.6.2 Straight Flume Studies 4-16
4.2.6.3 Annular Flume 4-18
4.2.6.4 In-Situ Methods 4-19

Table of Contents (continued)

Page

4.2.7 Critical Shear Stress and Erosion Rate Formulae 4-20
4.2.8 Discussion of Cohesive Soil Erosion Parameters Determined Through Experiment 4-22
4.2.9 Published Results of Erosion Parameters 4-23
4.3 Numerical Models of Cohesive Sediment Transport 4-31
4.3.1 One-Dimensional Models 4-31
4.3.2 Two-Dimensional Models 4-32
4.3.3 Three-Dimensional Models 4-33
4.3.4 Numerical Models of Contaminant Transport 4-34
4.4 Numerical Model GSTAR-1D 4-35
4.4.1 Conceptual Model 4-36
4.4.2 Active Layer Calculation 4-38
4.4.3 Consolidation 4-40
4.4.4 Bed Merge 4-41
4.4.5 Example Application 4-42
4.5 Summary 4-46
4.6 References 4-46

Chapter 5 – Sedimentation Modeling for Rivers and Reservoirs 5-1
5.1 Introduction 5-1
5.1.1 The Numerical Modeling Cycle 5-1
5.2 Mathematical Models 5-3
5.2.1 Three-Dimensional Models 5-3
5.2.2 Two-Dimensional Models 5-6
5.2.3 One-Dimensional Models 5-9
5.2.4 Bed Evolution 5-11
5.2.5 Auxiliary Equations 5-16
5.2.5.1 Flow Resistance 5-16
5.2.5.2 Sediment Transport 5-23
5.3 Numerical Solution Methods 5-26
5.3.1 Finite Difference Methods 5-27
5.3.2 Finite Element Methods 5-30
5.3.3 Finite Volume Methods 5-31
5.3.4 Other Discretization Methods 5-32
5.4 Modeling Morphologic Evolution 5-34
5.5 Reservoir Sedimentation Modeling 5-40

Table of Contents (continued)

Page

5.5.1 Reservoir Hydraulics 5-41
5.5.2 Sediment Transport in Reservoirs 5-44
5.5.3 Turbid Underflows 5-47
5.5.3.1 Plunge Point 5-48
5.5.3.2 Governing Equations 5-51
5.5.3.3 Additional Relationships 5-54
5.5.4 Difference Between Reservoirs and Other Bodies of Water 5-58
5.6 Data Requirements 5-59
5.7 One-Dimensional Model Comparison 5-62
5.8 Example: The GSTARS Models 5-62
5.8.1 Streamlines and Stream Tubes 5-64
5.8.2 Backwater Computations 5-65
5.8.3 Sediment Routing 5-67
5.8.4 Total Stream Power Minimization 5-73
5.8.5 Channel Side Slope Adjustments 5-74
5.8.6 Application Examples 5-75
5.9 Summary 5-82
5.10 References 5-83

Chapter 6 – Sustainable Development and Use of Reservoirs 6-1
6.1 Introduction 6-1
6.2 Sustainable Development and Use of Water Resources 6-1
6.3 Dynamic Adjustment of a River System 6-4
6.4 Planning 6-10
6.4.1 Background Information and Field Investigation 6-10
6.4.2 Basic Considerations 6-10
6.4.3 Sediment Control Measures 6-10
6.5 Design of Intakes 6-13
6.5.1 Location of Intakes 6-13
6.5.2 Types of Intakes for Sediment Control 6-14
6.6 Sediment Management for Large Reservoirs 6-16
6.7 Sediment Management for Small Reservoirs 6-17
6.7.1 Soil Conservation 6-17
6.7.2 Bypass of Incoming Sediment 6-18

Table of Contents (continued)

Page

6.7.3 Warping 6-18
6.7.4 Joint Operation of Reservoirs 6-18
6.7.5 Drawdown Flushing 6-19
6.7.6 Reservoir Emptying 6-19
6.7.7 Lateral Erosion 6-19
6.7.8 Siphoning Dredging 6-19
6.7.9 Dredging by Dredgers 6-20
6.7.10 Venting Density Current 6-20
6.7.11 Evaluation of Different Sediment Management Measures 6-20
6.8 Effective Management of Reservoir Sedimentation 6-21
6.9 Operational Rules 6-22
6.10 Cost of Sedimentation Prevention and Remediation 6-23
6.11 Reservoir Sustainability Criteria 6-24
6.12 Technical Tools 6-25
6.12.1 GSTARS 2.0/2.1 Models 6-25
6.12.2 GSTARS 3 Model 6-27
6.12.3 GSTAR-1D Model 6-28
6.12.4 GSTAR-W Model 6-28
6.12.5 Economic Model 6-28
6.13 Summary 6-29
6.14 References 6-30

Chapter 7 – River Processes and Restoration 7-1
7.1 Introduction 7-1
7.2 Conceptual Model 7-2
7.3 Data Collection, Analytical, and Numerical Modeling Tools 7-3
7.3.1 Data Collection Activities 7-3
7.3.2 Geomorphic Processes 7-6
7.3.2.1 Geology 7-10
7.3.2.2 Climate 7-10
7.3.2.3 Topography 7-11
7.3.2.4 Soils 7-11
7.3.2.5 Vegetation 7-11
7.3.2.6 Channel Morphology 7-12
7.3.2.7 Geomorphologic Mapping 7-13

Table of Contents (continued)

Page

7.3.2.8 Channel Geometry Analysis ... 7-15
7.3.2.9 Stream Classification ... 7-15
7.3.2.10 Channel Adjustments and Equilibrium ... 7-16
7.3.2.11 Geomorphic Summary ... 7-18
7.3.3 Disturbances Affecting the River Corridor ... 7-19
7.3.3.1 Dams ... 7-20
7.3.3.2 Diversions ... 7-21
7.3.3.3 Levees ... 7-22
7.3.3.4 Roads in the River Corridor ... 7-23
7.3.3.5 Bridges ... 7-23
7.3.3.6 Bank Protection ... 7-24
7.3.3.7 Removal of Vegetation and Woody Debris ... 7-25
7.3.3.8 Forestry Practices ... 7-25
7.3.3.9 Grazing (bank erosion) ... 7-26
7.3.3.10 Gravel Mining ... 7-26
7.3.3.11 Urbanization ... 7-26
7.3.3.12 Recreation ... 7-27
7.3.4 Hydrologic Analysis ... 7-27
7.3.4.1 Historical Discharge Data ... 7-27
7.3.4.2 Flood Frequency Analysis ... 7-28
7.3.4.3 Flow Duration Analysis ... 7-28
7.3.4.4 Ground Water Interaction ... 7-28
7.3.4.5 Channel Forming Discharge ... 7-29
7.3.5 Hydraulic Analysis and Modeling ... 7-30
7.3.5.1 Topographic Data Needed ... 7-30
7.3.5.2 Longitudinal Slope and Geometry Data ... 7-31
7.3.5.3 Physical and Numerical Models ... 7-31
7.3.6 Sediment Transport Analysis and Modeling ... 7-32
7.3.6.1 Sources of Upstream Sediment Supply ... 7-32
7.3.6.2 Total Stream Power ... 7-33
7.3.6.3 Incipient Motion ... 7-33
7.3.6.4 Sediment Particle Size Analysis ... 7-33
7.3.6.5 Sediment-Discharge Rating Curves ... 7-36
7.3.6.6 Reservoir Sediment Outflows ... 7-38
7.3.6.7 Scour and Degradation ... 7-38

Table of Contents (continued)

Page

7.3.6.8 Sediment Transport Equations 7-39
7.3.6.9 Sediment Considerations for Stable Channel Design 7-39
7.3.6.10 Evaluation and Potential Contaminants 7-41
7.3.7 Biologic Function and Habitat 7-41
7.4 Sediment Restoration Options 7-42
7.4.1 Goals and Objectives 7-42
7.4.2 Fully Assess the Range of Options 7-43
7.4.2.1 Sediment and Flow 7-43
7.4.2.2 Local Versus System-wide 7-43
7.4.2.3 Natural Versus Restrained Systems 7-44
7.4.2.4 Monitoring Versus Modification 7-45
7.4.3 Restoration Treatments 7-45
7.4.3.1 Restoration of the Historic Channel Migration Zone 7-45
7.4.3.2 Levee Setback and Removal 7-46
7.4.3.3 Roadway Setback 7-47
7.4.3.4 Lengthening Bridge Spans 7-47
7.4.3.5 Side Channel, Vegetation, and Woody Debris Recovery 7-48
7.4.3.6 Changes to Channel Cross Section or Sizing 7-48
7.4.3.7 Changes to Channel and Flood Plain Roughness 7-49
7.4.3.8 Bank Stabilization Concepts 7-49
7.4.3.9 Grade Control Structures 7-50
7.4.3.10 New Channel Design and Relocations 7-51
7.4.3.11 Special Flow Releases From Dams 7-52
7.4.4 Biologic Function and Habitat 7-53
7.4.4.1 Channel and Cross-Section Shape 7-53
7.4.4.2 Channel Banks 7-54
7.4.4.3 Channel Platform Characteristics 7-54
7.4.4.4 Changes in Channel Grade 7-54
7.4.4.5 Flow and Sediment Designs 7-55
7.4.5 Watershed Level Restoration 7-56
7.4.6 Uncertainty and Adaptive Management 7-56
7.5 Summary 7-57
7.6 References 7-58

Table of Contents (continued)

Page

Chapter 8 – Dam Decommissioning and Sediment Management 8-1

8.1 Introduction 8-1

8.2 Scope of Sediment Management Problems 8-2

8.3 Engineering Considerations of Dam Decommissioning 8-6

8.4 Sediment Management Alternatives 8-7

8.4.1 Integration of Dam Decommissioning and Sediment Management Alternatives 8-7

8.4.2 No Action Alternative 8-9

8.4.3 River Erosion Alternative 8-10

8.4.3.1 River Erosion Description 8-10

8.4.3.2 River Erosion Effects 8-12

8.4.3.3 Monitoring and Adaptive Management 8-13

8.4.4 Mechanical Removal Alternative 8-14

8.4.4.1 Sediment Removal Methods 8-15

8.4.4.2 Sediment Conveyance Methods 8-16

8.4.4.3 Long-Term Disposal 8-17

8.4.5 Stabilization Alternative 8-17

8.4.6 Comparison of Alternatives 8-19

8.5 Analysis Methods for River Erosion Alternative 8-20

8.5.1 Reservoir Erosion 8-21

8.5.1.1 Analytical Methods for Estimating Reservoir Erosion 8-23

8.5.1.2 Numerical Models 8-24

8.5.2 Downstream Impacts 8-26

8.5.2.1 Analytical Methods for Protecting Deposition Impacts 8-26

8.5.2.2 Numerical Modeling of Sediment Impacts 8-30

8.6 Summary 8-31

8.7 References 8-33

Chapter 9 – Reservoir Survey and Data Analysis 9-1

9.1 Introduction 9-1

9.2 Purpose of a Reservoir Survey 9-1

9.3 Sediment Hazards 9-3

9.4 Sediment Management 9-4

9.5 Frequency and Schedule of Surveys 9-5

9.6 Reservoir Survey Techniques 9-7

9.6.1 Shoreline Erosion 9-9

9.6.2 Data Density and Line Spacing 9-12

Table of Contents (continued)

Page

9.6.3 Cost of Conducting a Reservoir Survey 9-13
9.6.4 Selecting Appropriate Hydrographic Data Collection System and Software 9-13
9.7 Hydrographic Collection Equipment and Techniques 9-15
9.8 Global Positioning System 9-16
9.8.1 Absolute Positioning 9-17
9.8.2 Differential Positioning 9-18
9.8.3 Real-Time Kinematic GPS 9-21
9.8.4 GPS Errors 9-22
9.9 Horizontal and Vertical Control 9-23
9.9.1 Datums 9-24
9.10 Depth Measurements 9-24
9.10.1 Single Beam 9-24
9.10.2 Multibeam 9-29
9.10.3 Additional Sonar Methods 9-34
9.10.4 Single Beam Depth Records 9-35
9.11 Survey Accuracy and Quality 9-40
9.12 Survey Vessels 9-42
9.13 Survey Crew 9-43
9.14 Determination of Volume Deposits 9-43
9.14.1 Average-End-Area Method 9-45
9.14.2 Width Adjustment Method 9-45
9.14.3 Contour Method – Topographic Mapping 9-46
9.15 Final Results 9-46
9.15.1 Report 9-52
9.16 Reservoir Survey Terminology 9-56
9.17 Summary. 9-62
9.18 References 9-63

Appendix 1 – Notation

Appendix II – Conversion Factors

Appendix III – Physical Properties of Water

Author Index

Subject Index

Chapter 1
Introduction

Page

1.1 Background and Needs 1-1
1.2 Objectives 1-1
1.3 Manual Organization 1-2
1.4 Summary 1-5
1.5 References 1-5

Chapter 1
Introduction

by
Chih Ted Yang

1.1 Background and Needs

Surface erosion, sediment transport, scour, and deposition have been the subjects of study by engineers and geologists for centuries, due to their importance to economic and cultural development. Most ancient civilizations existed along rivers in order to use the water supply for irrigation and navigation. All rivers carry sediments, due to surface erosion from watersheds and bank erosion along the river. Our understanding of the dynamic equilibrium between sediment supply from upstream and a river's sediment transport capability is important to the success of river engineering design, operation, and maintenance.

Engineers built levees along rivers for flood control purposes. Reservoirs are built to ensure water supply and flood control. Canals are built for water supply and navigation. Sustainable use of these hydraulic structures depends on our understanding of the erosion and sedimentation processes and how to apply them to hydraulic designs. For example, soil conservation practice, check dams, sediment bypass devices, and sluicing are often used to reduce sediment inflow or remove sediments from a reservoir to prolong the useful life of a reservoir. The dynamic equilibrium, or regime, concept was used in the design of stable regime canals in India and Pakistan. More recently, computer models have been developed to simulate and predict the erosion and sediment transport, scour, and deposition processes.

There are many sediment transport books, such as those by Graf (1971), Yalin (1972), Simons and Sentürk (1977), Chang (1988), Julien (1995), and Yang (1996). These books were written mainly as university textbooks for teaching and research purposes. There is a gap between engineering and academic needs. Engineers often find it is difficult to apply erosion and sediment transport theories they have learned from the classroom to solve river engineering design problems. The American Society of Civil Engineers published the *Manuals and Reports of Engineering Practice No. 54 – Sedimentation Engineering* in 1935 (Vanoni, 1975). The erosion and sedimentation literature and methods summarized in that manual do not include those developed and used in the past thirty years. There is a need to develop and publish an erosion and sedimentation manual to summarize what we have learned in the past thirty years for the benefit of practicing engineers and geologists.

1.2 Objectives

Engineers in the Bureau of Reclamation's Sedimentation and River Hydraulics Group provide technical assistance and conduct studies to meet the needs of other Reclamation offices and of domestic and international water resources agencies. In addition to using the latest state-of-the-art technology, engineers in the Sedimentation and River Hydraulics Group often have to develop new technology, methods, and computer programs for solving erosion, sedimentation, and river hydraulic problems. All the authors of the *Erosion and Sedimentation Manual* are members of the Sedimentation and River Hydraulics Group. Information, computer programs, and materials included in the manual are based on proven technology existing in the literature and some of them were developed by the authors for solving practical engineering problems. The objectives of writing this manual are twofold: to summarize the authors' experience and knowledge and to share with the public what they have learned and used in solving erosion and sedimentation problems. The Erosion and Sedimentation Manual is intended for engineers with basic background and knowledge in open channel hydraulics, sediment transport, and river morphology. The manual can also be used as a reference book for university professors, graduate students, and researchers for solving practical engineering problems.

1.3 Manual Organization

This manual contains nine chapters and three appendices. Each chapter is self-contained, with an introduction, summary, and a list of references. Cross references are made to avoid duplications of materials in different chapters. Basic theories, concepts, and approaches in erosion, sediment transport, river morphology, computer modeling, and field survey are reviewed and summarized in the manual. Examples are used to illustrate how to use the methods and programs contained in the manual. Materials contained in each chapter and appendix are briefly summarized as follows.

Chapter 1 - Introduction

Chapter 1 describes the background, needs, and objectives of preparing and publishing this manual.

Chapter 2 - Erosion and Reservoir Sedimentation

Chapter 2 describes and evaluates empirical approaches based on the universal soil loss equation and its modified versions and the determination of sediment yield as a function of drainage area, drainage classification, or from direct measurements. The physically based approach is derived from the unit stream power theory for erosion and sediment transport and the minimum unit stream power or minimum stream power theory governing the river morphologic processes. Field data were used to compare the accuracy and applicability of the empirical and physically based approaches. The concept and approach used in developing the Generalized Sediment Transport model for Alluvial River Simulation (GSTARS) computer models GSTARS 2.1, GSTARS3, and GSTAR-W are summarized. It shows how a systematic approach based on consistent theories can be used to develop a model to simulate and predict the erosion and sediment transport, scour, and deposition processes in rivers and reservoirs in a watershed. This chapter ends with a summary of technology used in the determination of reservoir sediment trap efficiency, sediment density, and sediment distribution in a reservoir using conventional methods and the minimum unit stream power or minimum stream power methods.

Chapter 3: Noncohesive Sediment Transport

This chapter starts with the subject of incipient motion, followed by sediment transport functions based on regime, regression, probabilistic approaches, and deterministic approaches. Most of the commonly used sediment transport equations are summarized and compared. In addition to the conventional approaches, stream power, unit stream power, power balance, and gravitational power theories are summarized and compared. To address the impacts of fine sediment or wash load on sediment transport, the subject of nonequilibrium sediment transport is also included. This chapter ends with recommendations for selecting appropriate equations under different hydraulic and sediment conditions.

Chapter 4 - Cohesive Sediment Transport

The current level of understanding on cohesive sediment transport of fine matters is relatively primitive when compared with that of noncohesive sediment transport. This does not mean cohesive sediment transport is of less importance. Most pollutants are attached to and transported with fine sediments. The U.S. Environmental Protection Agency (2001) has identified sediment as the number one pollutant in the United States. This chapter summarizes cohesive sediment transport theories and experimental methods for determining erosion parameters. Computer models, especially the GSTARS3 and GSTAR-1D models, can be used to simulate the transport processes of cohesive sediments.

Chapter 5 - Sediment Modeling for Rivers and Reservoirs

This chapter starts with the numerical modeling cycle, followed by basic equations used in one- , two-, and three-dimensional models. Numerical solution methods, such as finite difference, finite element, and finite volume methods are introduced and compared. The stream tube concept and minimum total stream power theory are used in the development of the GSTARS. Examples of application of GSTARS 2.1 and GSTARS3 computer models are included to illustrate how the models can be applied to simulate and predict the sedimentation processes in rivers and reservoirs.

Chapter 6 – Sustainable Development and Use of Reservoirs

Sedimentation is a sure way to shorten the useful life of reservoirs. Due to environmental, political, social, economic, and geological considerations, sustainable development and use of reservoirs must be considered in the planning, design, construction, and operation of new reservoirs. For existing reservoirs, engineering methods should be developed and applied to prolong their useful life. This chapter provides a brief description of the planning process and design considerations for hydraulic structures to reduce sediment inflow to a reservoir and to sluicing sediment from a reservoir. Sediment management methods for large and small reservoirs are described and compared. Reservoir operation rules for different types of reservoirs are recommended, and sedimentation and prevention costs are included for engineers to consider.

GSTARS 2.1, GSTARS3, GSTAR-1D, GSTAR-W, and economic models are introduced as technical tools that are available for engineers to apply for analyzing and solving sedimentation problems.

Chapter 7 – River Processes and Restoration

A river is a dynamic system. Engineers must understand factors and principles governing river processes. Engineering design and construction of hydraulic structures without taking these factors and principles into consideration may not last long and may not serve the basic functions of rivers. Many existing hydraulic structures have been redesigned or modified in recent years to restore a river's basic functions. This chapter describes the geomorphic processes and possible disturbances affecting the river corridor. Analytical approaches for hydrologic, hydraulic, and sediment transport studies are summarized. Restoration options and treatments using structural and nonstructural measures are discussed and included in this chapter.

Chapter 8 – Dam Decommissioning and Sediment Management

More than 76,000 dams that are at least 6 feet high exist in the United States. While the great majority of these dams still provide beneficial use and function to the society, some of the dams may need to be decommissioned. Reasons for decommissioning include, but are not limited to, economics, dam safety and security, legal and financial liability, ecosystem restoration, and recreation considerations. This chapter describes reservoir sediment management problems and engineering considerations of dam decommissioning. Sediment management alternatives include no action, sediment removal by river erosion and by mechanical means, and stabilization. Special attention is paid to analysis methods for river erosion of reservoir sediments and their impacts on downstream river reaches.

Chapter 9 –Reservoir Survey and Data Analysis

Reservoir sedimentation is an ongoing natural depositional process that can remain below water and out of sight for a significant portion of the reservoir life, but lack of visual evidence does not reduce the potential impact. Reservoir sediment models have been developed for analyzing and solving sediment problems. Calibration and confirmation of these models can be achieved with accurate field data. This chapter presents methodology to measure reservoir bathymetry or topography with the goal of accurately updating reservoir sedimentation and storage capacity information in a timely and cost-efficient matter. Reclamation's Sedimentation and River Hydraulics Group continuously upgrades their technical procedures to reflect ever-changing technology, and the majority of the techniques provided are from experience gained. This chapter provides guidelines, techniques, and information for planning, collecting, analyzing, and reporting reservoir and river survey studies with the ultimate goals of preservation of the information and uniformity of collection and analysis.

Appendix I - Notation

Appendix II - Conversion Factors

Appendix III – Physical Properties of Water

Author Index

Subject Index

1.4 Summary

The *Erosion and Sedimentation Manual* provides a comprehensive coverage of subjects in nine chapters (i.e., introduction, erosion and reservoir sedimentation, noncohesive sediment transport, cohesive sediment transport, sediment modeling for rivers and reservoirs, sustainable development and use of reservoirs, river processes and restoration, dam decommissioning and sediment management, and reservoir surveys and data analysis). Each chapter is self-contained, with cross references of subjects that are discussed in different chapters of this manual. The manual also includes a list of commonly used notations used in the erosion and sedimentation literature, conversion factors between the Imperial and metric units, physical properties of water, and author and subject indexes for easy reference. Each chapter has a list of references for readers who would like to seek out more detailed information on specific subjects. The manual should serve as a useful book for researchers, university professors and graduate students, and engineers in solving erosion and sedimentation problems.

1.5 References

Chang, H.H. (1988). Fluvial Processes in River Engineering. John Wiley & Sons, Inc. (Reprint by Krieger Publishing Company, 1992.)

Graf, W.H. (1971). *Hydraulics of Sediment Transport,* The McGraw-Hill Companies, Inc.

Julien, P.Y. (1995). *Erosion and Sedimentation,* Cambridge University Press.

Simons, D.B., and Sentürk, F. (1977). *Sediment Transport Technology,* Water Resources Publications.

U.S. Environmental Protection Agency, Office of Wetlands, Oceans, and Watersheds (2001). *A Watershed Decade,* EPA 840-R-00-002, Washington, DC.

Vanoni, V.A., editor (1975). *ASCE Manuals and Reports on Engineering Practice–No. 54, Sedimentation Engineering.* American Society of Civil Engineers.

Yalin, M.S. (1972). *Mechanics of Sediment Transport,* Pergamon Press.

Yang, C.T. (1996). *Sediment Transport Theory and Practice,* The McGraw-Hill Companies, Inc. (Reprint by Krieger Publishing Company, 2003.)

Chapter 2
Erosion and Reservoir Sedimentation

Page

2.1 Introduction........ 2-1
2.2 Empirical Approach for Erosion Estimation........ 2-1
  2.2.1 Universal Soil Loss Equation........ 2-2
  2.2.2 Revised Universal Soil Loss Equation........ 2-8
  2.2.3 Modified Universal Soil Loss Equation........ 2-16
  2.2.4 Direct Measurement of Sediment Yield and Extension of Measured Data........ 2-17
  2.2.5 Sediment Yield as a Function of Drainage Area........ 2-18
  2.2.6 Sediment Yield Classification Procedure........ 2-19
2.3 Physically Based Approach for Erosion Estimates........ 2-19
2.4 Computer Model Simulation of Surface Erosion Process........ 2-29
  2.4.1 Total Maximum Daily Load of Sediment........ 2-34
  2.4.2 Generalized Sediment Transport Model for Alluvial River River Simulation (GSTARS)........ 2-36
  2.4.3 Rainfall-Runoff Relationship........ 2-38
  2.4.4 GSTAR-W Model........ 2-39
  2.4.5 Erosion Index Map........ 2-41
2.5 Example Case Studies........ 2-41
  2.5.1 Drainage Area Descriptions........ 2-41
  2.5.2 Example Computations of Sediment Yield........ 2-42
  2.5.3 Example Based on the RUSLE........ 2-42
  2.5.4 Example Based on Drainage Area........ 2-44
  2.5.5 Example Based on the Sediment Yield Classification Procedure........ 2-44
  2.5.6 Example Based on Unit Stream Power........ 2-46
    2.5.6.1 Flood Hydrology........ 2-46
    2.5.6.2 Application of the Sheet Erosion Equation........ 2-48
    2.5.6.3 Results........ 2-49
  2.5.7 Comparison of Different Approaches........ 2-54
2.6 Reservoir Sedimentation........ 2-57
  2.6.1 Reservoir Sediment Trap Efficiency........ 2-57
  2.6.2 Density of Deposited Sediment........ 2-60
  2.63 Sediment Distribution Within a Reservoir........ 2-64
  2.6.4 Delta Deposits........ 2-73
  2.6.5 Minimum Unit Stream Power and Minimum Stream Power Method........ 2-77
2.7 Summary........ 2-85
2.8 References........ 2-86

Chapter 2
Erosion and Reservoir Sedimentation

by
Timothy J. Randle, Chih Ted Yang, Joseph Daraio

2.1 Introduction

As a result of runoff from rainfall or snowmelt, soil particles on the surface of a watershed can be eroded and transported through the processes of sheet, rill, and gully erosion. Once eroded, sediment particles are transported through a river system and are eventually deposited in reservoirs, in lakes, or at sea. Engineering techniques used for the determination of erosion rate of a watershed rely mainly on empirical methods or field survey. This chapter reviews and summarizes these empirical methods.

During the 1997 19th Congress of the International Commission on Large Dams (ICOLD), the Sedimentation Committee (Basson, 2002) passed a resolution encouraging all member countries to (1) develop methods for the prediction of the surface erosion rate based on rainfall and soil properties, and (2) develop computer models for the simulation and prediction of reservoir sedimentation processes. Yang et al. (1998) outlined the methods that can be used to meet the goals of the ICOLD resolution. This chapter presents a physically based approach for erosion estimation based on unit stream power and minimum unit stream power theories. Details of the theories are given in Chapter 3 and in Yang's book, *Sediment Transport: Theory and Practice* (1996). This chapter also summarizes methods for the estimation of sediment inflow and distribution in a reservoir, based on empirical and computer model simulation.

2.2 Empirical Approach for Erosion Estimation

Sediment yield is the end product of erosion or wearing away of the land surface by the action of water, wind, ice, and gravity. The total amount of onsite sheet, rill, and gully erosion in a watershed is known as the gross erosion. However, not all of this eroded material enters the stream system. Some of the material is deposited as alluvial fans, along river channels, and across flood plains. The portion of the eroded material that is transported through the stream network to some point of interest is referred to as the sediment yield. Therefore, the amount of sediment inflow to a reservoir depends on the sediment yield produced by the upstream watershed. The factors that determine a watershed's sediment yield can be summarized as follows (Strand and Pemberton, 1982):

- Rainfall amount and intensity
- Soil type and geologic formation
- Ground cover
- Land use
- Topography
- Upland erosion rate, drainage network density, slope, shape, size, and alignment of channels

- Runoff
- Sediment characteristics—grain size, mineralogy, etc.
- Channel hydraulic characteristics

Most of the empirical approaches for the estimation of erosion rate are based on one of the following methods:

- Universal Soil Loss Equation (USLE) or its modified versions
- Sediment yield as a function of drainage area
- Sediment yield as a function of drainage characteristics

Empirical equations are developed using data collected from specific geographical areas; application of these equations should be limited to areas represented in the base data. Some investigators have attempted to revise or modify the USLE to apply it to areas other than the Central and Eastern United States.

2.2.1 Universal Soil Loss Equation

Soil erosion rates on cultivated land can be estimated by the use of the Universal Soil Loss Equation (Wischmeier and Smith, 1962, 1965, 1978). This method is based on statistical analyses of data from 47 locations in 24 states in the Central and Eastern United States. The Universal Soil Loss Equation is:

$$A = RKLSCP \tag{2.1}$$

where A = computed soil loss in tons/acre/year,
R = rainfall factor,
K = soil-erodibility factor,
L = slope-length factor,
S = slope-steepness factor,
C = cropping-management factor, and
P = erosion-control practice factor.

The rainfall factor R accounts for differences in rainfall intensity, duration, and frequency for different locations; that is, the average number of erosion-index units in a year of rain. Locational values of the R-factor can be obtained for the central and eastern parts of the United States from Figure 2.1. The R-factor thus obtained does not account for soil loss due to snowmelt and wind.

The soil-erodibility factor K is a measure of the intrinsic susceptibility of a given soil to soil erosion. It is the erosion rate per unit of erosion-index for a specific soil in cultivated, continuous fallow, on a 9-percent slope, 72.6 feet long. The K-factor values range from 0.7 for highly erodible loams and silt loams to less than 0.1 for sandy and gravelly soil with a high infiltration rate. Table 2.1 shows the K values for the Central and Eastern United States, recommended by Wischmeier and Smith (1965).

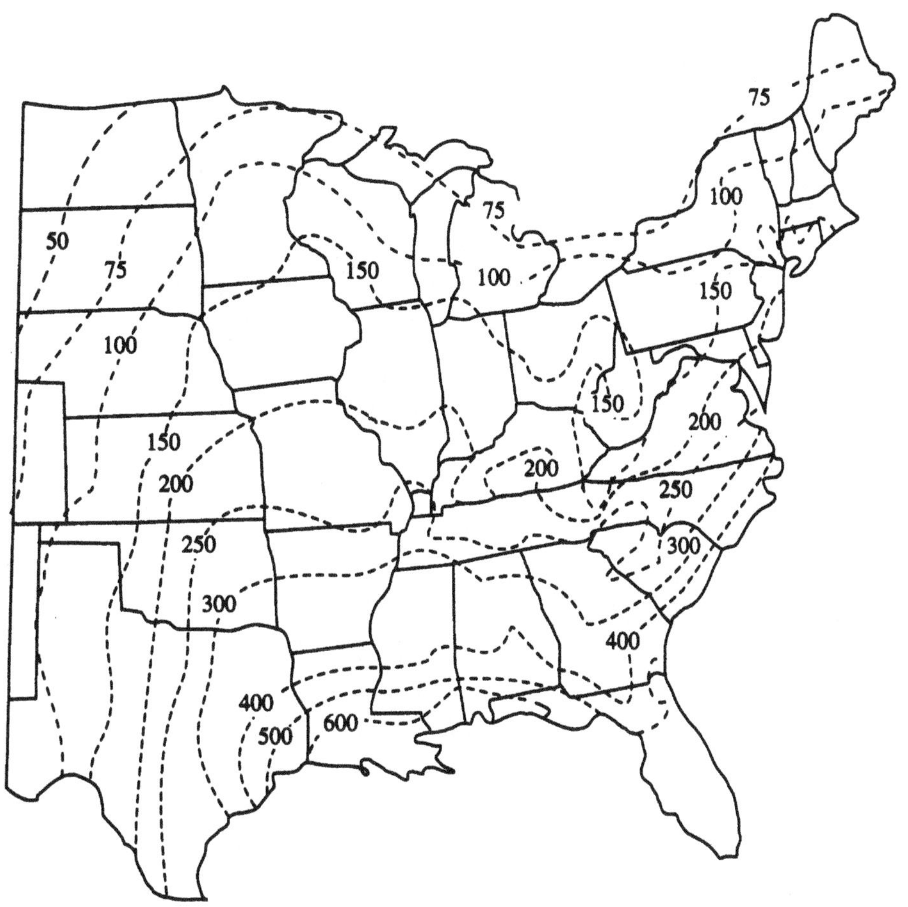

Figure 2.1. Isoerodent map of the R-factor values for the eastern portion of the United States (Wischmeier and Smith, 1965).

The slope-length factor L accounts for the increased quantity of runoff that occurs as distance from the top of the slope increases. It is the ratio of the soil loss from a given slope length to that from a 72.6-foot length, with all other conditions the same.

The slope-steepness factor S accounts for the increased velocity of runoff with slope steepness. It is the ratio of soil loss from a given slope steepness to that from a 9-percent slope. The effects of slope length and steepness are usually combined into one single factor; that is, the LS factor, which can be computed by:

$$LS = (\lambda / 72.6)^m (65.41 \sin^2 \theta + 4.56 \sin \theta + 0.065) \tag{2.2}$$

where λ = actual slope length in feet,
θ = angles of slope, and
m = an exponent with value ranging from 0.5 for slope equal to or greater than 5 percent to 0.2 for slope equal to or less than 1 percent.

Table 2.1. Relative erodibilities of key soils in the Central and Eastern United States (Wischmeier and Smith, 1965)

Soil	Location where evaluated	*K*-factor
Dunkirk silt loam	Geneva NY	0.69
Keene silt loam	Zanesville OH	0.48
Shelby loam	Bethany MO	0.41
Lodi loam	Blacksbury VA	0.39
Fayette silt loam	LaCrosse WI	0.38
Cecil sand clay loam	Watkinsville GA	0.36
Marshall silt loam	Clarinda IA	0.33
Ida silt loam	Castana IA	0.33
Mansic clay loam	Hays KS	0.32
Hagerstown silty clay loam	State College PA	0.31
Austin clay	Temple TX	0.29
Mexico silt loam	McCredie MO	0.28
Honeoye silt loam	Marcellus NY	0.28
Cecil sandy loam	Clemson SC	0.28
Ontario loam	Geneva NY	0.27
Cecil clay loam	Watkinsville GA	0.26
Boswell fine sandy loam	Tyler TX	0.25
Cecil sandy loam	Watkinsville GA	0.23
Zaneis fine sandy loam	Guthrie OK	0.22
Tifton loamy sand	Tifton GA	0.10
Freehold loamy sand	Marlboro NJ	0.08
Bath flaggy loam	Arnot NY	0.05
Albia gravelly loam	Beemerville NJ	0.03

Figure 2.2 expresses Equation (2.2) graphically. The results in Figure 2.2 were later extended to a slope length of 1,000 feet as shown in Table 2.2 (Wischmeier and Smith, 1978).

The cropping-management factor C accounts for the crop rotation used, tillage method, crop residue treatment, productivity level, and other agricultural practice variables. It is the ratio of soil loss from a field with given cropping and management practices to the loss from the fallow conditions used to evaluate the K-factor. The C-factor for an individual crop varies with the stage of crop growth, as shown in Table 2.3.

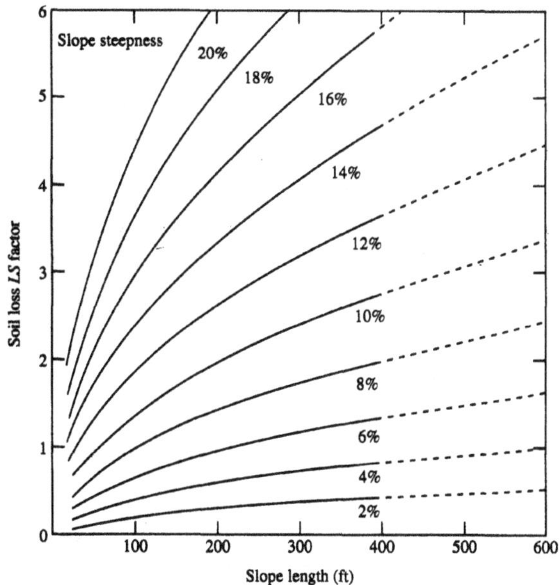

Figure 2.2. Topographic-effect graph used to determine *LS*-factor values for different slope-steepness—slope-length combinations (Wischmeier and Smith, 1965).

Table 2.2. Values of the topographic factor *LS* for specific combinations of slope length and steepness (Wischmeier and Smith, 1978)

Percent slope	Slope length (feet)											
	25	50	75	100	150	200	300	400	500	600	800	1,000
0.2	0.060	0.069	0.075	0.080	0.086	0.092	0.099	0.105	0.110	0.114	0.121	0.126
0.5	0.073	0.083	0.090	0.096	0.104	0.110	0.119	0.126	0.132	0.137	0.145	0.152
0.8	0.086	0.098	0.107	0.113	0.123	0.130	0.141	0.149	0.156	0.162	0.171	0.179
2	0.133	0.163	0.185	0.201	0.227	0.248	0.280	0.305	0.326	0.344	0.376	0.402
3	0.190	0.233	0.264	0.287	0.325	0.354	0.400	0.437	0.466	0.492	0.536	0.573
4	0.230	0.303	0.357	0.400	0.471	0.528	0.621	0.697	0.762	0.820	0.920	1.01
5	0.268	0.379	0.464	0.536	0.656	0.758	0.928	1.07	1.20	1.31	1.52	1.69
6	0.336	0.476	0.583	0.673	0.824	0.952	1.17	1.35	1.50	1.65	1.90	2.13
8	0.496	0.701	0.859	0.992	1.21	1.41	1.72	1.98	2.22	2.43	2.81	3.14
10	0.685	0.968	1.19	1.37	1.68	1.94	2.37	2.74	3.06	3.36	3.87	4.33
12	0.903	1.28	1.56	1.80	2.21	2.55	3.13	3.61	4.04	4.42	5.11	5.71
14	1.15	1.62	1.99	2.30	2.81	3.25	3.98	4.59	5.13	5.62	6.49	7.26
16	1.42	2.01	2.46	2.84	3.48	4.01	4.92	5.68	6.35	6.95	8.03	8.98
18	1.72	2.43	2.97	3.43	4.21	3.86	5.95	6.87	7.68	8.41	9.71	10.9
20	2.04	2.88	3.53	4.08	5.00	5.77	7.07	8.16	9.12	10.0	11.5	12.9

Table 2.3. Relative erodibilities of several crops for different crop sequences and yield levels at various stages of crop growth (Wischmeier and Smith, 1965)

Crop sequence	Crop yields		Soil-loss ratio for crop stage period[1]					
	Meadow	Corn	*F*	1	2	3	*4L*	*4R*
	(tons)	(bu)	(%)	(%)	(%)	(%)	(%)	(%)
Continuous fallow	-	-	100	100	100	100	-	100
1st-yr corn after meadow	1 to 2	40	15	32	30	19	30	50
1st-yr corn after meadow	2 to 3	70	10	28	19	12	18	40
1st-yr corn after meadow	3 to 5	100	8	25	17	10	15	35
2nd-yr corn after meadow, *RdR*[2]	2 to 3	70	60	65	51	24	-	65
2nd-yr corn after meadow, *RdL*	2 to 3	70	32	51	41	22	26	-
2nd-yr corn after meadow, *RdL+WC*	2 to 3	70	20	37	33	22	15	-
Corn, continuous, *RdR*	-	60	80	85	60	30	-	70
Corn, continuous, *RdL*	-	75	36	63	50	26	30	-
Corn, continuous, *RdL+WC*	-	75	22	46	41	26	15	-
Corn after oats with legume interseeding	-	60	25	40	38	24	30	-
Cotton, 1st-yr after meadow	2	-	15	34	40	30	30	-
Cotton, 2nd-yr after meadow	2	-	35	65	68	46	42	-
Cotton, continuous	-	-	45	80	80	52	48	-
Small grain with meadow interseeding, prior-crop residues on surface:								
After 1-yr corn after meadow	2	70	-	30	18	3	2	-
After 2-yr corn after meadow	2	70	-	40	24	5	3	-
After 2-yr cotton after meadow	2	-	-	50	35	5	3	-
Small grain after 1-yr corn after meadow, corn residues removed	2	-	-	50	40	15	3	-
Small grain on plowed seedbed, *RdR*	-	-	65	70	45	5	3	-
Established grass and legume meadow	3	-	0.4	0.4	0.4	0.4	0.4	0.4

[1] Crop stage periods: *F* = fallow; 1 = first month after seeding; 2 = second month after spring seeding; 3 = maturing crop to harvest; *4L* = residues; and *4R* = stubble.

[2] *RdR* = residues removed; *RdL* = residues left; *WC* = grass and legume winter-cover seeding.

The seasonal distribution of rainstorms in different locations influences the amount of erosion over the course of the year. The fraction of average annual erosion that occurs up to any point in the year varies according to geographical location. Figure 2.3 shows two sample erosion-index distribution curves for two parts of the United States.

The erosion-control practice factor *P* accounts for the effects of conservation practices, such as contouring, strip-cropping, and terracing, on erosion. It is the ratio of soil loss with a given practice to soil loss with straight-row farming parallel to the slope. For example, soil loss may be reduced by 50 percent on a 2- to 7-percent slope as a result of contouring. However, contouring becomes less effective with increasing slope. For steep slopes, terracing is a more effective conservation practice. Table 2.4 provides some suggested values of *P* based on recommendations of the U.S. Environmental Protection Agency and the Natural Resources Conservation Service (formerly the Soil Conservation Service).

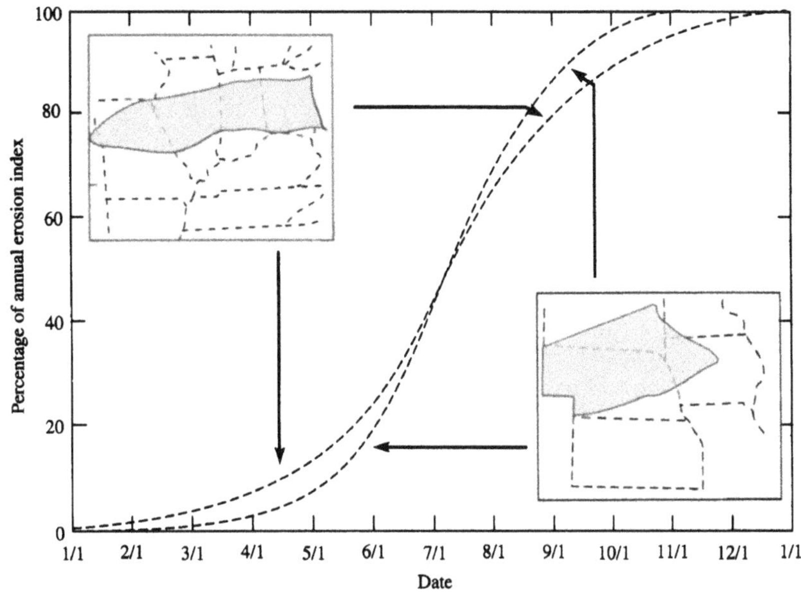

Figure 2.3. Erosion-index distribution curves for two sections of the United States (Wischmeier and Smith, 1965).

Table 2.4. Suggested *P* values for the erosion-control factor

Land slope (%)	Contouring[1]	Contour[2] furrows or pits	Contour ditches (wide spacing)
2.0 to 7	25.0 to 30	0.90	0.40
8.0 to 12	0.50	1.0	0.45
13.0 to 18	0.60	0.25	0.65
19.0 to 24	0.80	0.30	Factor values for this practice are not established.

[1] Topsoil spreading, tillage, and seeding on the contour. Contour limits–2%, 400 ft; 8%, 200 ft; 10%, 100 ft; 14 to 30%, 60 ft. The effectiveness of contouring beyond these limits is speculative.

2 Estimating values for surface manipulation of reclaimed land disturbed by surface mining. Furrows or pits installed on the contour. Spacing between furrows 40 to 60 inches with a minimum 6-inch depth. Pit spacing depends on pit size, but generally the pits should occupy 50% of the surface area.

The estimated soil loss from Equation (2.1) is the average value for a typical year, and the actual loss for any given year may be several times more or less than the average rate. It should also be noted that the computed soil loss gives the estimated soil erosion rates, based upon plot-sized areas of upland. It does not account for sediment detention due to vegetation, flat areas, or low areas. In the estimation of sediment inflow to a reservoir, the effects of rill, gully, and riverbank erosion and other sources, or erosion and deposition between upland and the reservoir, should also be considered. Another limitation of the use of the Universal Soil Loss Equation is that the R values given in Figure 2.1 do not include the western portion of the United States and other countries. Because it is an empirical equation, and the fact that the factors are based on agriculture practices in the United States, the application of the USLE is mainly limited to the Central and Eastern United States, even though successful examples of application can be found in other countries. Consequently, Equation (2.1) cannot be used directly in the Western United States or other countries without further studies of all the factors used in that equation.

Example 2.1 Determine the annual amount of soil loss from a contouring upland farm in central Illinois in the United States. The farmland has a size of 800 acres, the soil is in a silt loam, and the slope length is 400 feet with a slope steepness of 4 percent. The soil is covered with matured grass (Yang, 1996).

Solution:

From Figure 2.1, $R = 200$
From Table 2.1, $K = 0.33$
From Figure 2.2 or Table 2.2, $LS = 0.697$
From Table 2.3, $C = 0.004$
From Table 2.4, $P = 0.5$
$A = RKLSCP = 200 \times 0.33 \times 0.697 \times 0.004 \times 0.5 = 0.092$ tons/acre/year
Total annual loss of soil = 0.092 x 800 = 73 tons

2.2.2 Revised Universal Soil Loss Equation

Continued research and a deeper understanding of the erosion process prompted some needed revisions to the USLE. The Revised Universal Soil Loss Equation (RUSLE) retains the basic structure of the USLE, Equation (2.1). However, significant changes to the algorithms used to calculate the factors have been made in the RUSLE (Renard et al., 1994). The R factor has been expanded to include the Western United States (Figure 2.4) and corrections made to account for rainfall on ponded water. The K factor has been made time varying, and corrections were made for rock fragments in the soil profile. Slope length and steepness factors LS have been revised to account for the relation between rill and interrill erosion. The C factor no longer represents seasonal soil-loss ratios; it now represents a continuous function of prior land use PLU, surface cover SC, crop canopy CC, surface roughness SR, and soil moisture SM. The factor P has been expanded to include conditions for rangelands, contouring, stripcropping, and terracing. Additionally, seasonal variations in K, C, and P are accounted for by the use of climatic data, including twice monthly distributions of EI_{30} (product of kinetic energy of rainfall and 30-minute precipitation intensity) (Renard et al., 1996). The RUSLE factors are distinguished from the USLE factors by the subscript R. The majority of the information in this section is from Renard et al. (1996).

Figure 2.4. R_R isoerodent map of the Western United States. Units are hundreds (ft tonf in)/(ac hr yr). (From Renard et al., 1996).

Determination of the rainfall-runoff erosivity R factor in the USLE and RUSLE is made by use of the EI_{30} parameter, where E is the total storm energy and I_{30} is the maximum 30-minute rainfall intensity for the storm. The average EI_{30} is used to establish the isoerodent maps for the R factor.

An empirical relationship for calculating the kinetic energy of rainfall, used in calculating E, is used in the RUSLE. Isoerodent maps of the U.S. were updated for the RUSLE using

$$ke = 916 + 331\log_{10} i, \qquad i \leq 3 \text{ in h}^{-1} \tag{2.3}$$

$$ke = 1074, \qquad i > 3 \text{ in h}^{-1} \tag{2.4}$$

where ke = kinetic energy (ft ton acre^{-1} in^{-1}), and
i = rainfall intensity (in h^{-1}).

However, it is recommended by Renard et al. (1996) in the RUSLE handbook that the equation determined by Brown and Foster (1987)

$$ke_m = 0.29[1 - 0.72(e^{-0.05 i_m})] \tag{2.5}$$

where ke_m = kinetic energy of rainfall (MJ ha^{-1} mm^{-1} of rainfall), and
i_m = rainfall intensity (mm h^{-1}),

should be used for all calculations of the R factor. The kinetic energy of an entire storm is multiplied by the maximum 30-minute rainfall intensity I_{30} for that storm to get the EI_{30}.

An adjustment factor R_c is used to account for the protection from raindrops as a result of ponded water:

$$R_c = e^{(-0.49[y-1])} \tag{2.6}$$

where y = depth of flow or ponded water.

This adjustment in R is most important on land surfaces with little or no slope. Figure 2.5 shows the updated isoerodent map of the Eastern United States.

Corrections to the K factor have been made in the RUSLE to account for rock fragments in the soil matrix. Rock fragments present on the soil surface may act as an armoring layer causing a reduction in erosion and are accounted for by the C factor. Rock fragments present in the soil matrix have an effect on infiltration rates and hydraulic conductivity and, therefore, are accounted for with the K_R factor. The rate of reduction in saturated hydraulic conductivity resulting from the presence of rock fragments is given by:

$$K_b = K_f(1 - R_w) \tag{2.7}$$

where K_b = saturated hydraulic conductivity of the soil with rock fragments,
K_f = saturated hydraulic conductivity of the fine fraction of soil, and
R_w = percentage by weight of rock fragments > 2 mm.

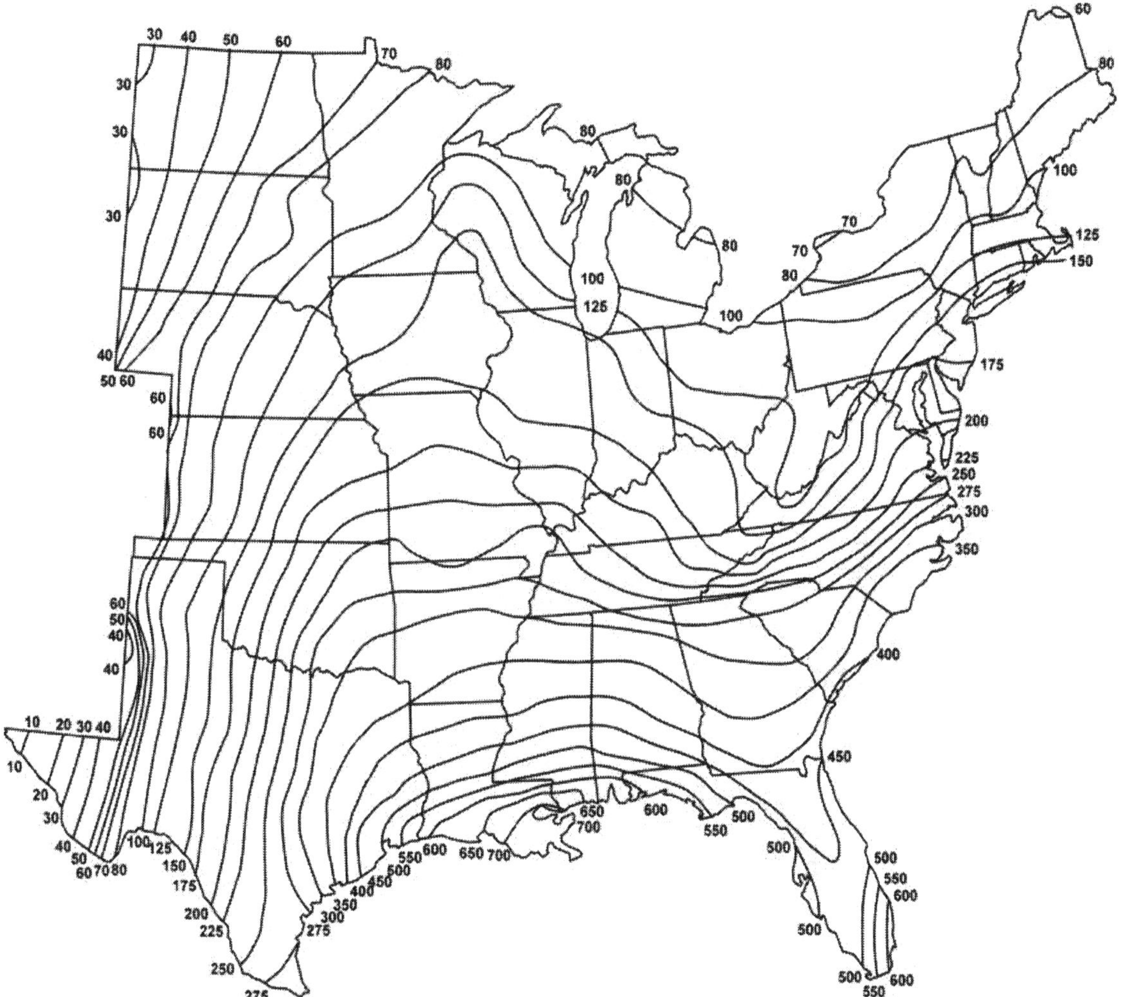

Figure 2.5. Adjusted R_R-factor isoerodent map of the Eastern United States. Units are hundreds (ft tonf in)/(ac hr yr). (From Renard et al., 1996).

An increase in rock fragments in the soil results in a corresponding decrease in the saturated hydraulic conductivity of the soil, thus leading to greater erosion potential and higher K_R factor values. Soil permeability classes that include the effects of rock fragments do not receive an adjustment of the K_R factor.

Additional changes to the K factor consist of the inclusion of seasonal effects as a result of soil freezing, soil texture, and soil water. Soil freezing and thawing cycles tend to increase the soil erodibility K factor by changing many soil properties, including soil structure, bulk density, hydraulic conductivity, soil strength, and aggregate stability. The occurrence of many freeze-thaw cycles will tend to increase the K factor, while the value of the soil erodibility factor will tend to decrease over the length of the growing season in areas that are not prone to freezing periods. An average annual value of K_R is estimated from:

$$K_{av} = \sum (EI_i)K_i / 100 \tag{2.8}$$

where EI_i = EI_{30} index at any time (calendar days),

$$K_i = K_{max}(K_{min} / K_{max})^{(t_i - t_{max})/\Delta t} \tag{2.9}$$

where
K_i = soil erodibility factor at any time (t_i in calendar days),
K_{max} = maximum soil erodibility factor at time t_{max},
K_{min} = minimum soil erodibility factor at time t_{min}, and
Δt = length of the frost-free period or growing period.

Figure 2.6 gives two examples of the variation in K_i with time for two soil types in two different climates. Table 2.5 gives some initial estimates of K_R for further use in the RUSLE computer program. The new K_R factor is designed to provide a more accurate yearly average value for K_i; e.g., for similar soils in different climates. Additionally, it allows for the RUSLE to be applied at smaller time scales, though it still does not allow for single event erosion modeling.

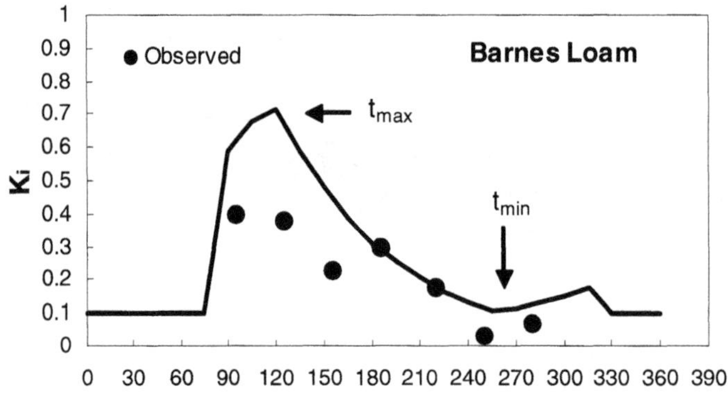

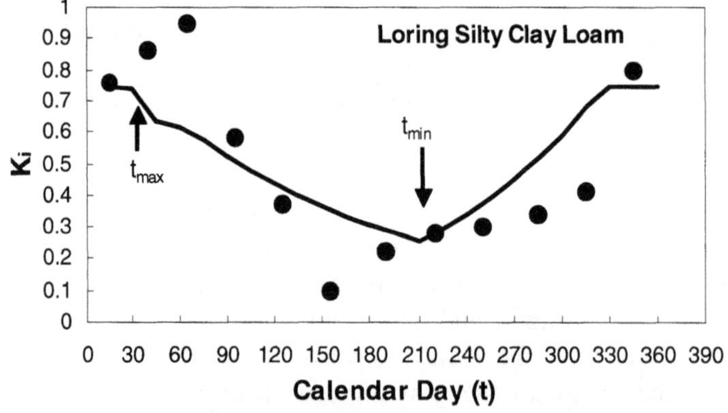

Figure 2.6 Relationship of K_i to calendar days for a Barnes loam soil near Morris, Minnesota, and a Loring silty clay loam soil near Holly Springs, Mississippi. K is given in U.S. customary units (from Renard et al., 1996).

Table 2.5. Initial K_R values for a variety of soil types in the Central and Eastern United States (Renard et al., 1996)

Soil type[1]	Location	Family	Period	Slope (%)	Length (ft)	K $\frac{ton}{acre\ eros.index}$
Bath sil.	Arnot, NY	Typic Fragiochrept	1938-45	19	72.6	0.05
Ontario l.	Geneva, NY	Glossoboric Hapludalf	1939-46	8	72.6	0.27
Cecil sl.	Clemson, SC	Typic Hapludalf	1940-42	7	180.7	0.28
Honeoye sil.	Marcellus, NY	Glossoboric Hapludalf	1939-41	18	72.6	0.28
Hagerstown sicl.	State College, PA	Typic Hapludalf	[2]NA	NA	NA	0.31
Fayette sil.	LaCrosse, WI	Typic Hapludalf	1933-46	16	72.6	0.38
Dunkirk sil.	Geneva, NY	Glossoboric Hapludalf	1939-46	5	72.6	0.69
Shelby l.	Bethany, MO	Typic Arguidoll	1931-40	8	72.6	0.53
Loring sicl.	Holly Springs, MS	Typic Fraguidalf	1963-68	5	72.6	0.49
Lexington sicl.	Holly Springs, MS	Typic Paleudalf	1963-68	5	72.6	0.44
Marshall sil.	Clarinda, IA	Typic Hapludoll	1933-39	9	72.6	0.43
Tifton ls.	Tifton, GA	Plinthic Paleudult	1962-66	3	83.1	[3]n.c.
Caribou grav. l.	Presque Isle, ME	Alfic Haplorthod	1962-69	8	72.6	n.c.
Barnes l.	Morris, MN	Udic Haploboroll	1962-70	6	72.6	0.23
Ida sil.	Castana, IA	Typic Udorthent	1960-70	14	72.6	0.27
Kenyon sil.	Independence, IA	Typic Hapludoll	1962-67	4.5	72.6	n.c.
Grundy sicl.	Beaconsfield, IA	Aquic Arguidoll	1960-69	4.5	72.6	n.c.

[1]si l. = silt loam, l. = loam, sl. = sandy loam, sicl. = silty clay loam, ls. = loamy sand, grav. l. = gravelly loam

[2]NA = Not available

[3]n.c. = Not calculated. However, soil-loss data for K-value computations are available from National Soil Erosion Laboratory, West Layfayette, Indiana

The slope length factor L is derived from plot data that indicate the following relation:

$$L = (\lambda / 72.6)^m \tag{2.10}$$

where λ = horizontal projection of the slope length, and
72.6 = RUSLE plot length in feet,

$$m = \beta / (1 + \beta) \tag{2.11}$$

where β = ratio of rill to interrill erosion.

The value of β when the soil is moderately susceptible to rill and interrill erosion is given by:

$$\beta = (\sin\theta / 0.0896) / [3.0(\sin\theta)^{0.8} + 0.56] \tag{2.12}$$

where θ = slope angle.

The parameter m in the RUSLE is a function of β (Equation 2.11). The newly defined L factor is combined with the original S factor to obtain a new LS_R factor. Values of m are in classes of low, moderate, and high, and tables are available in the RUSLE handbook for each of these classes to obtain values for LS_R. Table 2.6 gives an example of the new LS_R factor values for soils with low rill erosion rates. (Table 2.13 gives an example of LS_R values for soils with a high ratio of rill to interrill erosion.)

Table 2.6. Values of the topographic LS_R factor for slopes with a low ratio of rill to interrill erosion[1] (Renard et al., 1996)

	Horizontal slope length (ft)											
Slope (%)	25	50	75	100	150	200	250	300	400	600	800	1000
0.2	0.05	0.05	0.05	0.05	0.05	0.05	0.05	0.05	0.05	0.05	0.05	0.05
0.5	0.08	0.08	0.08	0.09	0.09	0.09	0.09	0.09	0.09	0.09	0.09	0.09
1.0	0.13	0.13	0.14	0.14	0.15	0.15	0.15	0.15	0.16	0.16	0.17	0.17
2.0	0.21	0.23	0.25	0.26	0.27	0.28	0.29	0.30	0.31	0.33	0.34	0.35
3.0	0.29	0.33	0.36	0.38	0.40	0.43	0.44	0.46	0.48	0.52	0.55	0.57
4.0	0.36	0.43	0.46	0.50	0.54	0.58	0.61	0.63	0.67	0.74	0.78	0.82
5.0	0.44	0.52	0.57	0.62	0.68	0.73	0.78	0.81	0.87	0.97	1.04	1.10
6.0	0.50	0.61	0.68	0.74	0.83	0.90	0.95	1.00	1.08	1.21	1.31	1.40
8.0	0.64	0.79	0.90	0.99	1.12	1.23	1.32	1.40	1.53	1.74	1.91	2.05
10.0	0.81	1.03	1.19	1.31	1.51	1.67	1.80	1.92	2.13	2.45	2.71	2.93
12.0	1.01	1.31	1.52	1.69	1.97	2.20	2.39	2.56	2.85	3.32	3.70	4.02
14.0	1.20	1.58	1.85	2.08	2.44	2.73	2.99	3.21	3.60	4.23	4.74	5.18
16.0	1.38	1.85	2.18	2.46	2.91	3.28	3.60	3.88	4.37	5.17	5.82	6.39
20.0	1.74	2.37	2.84	3.22	3.85	4.38	4.83	5.24	5.95	7.13	8.10	8.94
25.0	2.17	3.00	3.63	4.16	5.03	5.76	6.39	6.96	7.97	9.65	11.04	12.26
30.0	2.57	3.60	4.40	5.06	6.18	7.11	7.94	8.68	9.99	12.19	14.04	15.66
40.0	3.30	4.73	5.84	6.78	8.37	9.71	10.91	11.99	13.92	17.19	19.96	22.41
50.0	3.95	5.74	7.14	8.33	10.37	12.11	13.65	15.06	17.59	21.88	25.55	28.82
60.0	4.52	6.63	8.29	9.72	12.16	14.26	16.13	17.84	20.92	26.17	30.68	34.71

[1] Such as for rangeland and other consolidated soil conditions with cover (applicable to thawing soil where both rill and interrill erosion are significant).

The new cover-management factor C_R is based on a standard condition where a soil loss ratio SL_R is estimated relative to the reference condition (an area under clean-tilled continuous fallow). The SL_R is time variable, and values for SL_R are calculated every 15 days over the course of the year, based on the assumption that the important parameters remain constant over this time period. However, if, for example, a management operation changes in this time period, two values of SL_R are calculated for the 15-day time period. Soil Loss Ratio is calculated using the following relation:

$$SL_R = PLU \cdot CC \cdot SC \cdot SR \cdot SM \qquad (2.13)$$

where
- PLU = prior-land-use subfactor,
- CC = canopy-cover subfactor,
- SC = surface-cover subfactor,
- SR = surface-roughness subfactor, and
- SM = soil-moisture subfactor.

(See Renard et al., 1996, for details on calculating SL_R.) Once the values of SL_R are calculated for each time period, they are multiplied by the percentage of annual EI_{30} that occurs in that same time period and summed over the entire time period of investigation. This provides a new C_R factor for the RUSLE.

The supporting practices factor P is refined in the RUSLE and includes the effects of contouring, including tillage and planting on or near contours, stripcropping, terracing, subsurface drainage, and also includes rangeland conditions. Values for the new P_R factor are the least reliable of all the factors in the RUSLE (Renard et al., 1994); therefore, the physically-based model CREAMS (Kinsel, 1980) is used to supplement empirical information used in the RUSLE. The effects of various practices were analyzed using the model and represented as P_R subfactors that are then used to calculate an overall P_R factor. If a variety of supporting practices are present on a particular plot of land, the P_R subfactors are used to calculate an overall P_R factor and then used in the RUSLE. Calculation of the revised P_R factor, along with the calculation of all other factors as revised in the RUSLE, is facilitated by the use of a computer program, which is available at http://www.sedlab.olemiss.edu/rusle/. Use of the RUSLE would not be possible without it.

Example 2.2 Using the information given in example 2.1, in addition to the following information, determine the amount of annual soil loss using the RUSLE. Soil is dominated by interrill erosion with little or no rill erosion, 2% rock cover, no residual vegetative cover, 4-inch contour ridges, mature Bahiagrass, mechanically disturbed at harvest time.

Solution: The K_R, C_R, and P_R factors must be calculated using the RUSLE 1.06b program (download from http://www.sedlab.olemiss.edu/rusle/download.html).

From Figure 2.5, $R_R = 175$.
From Table 2.5, initial K_R value is 0.43, and using the RUSLE program (use city code 13001), $K_R = 0.38$.
From Table 2.6, $LS_R = 0.67$.
In the RUSLE program, select time invariant average annual value for C_R, determine effective root mass from Table 2.7, $C_R = 0.007$.
In the RUSLE program, select the frequent-disturbance option $P_R = 0.295$.

$A = R_R K_R LS_R C_R P_R = 175 \text{ x } 0.385 \text{ x } 0.67 \text{ x } 0.011 \text{ x } 0.295 = 0.146$ tons/acre/year

Total annual soil loss = 0.146 x 800 = 117 tons/year, which is greater than the 73 tons/year computed by the original USLE.

Table 2.7. Typical values of parameters required to estimate the C_R factor with the RUSLE computer program (Renard et al., 1996)

Common Name	Root mass in top 4 in (lbs acre-1)	Canopy cover just prior to harvest (%)	Effective fall height (ft)	Average annual yield (tons acre-1)
Grasses:				
Bahiagrass	1,900	95	0.1	4
Bermudagrass, coastal	3,900	100	0.2	8
Bermudagrass, common	2,400	100	0.1	3
Bluegrass, Kentucky	4,800	100	0.1	3
Brome grass, smooth	4,500	100	0.1	5
Dallisgrass	2,500	100	0.1	3
Fescue, tall	7,000	100	0.1	5
Orchardgrass	5,900	100	0.1	5
Timothy	2,900	95	0.1	5
Legumes:				
Alfalfa	3,500	100	0.2	6
Clover, ladino	1,400	100	0.2	3
Clover, red	2,100	100	0.1	4
Clover, sweet	1,200	90	2.0	2
Clover, white	1,900	100	0.1	2
Lespedeza, sericea	1,900	100	0.5	3
Trefoil, birdsfoot	2,400	100	0.3	4

These values are for mature, full pure stands on well-drained nonirrigated soils with moderate-to-high available water-holding capacity. These values hold for species shown only within their range of adaptation. Except for biennials, most forages do not attain a fully-developed root system until end of second growing season. Root mass values listed can be reduced by as much as half on excessively drained or shallow soils and in areas where rainfall during growing season is less than 18 in.

2.2.3 Modified Universal Soil Loss Equation

Williams (1975) modified the USLE to estimate sediment yield for a single runoff event. On the basis that runoff is a superior indicator of sediment yield than rainfall—i.e., no runoff yields no sediment, and there can be rainfall with little or no runoff—Williams replaced the *R* (rainfall erosivity) factor with a runoff factor. His analysis revealed that using the product of volume of runoff and peak discharge for an event yielded more accurate sediment yield predictions, especially for large events, than the USLE with the *R* factor. The Modified USLE, or MUSLE, is given by the following (Williams, 1975):

$$S = 95(Qp_p)^{0.56} KLSCP \tag{2.14}$$

where S = sediment yield for a single event in tons,
Q = total event runoff volume (ft^3),
p_p = event peak discharge (ft^3 s^{-1}), and
K, LS, C, and P = USLE parameters (Equation 2.1).

The comparison with the USLE was done by estimating the average annual soil loss with the USLE and comparing it to the annual soil loss calculated for each event over the course of the year using the MUSLE. The MUSLE has been tested (Williams, 1981; Smith et al., 1984) and found to perform satisfactorily on grassland and some mixed use watersheds. However, the utility of the MUSLE depends a great deal upon the accuracy of the hydrologic inputs.

Example 2.3 Using the same information from example 2.1, determine the sediment yield from a storm with a total runoff volume of 120 ft^3 and a peak discharge of 5 cfs.

Solution: From example 2.1, $K = 0.33$
$LS = 0.697$
$C = 0.004$
$P = 0.5$

$$(Qp_p)^{0.56} = 120 \text{ x } 5 = 600^{0.56} = 36$$

$$S = 95(Qp_p)^{0.56} KLSCP = 95 \text{ x } 36 \text{ x } 0.33 \text{ x } 0.697 \text{ x } 0.004 \text{ x } 0.5 = 1.57 \text{ tons}$$

In order to obtain an estimate of the annual soil loss from the MUSLE, soil loss from each event throughout the year needs to be calculated.

While the USLE, RUSLE, and MUSLE have met with practical success as an aid for conservation management decisions and the reduction of soil erosion from agricultural lands, they are not capable of simulating soil erosion as a dynamic process distributed throughout a watershed and changing in time. Although the MUSLE can estimate soil loss from a single event, neither it nor the USLE and RUSLE can estimate detachment, entrainment, transport, deposition, and redistribution of sediment within the watershed and are of limited application.

2.2.4 Direct Measurement of Sediment Yield and Extension of Measured Data

The most accurate method for determining the long-term sediment yield from a watershed is by direct measurement of sediment deposition in a reservoir (Blanton, 1982) or by direct measurement of streamflow, suspended sediment concentration, and bedload. If long-term records are available, then daily and average annual sediment loads can be computed. The average annual sediment load can then be used to estimate the long-term sediment yield. However, long-term measurements of river discharge are not always available. Long-term measurements of suspended sediment concentration are not commonly available, and long-term measurements of bedload are rare.

In the absence of long-term streamflow measurements for the site of interest, it may be possible to extend short-term measurements by empirical correlation with records from another stream gauge in the watershed or from a nearby watershed with similar drainage characteristics.

A short-term record of suspended sediment concentrations can be extended by correlation with streamflow. A power equation of the form, $C = aQ^b$, is most commonly used for regression analysis, where C is the sediment concentration, Q is the rate of streamflow, and a and b are regression coefficients. The relationship between streamflow and suspended sediment concentration can change with grain size, from low flows to high flows, from season to season, and from year to year. Therefore, enough measurements of suspended sediment concentration and streamflow are necessary to ensure that the regression equation is applicable over a wide range of streamflow conditions, seasons, and years.

A single regression equation may produce an acceptable correlation over a narrow range of conditions. However, separate regression equations may be necessary to achieve satisfactory correlations over a wide range of conditions. For example, the suspended sediment concentrations could be divided into wash load and bed-material load to develop separate regression equations for each. The data could also be sorted by streamflow to develop separate regression equations for low, medium, and high flows. The data may need to be sorted by season to develop separate regression equations for the winter and spring flood seasons. If enough data were available, a portion of the data could be used for the regression analysis, so that the remaining portion could be used for verification.

A short-term record of bedload measurements could be extended in the same manner as that described for the suspended sediment concentrations. If no bedload measurements were available, then bedload could be estimated as a percentage of the suspended sand load (typically 2 to 15%) or computed using one of many predictive equations (see Chapter 3, *Non-Cohesive Sediment Transport*). Strand and Pemberton (1982) presented a guide for estimating the ratio of bedload to suspended sediment load (Table 2.8). Table 2.8 presents five conditions that estimate the ratio of bedload to suspended sediment load as a function of the streambed material size, the fraction of the suspended load that is sand, and the suspended sediment concentration during floods. A bedload measurement program should be considered if the bedload could be more than 10 percent of the suspended sediment load.

Table 2.8. Bedload adjustment

Streambed material	Fraction of suspended sediment load that is sand (%)	Suspended sediment concentration (ppm)	Ratio of bedload to suspended sediment load
Sand	20–50	< 1,000	25–150
Sand	20–50	1,000–7,500	10–35
Sand	20–50	> 7,500	5
Compacted clay, gravels, cobbles, or boulders	< 25	Any	5–15
Clay and silt	Near 0	Any	< 2

2.2.5 Sediment Yield as a Function of Drainage Area

Empirical sediment yield equations can be developed strictly as a function of drainage area based on reservoir sediment survey data. For example, Strand (1975) developed the following empirical equation for Arizona, New Mexico, and California:

$$Q_s = 2.4\,A_d^{-0.229} \tag{2.15}$$

where Q_s = sediment yield in ac-ft/mi^2/yr, and
A_d = drainage area in mi^2.

Strand and Pemberton (1982) developed a similar empirical equation for the semiarid climate of the Southwestern United States:

$$Q_s = 1.84\,A_d^{-0.24} \tag{2.16}$$

This same approach can be used to develop equations for other regions.

2.2.6 Sediment Yield Classification Procedure

The Pacific Southwest Inter-Agency Committee (1968) developed a sediment yield classification procedure that predicts sediment yield as a function of nine individual drainage basin characteristics. These include surface geology, soils, climate, runoff, topography, ground cover, land use, upland erosion, and channel erosion. Each drainage basin characteristic is given a subjective numerical rating based on observation and experience. Table 2.9 presents the drainage basin characteristics considered by this method and their possible ratings. The sum of these ratings determines the drainage basin classification and the annual sediment yield per unit area (Table 2.10).

2.3 Physically Based Approach for Erosion Estimates

The minimum energy dissipation rate theory states that when a dynamic system reaches its equilibrium condition, its rate of energy dissipation is at a minimum (Yang and Song, 1986, and Yang, 1996). The minimum value depends on the constraints applied to the system. The rate of energy dissipation per unit weight of water is:

$$dY/dt = (dx/dt)\,(dY/dx) = VS = \text{unit stream power} \tag{2.17}$$

where Y = potential energy per unit weight of water,
t = time,
x = reach length,
dx/dt = velocity V, and
dY/dx = energy or water surface slope S.

For the equilibrium condition, the unit stream power VS will be at a minimum, subject to the constraints of carrying a given amount of water and sediment.

Table 2.9. List of drainage basin characteristics and possible range of numerical ratings (modified from Pacific Southwest Interagency Committee, Water Management Subcommittee, 1968)

Drainage basin characteristics	Sediment yield levels		
	High rating	Moderate rating	Low rating
Surface geology	10: marine shales and related mudstones and siltstones	5: rocks of medium hardness moderately weathered and fractured	0: massive hard formations
Soils	10: fine textured and easily dispersed or single grain salts and fine sands	5: medium textured, occasional rock fragments, or caliche crusted layers	0: frequent rock fragments, aggregated clays, or high organic content
Climate	10: frequent intense convective storms	5: infrequent convective storms, moderate intensity	0: humid climate with low intensity rainfall, arid climate with low intensity rainfall, or arid climate with rare convective storms
Runoff	10: high flows or volume per unit area	5: moderate flows or runoff volume per unit area	0: low flows or volume per unit area or rare runoff events
Topography	20: steep slopes (in excess of 30%), high relief, little or no flood plain development	10: moderate slopes (about 20%), moderate flood plain development	0: gentle slopes (less than 5%), extensive flood plain development
Ground cover	10: ground cover less than 20%, no rock or organic litter in surface soil	0: ground cover less than 40%, noticeable organic litter in surface soil	-10: area completely covered by vegetation, rock fragments, organic litter with little opportunity for rainfall to erode soil
Land use	10: more than 50% cultivated, sparse vegetation, and no rock in surface soil	0: less than 25% cultivated, less than 50% intensively grazed	-10: no cultivation, no recent logging, and only low intensity grazing, if any
Upland erosion	25: rill, gully, or landslide erosion over more than 50% of the area	10: rill, gully, or landslide erosion over about 25% of area	0: no apparent signs of erosion
Channel erosion	25: continuous or frequent bank erosion, or active headcuts and degradation in tributary channels	10: occasional channel erosion of bed or banks	0: wide shallow channels with mild gradients, channels in massive rock, large boulders, or dense vegetation or artificially protected channels

Table 2.10. Drainage basin sediment yield classification (Randle, 1996)

Drainage basin classification number	Total rating	Annual sediment yield (ac-ft/mi^2)
1	> 100	> 3
2	75 to 100	1.0 to 3.0
3	50 to 75	0.5 to 1.0
4	25 to 50	0.2 to 0.5
5	0 to 25	<0.2

Sediment transport rate is directly related to unit stream power (Yang, 1996). The basic form of Yang's (1973) unit stream power equation for sediment transport is:

$$\log C = I + J \log(VS/\omega - V_{cr}S/\omega) \tag{2.18}$$

where C = sediment concentration,
I, J = dimensionless parameters reflecting flow and sediment characteristics that are determined from regression analysis,
V = flow velocity,
S = energy or water surface slope of the flow,
ω = sediment particle fall velocity, and
V_{cr} = critical velocity required for incipient motion.

The unit stream power theory stems from a general concept in physics that the rate of energy dissipation used in transporting material should be related to the rate of material being transported. The original concept of unit stream power, or rate of potential energy dissipation per unit weight of water, was derived from a study of river morphology (Yang, 1971). The river systems observed today are the cumulative results of erosion and sediment transport. If unit stream power can be used to explain the results of erosion and sediment transport, it should be able to explain the process of erosion and sediment transport. The relationships between unit stream power and sediment transport in open channels and natural rivers have been addressed in many of Yang's publications. This section addresses the relationship between unit stream power and surface erosion.

For laminar flow over a smooth surface, the average flow velocity can be expressed by Horton et al. (1934):

$$V = (gSR^2)/(3\nu) \tag{2.19}$$

where V = average flow velocity,
S = slope,
g = gravitational acceleration,
R = hydraulic radius, which can be replaced by depth for sheet flow, and
ν = kinematic viscosity.

The shear velocity is:

$$U_* = \sqrt{gRS} \tag{2.20}$$

From Equations (2.19) and (2.20)

$$\frac{VS}{U_*^4} = \frac{1}{3g\nu} \tag{2.21}$$

In other words, the ratio between the unit stream power and the fourth power of the shear velocity is a constant for a fluid of a given viscosity.

For laminar flow over a rough bed, the grain shear stress can be expressed by:

$$\tau' = \frac{1}{8}\rho F' V^2 \tag{2.22}$$

where ρ = density of fluid, and
F' = a parameter.

Savat (1980) found:

$$F' = \frac{K}{R_e} \tag{2.23}$$

where K = a constant with a theoretical value of 24, and
R_e = Reynolds number.

From Equations (2.22) and (2.23):

$$\tau' = \frac{K\mu V}{8R} \tag{2.24}$$

where μ = dynamic viscosity.

Govers and Rauws (1986) assumed that:

$$R = \sqrt{\frac{3\nu V}{gS}} \tag{2.25}$$

then

$$\frac{VS}{U_*'} = \frac{192}{K^2 g\nu} \tag{2.26}$$

where U_*' = grain shear velocity.

Equations (2.21) and (2.26) indicate that the relationship between unit stream power and shear velocity due to grain roughness for sheet flows is well defined, regardless of whether the surface is smooth or rough. Figure 2.7 shows the relationship between sheet sediment concentration and grain shear velocity by Govers and Rauws (1986), based on data collected by Kramer and Meyer (1969), Rauws (1984), and Govers (1985). When Govers and Rauws (1986) replotted the same data, as shown in Figure 2.8, they showed a much better-defined relationship between sediment concentration and unit stream power. Figure 2.9 shows an example of comparison between measured and predicted sediment concentration based on unit stream power.

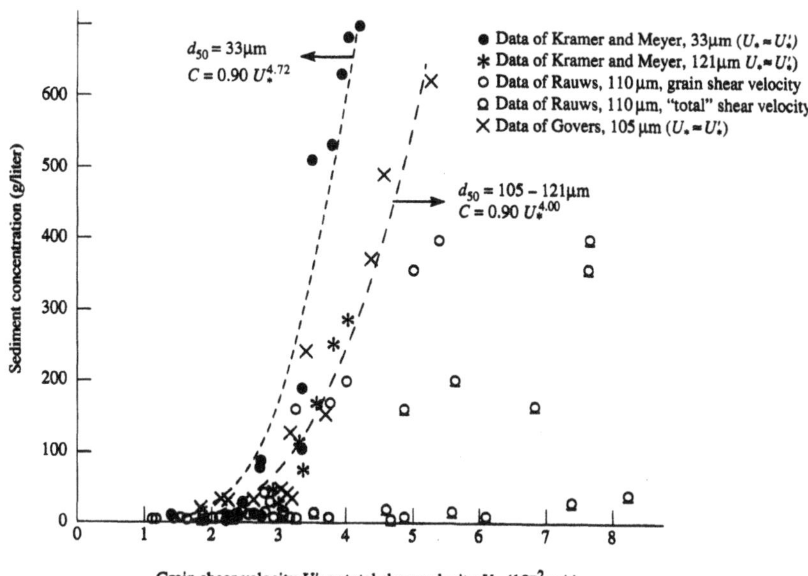

Figure 2.7. Relationship between sheet and rill flow sediment concentration and grain shear velocity (Govers and Rauws, 1986).

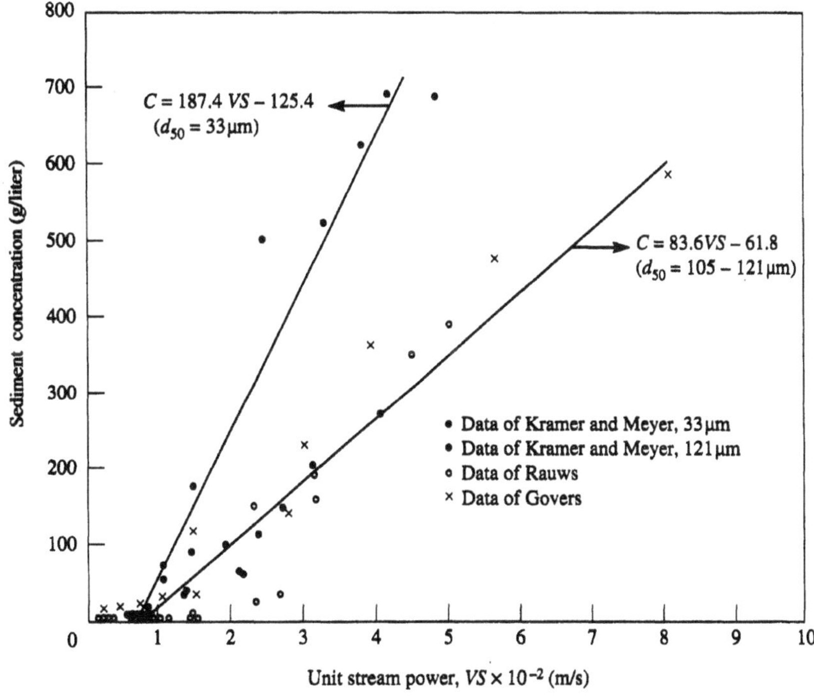

Figure 2.8. Relationship between sheet and rill flow sediment concentrations and unit stream power (Govers and Rauws, 1986).

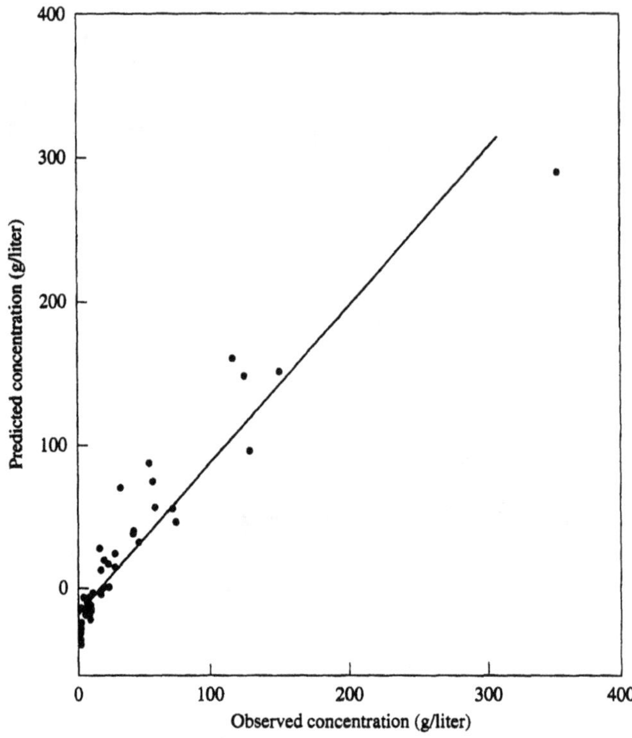

Figure 2.9. Comparison between measured and predicted sediment from surface erosion (Govers and Rauws, 1986).

Moore and Burch (1986) tested the direct application of Equation (2.18) to sheet and rill erosion. They reported from experimental results that:

$$I = 5.0105 \pm 0.0443 \tag{2.27}$$

$$J = 1.363 \pm 0.030 \tag{2.28}$$

Velocity was computed from Manning's equation, because it is difficult to measure for sheet flow, and they expressed unit stream power as:

$$VS = \left(\frac{Q}{B}\right)^{0.4} \frac{S^{1.3}}{n^{0.6}} \tag{2.29}$$

where Q = water discharge,
B = width of flow,
S = slope, and
n = Manning's roughness coefficient.

Similarly, the unit stream power for rill flow can be expressed by:

$$VS = \left(\frac{Q}{J}\right)^{0.25} \frac{S^{1.375}}{n^{0.75}} W \tag{2.30}$$

where J = number of rills crossing the contour element B, and
W = rill shape factor = (width/depth) $^{0.5}$.

It can be shown that for parabolic rills:

$$W = \sqrt{\frac{(1.5)^{0.5} a^{1.5}}{1.5a^2 + 4}} \tag{2.31}$$

for trapezoidal rills:

$$W = \left[\frac{(a+z)}{a + 2\sqrt{Z^2+1}}\right]^{0.5} \tag{2.32}$$

where a = rill width-depth ratio, and
Z = rill side slope.

Figure 2.10 shows the relationships among W, a, and Z for rills of different shapes. Figure 2.10 shows that when the width-depth ratio is greater than 2, the geometry has little impact on the value of the shape factor. Moore and Burch assumed that most natural rills can be approximated by a rectangular rill in the computation of W when a is greater than 2 or 3.

Yang's (1973) original unit stream power equation was intended for open channel flows. His dimensionless critical unit stream power required at incipient motion may not be directly applicable to sheet and rill flows. For sheet and rill flows with very shallow depth, Moore and Burch found that the critical unit stream power required at incipient motion can be approximated by a constant:

$$\frac{V_{cr}S}{v} = 4105 \text{ m}^{-1} \tag{2.33a}$$

or

$$V_{cr}S = 0.002 \text{ m/s} \tag{2.33b}$$

as shown in Figure 2.11. In Equation (2.33a), v = kinematic viscosity of water.

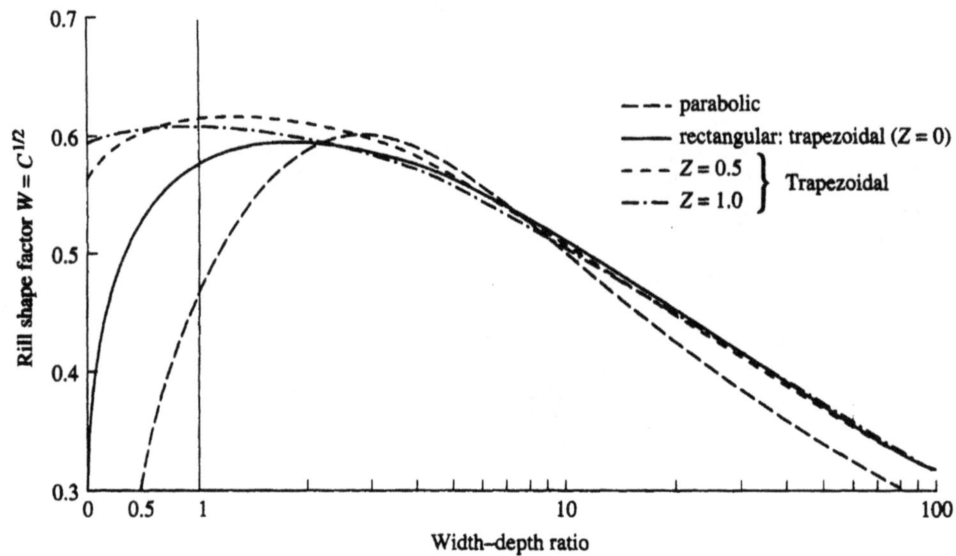

Figure 2.10. Relationship between rill shape factor and width-depth ratio for parabolic, rectangular, and trapezoidal rills (Moore and Burch, 1986).

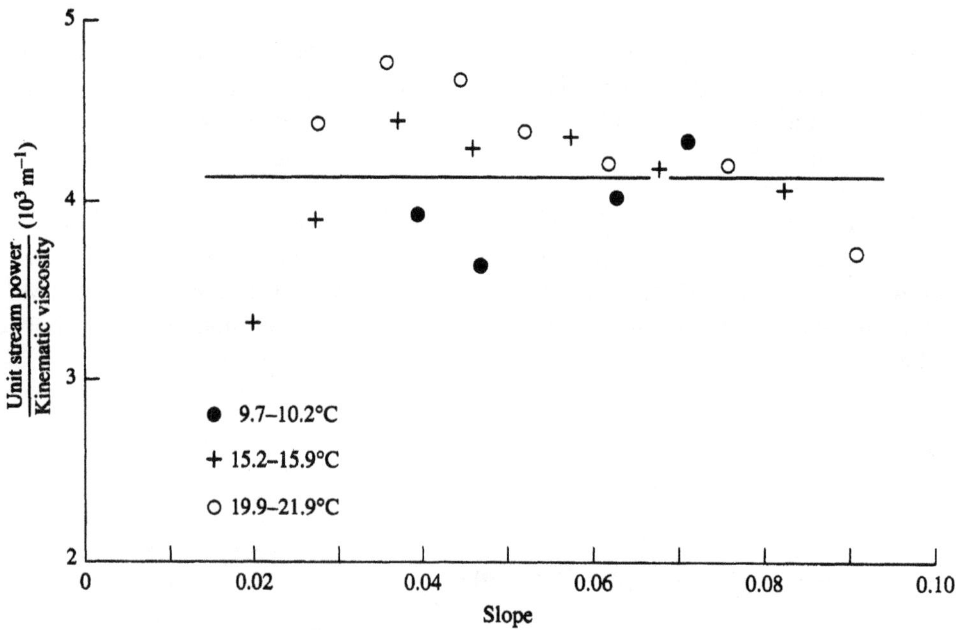

Figure 2.11. Relationship between the ratio of critical unit stream power and kinematic viscosity and the surface slope (Moore and Burch, 1986).

Moss et al. (1980) noted that sheet flow occurred initially, but as soon as general sediment motion ensued, the plane bed revealed its instability and rill cutting began. In accordance with the theory of minimum energy dissipation rate (Yang and Song, 1986, 1987; Yang et al., 1981) a rectangular channel with the least energy dissipation rate or maximum hydraulic efficiency should have a width-depth ratio of 2. For $a = 2$ and $Z = 0$, unit stream power for rill erosion can be computed by Equations (2.30) and (2.32). The number of rills generated by flow ranges from 1.5 at $Q = 0.0015\ m^3/s$ to 7 at $Q = 0.0003\ m^3/s$. Substituting the unit stream power thus obtained and a constant critical unit stream power of 0.002 m/s required at incipient motion, the sediment concentration due to sheet and rill erosion in the sand size range can be computed directly from Yang's 1973 equation. Yang's 1973 equation was intended for the movement of sediment particles in the ballistic or colliding region instead of the individual jump or saltation region. The comparisons shown in Figure 2.12 by Moore and Burch indicate that the rate of surface erosion can be accurately predicted by the unit stream power equation when the movement of sediment particles is in the ballistic dispersion region. The numbers shown in Figure 2.12 are sediment concentrations in parts per million by weight.

Yang's 1973 equation should not be applied to soils in the clay or fine silt size range directly because the terminal fall velocities of individual small particles are close to zero. In this case, the effective size of the aggregates of the eroded and transported materials should be used. The effective size increases with increasing flow rate and unit stream power. The estimated terminal fall velocities of these fine particles in water should also be adjusted for differences in the measured aggregate densities. For example, after these adjustments, effective particle diameters of aggregate size of the Middle Ridge clay loam and Irving clay for inter-rill and rill flow were determined to be 0.125 mm and 0.3 mm, respectively. With these effective diameters and a constant critical unit stream power of 0.002 m/s at incipient motion, Yang's 1973 equation can also be used for the estimation of surface erosion rate in the clay size range. Figure 2.13 shows that observed clay concentrations and predicted clay concentrations by Yang's (1973) equation using effective diameter of the clay aggregate, are in close agreement.

Combining Equations (2.18), (2.27), (2.28), and (2.29) yields the following equation for sheet erosion:

$$\log C = 5.0105 + 1.363 \log [\{(Q/B)^{0.4} S^{1.3} / n^{0.6} - 0.002\}/\omega] \qquad (2.34)$$

Similarly, the equation for rill erosion becomes:

$$\log C = 5.0105 + 1.363 \log [\{(Q/J)^{0.25} (S^{1.375} / n^{0.75})W - 0.002\}/\omega] \qquad (2.35)$$

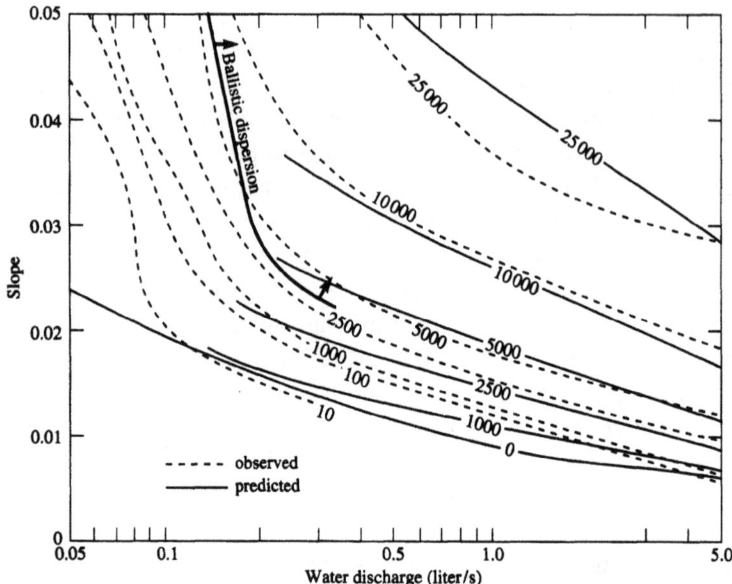

Figure 2.12. Comparison between observed and predicted sediment concentrations in ppm by weight from Yang's unit stream power equation with a plane bed composed of 0.43 mm sand (Moore and Burch, 1986).

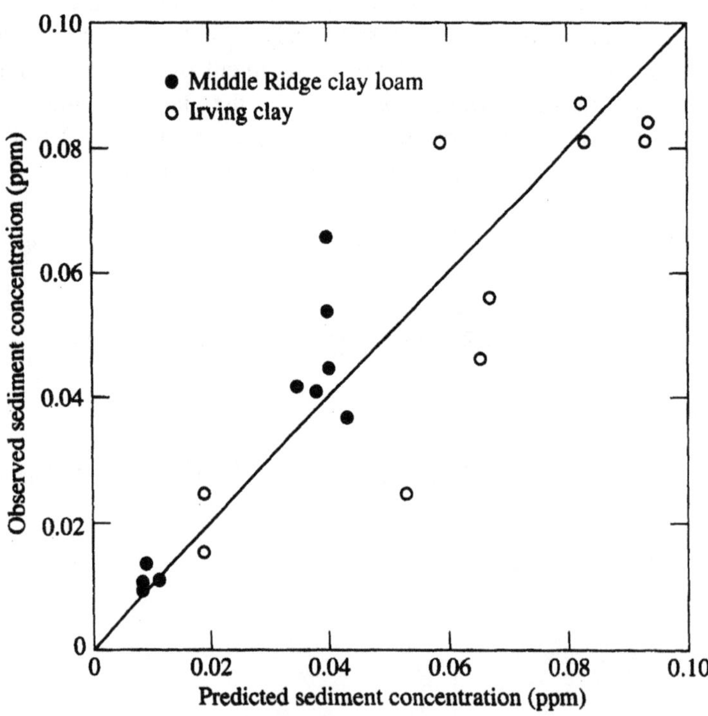

Figure 2.13. Comparison between observed and predicted clay concentrations from Yang's unit stream power equation (Moore and Burch, 1986).

The derivations and comparisons shown in this section confirm that with minor modifications, Yang's (1973) unit stream power equation can be used as a rational tool for the prediction of sheet and rill erosion rate, given the water discharge, surface roughness and slope, and median particle size or effective particle size and its associated fall velocity. This suggests that a rational method based on rainfall-runoff and the unit stream power relationship can be developed to replace the empirical Universal Soil Loss Equation for the prediction of soil loss due to sheet and rill erosion. It has been shown in the literature that Yang's unit stream power equations can be used to determine the rate of sediment transport in small and large rivers with accuracy. It is now possible to use the unit stream power theory to determine the total rate of sediment yield and transport from a watershed regardless of whether the sediment yield particles are transported by sheet, rill, or river flows. By doing so, the actual amount of sediment entering a reservoir can be determined by a consistent and rational method.

2.4 Computer Model Simulation of Surface Erosion Process

A multitude of computer models have been developed for various applications that utilize a wide array of techniques to simulate soil erosion within a watershed. Erosion models have been developed for different purposes including:

- Predictive tools for assessing soil loss for conservation planning, regulation, and soil erosion inventories.
- Predictive tools to assess where and when within a watershed soil erosion may be a problem.
- Research tools to better understand the erosion process (Nearing et al., 1994).

Watershed erosion models can be grouped into several categories:

- Empirically based, or derived, erosion models such as the USLE (Wischmeier and Smith, 1978) and the RUSLE (Renard et al., 1996).
- Physically based models, such as the Water Erosion Prediction Project (WEPP) (Nearing et al., 1989), Areal Non-point Source Watershed Environmental Response Simulation (ANSWERS) (Beasley et al., 1980), Chemicals, Runoff, and Erosion from Agricultural Management Systems (CREAMS) (Kinsel, 1980), Kinematic runoff and Erosion model (KINEROS) (Woolhiser et al., 1990), European Soil Erosion Model (EUROSEM) (Morgan et al., 1998), and Système Hydrologique Européen Sediment model (SHESED) (Wicks and Bathurst, 1996).
- Mixed empirical and physically based models, such as Cascade of Planes in Two Dimensions (CASC2D) (Johnson et al., 2000; Ogden and Julien, 2002), Agricultural Non- Point Source Pollution model (AGNPS) (Young et al., 1989), and Gridded Surface Subsurface Hydrological Analysis model (GSSHA) (Downer, 2002).

- GIS and Remote Sensing techniques that utilize one of the previously listed erosion models (Jürgens and Fander, 1993; Sharma and Singh, 1995; Mitasova et al., 2002).

Links to many other soil erosion models can be found on the World Wide Web at:

http://www3.bae.ncsu.edu/Regional-Bulletins/Modeling-Bulletin/
http://soilerosion.net/doc/models_menu.html

The performance of a given watershed-scale erosion model is best assessed within the context of its intended use. For instance, lumped empirical models of soil erosion, such as the USLE, are limited primarily to average sediment yield over a basin with the same characteristics as basins used in the model's development and cannot be used to assess spatial variability of erosion or to dynamically model the erosion process. Where applicable, USLE and RUSLE have been used with a good deal of success in assessing average yearly soil loss and in guiding land use and management decisions.

Distributed process-based models have been developed for a variety of different reasons (i.e., to assess and manage non-point source pollution; to explicitly model soil erosion; to model drainage basin evolution) and to be applied at a variety of different spatial and temporal scales with varying degrees of success. Synopsis of published reviews from applications of some of the available models are included below:

- An evaluation of WEPP in comparison to the USLE and the RUSLE indicates that WEPP predicts soil loss (kg/m^2) almost as well as the USLE and RUSLE at many sites, worse on others, and better on a few (Tiwari et al., 2000). Model efficiency, based on the Nash-Sutcliffe coefficient, ranged from -10.54 to 0.85 for the USLE and from -37.74 to 0.94 for WEPP. The Nash-Sutcliffe coefficient provides a measure of a model's performance over the course of an event as compared to the mean discharge for the event (Nash and Sutcliffe, 1970). A model efficiency of 1.0 represents a perfect fit of the model to observed values. Negative values indicate that use of the average (USLE in the evaluation of WEPP) is a better predictor than the model. The measured performance of WEPP compared to the USLE is considered a success given that the USLE performance at these sites is good and that the sites are where USLE parameters had been determined.

- A comparison by Bingner et al. (1989) of several erosion models applied to watersheds in Mississippi revealed that no model simulated sediment yield well on a consistent basis, though results are satisfactory to aid in management practice decisions. The models that were compared included CREAMS and the Simulator for Water Resources in Rural Basins (SWRRB) (Williams et al., 1985), the Erosion-Productivity Impact Calculator (EPIC) (Williams et al., 1984), ANSWERS, and AGNPS. For example, simulated results are within 50% of observed values for SWRRB (a modification of CREAMS) and AGNPS on one watershed and within 30% of observed values on another watershed. Error was as high as 500% for some models, and as low as 20% for others. The input parameters varied for each of these models and each model was developed for different applications. For example, ANSWERS and AGNPS are designed as single-event models on large watersheds (up to 10,000 ha), EPIC is designed for small watersheds (~1 ha), and CREAMS is designed for field sized watersheds (Bingner et al., 1989). A model like ANSWERS that did not perform as well on watersheds in Mississippi may require more updates of parameters, in which case its performance would be improved.

- Wicks and Bathurst (1996) show that SHESED does well at predicting sediment volume over the course of a snowmelt season but simulations at smaller time and spatial scales are less successful.

- EUROSEM, a single event model, was tested by Parsons and Wainwright (2000) on a watershed in Arizona. The hydrologic component did very poorly with good results for only the last 10 minutes of the simulation. Though EUROSEM underestimated runoff, soil erosion was overestimated by an order of magnitude. In order to obtain reasonable results, changing a measured parameter well beyond its recommended value was required.

- Kothyari and Jain (1997) used GIS techniques in combination with the USLE to estimate watershed-scale sediment yield. Performance of this model was adequate to poor with error in the range of 0.65-6.60 (ratio of observed sediment yield to simulated), which is in the range seen with physically based models.

While erosion models have been widely tested and evaluated (Mitchell et al., 1993; Wu et al., 1993; Smith et al., 1995; Bingner, 1996; Bouraoui and Dillaha, 1996; Zhang et al., 1996; Folly et al., 1999; Schröder, 2000; Tiwari et al., 2000; Ogden and Heilig, 2001; Kirnak, 2002), it is difficult to objectively compare the performance of these models to each other. That is, determining "the best" model depends on the watershed characteristics and the purpose of the investigation. Additionally, physically based models vary in the degree to which they represent the physical processes of erosion (Wu et al., 1993). For instance, models such as CREAMS, WEPP, and EUROSEM explicitly and separately account for erosion in interrill areas and rills, whereas models such as ANSWERS, CASC2D, and GSSHA lump rill and interrill erosion into a single process. However, if success in watershed-scale erosion modeling is defined by accuracy in prediction of sediment discharge at a watershed outlet, the following general comments may be made.

Watershed-scale erosion models tend to be less accurate for event-scale prediction of sediment yield than for average soil loss per year, per month, or over a number of events. It is likely that spatial variability and the random nature of the erosion process are at least partially responsible for inaccuracies on small time scales. Over longer time scales, these effects tend to average out; hence, the increased accuracy in model prediction over longer time periods. There is a tendency for models to overpredict erosion for small runoff conditions and underpredict erosion for larger runoff conditions (Nearing, 1998). This is the case with the USLE, RUSLE, WEPP, and several other models. However, Ogden and Heilig (2001) report overprediction on large events for CASC2D.

Many models utilize a simple relation between soil particle detachment, interrill erosion, and rainfall intensity or kinetic energy of rainfall, $D_r = f(i)$, where D_r is the rate of soil detached by rainfall (kg/m^2/s) and i is rainfall intensity (mm/hr). Parsons and Gadian (2000) question the validity of such a simple relation and point out that lack of a clear-cut relation brings much uncertainty into modeling soil erosion. A great deal of error may be introduced into a model as a result. For instance, Daraio (2002) introduced a simple relation between kinetic energy of rainfall and soil particle detachment to GSSHA. Some improvement in model performance was seen on smaller scales in dynamic modeling of erosion, but the model performed better without the rainfall detachment term on larger spatial scales. There is a need to better understand the relationship between rainfall intensity, raindrop size distribution, kinetic energy of rainfall, and soil erosion and incorporate this understanding into erosion models.

Two-dimensional models provide a more accurate prediction of spatial distributions of sediment concentrations than one-dimensional models, but there is little difference between one- and two-dimensional models at predicting total sediment yield at a defined outlet (Hong and Mostaghimi, 1997). This is expected, given the success of lumped empirical models at sediment yield prediction. The complexity of flow on overland surfaces and the redistribution of sediment that occurs in such a flow regime can be more accurately modeled in two dimensions than in one dimension.

Understanding and predicting redistribution of sediment through detachment and deposition that results from variations in micro-topography on upland surfaces represents a major challenge in erosion modeling. There is also a need to better understand the relationship between interrill and rill flow; i.e., what is the relative contribution of sediment from interrill areas (raindrop impact) relative to rill areas. For instance, Ziegler et al. (2000) found that raindrop impact contributed from 38-45% of total sediment from erosion on unpaved roads. The application of this result to upland erosion is not clear, and there is a lack of information on this topic in the literature. These general deficiencies must be remedied in order to meet the need for more accurate erosion modeling.

Only a few erosion models have been developed for the purpose of dynamically simulating suspended sediment concentrations and to estimate Total Maximum Daily Load (TMDL) of sediment (see Section 2.4.1) in watersheds. One such model is CASC2D. The soil erosion component of CASC2D has been developed for the purpose of dynamically simulating suspended sediment concentrations with the aim of assessing the TMDL of sediment (Ogden and Heilig, 2001). It uses modifications to the semi-empirical Kilinc and Richardson (1973) equation that estimates sediment yield as a function of the unit discharge of water and the slope of the land surface. This function is further modified by three of the six parameters from the empirical USLE. The relation is given by the following equation (Johnson et al., 2000):

$$q_s = 25500 q^{2.035} S_f^{1.664} \left(\frac{KCP}{0.15} \right) \tag{2.36}$$

where q_s = sediment unit discharge (tons/m/s),
q = unit discharge of overland flow (m^2/s) (calculated within the overland flow component of CASC2D),
S_f = friction slope, and
K ,C, and P = USLE parameters shown in Equation (2.1).

The factors K, C, and P are calibrated with constraints determined by values reported in the literature; e.g., as found in the RUSLE Handbook (Renard et al., 1996). These empirical factors have been derived as representing annual averages of soil loss, and use of them in an event-based dynamic model is problematic.

While the hydrologic component of CASC2D performs very well (Senarath et al., 2000), the overall performance of the erosion component of CASC2D is poor and there are several formulation areas in need of improvement (Ogden and Heilig, 2001). That is, major changes in the method of development are needed, such as using a purely process-based equation, rather than a semi-empirical equation, to simulate erosion. The sediment volume is underestimated by the model by up to 85%, and peak discharge is underestimated by an order of magnitude on internal sub-basins for the calibration event. Simulated sediment volume on a non-calibration event varied from 7 to 77% of observed volumes, and peak sediment discharge varied from 37 to 88% of observed values. The model grossly overestimated sediment yield on a heavy rainfall event, up to 360% error. The model does not reliably estimate sediment yield, nor does it dynamically model soil erosion accurately. GSSHA has been developed directly from CASC2D, and the erosion component in GSSHA is identical to the one in CASC2D. The preliminary indication, based upon an attempt to improve the erosion modeling capabilities of GSSHA (Daraio, 2002), is that Equation (2.36) is not a good predictor of erosion rates. It is likely that a new erosion algorithm and a new set of equations are needed to improve the model, including the addition of rill modeling capabilities. Currently, the erosion component of GSSHA and CASC2D is in its development phase and should not be used as a tool for determining the TMDL of sediment.

The GSTARS 2.1 and GSTARS3 models (Yang and Simões, 2000, 2002) were developed to simulate and predict river morphological changes as a result of human activities or natural events (see Section 2.4.2). The GSTARS models have broad capabilities and have had success in modeling sediment transport and deposition within rivers, lakes, and reservoirs. The inclusion of upland erosion capabilities has been proposed to be added to the GSTARS models. The addition of upland erosion capabilities to the GSTARS models would represent a comprehensive watershed model (GSTAR-W) that utilizes a systematic, consistent, and well-proven theoretical approach. The GSTARS models would apply the unit stream power theory (Yang, 1973, 1979) and the minimum energy dissipation rate theory (Yang and Song, 1987) towards modeling soil erosion resulting from rainfall and runoff on land surfaces.

Sediment yield from upland areas has been shown to be strongly related to unit stream power (Yang, 1996). Sediment concentrations in overland flow also show a good relationship with unit stream power (Nearing et al., 1997). Additionally, unit stream power has been shown to be superior to other relations at predicting erosion of loose sediment on soils over a wide variety of conditions (Nearing et al., 1997; Yang, 1996; Hairsine and Rose, 1992; Govers and Rauws, 1986; Moore and Burch, 1986). The ICOLD Sedimentation Committee report also confirmed that unit stream power is a good parameter for sedimentation studies.

While the erosion component of GSTAR-W is in its early stages of development, the fact that the model was developed as a process-based model to simulate sediment transport and river morphology gives it a great advantage over empirical and semi-empirical soil erosion models. Additionally, the integrated approach being taken in developing the erosion modeling capabilities of GSTAR-W is much more promising than current process-based approaches that have met with limited success.

In addition to the need for continued model development, there are some inherent difficulties to erosion modeling. Physically based models tend to require a relatively large number of calibrated parameters. This creates the need for good quality data sets, and also sets further limits on the applicability of such models. That is, it is not advisable to use a model in a watershed that does not have the requisite data. The most important parameters for process-based models are rainfall parameters (e.g., duration, intensity) and infiltration parameters (e.g., hydraulic conductivity). Poor quality input data can lead to large errors in erosion modeling. Additionally, soil erosion models are built upon the framework of hydrologic models that simulate the rainfall-runoff process. Any error that exists in the hydrologic model will be propagated with the error from the soil erosion model. However the error introduced from the simulated runoff is generally much less than the error from the simulation of erosion (Wu et al., 1993).

Due to the complexity of the surface erosion process, computer models are needed for the simulation of the process and the estimation of the surface erosion rate. The need for the determination of TMDL of sediment in a watershed also requires a process-based comprehensive computer model. The following five sections will describe the approaches used for developing a comprehensive, systematic, dynamic, and process-based model (Yang, 2002).

2.4.1 Total Maximum Daily Load of Sediment

The 1977 Clean Water Act (CWA) passed by the United States Congress sets goals and water quality standards (WQS) to "restore and maintain the chemical, physical, and biological integrity of the Nation's waters." The CWA also requires states, territories, and authorized tribes to develop lists of impaired waters. These impaired waters do not meet the WQS that states have set for them, even after point sources of pollution have installed the required level of pollution control technology. The law requires that states establish priority rankings for waters on the list and develop TMDLs for these waters. A TMDL specifies the maximum amount of point and non-point pollutant that a water body can receive and still meet the water quality standard. By law, the Environmental Protection Agency (EPA) must approve or disapprove state lists and TMDLs. If a submission is inadequate, the EPA must establish the list or the TMDL (U.S. Environmental Protection Agency, 2001).

A TMDL consists of three elements—total point source waste loads, total non-point source loads, and a margin of safety to account for the uncertainty of the technology needed for the determination of allowable loads. TMDLs are a form of pollution budget for pollutant allocations in a watershed. In the determination of TMDLs, seasonal and spatial variations must be taken into consideration. The EPA is under court order or consent decrees in many states to ensure that TMDLs are established by either the state or the EPA.

Table 2.11 is a U.S. Environmental Protection Agency (2000) list of causes of impairments by pollutants. Sediment is clearly the number one pollutant that causes water to be impaired. It should be noted that sediment type impacts have been combined with siltation, turbidity, suspended solids, etc.

Table 2.11. Causes of impairments

Pollutant	Number of times named as cause
Sediments	6,502
Nutrients	5,730
Pathogens	4,884
Metals	4,022
Dissolved oxygen	3,889
Other habitat alterations	2,163
pH	1,774
Temperature	1,752
Biologic impairment	1,331
Fish consumption advisories	1,247
Flow alterations	1,240
Pesticides	1,097
Ammonia	781
Legacy	546
Unknown	527
Organic	464

Non-point source pollution is the largest source of water pollution problems. It is the main reason that 40 percent of the assessed water bodies in the United States are unsafe for basic uses such as fishing or swimming. Most sediment in rivers, lakes, reservoirs, wetlands, and estuaries come from surface erosion in watersheds and bank erosion along rivers as non-point source pollutants.

When a tributary with heavy sediment load meets the main stem of a river, the sediment load from the tributary can be treated as a point source of input to the main stem. Similarly, sediments caused by landslides or produced at a construction site can also be treated as a point source of input to a stream. A comprehensive approach for the determination of TMDL of sediment from point sources and non-point sources should be an integrated approach for the whole river basin or watershed under consideration.

Sediments can be divided into fine and coarse. Rivers transport fine sediments mainly as suspended load. Fine sediments often carry various forms of agrochemical and other pollutants. Consequently, fine sediments can have significant impacts on water quality. Rivers transport coarse sediments mainly as bedload. A good quality of coarse sediments or gravel is essential for fish spawning. A comprehensive model for sediment TMDL should have the capabilities of integrating watershed sheet, rill, and gully erosion; sediment transport, scour and deposition in tributaries and rivers; and, finally, sediment deposition in lakes, reservoirs, wetlands, or at sea. It should be a process-oriented model based on sound theories and engineering practice in hydrology, hydraulics, and sediment transport. The model should be applicable to a wide range of graded materials with hydraulic conditions ranging from subcritical to supercritical flows. A geographic information system (GIS) or other technology should be used to minimize the need of field data for model calibration and application.

Different authors have proposed different sediment transport formulas. Sediment transport concentrations or loads computed by different formulas for a given river may differ significantly from each other and from measurements. Chapters 3 and 4 address the subjects of sediment transport for non-cohesive and cohesive materials, respectively.

2.4.2 Generalized Sediment Transport Model for Alluvial River Simulation (GSTARS)

The sediment concentration or load computed by a formula is the equilibrium sediment transport rate without scour nor deposition. Natural rivers constantly adjust their channel geometry, slope, and pattern in response to changing hydrologic, hydraulic, and geologic conditions and human activities to maintain dynamic equilibrium. To simulate and predict this type of dynamic adjustment, a sediment routing model is needed. An example of this type of model is the Reclamation's GSTARS 2.1 model (Yang and Simões, 2000). GSTARS 2.1 uses the stream tube concept in conjunction with the theory of minimum energy dissipation rate, or its simplified theory of minimum stream power, to simulate and predict the dynamic adjustments of channel geometry and profile in a semi-three-dimensional manner.

Figure 2.14 demonstrates the capability of GSTARS 2.1 to simulate and predict the dynamic adjustments of channel width, depth, and shape downstream of the unlined emergency spillway of Lake Mescalero in New Mexico. Figure 2.14 shows that the predicted results with optimization based on the theory of minimum stream power can more accurately simulate and predict the dynamic adjustments of channel shape and geometry than the simulation without the optimization options. Figure 2.14 also shows that the process of channel bank erosion can be simulated and predicted fairly accurately.

GSTARS3 (Yang and Simões, 2002) is an enhanced version of GSTARS 2.1 to simulate and predict the sedimentation processes in lakes and reservoirs. It can simulate and predict the formation and development of deltas, sedimentation consolidation, and changes of reservoir bed profiles as a result of sediment inflow in conjunction with reservoir operation.

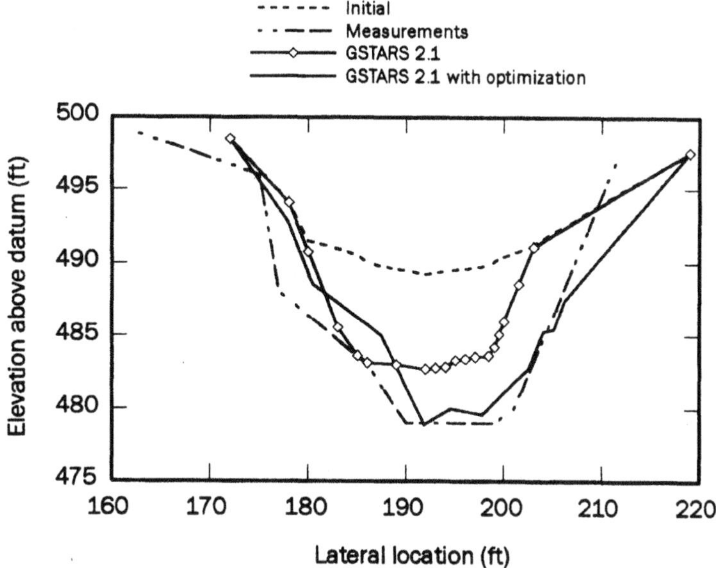

Figure 2.14. Comparison of results produced by GSTARS 2.1 and survey data for runs with and without width changes, due to stream power minimization (Yang and Simões, 2000).

Figure 2.15 shows an example of comparison between the predicted and observed delta formation (Swamee, 1974) in a laboratory flume. Figure 2.16 shows a comparison between the measured and simulated bed profiles using GSTARS3 for Tarbela Reservoir. GSTARS 2.1 and GSTARS3 enable us to simulate and predict the evolution of a river system with sediment from a tributary as a point source of sediment input and bank and bed erosion along a river reach as non-point source inputs to a river system. Reservoirs, lakes, and wetlands in a watershed can be considered as sinks for sediments.

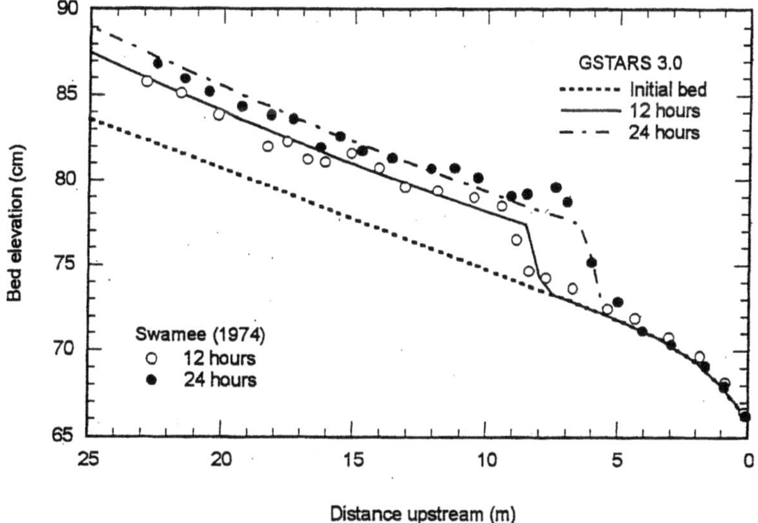

Figure 2.15. Comparison of experiments with simulations of reservoir delta development for two time instants (Yang and Simões, 2002).

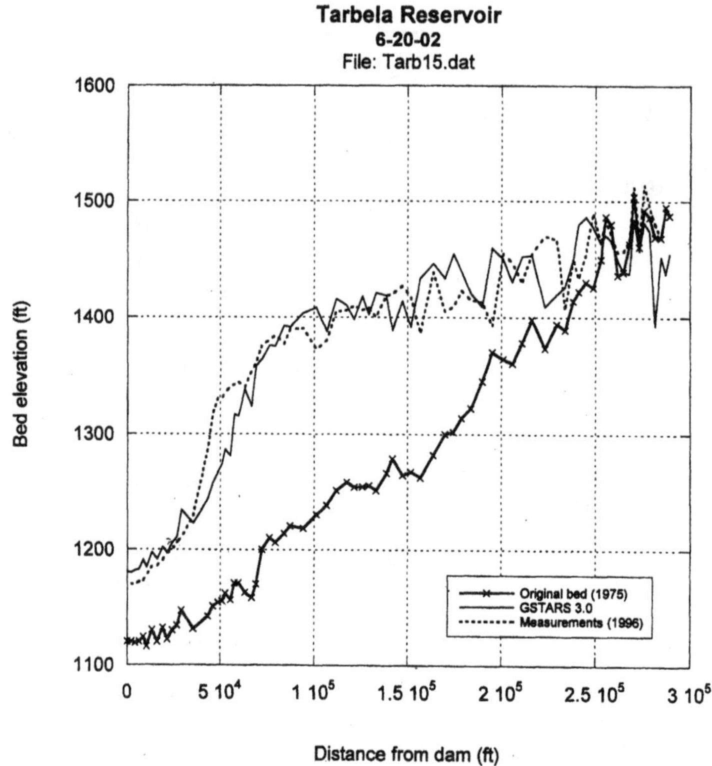

Figure 2.16. Comparison between measured and simulated bed profiles by GSTARS3 for the Tarbela Reservoir in Pakistan (Yang and Simões, 2001).

2.4.3 Rainfall-Runoff Relationship

Rainfall intensity, duration, and distribution in a watershed with given geologic and surface cover conditions will determine the surface runoff. Once the surface runoff is given, sheet, rill, and gully erosion rate of a watershed can be computed. *Computer Models of Watershed Hydrology* (Singh, 1995) summarizes some of the rainfall-runoff models. Some of these models also have certain abilities to simulate sheet erosion rates of a watershed. However, none of the existing models are based on a unified approach for the determination of erosion, sediment transport, and deposition in a watershed as described in this chapter. These models include, but are not limited to, the Precipitation-Runoff Modeling System (PRMS) by Leavesley et al. (1983) and the Hydrological Simulation Program—FORTRAN (HSPF) by Johanson et al. (1984). These models are modular, interactive programs. Input data include meteorologic, hydrologic, snow, and watershed descriptions. The outputs are runoff hydrographs, including maximum discharge, flow volume, and flow duration. Figure 2.17 is a schematic diagram of the PRMS model. The output information of these types of models can be used as part of the input information needed for a river sediment routing model such as Reclamation's GSTARS 2.1 and GSTARS3 computer models. Due to the complexity of sheet, rill, and gully erosion, a new model GSTAR-W needs to be developed and tested.

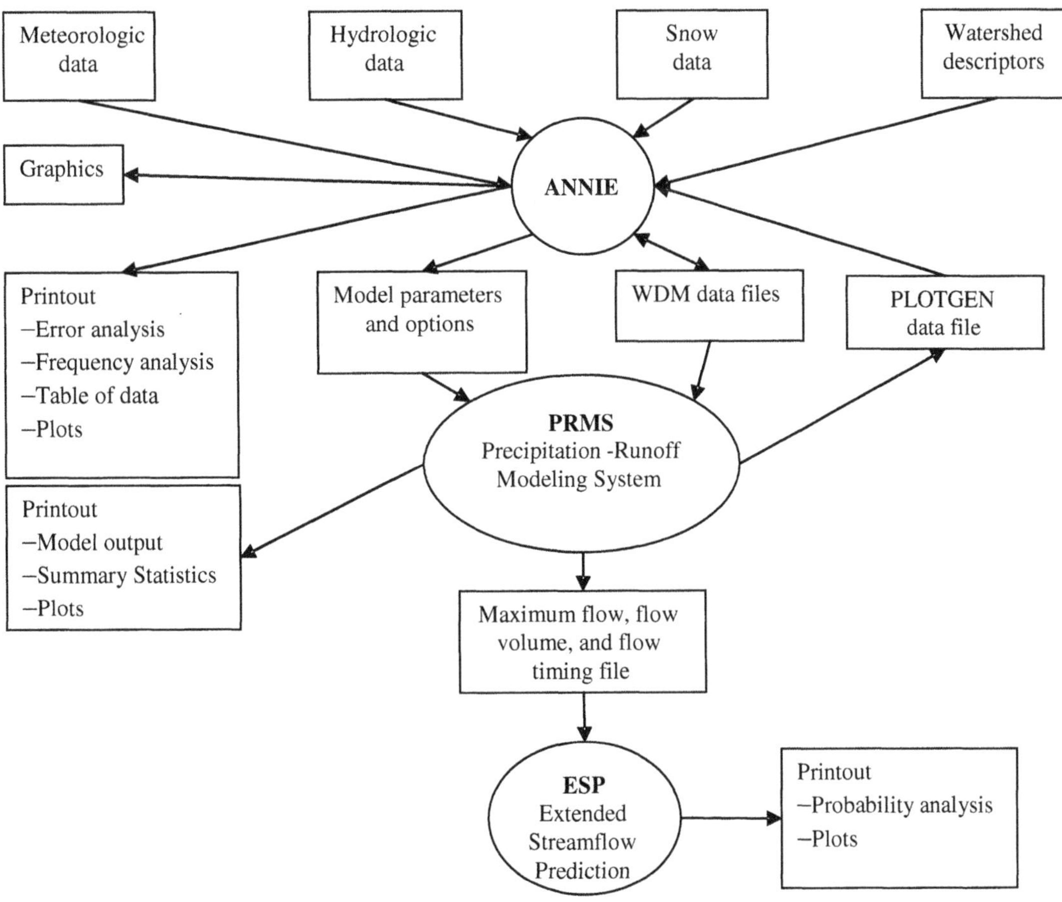

Figure 2.17. Flowchart of the PRMS model (Leavesley et al., 1993).

2.4.4 GSTAR-W Model

The loss of topsoil due to surface erosion not only can cause environmental problems, but it can also have adverse impacts on the agricultural productivity of a watershed. The United Nations Atomic Energy Agency has organized a 5-year international effort to determine surface erosion using radio isotopes as tracers. China has selected one watershed to test this fingerprinting technology. Reclamation will use the field data collected under different hydrologic, geologic, topographic, and sediment conditions for the calibration of GSTAR-W. Field data on rainfall-runoff relationships exist in the literature and will also be used for the calibration of GSTAR-W. GIS and other technology will be used to collect information on watershed topography, ground cover, and land use. With the calibrated and verified GSTAR-W and the already tested GSTARS 2.1 and GSTARS3 models, we can simulate and predict the sheet, rill, and gully erosion of a watershed as well as the river morphologic processes of bank and bed erosion, sediment transport, and depositions in rivers, lakes, reservoirs, and wetlands in a given watershed.

The EPA and other agencies can also use these integrated models to assess the impacts on TMDL of sediment due to a change of land use or other human activities. These models can become useful management tools for the selection of an optimum plan of action and the allocation of sediment TMDLs. Figure 2.18 is a decisionmaking flowchart of the integrated processes. It should be pointed out that computed hydraulic parameters from this integrated model can also be used for the determination of TMDLs of other pollutants in a watershed.

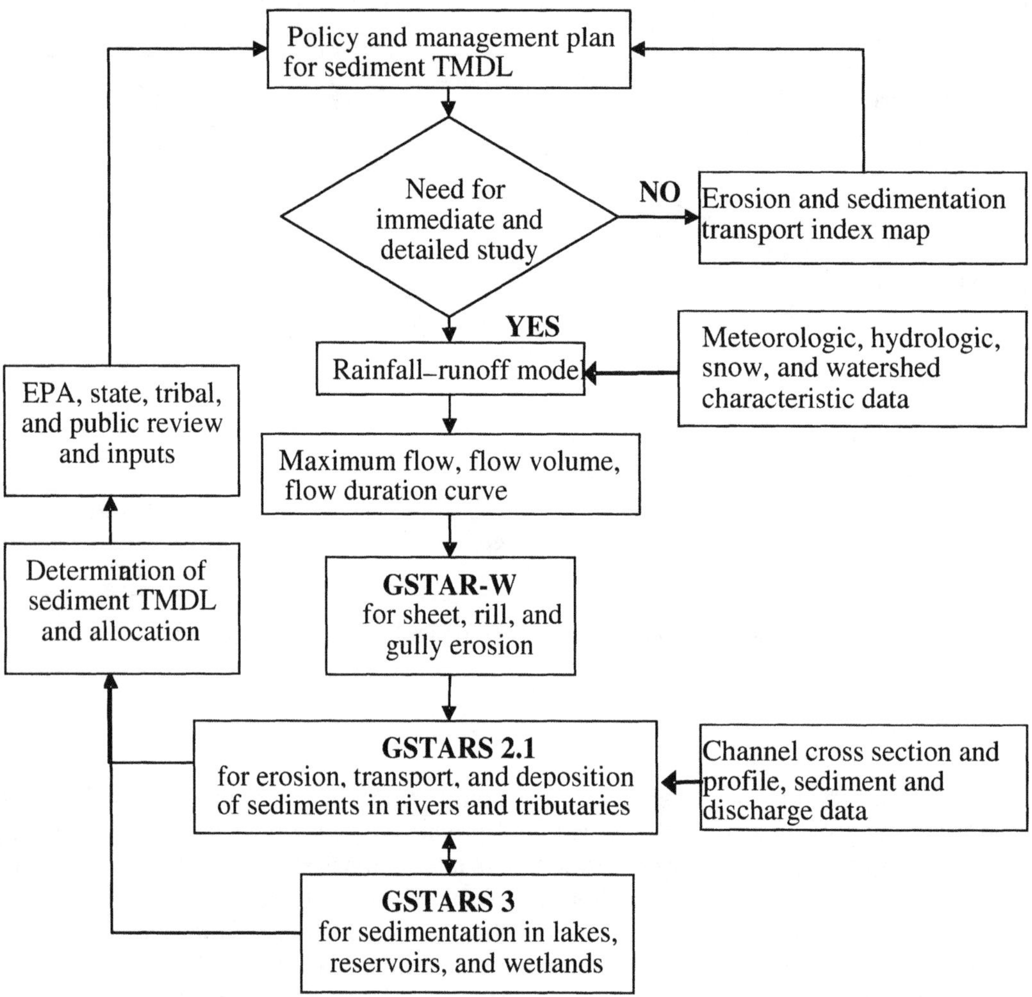

Figure 2.18. Sediment TMDL study and decision making flowchart.

2.4.5 Erosion Index Map

The amount of work and associated costs to determine TMDL of sediment are huge and will not be accomplished in a few years. A need exists to develop erosion index maps to identify areas that need immediate attention and possible remedial measures. It has been shown that the dimensionless unit stream power VS/ω is the most important parameter for the determination of erosion and sediment transport rates. GIS or topographic map information can be used for the estimation of slope S of a watershed. The average velocity V can be computed from Manning's formula for a given discharge Q, sediment size d, and surface roughness n. The fall velocity ω is proportional to the square root of particle diameter d. Thus, a preliminary estimation of VS/ω can be made, and erosion index maps can be developed based on the distribution of VS/ω. These maps may need to be modified with ground cover data which can also be obtained from GIS and other sources of information. Thus, a modified erosion index GVS/ω should be used. G is the ground cover factor with a value between 0 for paved surfaces and 1 for surfaces with no ground cover.

2.5 Example Case Studies

The methods described in this chapter are applied to an example case study in southwestern Arizona, where three small reservoirs have been proposed. The volume of sediment that would be expected to settle in these reservoirs over a 100-year period is computed for these examples. The study area is near the downstream end of the Colorado River basin in one of the driest desert regions of North America (Olmstead et al., 1973). The frequency of rainfall events that actually produce runoff is expected to be one to two times per year.

The first reservoir would have a storage capacity of 16,400 acre-feet and an average depth over 40 feet. The second reservoir would have a storage capacity of 8,700 acre-feet and an average depth of 21 feet. The third reservoir would have a storage capacity of 11,000 acre-feet and an average depth of 12 feet. Three separate drainage areas have been delineated for the first two reservoirs, and two separate drainage areas have been identified for the third reservoir.

2.5.1 Drainage Area Descriptions

The drainage areas for the proposed reservoir sites are desert foothills with steep to very steep terrain. The surface topography is composed of jagged rock, gravel, and sand. Little vegetation grows in these basins, except for sparsely spaced desert brush and clumps of grass. Stream bottoms are steep and sandy. The drainage areas are capable of producing flash flood conditions and large sediment volumes in the event of intense rainfall. Table 2.12 summarizes drainage basin characteristics for each of the proposed reservoir sites.

Since the reservoirs would normally be operated to completely contain runoff from local storms, they are expected to contain all of the sediment. Therefore, the trap efficiency for each reservoir is assumed to be 100 percent.

Table 2.12. Reservoir drainage basin characteristics

Proposed reservoir	Drainage basin	Drainage area (mi^2)	Drainage length (mi)	Average length (mi)	Drainage slope (%)
Reservoir 1	Unnamed Wash East	8.5	6.77	1.26	2.08
	Mission Wash	7.1	7.20	0.99	2.92
	Mission Wash East	2.2	1.50	1.47	6.00
Reservoir 2	Picacho Wash	43.7	16.20	2.70	0.90
	Unnamed Wash	30.2	12.60	2.40	1.39
	Picacho Wash East	2.0	3.44	0.58	1.43
Reservoir 3	Upper drainage area	9.72	2.86	3.40	2.67
	Lower drainage area	10.1	4.31	2.34	2.25

2.5.2 Example Computations of Sediment Yield

No stream gauges of flow or sediment load exist on any of the drainage areas, so the 100-year sediment yield cannot be based on direct measurements. The following four methods were used to estimate the amount of sediment inflow to the proposed reservoirs after 100 years (Randle, 1998):

- The Revised Universal Soil Loss Equation.
- Empirical equations that predict sediment yields as a function of drainage area.
- A sediment yield classification procedure that predicts sediment yields as a function of nine drainage basin characteristics.
- Unit stream power theory for sheet erosion that predicts sediment yields as a function of the runoff rate, velocity, drainage slope, and sediment particle characteristics.

2.5.3 Example Based on the RUSLE

Figure 2.4 gives a R_R value for southeastern Arizona that varies from 10 to 30. Some supplemental information is needed on the soil type and cover of the area, so a site assessment would be necessary before the RUSLE can be applied. The following is assumed to be the case for all sites. The soils are sandy with little or no structure, 80% silt and sand, no clay, and no organic matter. Assuming slow to moderate permeability with 20% coarse fragments > 3 inches and 80% less than 2 mm, calculation of K_R using the RUSLE program yields a value of 0.65. The area is a desert shrub habitat with approximately 10% canopy cover and 10% rock and residue cover; calculation of C_R using the RUSLE program yields a value of 0.11 because the area is highly susceptible to rill erosion. Table 2.13 is used to estimate the LS_R factor. The LS_R values shown in Table 2.14 were obtained by assuming the maximum slope length. The P_R factor is equal to 1 because no management practices are used. The sediment yield results are shown in Table 2.14.

Table 2.13. Values of the topographic LS_R factor for slopes with a high ratio of rill to interrill erosion (Renard et al., 1996)

	Horizontal slope length (ft)											
Slope (%)	25	50	75	100	150	200	250	300	400	600	800	1000
0.2	0.05	0.05	0.05	0.05	0.05	0.06	0.06	0.06	0.06	0.06	0.06	0.06
0.5	0.07	0.08	0.08	0.09	0.09	0.10	0.10	0.10	0.11	0.12	0.12	0.13
1.0	0.10	0.13	0.14	0.15	0.17	0.18	0.19	0.20	0.22	0.24	0.26	0.27
2.0	0.16	0.21	0.25	0.28	0.33	0.37	0.40	0.43	0.48	0.56	0.63	0.69
3.0	0.21	0.30	0.36	0.41	0.50	0.57	0.64	0.69	0.80	0.96	1.10	1.23
4.0	0.26	0.38	0.47	0.55	0.68	0.79	0.89	0.98	1.14	1.42	1.65	1.86
5.0	0.31	0.46	0.58	0.68	0.86	1.02	1.16	1.28	1.51	1.91	2.25	2.55
6.0	0.36	0.54	0.69	0.82	1.05	1.25	1.43	1.60	1.90	2.43	2.89	2.30
8.0	0.45	0.70	0.91	1.10	1.43	1.72	1.99	2.24	2.70	3.52	4.24	4.91
10.0	0.57	0.91	1.20	1.46	1.92	2.34	2.72	3.09	3.75	4.95	6.03	7.02
12.0	0.71	1.15	1.54	1.88	2.51	3.07	3.60	4.09	5.01	6.67	8.17	9.57
14.0	0.85	1.40	1.87	2.31	3.09	3.81	4.48	5.11	6.30	8.45	10.40	12.23
16.0	0.98	1.64	2.21	2.73	3.68	4.56	5.37	6.15	7.60	10.26	12.69	14.96
20.0	1.24	2.10	2.86	3.57	4.85	6.04	7.16	8.23	10.24	13.94	17.35	20.57
25.0	1.56	2.67	3.67	4.59	6.30	7.88	9.38	10.81	13.53	18.57	23.24	27.66
30.0	1.86	3.22	4.44	5.58	7.70	9.67	11.55	13.35	16.77	23.14	29.07	34.71
40.0	2.41	4.24	5.89	7.44	10.35	13.07	15.67	18.17	22.95	31.89	40.29	48.29
50.0	2.91	5.16	7.20	9.13	12.75	16.16	19.42	22.57	28.60	39.95	50.63	60.84
60.0	3.36	5.97	8.37	10.63	14.89	18.92	22.78	26.51	33.67	47.18	59.93	72.15

Such as for freshly prepared construction and other highly disturbed soil condition with little or no cover (not applicable to thawing soil).

Table 2.14. Sediment yield estimates based on the empirical RUSLE

				$R_R = 10$		$R_R = 30$	
Proposed Reservoir	Drainage Basin	LS_R	Area (mi^2)	Average annual sediment yield (acre-ft/mi^2)	100-year sediment yield (acre-ft)	Average annual sediment yield (acre-ft/mi^2)	100-year sediment yield (acre-ft)
Reservoir 1	Unnamed Wash East	0.69	8.5	0.24	206	0.73	619
	Mission Wash	1.23	7.1	0.43	307	1.30	922
	Mission Wash East	3.3	2.2	1.16	256	3.48	767
	Total		17.8		769		2,308
Reservoir 2	Picacho Wash	0.24	43.7	0.08	369	0.25	1,108
	Unnamed Wash	0.43	30.2	0.15	457	0.45	1,371
	Picacho Wash East	0.44	2.0	0.15	31	0.46	93
	Total		75.9		857		2,572
Reservoir 3	Upper drainage area	1	9.72	0.35	342	1.06	1,026
	Lower drainage area	0.85	10.1	0.30	302	0.90	907
	Total		19.82		644		1,933

2.5.4 Example Based on Drainage Area

The empirical sediment yield equations for Arizona, New Mexico, and California (Equation 2.15) and for the semiarid climate of the southwestern United States (Equation 2.16) were applied to the drainage areas listed in Table 2.12. Table 2.15 presents the predicted sediment yields, using both of these equations, for each of the proposed reservoirs.

Table 2.15. Sediment-yield estimates based on empirical equations as a function of drainage area

			Equation for New Mexico, Arizona, and California $Q_s = 2.4\,A^{-0.229}$		Equation for the southwestern United Sates $Q_s = 1.84\,A^{-0.24}$	
Proposed reservoir	Drainage Basin	Area (mi^2)	Average annual sediment yield (acre-ft/mi^2)	100-year sediment yield (acre-ft)	Average annual sediment yield (acre-ft/mi^2)	100-year sediment yield (acre-ft)
Reservoir 1	Unnamed Wash East	8.5	1.47	1,250	1.10	936
	Mission Wash	7.1	1.53	1,090	1.15	816
	Mission Wash East	2.2	2.00	440	1.52	335
	Total	17.8		3,000		2,000
Reservoir 2	Picacho Wash	43.7	1.01	4,420	0.74	3,250
	Unnamed Wash	30.2	1.10	3,320	0.81	2,450
	Picacho Wash East	2.0	2.05	410	1.56	310
	Total	75.9		8,000		6,000
Reservoir 3	Upper drainage area	9.72	1.42	1,380	1.07	1,030
	Lower drainage area	10.1	1.41	1,430	1.06	1,070
	Total	19.8		3,000		2,000

These 100-year sediment volume estimates are computed only as a function of drainage area and do not consider site-specific characteristics of the drainage basin or individual runoff events. The sediment yields from the two equations are different but they are of the same order of magnitude. For the second reservoir, the sediment yield estimates are 70 to 90 percent of the proposed reservoir storage capacity. This would indicate that more detailed investigations are warranted.

2.5.5 Example Based on the Sediment Yield Classification Procedure

The sediment yield classification procedures presented in Tables 2.9 and 2.10 were applied to all of the proposed reservoir drainage basins as a whole. All ratings were based on field inspection. Table 2.16 presents the estimated ratings for each drainage basin characteristic. These drainage basins are a class 3 sediment yield based on the ratings for each of the nine drainage basin characteristics. Applying an annual sediment yield of 0.5 to 1.0 acre-ft/mi^2 to each of the separate drainage basins provides an estimate of the 100-year sediment volume (Table 2.17). The ranges of sediment yields computed using this method tend to be less than those computed as a function of drainage area, but they are of the same order of magnitude.

Table 2.16. Estimated numerical ratings for the proposed reservoir drainage basins

Drainage basin characteristics	Possible ratings		Estimated sediment yield rating	Estimate description
Surface geology	0 - 10	5	Moderate	Varies from hard, dense crystalline rocks to unconsolidated alluvium and windblown sand.
Soils	0 - 10	0	Low	Surface material is sand, rock fragments, and bedrock outcrops.
Climate	0 - 10	0	Low	Arid climate with rare convective storms.
Runoff	0 - 10	0	Low	On average, only 1 to 2 storms per year that produce runoff.
Topography	0 - 20	20	High	Desert foothill terrain that is steep to very steep and dissected Piedmont slopes.
Ground cover	-10 - 10	10	High	Little vegetation, except for sparsely spaced desert brush and grass.
Land use	-10 – 10	-10	Low	There is no cultivation or grazing.
Upland erosion	0 – 25	10	Moderate	Upland mountains and hills are composed of older, more consolidated rocks and are estimated to have moderate erosion rates.
Channel erosion	0 – 25	25	High	Erosion on the dissected Piedmont slopes is the dominant process today and has been for the past several thousand years. Desert pavement is generally conspicuous.
Total rating		60	Class 3	0.5 to 1.0 ac-ft/mi^2/yr.

Table 2.17. 100-year sediment yield estimates

Proposed reservoir	Drainage basin	Area (mi^2)	100-year sediment yield (ac-ft)	
			at 0.5 ac-ft/mi^2/yr	at 1.0 ac-ft/mi^2/yr
Reservoir 1	Unnamed Wash East	8.5	425	850
	Mission Wash	7.1	355	710
	Mission Wash East	2.2	110	220
	Total	17.8	900	2,000
Reservoir 2	Picacho Wash	43.7	2,185	4,370
	Unnamed Wash	30.2	1,510	3,020
	Picacho Wash East	2.0	100	200
	Total	75.9	4,000	8,000
Reservoir 3	Upper drainage area	9.7	485	970
	Lower drainage area	10.1	505	1,010
	Total	19.8	1,000	2,000

2.5.6 Example Based on Unit Stream Power

The physically based equation for sheet erosion (Yang, 1996; Moore and Burch, 1986) was combined with a flood hydrology analysis to compute the 100-year sedimentation volume for each reservoir drainage basin. Required input for this procedure includes the following data:

- Flood hydrology analysis, including magnitude, duration, and frequency.
- Manning's roughness coefficient.
- Drainage slope and average width.
- Sediment particle characteristics including size, fall velocity, and incipient motion velocity.

2.5.6.1 Flood Hydrology

Flood hydrographs were computed for return periods of 100, 50, 25, 10, and 5 years. These hydrographs were determined from regional rainfall data because no stream gauge measurements existed. The flood hydrographs were computed with a 5-minute time step with total durations ranging from 2.3 to 14.1 hours. Table 2.18 lists the peak discharge values for each of these rainfall-runoff floods. Table 2.19 is an annual flow-duration table for the flood return periods of 100, 50, 25, 10, 5, 2, and 1 years.

Table 2.18. Peak discharge values computed for each drainage basin

Reservoir	Drainage basin	Peak flood discharge (ft^3/s)				
		100-year flood	50-year flood	25-year flood	10-year flood	5-year flood
Reservoir 1	Unnamed Wash East	3,840	2,830	2,540	1,130	652
	Mission Wash	3,410	2,510	1,770	1,010	579
	Mission Wash East	1,930	1,420	996	569	324
Reservoir 2	Picacho Wash	7,740	5,690	4,010	2,280	1,310
	Unnamed Wash	6,620	4,870	3,430	1,950	1,130
	Picacho Wash East	1,840	1,350	948	540	315
Reservoir 3	Upper drainage area	8,310	NA	NA	NA	NA
	Lower drainage area	4,460	3,280	2,310	1,320	755

On average, one to two rainfall events that produce runoff are expected to occur each year over the drainage basins of the proposed reservoirs. Over a 100-year period, this would amount to between 100 and 200 runoff events. From Table 2.19, the total number of annual floods is assumed to be 100 over a 100-year period, because this procedure only accounts for the largest flood from each year. For example, a 5-year flood would not be counted if it occurred in a year in which the peak discharge for that year was greater (i.e., had a return period of 10, 25, 50, or 100 years). Therefore, the annual series was transformed into a partial duration series using the method described by Linsley et al. (1975) and William Lane (Hydraulic Engineer, Reclamation, Denver, Colorado, personal communication). The partial series is made up of all floods above some selected base value. The base value is chosen so that not more than a certain number of floods N are included for each year. The partial series can then indicate the probability of floods being equaled or exceeded N times per year.

Table 2.19. Annual flow duration table

Flood return period	Number of times that the flood is equaled or exceeded during 100 years	Number of annual floods
100	1	1
50	2	1
25	4	2
10	10	6
5	20	10
2	50	30
1	100	50
Total		100

$$P_{annual} = 1/T_{partial} = 1 - (1 - P_{partial}/T_{annual})^{N} \qquad (2.37)$$

where P_{annual} = the annual probability of floods being equaled or exceeded once per year,
T_{annual} = the annual return period, in years, associated with the annual probability,
$P_{partial}$ = the probability of floods being equaled or exceeded N times per year, and
N = the number of floods per year.

Solving for $P_{partial}$, Equation (2.37) can be expressed as:

$$P_{partial} = 1/T_{partial} = N\,[1 - (1 - 1/T_{annual})^{1/N}] \qquad (2.38)$$

where $T_{partial}$ = the partial series return period, in years, associated with $P_{partial}$

The partial duration series was computed for a range of return periods, assuming no more than two floods per year for a total of 200 floods over a 100-year period (Table 2.20). The largest flood considered was that associated with the 200-year return period because, of all the floods exceeding the 100-year flood, half would be greater than the 200-year flood. From the partial duration series (last column of Table 2.20), the number of floods expected to occur during a 100-year period was used for the computation of the 100-year sediment volume.

Table 2.20. Transformation of annual flood series to a partial duration series, assuming no more than two floods per year

Annual peak flood series				Partial duration series		
Flood return period (yr)	Number of times exceeded in 100 yr	Number of floods in 100 yr	Flood return period (yr)	Flood return period (yr)	Number of times exceeded in 100 yr	Number of floods in 100 yr
>200.00	<0.500	1.000	200.00	99.505	<0.501	1.003
100.00	1.000	0.111	95.00	94.749	1.003	0.112
90.00	1.111	0.139	85.00	84.749	1.114	0.140
80.00	1.250	0.179	75.00	74.749	1.254	0.180
70.00	1.429	0.238	65.00	64.749	1.434	0.240
60.00	1.667	0.333	55.00	54.749	1.674	0.336
50.00	2.000	0.500	45.00	44.749	2.010	0.506
40.00	2.500	0.833	35.00	34.748	2.516	0.846
30.00	3.333	0.667	27.50	27.248	3.362	0.679
25.00	4.000	1.000	22.50	22.247	4.041	1.023
20.00	5.000	1.667	17.50	17.246	5.064	1.718
15.00	6.667	3.333	12.50	12.245	6.782	3.482
10.00	10.000	4.286	8.50	8.242	10.263	4.573
7.00	14.286	5.714	6.00	5.738	14.836	6.279
5.00	20.000	5.000	4.50	4.234	21.115	5.680
4.00	25.000	25.000	3.00	2.720	26.795	31.784
2.00	50.000	16.667	1.75	1.445	58.579	25.951
1.50	66.667	10.256	1.40	1.073	84.530	19.393
1.30	76.923	13.986	1.20	0.839	103.923	35.775
1.10	90.909	9.091	1.05	0.608	139.698	60.302
1.00	100.000	9.091	1.05	0.608	200.000	60.302
Total		100.00		Total		200.00

2.5.6.2 Application of the Sheet Erosion Equation

For each drainage basin, the sediment concentration was computed using Equation (2.18) for each 5-minute discharge of a given flood hydrograph. The sediment inflow volumes to each reservoir were computed for the 5-, 10-, 25-, 50-, and 100-year flood hydrographs. The sediment volumes corresponding to the return periods listed in Table 2.20 (5th column) were computed from a regression equation (specific to each drainage basin). The regression equations were computed from the logarithms of the sediment volumes (dependent variable) and the logarithms of the corresponding flood return periods (independent variable). The 100-year sediment volume was computed by accumulating the products of the sediment inflow volume, corresponding to a given range of floods, and the number of times floods in that range are expected to occur during a 100-year period (see Table 2.20, last column).

The Manning's *n* roughness coefficient in Equation (2.29) was assumed to be a constant of 0.030. Table 2.12 lists the drainage area and length, and slope *S* for each drainage basin. The drainage width was computed as the ratio of the drainage length to area. Sediment load was computed from each concentration, and a bulk density of 70 lbs/ft^3 was assumed to convert the sediment load to volume.

The soils of the drainage basins are primarily sand and coarser size material. The median sediment particle size was assumed to be within sand-size range (0.06 mm to 2.0 mm) but the size was not precisely known for any of the drainage basins. Therefore, 100-year sediment volumes were computed assuming a range of sand sizes. Table 2.21 lists the particle sizes and fall velocities used in the analysis.

Table 2.21. Sediment particle sizes and fall velocities (U.S. Committee on Water Resources, Subcommittee on Sedimentation, 1957)

Sediment particle size (mm)	Sediment particle fall velocity (cm/s)
0.06	0.25
0.1	0.60
0.2	1.81
0.5	5.72
1.0	11.4
2.0	19.0

2.5.6.3 Results

Tables 2.22 through 2.31 present summary results of the unit stream power procedure for each reservoir, drainage basin, flood, and assumed sediment particle size. Except for the 100-year flood, a flood- hydrology analysis was not completed for the upper drainage area.

The average 100-year sediment yield (per unit area) for the upper and lower drainage areas was assumed to be equal (Table 2.30). The 100-year sediment volume for the canal drainage area was computed by multiplying the average 100-year sediment yield per unit area by the canal drainage area of 9.72 mi^2 (see Table 2.31).

Sediment concentration was found to be sensitive to particle size. In some cases, the range of possible particle sizes could be reduced by examination of the computed peak sediment concentrations. A maximum concentration limit of 300,000 ppm was applied for the sand sizes of the study area. This is a reasonable limit, based on other streams, where long-term measurements exist. For example, the maximum mean daily concentration of record for the Rio Puerco near Bernardo, New Mexico (a major sediment-producing tributary of the Rio Grande) is 230,000 ppm.

Table 2.22. Reservoir 1, Unnamed Wash east drainage sediment yield estimates

Recurrence interval (yr)	Peak discharge (ft^3/s)	Sediment volume (ac-ft)					
		0.06 mm	0.1 mm	0.2 mm	0.5 mm	1 mm	2 mm
100	3,840	211.00	64.10	14.20	2.96	1.16	0.58
50	2,830	126.00	38.10	8.45	1.76	0.69	0.34
25	2,540	105.00	31.90	7.07	1.47	0.58	0.29
10	1,130	25.80	7.83	1.74	0.36	0.14	0.07
5	652	9.77	2.96	0.66	0.14	0.05	0.03
100-yr volume[1]			577.00	128.00	26.70	10.50	5.23

Recurrence interval (yr)	Peak discharge (ft^3/s)	Peak sediment concentration (ppm)					
		0.06 mm	0.1 mm	0.2 mm	0.5 mm	1 mm	2 mm
100	3,844	1,140,000	345,000	76,600	16,000	6,240	3,110
50	2,827	939,000	285,000	63,200	13,200	5,150	2,570
25	2,537	876,000	266,000	59,000	12,300	4,800	2,390
10	1,131	516,000	157,000	34,800	7,240	2,830	1,410
5	652	351,000	106,000	23,600	4,930	1,930	959

[1] See appendix for computation details.

Table 2.23. Reservoir 1, Mission Wash sediment yield estimates

Recurrence interval (yr)	Peak discharge (ft^3/s)	Sediment volume (ac-ft)					
		0.06 mm	0.1 mm	0.2 mm	0.5 mm	1 mm	2 mm
100	3,410	431.00	131.00	29.00	6.04	2.36	1.18
50	2,510	260.00	78.80	17.50	3.65	1.42	0.71
25	1,770	146.00	44.20	9.81	2.04	0.80	0.40
10	1,010	56.90	17.20	3.83	0.80	0.31	0.16
5	579	22.50	6.82	1.51	0.32	0.12	0.06
100-yr volume				256.00	53.30	20.80	10.40

Recurrence interval (yr)	Peak discharge (ft^3/s)	Peak sediment concentration (ppm)					
		0.06 mm	0.1 mm	0.2 mm	0.5 mm	1 mm	2 mm
100	3,410	2,350,000	713,000	158,000	33,000	12,900	6,420
50	2,510	1,960,000	595,000	132,000	27,500	10,800	5,360
25	1,770	1,590,000	482,000	107,000	22,300	8,720	4,340
10	1,010	1,130,000	341,000	75,800	15,800	6,170	3,080
5	579	795,000	241,000	53,500	11,200	4,360	2,170

Table 2.24. Reservoir 1, Mission Wash East sediment yield estimates

Recurrence interval (yr)	Peak discharge (ft³/s)	Sediment volume (ac-ft)					
		0.06 mm	0.1 mm	0.2 mm	0.5 mm	1 mm	2 mm
100	1,930	249.00	75.60	16.80	3.50	1.37	0.68
50	1,420	150.00	45.40	10.10	2.10	0.82	0.41
25	996	81.90	24.80	5.51	1.15	0.45	0.22
10	569	31.80	9.65	2.14	0.45	0.17	0.09
5	324	12.30	3.74	0.83	0.17	0.07	0.03
100-yr volume				144.00	30.00	11.70	5.85

Recurrence interval (yr)	Peak discharge (ft³/s)	Peak sediment concentration (ppm)					
		0.06 mm	0.1 mm	0.2 mm	0.5 mm	1 mm	2 mm
100	1,930	5,240,000	1,590,000	353,000	73,500	28,700	14,300
50	1,420	4,400,000	1,330,000	296,000	61,800	24,100	12,000
25	996	3,590,000	1,090,000	242,000	50,400	19,700	9,810
10	569	2,590,000	786,000	175,000	36,400	14,200	7,080
5	324	1,860,000	564,000	125,000	26,100	10,200	5,080

Table 2.25. Reservoir 1, total 100-year sediment volume

Total 100-year sediment volume (ac-ft)[1]					
0.06 mm	0.1 mm	0.2 mm	0.5 mm	1 mm	2 mm
1,000	1,000	500	100	40	20

[1] The total includes a sediment yield computed from a larger particle size. A larger particle size was used because the maximum probable concentration of 300,000 ppm was exceeded.

Table 2.26. Reservoir 2, Picacho Wash sediment yield estimates

Recurrence interval (yr)	Peak discharge (ft³/s)	Sediment volume (ac-ft)					
		0.06 mm	0.1 mm	0.2 mm	0.5 mm	1 mm	2 mm
100	7,740	101.00	30.70	6.82	1.42	0.56	0.28
50	5,690	54.10	16.40	3.64	0.76	0.30	0.15
25	4,010	25.20	7.64	1.70	0.35	0.14	0.07
10	2,280	6.29	1.91	0.42	0.09	0.03	0.02
5	1,310	1.05	0.32	0.07	0.01	0.01	0.00
100-yr volume		664.00	201.00	44.70	9.31	3.62	1.81

Table 2.26. Reservoir 2, Picacho Wash sediment yield estimates (continued)

Recurrence interval (yr)	Peak discharge (ft^3/s)	Peak sediment concentration (ppm)					
		0.06 mm	0.1 mm	0.2 mm	0.5 mm	1 mm	2 mm
100	7,740	156,000	47,300	10,500	2,190	855	426
50	5,690	118,000	35,900	7,970	1,660	648	323
25	4,010	83,500	25,300	5,620	1,170	458	228
10	2,280	42,500	12,900	2,860	597	233	116
5	1,310	16,400	4,990	1,110	231	90	45

Table 2.27. Reservoir 2, Unnamed Wash sediment yield estimates

Recurrence interval (yr)	Peak discharge (ft^3/s)	Sediment volume (ac-ft)					
		0.06 mm	0.1 mm	0.2 mm	0.5 mm	1 mm	2 mm
100	6,620	215.00	65.10	14.50	3.01	1.18	0.78
50	4,870	124.00	37.70	8.38	1.75	0.68	0.48
25	3,430	65.70	19.90	4.42	0.92	0.36	0.27
10	1,950	22.52	6.83	1.52	0.32	0.12	0.11
5	1,130	7.39	2.24	0.50	0.10	0.04	0.59
100-yr volume			484.00	107.00	22.40	8.76	4.36

Recurrence interval (yr)	Peak discharge (ft^3/s)	Peak sediment concentration (ppm)					
		0.06 mm	0.1 mm	0.2 mm	0.5 mm	1 mm	2 mm
100	6,620	452,000	137,000	30,400	6,340	2,480	1,240
50	4,870	364,000	110,000	24,500	5,110	2,000	995
25	3,430	282,000	85,400	19,000	3,950	1,540	769
10	1,950	180,000	54,600	12,100	2,520	986	492
5	1,130	110,000	33,500	7,440	1,550	605	302

Table 2.28. Reservoir 2, Picacho Wash East sediment yield estimates

Recurrence interval (yr)	Peak discharge (ft^3/s)	Sediment volume (ac-ft)					
		0.06 mm	0.1 mm	0.2 mm	0.5 mm	1 mm	2 mm
100	1,840	36.30	11.00	2.45	0.51	0.20	0.10
50	1,350	20.60	6.25	1.39	0.29	0.11	0.06
25	948	10.80	3.27	0.73	0.15	0.06	0.03
10	540	3.77	1.14	0.25	0.05	0.02	0.01
5	315	1.30	0.39	0.09	0.02	0.01	0.00
100-yr volume			81.40	18.10	3.76	1.47	0.74

Recurrence interval (yr)	Peak discharge (ft^3/s)	Peak sediment concentration (ppm)					
		0.06 mm	0.1 mm	0.2 mm	0.5 mm	1 mm	2 mm
100	1,840	529,000	160,000	35,600	7,420	2,900	1,450
50	1,350	428,000	130,000	28,800	6,010	2,350	1,170
25	948	333,000	101,000	22,400	4,670	1,820	909
10	540	217,000	65,800	14,600	3,040	1,190	593
5	315	138,000	41,800	9,270	1,930	755	376

Table 2.29. Reservoir 2, total 100-year sediment volume

Total 100-year sediment volume (ac-ft)[1]					
0.06 mm	0.1 mm	0.2 mm	0.5 mm	1 mm	2 mm
11,000	800	200	40	10	7

[1] The total includes a sediment yield computed from a larger particle size. A larger particle size was used, because the maximum probable concentration of 300,000 ppm was exceeded.

Table 2.30. Reservoir 3, reservoir drainage area sediment yield estimates

Recurrence interval (yr)	Peak discharge (ft^3/s)	Sediment volume (ac-ft)					
		0.06 mm	0.1 mm	0.2 mm	0.5 mm	1 mm	2 mm
100	4,460	192.00	58.30	12.90	2.70	1.05	0.53
50	3,280	115.00	34.80	7.73	1.61	0.63	0.31
25	2,310	62.70	19.00	4.22	0.88	0.34	0.17
10	1,320	23.50	7.14	1.58	0.33	0.13	0.06
5	755	8.63	2.62	0.58	0.12	0.05	0.02
100-yr volume			481.00	107.00	22.20	8.67	4.34
Average 100-yr sediment yield (ac-ft/mi2)			47.60	10.60	2.20	0.86	0.43

Table 2.30. Reservoir 3, reservoir drainage area sediment yield estimates (continued)

Recurrence interval	Peak discharge	Peak sediment concentration (ppm)					
(yr)	(ft³/s)	0.06 mm	0.1 mm	0.2 mm	0.5 mm	1 mm	2 mm
100	4,460	997,000	302,000	67,100	14,000	5,470	2,720
50	3,280	821,000	249,000	55,300	11,500	4,500	2,240
25	2,310	653,000	198,000	44,000	9,160	3,580	1,780
10	1,320	447,000	136,000	30,100	6,270	2,450	1,220
5	755	300,000	91,100	20,200	4,210	1,650	820

Table 2.31. Reservoir 3, total 100-year sediment volume

Drainage area	Total 100-year sediment volume[1] (ac-ft)					
	0.06 mm	0.1 mm	0.2 mm	0.5 mm	1 mm	2 mm
Upper drainage area		481	107	22.20	8.67	4.34
Lower drainage area		462	102	21.40	8.33	4.17
Total drainage area		900	200	40.00	20.00	9.00

[1] Computed by multiplying the average 100-year sediment yield per unit area (computed for the reservoir drainage area, see Table 2.24) by the canal drainage area of 9.72 mi^2.

2.5.7 Comparison of Different Approaches

Table 2.32 presents summary results from the three different methods. These results differ by two orders of magnitude for the low estimate and by up to one order of magnitude for the high estimate. Results from the RUSLE provided intermediate estimates as compared to the other empirical methods. The interpretation of the results from the RUSLE is complicated by the fact that the slope lengths in the basins are much greater than those used to determine the LS_R factor, and given the many assumptions made in determining the other factors.

Results from the empirical sediment yield equations consistently provide the largest estimates for the 100-year sediment volumes. These empirical equations are based on sedimentation measurements from reservoirs in Arizona, New Mexico, and California and from measurements of reservoirs throughout the southwestern United States. These reservoirs tend to be on drainage basins that have more annual rainfall than the proposed reservoirs in southwestern Arizona. Because the empirical equations are only a function of drainage area, they cannot take into account the drier and sandier conditions of the drainage areas. Therefore, the equations might be expected to overestimate the 100-year sediment yield.

Results from the sediment yield classification procedure provided the second highest sediment yield estimates. In a semi-quantitative fashion, this procedure takes into account many of the important variables affecting sediment yield from a drainage basin. The procedure is most

Table 2.32. Summary results of 100-year sediment volumes by four methods

Reservoir	Drainage area (mi ^{2})	Method	100-Year Sediment Volume (acre-ft) Low estimate	High estimate
Reservoir 1	17.8	Empirical RUSLE sediment estimate	800	2,300
		Empirical sediment yield equations	2,000	3,000
		Sediment yield classification procedure	900	2,000
		Unit stream power sheet erosion equation	20	1,000
Reservoir 2	75.9	Empirical RUSLE sediment estimate	900	2,600
		Empirical sediment yield equations	6,000	8,000
		Sediment yield classification procedure	4,000	8,000
		Unit stream power sheet erosion equation	7	1,000
Reservoir 3	19.8	Empirical RUSLE sediment estimate	600	1,900
		Empirical sediment yield equations	2,000	3,000
		Sediment yield classification procedure	1,000	2,000
		Unit stream power sheet erosion equation	9	900

sensitive to ratings for upland and channel erosion, topography, ground cover, and land use. In the case of the proposed reservoirs near Yuma, Arizona, the sediment yield ratings are high for the steep topography, sparse ground cover, and extensive erosion channel development. The ratings are low for the sandy soils and desert pavement, arid climate, and infrequent runoff. The procedure can be used to predict the relative difference in sediment yield between two or more drainage basins, but the procedure is still somewhat subjective when computing the actual sediment yield.

Results from the unit stream power sheet erosion equation provided the lowest sediment yield estimates. The sheet erosion equation accounts for the physical processes of erosion by taking into account the important variables of drainage slope, width, roughness, and sediment particle fall velocity and the runoff velocity, duration, and frequency. The drainage slope, S, runoff velocity, V, and sediment particle fall velocity, ω, are represented as dimensionless unit stream power (VS/ω), which Yang (1996) has shown to be applicable to a wide range of conditions. The sheet erosion equation is especially applicable to the drainage basins of the proposed reservoirs for the following reasons:

- The soils are mostly sand size, so particle cohesion can be ignored.
- Little or no vegetation exists to add cohesion to the sediment particles or complicate estimates of roughness.
- A reasonable estimate can be made for the total number of the runoff events over a 100-year period in this very arid climate.

- A maximum probable limit on sediment concentration can be applied to reduce the range of sediment particle sizes.

- The accuracy of the method could be improved if any of the following data were available:

  - A long-term record of rainfall, runoff, and sediment yield for each reservoir drainage basin.

  - A long-term record of rainfall, runoff, and sediment yield for a nearby and similar drainage basin for calibration purposes.

  - A sediment particle-size distribution of each drainage basin.

  - The areas of non-erodible and exposed bedrock of each drainage basin.

Results from the sheet erosion equation are believed to be the most accurate, because the most important variables are accounted for: dimensionless unit stream power and the magnitude, duration, and frequency of runoff events. Although results from this method are consistently lower than for the other two methods, 200 runoff events (over a 100-year period) are accounted for. Applying a maximum limit to the computed sediment concentration reduced the range of reasonable sediment particle sizes. The 100-year sediment volumes could only be greater if peak concentrations exceeded 300,000 ppm or if the runoff magnitudes or their frequency increased.

Sediment yield results for the assumed particle sizes of 1 and 2 mm were very low compared with smaller particle sizes and with the three other methods. The sediment particle size of 0.2 mm (fine sand) consistently provided the most reasonable high estimate in the sheet erosion equation for all drainage basins. Therefore, the results assuming a sediment particle size of 0.2 mm are used to represent the best estimate of the 100-year sediment volumes. Table 2.33 presents the 100-year sediment volume low estimates, high estimates, and best estimates for each proposed reservoir using the unit stream power sheet erosion equation.

Table 2.33. Low, high, and the best estimate of the 100-year sediment volume

Reservoir	Drainage area (mi^2)	Low estimate 100-year sedimentation volume (ac-ft)	High estimate 100-year sedimentation volume (ac-ft)	Best estimate 100-year sedimentation volume (ac-ft)
Reservoir 1	17.8	20	1,000	500
Reservoir 2	75.9	7	1,000	200
Reservoir 3	19.8	9	900	200

2.6 Reservoir Sedimentation

Rainfall, runoff, snowmelt, and river channel erosion provide a continuous supply of sediment that is hydraulically transported in rivers and streams. All reservoirs formed by dams on natural rivers are subject to some degree of sediment inflow and deposition. Because of the very low velocities in reservoirs, they tend to be very efficient sediment traps. Therefore, the amount of reservoir sedimentation over the life of the project needs to be predicted before the project is built. If the sediment inflow is large relative to the reservoir storage capacity, then the useful life of the reservoir may be very short. For example, a small reservoir on the Solomon River near Osborne, Kansas, filled with sediment during the first year of operation (Linsley and Franzini, 1979). If the inflowing sediments settle in the reservoir, then the clear water releases may degrade the downstream river channel (see Chapter 7, *River Processes and Restoration*).

There are several methods available for reducing reservoir sedimentation. These methods relate to the reservoir location and size, land use practices in the upstream watershed, and special considerations for the operation of the reservoir. In some cases, reservoirs can be operated for long-term sustainable use so that sedimentation eventually fills the reservoir (see Chapter 6, *Sustainable Development and Use of Reservoirs*).

Extensive literature exists on the subject of reservoir sedimentation. The book by Morris and Fan (1997), entitled *Reservoir Sedimentation Handbook* is an excellent reference and provides an extensive list of references.

2.6.1 Reservoir Sediment Trap Efficiency

The amount of sediment deposited within a reservoir depends on the trap efficiency. Reservoir trap efficiency is the ratio of the deposited sediment to the total sediment inflow and depends primarily upon the fall velocity of the various sediment particles, flow rate and velocity through the reservoir (Strand and Pemberton, 1982), as well as the size, depth, shape, and operation rules of the reservoir. The particle fall velocity is a function of particle size, shape, and density; water viscosity; and the chemical composition of the water and sediment. The rate of flow through the reservoir can be computed as the ratio of reservoir storage capacity to the rate of flow. The potential for reservoir sedimentation and associated problems can be estimated from the following six indicators:

- The reservoir storage capacity (at the normal pool elevation) relative to the mean annual volume of riverflow.
- The average and maximum width of the reservoir relative to the average and maximum width of the upstream river channel.
- The average and maximum depth of the reservoir relative to the average and maximum depth of the upstream river channel.

- The purposes for which the dam and reservoir are to be constructed and how the reservoir will be operated (e.g., normally full, frequently drawn down, or normally empty).

- The reservoir storage capacity relative to the mean annual sediment load of the inflowing rivers.

- The concentration of contaminants and heavy metals being supplied from the upstream watershed.

The ratio of the reservoir capacity to the mean annual streamflow volume can be used as an index to estimate the reservoir sediment trap efficiency. A greater relative reservoir size yields a greater potential sediment trap efficiency and reservoir sedimentation. Churchill (1948) developed a trap efficiency curve for settling basins, small reservoirs, flood retarding structures, semi-dry reservoirs, and reservoirs that are frequently sluiced.

Using data from Tennessee Valley Authority reservoirs, Churchill (1948) developed a relationship between the percent of incoming sediment passing through a reservoir and the sedimentation index of the reservoir (Figure 2.19). The sedimentation index is defined as the ratio of the period of retention to the mean velocity through the reservoir. The Churchill curve has been converted to a dimensionless expression by multiplying the sedimentation index by g, acceleration due to gravity.

The following description of terms will be helpful in using the Churchill curve:

Capacity—Capacity of the reservoir in the mean operating pool for the period to be analyzed in cubic feet.

Inflow—Average daily inflow rate during the study period in cubic feet per second.

Period of retention—Capacity divided by inflow rate.

Length—Reservoir length in feet at mean operating pool level.

Velocity—Mean velocity in feet per second, which is arrived at by dividing the inflow by the average cross-sectional area in square feet. The average cross-sectional area can be determined from the capacity divided by the length.

Sedimentation index—Period of retention divided by velocity.

Brune (1953) developed an empirical relationship for estimating the long-term reservoir trap efficiency for large storage or normal pond reservoir based on the correlation between the relative reservoir size and the trap efficiency observed in Tennessee Valley Authority reservoirs in the southeastern United States (see Figure 2.19). Using this relationship, reservoirs with the capacity to store more than 10 percent of the average annual inflow would be expected to trap between 75 and 100 percent of the inflowing sediment. Reservoirs with the capacity to store 1 percent of the average annual inflow would be expected to trap between 30 and 55 percent of the inflowing sediment. When the reservoir storage capacity is less than 0.1 percent of the average annual inflow, then the sediment trap efficiency would be near zero.

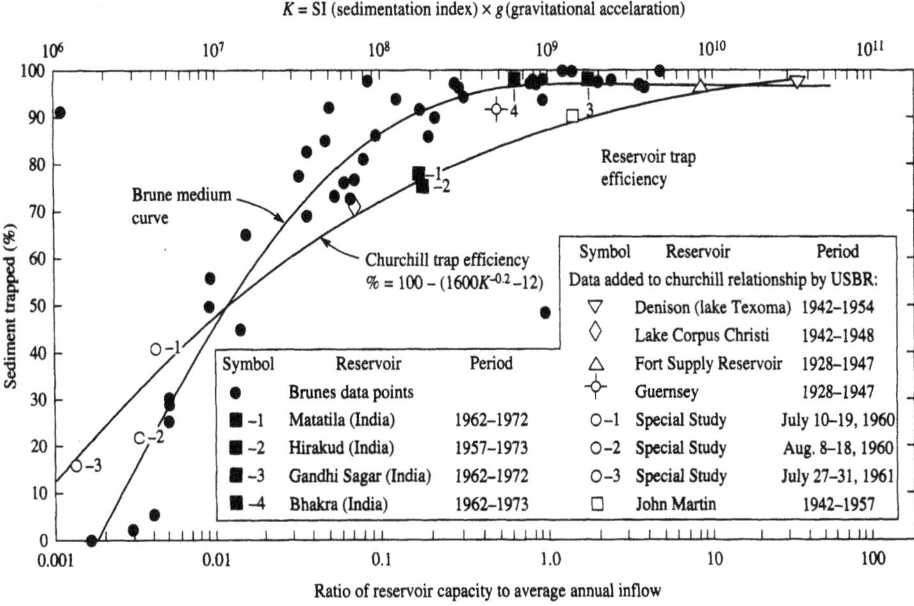

Figure 2.19. Trap efficiency curves (Churchill, 1948; Brune, 1953).

Figure 2.19 provides a good comparison of the Brune and Churchill methods for computing trap efficiencies using techniques developed by Murthy (1980). A general guideline is to use the Brune method for large storage or normal ponded reservoirs and the Churchill curve for settling basins, small reservoirs, flood retarding structures, semi-dry reservoirs, or reservoirs that are continuously sluiced. When the anticipated sediment accumulation is larger than 10 percent of the reservoir capacity, it is necessary that the trap efficiency be analyzed for incremental periods of the reservoir life.

The width and depth of the reservoir, relative to the width and depth of the upstream river channel, can also serve as indicators of reservoir sedimentation. Even if the reservoir capacity is small, relative to the mean annual inflow, a deep or wide reservoir may still trap some sediment.

The purposes for which a dam is constructed, along with legal constraints and hydrology, determine how the reservoir pool will be operated. The operation of the reservoir pool will influence the sediment trap efficiency and the spatial distribution and unit weight of sediments that settle within the reservoir. The reservoir trap efficiency of a given reservoir will be greatest if substantial portions of the inflows are stored during floods when the sediment concentrations are highest. If the reservoir is normally kept full (run of the river operation), floodflows pass through the reservoir and sediment trap efficiency is reduced. Coarse sediments would deposit as a delta at the far upstream end of the reservoir. When reservoirs are frequently drawn down, a portion of the reservoir sediments will be eroded and transported father downstream. Any clay-sized sediments that are exposed above the reservoir level will compact as they dry out (Strand and Pemberton, 1982).

Once sediment capacity is reached, the entire sediment load supplied by the upstream river channel is passed through the remaining reservoir. For example, the pool behind a diversion dam is typically filled with sediment within the first year or two of operation. For a large reservoir like Lake Powell, the average annual sediment inflow is 0.1 percent of the reservoir storage capacity.

If contaminants and heavy metals are transported into a reservoir, they will likely settle with the sediments in the reservoir. This may improve the water quality of the downstream river, but the water quality in the reservoir may degrade over time as the concentrations of contaminants and metals accumulate.

Once the estimated sediment inflow to a reservoir has been established, attention must be given to the effect the deposition of this sediment will have upon the life and daily operation of the reservoir (Strand and Pemberton, 1982). The mean annual sediment inflow, the trap efficiency of the reservoir, the ultimate density of the deposited sediment, and the distribution of the sediment within the reservoir all must be considered in the design of the dam.

Usually, to prevent premature loss of usable storage capacity, an additional volume of storage equal to the anticipated sediment deposition during the life of the reservoir is included in the original design. Reclamation has designed reservoirs to include sediment storage space whenever the anticipated sediment accumulation during the period of project economic analysis exceeds 5 percent of the total reservoir capacity (Strand and Pemberton, 1982). A 100-year period of economic analysis and sediment accumulation was used for those reservoirs. The allocated sediment space is provided to prevent encroachment on the required conservation storage space for the useful life of the project.

A schematic diagram of anticipated sediment deposition (Figure 2.20) shows the effect of sediment on storage. A distribution study with 100-year area and capacity curves similar to those shown on the left side of Figure 2.20 is needed whenever the 100-year sediment accumulation is more than 5 percent of the total reservoir capacity. In operational studies of a reservoir for determining the available water supply to satisfy projected water demands over the project life, an average can be used for the sediment accumulation during the economic life period. However, the total sediment deposition is used for design purposes to set the sediment elevation at the dam, to determine loss of storage due to sediment in any assigned storage space, and to help determine total storage requirements.

2.6.2 Density of Deposited Sediment

Samples of deposited sediments in reservoirs have provided useful information on the density of deposits. The density of deposited material in terms of dry mass per unit volume is used to convert total sediment inflow to a reservoir from a mass to a volume. The conversion is necessary when total sediment inflow is computed from a measured suspended and bed material sediment sampling program. Basic factors influencing density of sediment deposits in a reservoir are: (1) the manner in which the reservoir is operated; (2) the size of deposited sediment particles; and (3) the compaction or consolidation rate of deposited sediments.

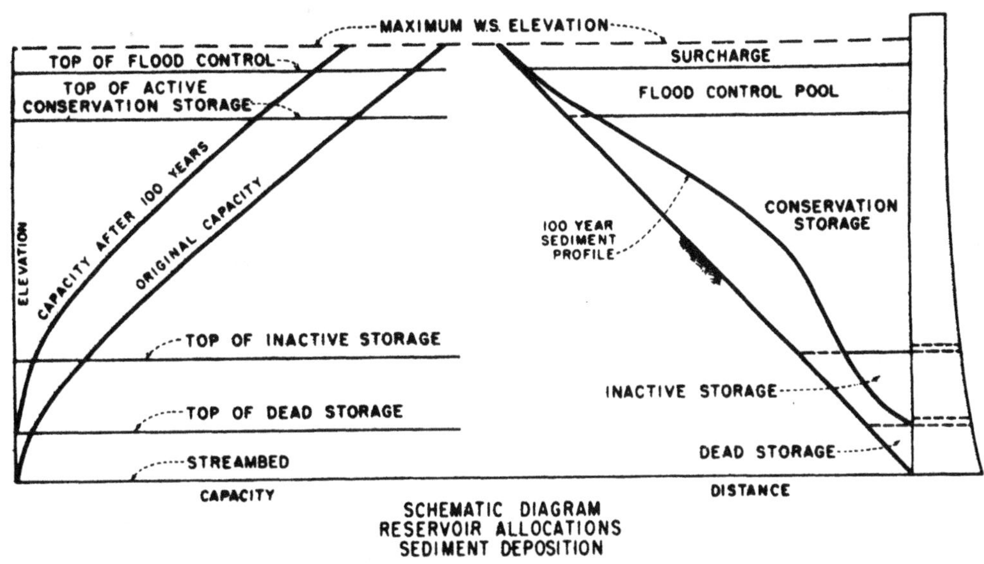

Figure 2.20. Schematic diagram of anticipated sediment deposition (Bureau of Reclamation, 1987).

The reservoir operation is probably the most influential of these factors. Sediments that have settled in reservoirs subjected to considerable drawdown are exposed to air for long periods and undergo a greater amount of consolidation. Reservoirs operating with a fairly stable pool do not allow the sediment deposits to dry out and consolidate to the same degree.

The size of the incoming sediment particles has a significant effect upon density. Sediment deposits composed of silt and sand will have higher densities than those in which clay predominates. The classification of sediment according to size as proposed by the American Geophysical Union (Vanoni, 1975) is as follows:

Sediment type	Size range in millimeters
Clay	Less than 0.004
Silt	0.004 to 0.062
Sand	0.062 to 2.0

The accumulation of new sediment deposits, on top of previously deposited sediments, changes the density of earlier deposits. This consolidation affects the average density over the estimated life of the reservoir, such as 100 years. Figure 2.21 shows a good example of consolidation of deposited sediments, taken from the report by Lara and Sanders (1970) for unit weights (densities) in Lake Mead at a sampling location with all clay-size material.

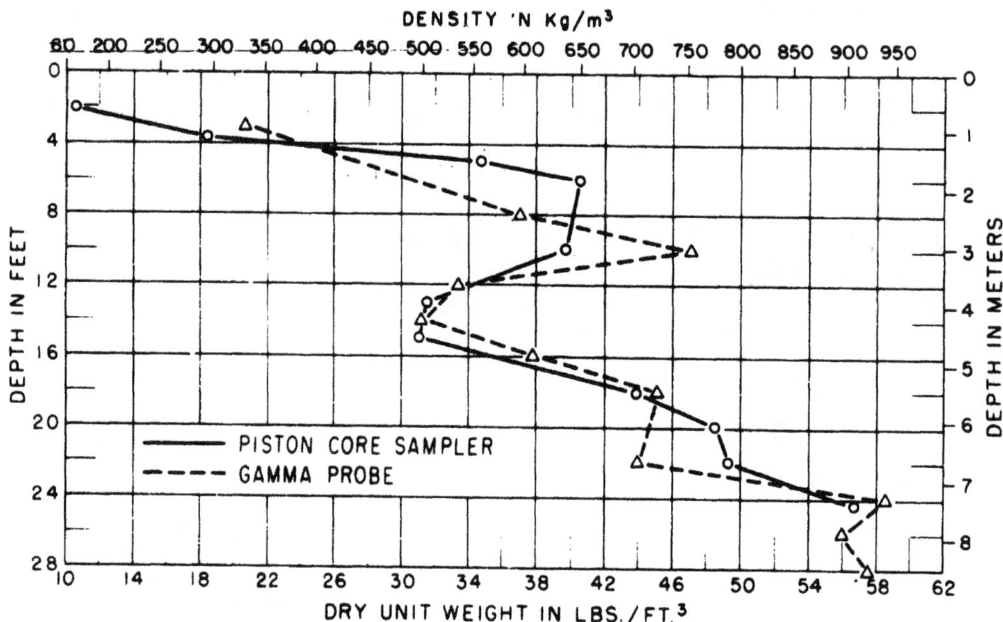

Figure 2.21. Comparison of densities on Lake Mead at location 5 (Bureau of Reclamation, 1987).

Three factors that should be taken into account in determining the density of deposited sediment are presented below. The influence of reservoir operation is the most significant because of the amount of consolidation or drying out that can occur in the clay fraction of the deposited material when a reservoir is subjected to considerable drawdown. The size of sediment particles entering the reservoir will also affect density, as shown by the variation in initial masses. Lara and Pemberton (1965) statistically analyzed some 1,300 samples for determining mathematical equations of variation of the unit weight of the deposits with the type of reservoir operation. Additional data on unit weight of deposited material from reservoir resurveys have supported the Lara and Pemberton (1965) equations (see Equation 2.39). The third factor is the years of operation of the reservoir.

Reservoir operations were classified according to operation as follows:

Operation	Reservoir operation
1	Sediment always submerged or nearly submerged
2	Normally moderate to considerable reservoir drawdown
3	Reservoir normally empty
4	Riverbed sediments upstream of reservoir

Selection of the proper reservoir operation number usually can be made from the operation study prepared for the reservoir.

Once the reservoir operation number has been selected, the density of the sediment deposits can be estimated using the following equation:

$$W = W_c p_c + W_m p_m + W_s p_s \tag{2.39}$$

where W = unit weight (lb/ft^3 or μg/m^3),
p_c, p_m, p_s = percentages of clay, silt, and sand, respectively, of the incoming sediment, and
W_c, W_m, W_s = unit weight of clay, silt, and sand, respectively, which can be obtained from the following tabulation:

Operation	Initial unit weight in lb/ft^3 (kg/m^3)		
	W_c	W_m	W_s
1	26 (416)	70 (1,20)	97 (1,50)
2	35 (561)	71 (1,40)	97 (1,50)
3	40 (641)	72 (1,50)	97 (1,50)
4	60 (961)	73 (1,70)	97 (1,50)

In determining the density of sediment deposits in reservoirs after a period of reservoir operation, it is recognized that part of the sediment will deposit in the reservoir in each of the T years of operation, and each year's deposits will have a different compaction time. Miller (1953) developed an approximation of the integral for determining the average density of all sediment deposited in T years of operation as follows:

$$W_T = W_o + 0.4343K \; \frac{T}{T-1}\left(\log_e T\right) - 1 \tag{2.40}$$

where W_T = average density after T years of reservoir operation,
W_o = initial unit weight (density) as derived from Equation (2.39), and
K = constant based on type of reservoir operation and sediment size analysis as obtained from the following table:

Reservoir operation	K for English units (metric units)		
	Sand	Silt	Clay
1	0	5.7	16
2	0	1.8	8.4
3	0	0	0

The K-factor of Equation (2.40) can be computed using Equation (2.41).

$$K = K_c p_c + K_m p_m + K_s p_s \tag{2.41}$$

where K_c, K_m, and K_s = the unit weight of clay, silt, and sand, respectively

As an example, the following data are known for a proposed reservoir with an operation number of 1 and a sized distribution of 23 percent clay, 40 percent silt, and 37 percent sand.

Then:

$$W = 26\ (0.23) + 70\ (0.40) + 97\ (0.37) = 6.0 + 28.0 + 35.9 = 70\ \text{lb/ft}^3\ (1120\ \text{kg/m}^3)$$

The 100-year average values to include compaction are computed as follows:

$$K = 16\ (0.23) + 5.7\ (0.40) + 0\ (0.37) = 3.68 + 2.28 + 0 = 5.96$$

$$W_{100} = 70 + 0.04343\ (5.96)\left[\frac{100}{99}(4.61) - 1\right] = 70 + 2.59\ (3.66) = 79\ \text{lb/ft}^3$$

This value may then be used to convert the initial weights (initial masses) of incoming sediment to the volume it will occupy in the reservoir after 100 years.

2.6.3 Sediment Distribution Within a Reservoir

The data obtained from surveys of existing reservoirs (U.S. Department of Agriculture, 1978) have been extensively used to develop empirical relationships for predicting sediment distribution patterns in reservoirs (Strand and Pemberton, 1982). Figures 2.22 and 2.23 illustrate the two most common techniques of showing sediment distribution, where sediment is distributed by depth and by longitudinal profile distance, respectively. Both methods clearly show that sediment deposition is not necessarily confined to the lower storage increments of the reservoir.

Sediment accumulation in a reservoir is usually distributed below the top of the conservation pool or normal water surface. However, if the reservoir has a flood control pool, and it is anticipated that the water surface will be held within this pool for significant periods of time, a portion of the sediment accumulation may be deposited within this pool. Figure 2.24 is a plot of data from 11 Great Plains reservoirs in the United States that may be used as a guide in estimating the portion of the total sediment accumulation that will deposit above the normal water surface. This plot should be regarded as a rough guide only, and the estimate obtained from it should be tempered with some judgment based upon the proposed reservoir operation and the nature of the incoming sediment. This curve is based on a limited amount of data and may be revised as more information becomes available.

The term "flood pool index" refers to the computed ratio of the flood control pool depth to the depth below the pool, multiplied by the percent of time the reservoir water surface will be within the flood control pool. This information for a proposed reservoir must be obtained from the reservoir operation study.

Once the quantity of sediment that will settle below the normal water surface has been established, the Empirical Area-Reduction Method may be used to estimate the distribution. This method was first developed from data gathered in the resurvey of 30 reservoirs and is described by Borland and Miller (1960) with revisions by Lara (1962). The method recognizes that distribution of sediment depends upon: (1) the manner in which the reservoir is to be operated; (2) the size of deposited sediment particle; (3) the shape of the reservoir; and (4) the volume of sediment deposited in the reservoir. The shape of the reservoir was adopted as the major criterion for development of empirically derived design curves for use in distributing sediment.

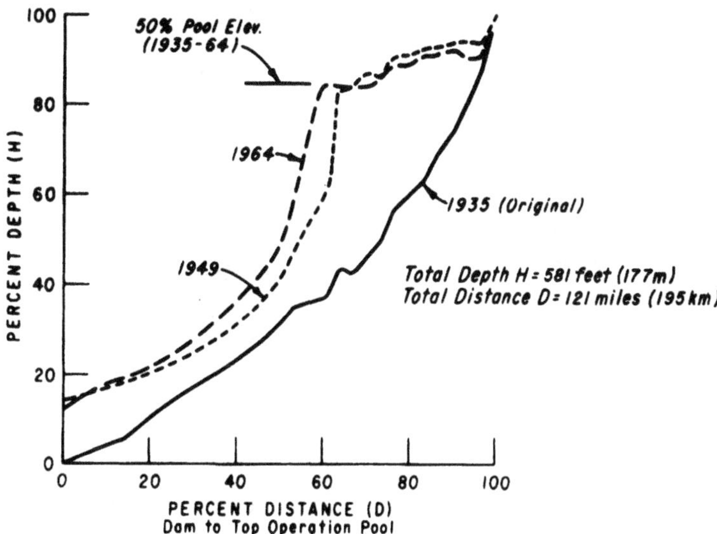

Figure 2.22. Sediment distribution from reservoir surveys of Lake Mead (Bureau of Reclamation, 1987).

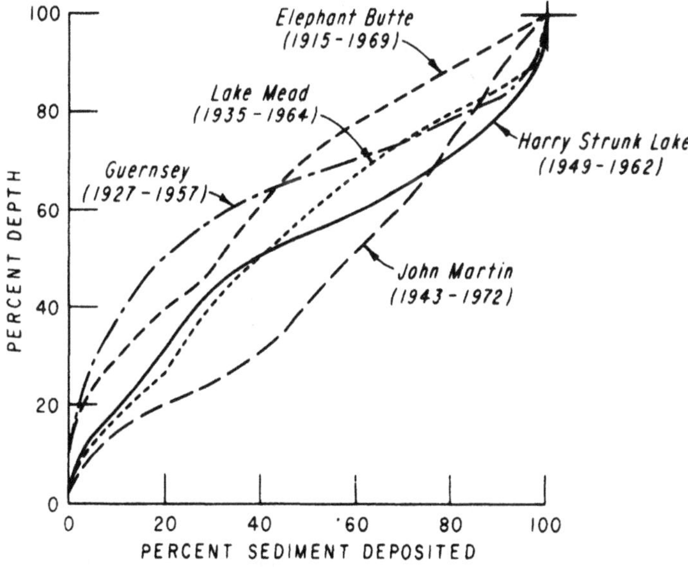

Figure 2.23. Sediment deposition profiles of several reservoirs (Bureau of Reclamation, 1987).

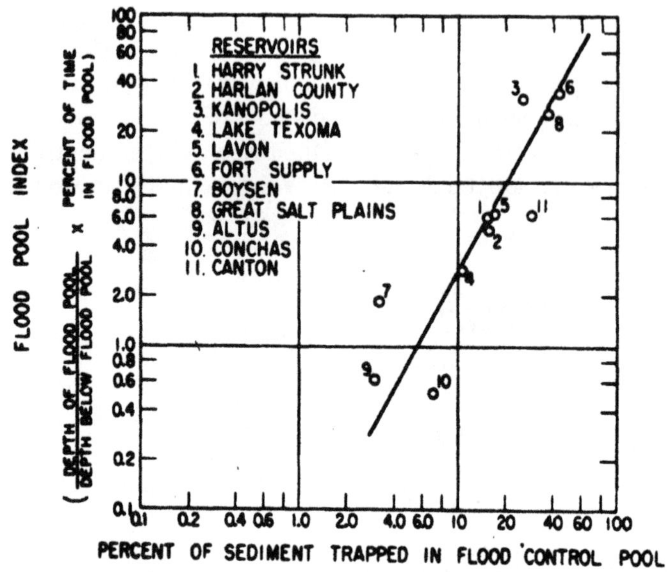

Figure 2.24. Sediment deposited in flood control pool (Bureau of Reclamation, 1987).

The design curve shown in Figure 2.25 can be used to predict reservoir sediment distribution as a function of depth. With equal weight applied to reservoir operation and shape, a weighted type distribution is selected from Table 2.34. In those cases where a choice of two weighted types are given, then a judicious decision can be made on whether the reservoir operation or shape of reservoir is more influential. The predominant size of reservoir sediment could be considered in this judgment of reservoir type from the following guidelines (see Figure 2.25):

Predominant size	Type
Sand or coarser	I
Silt	II
Clay	III

Table 2.34. Design type curve selection

Reservoir operation		Shape		
Class	Type	Class	Type	Weighted type
Sediment submerged	I	Lake	I	I
		Flood plain - foothill	II	I or II
		Hill and gorge	III	II
Moderate drawdown	II	Lake	I	I or II
		Flood plain - foothill	II	II
		Hill and gorge	III	II or III
Considerable drawdown	III	Lake	I	II
		Flood plain - foothill	II	II or III
		Hill and gorge	III	III
Normally empty	IV	All shapes		IV

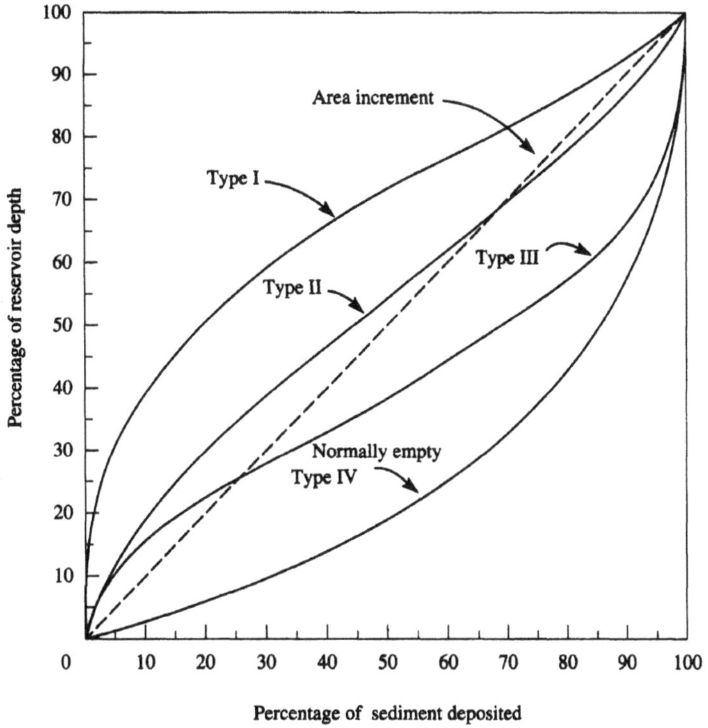

Figure 2.25. Sediment distribution design curves (Bureau of Reclamation, 1987).

Only for those cases with two possible type distributions should size of sediment be considered in selecting the design type curve. The size of sediments in most river systems is a mixture of clay, silt, and sand and has been found to be least important in selecting the design type curve from Figure 2.25.

Lara (1962) provides the detail on distributing sediment in a reservoir by the Empirical-Area Reduction Method. The appropriate design type curve is selected using the weighting procedure shown in Table 2.34.

The Area-Increment Method is based on the assumption that the area of sediment deposition remains constant throughout the reservoir depth. It is almost identical to the type II design curve (Figure 2.25) and is often used to estimate the new zero capacity elevation at the dam.

Strand and Pemberton (1982) give an example of a sediment distribution study for Theodore Roosevelt Dam, located on the Salt River in Arizona. Construction of the dam was completed in 1909, and a complete survey of the reservoir was made in 1981. The reservoir had an original total capacity of 1,530,500 acre-feet at elevation 2136 feet, the top of the active conservation pool. The purpose of this example is to: (1) compare the actual survey of 1981 with the distribution procedures; (2) show all of the steps involved in a distribution study; and (3) provide changes in capacity and projected sediment depths at the dam for 100, 200, and 300 years.

Table 2.35 gives the pertinent area-capacity data necessary to evaluate the actual 1981 survey and for use as a base in the distribution study. The total sediment accumulation in Theodore Roosevelt Lake, as determined from the 1981 survey, was 193,765 acre-feet. In the 72.4 years from closure of the dam in May 1909 until the survey in September 1981, the average annual sediment deposited was 2,676 acre-feet per year. The survey data from Table 2.35 were used to draw the sediment distribution design curve on Figure 2.26. To check the most appropriate design curve by the Empirical Area-Reduction Model, the volume of sediment accumulated in Theodore Roosevelt Lake from 1909 to 1981 was distributed by both a type II and III distribution, as shown in Figure 2.26. This comparison indicates that type II more closely resembles the actual survey. Figure 2.27 shows a plot of the area and capacity data from Table 2.35.

Table 2.35. Reservoir area and capacity data for Theodore Roosevelt Lake

	Original reservoir in 1909		1981 survey results	
Elevation (ft)	Area (acres)	Capacity (10^3 ac-ft)	Area (acres)	Capacity (10^3 ac-ft)
2136	17,785	1,530.5	17,337	1,336.7
2130	17,203	1,425.5	16,670	1,234.3
2120	16,177	1,258.5	15,617	1,072.4
2110	15,095	1,102.2	14,441	922.3
2100	14,104	956.5	13,555	782.6
2090	13,247	819.3	12,746	650.5
2080	11,939	693.3	11,331	530.0
2070	10,638	580.6	9,842	424.0
2060	9,482	479.9	8,230	333.8
2050	8,262	391.2	6,781	258.9
2040	7,106	314.6	5,569	197.6
2030	6,216	248.0	4,847	145.6
2020	5,286	190.3	4,212	100.3
2010	4,264	142.9	3,387	61.6
2000	3,544	103.8	2,036	35.0
1990	2,744	72.3	1,304	18.7
1980	1,985	48.9	903	7.6
1970	1,428	31.9	382	0.8
1960	1,020	19.7	[1]0	[1]0
1950	677	11.3		
1940	419	5.9		
1930	227	2.7		
1920	117	1.1		
1910	52	0.2		
1900	0	0		

[1] Sediment elevation at dam for 1981 survey is 1966 feet (599.2 m).

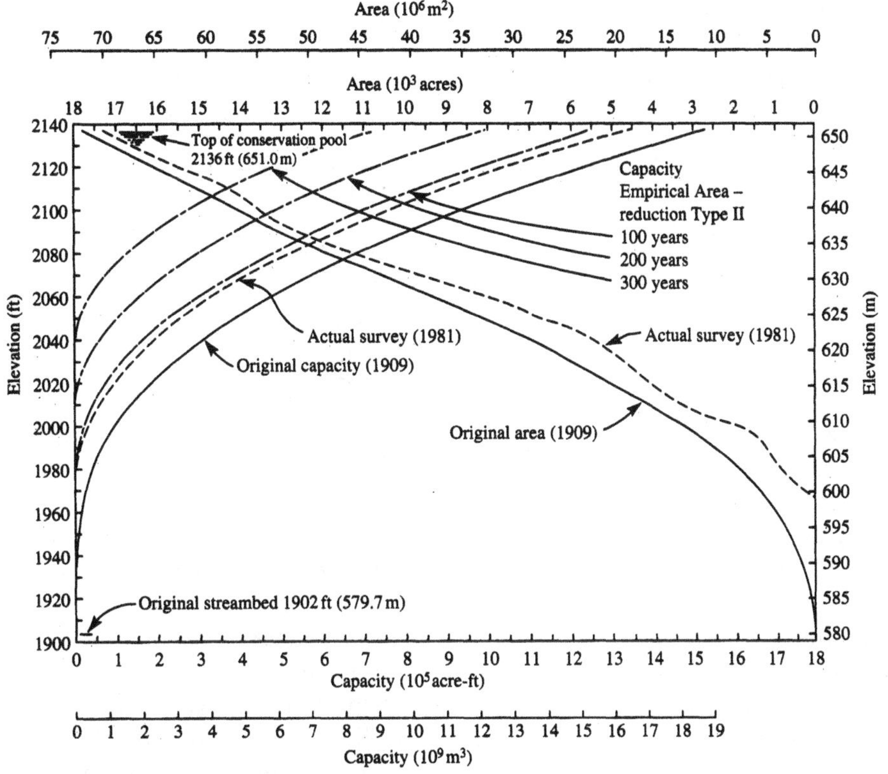

Figure 2.26. Area and capacity curves for Theodore Roosevelt Lake (Bureau of Reclamation, 1987).

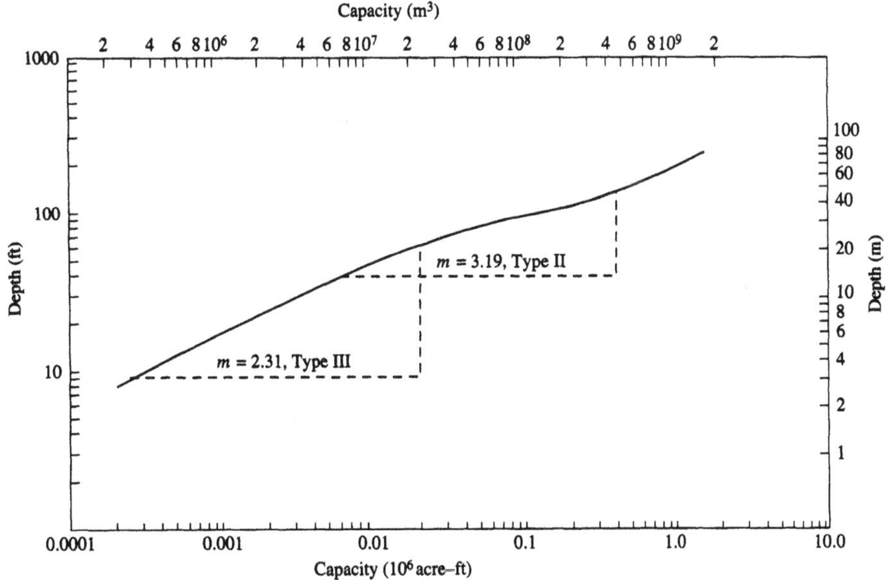

Figure 2.27. Depth versus capacity for Theodore Roosevelt Lake (Bureau of Reclamation, 1987).

The first step in the distribution study for the 100-, 200-, and 300-year period is a determination of the rate of sediment accumulation. In the case of Theodore Roosevelt Lake, the average annual rate determined from the 1981 survey and used for future projections, with the assumption that the compaction or unit weight of deposits will not change, is listed as follows:

Years	Sediment volume (acre-feet)
72.4 (1981)	193,765
100	267,600
200	535,200
300	802,800

No data existed on trap efficiency to apply to the above projections. The use of the rate from the 1981 survey results assumes that the trap efficiency for the first 72.4 years will remain the same through 300 years. In cases where sediment accumulation is determined from the total sediment load at a gauging station, then trap efficiency by use of Figure 2.19 and densities from Equations (2.39) and (2.40) are needed for computing the volume of sediment accumulation.

To complete this example, a logarithmic plot of the depth-capacity relationship for the original (1909) survey (Figure 2.27) for Theodore Roosevelt Lake, provided the shape factor for the reservoir type classification. Although the lower portion of the reservoir falls slightly in type III, the upper portion and overall slope indicate a type II classification. When assigning a type classification either for an existing reservoir or in distributing sediment on top of previous sediment deposits, it is important that the stage-capacity relationship only be plotted for the original survey. Studies have shown that a reservoir does not change type with continued sediment depositions. Once a reservoir has been assigned a type by shape, this classification will not change. However, it is possible that a change in reservoir operation could produce a new weighted type, as defined in Table 2.34.

The next step in the distribution study is computation of the elevation of sediment deposited at the dam. Table 2.36 shows a set of computations for determining the depth of sediment at the dam. The relative depth and a dimensionless function from the original area and capacity curves for Theodore Roosevelt Lake are computed as shown in Table 2.36 with the function:

$$F = \frac{S - V_h}{HA_h} \tag{2.42}$$

where

F = dimensionless function of total sediment deposition, capacity, depth, and area,

S = total sediment deposition,

V_h = reservoir capacity at a given elevation h,

H = original depth of reservoir, and

A_h = reservoir area at a given elevation h.

Table 2.36. Determination of elevation of sediment at Theodore Roosevelt Dam (Bureau of Reclamation, 1987)

	Year	Total sediment deposition (ac-ft)	Original depth of reservoir (ft)
1981 survey	72.4	193,765	234
	100	267,600	
	200	535,200	
	300	802,800	

Elevation (ft)	P relative depth	Original survey (1909) V_h capacity (ac-ft)	A_h area (ac)	HA_h 10^6 (ac-ft)	72.4 years $S\text{-}V_h$ (ac-ft)	F $\frac{S\text{-}V_h}{HA_h}$	100 years $S\text{-}V_h$ (ac-ft)	F $\frac{S\text{-}V_h}{HA_h}$	200 years $S\text{-}V_h$ (ac-ft)	F $\frac{S\text{-}V_h}{HA_h}$	300 years $S\text{-}V_h$ (ac-ft)	F $\frac{S\text{-}V_h}{HA_h}$
2080	0.761	693,315	11,939	2.79							109,485	0.0392
2070	0.718	580,590	10,638	2.49							222,210	0.0892
2060	0.675	479,928	9,482	2.22					55,272	0.0249	322,872	0.145
2050	0.632	391,207	8,262	1.93					143,993	0.0746	411,593	0.213
2040	0.590	314,623	7,106	1.66					220,577	0.133	488,177	0.294
2030	0.547	248,009	6,216	1.45			19,591	0.0135	287,191	0.198	554,791	0.383
2020	0.504	190,334	5,286	1.24			77,266	0.0623	344,866	0.278	612,466	0.494
2010	0.462	142,903	4,264	0.998	50,862	0.0510	124,697	0.125	392,297	0.393	659,897	0.661
2000	0.419	103,787	3,544	0.829	89,978	0.109	163,813	0.198	431,413	0.520	699,013	0.843
1990	0.376	72,347	2,744	0.642	121,418	0.189	195,253	0.304	462,853	0.721	730,453	1.138
1980	0.333	48,867	1,985	0.464	144,898	0.312	218,733	0.471	486,333	1.048	753,933	1.625
1970	0.291	31,935	1,428	0.334	161,830	0.485	235,665	0.706	503,265	1.507	770,865	2.308
1960	0.248	19,743	1,020	0.239	174,022	0.730	247,857	1.037	515,457	2.157	783,057	3.276
1950	0.205	11,328	677	0.158	182,437	1.155	256,272	1.622	523,872	3.316	791,472	5.009

A plot of the data points from Table 2.36 is superimposed on Figure 2.28 and the *p* value (relative depth) at which the line for any year crosses; the appropriate type curve will give the relative depth p_o equal to the new zero elevation at the dam. Figure 2.28 contains plotted curves of the full range of *F* values for all four reservoir types and the Area-Increment Method, as developed from the capacity and area design curves. For Theodore Roosevelt Dam, the intersect points for type II, as well as for the Area-Increment Method curves, gave sediment depths shown in Table 2.37. The Area-Increment Method is often selected because it will always intersect the *F* curve and, in many cases, gives a good check on the new zero capacity elevation at the dam. In the case of Theodore Roosevelt Dam, the 1981 survey had an observed elevation at the dam of 1966 feet (599.2 m), which was in better agreement with the Area-Increment Method value than any of the type curves. Data from Table 2.37 can be used to predict useful life of a reservoir or projection beyond the 300 years.

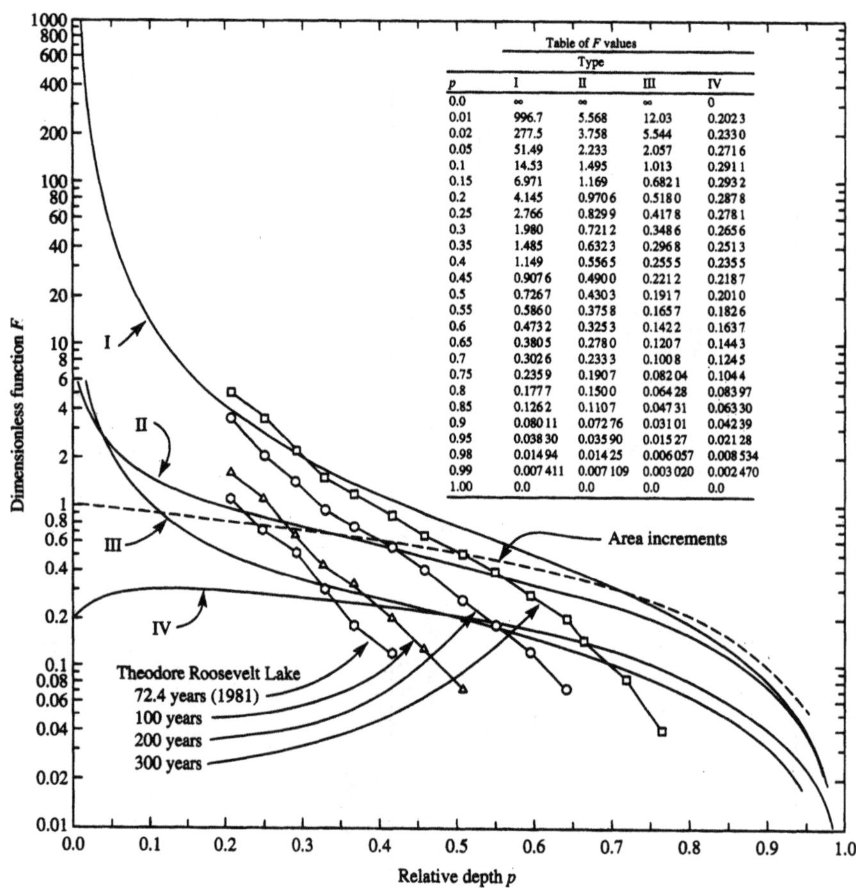

	Type			
p	I	II	III	IV
0.0	∞	∞	∞	0
0.01	996.7	5.568	12.03	0.2023
0.02	277.5	3.758	5.544	0.2330
0.05	51.49	2.233	2.057	0.2716
0.1	14.53	1.495	1.013	0.2911
0.15	6.971	1.169	0.6821	0.2932
0.2	4.145	0.9706	0.5180	0.2878
0.25	2.766	0.8299	0.4178	0.2781
0.3	1.980	0.7212	0.3486	0.2656
0.35	1.485	0.6323	0.2968	0.2513
0.4	1.149	0.5565	0.2555	0.2355
0.45	0.9076	0.4900	0.2212	0.2187
0.5	0.7267	0.4303	0.1917	0.2010
0.55	0.5860	0.3758	0.1657	0.1826
0.6	0.4732	0.3253	0.1422	0.1637
0.65	0.3805	0.2780	0.1207	0.1443
0.7	0.3026	0.2333	0.1008	0.1245
0.75	0.2359	0.1907	0.08204	0.1044
0.8	0.1777	0.1500	0.06428	0.08397
0.85	0.1262	0.1107	0.04731	0.06330
0.9	0.08011	0.07276	0.03101	0.04239
0.95	0.03830	0.03590	0.01527	0.02128
0.98	0.01494	0.01425	0.006057	0.008534
0.99	0.007411	0.007109	0.003020	0.002470
1.00	0.0	0.0	0.0	0.0

Figure 2.28. Curves to determine the depth of sediment at the dam (Bureau of Reclamation, 1987).

Table 2.37. Elevation of sediment at Theodore Roosevelt Dam
H = 234 ft

Years	Type II			Area increment		
	P_o	P_oH	Elevation (ft)	P_o	P_oH	Elevation (ft)
72.4 (1981)	0.23	54	1956	0.247	58	1960
100	0.284	66	1968	0.290	68	1970
200	0.418	98	2000	0.4	94	1996
300	0.553	129	2031	0.506	118	2020

The final step in the distribution study is to distribute a specified volume of sediment, which, for the example selected, involved the 72.4-, 100-, 200-, and 300-year volume in Theodore Roosevelt Lake by the type II design curve. Figure 2.26 shows the results of this distribution using procedures described by Lara (1962). Table 2.38 shows an example of the results for the 100-year distribution by use of the Empirical Area-Reduction Method and type II design curves. Although the example given is for type II, the equations for the relative sediment area a for each type are as follows (Bureau of Reclamation, 1987):

Type	Equation
I	$a = 5.074\, p^{1.85} (1 - p)^{0.35}$
II	$a = 2.487\, p^{0.57} (1 - p)^{0.41}$
III	$a = 16.967\, p^{1.15} (1 - p)^{2.32}$
IV	$a = 1.486\, p^{-0.25} (1 - p)^{1.34}$

where a = relative sediment area,
p = relative depth of reservoir measured from the bottom, and
p_o = relative depth at zero capacity.

2.6.4 Delta Deposits

Another phenomenon of reservoir sediment deposition is the distribution of sediment longitudinally as illustrated in Figure 2.22 for Lake Mead. The extreme upstream portion of the deposition profile is the formation of delta deposits. The major consequence of these delta deposits is the raising of the backwater elevations in the channel upstream from a reservoir. Therefore, the delta may cause a flood potential that would not be anticipated from pre-project channel conditions and proposed reservoir operating water surfaces. Predicting the delta development within a reservoir is a complex problem because of variables, such as operation of the reservoir, sizes of sediment, and hydraulics (in particular, the width of the upper reaches of the reservoir). Sediments deposited in the delta are continually being reworked into the downstream storage area at times of low reservoir stage and during extreme flood discharges.

Table 2.38. Theodore Roosevelt Lake; Type II reservoir sediment deposition study - empirical area reduction method. Sediment inflow = 267,600 acre-ft (Bureau of Reclamation, 1987)

	Original		Relative		Sediment		Revised	
Elevation ft	Area acres	Capacity acre-ft	Depth	Area	Area acres	Volume acre-ft	Area acres	Capacity acre-ft
2136.0	17,785.0	1,530,499	1.000	0.000	0.0	267,600	17,785.0	1,262,899
2130.0	17,203.0	1,425,512	0.974	0.546	699.1	265,503	15,503.9	1,160,009
2120.0	16,177.0	1,258,547	0.932	0.795	1018.8	256,914	15,158.2	1,001,633
2110.0	15,095.0	1,102,215	0.889	0.945	1210.3	245,768	13,884.7	856,447
2100.0	14,104.0	956,455	0.846	1.050	1344.8	232,993	12,759.2	723,462
2090.0	13,247.0	819,272	0.803	1.127	1443.6	219,051	11,803.4	600,221
2080.0	11,939.0	693,315	0.761	1.184	1516.9	204,248	10,422.1	489,067
2070.0	10,638.0	580,590	0.718	1.225	1570.0	188,814	9,068.0	391,776
2060.0	9,422.0	479,928	0.675	1.254	1606.3	172,293	7,875.7	306,996
2050.0	8,262.0	391,207	0.632	1.271	1628.0	156,761	6,634.0	234,446
2040.0	7,106.0	314,623	0.590	1.277	1636.5	140,438	5,469.5	174,185
2030.0	6,216.0	248,009	0.547	1.274	1632.8	124,092	4,583.2	123,917
2020.0	5,286.0	190,334	0.504	1.263	1617.6	107,840	3,668.4	82,494
2010.0	4,264.0	142,903	0.462	1.242	1591.0	91,797	2,673.0	51,106
2000.0	3,544.0	103,787	0.419	1.212	1553.1	76,076	1,990.9	27,711
1990.0	2,744.0	72,347	0.376	1.174	1503.8	60,792	1,240.2	11,555
1980.0	1,985.0	48,867	0.333	1.126	1443.0	46,057	542.0	2,810
1970.0	1,428.0	31,935	0.291	1.068	1381.5	31,935	46.5	33
1968.6	1,369.7	29,983	0.284	1.059	1369.7	29,983	0.0	0
1960.0	1,020.0	19,743	0.248	0.999	1020.0	19,743	0.0	0
1950.0	677.0	11,328	0.205	0.918	677.0	11,328	0.0	0
1940.0	419.0	5,893	0.162	0.821	419.0	5,893	0.0	0
1930.0	227.0	2,735	0.120	0.704	227.0	2,735	0.0	0
1920.0	117.0	1,059	0.077	0.558	117.0	1,059	0.0	0
1910.0	52.0	211	0.034	0.358	52.0	211	0.0	0
1902.0	0.0	0	0.000	.000	0.0	0	0.0	0

A delta study is needed for situations involving the construction of railroads or highway bridges in the delta area, defining inundated property such as urban areas or farmland, and design of protective structures to control inundation of property. The 100-year flood peak discharge is often used for inundation comparison in the flood plain, with the delta size over the life of the project to represent average conditions for the 100-year event.

An empirical procedure exists for the prediction of delta formation that is based upon observed delta deposits in existing reservoirs (Strand and Pemberton, 1982). Figure 2.29 shows a typical delta profile. It is defined by a topset slope, foreset slope, and a pivot point between the two slopes at the median reservoir operating level. The quantity of material to be placed in the delta is assumed to be equal to the volume of sand-size material or coarser (> 0.062 mm) entering the reservoir over the project life. A trial and error method, utilizing topographic data and volume computations by average end-area method, is used to arrive at a final delta location.

The topset slope of the delta is computed by one or more of several methods: (1) a statistical analysis of existing delta slopes that support a value equal to one-half of the existing channel slope (Figure 2.30); (2) topset slope from a comparable existing reservoir; or (3) zero bedload transport slope from bedload equations, such as those by Meyer-Peter and Müller (1948), Sheppard (1960), or Schoklitsch (1934). An example of the topset slope computed by the Meyer-Peter and Müller beginning transport equation for zero bedload transport is given by:

$$S_T = K \frac{\frac{Q}{Q_B}\left[\frac{n_s}{\left(d_{90}\right)^{1/6}}\right]^{3/2}}{D} d \qquad (2.43)$$

where

S_T = topset slope,

K = coefficient equal to 0.19,

Q/Q_B = ratio of total flow in ft^3/s to flow over bed of stream in ft^3/s (Q/Q_B is normally equal to 1). Discharge is referred to as dominant discharge and is usually determined by either channel bank full flow or as the 1.5-year flood peak,

d = diameter of bed material on topset slope, usually determined as weighted mean diameter in millimeters,

d_{90} = diameter of bed material for 90% finer than in millimeters,

D = maximum channel depth at dominant discharge in feet, and

n_s = Mannings roughness coefficient for the bed of the channel.

The Meyer-Peter and Müller equation, or any other equation selected for zero transport, will yield slope at which the bed material will no longer be transported, which must necessarily be true for the delta to form.

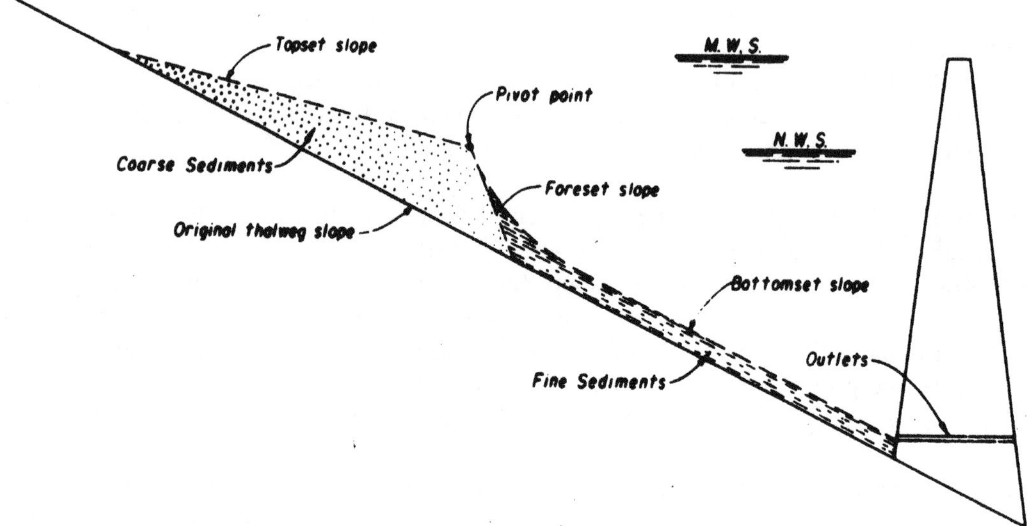

Figure 2.29. Typical sediment deposition profile (Bureau of Reclamation, 1987).

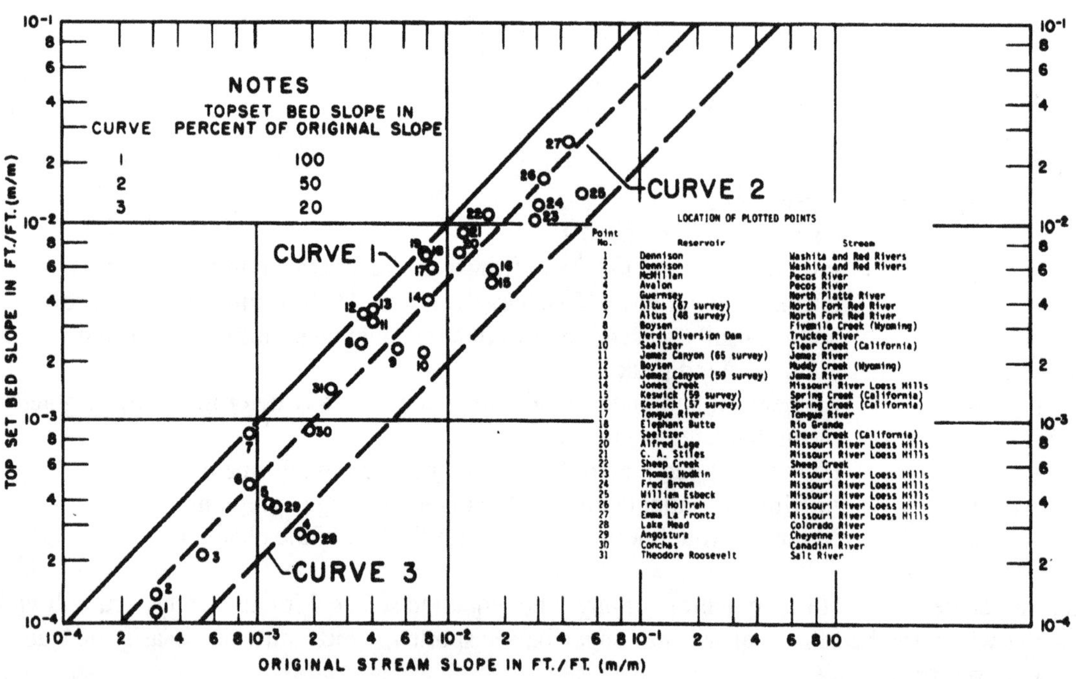

Point No.	Reservoir	Stream
1	Dennison	Washita and Red Rivers
2	Dennison	Washita and Red Rivers
3	McMillan	Pecos River
4	Avalon	Pecos River
5	Guernsey	North Platte River
6	Altus (67 survey)	North Fork Red River
7	Altus (48 survey)	North Fork Red River
8	Boysen	Fivemile Creek (Wyoming)
9	Verdi Diversion Dam	Truckee River
10	Saeltzer	Clear Creek (California)
11	Jemez Canyon (65 survey)	Jemez River
12	Boysen	Muddy Creek (Wyoming)
13	Jemez Canyon (59 survey)	Jemez River
14	Jones Creek	Missouri River Loess Hills
15	Keswick (59 survey)	Spring Creek (California)
16	Keswick (57 survey)	Spring Creek (California)
17	Tongue River	Tongue River
18	Elephant Butte	Rio Grande
19	Saeltzer	Clear Creek (California)
20	Alfred Lage	Missouri River Loess Hills
21	C. A. Stiles	Missouri River Loess Hills
22	Sheep Creek	Sheep Creek
23	Thomas Hodkin	Missouri River Loess Hills
24	Fred Brown	Missouri River Loess Hills
25	William Esbeck	Missouri River Loess Hills
26	Fred Hollrah	Missouri River Loess Hills
27	Emma La Frontz	Missouri River Loess Hills
28	Lake Mead	Colorado River
29	Angostura	Cheyenne River
30	Conchas	Canadian River
31	Theodore Roosevelt	Salt River

Figure 2.30. Topset slope versus original stream slope from existing reservoirs (Bureau of Reclamation, 1987).

The location of the pivot point between the topset and foreset slopes depends primarily on the operation of the reservoir and the existing channel slope in the delta area. If the reservoir is operated near the top of the conservation pool a large portion of the time, the elevation of the top of the conservation pool will be the pivot point elevation. Conversely, if the reservoir water surface has frequent fluctuations and a deeply entrenched inflow channel, a mean operating pool elevation should be used to establish the pivot point. In the extreme situation when a reservoir is emptied every year during the flood peak flows for sluicing sediment, the pivot point will be at the sluiceway.

As an initial guess, the upstream end of the delta is set at the intersection of the maximum water surface and the original streambed, and the topset slope is projected from that point to the anticipated pivot point elevation to begin the first trial computations of delta volume.

The average of foreset slopes observed in Reclamation reservoir resurveys is 6.5 times the topset slope. However, some reservoirs exhibit a foreset slope considerably greater than this; for example, Lake Mead's foreset slope is 100 times the topset. By adopting a foreset slope of 6.5 times the topset, the first trial delta fit can be completed.

The volume of sediment computed from the channel cross sections with the delta imposed on them should agree with the volume of sand size or larger material anticipated to come from the delta stream. The quantity of sediment in the delta above normal water surface elevation should also agree with that estimated to settle above the normal operating level, as shown in Figure 2.25. If the adjustment necessary to attain agreement is minor, it can usually be accomplished by a small change in the foreset slope. If a significant change in delta size is needed, the pivot point can be moved forward or backward in the reservoir, while maintaining the previously determined elevation of the point. The topset slope is then projected backward from the new pivot point location, and the delta volume is again computed. The intersection of the delta topset and the original streambed may fall above the maximum water surface elevation, a condition that has been observed in small reservoirs. The delta formation can also be determined from computer modeling (see Chapter 5, *Sedimentation Modeling for Rivers and Reservoirs*).

2.6.5 Minimum Unit Stream Power and Minimum Stream Power Method

Yang (1971) first derived the theory of minimum unit stream power for river morphology from thermodynamics. The theory states that for a closed and dissipative system under dynamic equilibrium:

$$\frac{dY}{dt} = \frac{dx}{dt}\frac{dY}{dx} = VS = \text{a minimum} \tag{2.44}$$

where Y = potential energy per unit weight of water,
x = distance,
V = average flow velocity,
S = slope,
VS = unit stream power, and
t = time.

The minimum value in Equation (2.44) depends on the constraints applied to the system. Yang (1976) later applied the theory to fluvial hydraulics computations. Yang (1976) and Yang and Song (1986, 1987) derived the theory of minimum energy dissipation rate from basic theories in fluid mechanics and mathematics. The theories of minimum stream power and minimum unit stream power are two of the special and simplified theories of the more general theory of minimum energy dissipation rate. These theories have been applied to solve a wide range of fluvial hydraulic problems.

Yang and Molinas (1982) showed that unit stream power can be obtained through the integration of the product of shear stress τ and velocity gradient du/dy, where u = time-averaged local velocity in an open channel flow. Consequently, minimization of VS is equivalent to minimization of $\tau\,(du/dy)$. Annandale (1987) called $\tau\,(du/dy)$ the applied unit stream power. It can be shown that minimization of applied unit stream power is equivalent to:

$$\sqrt{gDS} = \text{a minimum} = \text{a constant} \tag{2.45}$$

where g = gravitational acceleration, and
D = water depth.

Equation (2.45) can be used to determine the longitudinal bed profile of a reservoir in or near a stable condition. Annandale (1987) verified the validity of Equation (2.45) using data from Van Rhynereldpass Reservoir in South Africa, as shown in Figure 2.31. Figure 2.31 (b) indicates that a constant value of $\sqrt{gDS} = 6\times10^{-3}$ m/s can be used for the Van Rhynereldpass Reservoir. In order to apply Equation (2.45) to the determination of the longitudinal bed profile of a reservoir, it is assumed that the shear velocity of the river at the entrance of the reservoir remains constant through the reservoir. A modified backwater surface profile computation through the reservoir is then made by assuming two free surfaces; that is, water surface and bed surface. The bed surface is adjusted in the computation, such that Equation (2.45) is satisfied.

For most natural rivers and reservoirs, a more generalized theory of minimum stream power is applicable; that is:

$$QS = \text{a minimum} = \text{a constant} \tag{2.46}$$

where Q = water discharge, and
QS = stream power.

The minimum value in Equation (2.46) depends on the constraints applied to the system. Chang (1982) and Annandale (1984) found that when the stream power approaches a minimum value for relatively short reach:

$$\frac{dQ}{dx} \to 0 \tag{2.47}$$

The relationship between channel cross-sectional area A and wetted perimeter P becomes:

$$\frac{dA}{dx} = \frac{A}{Q}\frac{dQ}{dx} + \frac{A}{3P}\frac{dP}{dx} = \frac{A}{3P}\frac{dP}{dx} \tag{2.48}$$

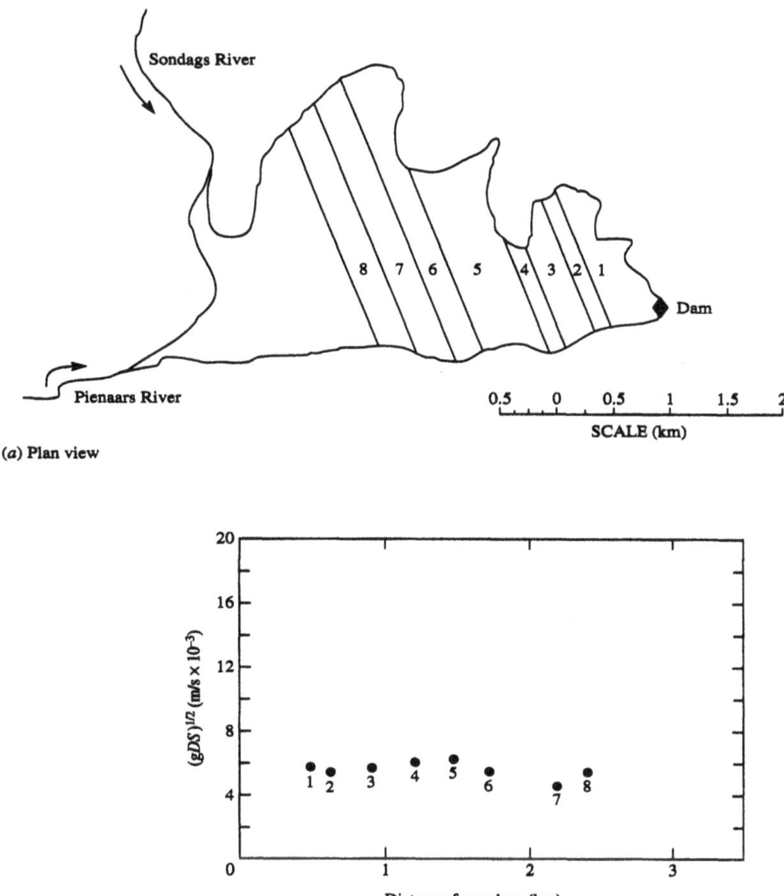

Figure 2.31. Relationship between shear velocity and distance for the Van Rhyneveld Reservoir (Annandale, 1987).

Equation (2.48) can be used to develop dimensionless cumulative mass curves as a function of dP/dx. Figure 2.32 shows the curves developed by Annandale (1987) based on data from 11 reservoirs in South Africa, where sediments are deposited below full supply level or below the crest of the spillway of a reservoir. Figure 2.33 shows the result by Annandale for sediments deposited above the full supply level. To illustrate the computational procedures based on a minimum unit stream power and minimum stream power theory, Yang (1996) summarized examples used by Annandale (1987) in the following example problems.

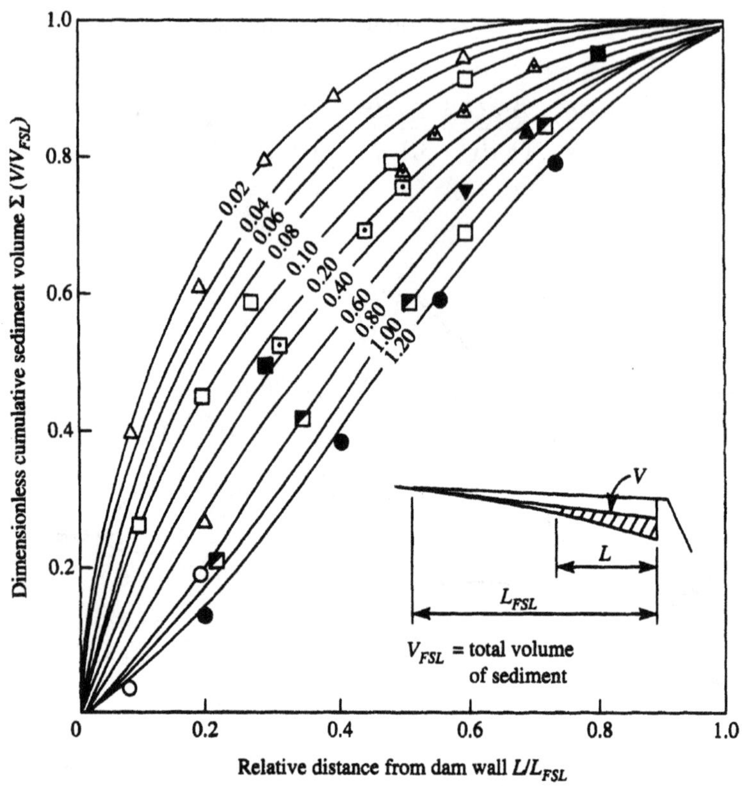

Figure 2.32. Dimensionless cumulative mass curves explaining sediment distribution below full supply level as a function of dP/dx for stable conditions (Annandale, 1987).

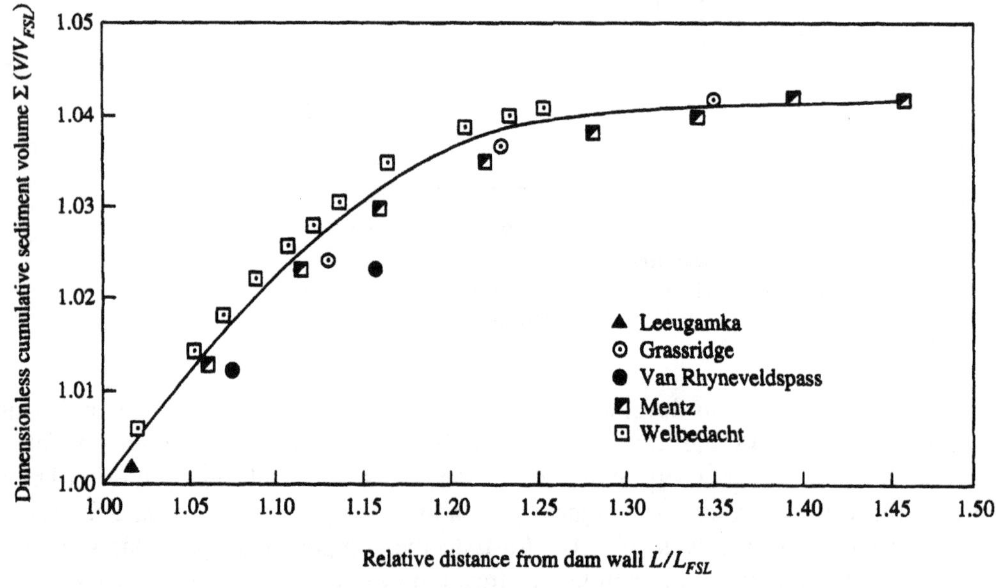

Figure 2.33. Sediment distribution above full supply level (Annandale, 1987).

Example 2.4 The river and reservoir schematic shown in Figure 2.34 below has the following properties: Width = B = 1 m; Flow depth in river = D = 1 m; Bed slope of river = S_0 = 0.002; Bed slope of reservoir = S_0' = 0.02; Manning's n = 0.03;

Hydraulic radius of river reach $= R = \dfrac{B \times D}{2D + B} = 0.333$ m;

Flow velocity in river $= V = \dfrac{R^{2/3} S^{1/2}}{n} = 0.717$ m/s;

Discharge = $Q = VA = 0.717$ m^3/s; Shear velocity in river reach = $\sqrt{gDS_0} = 0.14$ m/s.

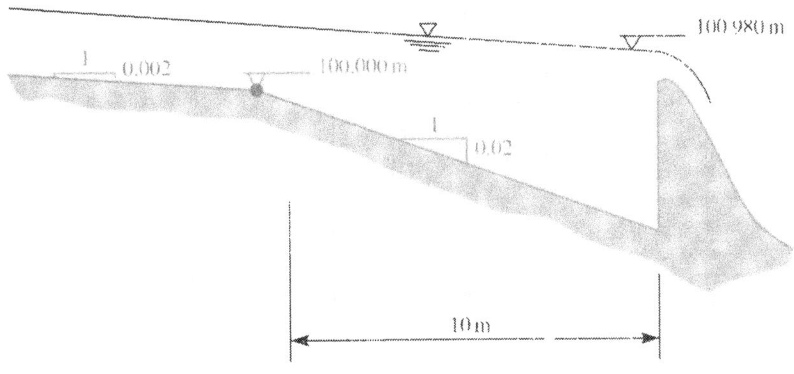

Figure 2.34. River and reservoir schematic for Example 2.4.

It is assumed that the river is in a dynamic, equilibrium condition. The constant value assumed for the shear velocity in the river is applicable for the reservoir and is therefore set at 0.14 m/s. Compute the reservoir bed surface profile based on minimum unit stream power theory.

Solution: Table 2.39 (Yang, 1996) summarizes the computation, based on Equation (2.45).

The final computed reservoir bed surface profile is:

Reach (m)	Bed level (m)
0	99.980
4	99.988
8	99.996
10	100.000

Table 2.39. Reservoir bed surface profile computation (Annandale, 1987; Yang, 1996)

Chainage (m) (1)	Stage (m) (2)	Assumed bed level (m) (3)	Flow depth (m) (4)	Area (m^2) (5)	$V^2/2g$ (m) (6)	Total head (m) (7)	P (m) (8)	R (m) (9)	$R^{4/3}$ ($m^{4/3}$) (10)	Friction slope S (m/m) (11)	Ave. S over reach (m/m) (12)	Reach length (m) (13)	h_f (m) (14)	Total head (m) (15)	$\sqrt{gDS}$ (m/s) (16)	Notes (17)
0	100.980	99.900	1.080	1.080	0.0225	101.002	3.160	0.341	0.239	0.0017					0.133	(i)
0	100.980	99.980	1.000	1.000	0.0262	101.006	3.000	0.333	0.231	0.0020					0.140	(ii)
4	101.000	99.990	1.010	1.010	0.0257	101.025	3.020	0.334	0.232	0.0020	0.0020	4	0.0078	101.014		(iii)
4	100.988	99.990	0.998	0.998	0.0263	101.014	2.996	0.333	0.230	0.0020	0.0020	4	0.0081	101.014	0.140	(iv)
4	100.988	99.988	1.000	1.000	0.0262	101.014	3.000	0.333	0.231	0.0020	0.0020	4	0.0080	101.014	0.140	(v)
8	101.100	99.990	1.110	1.110	0.0213	101.121	3.220	0.344	0.241	0.0016	0.0018	4	0.0071	101.021		(iii)
8	100.997	99.990	1.007	1.007	0.0259	101.022	3.014	0.334	0.231	0.0020	0.0020	4	0.0079	101.022	0.139	(iv)
8	100.996	99.996	1.000	1.000	0.0262	101.022	3.000	0.333	0.231	0.0020	0.0020	4	0.0080	101.022	0.140	(v)
10	101.000	100.00	1.000	1.000	0.0262	101.026	3.000	0.333	0.231	0.0020	0.0020	2	0.0040	101.026	0.1400	

Notes:

(i) Shear velocity $\neq$ 0.14 m/s, repeat calculation with new assumed bed level

(ii) Shear velocity = 0.14 m/s, proceed to next reach

(iii) Total head in column (7) $\neq$ total head in column (15), adjust water stage and repeat calculation

(iv) Total heads balance but shear velocity $\neq$ 0.140 m/s, adjust bed level and repeat calculation

Example 2.5 Figure 2.35(a) shows the plan view of Lake Mentz in South Africa. The estimated volume of sediment expected to be deposited in the reservoir is 129 x 10^6 m^3. Assume that the wetted perimeter can be replaced by reservoir width. Figure 2.35(b) shows the relationship between reservoir width and distance at full supply level. From Figure 2.35(b), $dP/dx = 0.8$. This value is used to select the curve from Figure 2.32 for sediment volume computation below full supply level. Sediment volume with L/L_{FSL} greater than 1.0, can be obtained from Figure 2.33. Table 2.40 summarizes the computations (Annandale, 1987; Yang, 1996).

Table 2.40. Lake Mentz reservoir sedimentation volume computation

Relative distance L/L_{FSL}	Actual distance (m)	Dimensionless cumulative sediment vol. $\Sigma(V/V_{FSL})$	Estimated cumulative sediment vol. (10^6 m^3)	Estimated sediment vol. sections (10^6 m^3)
0.0	0	0.00	0.00	0.00
0.1	1,200	0.08	0.92	9.92
0.2	2,400	0.20	24.81	14.89
0.3	3,600	0.36	44.65	19.84
0.4	4,800	0.50	62.02	17.37
0.5	6,000	0.60	74.42	12.40
0.6	7,200	0.70	86.83	12.41
0.7	8,400	0.82	101.71	14.88
0.8	9,600	0.90	111.63	9.92
0.9	10,800	0.95	117.84	6.21
1.0	12,000	1.00	124.04	6.20
1.1	13,200	1.02	126.52	2.48
1.2	14,400	1.03	127.76	1.24
1.3	15,600	1.04	129.00	1.24
1.4	16,800	1.04	129.00	0.00

The Sanmenxia Reservoir on the Yellow River in China has severe sedimentation problems. The project went through three phases of reconstruction to modify its operation since its completion. The modifications include reopening low level diversion tunnels and constructing side tunnels to sluice sediments. The operation rules also changed to releasing water with high sediment concentration during floods and storing water with lower sediment concentration after floods. Since these modifications, sediment inflow into and outflow from the reservoir is now in a state of dynamic equilibrium. During the long process of trial-and-error to determine the optimum reconstruction and modification of operation rules, the Yellow River Conservancy Commission (He et al., 1987) collected valuable data on scour and deposition in the reservoir. Yang (1996) used the Sanmenxia Reservoir data shown in Figure 2.36 to demonstrate the application of minimum stream power shown in Equation (2.46). Figure 2.36 shows that scour occurs when QS is greater than 0.3 m^3/s, while deposition occurs when QS is less than 0.3 m^3/s. The state of dynamic equilibrium can be maintained at $QS = 0.3$ m^3/s = a constant. These results indicate that the theory of minimum stream power, as stated in Equation (2.46), can be applied directly to the design and operation of a reservoir to maintain a dynamic equilibrium between sediment inflow and outflow. Figure 2.36 shows the actual measured values of QS x 10^{-4} m^3/s.

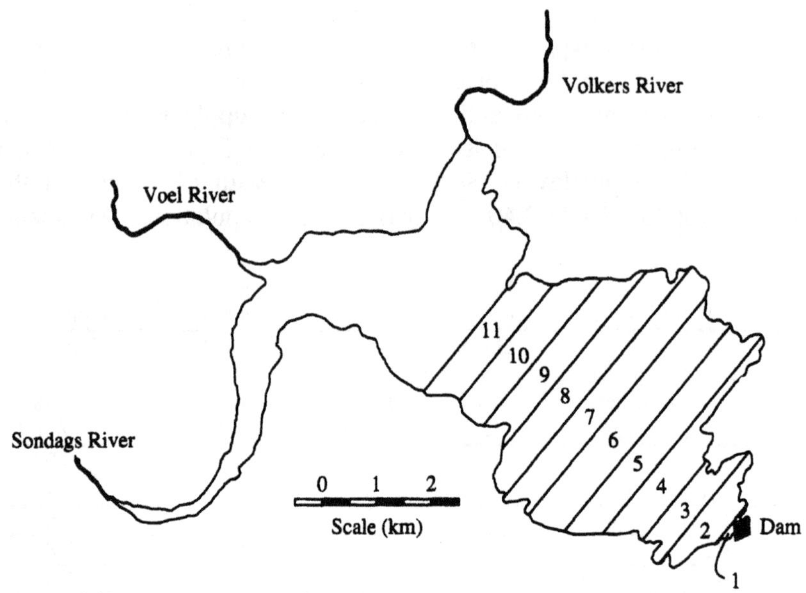

(*a*) Plan view

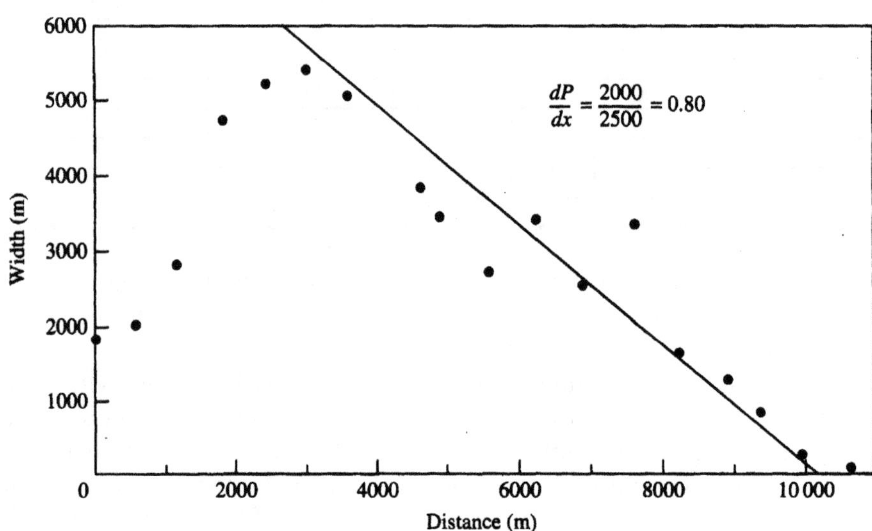

(*b*) Width–distance relationship

Figure 2.35. Plan view and width-distance relationship for Lake Metz in South Africa (Annandale, 1987).

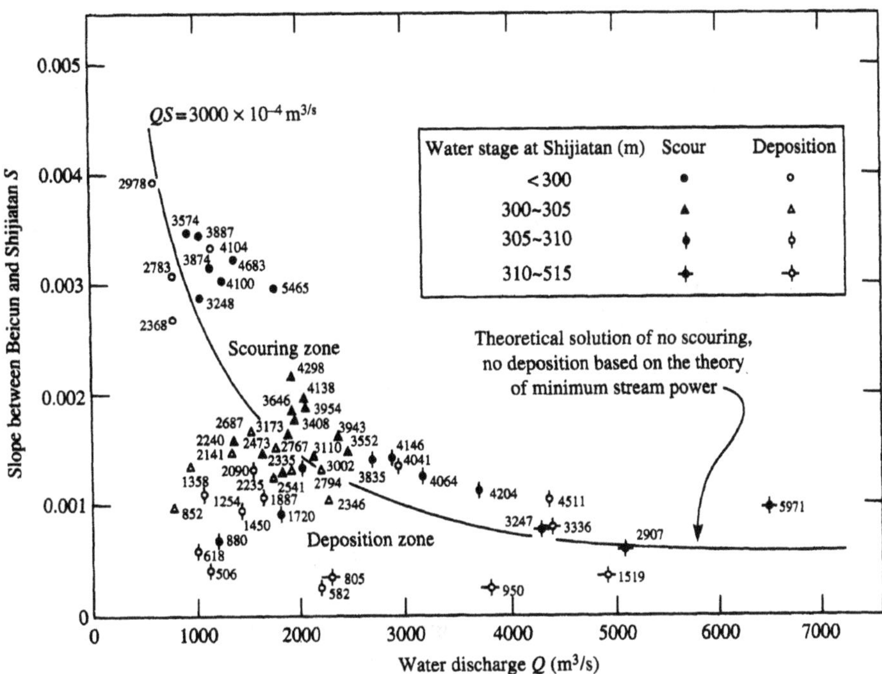

Figure 2.36. Determination of the Sanmenxia Reservoir scouring and deposition process based on the minimum stream power theory (Yang, 1996).

2.7 Summary

This chapter provides a detailed review and evaluation of empirical approaches used for the estimation of erosion rate or sediment yield. Empirical approaches include the use of the Universal Soil Loss Equation and its revised and modified versions, and direct measurement of sediment yield. Site-specific estimates of sediment yield from a watershed can also be computed from empirical equations based on drainage area or basin characteristics.

Theories of minimum energy dissipation rate and its simplified versions of minimum stream power or minimum unit stream power in conjunction with the unit stream power theory for sediment transport can be used as the basis for the computation of sheet, rill, and gully erosion rates. The GSTARS computer series can be used to systematically compute erosion rates and sediment transport, scour, and deposition in a watershed.

Reservoir sedimentation processes and computations based on empirical methods and analytical methods using minimum unit stream power and minimum stream power theories are included in this chapter. Example computations of erosion rates and reservoir sedimentation are used to illustrate the methods described in this chapter.

2.8 References

Annandale, G.W. (1984). "Predicting the Distribution of Deposited Sediment in Southern African Reservoirs," *IAHR Proceedings of the Symposium on Challenges in African Hydrology and Water Resources*, Harare, Zimbabwe.

Annandale, G. W. (1987). *Reservoir Sedimentation*, Elsevier Science Publishers, Amsterdam.

ASCE (1975). *Sedimentation Engineering*, American Society of Civil Engineering Manual No. 54, V.A. Vanoni, editor.

Basson, G. (2002). "Mathematical Modelling of Sediment Transport and Deposition in Reservoirs—Guidelines and Case Studies," *International Commission on Large Dams Sedimentation Committee Report.*

Beasley, D.B., L.F. Huggins, and E.J. Monke (1980). "ANSWERS: A model for watershed planning," *Transactions of the ASAE*, vol. 23, no. 4, pp. 938-944.

Bingner, R.L. (1996). "Runoff Simulated from Goodwin Creek Watershed Using SWAT," *Transactions of the ASAE*, vol 39, no. 1, pp. 85-89.

Bingner, R.L., C.E. Murphree, and C.K. Mutchler (1989). "Comparison of Sediment Yield Models on Watersheds in Mississippi," *Transactions of the ASAE*, vol. 32, no. 2, pp. 529-534.

Blanton, J.O., III (1982). "Procedures for Monitoring Reservoir Sedimentation," Bureau of Reclamation, Denver, Colorado.

Borland, W.M., and C.R. Miller (1960). "Distribution of Sediment in Large Reservoirs," *Transactions of the ASCE*, vol. 125, pp. 166-180.

Bouraoui, F.B., and T.A. Dillaha (1996). "ANSWERS-2000: Runoff and Sediment Transport Model," *Journal of Environmental Engineering*, vol. 122, no. 6, pp. 493-502.

Brown, L.C., and G.R. Foster (1987). "Storm Erosivity Using Idealized Intensity Distributions," *Transactions of the American Society of Agricultural Engineers*, vol. 30, no. 2, pp. 379-386.

Brune, G.M. (1953). "Trap Efficiency of Reservoirs," *Transactions of American Geophysical Union*, vol. 34, no. 3 pp. 407-418.

Bureau of Reclamation (1951). "Stable Channel Profiles," Hydraulic Laboratory Report no. Hyd. 325, Denver, Colorado.

Bureau of Reclamation (1987). *Design of Small Dams,* 3rd Edition, Denver, Colorado.

Chang, H.H. (1982). "Fluvial Hydraulics of Delta and Alluvial Fans," *Journal of the Hydraulics Division, ASCE*, vol. 108, no. HY11.

Churchill, M.A. (1948). Discussion of "Analysis and Use of Reservoir Sedimentation Data," by L.C. Gottschalk, *Proceedings, Federal Interagency Sedimentation Conference*, Denver, Colorado, pp. 139-140.

Daraio, J.A. (2002). "Assessing the Effects of Rainfall Kinetic Energy on Channel Suspended Sediment Concentrations for Physically-Based Distributed Modeling of Event-Scale Erosion," MS Thesis, Department of Civil and Environmental Engineering, University of Connecticut, Storrs, Connecticut.

Downer, C.W. (2002) "Identification and Modeling of Important Streamflow Producing Processes," Ph.D. Dissertation, Department of Civil and Environmental Engineering, University of Connecticut, Storrs, Connecticut.

Folly, A., J.N. Quinton, and R.E. Smith (1999). "Evaluation of the EUROSEM Model for the Catsop Watershed, The Netherlands," *Catena*, vol. 37, pp. 507-519.

Hong, S., and G.S. Govers (1985). "Selectivity and Transport Capacity of Thin Flows in Relation to Rill Generation," *Catena*, vol. 12, pp. 35-49.

Govers, G. (1985). "Selectivity and Transport of Thin Flows in Relation to Rill Erosion," *Catena*, vol. 12, pp. 35-49.

Govers, G., and G. Rauws (1986). "Transporting Capacity of Overland Flow on Plane and on Irregular Beds," *Earth Surface Processes and Landforms*, vol. 11, pp. 515-524.

Hairsine, P.B., and C.W. Rose (1992). "Modeling Water Erosion Due to Overland Flow Using Physical Principles 1, Sheet Flow," *Water Resources Research*, vol. 28, no. 1, pp. 237-243.

He, G., Z. Hua, and G. Wang (1987). "Regulation of Streamflow and Sediment Regimen by Sanmenxia Reservoir and Particularities of Scouring and Deposition," *Selected Papers of Researches on the Yellow River and Present Practice*, Yellow River Conservancy Commission, Zhengzou, China, pp. 166-175.

Hong, S., and S. Mostaghimi (1997). "Comparison of 1-D and 2-D Modeling of Overland Runoff and Sediment Transport," *Journal of the American Water Resources Association*, vol. 33, no. 5, pp. 1103-1116.

Horton, R.E., H.R. Leach, and R. Van Vlieit (1934). "Laminar Sheet-Flow," *Transactions of the American Geophysics Union*, vol. 2, pp. 393-404.

Johanson, R.C., J.C. Imhoff, J.L. Kittle, Jr., and A.S. Donigian (1984). *Hydrological Simulation Program CFORTRAN (HSPF): User Manual for Release 8.0*, EPA 600/3-84-066, Environmental Research Laboratory, U.S. EPA, Athens, Georgia.

Johnson, B.E., P.Y. Julien, D.K. Molnar, and C.C. Watson (2000). "The Two-Dimensional Upland Erosion Model CASC2D-SED," *Journal of the American Water Resources Association*, vol. 36, no. 1, pp. 31-42.

Jürgens, C., and M. Fander (1993). "Soil Erosion Assessment and Simulation by Means of SGEOS and Ancillary Digital Data," *International Journal of Remote Sensing*, vol. 14, no. 15, pp. 2847-2855.

Kilinc, M., and E.V. Richardson (1973). "Mechanics of Soil Erosion from Overland Flow Generated by Simulated Rainfall," Hydrology Paper no. 63, Colorado State University, Fort Collins, Colorado.

Kinsel, W.G. (1980). "CREAMS: A Field Scale Model for Chemicals, Runoff, and Erosion," in *Agricultural Management Systems*, U.S. Department of Agriculture, Conservation Report no. 26, 640 pp.

Kirnak, H. (2002). "Comparison of Erosion and Runoff Predicted by WEPP and AGNPS Models Using a Geographical Information System," *Turk J Agric For*, vol. 26, pp. 261-268.

Kothyari, U.C., and S.K. Jain (1997). "Sediment Yield Estimation Using GIS," *Hydrological Sciences Journal*, vol. 42, no. 6, pp. 833-843.

Kramer, L.A., and L.D. Meyer (1969). "Small Amount of Surface Runoff Reduces Soil Erosion and Runoff Velocity," *Transactions of the American Society of Agriculture Engineers*, vol. 12, pp. 638-648.

Lane, E.W. (1955). "The Importance of Fluvial Morphology in Hydraulic Engineering," *Journal of the Hydraulic Division, Proceedings, American Society of Civil Engineers*, vol. 81, paper 745, pp. 1-17.

Lane, E.W., and V.A. Koelzer (1943). "Density of Sediments Deposited in Reservoirs," in *A Study of Methods Used in Measurement and Analysis of Sediment Loads in Streams*, Report no. 9, Interagency Committee on Water Resources.

Lara, J.M. (1962). "Revision of the Procedure to Compute Sediment Distribution in Large Reservoirs," U.S. Bureau of Reclamation, Denver, Colorado.

Lara, J.M., and E.L. Pemberton (1965). "Initial Unit Weight of Deposited Sediments," *Proceedings, Federal Interagency Sedimentation Conference*, 1963, U.S. Agriculture Research Service Miscellaneous Publication, no. 970, pp. 818-845.

Lara, J.M., and H.I. Sanders (1970). "The 1963-64 Lake Mead Survey," U.S. Bureau of Reclamation, Denver, Colorado.

Leavesley, G.H., R.W. Lichty, B.M. Troutman, and L.G. Saindon (1983). *Precipitation-Runoff Modeling System—Users Manual*, USGS Water Resources Investigations Report 83-4238.

Linsley, R.K., and J.B. Franzini (1979). *Water-Resources Engineering*, McGraw-Hill Book Company, New York.

Linsley, R.K., Jr., M.A. Kohler, and J.L.H. Paulhus (1975). *Hydrology for Engineers*, McGraw-Hill Book Company, New York.

Meyer-Peter, E., and R. Müller (1948). "Formulas for Bed-load Transport," Meeting of the International Association for Hydraulic Structure Research, second meeting, Stockholm.

Miller, C.R. (1953). "Determination of the Unit Weight of Sediment for Use in Sediment Volume Computations," U.S. Bureau of Reclamation, Denver, Colorado.

Mitasova, H., L. Mitas, B. Brown, and D. Johnston (2002). "Distributed Hydrologic and Erosion Modeling for Watershed Management," online document http://skagit.meas.ncsu.edu/~helena/gmslab/courtcreek/wmb8.html.

Mitchell, J.K., B.A. Engel, R. Srinivasan, and S.S.Y. Wang (1993). "Validation of AGNPS for Small Watersheds using an Integrated AGNPS/GIS System," *Water Resources Bulletin*, vol. 29, no. 5, pp. 833-842.

Moore, I.D., and G.J. Burch (1986). "Sediment Transport Capacity of Sheet and Rill Flow: Application of Unit Stream Power Theory," *Water Resources Research*, vol. 22, no. 8, pp. 1350-1360.

Morgan, R.P.C., J.N. Quinton, R.E. Smith, G. Govers, J.W.A. Poesen, K. Auerswald, G. Chisci, D. Torri, and M.E. Styczen (1998). "The European Soil Erosion Model (EUROSEM): A Dynamic Approach for Predicting Sediment Transport from Fields and Small Catchments," *Earth Surface Process and Landforms*, vol. 23, pp. 527-544.

Morris, G.L., and J. Fan (1997). *Reservoir Sedimentation Handbook*, McGraw-Hill Book Company, New York, 758 pp.

Moss, A.J., P.H. Walker, and J. Hutka (1980). "Movement of Loose, Sandy Detritus by Shallow Water Flows: An Experimental Study," *Sedimentation Geology*, vol. 25, pp. 43-66.

Murthy, B.N. (1980). "Life of Reservoir," Technical Report No. 19, Central Board of Irrigation and Power, New Delhi.

Nash, J.E., and J.V. Sutcliffe (1970). "River Flow Forecasting through Conceptual Models Part ICA Discussion of Principles," *Journal of Hydrology*, vol. 10, pp. 282-290.

Nearing, M.A. 1998. "Why Soil Erosion Models Over-Predict Small Soil Losses and Under-Predict Large Soil Losses," *Catena*, vol. 32, pp. 15-22.

Nearing, M.A., L.J. Lane, and V.L. Lopes (1994). "Modeling Soil Erosion," in *Soil Erosion Research Methods*, Rattan Lal (ed.), St. Lucie Press, Delray Beach, Florida.

Nearing, M.A., G.R. Foster, L.J. Lane, and S.C. Finker (1989). "A Process-Based Soil Erosion Model for USDA-Water Erosion Prediction Project Technology," *Transactions of the American Society of Agricultural Engineers*, vol. 32, no. 5, pp. 1587-1593.

Nearing, M.A., L.D. Norton, D.A. Bulgakov, G.A. Larionov, L.T. West, and K.M. Dontsova (1997). "Hydraulics and Erosion in Eroding Rills," *Water Resources Research*, vol. 33, no. 4, pp. 865-876.

Ogden, F.L., and A. Heilig (2001). "Two-Dimensional Watershed Scale Erosion Modeling with CASC2D," Harmon, R.S., and Doe, W.W., III (eds.), in *Landscape Erosion and Evolution Modeling*, Kluwer Academic/Plenum Publishers, New York.

Ogden, F.L., and P.Y. Julien (2002). "CASC2D: A Two-Dimensional, Physically-Based, Hortonian, Hydrologic Model," in *Mathematical Models of Small Watershed Hydrology and Applications*, V.J. Singh, and D. Freverts, eds., Water Resources Publications, Littleton, Colorado.

Olmsted, F.H., O.J. Loeltz, and I. Burdge (1973). "Geohydrology of the Yuma Area, Arizona and California," *Water Resources of Lower Colorado River—Salton Sea Area*, Geological Survey Professional Paper 486-H, United States Government Printing Office, Washington DC.

Pacific Southwest Interagency Committee, Water Management Subcommittee (1968). "Factors Affecting Sediment Yield in the Pacific Southwest Area and Selection and Evaluation of Measures for Reduction of Erosion and Sediment Yield."

Pacific Southwest Interagency Committee (1968). "Factors Affecting Sediment Yield and Measures for the Reduction of Erosion and Sediment Yield."

Parsons, A.J., and A.M. Gadian (2000). "Uncertainty in Modelling the Detachment of Soil by Rainfall," *Earth Surface Processes and Landforms*, vol. 25, pp. 723-728.

Parsons, A.J., and J. Wainwright (2000). "A Process-Based Evaluation of a Process-Based Soil-Erosion Model," in *Soil Erosion: Application of Physically Based Models*, J. Schmidt, ed., Springer: Berlin, Germany.

Randle, T.J. (1996). "Lower Colorado River Regulatory Storage Study Sedimentation Volume After 100 Years for the Proposed Reservoirs Along the All American Canal and Gila Gravity Main Canal," U.S. Bureau of Reclamation, Technical Service Center, Denver, Colorado.

Randle, T.J. (1998). "Predicting Sediment Yield From Arid Drainage Areas" in *Proceedings of the First Federal Interagency Hydrologic Modeling Conference*, Interagency Advisory Committee on Water Data, Subcommittee on Hydrology, Las Vegas, Nevada, April 19-23, 1998.

Rauws, G. (1984). "De Transportcapaiteit van Afterfow Over Een Ruw Oppervla K: Laboratoriumexperimenten," Licentiaats thesis KUL, Fac. Wet., Leuven.

Renard, K.G., J.M. Laflen, G.R. Foster, and D.K. McCool (1994). "The Revised Universal Soil Loss Equation," in *Soil Erosion Research Methods,* Rattan Lal (ed.), St. Lucie Press, Delray Beach, Florida.

Renard, K.G., G.R. Foster, G.A. Weesies, D.K. McCool, and D.C. Yoder (1996). *Predicting Soil Erosion by Water: A Guide to Conservation Planning with the Revised Universal Soil Loss Equation*, U.S. Department of Agriculture, Agriculture Handbook 703, 384 pp.

Rubey, W.W. (1933). "Settling Velocities of Gravel, Sand, and Silt Particles," *American Journal of Science*, vol. 25, pp. 325-338.

Savat, J. (1980). "Resistance to Flow in Rough Supercritical Sheet Flow," *Earth Surface Processes and Landforms*, vol. 5, pp. 103-122.

Schoklitsch, A. (1934). "Der Gesuhiebetrieb und di Geschiebefracht," *Wasserkraft und Wasserwirtschaft*, vol. 29, no. 4, pp. 37-43.

Schröder, A. (2000). "WEPP, EUROSEM, E-2D: Results of Applications at the Plot Scale," in *Soil Erosion Application of Physically Based Models*, J. Schmidt (ed.), Springer, Berlin.

Senarath, S.U.S., F.L. Ogden, C.W. Downer, and H.O. Sharif (2000). On the Calibration and Verification of Two-Dimensional, Distributed, Hortonian, Continuous Watershed Models, *Water Resources Research*, vol. 36, no. 6, pp. 1495-1510.

Sharma, K.D., and S. Singh (1995). "Satellite Remote Sensing for Soil Erosion Modelling Using the ANSWERS Model," *Hydrological Sciences Journal*, vol. 40, no. 2, pp. 259-272.

Sheppard, J.R. (1960). "Investigation of Meyer-Peter, Müller Bed-load Formulas," U.S. Bureau of Reclamation, Denver, Colorado.

Singh, V.J. (1995). *Computer Models of Watershed Hydrology*, Water Resources Publication, Highlands Ranch, Colorado.

Smith, R.E., D.A. Goodrich, and J.N. Quinton (1995). "Dynamic Distributed Simulation of Watershed Erosion: KINEROS II and EUROSEM," *Journal of Soil and Water Conservation*, vol. 50, pp. 517-520.

Smith, S.J., J.R. Williams, R.G. Menzel, and G.A. Coleman (1984). "Prediction of Sediment Yield from Southern Plains Grasslands with the Modified Universal Soil Loss Equation," *Journal of Range Management*, vol. 37, no. 4, pp. 295-297.

Strand, R.I. (1975). "Bureau of Reclamation Procedures for Predicting Sediment Yield," in *Present and Prospective Technology for Predicting Sediment Yields and Sources*, Proceedings of the Sediment-Yield Workshop, USDA Sedimentation Laboratory, Oxford, Mississippi, November 28-30, 1972.

Strand, R.I., and E.L. Pemberton (1982). *Reservoir Sedimentation Technical Guidelines for Bureau of Reclamation*, U.S. Bureau of Reclamation, Denver, Colorado, 48 pp.

Swamee, P.K. (1974). "Analytic and Experimental Investigation of Streambed Variation Upstream of a Dam," Ph.D. Thesis, Department of Civil Engineering, University of Roorkee, India.

Tiwari, A.K., L.M. Risse, and M.A. Nearing (2000). "Evaluation of WEPP and its Comparison with USLE and RUSLE," *Transactions of the American Society of Agricultural Engineers*, vol. 43, no. 5, pp. 1129-1135.

U.S. Department of Agriculture (1978). "Sediment Deposition in U.S. Reservoirs, Summary of Data Reported Through 1975," Miscellaneous Publication No. 1362, Agriculture Research Service.

U.S. Environmental Protection Agency (2001). *A Watershed Decade*, EPA 840-R-00-002, Office of Wetlands, Oceans and Watersheds, Washington DC.

U.S. Interagency Committee on Water Resources (1957). Subcommittee on Sedimentation, "Some Fundamentals of Particle Size Analysis," Report no. 12.

U.S. Government Handbook (1978). "Chapter 3 - Sediment, National Handbook of Recommended Methods for Water-Data Acquisition."

U.S. Interagency Sedimentation Project, "A Study of Methods Used in Measurement and Analysis of Sediment Loads in Streams," Reports No. 1 through 14 and A through W, Subcommittee on Sedimentation, 1940 to 1981.

Vanoni, V.A. (1975). *Sedimentation Engineering*, ASCE Manual 54, ASCE, New York.

Wicks, J.M., and J.C. Bathurst (1996). "SHESED: A Physically Based, Distributed Erosion and Sediment Yield Component for the SHE Hydrological Modelling System," *Journal of Hydrology*, vol. 175, pp. 213-238.

Williams, J.R., A.D. Nicks, and J.G. Arnold (1985) "Simulator for Water Resources in Rural Basins," *Journal of Hydraulic Engineering, ASCE*, vol. 111, no. 6, pp. 970-986.

Williams, J.R., C.A. Jones, and P.T. Dyke (1984). "A modeling approach to determining the relationship between erosion and soil productivity," *Transaction of the American Society of Agricultural Engineers*, vol. 27, pp. 129-144.

Williams, J.R. (1975). "Sediment-Yield Prediction with Universal Equation Using Runoff Energy Factor," in *Present and Prospective Technology for Predicting Sediment Yields and Sources*, ARS-S-40, USDA-ARS.

Williams, J.R. (1981). "Testing the modified Universal Soil Loss Equation," in *Estimating Erosion and Sediment Yield on Rangelands*, USDA, ARM-W-26:157-164.

Wischmeier, W.H., and D.D. Smith (1962). "Soil-Loss Estimation as a Tool in Soil and Water Management Planning," Institute of Association of Scientific Hydrology, Publication No. 59, pp. 148-159.

Wischmeier, W.H., and D.D. Smith. (1965). *Predicting Rainfall—Erosion Losses from Cropland East of the Rocky Mountains*, U.S. Department of Agriculture, Agriculture Handbook No. 282, 48 p.

Wischmeier, W.H., and D.D. Smith (1978). *Predicting Rainfall Erosion Losses—Guide to Conservation Planning*, U.S. Department of Agriculture, Agriculture Handbook No. 537.

Woolhiser, D.A., R.E. Smith, and D.C. Goodrich (1990). *KINEROS, A Kinematic Runoff and Erosion Model: Documentation and User Manual*, U.S. Department of Agriculture, Agricultural Research Service, ARS-77, 130 p.

Wu, T.H., J.A. Hall, and J.V. Bonta (1993). "Evaluation of runoff and erosion models," *Journal of Irrigation and Drainage Engineering*, vol. 119, no. 2, pp. 364-382.

Yang, C.T. (1971). "Potential Energy and Stream Morphology," *Water Resources Research*, vol. 7, no. 2, pp. 311-322.

Yang, C.T. (1973). "Incipient Motion and Sediment Transport," *Journal of the Hydraulics Division, ASCE*, vol. 99, no. HY10, pp. 1679-1704.

Yang, C.T. (1976). "Minimum Unit Stream Power and Fluvial Hydraulics," *Journal of the Hydraulics Division, ASCE*, vol. 102, No. HY7, pp. 919-934.

Yang, C.T. (1979). "Unit Stream Power Equations for Total Load," *Journal of Hydrology*, vol. 40, pp. 123-138.

Yang, C.T. (1996). *Sediment Transport Theory and Practice*, The McGraw-Hill Companies, Inc., New York, 396 p. (reprint by Krieger Publishing Company, Malabar, Florida, 2003).

Yang, C.T. (2002). "Total Maximum Daily Load of Sediment," *Proceedings of the International Workshop on Ecological, Sociological, and Economic Implications of Sediment in Reservoirs*, April 8-10, Pestrum, Italy.

Yang, C.T., and A. Molinas (1982). "Sediment Transport and Unit Stream Power Function," *Journal of the Hydraulics Division, ASCE*, vol. 108, no. HY6, pp. 774-793.

Yang, C.T., and C.C.S. Song (1986). "Theory of Minimum Energy and Energy Dissipation Rate," *Encyclopedia of Fluid Mechanics*, Chapter 11, Gulf Publishing Company, Houston, Texas, pp. 353-399.

Yang, C.T., and C.C.S. Song (1987). "Theory of Minimum Rate of Energy Dissipation," *Journal of the Hydraulics Division*, ASCE, vol. 105, no. HY7, pp. 769-784.

Yang, C.T., and F.J.M. Simões (2000). *User's Manual for GSTARS 2.1 (Generalized Stream Tube model for Alluvial River Simulation Version 2.1)*, U.S. Bureau of Reclamation, Technical Service Center, Denver, Colorado.

Yang, C.T., and F.J.M. Simões (2002). *User's Manual for GSTARS3 (Generalized Sediment Transport model for Alluvial River Simulation Version 3)*, U.S. Bureau of Reclamation, Technical Service Center, Denver, Colorado.

Yang, C.T., C.C.S. Song, and M.J. Woldenberg (1981). "Hydraulic Geometry and Minimum Rate of Energy Dissipation," *Water Resources Research*, vol. 17, no. 4, pp. 1014-1018.

Yang, C.T., T.J. Randle, and S.K. Hsu (1998). "Surface Erosion, Sediment Transport, and Reservoir Sedimentation," *Proceedings of the Symposium on Modeling Soil Erosion, Sediment Transport, and Closely Related Hydrological Processes*, Vienna, International Association of Hydrologic Sciences Publication, no. 249, pp. 3-12.

Young, R.A., C.A. Onstad, D.D. Bosch, and W.P. Anderson (1989). "AGNPS: A Non-point-Source Pollution Model for Evaluating Agricultural Watersheds," *Journal of Soil and Water Conservation*, vol. 44, no. 2, pp. 168-172.

Zhang, X.C., M.A. Nearing, L.M. Risse, and K.C. McGregor (1996). "Evaluation of WEPP Runoff and Soil Loss Predictions Using Natural Runoff Plot Data," *Transactions of the American Society of Agricultural Engineers*, vol. 39, no. 3, pp. 855-863.

Ziegler, A.D., R.A. Sutherland, and T.W. Giambelluca (2000). "Partitioning Total Erosion on Unpaved Roads into Splash and Hydraulic Components: The Roles of Interstorm Surface Preparation and Dynamic Erodibility," *Water Resources Research.*, vol. 36, no. 9, pp. 2787-2791.

Chapter 3
Noncohesive Sediment Transport

Page

3.1 Introduction 3-1
3.2 Incipient Motion 3-1
3.2.1 Shear Stress Approach 3-2
3.2.2 Velocity Approach 3-7
3.3 Sediment Transport Functions 3-12
3.3.1 Regime Approach 3-12
3.3.2 Regression Approach 3-14
3.3.3 Probabilistic Approach 3-16
3.3.4 Deterministic Approach 3-17
3.3.5 Stream Power Approach 3-23
3.3.5.1 Bagnold's Approach 3-23
3.3.5.2 Engelund and Hansen's Approach 3-25
3.3.5.3 Ackers and White's Approach 3-25
3.3.6 Unit Stream Power Approach 3-28
3.3.7 Power Balance Approach 3-32
3.3.8 Gravitational Power Approach 3-34
3.4 Other Commonly Used Sediment Transport Functions 3-36
3.4.1 Schoklitsch Bedload Formula 3-36
3.4.2 Kalinske Bedload Formula 3-37
3.4.3 Meyer-Peter and Müller Formula 3-39
3.4.4 Rottner Bedload Formula 3-40
3.4.5 Einstein Bedload Formula 3-41
3.4.6 Laursen Bed-Material Load Formula 3-41
3.4.7 Colby Bed-Material Load Formula 3-42
3.4.8 Einstein Bed-Material Load Formula 3-44
3.4.9 Toffaleti Formula 3-44
3.5 Fall Velocity 3-45
3.6 Resistance to Flow 3-47
3.6.1 Einstein's Method 3-49
3.6.2 Engelund and Hansen's Method 3-54
3.6.3 Yang's Method 3-58
3.7 Nonequilibrium Sediment Transport 3-63
3.8 Comparison and Selection of Sediment Transport Formulas 3-63
3.8.1 Direct Comparisons with Measurements 3-64
3.8.2 Comparison by Size Fraction 3-73
3.8.3 Computer Model Simulation Comparison 3-77
3.8.4 Selection of Sediment Transport Formulas 3-83
3.8.4.1 Dimensionless Parameters 3-85
3.8.4.2 Data Analysis 3-86
3.8.4.3 Procedures for Selecting Sediment Transport Formulas 3-102
3.9 Summary 3-104
3.10 References 3-104

Chapter 3
Noncohesive Sediment Transport

by

Chih Ted Yang

3.1 Introduction

Engineers, geologists, and river morphologists have studied the subject of sediment transport for centuries. Different approaches have been used for the development of sediment transport functions or formulas. These formulas have been used for solving engineering and environmental problems. Results obtained from different approaches often differ drastically from each other and from observations in the field. Some of the basic concepts, their limits of application, and the interrelationships among them have become clear to us only in recent years. Many of the complex aspects of sediment transport are yet to be understood, and they remain among the challenging subjects for future studies.

The mechanics of sediment transport for cohesive and noncohesive materials are different. Issues relating to cohesive sediment transport will be addressed in chapter 4. This chapter addresses noncohesive sediment transport only. This chapter starts with a review of the basic concepts and approaches used in the derivation of incipient motion criteria and sediment transport functions or formulas. Evaluations and comparisons of some of the commonly used criteria and transport functions give readers general guidance on the selection of proper functions under different flow and sediment conditions. Some of the materials summarized in this chapter can be found in the book *Sediment Transport Theory and Practice* (Yang, 1996). Most noncohesive sediment transport formulas were developed for sediment transport in clear water under equilibrium conditions. Understanding sediment transport in sediment-laden flows with a high concentration of wash load is necessary for solving practical engineering problems. The need to consider nonequilibrium sediment transport in a sediment routing model is also addressed in this chapter.

3.2 Incipient Motion

Incipient motion is important in the study of sediment transport, channel degradation, and stable channel design. Due to the stochastic nature of sediment movement along an alluvial bed, it is difficult to define precisely at what flow condition a sediment particle will begin to move. Consequently, it depends more or less on an investigator's definition of incipient motion. They use terms such as "initial motion," "several grain moving," "weak movement," and "critical movement." In spite of these differences in definition, significant progress has been made on the study of incipient motion, both theoretically and experimentally.

Figure 3.1 shows the forces acting on a spherical sediment particle at the bottom of an open channel. For most natural rivers, the channel slopes are small enough that the component of gravitational force in the direction of flow can be neglected compared with other forces acting on a spherical sediment particle. The forces to be considered are the drag force F_D, lift force F_L, submerged weight W_S, and resistance force F_R. A sediment particle is at a state of incipient motion when one of the following conditions is satisfied:

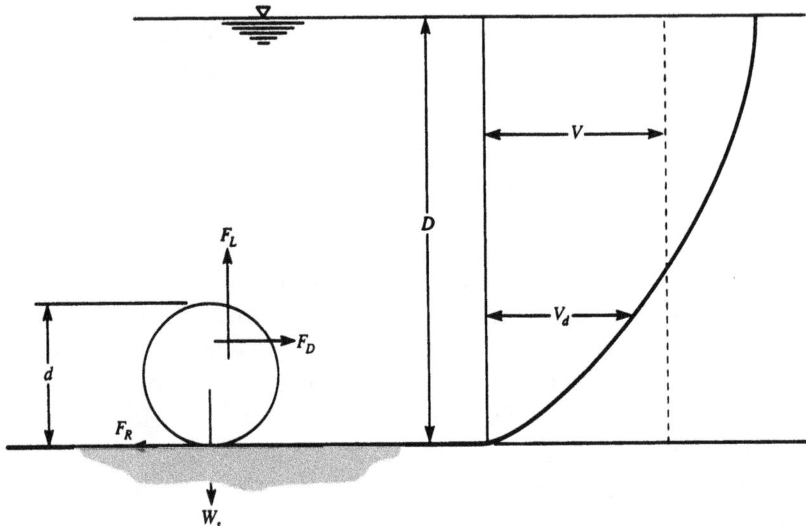

Figure 3.1. Diagram of forces acting on a sediment particle in open channel flow (Yang, 1973).

$$F_L = W_s \tag{3.1}$$

$$F_D = F_R \tag{3.2}$$

$$M_O = M_R \tag{3.3}$$

where M_O = overturning moment due to F_D and F_L, and
M_R = resisting moment due to F_L and W_s.

Most incipient motion criteria are derived from either a shear stress or a velocity approach.

3.2.1 Shear Stress Approach

One of the most prominent and widely used incipient motion criteria is the Shields diagram (1936) based on shear stress. Shields assumed that the factors in the determination of incipient motion are the shear stress τ, the difference in density between sediment and fluid ρ_s - ρ_f, the diameter of the particle d, the kinematic viscosity v, and the gravitational acceleration g. These five quantities can be grouped into two dimensionless quantities, namely,

$$d\frac{\left(\tau_c/\rho_f\right)^{1/2}}{v} = \frac{dU_*}{v} \tag{3.4}$$

and

$$\frac{\tau_c}{d\left(\rho_s-\rho_f\right)g} = \frac{\tau_c}{d\gamma\left[\left(\rho_s/\rho_f\right)-1\right]} \tag{3.5}$$

where ρ_s and ρ_f = densities of sediment and fluid, respectively,
γ = specific weight of water,
U_* = shear velocity, and
τ_c = critical shear stress at initial motion.

The relationship between these two parameters is then determined experimentally. Figure 3.2 shows the experimental results obtained by Shields and other investigators at incipient motion. At points above the curve, the particle will move. At points below the curve, the flow is unable to move the particle. It should be pointed out that Shields did not fit a curve to the data but showed a band of considerable width. Rouse (1939) first proposed the curve shown in Figure 3.2. Although engineers have used the Shields diagram widely as a criterion for incipient motion, dissatisfactions can be found in the literature. Yang (1973) pointed out the following factors and suggested that the Shields' diagram may not be the most desirable criterion for incipient motion.

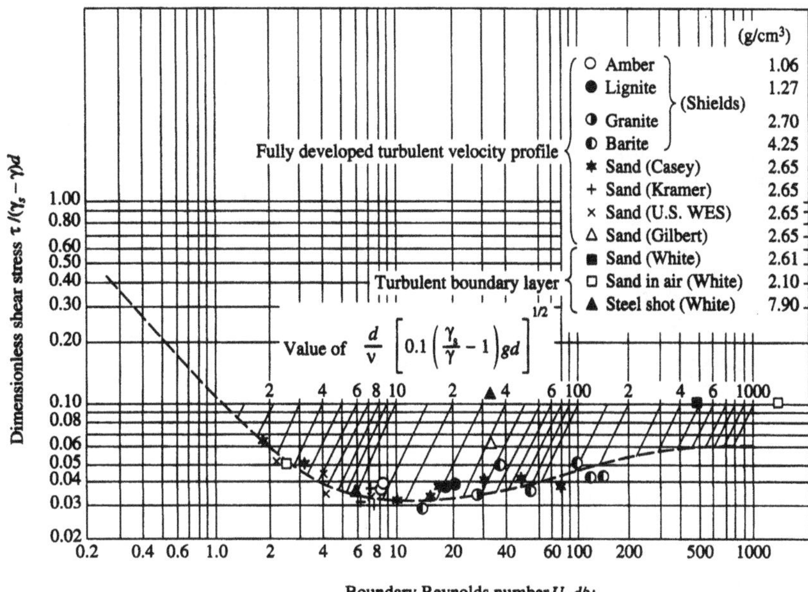

Figure 3.2. Shields diagram for incipient motion (Vanoni, 1975).

- The justification for selecting shear stress instead of average flow velocity is based on the existence of a universal velocity distribution law that facilitates computation of the shear stress from shear velocity and fluid density. Theoretically, water depth does not appear to be related directly to the shear stress calculation, while the main velocity is a function of water depth. However, in common practice, the shear stress is replaced by the average shear stress or tractive force $\tau = \gamma DS$, where γ is the specific weight of water, D is the water depth, and S is the energy slope. In this case, the average shear stress depends on the water depth.

- Although by assuming the existence of a universal velocity distribution law, the shear velocity or shear stress is a measure of the intensity of turbulent fluctuations, our present knowledge of turbulence is limited mainly to laboratory studies.

- Shields derived his criterion for incipient motion by using the concept of a laminar sublayer, according to which the laminar sublayer should not have any effect on the velocity distribution when the shear velocity Reynolds number is greater than 70. However, the Shields diagram clearly indicates that his dimensionless critical shear stress still varies with shear velocity Reynolds number when the latter is greater than 70.

- Shields extends his curve to a straight line when the shear velocity Reynolds number is less than three. This means that when the sediment particle is very small, the critical tractive force is independent of sediment size (Liu, 1958). However, White (1940) showed that for a small shear velocity Reynolds number, the critical tractive force is proportional to the sediment size.

- It is not appropriate to use both shear stress τ and shear velocity U_* in the Shields diagram as dependent and independent variables because they are interchangeable by $U_* = (\tau/\rho)^{1/2}$, where ρ is the fluid density. Consequently, the critical shear stress cannot be determined directly from Shields' diagram; it must be determined through trial and error.

- Shields simplified the problem by neglecting the lift force and considering only the drag force. The lift force cannot be neglected, especially at high shear velocity Reynolds numbers.

- Because the rate of sediment transport cannot be uniquely determined by shear stress (Brooks, 1955; Yang, 1972), it is questionable whether critical shear stress should be used as the criterion for incipient motion of sediment transport.

One of the objections to the use of the Shields diagram is that the dependent variables appear in both ordinate and abscissa parameters. Depending on the nature of the problem, the dependent variable can be critical shear stress or grain size. The American Society of Civil Engineers Task Committee on the Preparation of a Sediment manual (Vanoni, 1977) uses a third parameter

$$\frac{d}{\nu}\left[0.1\left(\frac{\gamma_s}{\gamma}-1\right)gd\right]^{1/2}$$

as shown in Figure 3.2. The use of this parameter enables us to determine its intersection with the Shields diagram and its corresponding values of shear stress. The basic relationship shown in Figure 3.2 has been tested and modified by different investigators. Figure 3.3 shows the results summarized by Govers (1987) in accordance with a modified Shields diagram suggested by Yalin and Karahan (1979).

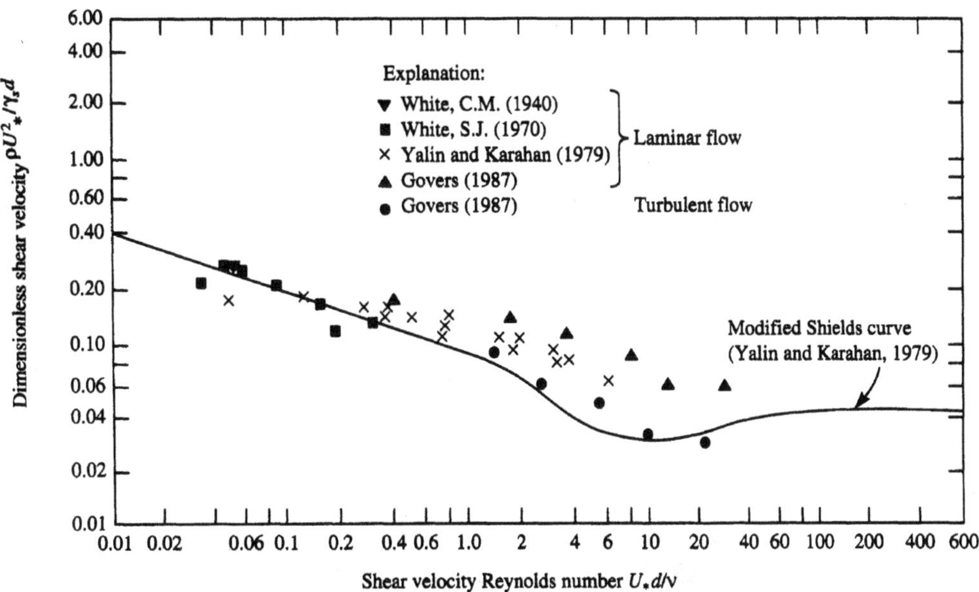

Figure 3.3. Modified Shields diagram (Govers, 1987).

Bureau of Reclamation (1987) developed some stable channel design criteria based on the critical shear stress required to move sediment particles in channels under different flow and sediment conditions. The critical tractive force can be expressed by:

$$\tau_c = \gamma DS \tag{3.6}$$

where τ_c = critical tractive force or shear stress (in lb/ft^2 or g/m^2),
γ = specific weight of water (= 62.4 lb/ft^3 or 1 ton/m^3), and
D = mean flow depth (in ft or m).

Figure 3.4 shows the relationship between critical tractive force and mean sediment diameter for stable channel design recommended by Bureau of Reclamation (1977).

Lane (1953) developed stable channel design curves for trapezoidal channels with different typical side slopes. These curves are based on maximum allowable tractive force and are shown in Figure 3.5. Figure 3.5(a) is for the channel sides, and Figure 3.5(b) is for the channel bottom. Figure 3.5 indicates that the maximum shear stress is about equal to γDS and $0.75\gamma DS$ for the bottom and the sides of the channel, respectively. Lane's study also shows that shear stress is zero at the corners.

The shear stress acting on the channel side at incipient motion is:

$$\tau_w = W_s \cos\theta \tan\phi \left(1 - \frac{\tan^2\theta}{\tan^2\phi}\right)^{1/2} \tag{3.7}$$

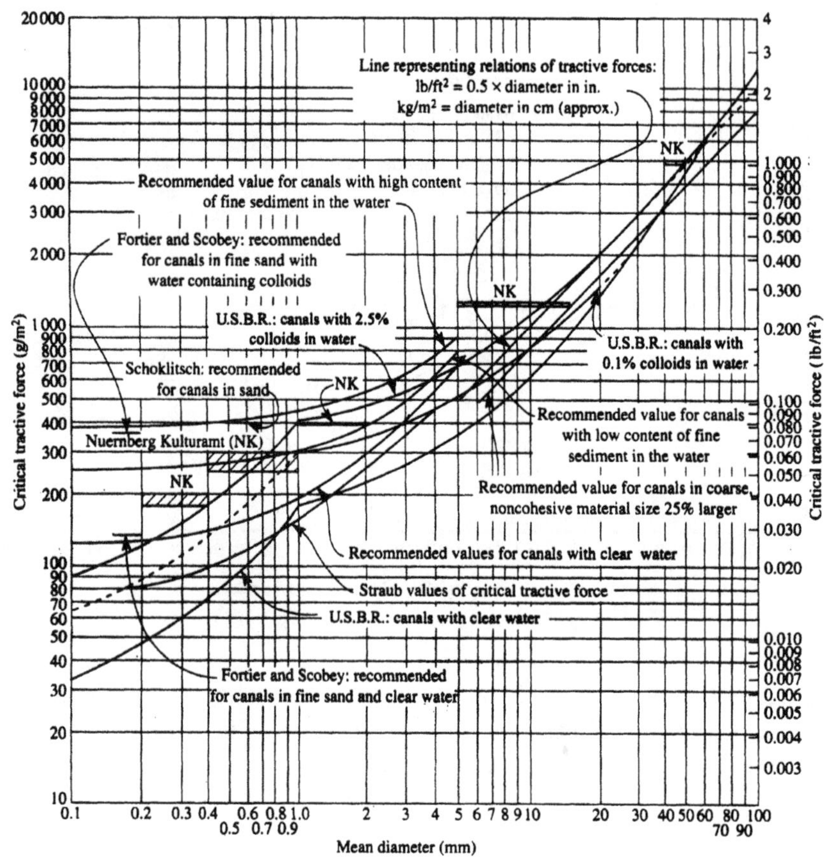

Figure 3.4. Tractive force versus transportable sediment size (Bureau of Reclamation, 1987).

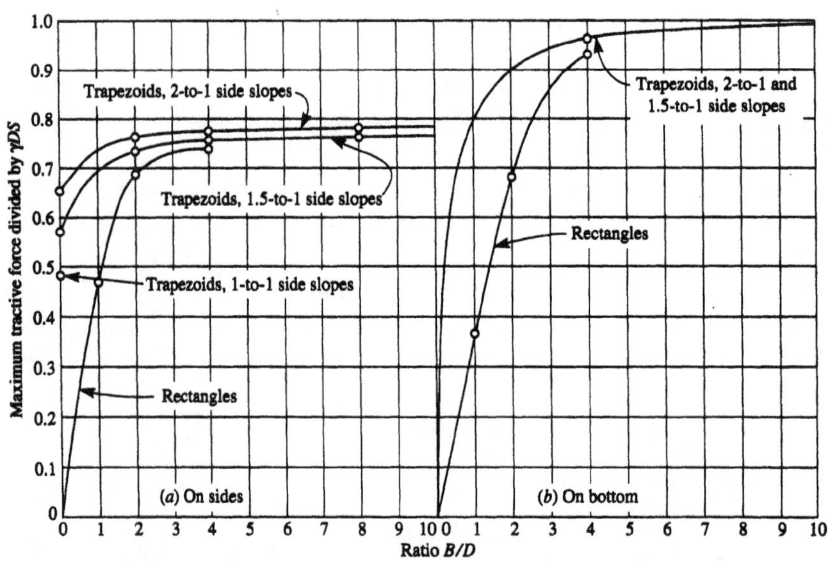

Figure 3.5. Maximum shear stress in a channel (Lane, 1953).

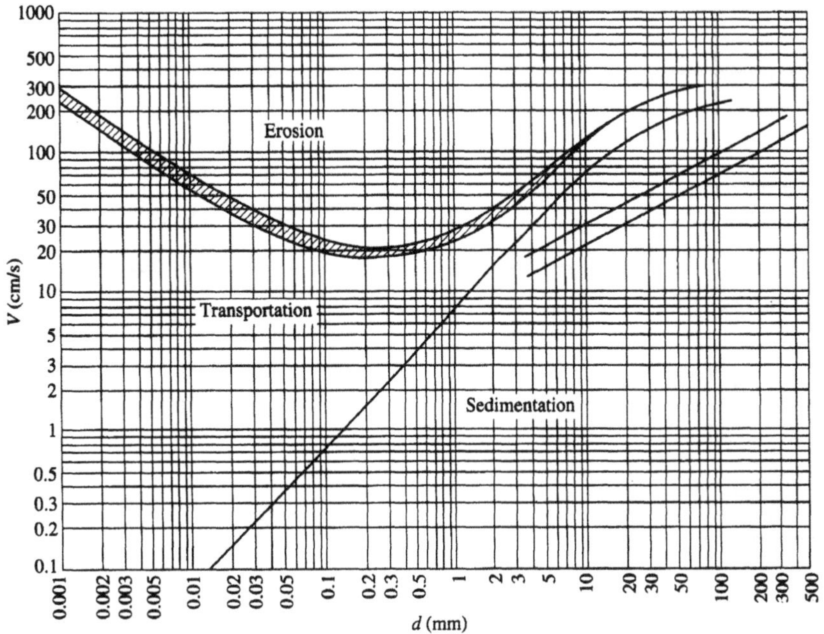

Figure 3.6. Erosion-deposition criteria for uniform particles (Hjulstrom, 1935).

At the bottom of a channel, θ = 0, and equation (3.7) becomes:

$$\tau_b = W_s \tan\phi \tag{3.8}$$

The ratio of limiting tractive forces acting on the channel side and channel bottom is:

$$K = \frac{\tau_w}{\tau_b} = \cos\theta\left(1 - \frac{\tan^2\theta}{\tan^2\phi}\right)^{1/2} \tag{3.9}$$

For stable channel design, the value of τ_b can be obtained from the Shields diagram as shown in Figure 3.2, or from Figure 3.4 for channels of different materials.

3.2.2 Velocity Approach

Fortier and Scobey (1926) made an extensive field survey of maximum permissible values of mean velocity in canals. Table 3.1 shows their permissible velocities for canals of different materials. Hjulstrom (1935) made detailed analyses of the movement of uniform materials on the bottom of channels. Figure 3.6 gives the relationship between sediment size and average flow velocity for erosion, transportation, and sedimentation. The American Society of Civil Engineers Sedimentation Task Committee (Vanoni, 1977) suggested the use of Figure 3.7 for stable channel design.

Table 3.1 Permissible canal velocities (Fortier and Scobey, 1926)

Original material excavated for canal (1)	Velocity* (ft/s) Clear water, no detritus (2)	Water transporting colloidal silts (3)	Water transporting noncolloidal silts, sands, gravels, or rock fragments (4)
Fine sand (noncolloidal)	1.50	2.50	1.50
Sandy loam (noncolloidal)	1.75	2.50	2.00
Silt loam (noncolloidal)	2.00	3.00	2.00
Alluvial silts when noncolloidal	2.00	3.50	2.00
Ordinary firm loam	2.50	3.50	2.25
Volcanic ash	2.50	3.50	2.00
Fine gravel	2.50	5.00	3.75
Stiff clay (very colloidal)	3.75	5.00	3.00
Graded, loam to cobbles, when noncolloidal	3.75	5.00	5.00
Alluvial silts when colloidal	3.75	5.00	3.00
Graded, silt to cobbles, when colloidal	4.00	5.50	5.00
Coarse gravel (noncolloidal)	4.00	6.00	6.50
Cobbles and shingles	5.00	5.50	6.50
Shales and hard pans	6.00	6.00	5.00

* For channels with depth of 3 ft or less after aging

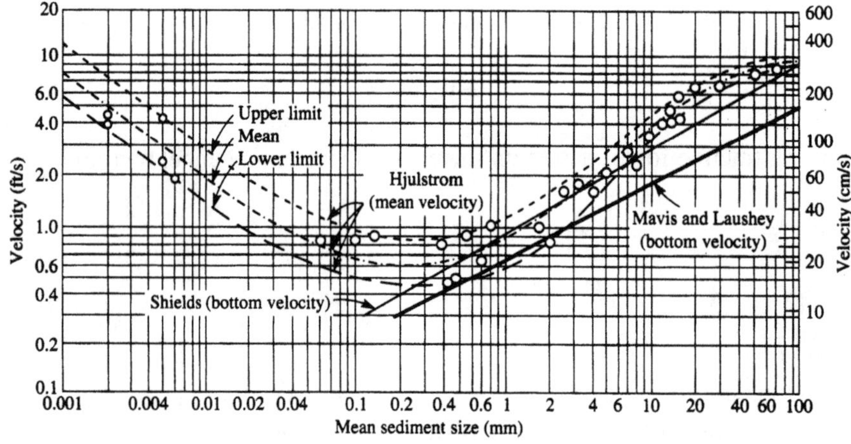

Figure 3.7. Critical water velocities for quartz sediment as a function of mean grain size (Vanoni, 1977).

Yang (1973) applied some basic theories in fluid mechanics to develop his incipient motion criteria. At incipient motion, the resistance force F_R in Figure 3.1 should be balanced by the drag force F_D.

It can be shown that:

$$F_D = \frac{\pi d^3}{6\psi_1}(\rho_s - \rho) g \left(\frac{V}{\omega}\right)^2 \left[\frac{B}{5.75\left[\log(D/d)-1\right]+B}\right]^2 \tag{3.10}$$

The lift force acting on the particle can be obtained as:

$$F_L = \frac{\pi d^3}{6\psi_1\psi_2}(\rho_s - \rho) g \left(\frac{V}{\omega}\right)^2 \left[\frac{B}{5.75\left[\log(D/d)-1\right]+B}\right]^2 \tag{3.11}$$

The submerged weight of the particle is:

$$W_S = \frac{\pi d^3}{6}(\rho_S - \rho)g \tag{3.12}$$

The resistant force:

$$\begin{aligned} F_D &= \psi_3 (W_S - F_L) \\ &= \frac{\psi_3 \pi d^3}{6}\left(\rho_S - \rho\right) g \left\{1 - \frac{1}{\psi_1\psi_2}\left(\frac{V}{\omega}\right)^2 \left[\frac{B}{5.75\left[\log(D/d)-1\right]+B}\right]^2\right\} \end{aligned} \tag{3.13}$$

where ψ_1, ψ_2, ψ_3 = coefficients,
ρ, ρ_S = density of water and sediment, respectively,
D = average flow depth,
D = sediment particle diameter,
ω = sediment particle fall velocity,
V = average flow velocity, and
B = roughness function.

Assume that the incipient motion occurs when $F_D = F_R$. From equations (3.10) and (3.13):

$$\frac{V_{cr}}{\omega} = \left[\frac{5.75\left[\log(D/d)-1\right]}{B}+1\right]\left(\frac{\psi_1\psi_2\psi_3}{\psi_2+\psi_3}\right)^{1/2} \tag{3.14}$$

where V_{cr} = average critical velocity at incipient motion, and
V_{cr}/ω = dimensionless critical velocity.

In the hydraulically smooth regime, B is a function of only the shear velocity Reynolds number U_*d/ν, that is,

$$B = 5.5 + 5.75 \log\frac{U_* d}{\nu}, \quad 0 < \frac{U_* d}{\nu} < 5 \tag{3.15}$$

where U_* = shear velocity, and
ν = kinematic viscosity of water.

Then equation (3.14) becomes:

$$\frac{V_{cr}}{\omega}=\left[\frac{\log(D/d)-1}{\log(U_*d/\nu)+0.956}+1\right]\left(\frac{\psi_1\psi_2\psi_3}{\psi_2+\psi_3}\right)^{1/2} \tag{3.16}$$

which is a hyperbola on a semilog plot between V_{cr}/ω and U_*d/ν. The relative roughness d/D should not have any significant influence on the shape of this hyperbola in the hydraulically smooth regime. In the completely rough regime, the laminar friction contribution can be neglected, and B is a function of only the relative roughness d/D, that is:

$$B=8.5,\qquad \frac{U_*d}{\nu}>70 \tag{3.17}$$

Then equation (3.14) becomes:

$$\frac{V_{cr}}{\omega}=\left[\frac{\log(D/d)-1}{1.48}+1\right]\left(\frac{\psi_1\psi_2\psi_3}{\psi_2+\psi_3}\right)^{1/2} \tag{3.18}$$

Equation (3.18) indicates that in the completely rough regime, the plot of V_{cr}/ω against U_*d/ν is a straight horizontal line. The position of this horizontal line depends on the value of the relative roughness, ψ_1, ψ_2, and ψ_3.

In the transition regime with the shear velocity Reynolds number between 5 and 70, protrusions extend partly outside the laminar sublayer. Both the laminar friction and turbulent friction contributions should be considered. In this case, B deviates gradually from equation (3.15) with increasing U_*d/ν. It is reasonable to expect that, basically, equation (3.16) is still valid, but with the relative roughness d/D playing an increasingly important role as U_*d/ν increases.

Yang (1973) used laboratory data collected by different investigators for the determination of coefficients in equations (3.16) and (3.18). The incipient motion criteria thus obtained are:

$$\frac{V_{cr}}{\omega}=\frac{2.5}{\log(U_*d/\nu)-0.06}+0.66,\qquad 1.2<\frac{U_*d}{\nu}<70 \tag{3.19}$$

and

$$\frac{V_{cr}}{\omega}=2.05,\qquad 70\le\frac{U_*d}{\nu} \tag{3.20}$$

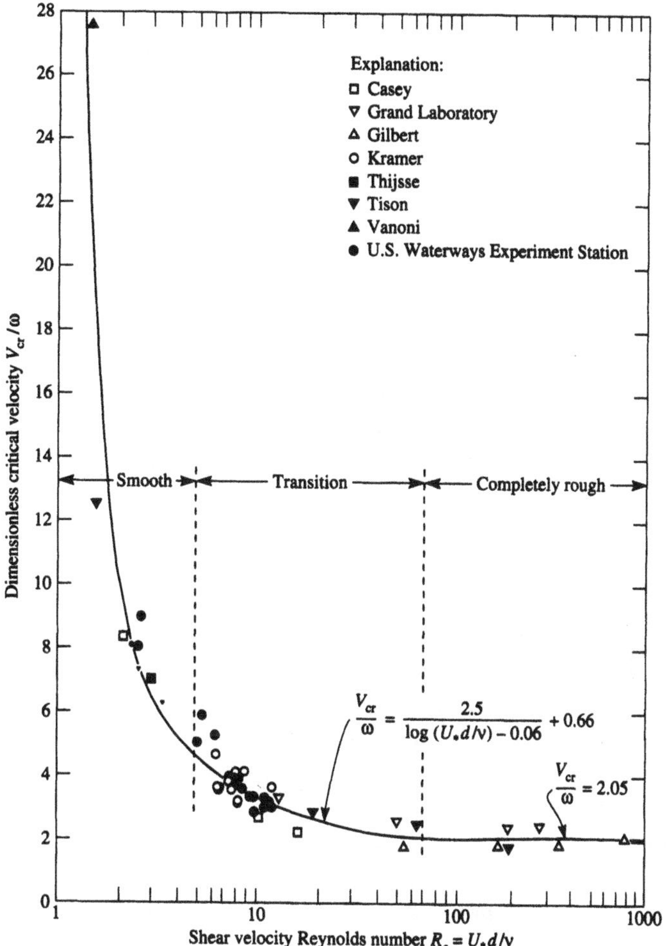

Figure 3.8. Relationship between dimensionless critical average velocity and Reynolds number (Yang, 1973).

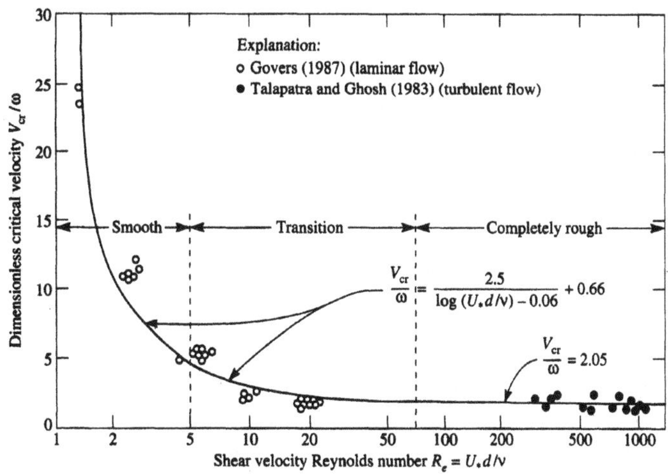

Figure 3.9. Verification of Yang's incipient motion criteria (Yang, 1996, 2003).

Equation (3.19) indicates that the relationship between dimensionless critical average flow velocity and Reynolds number follows a hyperbola when the Reynolds number is less than 70. When the Reynolds number is greater than 70, V_{cr}/ω becomes a constant, as shown in equation (3.20). Figure 3.8 shows comparisons between equations (3.19), (3.20), and laboratory data. Figure 3.9 summarizes independent laboratory verification of Yang's criteria by Govers (1987) and Talapatra and Ghosh (1983).

3.3 Sediment Transport Functions

The basic approaches used in the derivation of sediment transport functions or formulas are the regime, regression, probabilistic, and deterministic approaches. The basic assumptions, their limits of applications, and the theoretical basis of the above approaches and some of the more recent approaches based on the power concept are summarized herein.

3.3.1 Regime Approach

A regime channel is an alluvial channel in dynamic equilibrium without noticeable long-term aggradations, degradation, or change of channel geometry and profile. Some site-specific quantitative relationships exist among sediment transport rates or concentration, hydraulic parameters, and channel geometry parameters. The so-called "regime theory" or "regime equations" are empirical results based on long-term observations of stable canals in India and Pakistan. Blench (1969) summarizes the range of regime channel data as shown in Table 3.2. The regime equations obtained from the regime concept are mainly obtained from the regression analysis of regime canal data.

Different sets of regime equations have been proposed by different investigators, such as those by Blench (1969), Kennedy (1895), and Lacy (1929). According to Blench, applications of regime equations have the following limitations:

- Steady discharge.
- Steady bed-sediment discharges of too small an amount to appear explicitly in the equations.
- Duned sand bed with the particle size distribution natural in the sense of following log-normal distribution.
- Suspended load insufficient to affect the equations.
- Steep, cohesive sides that are erodible or depositable from suspension and behave as hydraulically smooth.
- Straightness in the plan, so that the smoothed, duned bed is level across the cross-section.
- Uniform section and slope.

- Constant water viscosity.

- Range of important parameters as shown in Table 3.2 or in whatever extrapolated range permits the same phase of flow.

Table 3.2 Regime canal data range (after Blench, 1969)

Particle size *d*, mm	0.10–0.60
Silt grading	log probability
Concentration per 10^5	0 to about 3
Suspended load	0–1%
Water temperature	50–86 °F
Channel sides material	clay, smooth
Width-depth ratio, *B/D*	4–30
V^2/D, ft/s^2	0.5–1.5
VB/v	10^6–10^8
Water discharge, *Q*, ft^3/s	1–10,000
Bed form	dunes
D/d	> 1,000

Specifically, the equations are unlikely to apply if the width-depth ratio falls below about 5, or the depth below about 400 millimeters.

The channel-forming discharge, or the dominant discharge, and sediment load or silt factors are the two most important factors to be considered in regime equations. The regime equations are useful engineering tools for stable canal design, especially for those in Pakistan and India. However, they have been subject to criticism for their lack of rational and physical rigor. No regime equations are given in this chapter. Readers who are interested in the application of regime equations should study the conditions under which these empirical equations were obtained. Applications of regime equations to conditions outside of the range of data used in deriving them could lead to erroneous results.

The concept of "regime" is similar to the concepts of "dynamic equilibrium" and "hydraulic geometry." Lacy's (1929) regime equation describing the relationships among channel slope *S*, water discharge *Q*, and silt factor f_s for sediment transport is:

$$S = 0.0005423 \frac{f_s^{5/3}}{Q^{1/6}} \tag{3.21}$$

Leopold and Maddock's (1953) hydraulic geometry relationships are:

$$W = aQ^b \tag{3.22}$$

$$D = cQ^f \tag{3.23}$$

$$V = kQ^m \tag{3.24}$$

where W = channel width,
D = channel depth,
V = average flow velocity,
Q = water discharge, and
a, b, c, f, k, m = site-specific constants.

Yang, et al. (1981) applied the unit stream power theory for sediment transport (Yang, 1973), the theory of minimum unit stream power (Yang, 1971, 1976; Yang and Song, 1979, 1986), and the hydraulic geometry relationships shown in equations (3.22) through (3.24) to derive the relationship between Q and S. They also assumed that:

$$S = iQ^j \tag{3.25}$$

where i, j = constants.

The theoretically derived j value is $-2/11$, which is very close to the empirical value of $-1/6$ shown in equation (3.21).

3.3.2 Regression Approach

Some researchers believe that sediment transport is such a complex phenomenon that no single hydraulic parameter or combination of parameters can be found to describe sediment transport rate under all conditions. Instead of trying to find a dominant variable that can determine the rate of sediment transport, they recommend the use of regressions based on laboratory and field data. The parameters used in these regression equations may or may not have any physical meaning relating to the mechanics of sediment transport.

Shen and Hung (1972) proposed the following regression equation based on 587 sets of laboratory data in the sand size range:

$$\log C_t = -107{,}404.45938164 + 324{,}214.74734085Y - 326{,}309.58908739Y^2 + 109{,}503.87232539Y^3 \tag{3.26}$$

where Y = $(VS^{0.57}/\omega^{0.32})^{0.00750189}$,
C_t = total sediment concentration in ppm by weight, and
ω = average fall velocity of sediment particles.

Before equation (3.26) was finally adopted by Shen and Hung, they performed a sensitivity analysis on the importance of different variables to the rate of sediment transport. Because laboratory data have limited range of variation of water depth, the sensitivity analysis indicated that the rate of sediment transport was not sensitive to changes in water depth. Consequently, water depth was eliminated from consideration. The dimensionally nonhomogeneous parameters used and the lack of ability to reflect the effect of depth change limit the application of equation (3.26) to laboratory flumes and small rivers with particles in the sand size range.

Karim and Kennedy (1990) used nonlinear, multiple-regression analyses to derive relations between flow velocity, sediment discharge, bed-form geometry, and friction factor of alluvial rivers. They used a total of 339 sets of river data and 608 sets of flume data in the analyses. The sediment discharge and velocity relationships adopted by them have the following general forms:

$$\log \frac{q_s}{\left(1.65 g d_{50}^3\right)^{1/2}} = A_0 + A_{ijk} \sum_i \sum_j \sum_k \log X_i \log X_j \log X_k \tag{3.27}$$

$$\log \frac{V}{\left(1.65 g d_{50}^3\right)^{1/2}} = B_0 + B_{pqr} \sum_p \sum_q \sum_r \log X_p \log X_q \log X_r \tag{3.28}$$

where q_s = volumetric total sediment discharge per unit width,
g = gravitational acceleration,
d_{50} = median bed-material particle diameter,
V = mean velocity,
A_0, A_{ijk}, B_0, and B_{pqr} = constants determined from regression analyses, and
X_i, X_j, X_k, X_p, X_q, and X_r = nondimensional independent variables.

The uncoupled relations recommended by Karim and Kennedy are:

$$\begin{aligned} \log \frac{q_s}{(1.65 g d_{50}^3)^{1/2}} = & -2.279 + 2.972 \log \frac{V}{(1.65 g d_{50})^{1/2}} \\ & + 1.060 \log \frac{V}{(1.65 g d_{50})^{1/2}} \log \frac{U_* - U_{*c}}{(1.65 g d_{50})^{1/2}} \\ & + 0.299 \log \frac{D}{d_{50}} \log \frac{U_* - U_{*c}}{(1.65 g d_{50})^{1/2}} \end{aligned} \tag{3.29}$$

and

$$\frac{V}{(1.65 g d_{50})^{1/2}} = 2.822 \left[\frac{q}{(1.65 g d_{50}^3)^{1/2}} \right]^{0.376} S^{0.310} \tag{3.30}$$

where q = water discharge per unit width,
S = energy slope,
V = average flow velocity,
U_* = bed shear velocity = $(gDS)^{1/2}$,
U_{*c} = Shields' value of critical shear velocity at incipient motion, and
D = water depth.

Equation (3.30) can be used for flows well above the incipient sediment motion. If it is necessary to take into account the bed configuration changes in the development of a friction or velocity predictor, equation (3.30) should be replaced by:

$$\frac{V}{(1.65gd_{50})^{1/2}} = 9.82\left[\frac{q}{(1.65gd_{50}^{3})^{1/2}}\right]^{0.216}\left(\frac{f}{f_0}\right)^{-0.164} \tag{3.31}$$

where f = the Darcy-Weisbach friction factor.

The grain roughness factor f_0 can be expressed as:

$$f_0 = \frac{8}{\left[6.25 + 2.5\ln(D/2.5d_{50})\right]^2} \tag{3.32}$$

The friction factor ratio f/f_0 in equation (3.31) can be computed as:

$$\frac{f}{f_o} = 1.20 + 8.92\left[0.08 + 2.24\left(\frac{\theta}{3}\right) - 18.13\left(\frac{\theta}{3}\right)^2 + 70.90\left(\frac{\theta}{3}\right)^3 - 88.33\left(\frac{\theta}{3}\right)^4\right] \quad \text{for } \theta \le 1.5 \tag{3.33a}$$

$$\frac{f}{f_o} = 1.20 \quad \text{for } \theta > 1.5 \tag{3.33b}$$

where

$$\theta = \frac{\tau_0}{1.65\gamma d_{50}} = \frac{DS}{1.65d_{50}} \tag{3.34}$$

and γ = specific weight of water.

Equations (3.29), (3.31), and (3.33) constitute a set of coupled sediment discharge friction, and bed-form relations. Yang (1996) summarized the interaction scheme for solving equations (3.29), (3.31), and (3.33) for a set of known values of q, S, and d_{50}.

A regression equation may give fairly accurate results for engineering purposes if the equation is applied to conditions similar to those from where the equation was derived. Application of a regression equation outside the range of data used for deriving the regression equation should be carried out with caution. In general, regression equations without a theoretical basis and without using dimensionless parameters should not be used for predicting sediment transport rate or concentration in natural rivers.

3.3.3 Probabilistic Approach

Einstein (1950) pioneered sediment transport studies from the probabilistic approach. He assumed that the beginning and ceasing of sediment motion can be expressed in terms of probability. He also assumed that the movement of bedload is a series of steps followed by rest periods. The average step length is 100 times the particle diameter. Einstein used the hiding correction factor and lifting correction factor to better match theoretical results with observed laboratory data.

In spite of the sophisticated theories used, the Einstein bedload transport function is not a popular one for engineering applications. This is partially due to the complex computational procedures required. However, the probabilistic approach developed by Einstein has been used as a theoretical basis for developing other transport functions, such as the method proposed by Toffaleti (1969).

Based on the mode of transport, total sediment load consists of bedload and suspended load. Total load can also be divided into measured and unmeasured load. The original Einstein function has been modified by others for the estimation of unmeasured load. The original Einstein function is a predictive function for sediment transport. The "modified Einstein method" is not a predictive function. The method can be used to estimate bedload or unmeasured load based on measured suspended load for the estimation of total load or total bed-material load. The method proposed by Colby and Hembree (1955) is one of the most commonly used modified Einstein methods for the computation of total bed-material load.

Application of the original Einstein method and the modified Einstein method is labor intensive. Unless necessary, these methods are not commonly used for solving engineering problems or used in a computer model for routing sediment. Yang (1996) provided detailed explanations of these methods with step-by-step computation examples for engineers to follow.

3.3.4 Deterministic Approach

The basic assumption in a deterministic approach is the existence of one-to-one relationship between independent and dependent variables. Conventional, dominant, independent variables used in sediment transport studies are water discharge, average flow velocity, shear stress, and energy or water surface slope. More recently, the use of stream power and unit stream power have gained increasing acceptance as important parameters for the determination of sediment transport rate or concentration. Other independent parameters used in sediment transport functions are sediment particle diameter, water temperature, or kinematic viscosity. The accuracy of a deterministic sediment transport formula depends on the generality and validity of the assumption of whether a unique relationship between dependent and independent variables exists. Deterministic sediment transport formulas can be expressed by one of the following forms:

$$q_s = A_1(Q - Q_c)^{B_1} \tag{3.35}$$

$$q_s = A_2(V - V_c)^{B_2} \tag{3.36}$$

$$q_s = A_3(S - S_c)^{B_3} \tag{3.37}$$

$$q_s = A_4(\tau - \tau_c)^{B_4} \tag{3.38}$$

$$q_s = A_5(\tau V - \tau_c V_c)^{B_5} \tag{3.39}$$

$$q_s = A_6(VS - V_c S_c)^{B_6} \tag{3.40}$$

where q_s = sediment discharge per unit width of channel,
Q = water discharge,
V = average flow velocity,
S = energy or water surface slope,
τ = shear stress,
τV = stream power per unit bed area,
VS = unit stream power,
$A_1, A_2, A_3, A_4, A_5, A_6, B_1, B_2, B_3, B_4, B_5, B_6$ = parameters related to flow and sediment conditions, and
c = subscript denoting the critical condition at incipient motion.

Yang (1972, 1983) used laboratory data collected by Guy et al. (1966) from a laboratory flume with 0.93-mm sand, as shown in Figure 3.10, as an example to examine the validity of these assumptions.

Figure 3.10(a) shows the relationship between the total sediment discharge and water discharge. For a given value of Q, two different values of q_t can be obtained. Field data obtained by Leopold and Maddock (1953) also indicate similar results. Some of Gilbert's (1914) data indicate that no correlation exists at all between water discharge and sediment discharge. Apparently, different sediment discharges can be transported by the same water discharge, and a given sediment discharge can be transported by different water discharges. The same sets of data shown in Figure 3.10(a) are plotted in Figure 3.10(b) to show the relationship between total sediment discharge and average velocity. Although q_t increases steadily with increasing V, it is apparent that for approximately the same value of V, the value of q_t can differ considerably, owing to the steepness of the curve. Some of Gilbert's (1914) data also indicate that the correlations between q_t and V are very weak. Figure 3.10(c) indicates that different amounts of total sediment discharges can be obtained at the same slope, and different slopes can also produce the same sediment discharge. Figure 3.10(d) shows that a fairly well-defined correlation exists between total sediment discharge and shear stress when total sediment discharge is in the middle range of the curve. For either higher or lower sediment discharge, the curve becomes vertical, which means that for the same shear stress, numerous values of sediment discharge can be obtained.

It is apparent from Figure 3-10(a-d) that more than one value of total sediment discharge can be obtained for the same value of water discharge, velocity, slope, or shear stress. The validity of the assumption that total sediment discharge of a given particle size could be determined from water discharge, velocity, slope, or shear stress is questionable.

Because of the basic weakness of these assumptions, the generality of an equation derived from one of these assumptions is also questionable. When the same sets of data are plotted on Figure 3.10(e), with stream power as the independent variable, the correlation improves. Further improvement can be made by using unit stream power as the dominant variable, as shown in Figure 3.10(f). This close correlation exists in spite of the presence of different bed forms, such as plane bed, dune, transition, and standing wave.

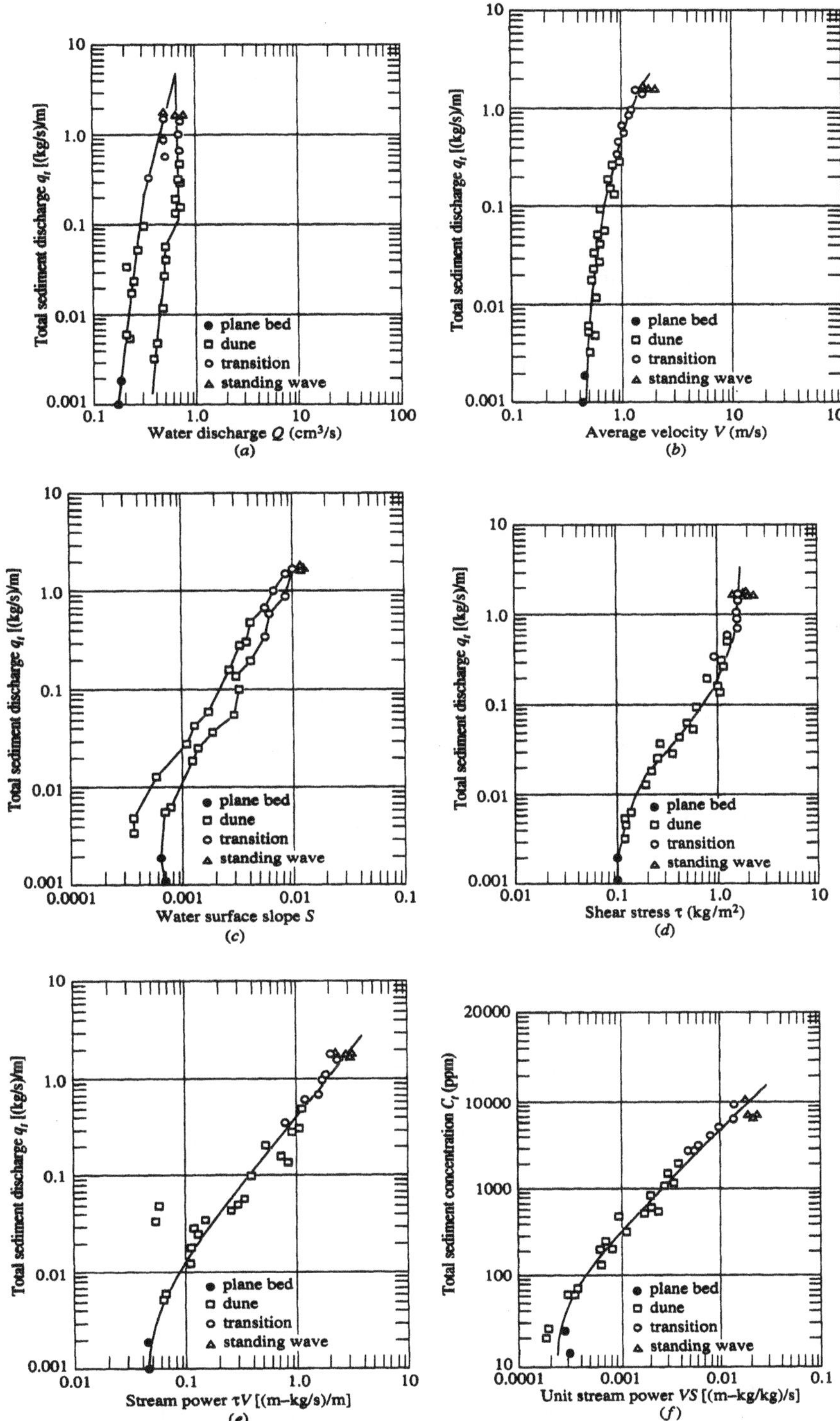

Figure 3.10. Relationships between total sediment discharge and (a) water discharge, (b) velocity, (c) slope, (d) shear stress, (e) stream power, and (f) unit stream power, for 0.93-mm sand in an 8-ft wide flume (Yang, 1972, 1983).

The close relationship between total sediment concentration and unit stream power exists not only in straight channels but also in those channels that are in the process of changing their patterns from straight to meandering, and to braided channels, as shown in Figure 3.11 (Yang, 1977). Schumm and Khan (1972) collected these data.

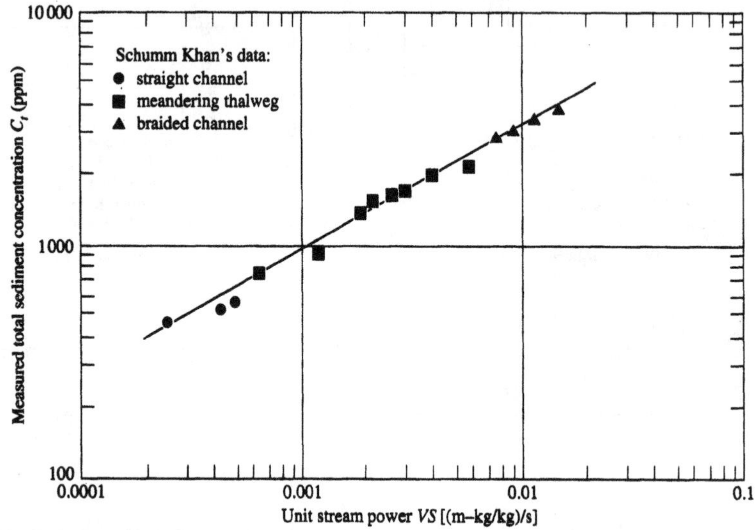

Figure 3.11. Relationship between total concentration and unit stream power during process of channel pattern development from straight to meandering, and to braided (Yang, 1977).

Vanoni (1978), among others, has confirmed the fact that unit stream power dominates sediment discharge or concentration. It is apparent from the results in Figure 3.12 that sediment concentration cannot be determined from relative roughness D/d_{50} and Froude number Fr. However, when the same data are plotted in Figure 3.13 using dimensionless unit stream power VS/ω as the dominant variable, the improvement is apparent.

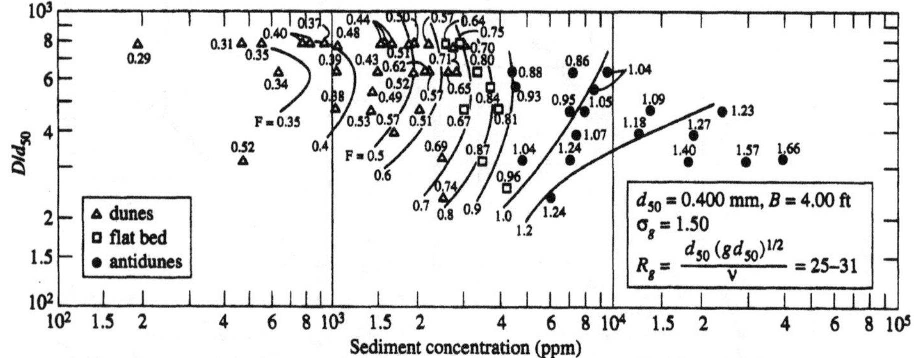

Figure 3.12. Plot of Stein's (1965) data as sediment discharge concentration against Froude number Fr (indicated by the number next to each data point) and the ratio of flow depth D to bed-sediment size d_{50} (Vanoni, 1978).

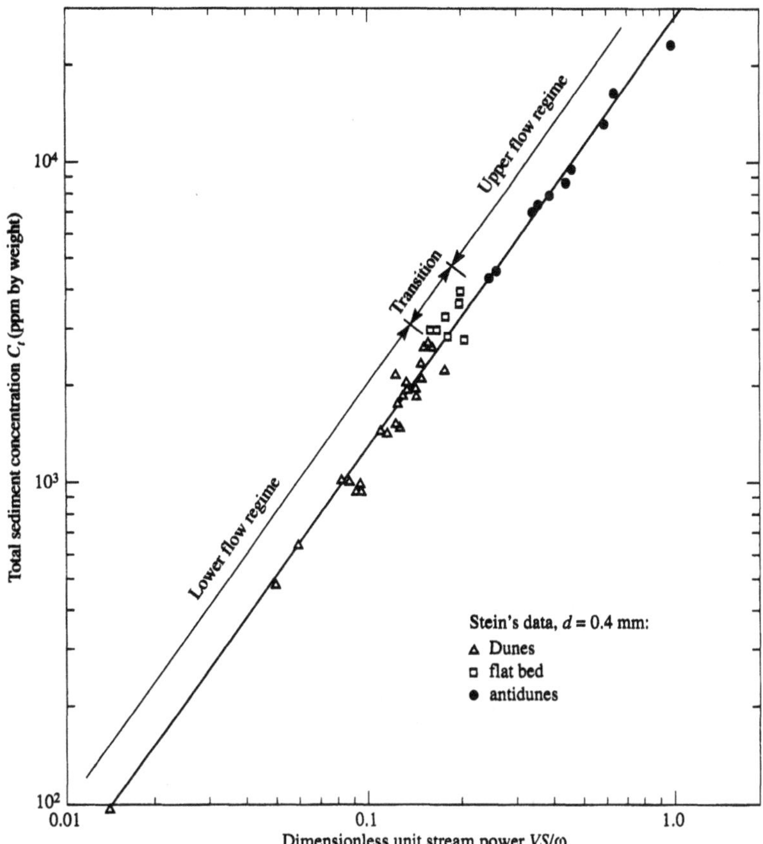

Figure 3.13. Relationship between sediment concentration and dimensionless unit stream power (Yang and Kong, 1991).

Many investigators believe that shear stress τ or stream power τV would be more suitable for the study of coarse material or bedload movement, because these parameters represent the force or power acting along the bed. Yang and Molinas (1982) have shown theoretically that bedload and suspended load, as well as total load, are directly related to unit stream power.

Yang (1983, 1984) used Meyer-Peter and Müller's (1948) gravel data to verify the theoretical finding that bedload can be more accurately determined by unit stream power than by shear stress or stream power. Figure 3.14 shows the loop effect when shear stress or stream power is used as the dominant variable. Gilbert's (1914) data (Figure 3.15) indicate that a family of curves exists between gravel concentration and shear stress or stream power, with water discharge as the third parameter. These results indicate that bedload may not be determined by using shear stress, stream power, or water discharge as the dominant variable. In each case, more than one value of gravel concentration can be obtained at a given value of shear stress, stream power, or water discharge. However, the well-defined strong correlation between gravel concentration and dimensionless unit stream power VS/ω shown in Figures 3.14 and 3.15 is apparent.

It can be concluded that, of all the parameters used in the determination of sediment transport rate, stream power and unit stream power have the strongest correlation with sediment transport rate or concentration. Based on the theoretical derivations and measured data, unit stream power VS or dimensionless unit stream power VS/ω are preferable to other parameters for the determination of sediment transport rate or concentration. The lack of well-defined strong correlation between sediment load or concentration and a dominant variable selected for the development of a sediment transport equation may be the fundamental reason for discrepancies between computed and measured results under different flow and sediment conditions.

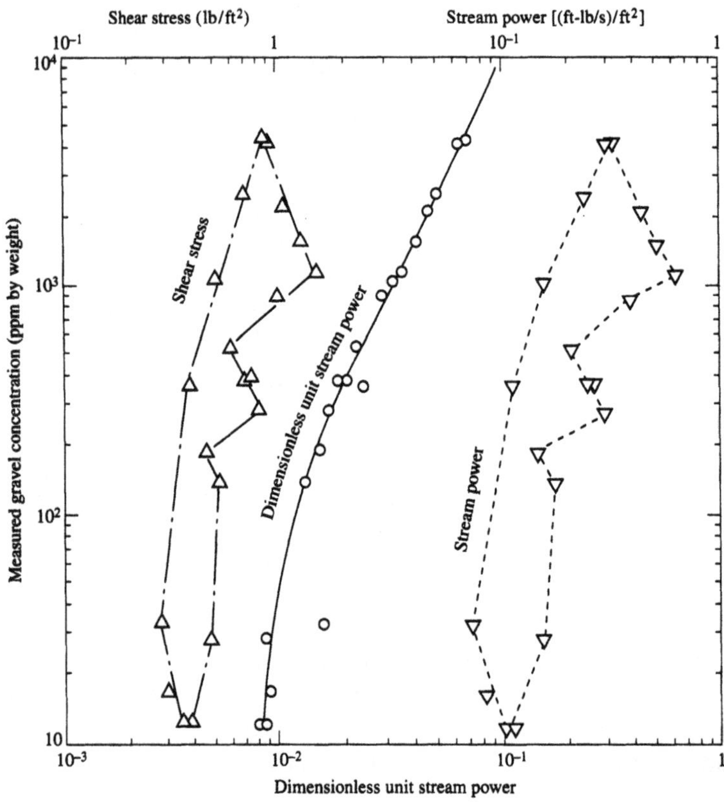

Figure 3.14. Relationship between dimensionless unit stream power, stream power, shear stress, and 5.12-mm gravel concentration measure by Meyer-Peter and Müller (Yang, 1984).

Yang (1996) summarized more detailed explanations and derivations. Due to the importance of stream power, unit stream power, and other power approaches to the determination of sediment transport rate or concentration, more detailed analyses will be made in the following sections.

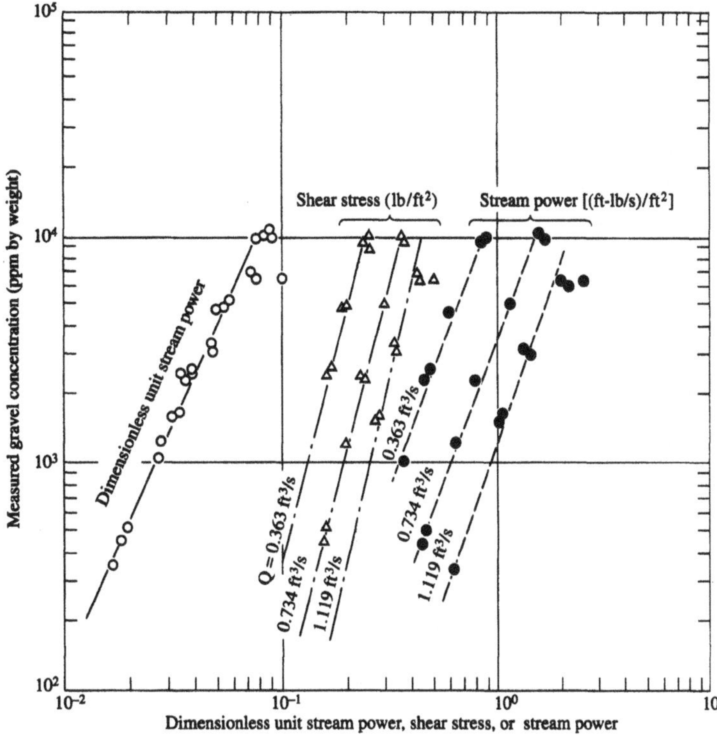

Figure 3.15. Relationship between dimensionless unit stream power, shear stress, stream power, and 4.94-mm gravel concentration measured by Gilbert from a 0.2-m flume (Yang, 1983, 1984).

3.3.5 Stream Power Approach

Bagnold (1966) introduced the stream power concept for sediment transport based on general physics. Engelund and Hansen (1972), and Ackers and White (1973) later used the concept as the theoretical basis for developing their sediment transport functions (Yang, 2002). These transport functions are summarized herein.

3.3.5.1 Bagnold's Approach

From general physics, the rate of energy used in transporting materials should be related to the rate of materials being transported. Bagnold (1966) defined stream power τV as the power per unit bed area which can be used to transport sediment. Bagnold's basic relationship is:

$$\frac{\gamma_s - \gamma}{\gamma} q_{bw} \tan\alpha = \tau V e_b \tag{3.41}$$

where γ_s and γ = specific weights of sediment and water, respectively,
q_{bw} = bedload transport rate by weight per unit channel width,
$\tan\alpha$ = ratio of tangential to normal shear force,

τ = shear force acting along the bed,
V = average flow velocity, and
e_b = efficiency coefficient.

In equation (3.41), the values of e_b and tan α were given by Bagnold in two separate figures. The rate of work needed in transporting the suspended load is:

$$\phi_s = \frac{\gamma_s - \gamma}{\gamma} q_{sw} \frac{\omega}{\bar{u}_s} \tag{3.42}$$

where q_{sw} = suspended load discharge in dry weight per unit time and width,
$\bar{u}_s$ = mean transport velocity of suspended load, and
ω = fall velocity of suspended sediment.

The rate of energy available for transporting the suspended load is:

$$\phi_s' = \tau V (1 - e_b) \tag{3.43}$$

Based on general physics, the rate of work being done should be related to the power available times the efficiency of the system; that is:

$$\frac{\gamma_s - \gamma}{\gamma} q_{sw} \frac{\omega}{\bar{u}_s} = \tau V (1 - e_b) e_s \tag{3.44}$$

where e_s = suspended load transport efficiency coefficient.

Equation (3.44) can be rearranged as:

$$\frac{\gamma_s - \gamma}{\gamma} q_{sw} = (1 - e_b) e_s \frac{\bar{u}_s}{\omega} \tau V \tag{3.45}$$

Assuming $\bar{u}_s = V$, Bagnold found $(1 - e_b)e_s = 0.01$ from flume data. Thus, the suspended load can be computed by:

$$\frac{\gamma_s - \gamma}{\gamma} q_{sw} = 0.01 \tau V^2 / \omega \tag{3.46}$$

The total load in dry weight per unit time and unit width is the sum of bedload and suspended load; that is, from equations (3.41) and (3.46):

$$q_t = q_{bw} + q_{sw} = \frac{\gamma}{\gamma_s - \gamma} \tau V \left(\frac{e_b}{\tan \alpha} + 0.01 \frac{V}{\omega} \right) \tag{3.47}$$

where q_t = total load [in (lb/s)/ft].

3.3.5.2 Engelund and Hansen's Approach

Engelund and Hansen (1972) applied Bagnold's stream power concept and the similarity principle to obtain a sediment transport function:

$$f'\phi = 0.1\theta^{5/2} \tag{3.48}$$

with

$$f' = \frac{2gSD}{V^2} \tag{3.49}$$

$$\phi = \frac{q_t}{\gamma_s}\left[\left(\frac{\gamma_s - \gamma}{\gamma}\right)gd^3\right]^{-1/2} \tag{3.50}$$

$$\theta = \frac{\tau}{(\gamma_s - \gamma)d} \tag{3.51}$$

where g = gravitational acceleration,
S = energy slope,
V = average flow velocity,
q_t = total sediment discharge by weight per unit width,
γ_s and γ = specific weights of sediment and water, respectively,
d = median particle diameter, and
τ = shear stress along the bed.

Strictly speaking, equation (3.48) should be applied to those flows with dune beds in accordance with the similarity principle. However, Engelund and Hansen found that it can be applied to the dune bed and the upper flow regime with particle size greater than 0.15 mm without serious deviation from the theory. Yang (2002) made step-by-step theoretical derivations to show that the basic form of Engelund and Hansen's transport function can be obtained from Bagnold's stream power concept. Yang (1966) also provided a numerical example on the application of Engelund and Hansen's transport function.

3.3.5.3 Ackers and White's Approach

Ackers and White (1973) applied dimensional analysis to express mobility and sediment transport rate in terms of some dimensionless parameters. Their mobility number for sediment transport is:

$$F_{gr} = U_*^n\left[gd\left(\frac{\gamma_s}{\gamma} - 1\right)\right]^{-1/2}\left[\frac{V}{\sqrt{32}\log(\alpha D/d)}\right]^{1-n} \tag{3.52}$$

where U_* = shear velocity,
n = transition exponent, depending on sediment size,
α = coefficient in rough turbulent equation (= 10),
d = sediment particle size, and
D = water depth.

They also expressed the sediment size by a dimensionless grain diameter:

$$d_{gr} = d\left[\frac{g(\gamma_s/\gamma - 1)}{\nu^2}\right]^{1/3} \tag{3.53}$$

where ν = kinematic viscosity.

A general dimensionless sediment transport function can then be expressed as:

$$G_{gr} = f\left(F_{gr}, d_{gr}\right) \tag{3.54}$$

with

$$G_{gr} = \frac{XD}{d\gamma_s/\gamma}\left(\frac{U_*}{V}\right)^n \tag{3.55}$$

where X = rate of sediment transport in terms of mass flow per unit mass flow rate; i.e., concentration by weight of fluid flux.

The generalized dimensionless sediment transport function can also be expressed as:

$$G_{gr} = C\left(\frac{F_{gr}}{A} - 1\right)^m \tag{3.56}$$

Ackers and White (1973) determined the values of A, C, m, and n based on best-fit curves of laboratory data with sediment size greater than 0.04 mm and Froude number less than 0.8. For the transition zone with $1 < d_{gr} \leq 60$,

$$n = 1.00 - 0.56 \log d_{gr} \tag{3.57}$$

$$A = 0.23 d_{gr}^{-1/2} + 0.14 \tag{3.58}$$

For coarse sediment, $d_{gr} > 60$:

$$n = 0.00 \tag{3.59}$$

$$A = 0.17 \tag{3.60}$$

$$m = 1.50 \tag{3.61}$$

$$C = 0.025 \tag{3.62}$$

For the transition zone:

$$m = \frac{9.66}{d_{gr}} + 1.34 \tag{3.63}$$

$$\log C = 2.86 \log d_{gr} - (\log d_{gr})^2 - 3.53 \tag{3.64}$$

The procedure for the computation of sediment transport rate using Ackers and White's approach is summarized as follows:

1. Determine the value of d_{gr} from known values of d, g, γ_s/γ, and v in equation (3.53).

2. Determine values of n, A, m, and C associated with the derived d_{gr} value from equations (3.57) through (3.64).

3. Compute the value of the particle mobility F_{gr} from equation (3.52).

4. Determine the value of G_{gr} from equation (3.56), which represents a graphical version of the new sediment transport function.

5. Convert G_{gr} to sediment flux X, in ppm by weight of fluid flux, using equation (3.55).

Although it is not apparent from the above procedures, Yang (2002) provided step-by-step derivations to show that Ackers and White's basic transport function can be derived from Bagnold's stream power concept.

The original Ackers and White formula is known to overpredict transport rates for fine sediments (smaller than 0.2 mm) and for relatively coarse sediments. To correct that tendency, a revised form of the coefficients was published in 1990 (HR Wallingford, 1990). Table 3.3 gives the comparison between the original and revised coefficients.

Reclamation's computer models GSTARS 2.1 (Yang and Simões, 2000) and GSTARS3 (Yang and Simões, 2002) allow users to select either the 1973 or the 1990 values in their application of the Ackers and White sediment transport function.

Table 3.3. Coefficients for the 1973 and 1990 versions of the Ackers and White transport function

1973	1990
$1 < d_{gr} \le 60$ $A = 0.23 d_{gr}^{-1/2} + 0.14$	$A = 0.23 d_{gr}^{-1/2} + 0.14$
$\log C = -3.53 + 2.86 \log d_{gr} - (\log d_{gr})^2$	$\log C = -3.46 + 2.79 \log d_{gr} - 0.98 (\log d_{gr})^2$
$m = 9.66\, d_{gr}^{-1} + 1.34$	$m = 6.83\, d_{gr}^{-1} + 1.67$
$n = 1.00 - 0.56 \log d_{gr}$	$n = 1.00 - 0.56 \log d_{gr}$
$d_{gr} > 60$ $A = 0.17$	$A = 0.17$
$C = 0.025$	$C = 0.025$
$m = 1.50$	$m = 1.78$
$n = 0$	$n = 0$

3.3.6 Unit Stream Power Approach

The rate of energy per unit weight of water available for transporting water and sediment in an open channel with reach length x and total drop of Y is:

$$\frac{dY}{dt} = \frac{dx}{dt}\frac{dY}{dx} = VS \tag{3.65}$$

where V = average flow velocity, and
S = energy or water surface slope.

Yang (1972) defines unit stream power as the velocity-slope product shown in equation (3.65). The rate of work being done by a unit weight of water in transporting sediment must be directly related to the rate of work available to a unit weight of water. Thus, total sediment concentration or total bed-material load must be directly related to unit stream power. While Bagnold (1966) emphasized the power applies to a unit bed area, Yang (1972, 1973) emphasized the power available per unit weight of water to transport sediments.

To determine total sediment concentration, Yang (1973) considered a relation between the relevant variables of the form

$$\Phi(C_t, VS, U_*, \nu, \omega, d) = 0 \tag{3.66}$$

where C_t = total sediment concentration, with wash load excluded (in ppm by weight):
VS = unit stream power,
U_* = shear velocity,
V = kinematic viscosity,
ω = fall velocity of sediment, and
d = median particle diameter.

Using Buckingham's π theorem and the analysis of laboratory data, C_t in equation (3.66) can be expressed in the following dimensionless form:

$$\log C_t = I + J \log\left(\frac{VS}{\omega} - \frac{V_{cr}S}{\omega}\right) \tag{3.67}$$

where $V_{cr}S/\omega$ = critical dimensionless unit stream power at incipient motion.

I and J in equation (3.67) are dimensionless parameters reflecting the flow and sediment characteristics, that is:

$$I = a_1 + a_2 \log\frac{\omega d}{v} + a_3 \log\frac{U_*}{\omega} \tag{3.68}$$

$$J = b_1 + b_2 \log\frac{\omega d}{v} + b_3 \log\frac{U_*}{\omega} \tag{3.69}$$

where $a_1, a_2, a_3, b_1, b_2, b_3$ = coefficients.

Yang (1973) used 463 sets of laboratory data for the determination of coefficients in equations (3.68) and (3.69). The dimensionless unit stream power equation for sand transport thus obtained is:

$$\begin{aligned}\log C_{ts} &= 5.435 - 0.286 \log\frac{\omega d}{v} - 0.457 \log\frac{U_*}{\omega} \\ &+ \left(1.799 - 0.409 \log\frac{\omega d}{\omega} - 0.314 \log\frac{U_*}{\omega}\right)\log\left(\frac{VS}{\omega} - \frac{V_{cr}S}{\omega}\right)\end{aligned} \tag{3.70}$$

where C_{ts} = total sand concentration in ppm by weight.

The critical dimensionless unit stream power $V_{cr}S/\omega$ is the product of dimensionless critical velocity $V_{cr}S/\omega$ shown in equations (3.19) and (3.20) and the energy slope S. Yang and Molinas (1982) made a step-by-step derivation to show that sediment concentration is indeed directly related to unit stream power, based on basic theories in fluid mechanics and turbulence. They showed that the vertical sediment concentration distribution is directly related to the vertical distribution of turbulence energy production rate; that is:

$$\frac{\bar{C}}{\bar{C}_a} = \left[\frac{\tau_{xy} d\bar{U}_x/dy}{\left(\tau_{xy} d\bar{U}_x/dy\right)_{y=a}}\right]^{Z_1} \tag{3.71}$$

where $\bar{C}, \bar{C}_a$ = time-averaged sediment concentration at a given cross-section and at a depth a above the bed, respectively,
τ_{xy} = turbulence shear stress,
dU_x/dy = velocity gradient,
$\tau_{xy}\, dU_x/dy$ = turbulence energy production rate,
Z_1 = $\omega/k\beta U_*$,
ω = sediment particle fall velocity,
β = coefficient,
k = von Karman constant, and
U_* = shear velocity.

Figure 3.16 shows comparisons between measured and theoretical results from equation (3.71). This confirmation is independent from the selection of reference elevation a.

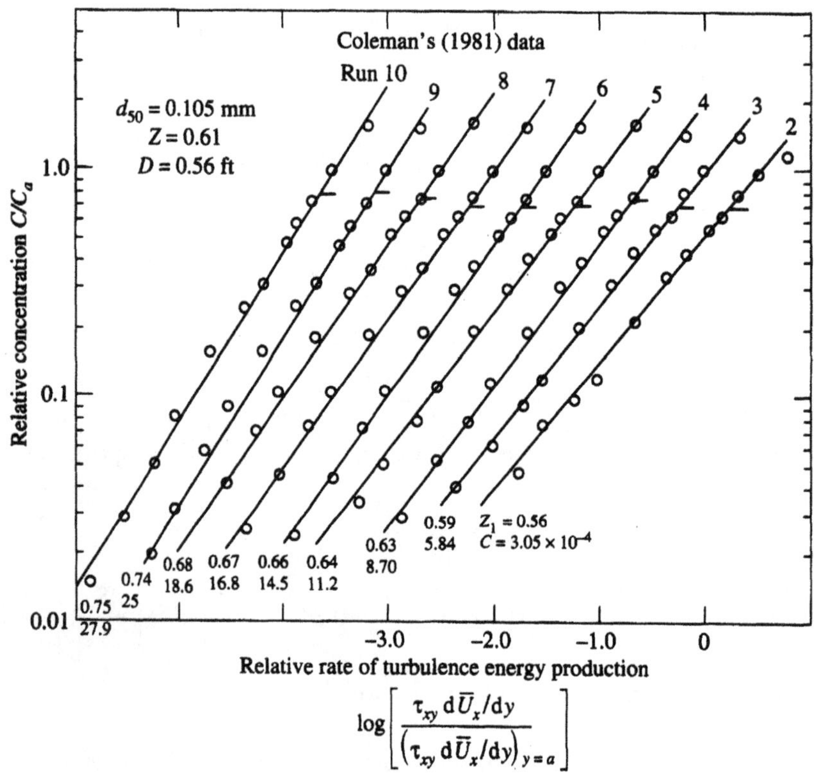

$$\log\left[\frac{\tau_{xy}\, d\bar{U}_x/dy}{\left(\tau_{xy}\, d\bar{U}_x/dy\right)_{y=a}}\right]$$

Figure 3.16. Comparison between theoretical and measured suspended sediment concentration distributions (Yang, 1985).

For sediment concentration higher than about 100 ppm by weight, the need to include incipient motion criteria in a sediment transport equation decreases. Yang (1979) introduced the following dimensionless unit stream power equation for sand transport with concentration higher than 100 ppm:

$$\log C_{ts} = 5.165 - 0.153 \log \frac{\omega d}{\nu} - 0.297 \log \frac{U_*}{\omega} + \left(1.780 - 0.360 \log \frac{\omega d}{\nu} - 0.480 \log \frac{U_*}{\omega}\right) \log \frac{VS}{\omega} \tag{3.72}$$

Yang (1984) extended his dimensionless unit stream power equation for sand transport to gravel transport by calibrating the coefficients in equations (3.68) and (3.69) with gravel data. The gravel equation thus obtained is:

$$\log C_{tg} = 6.681 - 0.633 \log \frac{\omega d}{\nu} - 4.816 \log \frac{U_*}{\omega} + \left(2.784 - 0.305 \log \frac{\omega d}{\nu} - 0.282 \log \frac{U_*}{\omega}\right) \log \left(\frac{VS}{\omega} - \frac{V_{cr}S}{\omega}\right) \tag{3.73}$$

where C_{tg} = total gravel concentration in ppm by weight.

The incipient motion criteria given in equations (3.19) and (3.20) should be used for equation (3.73).

Most of the sediment transport equations were developed for sediment transport in rivers where the effect of fine or wash load on fall velocity, viscosity, and relative density can be ignored. The Yellow River in China is known for its high sediment concentration and wash load. The relationship between fall velocity of sediment in clear water and that of a sediment-laden flow of the Yellow River is:

$$\omega_m = \omega \left(1 - C_v\right)^{7.0} \tag{3.74}$$

where ω and ω_m = sediment particle fall velocities in clear water and in sediment-laden flow, respectively, and

C_v = suspended sediment concentration by volume, including wash load.

The kinematic viscosity of the sediment-laden Yellow River is:

$$\nu_m = \frac{\rho}{\rho_m} - e^{5.06 C_v} \nu \tag{3.75}$$

where ρ and ρ_m = specific densities of water and sediment-laden flow, respectively, and:

$$\rho_m = \rho + \left(\rho_s - \rho\right) C_v \tag{3.76}$$

where ρ_s = specific density of sediment particles.

If sediments are transported in a sediment-laden flow with high concentrations of fine materials, it can be shown that:

$$C_v = (1 - e_b)\frac{\gamma_m}{\gamma_s - \gamma_m}\left(\frac{VS}{\omega_m}\right) \tag{3.77}$$

where γ and γ_m = specific weights of sediment and sediment-laden flow, respectively, and
e_b = efficient coefficient for bedload.

It can be seen from equation (3.77) that when the unit stream power concept is applied to the estimation of sediment transport in sediment-laden flows, a modified dimensionless unit stream power $[\gamma_s/(\gamma_s - \gamma_m)]VS/\omega_m$ should be used. The modified Yang's unit stream power formula (Yang et al., 1996) for a sediment-laden river, such as the Yellow River, becomes:

$$\begin{aligned}\log C_{ts} &= 5.165 - 0.153\log\frac{\omega_m d}{v_m} - 0.297\log\frac{U_*}{\omega_m} \\ &+ \left(1.780 - 0.360\log\frac{\omega_m d}{v_m} - 0.480\log\frac{U_*}{\omega_m}\right)\log\left(\frac{\gamma_m}{\gamma_s - \gamma_m}\frac{VS}{\omega_m}\right)\end{aligned} \tag{3.78}$$

It should be noted that the coefficients in equation (3.78) are identical to those in equation (3.72). However, the values of fall velocity, kinematic viscosity, and relative specific weight are modified for sediment transport in sediment-laden flows with high concentrations of fine suspended materials.

It has been the conventional assumption that wash load depends on supply and is not a function of the hydraulic characteristics of a river. Yang (1966) demonstrated that the conjunctive use of equations (3.72) and (3.78) can determine not only bed-material load but also wash load in a sediment-laden river. Yang and Simões (2005) made a systematic and thorough analysis of 1,160 sets of data collected from 9 gauging stations along the Middle and Lower Yellow River. They confirmed that the method suggested by Yang (1996) can be used to compute wash load, bed-material load, and total load in the Yellow River with accuracy.

3.3.7 Power Balance Approach

Pacheco-Ceballos (1989) derived a sediment transport function based on power balance between total power available and total power expenditure in a stream; that is:

$$P = P_1 + P_s + P_b + P_2 \tag{3.79}$$

where P = total power available per unit channel width,
P_1 = power expenditure per unit width to overcome resistance to flow,
P_s = power expenditure per unit width to transport suspended load,
P_b = power expenditure per unit width to transport bedload, and
P_2 = power expenditure per unit width by minor or other causes which will not be considered hereinafter.

According to Bagnold (1966):

$$P = \tau_0 V = \rho g D S V \tag{3.80}$$

where ρ = density of water,
g = gravitational acceleration, and
D = average depth of flow.

According to Einstein and Chien (1952):

$$P_s = (\rho_s - \rho) g \frac{Q_s \omega}{BV} \tag{3.81}$$

where ρ_s = density of sediment,
Q_s = suspended load,
ω = fall velocity of sediment, and
B = channel width.

Accounting to the power concept and balance of acting force,

$$P_b = g Q_b \frac{\rho_s - \rho}{B} \tan \phi \tag{3.82}$$

where: Q_s = bedload, and
$\tan \phi$ = angle of repose of sediments.

If it is assumed that a certain portion of the available power is used to overcome resistance to flow, then:

$$P_1 = K_0 P = K_0 \rho g S Q / B \tag{3.83}$$

where K_0 = proportionality factor, and
Q = water discharge.

Substituting equations (3.80) through (3.83) into equation (3.79) yields:

$$K = \frac{V Q_b \tan \phi + \omega Q_s}{QVS} \tag{3.84}$$

where

$$K = \frac{(1 - K_0)\rho}{\rho_s - \rho} \tag{3.85}$$

The total sediment concentration can be expressed in the following general form:

$$C_t = \frac{KVS}{K''V\tan\phi + (1 - K'')\omega} = K'VS \tag{3.86}$$

where
K'' = ratio between bedload and total load,
K' = parameter,
C_t = total sediment concentration, and
VS = Yang's unit stream power.

When $K''= 1$, equation (3.86) becomes a bedload equation; that is:

$$C_b = \frac{KVS}{V\tan\phi} \tag{3.87}$$

When $K''= 0$, equation (3.86) becomes a suspended-load equation, that is:

$$C_s = \frac{KVS}{\omega} \tag{3.88}$$

Thus, the analytical derivation by Pacheco-Ceballos (1989) based on power balance shows that bedload, suspended-load, and total-load concentrations are all functions of unit stream power. It should be pointed out that K is not a constant. The K value given by Pacheco-Ceballos is:

$$K = \frac{\rho_m}{\Delta\rho D b_f}\left(\frac{aV_b}{V} + ea_s\right) \tag{3.89}$$

where
ρ_m = density of water and sediment mixture,
$\Delta\rho$ = $(\rho_s - \rho)/\rho$,
a and a_s = thicknesses of bed layer and suspended layer, respectively,
e = dimensionless coefficient,
D = average depth of flow,
b_f = bed form shape factor, and
V_b = bottom velocity.

3.3.8 Gravitational Power Approach

Velikanov (1954) derived his transport function from the gravitational power theory. He divided the rate of energy dissipation for sediment transport into two parts. These are the power required to overcome flow resistance and the power required to keep sediment particles in suspension against the gravitational force. Velikanov's basic relationship can be expressed as:

$$\rho g(1 - C_{vy})V_y S = \rho V_y \frac{d\left[(1 - C_{vy})\overline{u_x u_y}\right]}{dy} + g(\rho_s - \rho)C_{vy}(1 - C_{vy})\omega \tag{3.90}$$

(I) (II) (III)

where C_{vy} = time-averaged sediment concentration at a distance y above the bed (in % by volume),
V_y = time averaged flow velocity at a distance y above the bed,
u_x and u_y = fluctuating parts of velocity in the x and y directions, respectively,
ρ_s and ρ = densities of sediment and water, respectively, and
g = gravitational acceleration.

Equation (3.90) has the following physical meaning:

(I) = effective power available per unit volume of flowing water,

(II) = rate of energy dissipation per unit volume of flow to overcome resistance, and

(III) = rate of energy dissipation per unit volume of flow to keep sediment particles in suspension.

Assuming that the sediment concentration is small, integration of equation (3.90) over the depth of flow, D, yields:

$$gVSD = \frac{f_0 V^3}{8} + \frac{\rho_s - \rho}{\rho} gD\omega C_v \tag{3.91}$$

where C_v = average sediment concentration by volume.

Equation (3.91) shows that sediment concentration by volume is a function of unit stream power.

The Darcy-Weisbach resistance coefficients with and without sediment can be expressed, respectively, as:

$$f = \frac{8gDS}{V^2} \text{ for } C_v \neq 0 \tag{3.92}$$

$$f_0 = \frac{8gDS_0}{V^2} \text{ for } C_v = 0 \tag{3.93}$$

where S and S_0 = energy slopes with and without sediment, respectively, and
C_v = time-averaged sediment concentration (in % by volume).

It can be shown that Velikanov's equation can be expressed in the following general form:

$$C_v = K \frac{V^3}{gD\omega} \tag{3.94}$$

where K = a coefficient to be determined from measured data.

Several Chinese researchers have used Velikanov's gravitational power theory as the theoretical basis for the derivation of sediment transport equations. For example, Dou (1974) suggested that the rate of energy dissipation used by flowing water to keep sediment particles in suspension should be equal to that used by sediment particles in suspension, and proposed the following equation:

$$C_t = K_2 \frac{V^3}{gD\omega} \tag{3.95}$$

where K_2 = a variable to be determined, and
C_t = total sediment concentration.

Zhang (1959) assumed that the rate of energy dissipation used in keeping sediment particles in suspension should come from turbulence instead of the effective power available from the flow. He also considered the damping effect and believed that the existence of suspended sediment particles could reduce the strength of turbulence. Zhang's equation for sediment transport is:

$$C_t = K_3 \left(\frac{V^3}{gR\omega} \right)^m \tag{3.96}$$

where K_3 and m = parameters related to sediment concentration, and
R = hydraulic radius.

Yang (1996) gave a detailed comparison of transport functions based on gravitational and unit stream power approaches.

3.4 Other Commonly Used Sediment Transport Functions

Engineers have used sediment transport functions, formulas, or equations obtained from different approaches described in section 3.3 for solving engineering and river morphological problems. In addition to those proposed by Bagnold (1966), Ackers and White (1973), Engelund and Hansen (1967), and by Yang (1973, 1979, 1984) described previously, other commonly used transport formulas are summarized herein. Yang (1996) has published more detailed descriptions of the commonly used formulas, their theoretical basis, and their limits of application. Stevens and Yang (1984) published computer programs for 13 commonly used sediment transport formulas for PC application. They are given in Yang's book (1996, 2003).

3.4.1 Schoklitsch Bedload Formula

Schoklitsch (1934) developed a bedload formula based mainly on Gilbert's (1914) flume data with median sediment sizes ranging from 0.3 to 5mm. The Schoklitsch formula for unigranular material is:

$$G_s = \frac{86.7}{\sqrt{D}} S^{3/2}(Q - Wq_0) \tag{3.97}$$

where:

$$q_0 = \frac{0.00532D}{S^{4/3}} \tag{3.98}$$

where G_s = the bedload discharge, in lb/s,
D = the mean grain diameter, in in.,
S = the energy gradient, in ft per ft,
Q = the water discharge in ft^3/s,
W = the width, in ft, and
q_0 = the critical discharge, in ft^3/s per ft of width.

The formula can be applied to mixtures by summing the computed bedload discharges for all size fractions. The discharge for each size fraction is computed using the mean diameter and the fraction of the sediment in the sized fraction. Converting the equation for use with mixtures and changing the grain diameter from inches to feet and the bedload discharge from pounds to pounds per foot of width gives:

$$g_s = \sum_{i=1}^{n} i_b \frac{25}{\sqrt{D_{si}}} S^{3/2} \left(q - q_0\right) \tag{3.99}$$

where:

$$q_0 = \frac{0.0638 D_{si}}{S^{4/3}} \tag{3.100}$$

where g_s = the bedload discharge, in lb/s per ft of width,
i_b = the fraction, by weight, of bed material in a given size fraction,
D_{si} = the mean grain diameter, in ft, of sediment in size fraction *i,*
Q = the water discharge, in ft^3/s per ft of width,
q_0 = the critical discharge, in ft^3/s per ft of width, for sediment of diameter D_{si}, and,
n = the number of size fractions in the bed-material mixture.

3.4.2 Kalinske Bedload Formula

The formula developed by Kalinske (1947) for computing bedload discharge of unigranular material is based on the continuity equation, which states that the bedload discharge is equal to the product of the average velocity of the particles in motion, the weight of each particle, and the number of particles. The average particle velocity is related to the ratio of the critical shear to the total shear. The formula is:

$$g_s = \sum_{i=1}^{n} U_* \gamma_s D_{si} P_i 7.3 \left(\frac{\overline{U_g}}{\overline{U}}\right) \tag{3.101}$$

where:

$$U_* = \frac{\sqrt{\tau_0}}{\rho} \tag{3.102}$$

$$\frac{\bar{U}_g}{\bar{U}} = f\left(\frac{\tau_{ci}}{\tau_0}\right) \tag{3.103}$$

$$\tau_{ci} = 12 D_{si} \tag{3.104}$$

$$P_i = \frac{0.35}{m}\left(\frac{i_b}{D_{si}}\right) \tag{3.105}$$

where g_s = the bedload discharge in lb/s per ft of width,
n = the number of size fractions in the bed-material mixture,
U_* = the shear velocity in ft/s,
g_s = the specific weight of the sediment in lb/ft^3,
D_{si} = the mean grain diameter in ft of sediment in size fraction i,
P_i = the proportion of the bed area occupied by the particles in size fraction i,
$\bar{U}_g$ = the average velocity, in ft/s, of particles in size fraction i,
$\bar{U}$ = the mean velocity of flow, in ft/s, at the grain level,
τ_0 = the total shear at the bed, in lb/ft^2, which equals $62.4dS$,
d = the mean depth in ft,
S = the energy gradient in ft per ft,
ρ = the density of water in slugs per ft^3,
f = denotes function of,
τ_{ci} = the critical tractive force in lb/ft^2,
m = the summation of values of i_b/D_{si} for all size fractions in the bed-material mixture, and
i_b = the fraction, by weight, of bed material in a given size fraction.

Using the values of 165.36 for γ_s and 1.94 for ρ, the formula is:

$$g_s = 25.28\sqrt{\tau_0}\sum_{i=1}^{n}\tau_{ci}\frac{\frac{i_b}{D_{si}}}{m}\left(\frac{\bar{U}_g}{\bar{U}}\right) \tag{3.106}$$

Figure 3.17 shows values of τ_{ci}/τ_0.

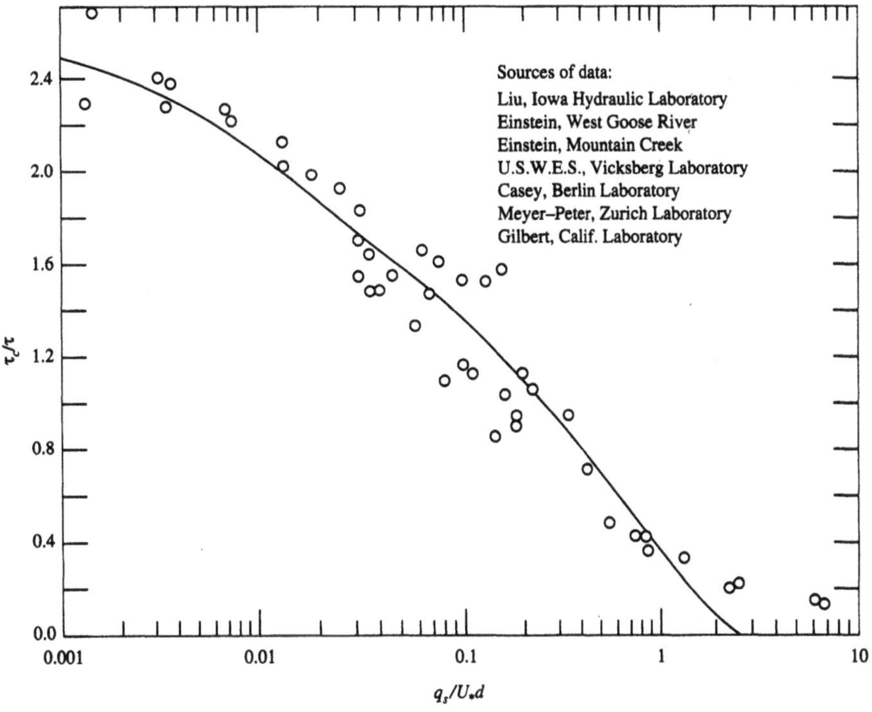

Figure 3.17. Kalinske's bed-load equation (Kalinske, 1947).

3.4.3 Meyer-Peter and Müller Formula

Meyer-Peter and Müller (1948) developed an empirical formula for the bedload discharge in natural streams. The original form of the formula in metric units for a rectangular channel is:

$$\gamma\frac{Q_s}{Q}\left(\frac{\mathrm{K}_s}{\mathrm{K}_r}\right)^{3/2} dS = 0.047\gamma_s' D_m + 0.25\left(\frac{\gamma}{\mathrm{g}}\right)^{1/3} g_s^{2/3} \tag{3.107}$$

in which:

$$D_m = \sum_{i=1}^{n} D_{si} i_b \tag{3.108}$$

where γ = the specific weight of water and equals 1 t/m^3,
Q_s = that part of the water discharge apportioned to the bed in l/s,
Q = the total water discharge in L/s,
K_s = Strickler's coefficient of bed roughness, equal to 1 divided by Manning's roughness coefficient n_s,
K_r = the coefficient of particle roughness, equal to $26/D_{90}^{1/6}$,
D_{90} = the particle size, in m, for which 90% of the bed mixture is finer,
d = the mean depth in m,
S = the energy gradient in m per m,

γ_s = the specific weight of sediment underwater, equal to 1.65 t/m^3 for quartz,
D_m = the effective diameter of bed-material mixture in m,
g = the acceleration of gravity, equal to 9.815 m/s^2,
g_s = the bedload discharge measured underwater in t/s per m of width,
n = the number of size fractions in the bed material,
D_{si} = the mean grain diameter, in m, of the sediment in size fraction i, and
i_b = the fraction, by weight, of bed material in a given size fraction.

Converting the formula to English units gives:

$$g_s = \left[0.368 \frac{Q_s}{Q} \left(\frac{D_{90}^{1/6}}{n_s} \right)^{3/2} dS - 0.0698 D_m \right]^{3/2} \tag{3.109}$$

where g_s = the bedload discharge for dry weight, in lb/s per ft of width,
Q, Q_s = sediment and water discharges, respectively, in ft^3/s,
D_{90}, D_m = sediment particle diameter at which 90% of the material, by weight, is finer and mean particle diameter, respectively,
d = water depth in ft, and
n_s = Manning's roughness value for the bed of the stream.

3.4.4 Rottner Bedload Formula

Rottner (1959) developed an equation to express bedload discharge in terms of the flow parameters based on dimensional considerations and empirical coefficients. Rottner applied a regression analysis to determine the effect of a relative roughness parameter D_{90}/d. Rottner's equation is dimensionally homogenous, so that it can be presented directly in English units:

$$g_s = \gamma_s \left[\left(S_g - 1 \right) g d^3 \right]^{1/2} \left\{ \frac{V}{\sqrt{\left(S_g - 1 \right) g d}} \left[0.667 \left(\frac{D_{50}}{d} \right)^{2/3} - 0.14 \right] - 0.778 \left(\frac{D_{50}}{d} \right) \right\} \tag{3.110}$$

where g_s = the bedload discharge in lb/ft of width,
γ_s = the specific weight of sediment in lb/ft^3,
S_g = the specific gravity of the sediment,
g = the acceleration of gravity in ft/s^2,
d = the mean depth in ft,
V = the mean velocity in ft/s, and
D_{50} = the particle size, in ft, at which 50% of the bed material by weight is finer.

In this derivation, wall and bed form effects were excluded. Rottner stated that his equation may not be applicable when small quantities of bed material are being moved.

3.4.5 Einstein Bedload Formula

The bedload function developed by Einstein (1950) is derived from the concept of probabilities of particle motion. Due to the complexity of the bedload function, a description of the procedure will not be presented here. Interested readers should refer to Einstein's original paper or the summary published by Yang (1996).

3.4.6 Laursen Bed-Material Load Formula

The equation developed by Laursen (1958) to compute the mean concentration of bed-material discharge is based on empirical relations:

$$\bar{C} = \sum_{i=1}^{n} i_b \left(\frac{D_{si}}{d}\right)^{7/6} \left(\frac{\tau_o}{\tau_c} - 1\right) f\left(\frac{U_*}{\omega_i}\right) \qquad (3.111)$$

where:

$$\tau_0' = \frac{\rho V^2}{58}\left(\frac{D_{50}}{d}\right)^{1/3} \qquad (3.112)$$

$$\tau_c = Y_c \rho g (S_g - 1) D_{si} \qquad (3.113)$$

where $\bar{C}$ = the concentration of bed-material discharge in % by weight,
n = the number of size fractions in the bed material,
i_b = the fraction, by weight, of bed material in a given size fraction,
D_{si} = the mean grain diameter, in ft, of the sediment in size fraction i,
d = the mean depth in ft,
τ_0' = Laursen's bed shear stress due to grain resistance,
τ_c = critical shear stress for particles of a size fraction,
f = denotes function of,
U_* = the shear velocity in ft/s,
ω_i = the fall velocity, in ft/s, of sediment particles of diameter D_{si},
ρ = the density of water in slugs per ft^3,
V = the mean velocity in ft/s,
D_{50} = the particle size, in ft, at which 50% of the bed material, by weight, is finer,
Y_c = a coefficient relating critical tractive force to sediment size,
g = acceleration of gravity in ft/s^2, and
S_g = the specific gravity of sediment.

The density ρ has been introduced into the original τ_0' equation presented by Laursen so that the equation is dimensionally homogeneous, and Laursen's coefficient has been changed accordingly. Substituting for τ_0' and τ_c in equation (3.111) and converting $\bar{C}$ to C gives:

$$C = 10^4 \sum_{i=1}^{n} i_b \left(\frac{D_{si}}{d}\right)^{7/6} \left[\frac{V^2}{58 Y_c D_{si}(S_g - 1)gd}\left(\frac{D_{50}}{d}\right)^{1/3} - 1\right] f\left(\frac{U_*}{\omega_i}\right) \tag{3.114}$$

where C = the concentration of bed-material discharge, in parts per million by weight.

Figure 3.18 shows values of $f(U_*/\omega_i)$.

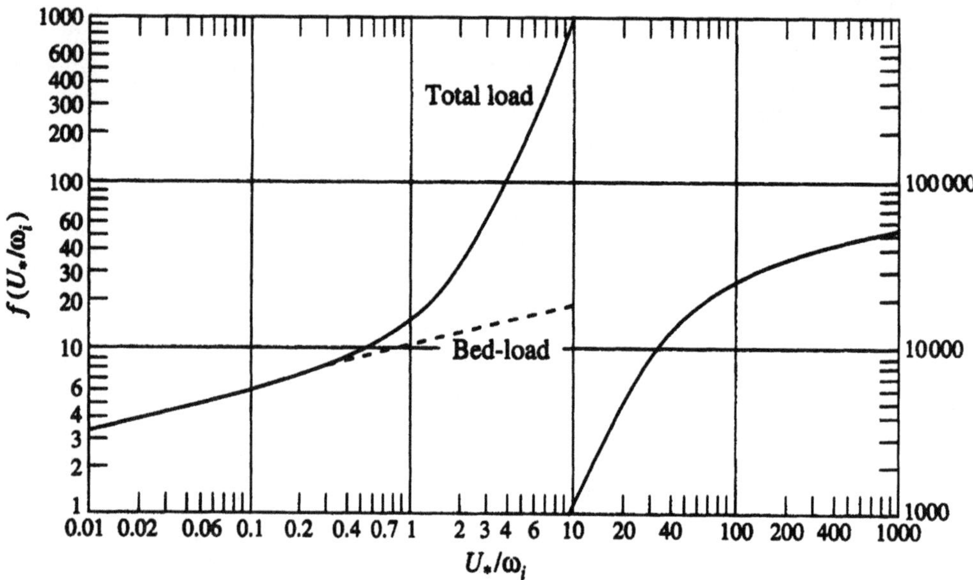

Figure 3.18 Function $f(U_*/\omega_i)$ in Laursen's approach (Laursen, 1958).

3.4.7 Colby Bed-Material Load Formula

Colby (1964) presented a graphical method to determine the discharge of sand-size bed material that ranged from 0.1 to 0.8 mm. The bed-material discharge g_s, in lb/s/ft of width, at a water temperature of 15.6 degrees Celsius (°C) (Colby's 1964 fig. 6) is:

$$g_s = A(V - V_c)^B 0.672 \tag{3.115}$$

where:

$$V_c = 0.4673 d^{0.1} D_{50}^{0.33} \tag{3.116}$$

where V = the mean velocity in ft/s,
V_c = the critical velocity in ft/s,
D = the mean depth in ft,
d_{50} = the practical size, in mm, at which 50% of the bed material by weight is finer,
A = a coefficient, and
B = an exponent.

Colby developed his graphical solutions for total load mainly from laboratory and field data using Einstein's (1950) bedload function as a guide. His graphical solutions are shown in Figures 3.19 and 3.20. The required information in Colby's approach comprises the mean flow velocity V, average depth D, median particle diameter d_{50}, water temperature T, and fine sediment concentration C_f. The total load can be computed by the following procedure:

Step 1: with the given V and d_{50}, determine the uncorrected sediment discharge q_{ti} for the two depths shown in Figure 3.19 that are larger and smaller than the given depth D, respectively.

Step 2: interpolate the correct sediment discharge q_{ti} for the given depth D on a logarithmic scale of depth versus q_{ti}.

Step 3: with the given depth D, median particle size d_{50}, temperature T, and fine sediment concentration C_f, determine the correction factors k_1, k_2, and k_3 from Figure 3.20.

Step 4: the total sediment discharge (in ton/day/ft of channel width), corrected for the effect of water temperature, fine suspended sediment, and sediment size, is:

$$q_t = [1 + (k_1 k_2 - 1) 0.01 k_3] q_{ti} \tag{3.117}$$

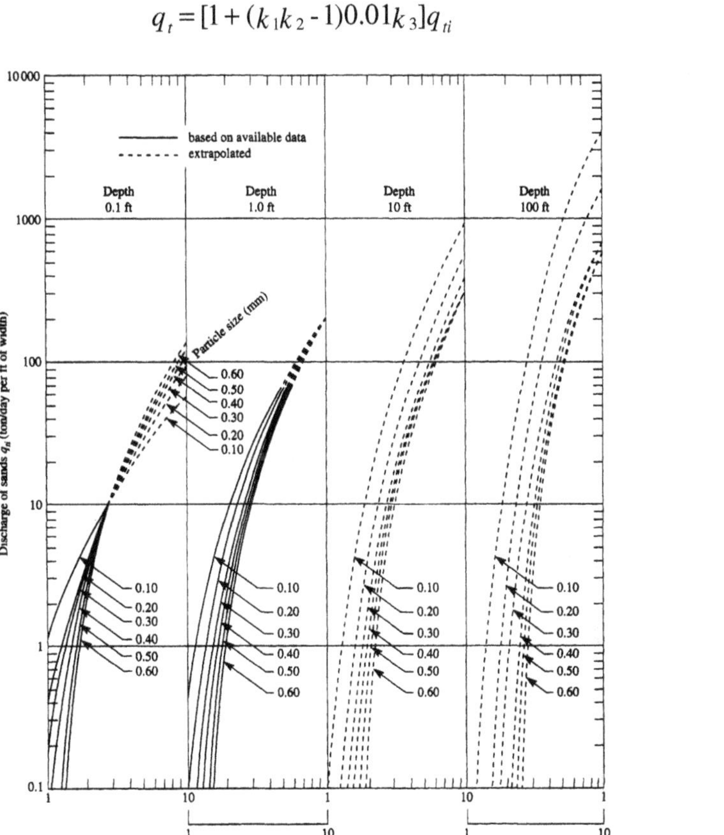

Figure 3.19 Relationship of discharge of sands to mean velocity for six median sizes of bed sands, four depths of flow, and a water temperature of 60 °F (Colby, 1964).

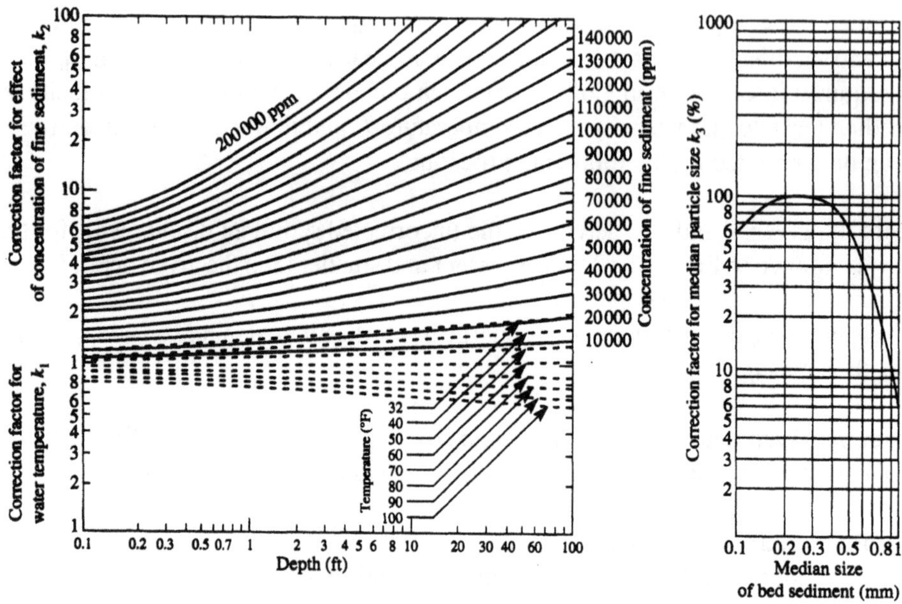

Figure 3.20. Approximate effect of water temperature and concentration of fine sediment on the relationship of discharge of sands to mean velocity (Colby, 1964).

From Figure 3.20, $k_1 = 1$ for $T = 60$ °F, $k_2 = 1$ where the effect of fine sediment can be neglected, and $k_3 = 100$ when the median particle size is in the range of 0.2 to 0.3 mm. Because of the range of data used in the determination of the rating curves shown in Figures 3.19 and 3.20, Colby's approach should not be applied to rivers with median sediment diameter greater than 0.6 mm and depth greater than 3 m.

3.4.8 Einstein Bed-Material Load Formula

Einstein (1950) presented a method to combine his computed bedload discharges with a computed suspended bed-material discharge to yield the total bed-material discharge. A complete description of the complex procedure will not be presented here. Interested readers should follow the original Einstein paper or the summary made by Yang (1996) to apply the Einstein bed-material formula.

3.4.9 Toffaleti Formula

The procedure to determine bed-material discharge developed by Toffaleti (1968) is based on the concepts of Einstein (1950) with three modifications:

1. Velocity distribution in the vertical is obtained from an expression different from that used by Einstein;

2. Several of Einstein's correction factors are adjusted and combined; and

3. The height of the zone of bedload transport is changed from Einstein's two grain diameters.

Toffaleti defines his bed-material discharge as total river sand discharge, even though he defines the range of bed-size material from 0.062 to 16 mm. The complex procedures in the Toffaleti formula will not be presented here. Interested readers should follow the original Toffaleti procedures or the summary by Yang (1996) to apply the procedures.

3.5 Fall Velocity

Sediment particle fall velocity is one of the important parameters used in most sediment transport functions or formulas. Depending on the sediment transport functions used and sediment particle size in a particular study, different methods have been developed for the computation of sediment particle fall velocity. Some of the commonly used methods for fall velocity computation are summarized herein.

When Toffaleti's equation is used, Rubey's (1933) formula should be employed; that is:

$$\omega_s = F\sqrt{dg(G-1)} \tag{3.118}$$

where

$$F = \left[\frac{2}{3} + \frac{36\nu^2}{gd^3(G-1)}\right]^{1/2} - \left[\frac{36\nu^2}{gd^3(G-1)}\right]^{1/2} \tag{3.119}$$

for particles with diameter, d, between 0.0625 mm and 1 mm, and where $F = 0.79$ for particles greater than 1 mm. In the above equations, ω_s = fall velocity of sediments; g = acceleration due to gravity; G = specific gravity of sediment = 2.65; and ν = kinematic viscosity of water. The viscosity of water is computed from the water temperature, T, using the following expression:

$$\nu = \frac{1.792\times10^{-6}}{1.0 + 0.0337T + 0.000221T^2} \tag{3.120}$$

with T in degrees Centigrade and ν in m^2/s.

When any of the other sediment transport formulas are used, the values recommended by the U.S. Interagency Committee on Water Resources Subcommittee on Sedimentation (1957) are used (Figure 3.21). Yang and Simões (2002) use a value for the Corey shape factor of SF = 0.7, for natural sand in their computer model GSTARS3, where:

$$SF = \frac{c}{\sqrt{ab}} \tag{3.121}$$

where a, b, and c = the length of the longest, the intermediate, and the shortest mutually perpendicular axes of the particle, respectively.

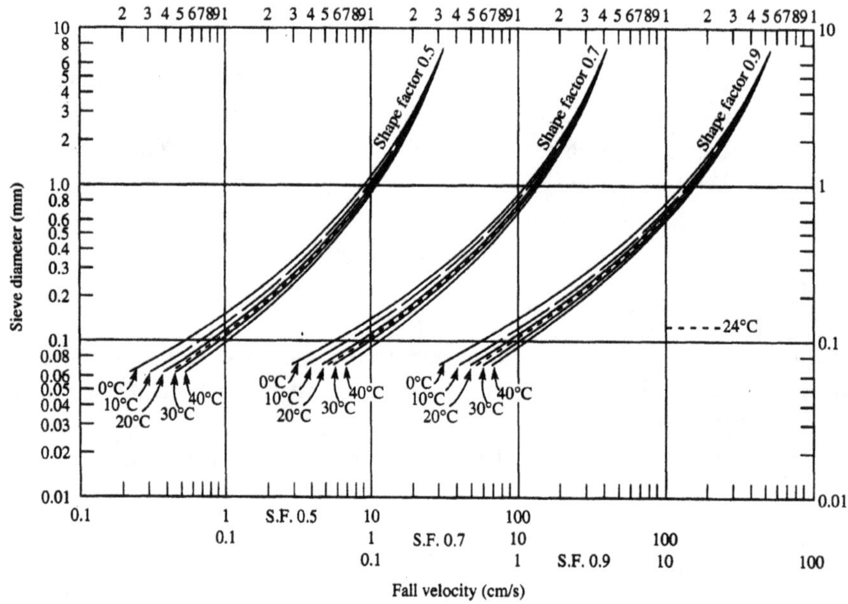

Figure 3.21.—Relation between particle sieve diameter and its fall velocity according to the U.S. Interagency Committee on Water Resources Subcommittee on Sedimentation (1957).

Yang and Simões (2002) also used the following approximations for the computation of fall velocities. For particles with diameters greater than 10 mm, which are above the range given in Figure 3.21, the following formula is used:

$$\omega_s = 1.1\sqrt{(G - 1)gd} \tag{3.122}$$

For particles in the silt and clay size ranges, namely with diameters between 1 and 62.5 µm, the sediment fall velocities are computed from the following equations:

unhindered settling:

$$\omega_s = \frac{(G - 1)gd^2}{18v} \text{ for } C < C_1 \tag{3.123}$$

flocculation range:

$$\omega_s = MC^N \text{ for } C_1 \le C \le C_2 \tag{3.124}$$

hindered settling:

$$\omega_s = \omega_0(1 - kC)^l \text{ for } C > C_2 \tag{3.125}$$

where ω_0 is found by equations (3.124) and (3.125) at C = C_2, that is:

$$\omega_0 = \frac{MC_2^N}{(1 - kC_2)^l} \tag{3.126}$$

and k, l, M, and N are site-specific constants supplied by the user; figure 3.22 shows fall velocities in flocculation range for different natural conditions. The expression $\omega_s = 1.0C^{1.0}$ represents the average values with ω_s in mm/s and C in kg/m^3.

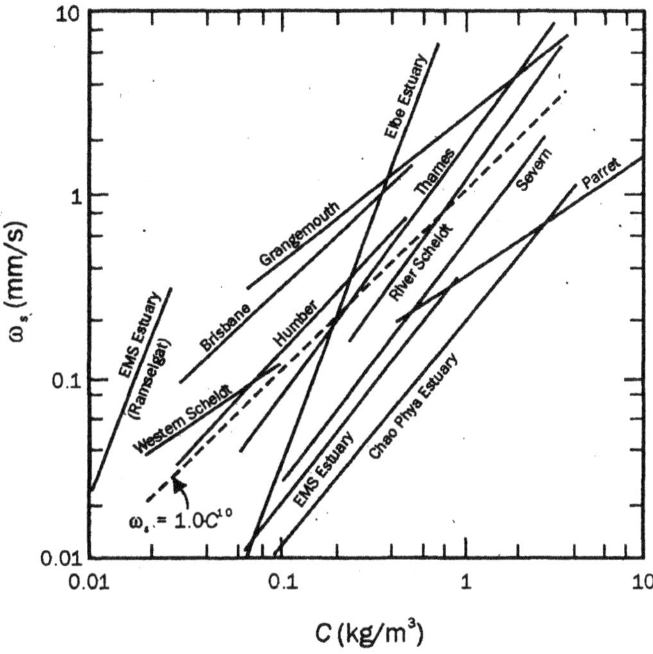

Figure 3.22. Variability of the parameters M and N of eq. (3.124) for several well known rivers and estuaries (Yang and Simões, 2002).

3.6 Resistance to Flow

For a steady, uniform, open channel flow of constant width W without sediment, the water depth D and velocity V can be determined for a given discharge Q and channel slope S by using the continuity equation:

$$Q = WDV \tag{3.127}$$

and a friction equation, such as the Darcy-Weisback formula:

$$V = \sqrt{\frac{8gRS}{f}} \tag{3.128}$$

where V = average flow velocity,
g = gravitational acceleration,

R = hydraulic radius,
S = water surface or energy slope, and
f = Darcy-Weisbach friction factor.

For fluid hydraulics with sediment transport, the total roughness for resistance to flow consists of two parts. If equation (3.128) is used:

$$f = f' + f'' \tag{3.129}$$

where f' = Darcy-Weisbach friction factor due to grain roughness, and
f'' = Darcy-Weisbach friction factor due to form roughness on the existence of bed forms.

Figure 3.23 is based on the data collected by Guy et al. (1966) in a laboratory flume with 0.19-mm sand. Figure 3.23 shows that f' is a constant, but the f'' value depends on the bed form, such as plane bed, ripple, dune, transition, antidune, and chute-pool. Although empirical methods exist for the determinations of bed forms, no consistent result can be obtained from empirical methods. Consequently, the Darcy-Weisbach friction factor f or the Manning's coefficient n cannot be assumed as a given constant in an alluvial channel with sediment. Assume that sediment concentration can be determined by the following function:

$$C_t = \Phi\,(V, S, D, d, \nu, \omega) \tag{3.130}$$

where C_t = total sediment concentration, in parts per million by weight,
d = median sieve diameter of bed material,
ν = kinematic viscosity of water, and
ω = terminal fall velocity of sediment.

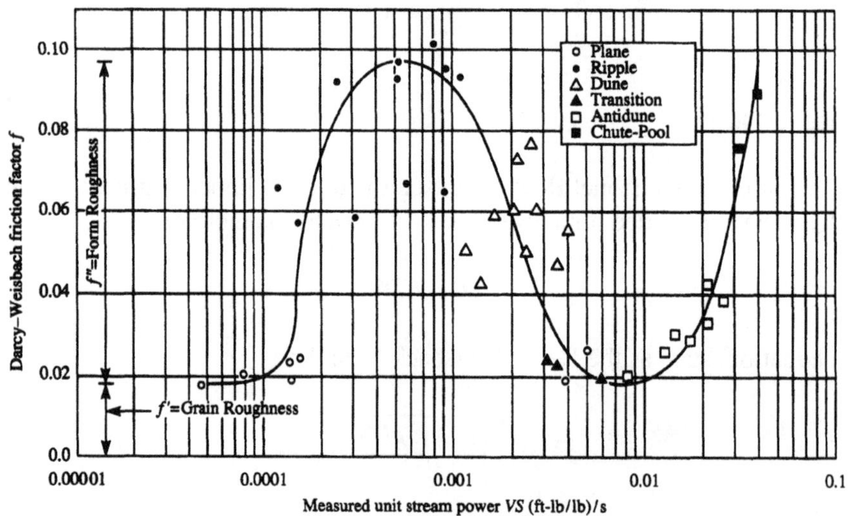

Figure 3.23 Variation of friction factor with bed form and measured unit stream power (Yang, 1996).

Because the f value of an alluvial channel cannot be predicted with confidence, we have equations (3.127) and (3.130) with three unknown, namely *V, D,* and *S*. Thus, fluvial hydraulics is still basically indeterminate despite the significant progress made in the past decades.

Due to the site-specific nature of empirical methods, they will not be introduced here. The following sections will introduce only analytical methods for the determination of resistance to flow or the roughness coefficient, or the determination of *V, S, D* without prior knowledge of the roughness coefficient.

3.6.1 Einstein's Method

Einstein (1950) expressed the resistance due to grain roughness by:

$$\frac{V}{U'_*} = 5.75 \log\left(12.27\frac{R'}{k_s}x\right) \tag{3.131}$$

where U'_* = shear velocity due to skin friction or grain roughness = $(gR'S)^{1/2}$,
R' = hydraulic radius due to skin friction,
k_s = equivalent grain roughness = d_{65},
x = a function of k_s/δ, and
δ = boundary layer thickness, which can be expressed as:

$$\delta = \frac{11.6\nu}{U'_*} \tag{3.132}$$

where ν = kinematic viscosity.

Figure 3.24 shows the relationship between x and k_s/δ suggested by Einstein (1950). With the given values of V, d_{65}, and x determined from Figure 3.24, equation (3.131) can be used to compute the value of R'. Einstein (1950) suggested that:

$$\frac{V}{U''_*} = \phi(\psi') \tag{3.133}$$

where

$$\psi' = \frac{\gamma_s - \gamma}{\gamma}\frac{d_{35}}{SR'} \tag{3.134}$$

The functional relationship between V/U''_* and ψ' was determined from field data by Einstein and Barbarossa (1952) as shown in Figure 3.25.

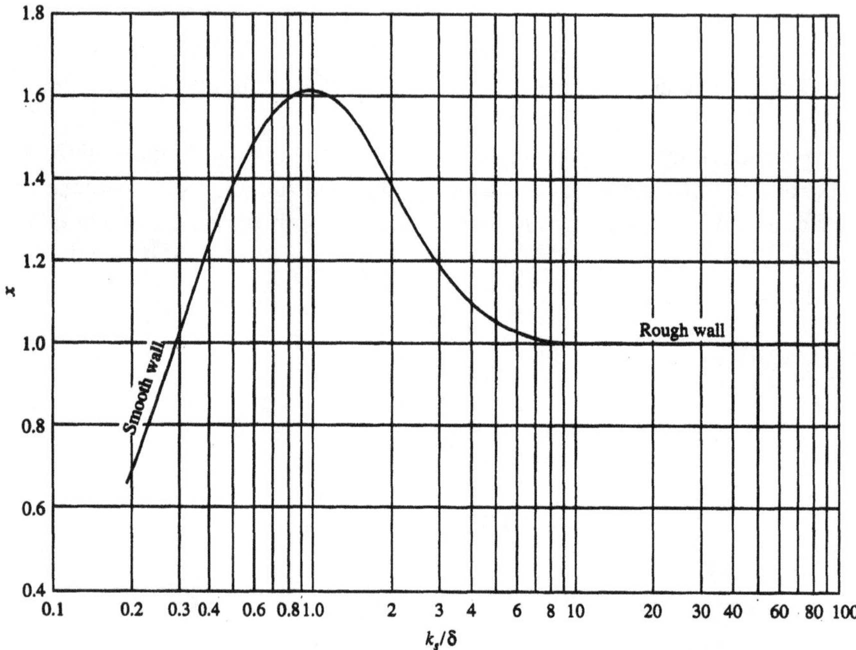

Figure 3.24. Correction factor in the logarithmic velocity distribution (Einstein, 1950).

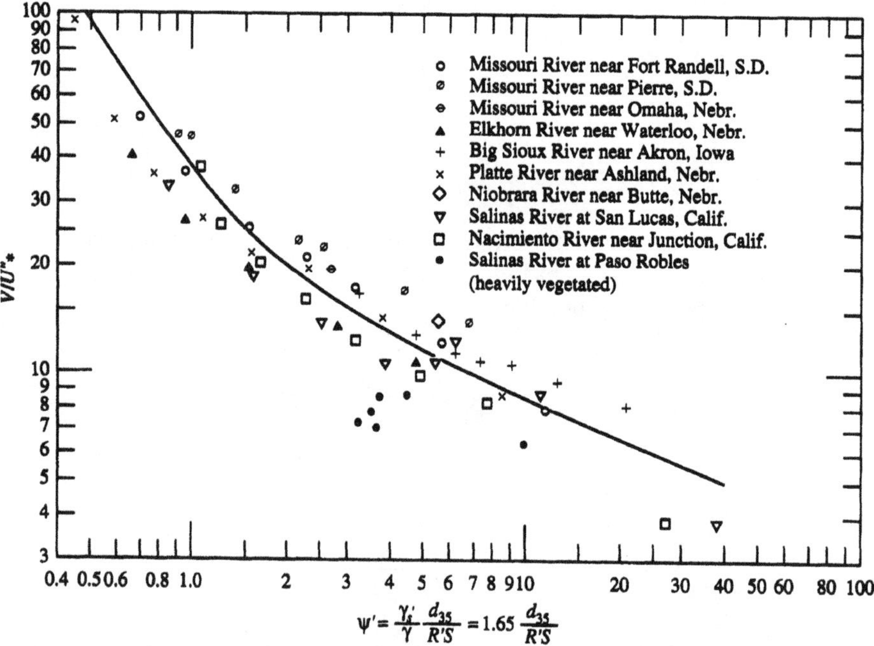

Figure 3.25. Friction loss due to channel irregularities as a function of sediment transport rate (Einstein and Barbarossa, 1952).

Einstein and Barbarossa (1952) suggested the following procedures for the computation of total hydraulic radius due to grain and form roughness when the water discharge is given, or vice versa.

Case A. Determine R with given Q

Step 1: Assume a value of R'.

Step 2: Apply equation (3.131) and Figure 3.24 to determine V.

Step 3. Compute ψ' using equation (3.134) and the corresponding value of V/U''_* from Figure 3.25.

Step 4: Compute U'_* and the corresponding value of R''.

Step 5: Compute $R = R' + R''$ and the corresponding channel cross-sectional area A.

Step 6: Verify using the continuity equation $Q = VA$. If the computed Q agrees with the given Q, the problem is solved. Otherwise, assume another value of R' and repeat the procedure until agreement is reached between the computed and the given Q.

Case B. Determine Q with given R. The five first steps are identical to those for case A. After the R value has been computed, it is compared with the given value of R. If these values agree, the problem is solved, and $Q = VA$. If not, the computation procedures will be repeated by assuming different values of R' until the computed R agrees with the given R. Yang (1996) gave the following examples, using Einstein's method.

Example 3.1 Given the following data, determine the flow depth D for the channel shown using the Einstein procedures:

$Q = 40\ \text{m}^3/\text{s}$, $B = 5$ m
$\nu = 10^{-6}\ \text{m}^2/\text{s}$, $S = 0.0008$
Specific gravity of sand = 2.65
$d_{35} = 0.3$ mm, $d_{65} = 0.9$ mm

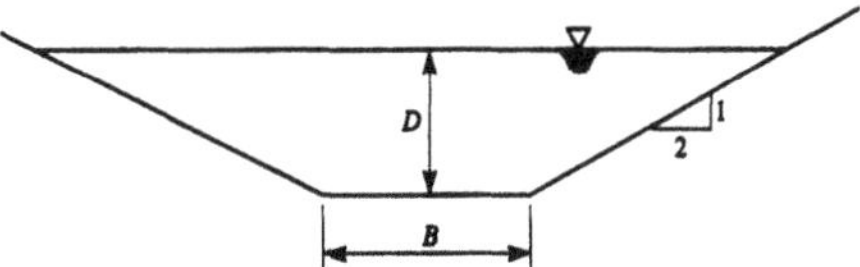

Solution:
(a) Assume R'.
(b) Determine velocity from equation (3.12):

$$V = 5.75 U'_* \log\left(12.27 \frac{R'}{k_s} x\right)$$

The equivalent sand roughness k_s may be taken as equal to $d_{65} = 0.0009$ m, and shear velocity U'_* is

$$U'_* = (gR'S)^{1/2} = 0.089(R')^{1/2}$$

The correction factor x is a function of k_s/δ, and may be read from Figure 3.24. The laminar sublayer thickness δ can be estimated from equation (3.132); that is,

$$\delta = 11.6\frac{v}{U'_*} = \frac{11.6(10^{-6})}{0.089(R')^{1/2}} = \frac{1.31\times10^{-4}}{(R')^{1/2}}$$

so

$$\frac{k_s}{\delta} = \frac{0.0009(R')^{1/2}}{1.31\times10^{-4}} = 6.87(R')^{1/2}$$

Substituting for U'_* and k_s, the velocity can be estimated from:

$$V = 0.509(R')^{1/2}\log(13.633R')^{1/2}$$

(c) Compute ψ' from equation (3.134),

$$\psi' = (2.65-1)\frac{d_{35}}{SR'} = 1.65\frac{0.0003}{0.0008R'} = \frac{0.619}{R}$$

and determine V/U'_* from Figure 3.25.

(d) Compute U'_* and R'' from:

$$U''_* = \left(\frac{V}{U''_*}\right)^{-1} V$$

$$R'' = \frac{(U''_*)^2}{gS} = \frac{(U''_*)^2}{0.0078}$$

(e) Determine $R = R' + R''$ and the corresponding depth D and area A.

(f) Determine $Q = AV$, and reiterate if necessary.

The determination of depth and area from the hydraulic radius may be facilitated by developing curves relating these variables. The relations may be expressed as:

$$A = 5D + 2D^2$$

$$R = \frac{5D + 2D^2}{5 + 4.47D}$$

Assuming values of *D*, the relationship between *D*, *A*, and *R* can be computed from the above two equations as follows:

D	A	R
0.6	3.72	0.484
0.8	5.28	0.616
1.0	7.00	0.737
1.2	8.88	0.857
1.5	12.00	1.025
2.0	18.00	1.290

The following is a tabulation of the solution procedure:

R' (m)	$\frac{k_s}{\delta}$	x	V (m/s)	ψ'	$\frac{V}{U'_*}$	U'_* (m/s)	R'' (m)	R (m)	A (m^2)	Q (m^3/s)
0.50	4.86	1.06	1.39	1.24	31	0.045	0.26	0.76	7.0	9.7
0.20	3.07	1.18	0.798	3.10	15	0.053	0.36	0.56	4.5	3.6
1.00	6.87	1.02	2.11	0.619	75	0.028	0.10	1.10	14.0	29.5
1.20	7.53	1.01	2.35	0.516	97	0.024	0.08	1.28	18.0	42.3
1.15	7.37	1.01	2.29	0.538	90	0.025	0.08	1.23	16.5	37.8
1.17	7.43	1.01	2.32	0.529	93	0.025	0.08	1.25	17.0	39.4
1.18	7.46	1.01	2.33	0.525	94	0.025	0.08	1.26	17.5	40.8

For $Q = 40$ m^3/s, $R = 1.254$ m
The corresponding water depth is $D = 1.93$ m.

Example 3.2 Use the fluid and sediment properties given in example 3.1 and the flow depth determined there; compute the water discharge using the Einstein procedure.

Solution: Use the same procedure as outlined for example 3.1, but reiterate until the computed *R* agrees with the actual *R*; then determine the discharge $Q = AV$.

The following is a tabulation of the solution procedure:

R' (m)	$\frac{k_s}{\delta}$	x	V (m/s)	ψ'	$\frac{V}{U'_*}$	U'_* (m/s)	R'' (m)	R (m)
1.17	7.43	1.01	2.32	0.529	93	0.025	0.08	1.25
1.18	7.46	1.01	2.33	0.525	94	0.025	0.08	1.26

For $R = 1.254$ m, $V = 93.4 \times 0.025 = 2.335$ m/s
Channel cross-sectional area: $A = 5(1.93) + 2(1.93)^2 = 17.10$ m^2
Discharge: $Q = 17.10\,(2.335) = 39.9$ m^3/s ≈ 40 m^3/s

3.6.2 Engelund and Hansen's Method

Engelund and Hansen (1966) expressed the energy loss or frictional slope due to bed form as:

$$S'' = \frac{\Delta H''}{L} = \frac{q^2}{2gL}\left(\frac{1}{D - \frac{1}{2}A_m} - \frac{1}{D + \frac{1}{2}A_m}\right) = \frac{V^2}{2gL}\left(\frac{A_m}{D}\right)^2 \tag{3.135}$$

where $\Delta H''$ = frictional loss due to bed forms of wave length L,
q = flow discharge per unit width,
D = mean depth, and
A_m = amplitude of sand waves.

The total shear stress can also be expressed as:

$$\tau = \gamma R\,(S' + S'') \tag{3.136}$$

or

$$\frac{\tau}{\gamma R} = \frac{\tau'}{\gamma R} + S'' \tag{3.137}$$

Substituting equation (3.135) for S'' into equation (3.137) and assuming $R = D$ for a wide open channel,

$$\frac{\tau}{\gamma D} = \frac{\tau'}{\gamma D} + \frac{V^2}{2gL}\left(\frac{A_m}{D}\right)^2 \tag{3.138}$$

Let

$$\theta = \frac{DS}{[(\rho_s/\rho) - 1]d} \tag{3.139}$$

$$\theta' = \frac{D'S}{[(\rho_s/\rho) - 1]d} \tag{3.140}$$

and

$$\theta'' = \frac{1}{2}F_r^2 \frac{A_m^2}{[(\rho_s/\rho) - 1]dL} \tag{3.141}$$

where ρ_s and ρ = densities of sediment and water, respectively,
D and D' = water depth and corresponding depth due to grain roughness, respectively,
d = sediment particle size, and
Fr = Froude number = $V/(gD)^{1/2}$

From equations (3.139), (3.140), and (3.141):

$$\theta = \theta' + \theta'' \tag{3.142}$$

This relation was proposed by Engelund and Hansen (1967). For narrow channels, D and D' should be replaced by R and R' in equations (3.138) to (3.140). Figure 3.26 shows the relationship between θ and θ' for different bed forms. For the upper flow region, it can be assumed that form drag is not associated with the flow and $\theta = \theta'$. Figure 3.26 can be applied to the determination of a stage-discharge relationship by the following procedures:

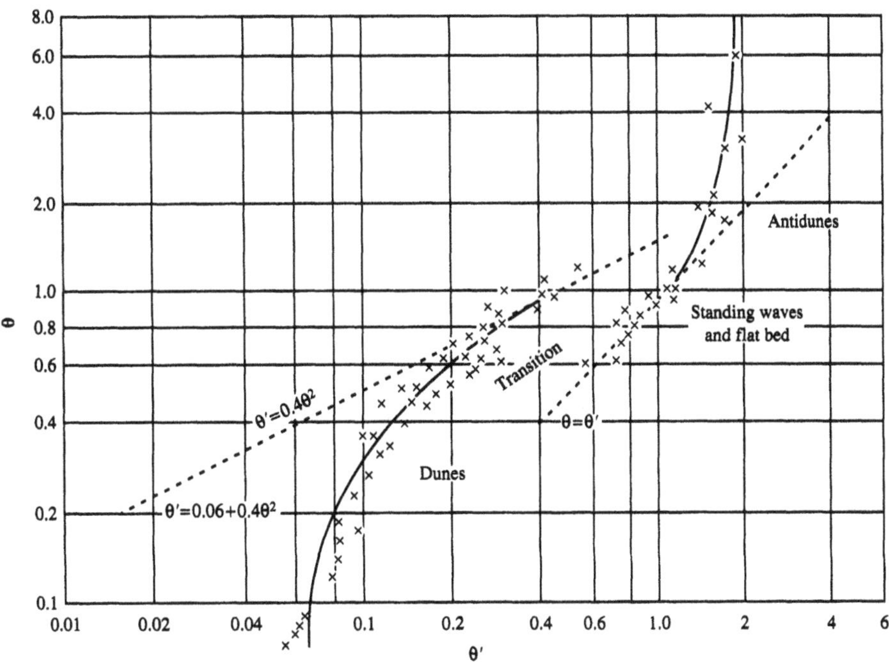

Figure 3.26. Flow resistance relationship (Engelund and Hansen, 1967).

Step 1: Determine S and D from a field survey of slope and channel cross-section.

Step 2: Compute θ from equation (3.139) for the given sediment size d.

Step 3: Determine θ' from Figure (3.26) with θ from step 2.

Step 4: Compute D' from equation (3.140).

Step 5: Compute V from equation (3.141).

Step 6: Determine the channel cross-sectional area A corresponding to the D value selected in step 1.

Step 7: Compute $Q = AV$. The stage-discharge relationship can be determined by selecting different D values and repeating the process.

Yang (1996) gave the following example using Engelund and Hansen's method.

Example 3.3 For the fluid and sediment properties and channel cross-section given in example 3.1, obtain the stage-discharge relationship using the procedure proposed by Engelund and Hansen.

Solution:

(a) Assume a depth of flow D.

(b) Compute θ for given R, S, and d from equation (3.139).

$$\theta = \frac{RS}{(\rho_s / \rho - 1)d}$$

For this analysis, the slope will be assumed equal to S_o (uniform flow), and the sediment size d will be assumed equal to:

$$d = ½(d_{35} + d_{65}) = ½(0.3 + 0.9) = 0.6 \text{ mm}$$

The hydraulic radius R may be determined from the assumed depth as:

$$R = \frac{5D + 2D^2}{5 + 2D\sqrt{5}}$$

Substituting:

$$\theta = \frac{0.0008(5D + 2D^2)}{1.65(0.0006)(5 + 2D\sqrt{5})} = 0.808\frac{5D + 2D^2}{5 + 2D\sqrt{5}}$$

(c) Determine θ' from Figure 3.26.

(d) Compute R' from:

$$R' = \frac{\theta'(\rho_s / \rho - 1)d}{S} = \frac{\theta'(1.65)(0.0006)}{0.0008} = 1.24\theta'$$

(e) Compute the velocity V from equation (3.131).

$$V = 5.75U'_* \log\left(12.27\frac{R'}{k_s}x\right)$$

The shear velocity $U_*' = (gR'S)^{1/2} = [9.81(0.0008)R']^{1/2} = 0.089(R')^{1/2}$. The equivalent sand roughness k_s may be taken as equal to d_{65} = 0.9 mm, and the correction factor x may be determined from Figure 3.24. A necessary parameter for the use of Figure 3.24 is k_s/δ, which can be computed from equation (3.132)

$$\frac{k_s}{\delta} = \frac{k_s U_*'}{11.6\nu} = \frac{0.0009(0.089)(R')^{1/2}}{11.6(10^{-6})} = 6.87(R')^{1/2}$$

(f) Compute the cross-sectional area A from

$$A = 5D + 2D^2$$

(g) Determine the discharge Q by continuity as

$$Q = AV$$

This procedure can be repeated for various values of D. Computations are shown in the table below.

The stage-discharge relationship for example 3.3 is shown below:

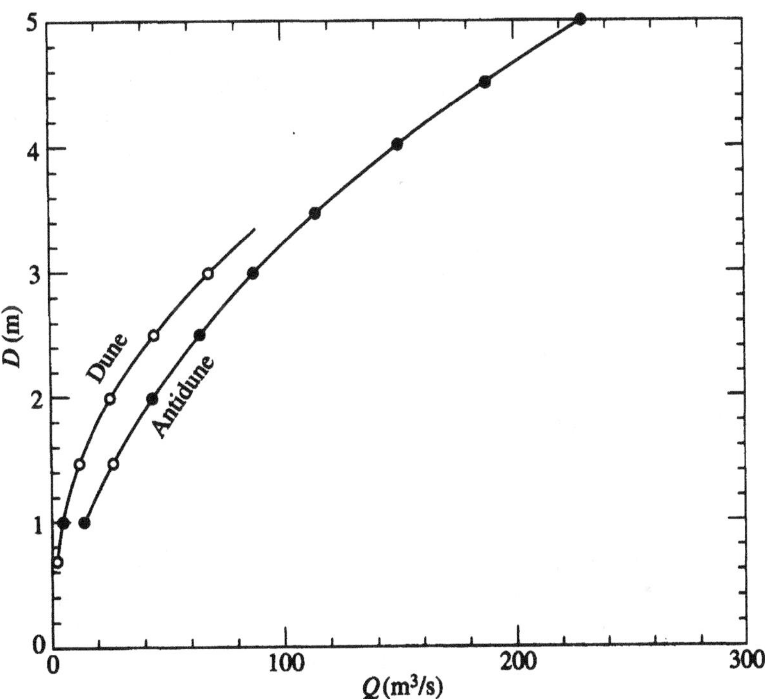

D (m)	R (m)	θ	θ′	R' (m)	k_s/δ	x	U' (m/s)	V (m/s)	Q (m^3/s)
0.5	0.415	0.335	0.12	0.15	2.7	1.22	0.034	0.663	2.0
1.0	0.739	0.597	0.18	0.22	3.2	1.05	0.042	0.845	5.9
			(0.59)	(0.73)	(5.9)	(1.02)	(0.076)	(1.76)	(12.3)
1.5	1.02	0.828	0.28	0.35	4.0	1.10	0.052	1.10	13.2
			(0.80)	(0.99)	(6.8)	(1.01)	(0.088)	(2.09)	(25.1)
2.0	1.29	1.04	0.50	0.62	5.4	1.02	0.070	1.58	28.4
			(1.0)	(1.24)	(7.7)	(1.00)	(0.099)	(2.41)	(43.4)
2.5	1.55	1.25	0.66	0.82	6.2	1.00	0.080	1.86	46.5
			(1.2)	(1.49)	(8.4)	(1.00)	(0.108)	(2.68)	(67.0)
3.0	1.79	1.45	0.87	1.08	7.1	1.00	0.092	2.20	72.6
			(1.25)	(1.55)	(8.6)	(1.00)	(0.110)	(2.74)	(90.4)
3.5	2.03	1.64	(1.3)	(1.61)	8.7	1.00	0.112	2.80	118
4.0	2.27	1.84	(1.37)	(1.70)	9.0	1.00	0.116	2.91	151
4.5	2.51	2.03	(1.43)	(1.77)	9.1	1.00	0.118	2.97	187
5.0	2.74	2.21	(1.5)	(1.86)	9.4	1.00	0.121	3.06	230

Values in parenthesis are for the upper flow regime or antidune.

3.6.3 Yang's Method

The theory of minimum rate of energy dissipation (Yang, 1976; Yang and Song, 1979, 1984) states that when a dynamic system reaches its equilibrium condition, its rate of energy dissipation is a minimum. The minimum value depends on the constraints applied to the system. For a uniform flow of a given channel width, where the rate of energy dissipation due to sediment transport can be neglected, the rate of energy dissipation per unit weight of water is:

$$\frac{dY}{dt} = \frac{dx}{dt}\frac{dY}{dx} = VS = \text{unit stream power} \tag{3.143}$$

where Y = potential energy per unit weight of water.

Thus, the theory of minimum unit stream power requires that:

$$VS = V_m S_m = \text{a minimum} \tag{3.144}$$

subject to the given constraints of carrying a given amount of water discharge Q and sediment concentration C_t of a given size d. The subscript m denotes the value obtained with minimum unit stream power. Utilization of equation (3.144) in conjunction with equations (3.127) and (3.130) can give a solution for the three unknowns, V, D, and S, without any knowledge of the total roughness. The procedures by Yang (1973) for the determination of Manning's coefficient based on his dimensionless unit stream power formula (Yang, 1973) are as follows.

Step 1: Assume a value of the depth D.

Step 2: For the values of Q, C_{ts}, W, d, ω, and ν, solve equations (3.127) and (3.70) for V and S.

Step 3: Compute the unit stream power as the product of V and S.

Step 4: Select another D and repeat the steps.

Step 5: Compare all the computed VS values and select the one with minimum value as the solution in accordance with equation (3.144).

Step 6: Once VS has been determined, the corresponding values of V, S, and D can be computed from equations (3.127) and (3.70). Manning's coefficient can be computed from Manning's formula without any knowledge of the bed form.

Figure 3.27 shows an example of the relationship between generated unit stream power $V_i S_i$ and water depth D_i. The minimum unit stream power $V_m S_m$ determined is in close agreement with the measured unit stream power VS. Figure 3.28 shows examples of comparisons between measured and computed results from the above procedure. The subscript m in Figure 3.27 denotes the value obtained using equation (3.144). In the above procedures, it is assumed that sediment transport equations used are accurate in predicting the total bed-material concentration. If the measured concentration is significantly different from the computed one, the agreement may not be as good as those shown in Figure 3.27.

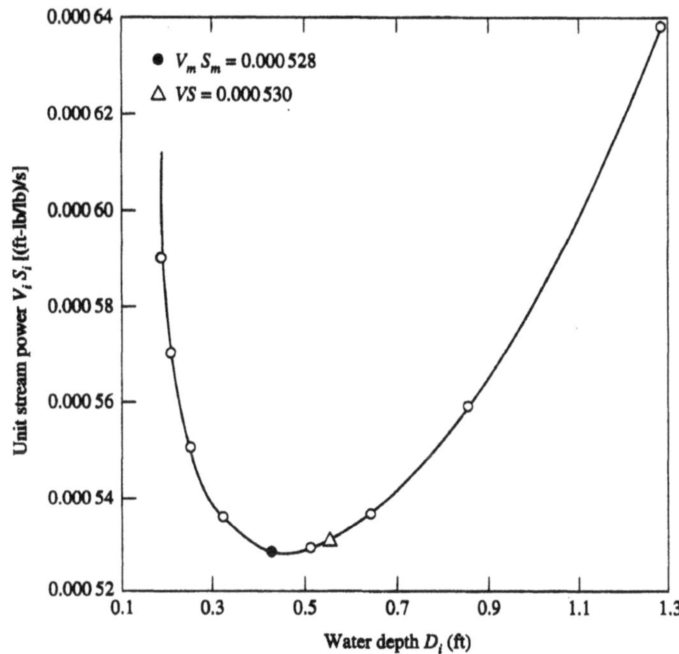

Figure 3.27. Relationship between unit stream power and water depth with 0.19-mm sand in a laboratory flume (Yang, 1976).

Parker (1977), in his discussion of Yang's paper (1976), compared resistance relationships obtained from the theory of minimum unit stream power and those from extensive actual data fitting. Figure 3.28 shows Parker's comparison. These results suggest that the theory of minimum unit stream power can provide a simple theoretical tool for the determination of roughness of alluvial channels, at least for the lower flow region, where the sediment transport rate is not too high, and the rate of energy dissipation due to sediment transport can be neglected. As the sediment concentration or the Froude number increases, the rate of energy dissipation can no longer be neglected, and the accuracy of Yang's method decreases.

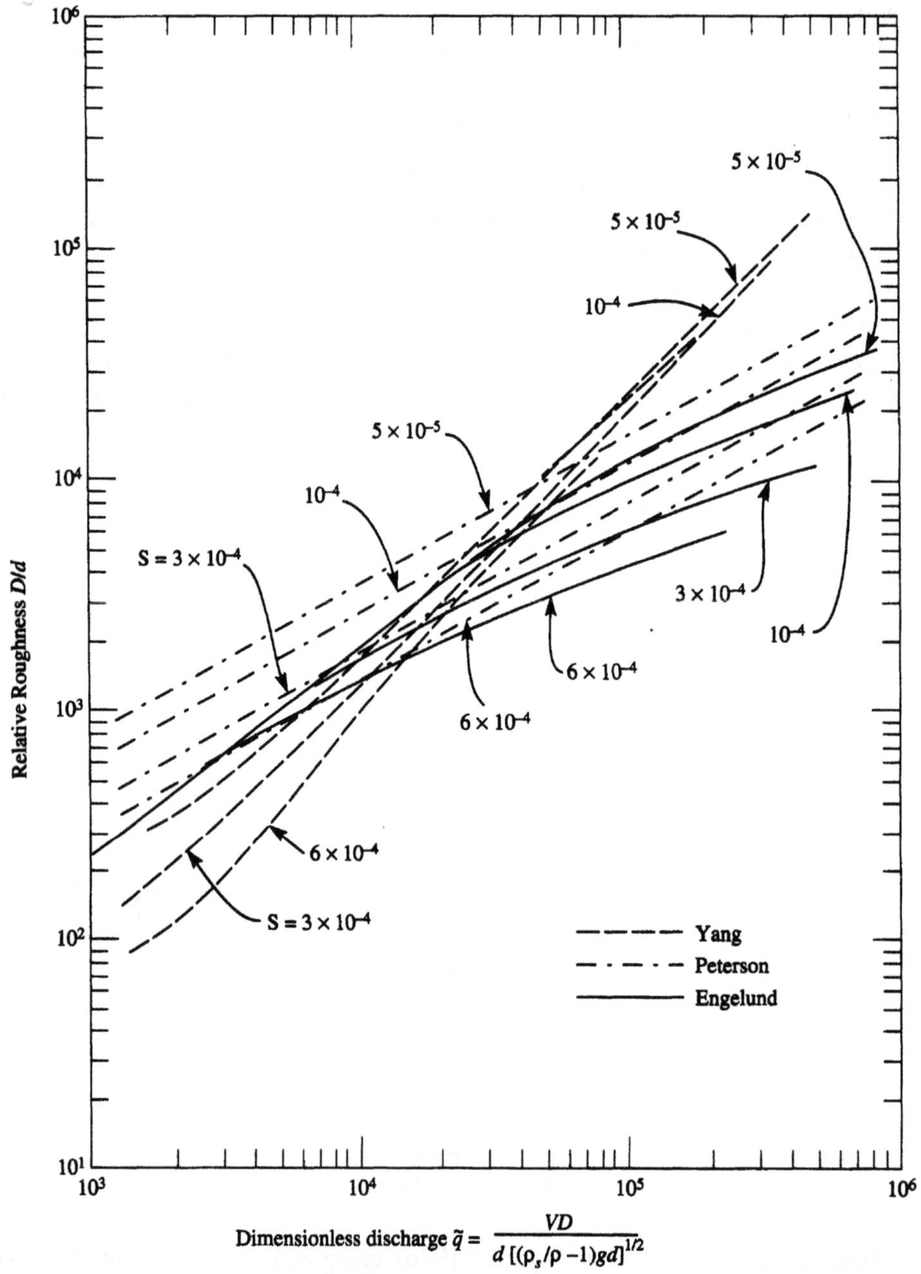

Figure 3.28. Comparisons between relative roughness determined from the theory of minimum unit stream power and those obtained by Peterson and Engelund (Parker, 1977).

Other sediment transport formulas can also be used in Yang's method as long as the formula can accurately estimate sediment load or concentration at the study site. The following example is given to illustrate the application of this method (Yang, 1996):

Example 3.4 The following data were collected from Rio Grande River Section F with width of 370 ft near Bernalillo, New Mexico.

$$d_{50} = 0.31 \text{ mm} \quad V = 3.2 \text{ ft/s} \quad D = 2.41 \text{ ft} \quad S = 0.00076 \quad T = 21.1\ ^\circ\text{C}$$

Determine Manning's roughness coefficient using the minimum unit stream power theory and Yang's (1973) unit stream power equation.

Solution: The computed sediment concentration from equation (3.70) is 517 ppm by weight. The following table summarizes the minimum unit stream power computation:

D_i (ft)	V_i (ft/s)	S_i	$V_i\, S_i$ [(ft-lb/lb)/s]
3.51	2.2	0.001114	0.002451
3.08	2.5	0.000977	0.002443
2.75	2.8	0.000870	0.002435
2.49	3.1	0.000784	0.002431
2.27	3.4	0.000715	0.002430 (min)
2.08	3.7	0.000657	0.002432
1.93	4.0	0.000608	0.002433
1.79	4.3	0.000566	0.002434
1.71	4.5	0.000541	0.002435

The minimum unit stream power $V_m\, S_m$ = 0.002430 (ft-lb/lb)/s, which is close to the measured unit stream power VS = 0.002432 (ft-lb/lb)/s. The corresponding values of depth, velocity, and slope are:

$$D_m = 2.27 \text{ ft} \quad V_m = 3.4 \text{ ft/s} \quad S_m = 0.000715$$

Manning's roughness coefficient with minimum unit stream power is:

$$n_m = \frac{1.49}{V_m} D_m^{2/3} S_m^{1/2} = \frac{1.49}{3.4}(2.27)^{2/3}(0.000715)^{1/2} = 0.0203$$

The actual n value based on the measured V, S, and D is:

$$n = \frac{1.49}{3.2}(2.41)^{2/3}(0.00076)^{1/2} = 0.0231$$

Figure 3.29 summarizes the comparisons between computed values based on Yang's methods and measurements from two river stations of the Rio Grande.

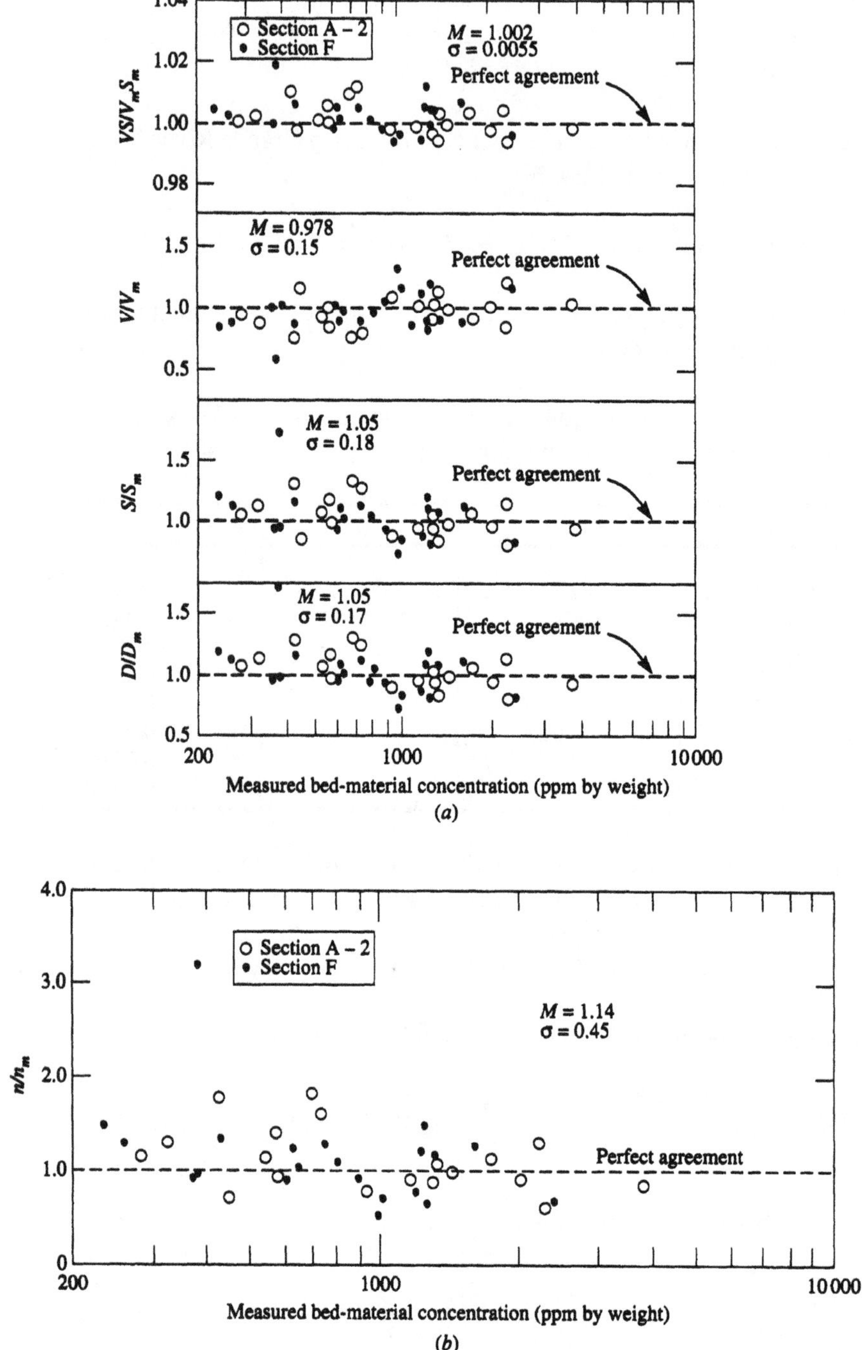

Figure 3.29. Comparisons between measured data from the Rio Grande and computed values from the theory of minimum unit stream power: (a) hydraulic parameters; (b) Manning's roughness coefficient (Yang and Song, 1979).

3.7 Nonequilibrium Sediment Transport

Most of the sediment transport functions were derived under the equilibrium condition with no scour nor deposition. The computed sediment load or concentration in a river from a sediment transport function is the river's sediment-carrying capacity.

When the wash load or concentration of fine material is high, a transport function should be modified by taking the effects of wash load into consideration before its application. An example of this type of modification is the modified dimensionless unit stream power formula proposed by Yang et al. (1996) as shown in equation (3.78). When a sediment transport function is used in a computer model for sediment routing, we also assume equilibrium sediment transport. Under this condition, if a river's sediment-carrying capacity determined from a sediment transport function is different from the sediment supply rate from upstream, scour or deposition would occur instantaneously. This assumption is valid for sand or coarse materials. For fine materials, the concept of nonequilibrium sediment transport should be applied. Based on the analytical solution of the convection-diffusion equation, Han (1980) proposed the following equation for the determination of nonequilibrium sediment transport rate:

$$C_i = C_{t,i} + \left(C_{i-1} - C_{t,i-1}\right)\exp\left\{-\frac{\alpha\omega_s\Delta x}{q}\right\} + \left(C_{t,i-1} - C_{t,i}\right)\left(\frac{q}{\alpha\omega_s\Delta x}\right)\left[1-\exp\left\{-\frac{\alpha\omega_s\Delta x}{q}\right\}\right] \tag{3.145}$$

where C = sediment concentration,
C_t = sediment-carrying capacity, computed from an equilibrium sediment transport function,
q = discharge of flow per unit width,
Δx = reach length,
ω_s = sediment fall velocity,
i = cross-section index (increasing from upstream to downstream), and
α = a dimensionless parameter.

Equation (3.145) is employed for each of the particle size fractions in the cohesiveless range; that is, with diameter greater than 62.5µm. The parameter α is a recovery factor. Han and He (1990) recommended a value of 0.25 for deposition and 2.0 for entrainment. Although equation (3.145) was derived for suspended load, its application to bed-material load is reasonable. Yang and Simões (2002) gave more detailed analysis on the use of equation (3.145) for sediment routing.

3.8 Comparison and Selection of Sediment Transport Formulas

The selection of appropriate sediment transport formulas under different flow and sediment conditions are important to sediment transport and river morphologic studies. Computed sediment load or concentration from different sediment transport formulas can give vastly different results from each other and from field measurement. Consequently, engineers must compare the accuracies and limits of application of different formulas before their final selection. Comparisons of accuracies of sediment formulas were published by Schulits and Hill (1968), White et al. (1975), Yang (1976, 1979, 1984, 1996), Alonso (1980), Brownlie (1981), Yang and Molinas (1982), ASCE (1982), Vetter

(1989), German Association for Water and Land Improvement (1990), Yang and Wan (1991), and Yang and Huang (2001). The comparisons were made directly based on measured results or indirectly based on simulated results of a computer model.

3.8.1 Direct Comparisons with Measurements

Vanoni (1975) compared the computed sediment discharges from different equations with the measured results from natural rivers. Yang (1977) replotted his comparisons. The total measured sediment load does not include wash load. Figures 3.30 and 3.31 show these comparisons. With the exception of Yang's (1973) unit stream power equation, the results in Figures 3.30 and 3.31 are obtained directly from Vanoni's (1971) comparisons.

Figure 3.30 shows a comparison between computed and measured results by Colby and Hembree (1955) from the Niobrara River near Cody, Nebraska. Among the 14 equations, computed results from Yang's (1973) unit stream power equation give the best agreement with measurements. Colby's, Lauren's and, Toffaleti's equations and Einstein's bedload function can all provide reasonable estimates of the total sediment discharge form the Niobrara River. Figure 3.31 shows that Yang's (1973) unit stream power equation is the only one that can provide a close estimate of the total sediment discharge in Mountain Creek. The Schoklitsch equation ranks second in accuracy in this case.

White, Milli, and Crabe (1975) reviewed and compared sediment transport theories. They reviewed and compared most of the available equations at that time, with the exception of Yang's (1973) and Shen and Hung's (1972) equations. Their comparison was based on over 1,000 flume experiments and 260 field measurements. They excluded data with Froude numbers greater than 0.8. They used two dimensionless parameters for comparison purposes: the dimensionless particle size D_{gr} and the discrepancy ratio. The latter is defined as the ratio between calculated and measured sediment loads. D_{gr} is defined as:

$$D_{gr} = \left[\frac{g(\rho_s / \rho - 1)}{v^2} \right]^{1/3} d \tag{3.146}$$

where g = gravitational acceleration,
ρ_s and ρ = densities of sediment and water, respectively,
v = kinematic viscosity of water, and,
d = particle diameter.

Comparisons made by White et al. (1975) indicated that Ackers and White's (1973) equation is the most accurate, followed by Engelund and Hansen's (1972), Rottner's (1959), Einstein's (1950), Bishop, Simons, and Richardson's (1965), Toffaleti's (1969), Bagnold's (1966), and Meyer-Peter and Müller's (1948) equations.

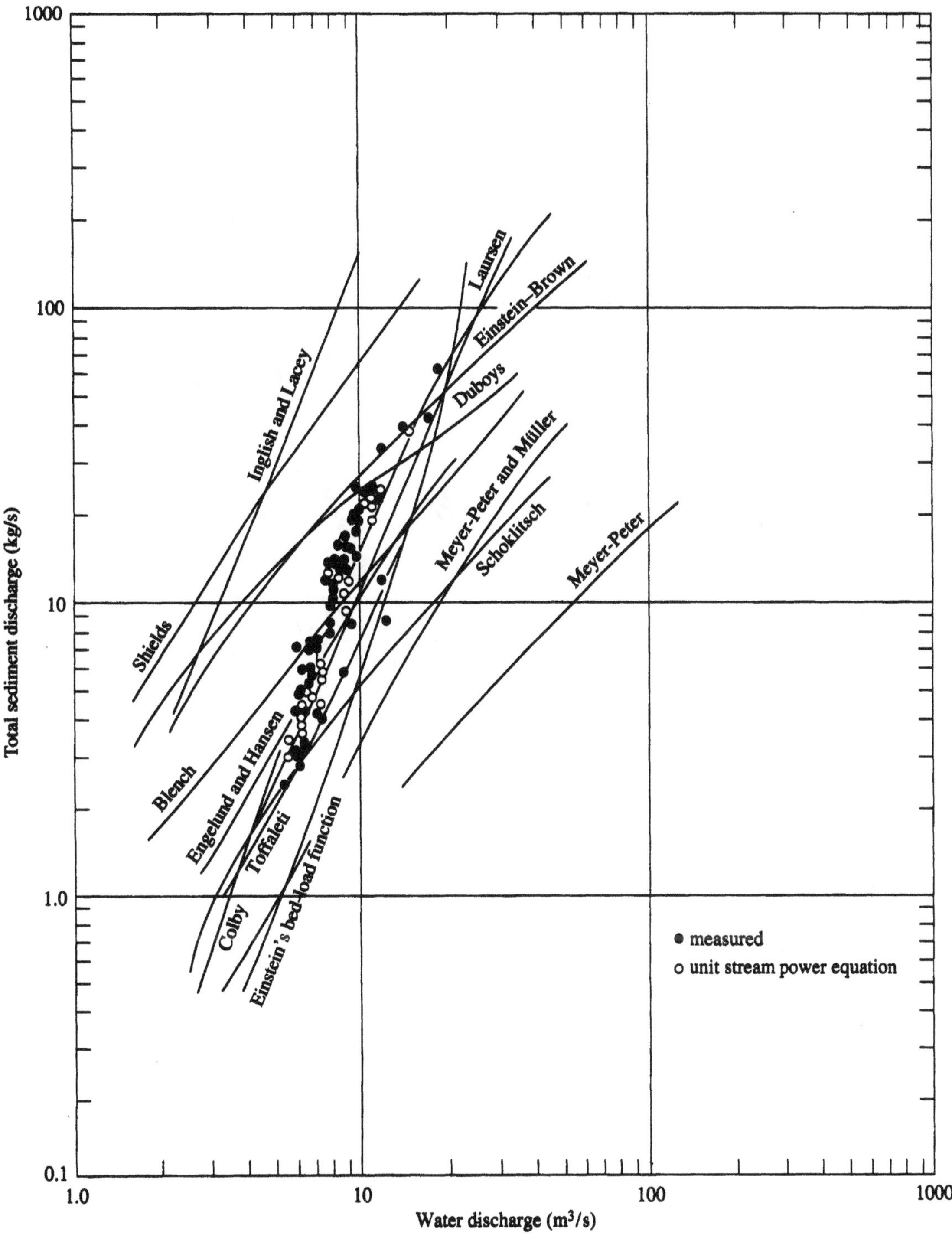

Figure 3.30. Comparison between measured total sediment discharge of the Niobrara River near Cody, Nebraska, and computed results of various equations (Yang, 1977).

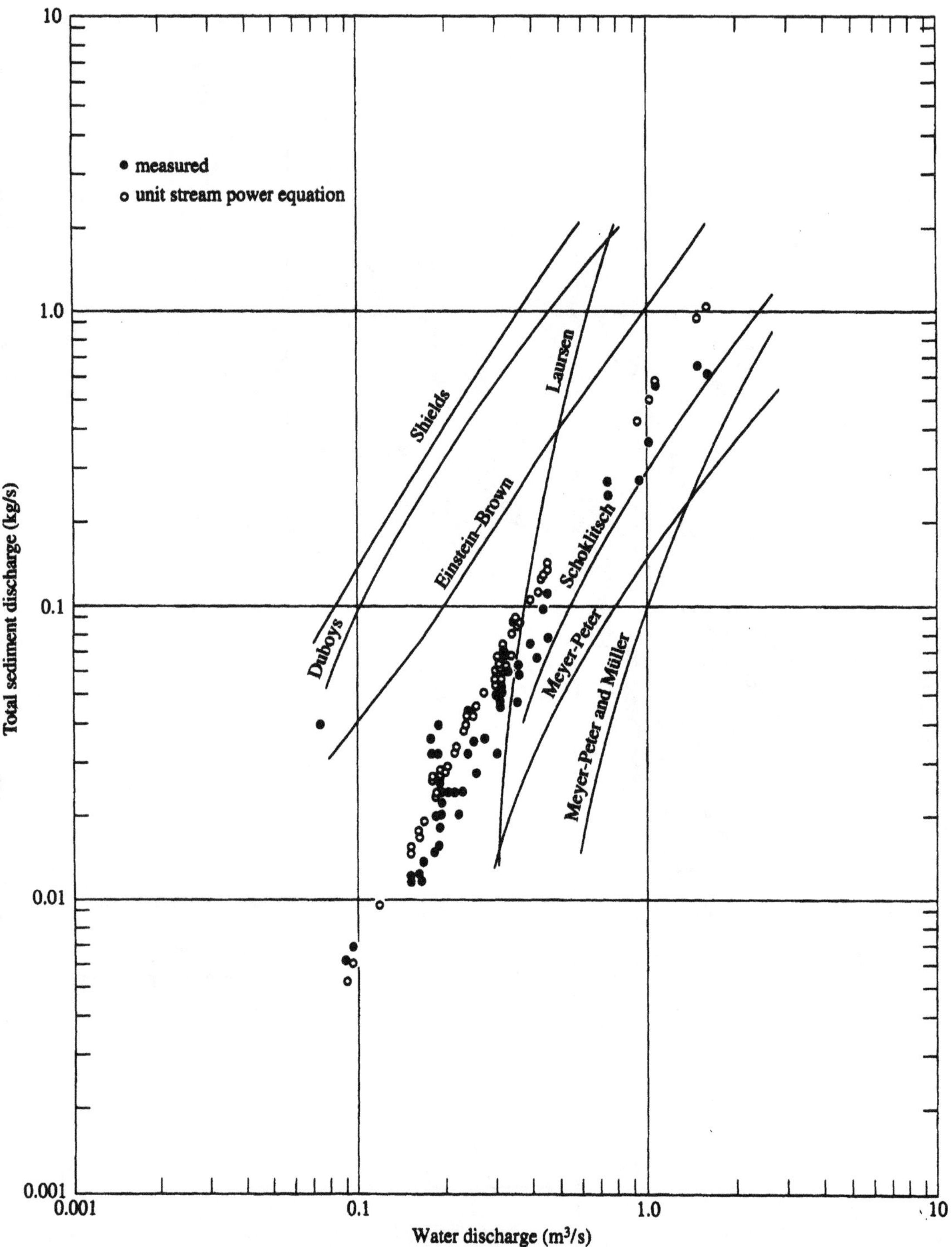

Figure 3.31. Comparison between measured total sediment discharge of Mountain Creek at Greenville, South Carolina, and computed results of various equations (Yang, 1977).

Yang (1976) made a similar analysis of 1,247 sets of laboratory and river data and discussed the results of White et al. (1975). Because the data used for comparison by Yang and by White et al. are basically the same, Table 3.4 combines the comparisons to give a relative rating of different sediment transport equations.

Table 3.4. Summary of accuracies of different equations (Yang, 1976)

Equation	Percent of data with discrepancy ratio between ½ and 2
Yang (1973)	91
Shen and Hung* (1972)	85
Ackers and White (1973)	68
Engelund and Hansen (1972)	63
Rottner (1959)	56
Einstein (1950)	46
Bishop et al. (1965)	39
Toffaleti (1969)	37
Bagnold (1966)	22
Meyer-Peter and Müller (1948)	10

* Should not be applied to large rivers

Alonso (1980) and Alonso et al. (1982) made systematic and detailed evaluations of sediment transport equations. The equations they evaluated cover wide ranges of sediment size, from very fine to very coarse. Among the 31 transport equations initially considered by Alonso (1980), only 8 received detailed comparison and evaluation. Some of the equations were not included for detailed evaluation by Alonso because they have not received extensive application. Others, such as Toffaleti's (1969) and the modified Einstein (Hubbell and Matejke, 1955) methods, are too complicated or require knowledge of the concentration of the measured suspended load and, therefore, not suitable for hydrologic or engineering simulation. Table 3.5 shows the results of the comparison by Alonso (1980) for sand transport. The MPME method, as shown in Table 3.5, estimates the total load by adding the bedload predicted by Meyer-Peter and Müller (1948) formulas to the suspended load computed by Einstein's (1950) procedures.

Alonso limited his comparisons of field data to those where the total bed-material load can be measured by special facilities. Thus, uncertainties in the unmeasured load do not exist. Table 3.5 indicates that Yang's (1973) equation has an average error of 1 percent for both field and flume data. When the depth-particle diameter ratio D/d is less then 70, the flow is shallow, and surface wave effects become important. In this range, most sediment formulas may fail because they do not account for interactions with free surface waves.

Table 3.6 provides a summary rating of selected sediment transport formulas by the American Society of Civil Engineers (ASCE, 1982). The German Association for Water and Land Improvement (1990) published similar ratings.

Table 3.5. Analysis of discrepancy ratio distributions of different transport formulas (Alonso, 1980)

Formula	Number of tests	Ratio between predicted and measured load				Percentage of tests with ratio between ½ and 2
		Mean	95% confidence limits of the mean		Standard deviation	
Field data						
Ackers and White (1973)	40	1.27	1.05	1.48	0.68	87.8
Engelund and Hansen (1972)	40	1.46	1.28	1.64	0.56	82.9
Laursen (1958)	40	0.65	0.49	0.80	0.48	56.1
MPME**	40	0.83	0.50	1.15	1.02	58.5
Yang (1973)	40	1.01	0.89	1.13	0.39	92.7
Bagnold (1966)	40	0.39	0.31	0.47	0.26	32.0
Meyer-Peter and Müller (1948)	40	0.24	0.22	0.27	0.09	0
Yalin (1963)	40	2.59	2.08	3.11	1.62	46.3
Flume data with $D/d \geq 70$						
Ackers and White (1973)	177	1.34	1.24	1.54	1.29	73.0
Engelund and Hansen (1972)	177	0.73	0.63	0.83	0.68	51.1
Laursen (1958)	177	0.81	0.73	0.88	0.51	71.4
MPME**	177	3.11	2.95	3.52	2.75	42.1
Yang (1973)	177	0.99	0.93	1.08	0.60	79.8
Bagnold (1966)	177	0.85	0.81	1.22	2.50	20.8
Meyer-Peter and Müller (1948)	177	0.40	0.39	0.47	0.49	18.5
Yalin (1963)	177	1.62	1.38	2.23	4.08	32.6
Flume data with $D/d \leq 70$						
Ackers and White (1973)	48	1.12	0.93	1.28	0.52	89.6
Engelund and Hansen (1972)	48	0.75	0.59	0.90	0.50	66.7
Laursen (1958)	48	1.04	0.76	1.32	0.99	79.2
MPME**	48	1.34	1.04	1.64	1.04	66.7
Yang (1973)	48	0.90	0.79	1.05	0.51	85.4
Bagnold (1966)	48	1.53	1.46	1.87	1.14	45.8
Meyer-Peter and Müller (1948)	48	1.03	1.00	1.27	0.83	72.9
Yalin (1963)	48	1.92	1.45	2.41	1.65	64.6

* MPME = Meyer-Peter and Müller's (1948) formula for bedload and Einstein's (1950) formula for suspended load.

Table 3.6. Summary of rating of selected sediment transport formulas (ASCE, 1982)

Formula number (1)	Reference (2)	Type (3)	Comments (4)
1	Ackers and White (1973)	Total load	rank* = 3
2	Engelund and Hansen (1967)	Total load	rank = 4
3	Laursen (1958)	Total load	rank = 2
4	MPME	Total load	rank = 6
5	Yang (1973)	Total load	rank = 1, best overall predictions
6	Bagnold (1966)	Bedload	rank = 5
7	Meyer-Peter and Müller (1948)	Bedload	rank = 7
8	Yalin (1963)	Bedload	rank = 8

* Based on mean discrepancy ratio (calculated over observed transport rate) from 40 tests using field data and 165 tests using flume data

Direct comparisons between measured and computed results from different sediment transport equations indicate that, on the average, Yang's (1973) dimensionless unit stream power equation is more accurate than others for sediment transport in the sand size range. Figure 3.32 shows a summary comparison between measured bed-material discharge from six river stations and computed results from Yang's (1973) equation.

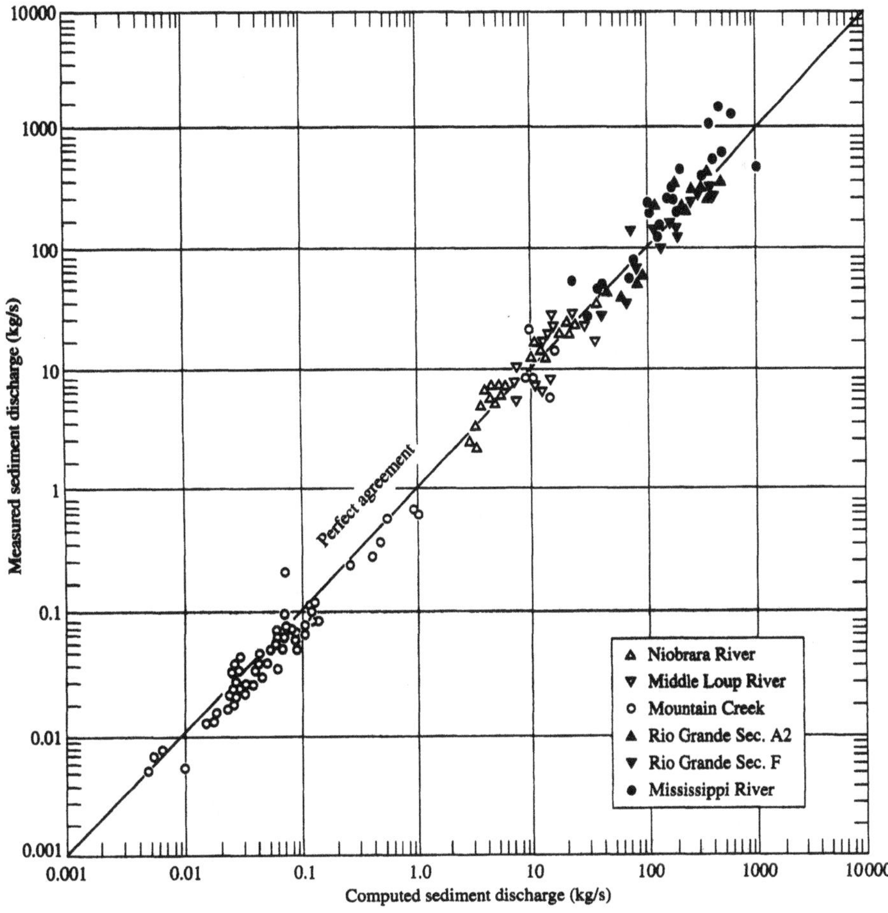

Figure 3.32. Comparison between measured total bed-material discharge from six river stations and computed results from Yang's (1973) equation (Yang, 1979, 1980).

The results shown in Table 3.7 indicate that the average mean discrepancy ratio of Yang's (1973) equation for 1,247 sets of laboratory and river data is 1.03. This means that, on the average, Yang's (1973) equation has an error of 3 percent. Figure 3.33 shows that the distributions of discrepancy ratio of Yang's (1973) equation for both laboratory and river data follow normal distributions. This means that no systematic error exists in Yang's (1973) equation. The reason that computed loads for natural rivers are generally higher than measurements is that Yang's (1973) equation includes loads in the unmeasured zone, while for most natural rivers, loads in the unmeasured zone are not included in the measurements.

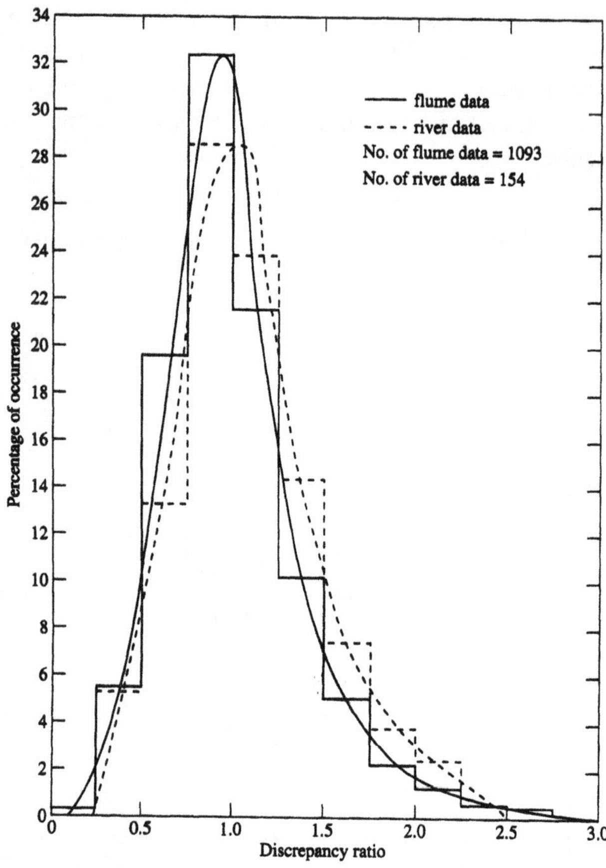

Figure 3.33. Distribution of discrepancy ratio of Yang's (1973) equation (Yang, 1977).

Table 3.7. Summary of accuracy of Yang's (1973) equation (Yang, 1977)

	Discrepancy ratio							
	Max. (1)	Mean (2)	Min. (3)	0.75–1.25 (4)	0.5–1.50 (5)	0.25–1.75 (6)	0.5–2.0 (7)	No. of data sets (8)
Sand in laboratory flumes	2.05	1.02	0.57	54%	84%	94%	91%	1,093
Sand in rivers	1.92	1.08	0.47	53%	80%	93%	92%	154
All data	2.03	1.03	0.56	54%	83%	94%	91%	1,247

Most of the comparisons of accuracy of equations were made for data collected in the sand size range. For coarser materials, sediments mainly travel as bedload or in the unmeasured zone. No reliable instrument can be used to measure bedload in most natural rivers under normal conditions. Thus, comparisons can be made only for laboratory flume data, where bedload can be measured by special equipment. Figure 3.34 shows an example of a comparison of four equations. It indicates that equations of Yang (1984), Engelund and Hansen (1972), Ackers and White (1973), and Meyer-Peter and Müller (1948) are all reasonably accurate for Gilbert's (1914) 7.0- mm gravel data collected from a laboratory flume. However, with the exception of Yang's (1984) gravel equation, the agreement between measured (Cassie, 1935) and computed results shown in Figure 3.35 is poor. This is due to the lack of generality of the assumptions used in the development of these equations, as explained in section 3.3.4.

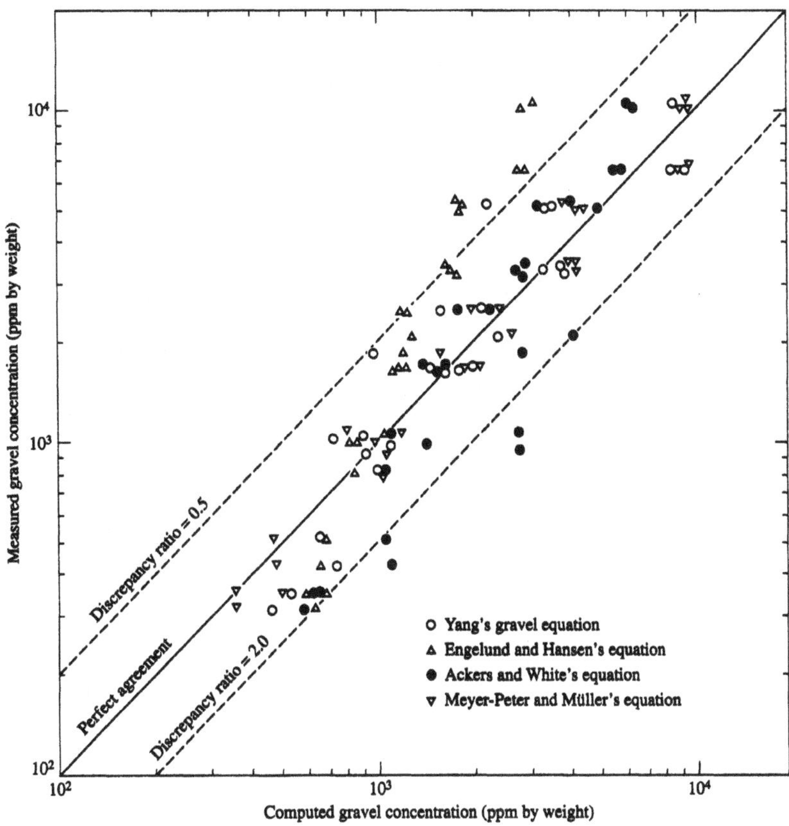

Figure 3.34. Comparison between 7.01-mm gravel concentration measured by Gilbert (1914) and results computed using different equations (Yang, 1984).

Among deterministic sediment transport equations, the modified Yang's (1996) unit stream power equation (3.78) is the one that can be applied to flows with high concentration of wash load. Yang et al. (1996) compared the computed results from equation (3.78) and 580 sets of measured data from 9 gauging stations along the Middle and Lower Yellow River. Their comparisons have an averaged discrepancy ratio of 1.0034 and a standard deviation of 1.6692. Figure 3.36 shows their comparisons. The slope of the Middle and Lower Yellow River is very flat. The flatter the slope, the higher the percentage error of measurement that can be caused by water surface fluctuation.

Figure 3.37 shows a comparison between the computed and measured results from the Yellow River, excluding 112 sets of data with slope less than 0.0001 from a total of 580 sets of data. The improvement shown in Figure 3.37 over that in Figure 3.36 is apparent. Thus, in a comparison with field data, the possibility of having measurement errors should not be overlooked.

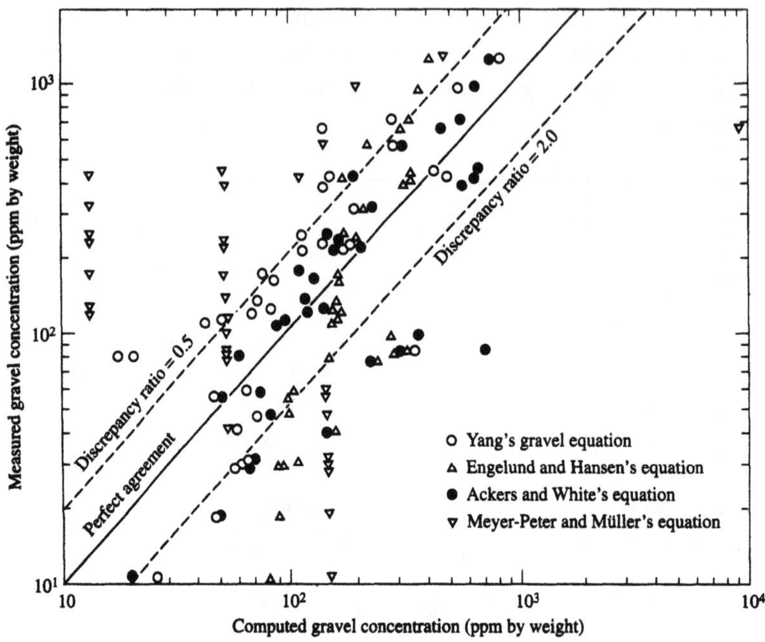

Figure 3.35. Comparison between 2.46-mm gravel concentration measured by Cassie (1935) and results computed using different equations (Yang, 1984).

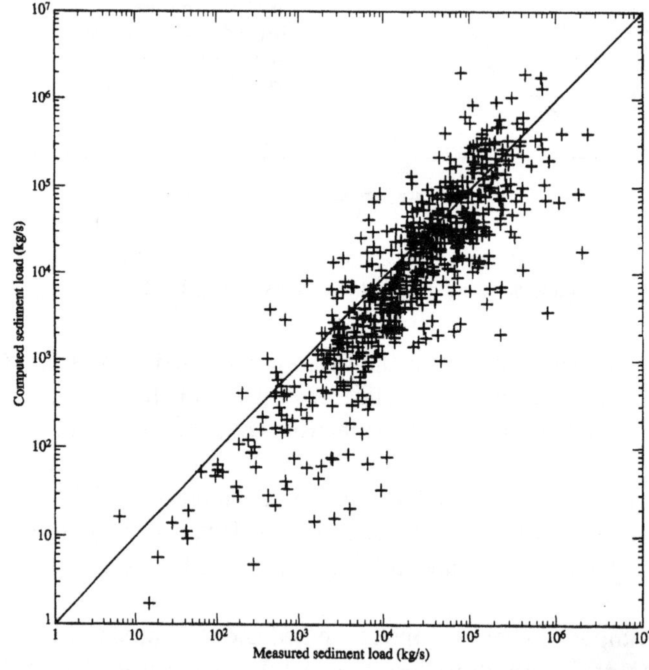

Figure 3.36. Comparison between computed and measured results based on the modified Yang's unit stream power formula, equation (3.78), and measurements from the Yellow River with sediment diameter larger than 0.01 mm.

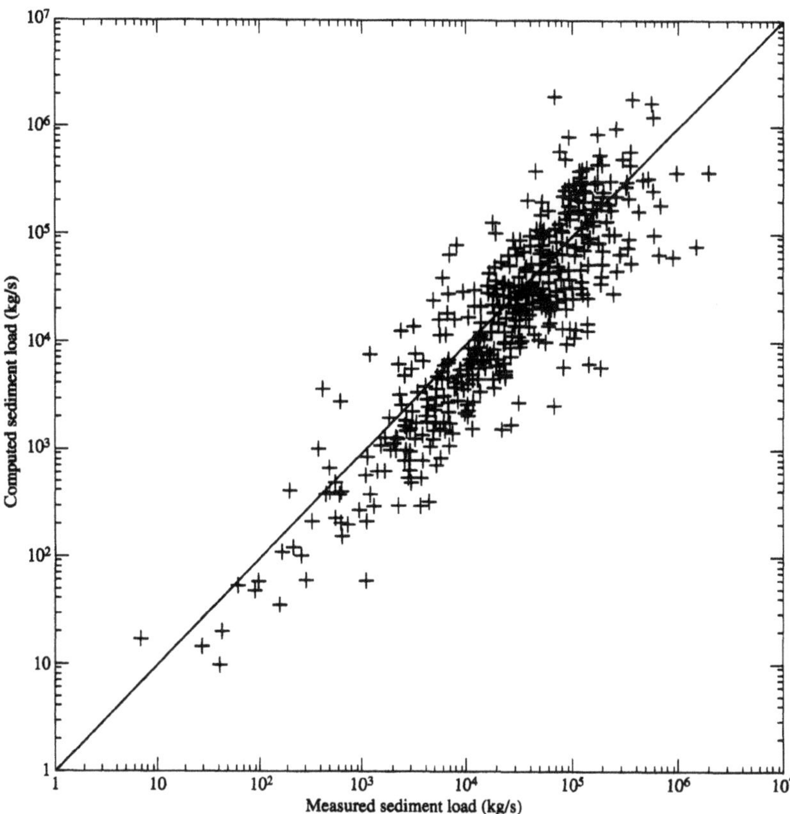

Figure 3.37. Comparison between computed and measured results based on the modified Yang's unit stream power formula, equation (3.78), and measurements from the Yellow River with sediment diameter larger than 0.01 mm and slope greater than 0.0001.

Equation (3.70) was developed as a predictive equation for sand transport. Figure 3.38 indicates that equation (3.70) can be used to predict sediment transport rate in the clay-size range if the effective diameter of clay aggregate is used. The scattering shown in Figure 3.38 was mainly due to the fact that different numbers of fine particles are bunched together to form clay aggregate of different effective diameters. Moore and Burch (1986) applied equation (3.70) in conjunction with the theory of minimum unit stream power for the determination of surface and rill erosion rate. Figure 3.39 indicates that equation (3.70) can accurately predict surface and rill erosion rate, especially if soil particles are in the ballistic dispersion mode when most sediment particles are being eroded (see Chapter 2, Erosion and Reservoir Sedimentation).

3.8.2 Comparison by Size Fraction

Not all sediment particles move at the same rate under a given flow condition when the particle sizes are not uniform. Yang and Wan (1991) made detailed comparisons of formulas based on size fraction.

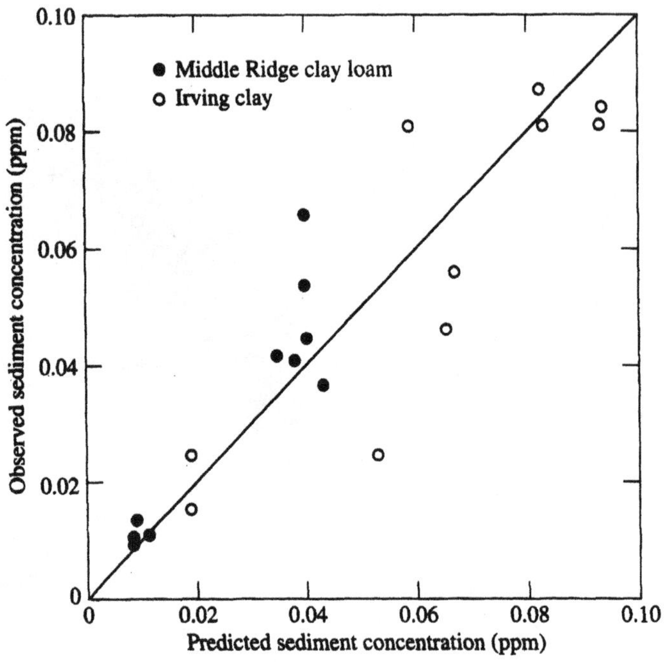

Figure 3.38. Comparison between observed and predicted clay concentrations from Yang's unit stream power (Moore and Burch, 1986).

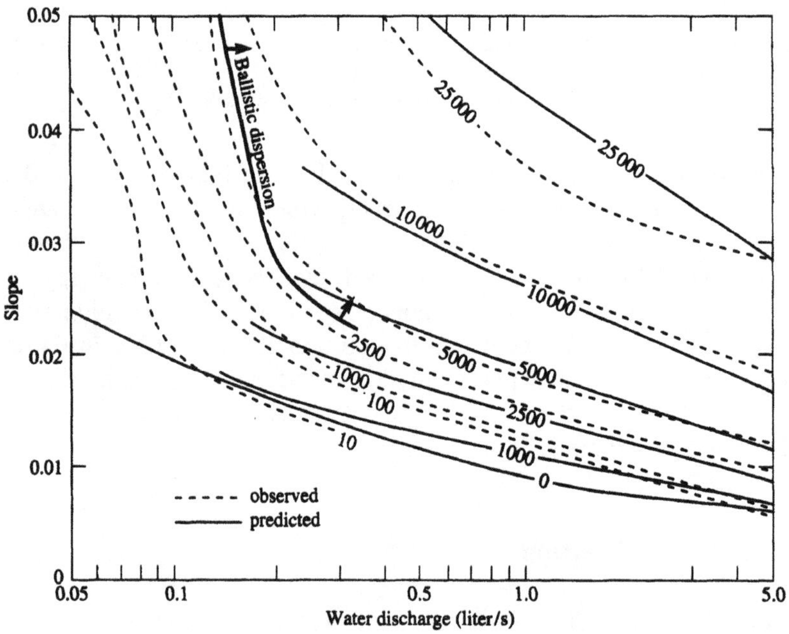

Figure 3.39. Comparison between observed and predicted sediment concentrations in ppm, by weight, from Yang's unit stream power equation with a plane bed composed of 0.43-mm sand (Moore and Burch, 1986).

They defined the discrepancy ratio γ_1 as the ratio between the median particle diameter d_c in transportation, computed by a formula, and actually measured particle diameter d_m in transportation; that is,

$$\gamma_1 = \frac{d_c}{d_m} \tag{3.147}$$

They also defined the discrepancy ratio γ_2 as the ratio between the median particle diameter d_c in transportation, computed by a formula, and d_{50} of the original bed materials on the alluvial bed; that is,

$$\gamma_2 = \frac{d_c}{d_{50}} \tag{3.148}$$

Most sediment transport equations were originally developed for fairly uniform bed materials. When they are applied to nonuniform materials, the total sediment concentration can be computed by size fraction (Yang, 1988):

$$C = \sum_{i=1}^{j} p_i C_i \tag{3.149}$$

where:
j = total number of size fractions in the computation,
p_i = percentage of material available in size i, and
C and C_i = total concentration and concentration for size i computed from an equation, respectively.

The discrepancy ratio γ_1 should give an indication of the average accuracy of a formula in predicting the size distribution of bed materials in transportation. The discrepancy ratio γ_2 should give us an indication of the reasonableness of a formula in predicting the effect of the sorting process or the reduction of average particle size in the transport process. The results shown in Table 3.8 indicate that Yang's (1973) fraction formula has the best overall discrepancy ratio of 0.95. These results also show that the γ_1 value for Yang's (1973) fraction formula is not very sensitive to variations in Froude number. They suggest that Yang's fraction formula can be used with accuracy to predict size distribution of bed materials in transportation. This study also indicates that the median sizes of bed materials in transportation predicted by Laursen (1958) and Toffaleti (1968) are too small, while those predicted by Einstein (1950) are too large.

Table 3.9 indicates that, with the exception of Einstein's formula, bed materials in transportation computed by Laursen (1958), Yang (1973) by size fraction, and Toffaleti (1968) are finer than the original bed materials on the bed, which is consistent with the sorting phenomena. This sorting process explains why bed-material size should decrease in the downstream direction. The measured γ_2 value based on Yang's (1973) fraction formula changes very little, and Table 3.9 shows an average value of 0.77. The γ_2 values of Einstein's (1950) formula are greater than unity for all flow conditions, which means that the materials in transportation computed using Einstein's formula are coarser than the original bed materials, and the bed-material size would increase in the downstream direction, which is not reasonable. Yang and Wan's results suggest that Einstein's hiding and lifting factors may overcorrect the effect of nonuniformity of bed-material size on transport of graded bed materials. Einstein's assumption, that the average step length of 100 particle diameters implies that larger particles would have longer step length, is also erroneous.

Table 3.8. Comparison between computed and measured bed-material sizes in transportation (Yang and Wan, 1991)

Formula	Discrepancy ratio γ_1					Number of data sets
	Mean	Percentage of data in the range			Standard deviation	
		0.75–1.25	0.5–1.5	0.25–1.75		
F_r = 0.20–0.30						
Laursen	0.86	79	100	100	0.15	19
Yang (fraction)	1.09	79	100	100	0.18	19
Einstein	2.64	0	0	0	0.84	19
Toffaleti	0.77	42	100	100	0.19	19
F_r = 0.30–0.50						
Laursen	0.82	60	97	100	0.19	117
Yang (fraction)	1.03	85	93	100	0.21	117
Einstein	1.94	22	43	50	0.86	117
Toffaleti	0.61	32	59	85	0.28	117
F_r = 0.50–1.00						
Laursen	0.78	60	88	95	0.23	86
Yang (fraction)	0.91	80	92	99	0.25	86
Einstein	1.41	55	77	84	0.64	86
Toffaleti	0.55	22	51	83	0.27	86
F_r = 1.00–2.00						
Laursen	0.73	48	89	100	0.16	83
Yang (fraction)	0.85	66	99	100	0.18	83
Einstein	0.99	86	99	100	0.18	83
Toffaleti	0.53	22	47	94	0.23	83
All data						
Laursen	0.79	58	92	99	0.19	305
Yang (fraction)	0.95	78	95	100	0.21	305
Einstein	1.58	47	65	70	0.61	305
Toffaleti	0.58	27	65	87	0.26	305

Table 3.9. Comparison between computed and bed-material size in transportation and measured original bed-material size (Yang and Wan, 1991)

Formula	Discrepancy ratio γ_2					Number of data sets
	Mean	Percentage of data in the range			Standard deviation	
		0.75–1.25	0.5–1.5	0.25–1.75		
F_r = 0.20–0.30						
Laursen	0.66	26	84	100	0.14	19
Yang (fraction)	0.81	89	100	100	0.08	19
Einstein	1.98	0	26	37	0.54	19
Toffaleti	0.59	26	74	100	0.16	19
Measured value	0.77	58	95	100	0.10	19
F_r = 0.30–0.50						
Laursen	0.61	17	87	91	0.16	117
Yang (fraction)	0.76	80	91	91	0.18	117
Einstein	1.34	51	70	87	0.39	117
Toffaleti	0.43	8	40	79	0.20	117
Measured value	0.77	56	91	96	0.17	117
F_r = 0.50–1.00						
Laursen	0.63	37	79	91	0.19	86
Yang (fraction)	0.73	70	88	93	0.20	86
Einstein	1.06	90	97	100	0.17	86
Toffaleti	0.43	13	40	78	0.21	86
Measured value	0.77	79	91	99	0.17	86
F_r = 1.00–2.00						
Laursen	0.65	17	93	100	0.10	83
Yang (fraction)	0.76	61	100	100	0.08	83
Einstein	0.90	80	100	100	0.14	83
Toffaleti	0.46	5	36	94	0.15	83
Measured value	0.77	70	98	100	0.17	83
All data						
Laursen	0.63	23	86	94	0.15	305
Yang (fraction)	0.75	73	93	95	0.15	305
Einstein	1.18	67	83	91	0.27	305
Toffaleti	0.45	10	41	84	0.19	305
Measured value	0.77	66	93	98	0.17	305

3.8.3 Computer Model Simulation Comparison

Computer models have been increasingly used to predict or simulate the scour and deposition procedures of a river due to artificial or natural causes. The simulated results are sensitive to the selection of sediment transport equations used in the computer model. Therefore, the agreement between the measured and simulated results from a sediment transport equation is an indication of the accuracy of that equation. One of the most commonly used one-dimensional sediment routing models is HEC-6, developed by the U.S. Army Corps of Engineers (1977, 1993).

The U.S. Army Corps of Engineers (1982) applied the HEC-6 model to the study of scour and deposition process along several rivers due to engineering constructions. The sediment transport equations included in HEC-6 that were selected by the Los Angeles District of the Corps for comparison included those of Yang (1973, 1984), Toffaleti (1969), Laursen (1958), and DuBoys (1879). After a thorough comparison of all the transport equations available in HEC-6, Yang's (1973) equation was selected.

The Los Angeles District gave the following reasons:

> This function was selected because of (1) previous successful application in sediment studies performed on similar streams in southern California by the Los Angeles District, (2) the conclusions reported in a study conducted by the U.S. Department of Agriculture, and (3) comparison with the results from other transport functions applied in this study.

Before the Corps finally selected Yang's (1973) equation, sensitivity tests of the results using different transport functions in HEC-6 were made. These tests reached the following conclusions:

> Of the four functions applied, the Toffaleti transport capacity was found to be much less than the others. The result has reasonably small changes in computed bed elevations. The Duboys equation produced trends opposite from those predicted in the preliminary analysis indicated in table 1 (U.S. Army Corps of Engineers, 1982). Likewise, the Laursen function produced trends in the middle reach that were opposite from those predicted, and moreover, indicated unreasonably high deposition in the downstream reach. By contrast, the Yang equation produced trends that agreed well with the preliminary analysis throughout the study reach with the exception of the very downstream end, as was previously discussed (due to the lack of reliable estimation of Manning's *n* value). Thus, even though the computed changes in bed elevation were found to be very sensitive to different functions, the Yang equation clearly yielded the most reasonable results of the four functions incorporated into the HEC-6 program. For this reason and for the reasons discussed previously, it was concluded that the Yang function is the most appropriate to use in simulating sediment transport in the San Luis Rey River.

Figures 3.40 and 3.41 are two examples of comparisons made by the Corps of Engineers, Los Angeles District. These data are in the sand-size range. The comparisons indicate that generally good correlation between the observed and reconstituted bed profiles was obtained from the HEC-6 model using Yang's (1973) equation.

The HEC-6 computer model is a one-dimensional model for water and sediment routing. The bed elevation adjustment is parallel to the original bed without any variation in the lateral direction. The Bureau of Reclamation's GSTARS (Molinas and Yang, 1986) is a generalized stream tube model for alluvial river simulation. GSTARS can simulate the hydraulic conditions in a semi-two-dimensional manner, and channel geometry change in a semi-three-dimensional manner. Figure 3.42 shows a three-dimensional plot of the variation of computed scour pattern at the Mississippi River Lock and Dam No. 26 replacement site. Yang's 1973 sand formula and his 1984 gravel formula were used in the GSTARS simulation. Figure 3.43 shows the comparison between measured and computed results based on GSTARS. Because GSTARS cannot simulate secondary flow and eddies, a simplified assumption of a straight line extension of the cofferdam, as shown in Figure 3.43(b), was adopted. Despite this simplification, Figure 3.43 shows that the scour patterns predicted by GSTARS using Yang's sand and gravel formulas agree very well with measured results.

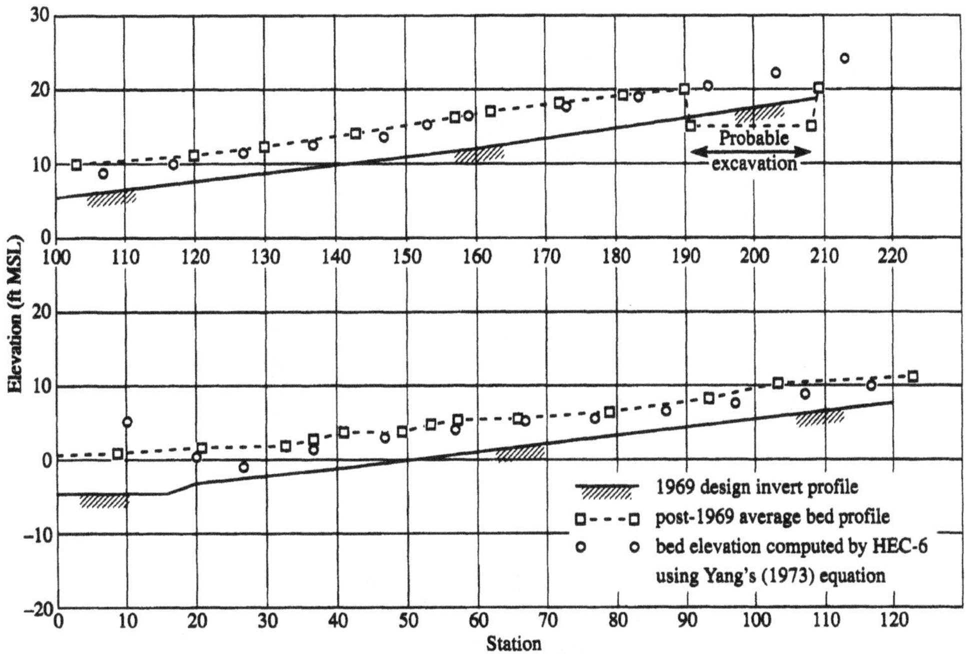

Figure 3.40. Reconstituted bed profiles of the Lower Santa Ana River after the 1969 flood, using Yang's (1973) equation (U.S. Army Corps of Engineers, 1982).

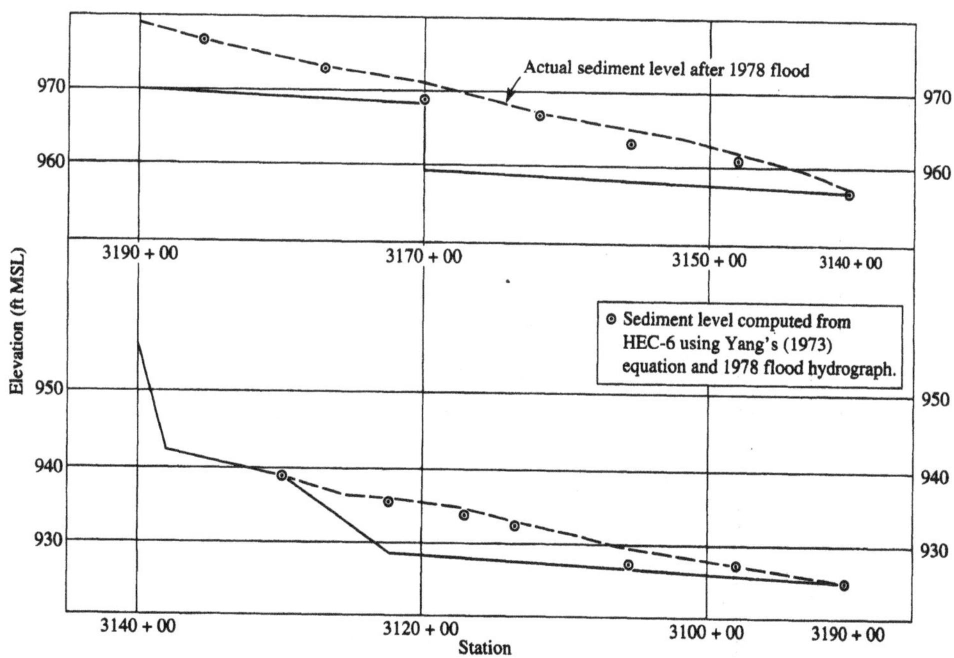

Figure 3.41. Reconstituted bed profiles of the Upper Santa Ana River after the 1978 flood, using Yang's (1973) equation (U.S. Army Corps of Engineers, 1982).

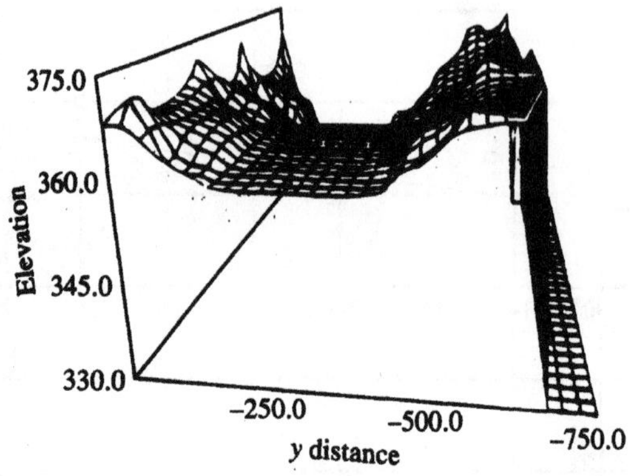

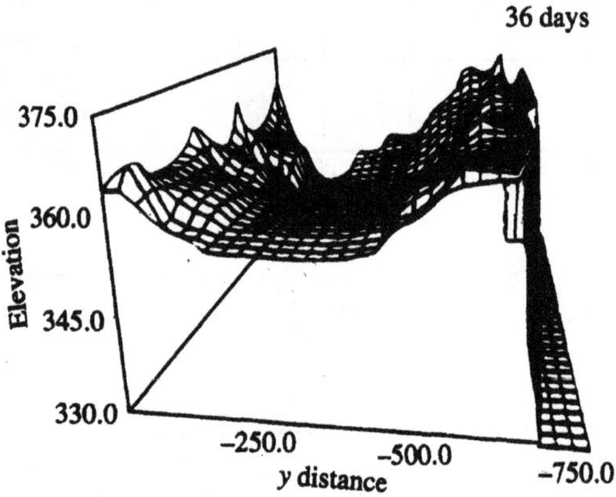

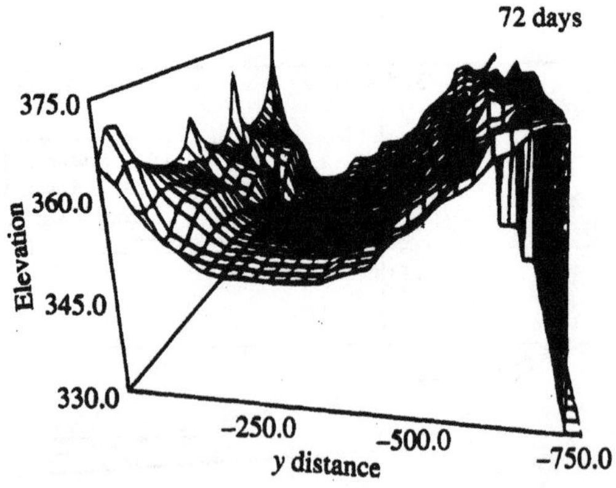

Figure 3.42. Three-dimensional plot of the variation of computed scour pattern at the Mississippi River Lock and Dam No. 26 replacement site (Yang et al., 1989).

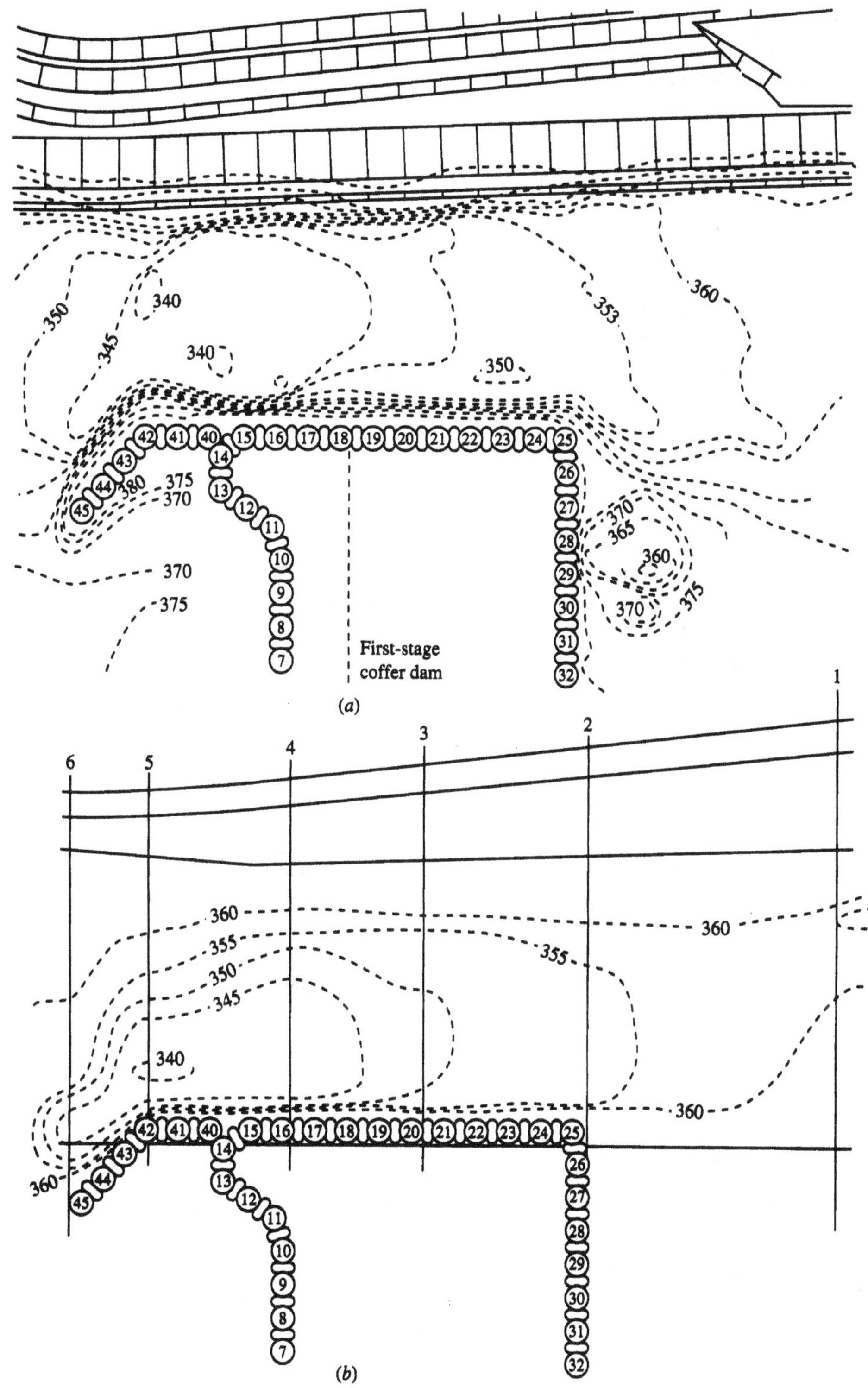

Figure 3.43. Scour pattern (a) measured and (b) computed, based on the flow condition of April 1,1982, at the Mississippi River Lock and Dam No. 26 replacement site (Yang et al., 1989).

The GSTARS computer model series has evolved through different revised and improved versions since its original release in 1986. They are GSTARS 2.0 (Yang et al., 1998), GSTARS 2.1 (Yang and Simões, 2000), and GSTARS3 (Yang and Simões, 2002). Information on these programs can be found by accessing website: http://www.usbr.gov/pmts/sediment. One of the important features of all the GSTARS models is the ability to simulate and predict channel width adjustments based on the theory of minimum energy dissipation rate (Yang, 1976; Yang and Song, 1979, 1984) or its simplified version of minimum stream power. Figure 3.44 compares the measured and predicted channel cross-sectional change of the unlined emergency spillway downstream from Lake Mescalero in New Mexico. The computation was based on Yang's sand (1973) and gravel (1984) formulas using GSTARS 2.1. It is apparent that the use of the optimization option based on the theory of minimum stream power can more accurately predict and simulate channel geometry changes. It is also apparent that the accuracy of simulated results depends not only on the selection of a sediment transport formula, but also on the capability and limits of application of the computer model used in the simulation.

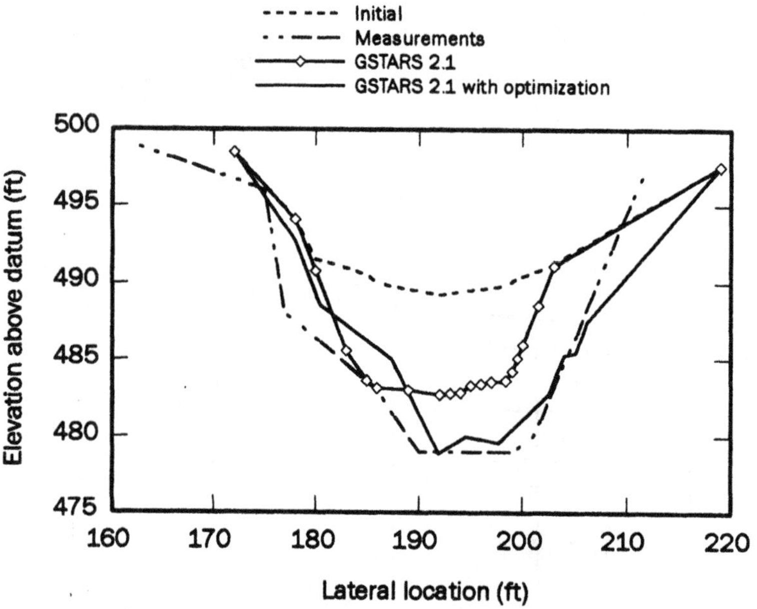

Figure 3.44. Comparison of results produced by GSTARS 2.1 and survey data for runs with and without width changes due to stream power minimization (Yang and Simões, 1998).

It is difficult to determine the accuracy and applicability of a bedload or gravel transport formula directly when it is applied to a natural river. This is because of the limitation of existing sampling methods. Chang (1991, 1994) developed a method for selecting a gravel transport formula based on the measured changes in stream morphology instead of site-specific gravel transport data.

The measured scour at the Highway No. 32 bridge crossing Stony Creek in Glen County, California ,is 77.6 m^2 in cross-sectional area. The simulated values based on Meyer-Peter and Müller's (1948), Parker's (1990), Yang's (1984), and Engelund and Hansen's formulas are 58.5, 79.9, 75.2, and 143.1 m^2, respectively. The measured deposition at station 46200 is 150 m^2 in cross-sectional area.

The simulated values based on Meyer-Peter and Müller's, Parker's, Yang's, and Engelund and Hansen's formulas are 63, 155, 149, and 273 m^2, respectively. These results indicate that the gravel formulas of Yang and Parker can accurately predict the scour and deposition process. Engelund and Hansen's formula produced a higher transport rate, while Meyer-Peter and Müller's produced a lower transport rate than the measurements.

Although the depositions simulated using Parker's formula and Yang's (1984) formula are similar, Yang's showed a more uniform distribution of deposition along the channel and correlated better with measurement (Chang, 1991). For this reason, Chang (1991, 1994) adopted Yang's formula for the Stony Creek morphological study. Figures 3.45 and 3.46 show examples of Chang's simulation results using Yang's (1984) formula.

3.8.4 Selection of Sediment Transport Formulas

The ranking of the accuracy of formulas in the published comparisons is not consistent, mainly because they were based on different sets of data. Some of the comparisons are not strictly valid, because data outside of the range of application recommended by the authors of the formulas were used in the comparison. Although no lack of data for comparison exists, the accuracies of data, especially field data, may be questionable.

Yang and Huang (2001) published a comprehensive comparison of 13 sediment transport formulas to determine their limits of application. Published, reliable data by different authors were used to give unbiased comparisons. Different amounts of data were used for different formulas because only the data within the applicable range of a formula are used to test its accuracy. Dimensionless parameters were used to determine the sensitivities of formulas to these parameters.

Stevens and Yang (1989) published FORTRAN and BASIC computer programs for 13 commonly used sediment transport formulas in river engineering. Yang's 1996 book, *Sediment Transport Theory and Practice*, includes the complete source codes in both FORTRAN and BASIC and a floppy diskette of the programs. The 13 formulas are those proposed by Schoklitsch (1934), Kalinske (1947), Meyer-Peter and Müller (1948), Einstein (1950) for bedload, Einstein (1950) for bed-material load, Laursen (1958), Rottner (1959), Engelund and Hansen (1967), Toffaleti (1968), Ackers and White (1973), Yang (1973) for sand transport with incipient motion criteria, Yang (1979) for sand transport without incipient motion criteria, and Yang (1984) for gravel transport. Yang and Huang (2001) selected these formulas, because the computer program used in comparison is readily available to the public. Many of these formulas have been incorporated in sediment transport models, such as the U.S. Army Corps of Engineers' HEC-6 computer model, Scour and Deposition in Rivers and Reservoirs (1993), and the Bureau of Reclamation's Generalized Stream Tube Model for Alluvial River Simulation (GSTARS) by Molinas and Yang (1986) and its revised and improved versions of GSTARS2 (Yang, et al., 1998), GSTARS 2.1 (Yang and Simões, 2000), and GSTARS3 (Yang and Simões, 2002).

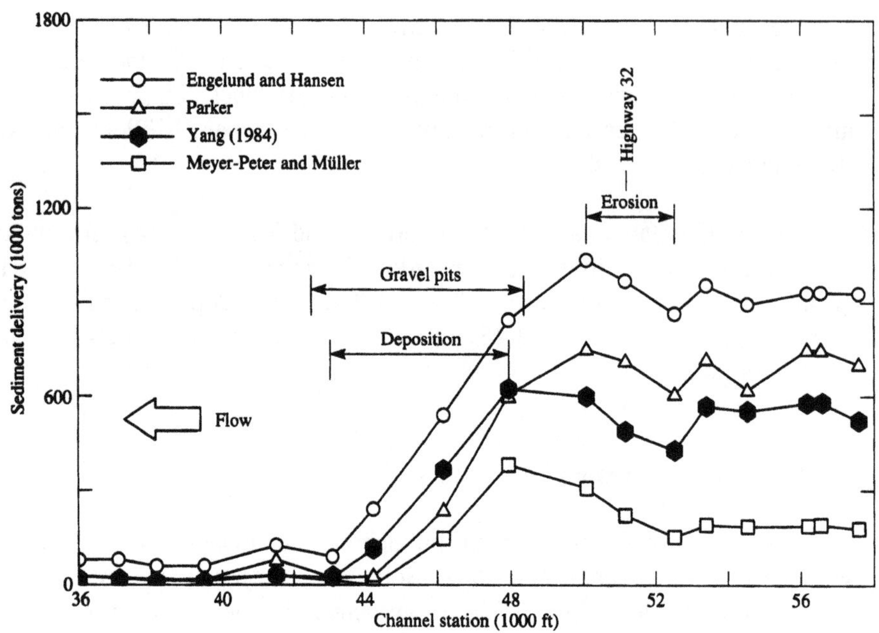

Figure 3.45. Spatial variations of the Stony Creek sediment delivery by the 1978 flood based on four sediment-transport formulas (Chang, 1994).

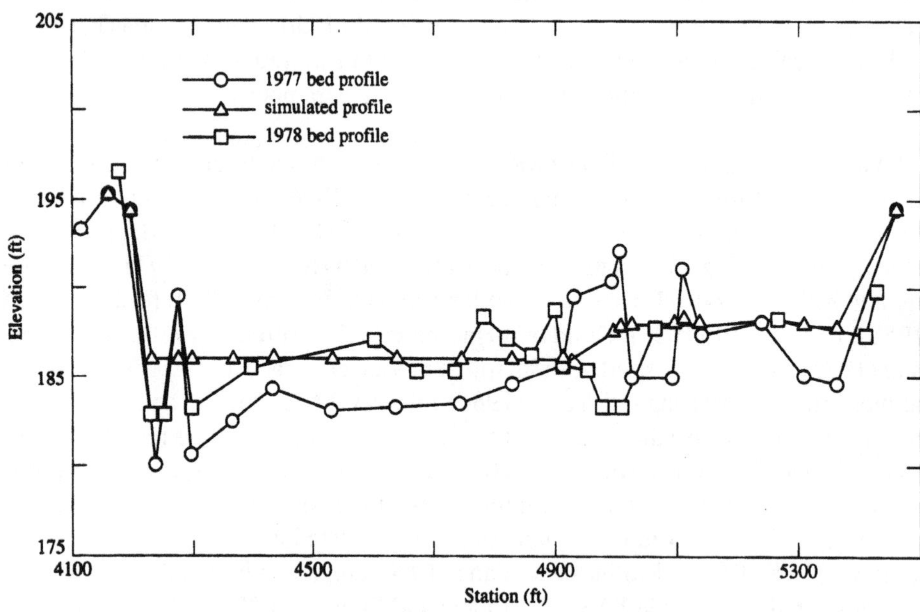

Figure 3.46. Measured cross-sectional changes at Stony Creek section 52400 and those simulated based on Yang's (1984) formula (courtesy of Chang).

3.8.4.1 Dimensionless Parameters

The accuracy of a sediment transport formula may vary with varying flow and sediment conditions. To determine the sensitivities of a transport formula to varying flow and sediment conditions, Yang and Huang (2001) selected seven dimensionless parameters for comparison. They are dimensionless particle diameter, relative depth, Froude number, relative shear stress, dimensionless unit stream power, sediment concentration, and discrepancy ratio.

Different transport formulas were developed for sediment transport in different size ranges. The dimensionless particle diameter used in the comparisons is defined as:

$$D_* = d\left[\frac{\gamma_s - \gamma}{\gamma} g / v^2\right]^{1/3} \tag{3.150}$$

where:
d = sediment particle diameter,
γ_s, γ = specific weight of sediment and water, respectively,
g = gravitational acceleration, and
v = kinetic viscosity of water.

The relative depth is defined as the ratio between average water depth D and sediment particle diameter d. The inverse of relative depth is the relative roughness, which has been considered by many investigators as an important parameter for the determination of sediment transport rate and resistance to flow. One major difference between laboratory and river data is that the former have a much smaller value of relative depth. If the relative depth is small, say less than 50, the water surface wave and the size of bed form may affect accuracy of measurements.

Froude number is one of the most important parameters for open channel flow studies. Most sediment transport formulas were developed for subcritical flows.

Relative shear velocity is defined as the ratio between shear velocity U_* and sediment particle fall velocity ω. Many researchers consider U_*/ω as an index of flow intensity for sediment transport. For example, Julien (1995) believes that there is no sediment movement if $U_*/\omega < 0.2$; sediment transport is in the form of bedload if $0.2 < U_*/\omega < 0.4$; sediment transport is in the form of both bedload and suspended load if $0.4 < U_*/\omega < 2.5$; sediment transport is in the form of suspended load if $U_*/\omega > 2.5$.

Yang (1973) defined the dimensionless unit stream power as VS/ω, where V = cross-sectional average flow velocity; S = energy or water surface slope; and ω = sediment particle fall velocity. Yang (1973, 1996) considered VS/ω the most important parameter for the determination of sediment concentration or sediment transport rate.

Sediment concentration is defined as the ratio between sediment transport rate and water discharge by weight.

Discrepancy ratio is defined as the ratio between computed sediment concentration and measured sediment concentration; that is,

$$R = C_c/C_m \tag{3.151}$$

where C_c = computed sediment concentration in parts per million by weight, and
C_m = measured total bed-material concentration in parts per million by weight.

The average discrepancy ratio is defined as:

$$\bar{R} = \frac{\sum_{i=1}^{j} R_i}{j} \tag{3.152}$$

where i = data set number, and
j = total number of data used in the comparison.

3.8.4.2 Data Analysis

A total of more than 6,200 sets of sediment transport and hydraulic data were available to Yang and Huang (2001) for preliminary comparison and analysis. One of the difficulties in the selection of data for final comparison and analysis is the determination of accuracies of data published by different investigators. The following criteria were used to eliminate data of questionable accuracy:

- Only those data published by an investigator with more than 50 percent in a range of discrepancy ratio between 0.5 and 2, based on two or more of the 13 formulas, were included. Data with less than 10 sets were excluded. A total of 3,391 sets of data met this requirement. These data were compiled by Yang (2001).

- To avoid the uncertainties related to incipient motion, measured sediment concentrations less than 10 ppm, by weight, were excluded.

- Most of the laboratory data were fairly uniform in size. The median particle diameter was used for all sediment transport formula computations. The gradation coefficient is defined as:

$$\sigma = \frac{1}{2}\left(\frac{d_{84.1}}{d_{50}} + \frac{d_{50}}{d_{15.9}}\right) \tag{3.153}$$

where $d_{15.9}$, d_{50}, $d_{84.1}$ = sediment particle size corresponding to 15.9%, 50%, and 84.1% finer, respectively.

Data with $\sigma \geq 2.0$ were excluded from further analysis.

- To avoid the inclusion of wash load, data with median particle diameters of less than 0.0625 mm were excluded.

All the laboratory data had to be collected under steady equilibrium conditions. Natural river sediment and hydraulic data had to be collected within a day, and flow conditions had to be fairly steady to ensure a close relationship between sediment and flow conditions for a given set of river data.

Based on the above criteria, a total of 3,225 sets of laboratory data and 166 sets of river data were selected for final analysis and comparison. Table 3.10 summarizes these data.

Some of the transport formulas were intended for sand transport and some for gravel transport. The second step of comparison was to determine the range of application of sediment particle size based on discrepancy ratio for each formula. Table 3.11 shows the results. Based on the results shown in Table 3.11, Table 3.12 gives the ranges of application of the 13 formulas. Yang and Huang (2001) used only those data within the range of application of each formula as shown in Table 3.12 for further comparison and analysis.

Table 3.13 summarizes the sensitivity of the accuracy of formulas as a function of relative depth. The relatively large variations of discrepancy ratio for 13 formulas with $4 < D/d < 50$ suggest that the influences of water surface wave and bed form may be significant. If we exclude the data with $4 < D/d < 50$, Yang's 1979 sand formula is least sensitive to the variation of relative depth, followed by Yang's 1973 sand formula, and Yang's 1984 gravel formula. The Rottner formula and the Kalinske formula are the most sensitive. The Ackers and White formula has a tendency to overestimate sediment concentration with increasing flow depth, while the Engelund and Hansen formula has the reverse tendency.

Table 3.14 and Figure 3.47 summarize the sensitivity of the accuracy of formulas as a function of Froude number. The Rottner formula is most sensitive to the variation of Froude number, followed by Einstein's bedload and bed-material load formulas and the Kalinske formula. Yang's 1979 and 1973 sand formulas are least sensitive to the variation of Froude number. Table 3.14 shows that Yang's 1973, 1979, and 1984 formulas can be applied to subcritical, supercritical, and transitional flow regimes, while other formulas should be applied to subcritical flow only. Table 3.15 summarizes the sensitivity of the accuracy of formulas as a function of relative shear velocity. The Rottner and Kalinske formulas are most sensitive to the variation of relative shear velocity. Yang's 1973, 1979, and 1984 formulas are least sensitive to the variation of relative shear velocity.

Yang considered the dimensionless unit stream power to be the most important parameter in his 1973, 1979, and 1984 formulas. Table 3.16 shows that Yang's three formulas consistently and reliably predict sediment concentration or transport rates. The formulas by Ackers and White and by Engelund and Hansen also can give accurate estimation of sediment concentration or load for a wide range of dimensionless unit stream power. The least reliable ones are the Rottner, Kalinske, and Einstein's bedload and bed-material load formulas. While the Kalinske and Laursen formulas consistently overestimate sediment concentration and transport rate, the Meyer-Peter and Müller formula consistently underestimates sediment concentration and transport rate.

Table 3.17 and Figure 3.48 summarize the accuracies of transport equations as a function of measured sediment concentration. Accuracy apparently increases for all formulas when the measured sediment concentration is greater than 100 ppm by weight. This may be related to the fact that it is more difficult to measure accurately when the concentration is low. If we limit our comparisons with concentration greater than 100 ppm by weight, the most accurate formulas are those proposed by Yang in 1973, 1979, and 1984. The Ackers and White and the Engelund and Hansen formulas can also give reasonable estimates. The least accurate ones are the Kalinske, Rottner, Einstein bedload and bed-material load, Taffoletti, and the Meyer-Peter and Müller formulas.

The difference between Yang's 1973 and 1979 formulas is that the 1973 formula includes incipient motion criteria, while the 1979 formula does not have incipient motion criteria. Consequently, the 1973 formula should be used where measured total bed-material concentration is less than 100 ppm by weight. The 1979 formula should give slightly more accurate results at high concentrations because the uncertainty and the importance of incipient motion criteria decrease with increasing sediment concentration. Tables 3.18 and 3.19 and Figure 3.49 summarize the comparison between Yang's 1973 and 1979 formulas. It is apparent that the 1973 formula should be used where total bed-material concentration is less than 100 ppm by weight, while the 1979 formula is slightly more accurate where the concentration is greater than 100 ppm by weight.

The Meyer-Peter and Müller and the 1984 Yang formulas should be used for bed materials in the very coarse sand to coarse gravel range. Figure 3.50 shows that the 1984 Yang formula gives more reasonable prediction than the Meyer-Peter and Müller formula.

Table 3.19 summarizes the recommended ranges of application and the accuracy of 13 formulas. It is apparent that formulas based on energy dissipation rate either directly or indirectly, such as those by Yang, Ackers and White, and Engelund and Hansen, outperform those based on other approaches. The Einstein transport functions were based on probability concepts. In spite of the sophisticated theories and the complicated computational procedures used, Einstein's bedload and bed-material transport formulas are less accurate than others for engineering applications. This is mainly due to the lack of generality of Einstein's assumptions, such as step length, hiding factor, and lifting factor (Yang and Wan, 1991). Einstein's formulas should not be used in any computer model if sediment routing based on size fractions is performed. Yang and Wan (1991) pointed out that if computation is based on size fraction using Einstein's formulas, sediment in transportation would be coarser than the original bed-material gradation, and coarser materials would be transported further in the downstream direction at a higher rate than the finer materials.

The Rottner formula is a regression equation without much theoretical basis. The results shown in Table 3.19 indicate that the Rottner formula is less reliable than others based on discrepancy ratio. Formulas purely based on regression analysis should not be applied to places other than where the data were used in the original regression analyses.

Table 3.19 also indicates that the classical approach based on shear stress, such as the Kalinske and the Meyer-Peter and Müller formulas, is less accurate than those based on the energy dissipation rate theories used by Yang directly and by Ackers and White, and Engelund and Hansen indirectly. Yang's approach was based on his unit stream power theory, while Ackers and White and Engelund and Hansen applied Bagnold's (1966) stream power concept to obtain their transport functions (Yang, 1996, 2002).

Table 3.10. Summary of basic data (Yang and Huang, 2001)

Author	D_J	s	D/d	Fr	U_J/ω	VS/ω	C	N
Ansely (1963)	5.83	1.33	58.9-157.0	2.301-3.362	2.042-3.446	1.0312-2.2163	29576-198664	26
Chyn (1935)	19.5-21.0	1.23-1.58	59.4-106.0	0.514-0.764	0.261-0.440	0.0043-0.0152	123-751	22
MacDougal (1933)	16.5-31.5	1.29-1.71	29.6-190.3	0.433-0.799	0.218-0.507	0.0038-0.0212	123-1237	74
USACE (1935)	4.50-12.5	1.31-1.94	46.6-1021	0.253-0.735	0.208-3.260	0.0043-0.00786	10-833	279
USACE (1936)	23.8	1.44	30.5-206.9	0.324-0.674	0.206-0.506	0.0032-0.0102	16-379	101
Sato et al. (1958)	26.0-114.5	1.00	60.2-421.1	0.189-0.754	0.210-0.626	0.0010-0.0115	10-500	219
Casey (1935)	61.5	1.16	11.0-89.1	0.425-0.880	0.179-0.286	0.0034-0.0173	10-960	36
Meyer-Peter and Müller (1948)	130.3-716.3	1.00	11.1-47.7	0.623-1.414	0.222-0.440	0.0092-0.0787	10-7000	51
Graf and Suszka (1987)	307.5-587.5	1.23-1.24	3.99-20.9	0.772-1.264	0.205-0.293	0.0114-0.0552	12-2910	101
Song et al. (1998)	307.5	1.37	6.84-17.1	0.698-0.991	0.227-0.288	0.0113-0.0316	11-2519	48
							Total	3225
(b) River data								
Colby and Hembree (1955)	7.08	1.76	1465-2036	0.304-0.535	1.763-3.294	0.0205-0.0716	392-2220	25
Hubbell and Matejka (1959)	4.50-6.00	1.58-2.54	1365-2019	0.326-0.723	2.165-4.425	0.0263-0.0919	632-2440	15
Nordin (1964)	4.75-9.75	1.44-1.89	1107-5045	0.258-0.735	1.055-3.607	0.0112-0.0591	260-3787	42
Jordan (1965)	4.75-19.5	1.43-1.98	9735-45078	0.100-0.158	0.710-4.579	0.0005-0.0064	13.1-226	23
Einstein (1944)	25.0	1.84	61.0-399.3	0.394-0.497	0.251-0.710	0.0047-0.0106	40-664	61
							Total	166

Total number of laboratory and river data = 3,391

Note: D_J = dimensionless diameter; σ = gradation; D/d = relative depth; Fr = Froude number; U_J/ω = ratio of shear velocity to fall velocity; VS/ω = dimensionless unit stream power; C = Concentration (ppm by weight); N = number of data set

Table 3.10. Summary of basic data (Yang and Huang, 2001)

Author	D_l	s	D/d	Fr	U_l/ω	VS/ω	C	N
(a) Laboratory data								
Gilbert (1914)	7.63-175.3	1.06-1.34	5.74-295.9	0.292-3.540	0.240-1.998	0.0057-0.6628	77-35340	886
Guy et al. (1996)	4.75-30.0	1.25-1.67	109.2-1701	0.220-1.698	0.235-7.236	0.0014-0.6533	10-50000	272
Willis et al. (1972)	2.50	1.30	1036-3780	0.218-1.005	4.217-10.427	0.0167-0.3810	87-19400	96
Willis (1979)	13.5	1.12	191.9-276.6	0.272-1.155	0.437-1.276	0.0035-0.1248	15-6670	32
Willis (1983a)	13.8	1.60	698.3-2810	0.163-0.643	0.776-2.392	0.0024-0.0693	27-4620	42
Willis (1983b)	13.8	1.60	310.0-642.9	0.284-1.159	0.395-1.533	0.0021-0.1603	61-6180	27
Barton and Lin (1955)	4.50	1.26	508.0-1321	0.161-0.872	1.428-3.428	0.0119-0.1141	19-3776	28
Stein (1965)	10.0	1.50	228.6-777.2	0.243-1.664	0.747-2.467	0.0045-0.3118	93-39293	57
Nordin (1976)	6.25	1.44	951.0-3438	0.222-1.128	1.308-3.722	0.0041-0.2744	18-17200	45
Foley (1975)	7.25	1.37	102.0-162.9	0.656-1.375	0.953-1.554	0.0393-0.2193	845-11693	12
Taylor (1971)	5.70	1.52	346.2-701.8	0.278-0.988	1.106-2.653	0.0111-0.1146	14-2270	13
Williams (1970)	33.8	1.20	20.1-164.8	0.343-3.504	0.216-1.490	0.0020-0.5207	10-34575	175
Kennedy (1961)	5.83-13.7	1.14-1.47	41.1-465.7	0.499-1.964	0.639-4.137	0.0355-0.7779	490-58500	41
Brooks (1957)	2.20-3.63	1.11-1.17	325.8-983.7	0.274-0.799	2.545-8.507	0.0425-0.2759	190-5300	21
Vanoni and Brooks (1957)	3.43	1.38	527.3-1230	0.252-0.810	2.061-4.377	0.0078-0.1613	37-3000	14
Nomicos (1956)	3.80	1.76	483.3-508.7	0.287-0.956	2.246-3.755	0.0323-0.2136	300-5600	12
Laursen (1958)	2.75	1.20	692.7-2757	0.243-0.863	4.440-6.626	0.0224-0.1580	140-5150	16
Davis (1971)	3.75	1.17	508.0-2032	0.190-0.623	2.083-3.844	0.0073-0.1024	11-1760	70
Pratt (1970)	12.0	1.11	159.4-956.5	0.210-0.502	0.407-1.074	0.0016-0.0195	12-560	29
Singh(1960)	15.5	1.16	23.6-329.4	0.313-1.244	0.269-0.954	0.0041-0.1355	19-9200	286
Znamenskaya (1963)	20.0	1.60	62.5-254.9	0.422-1.213	0.298-0.862	0.0055-0.0478	126-3000	26
Straub (1954)	4.78	1.40	218.6-1232	0.399-1.299	1.800-2.626	0.0222-0.2788	423-12600	18
Krishnappan and Engel (1988)	30.0	1.00	118.1-137.9	0.459-0.765	0.283-0.745	0.0040-0.0451	88-2087	15
Wang et al. (1998)	2.78	1.94	845.8-1229	0.329-1.128	6.894-13.716	0.1045-0.9641	13750-118180	35

Table 3.11. Applicability test of formulas according to dimensionless diameter D_I (all data) (Yang and Huang, 2001)

Author of formula	D_I = 1.56-6.25 (d = 0.0625-0.25mm)			D_I = 6.25-20.0 (d = 0.25-0.8mm)			D_I = 20.0-50.0 (d = 0.8-2.0mm)			D_I = 50.0-720.0 (d = 2.0-28.8mm)			N_T
	R			R			R			R			
	$\overline{R}$	0.5-2.0	N	$\overline{R}$	0.5-2.0	N	$\overline{R}$	0.5-2.0	N	$\overline{R}$	0.5-2.0	N	
Ackers and White (1973)	1.31	77%	505	1.06	95%	1700	1.07	89%	491	1.26	74%	535	3231
Einstein (1950)	0.23	30%	505	1.38	52%	1703	1.77	53%	523	2.45	25%	553	3284
Einstein (1950)	0.55	46%	505	1.42	64%	1703	1.83	52%	523	2.49	21%	553	3284
Engelund and Hansen (1967)	0.87	82%	505	1.22	88%	1703	1.31	83%	523	1.63	72%	553	3284
Kalinske (1947)	1.23	49%	505	1.88	33%	1703	3.62	9%	523	5.84	4%	553	3284
Laursen (1958)	1.26	82%	495	1.29	85%	1690	1.48	67%	491	2.11	43%	473	3149
Meyer-Peter and Müller (1948)	0.16	11%	502	0.61	60%	1617	0.44	36%	374	0.58	63%	308	2801
Rottner (1959)	0.63	58%	505	1.84	47%	1703	3.77	11%	523	8.34	3%	553	3284
Schoklitsch (1934)	0.43	39%	488	0.82	83%	1242	1.25	73%	224	1.31	85%	284	2238
Toffaleti (1968)	0.21	26%	505	0.38	35%	1703	0.79	54%	523	1.68	48%	553	3284
Yang (sand) (1973)	1.06	90%	505	1.04	93%	1703	1.24	86%	523	9.86	6%	528	3259
Yang (sand) (1979)	0.99	94%	505	1.01	96%	1703	1.21	85%	523	8.85	7%	528	3259
Yang (gravel) (1984)	0.03	1%	505	0.29	24%	1703	0.66	53%	523	0.89	81%	528	3259

Note: R = discrepancy ratio; $\overline{R}$ = average discrepancy ratio; N = number of data sets; N_T = total number of data

Table 3.12. Range of application of median sediment particle size (Yang and Huang, 2001)

Author of formula	Median particle diameter (mm)
Ackers and White (1973)	0.065–32 (coarse silt–coarse gravel)
Einstein Bedload (1950)	0.25–32 (medium sand–coarse gravel)
Einstein Bed material (1950)	0.0625–32 (coarse silt–coarse gravel)
Engelund and Hansen (1967)	0.0625–32 (coarse silt–coarse gravel)
Kalinske (1947)	0.0625–2 (coarse silt–coarse sand)
Laursen (1958)	0.0625–2 (coarse silt–coarse sand)
Meyer-Peter and Müller (1948)	2.0–32 (very coarse sand–coarse gravel)
Rottner (1959)	0.0625–2 (coarse silt–very coarse sand)
Schoklitsch (1934)	0.25–32 (median sand–very coarse gravel)
Toffaleti (1968)	0.25–32 (median sand–coarse gravel)
Yang (sand) (1973)	0.0625–2.0 (coarse silt–very coarse sand)
Yang (sand) (1979)	0.0625–2.0 (coarse silt–very coarse sand)
Yang (gravel) (1984)	2.0–32 (very coarse sand–coarse gravel)

Table 3.13. Applicability test of formulas according to relative depth D/d (using applicable data) (Yang and Huang, 2001)

Author of formula	D/d = 4.0–50			D/d = 50–200			D/d = 200–1000			D/d = 1000–50,000			N_T
	R			R			R			R		N	
	$\overline{R}$	0.5–2.0	N	$\overline{R}$	0.5–2.0	N	$\overline{R}$	0.5–2.0	N	$\overline{R}$	0.5–2.0		
Ackers and White (1973)	1.27	75%	589	1.08	94%	1561	1.05	90%	646	1.28	79%	436	3232
Einstein (bedload) (1950)	2.10	32%	624	1.66	52%	1521	1.46	50%	448	0.76	46%	186	2779
Einstein (bed material) (1950)	2.17	31%	624	1.60	52%	1577	1.41	62%	647	0.55	68%	436	3284
Engelund and Hansen (1967)	1.68	73%	624	1.23	85%	1577	1.17	91%	647	0.82	83%	436	3284
Kalinske (1947)	3.76	11%	289	2.20	28%	1385	1.65	38%	621	1.28	46%	436	2731
Laursen (1958)	1.74	68%	266	1.31	81%	1356	1.23	84%	618	1.22	86%	436	2676
Meyer-Peter and Müller (1948)	0.63	71%	136	0.52	55%	150	0.68	68%	22	-	-	0	308
Rottner (1959)	4.46	9%	289	2.06	33%	1385	1.57	59%	621	0.70	69%	436	2731
Schoklitsch (1934)	1.25	81%	237	1.02	86%	931	0.74	80%	401	0.71	68%	181	1750
Toffaleti (1968)	1.56	49%	624	0.52	42%	1521	0.37	32%	448	0.32	30%	186	2779
Yang (sand) (1973)	1.24	86%	289	1.10	90%	1385	1.06	93%	621	1.02	95%	436	2731
Yang (sand) (1979)	1.25	85%	289	1.04	93%	1385	1.01	96%	621	1.00	97%	436	2731
Yang (gravel) (1984)	0.83	79%	264	0.96	83%	238	0.82	84%	26	-	-	0	528

Note: R = discrepancy ratio; $\overline{R}$ = average discrepancy ratio; N = number of data sets; N_T = total number of data

Table 3.14. Applicability test of formulas according to Froude number Fr (using applicable data) (Yang and Huang, 2001)

Author of formula	$Fr = 0.10–0.40$			$Fr = 0.40–0.80$			$Fr = 0.80–1.20$			$Fr = 1.20–3.60$			N_T
	R			R			R			R		N	
	$\overline{R}$	0.5–2.0	N	$\overline{R}$	0.5–2.0	N	$\overline{R}$	0.5–2.0	N	$\overline{R}$	0.5–2.0		
Ackers and White (1973)	1.09	88%	641	1.08	94%	1349	1.11	84%	644	1.33	78%	597	3231
Einstein (bedload) (1950)	1.12	62%	421	2.23	42%	1237	1.93	49%	564	0.56	44%	557	2779
Einstein (bed material) (1950)	0.88	47%	647	1.90	50%	1387	2.22	49%	653	0.63	66%	597	3284
Engelund and Hansen (1967)	1.61	80%	647	1.27	83%	1387	1.14	87%	653	0.93	85%	597	3284
Kalinske (1947)	1.44	44%	639	1.91	35%	1162	2.34	24%	424	3.13	13%	506	2731
Laursen (1958)	1.39	69%	611	1.33	84%	1138	1.32	88%	421	1.21	84%	506	2676
Meyer-Peter and Müller (1948)	-	-	-	0.68	72%	94	0.54	60%	174	0.52	55%	40	308
Rottner (1959)	0.51	31%	659	3.25	39%	1142	2.48	44%	424	0.64	62%	506	2731
Schoklitsch (1934)	1.29	80%	47	1.16	85%	611	0.87	82%	537	0.78	79%	555	1750
Toffaleti (1968)	0.34	32%	421	0.55	40%	1237	0.76	47%	564	1.32	45%	557	2779
Yang (sand) (1973)	1.18	88%	659	1.07	91%	1142	1.04	92%	424	1.02	95%	506	2731
Yang (sand) (1979)	1.14	90%	659	1.03	93%	1142	1.01	96%	424	0.99	97%	506	2731
Yang (gravel) (1984)	0.74	75%	8	0.86	79%	216	0.91	82%	263	0.94	87%	41	528

Note: R = discrepancy ratio; $\overline{R}$ = average discrepancy ratio; N = number of data sets; N_T = total number of data

Table 3.15. Applicability test of formulas according to relative shear velocity U_1 $/\omega$ (using applicable data) (Yang and Huang, 2001)

Author of formula	U_1 $/\omega$ = 0.18-0.40			U_1 $/\omega$ = 0.40-1.00			U_1 $/\omega$ = 1.00-2.50			U_1 $/\omega$ = 2.50-15.00			N_T
	R			R			R			R			
	$\overline{R}$	0.5-2.0	N	$\overline{R}$	0.5-2.0	N	$\overline{R}$	0.5-2.0	N	$\overline{R}$	0.5-2.0	N	
Ackers and White (1973)	1.30	80%	1030	0.97	96%	1237	1.06	90%	552	1.32	80%	412	3231
Einstein (bedload) (1950)	2.00	35%	1081	1.42	58%	1229	1.53	43%	461	0.65	45%	28	2799
Einstein (bed material) (1950)	2.07	33%	1081	1.47	57%	1239	1.33	74%	552	0.57	58%	412	3284
Engelund and Hansen (1967)	1.64	76%	1081	1.08	89%	1239	1.11	86%	552	0.92	84%	412	3284
Kalinske (1947)	3.38	11%	640	1.97	31%	1127	1.51	40%	552	1.21	52%	412	2731
Laursen (1958)	1.49	74%	601	1.28	82%	1115	1.26	84%	548	1.25	85%	412	2676
Meyer-Peter and Müller (1948)	0.60	65%	212	0.55	60%	93	-	-	-	-	-	-	305
Rottner (1959)	3.54	13%	640	2.20	44%	1127	0.82	73%	552	0.55	41%	412	2731
Schoklitsch (1934)	1.27	82%	372	0.91	85%	910	0.80	77%	441	0.63	65%	27	1750
Toffaleti (1968)	1.12	49%	1081	0.48	38%	1229	0.39	30%	461	0.30	28%	28	2799
Yang (sand) (1973)	1.17	88%	640	1.06	92%	1127	1.05	93%	552	1.01	94%	412	2731
Yang (sand) (1979)	1.14	90%	640	1.03	94%	1127	1.01	96%	552	0.98	95%	412	2731
Yang (gravel) (1984)	0.88	80%	386	0.92	84%	142	-	-	-	-	-	-	528

Note: R = discrepancy ratio; $\overline{R}$ = average discrepancy ratio; N = number of data sets; N_T = total number of data

Table 3.16. Applicability test of formulas according to dimensionless unit stream power VS/ω (using applicable data) (Yang and Huang, 2001)

Author of formula	VS/ω = 0.0005-0.10			VS/ω = 0.10-0.02			VS/ω = 0.05-0.10			VS/ω = 0.10-2.50			N_T
	R			R			R			R			
	$\overline{R}$	0.5-2.0	N	$\overline{R}$	0.5-2.0	N	$\overline{R}$	0.5-2.0	N	$\overline{R}$	0.5-2.0	N	
Ackers and White (1973)	1.18	87%	847	1.09	90%	1141	1.02	94%	505	1.23	81%	738	3231
Einstein (bedload) (1950)	2.21	43%	897	1.69	43%	1105	1.39	60%	361	0.67	54%	416	2779
Einstein (bed material) (1950)	2.22	42%	897	1.70	49%	1144	1.38	67%	505	0.54	59%	738	3284
Engelund and Hansen (1967)	1.57	73%	897	1.23	87%	1144	1.18	90%	505	0.94	87%	738	3284
Kalinske (1947)	3.63	11%	513	2.15	29%	986	1.57	36%	494	1.30	46%	738	2731
Laursen (1958)	1.65	72%	476	1.25	83%	971	1.27	82%	491	1.23	84%	738	2676
Meyer-Peter and Müller (1948)	0.63	68%	176	0.53	59%	121	0.46	37%	8	-	-	-	305
Rottner (1959)	4.18	11%	513	2.17	41%	986	1.44	62%	494	0.58	52%	738	2731
Schoklitsch (1934)	1.29	83%	121	1.09	87%	904	0.82	82%	314	0.66	71%	411	1750
Toffaleti (1968)	1.31	47%	897	0.50	40%	1105	0.37	37%	361	0.31	32%	416	2779
Yang (sand) (1973)	1.21	85%	513	1.08	91%	986	1.05	92%	494	1.02	95%	738	2731
Yang (sand) (1979)	1.22	84%	513	1.02	96%	986	1.01	95%	494	0.98	97%	738	2731
Yang (gravel) (1984)	0.85	78%	334	0.96	86%	1891	0.92	91%	11	-	-	-	2236

Note: R = discrepancy ratio; $\overline{R}$ = average discrepancy ratio; N = number of data sets; N_T = total number of data

Table 3.17. Applicability test of formulas according to sediment concentration C (using applicable data) (Yang and Huang, 2001)

Author of formula	C = 10.0-100 ppm			C = 100-1000 ppm			C = 1000-10,000 ppm			C = 10,000-20,000 ppm			N_T
	R			R			R			R			
	$\overline{R}$	0.5-2.0	N	$\overline{R}$	0.5-2.0	N	$\overline{R}$	0.5-2.0	N	$\overline{R}$	0.5-2.0	N	
Ackers and White (1973)	1.22	78%	480	1.14	87%	1211	1.05	94%	1152	1.26	85%	388	3231
Einstein (bedload) (1950)	2.39	28%	505	1.49	47%	1185	1.58	54%	993	0.77	57%	116	2779
Einstein (bed material) (1950)	2.44	24%	521	1.51	55%	1223	1.49	61%	1152	0.50	54%	388	3284
Engelund and Hansen (1967)	1.55	74%	521	1.39	84%	1223	1.09	88%	1152	0.88	82%	388	3284
Kalinske (1947)	4.72	7%	204	2.28	28%	1079	1.71	36%	1060	1.24	41%	388	2731
Laursen (1958)	1.86	71%	178	1.34	80%	1052	1.20	84%	1058	1.34	81%	388	2676
Meyer-Peter and Müller (1948)	0.66	73%	77	0.59	64%	109	0.53	57%	112	0.57	61%	7	305
Rottner (1959)	4.25	12%	204	2.19	35%	1079	1.82	46%	1060	0.68	67%	388	2731
Schoklitsch (1934)	1.29	81%	96	1.11	85%	662	0.84	83%	878	0.66	58%	114	1750
Toffaleti (1968)	1.49	49%	505	0.66	42%	1185	0.42	37%	993	0.32	28%	116	2799
Yang (sand) (1973)	1.28	85%	204	1.09	89%	1079	1.06	93%	1060	1.02	95%	388	2731
Yang (sand) (1979)	1.30	83%	204	1.05	92%	1079	1.01	96%	1060	0.99	97%	388	2731
Yang (gravel) (1984)	0.78	76%	203	0.91	83%	181	1.03	87%	137	0.91	86%	7	528

Note: R = discrepancy ratio; $\overline{R}$ = average discrepancy ratio; N = number of data sets; N_T = total number of data

Table 3.18. Comparison of equations of Yang (1973) and Yang (1979) for sand transport (Yang and Huang, 2001)

Author of formula	C = 10.0–40.0 ppm			C = 40.0–70.0 ppm			C = 70.0–100.0 ppm		
	Discrepancy ratio		No. of data sets	Discrepancy ratio		No. of data sets	Discrepancy ratio		No. of data sets
	Mean	0.5–2.0		Mean	0.5–2.0		Mean	0.5–2.0	
Yang (1973)	1.46	80%	37	1.30	84%	58	1.21	87%	109
Yang (1979)	1.52	73%	37	1.33	80%	58	1.21	88%	109
	C = 100–1000 ppm			C = 1000–10,000			C = 10,000–200,000		
Yang (1973)	1.09	89%	1079	1.06	93%	1060	1.02	95%	388
Yang (1979)	1.09	92%	1079	1.01	96%	1060	0.99	97%	388

Table 3.19. Summary of comparison of accuracy of formulas in their applicable ranges (Yang and Huang, 2001)

Author of formula	Discrepancy ratio: Mean	Discrepancy ratio: Data between 0.5 and 2.0	No. of data sets
For coarse silt to very coarse sand, d_{50} = 0.0625–2 mm			
Yang (1979)	1.04	94%	2731
Yang (1973)	1.08	91%	2731
Ackers and White (1973)	1.11	90%	2696
Engelund and Hansen (1967)	1.17	93%	2731
Laursen (1958)	1.32	81%	2676
Einstein (bed material) (1950)	1.34	58%	2731
Rottner (1959)	1.99	42%	2731
Kalinske (1947)	2.09	31%	2731
For medium sand to coarse gravel, d_{50} = 0.25–32 mm			
Schoklitsch (1934)	0.85	82%	1750
Toffaleti (1968)	0.72	41%	2779
Einstein (bedload) (1950)	1.67	47%	2779
For very coarse sand to coarse gravel, d_{50} = 2–32 mm			
Yang (1984)	0.89	81%	528
Meyer-Peter and Müller (1948)	0.58	63%	308
For coarse silt to coarse gravel, d_{50} = 0.0625–32 mm			
Yang (1979) and Yang (1984)	1.02	91%	3259
Yang (1973) and Yang (1984)	1.05	89%	3259
Ackers and White (1973)	1.13	88%	3231
Engelund and Hansen (1967)	1.25	84%	3284
Einstein (bed-material) (1950)	1.53	52%	3284

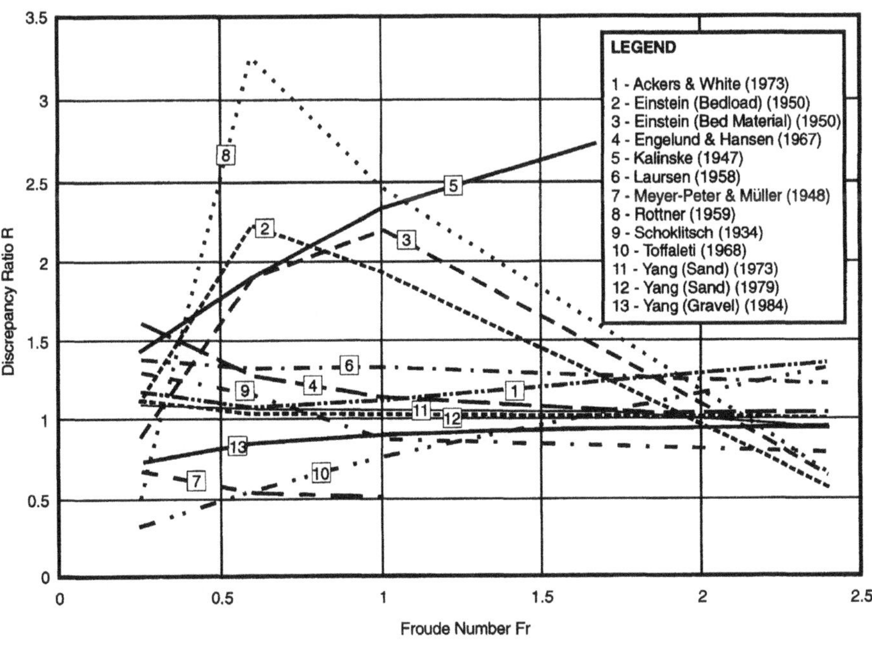

Figure 3.47. Comparison of discrepancy ratio based on Froude number.

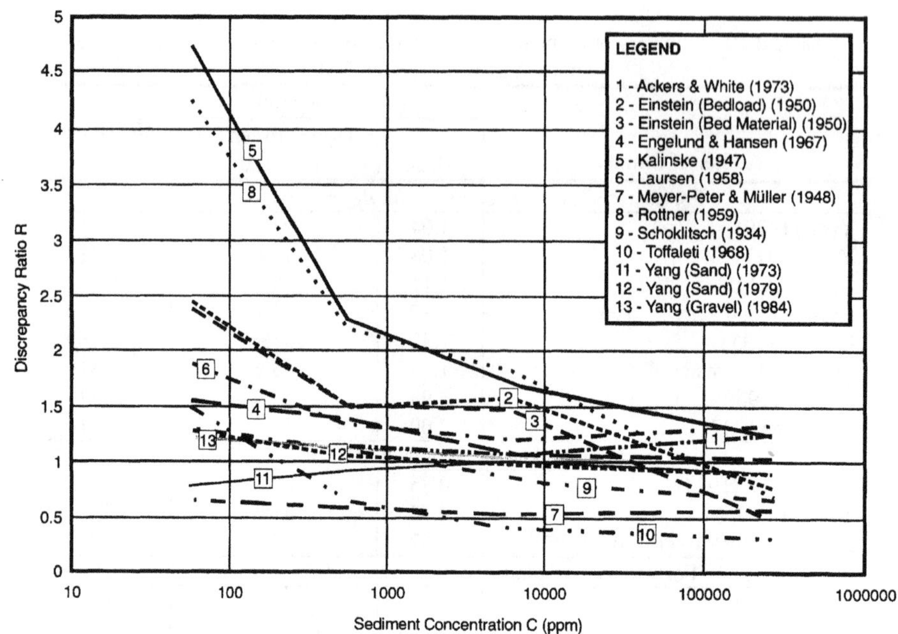

Figure 3.48. Comparison of discrepancy ratio based on concentration.

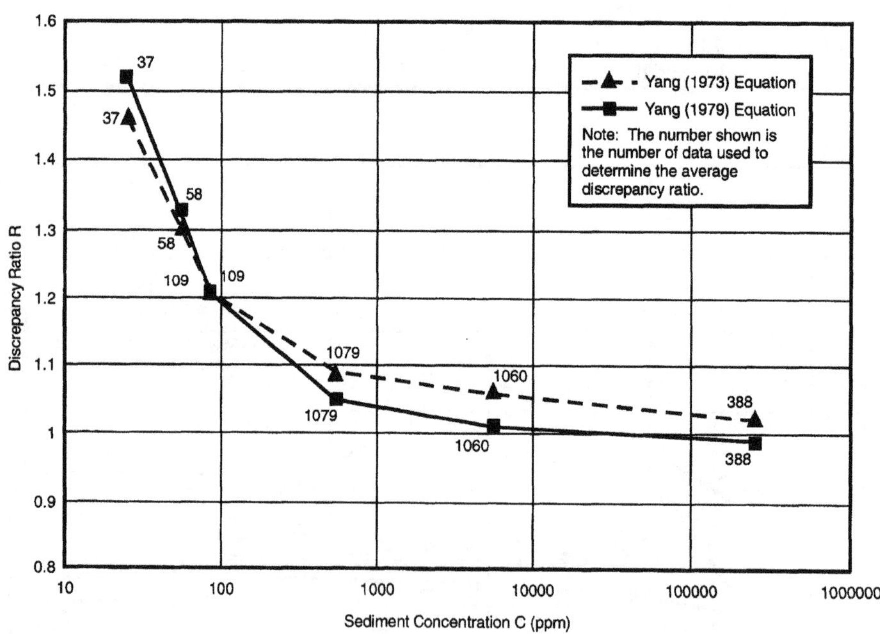

Figure 3.49. Comparison of equations of Yang (1973) and Yang (1979) for sand transport.

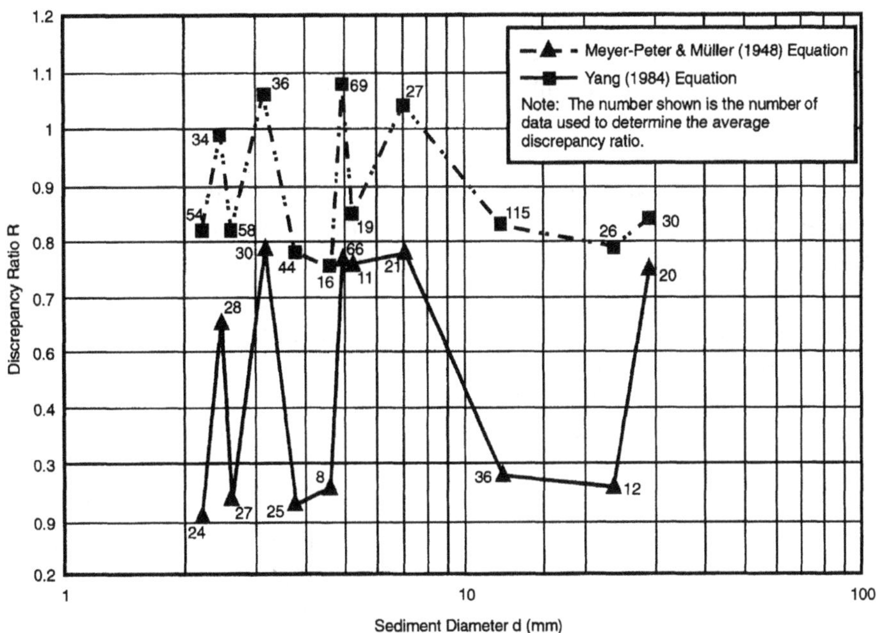

Figure 3.50. Comparison of equations of Meyer-Peter and Müller (1948) and Yang (1984) for gravel transport.

Most of the river sediment transport studies involve sediments in the coarse silt to coarse gravel size range. Table 3.19 indicates that the priority of selection should be Yang (1979) for $d_{50} < 2$ mm plus Yang (1984) for $d_{50} > 2$ mm, followed by Yang (1973) for $d_{50} < 2$ mm plus Yang (1984) for $d_{50} > 2$ mm, and then followed by Ackers and White (1973) and Engelund and Hansen (1967). If the local conditions on the range of variations of dimensionless particle diameter, relative depth, Froude number, relative shear velocity, dimensionless unit stream power, and measured bed-material load concentration are available, Tables 3.11 to 3.19 should be used as references to finalize the selection of the most appropriate formula for engineers to use.

The analyses by Yang and Huang (2001) reached the following conclusions:

- Sediment transport formulas based on energy dissipation rate or the power concept are more accurate than those based on other concepts. Yang's (1973, 1979, 1984) formulas were derived directly from the unit stream power theory, while the formulas by Engelund and Hansen (1967) and by Ackers and White (1973) were obtained indirectly from Bagnold's (1966) stream power concept.

- Among the 13 formulas compared, Yang's 1973, 1979, and 1984 formulas are the most robust, and their accuracies are least sensitive to the variation of relative depth, Froude number, dimensionless shear velocity, dimensionless unit stream power, and sediment concentration.

- With the exception of Yang's (1973, 1979, and 1984) dimensionless unit stream power formulas and Engelund and Hansen's (1967) formula, the application of other sediment transport formulas should be limited to subcritical flows.

- Engineers should use Table 3.19 as a reference for the preliminary selection of appropriate formulas for different size ranges of sediment particle diameter. Tables 3.13 to 3.17 should be used to determine whether a formula is suitable for a given range of dimensionless parameters before the final selection of formula is made.

- Yang's 1973 and 1979 sand transport formulas have about the same degree of accuracy. However, the 1973 formula with incipient motion criteria is slightly more accurate when the sand concentration is less than 100 ppm, while the 1979 formula without incipient motion criteria is slightly more accurate for concentrations higher than 100 ppm.

- The Einstein bed-material load (1950) and bedload (1950) formulas and those by Toffaleti (1958) and Meyer-Peter and Müller are not as accurate as those formulas based on the power approach. Some engineers use the Meyer-Peter and Müller formula for bedload, and the Einstein bed-material or Toffaleti formula for suspended load for the estimate of total load. This kind of combined use may not be justified from a theoretical point of view nor from the accuracies of these equations based on the results shown in this chapter.

3.8.4.3 Procedures for Selecting Sediment Transport Formulas

No perfect assumption exists that can be used to derive a sediment transport formula. However, the generalities of these assumptions do differ. Based on the majority of published data, it appears that unit stream power dominates the rate of sediment transport or sediment concentration more than any other variable. Even if perfect assumptions could be found and used in the derivation of a formula, the coefficients in the formula would still have to be determined by comparing the mathematical model and measured data. Thus, the applicability of a formula depends not only on the assumptions and theories used in its derivation, but also on the range of data used for the determination of the coefficients in the formula. Sediment discharge in natural rivers depends not only on the independent variables mentioned in previous sections, but also on the gradation and shape factor of sediment, the percentage of bed surface covered by coarse material, the availability of bed material for transport, variations in the hydrologic cycle, the rate of supply of fine material or wash load, the water temperature, the channel pattern and bed configuration, the strength of turbulence, etc. Because of the tremendous uncertainties involved in estimating sediment discharge at different flow and sediment conditions under different hydrologic, geologic, and climatic constraints, it is extremely difficult, if not impossible, to recommend one formula for engineers and geologists to use in the field under all circumstances (Yang, 1996). The following procedures are based on the recommendations made by Yang (1977, 1980, 1996) with minor modifications.

Step 1: Determine the kind of field data available or measurable within the time, budget, and staffing limits.

Step 2: Examine all the formulas and select those with measured values of independent variables determined from step 1.

Step 3: Compare the field situation and the limitations of formulas selected in step 2. If more than one formula can be used, calculate the rate of sediment transport by these formulas and compare the results.

Step 4: Decide which formulas can best agree with the measured sediment load, and use these to estimate the rate of sediment transport at those flow conditions when actual measurements are not possible.

Step 5: In the absence of measured sediment load for comparison, the following formulas or procedures should be considered:

- Use Meyer-Peter and Müller's formula when the bed material is coarser than 5 mm;
- Use Einstein's bedload transport function when bedload is a significant portion of the total load;
- Use Toffaleti's formula for large sand-bed rivers;
- Use Colby's formula for rivers with depth less than 10 ft;
- Use Shen and Hung's regression formula for laboratory flumes and very small rivers;
- Use Karim and Kennedy's regression formula for natural rivers with a wide range of variations of flow and sediment conditions;
- Use Yang's (1973) formula for sand transport in laboratory flumes and natural rivers;
- Use Yang's (1979) formula for sand transport when the critical unit stream power at incipient motion can be neglected;
- Use Yang's (1984) or Parker's (1990) gravel formulas for bedload or gravel transport;
- Use the modified Yang (1996) formula for nonequilibrium, high-concentration flows when wash load or concentration of fine material is high;
- Use Ackers and White's or Engelund and Hansen's formula for the subcritical flow condition in the lower flow regime;
- Use Yang's formulas (1973, 1979, 1984) for subcritical, transition, and supercritical flow conditions in the lower and upper flow regimes;
- Use Laursen's formula for laboratory flumes and shallow rivers with fine sand or coarse silt;
- Use Meyer-Peter and Müller's formula for bedload and the modified Einstein's formula for suspended load to obtain total load;
- A regime or regression formula can be applied to a river only if the flow and sediment conditions are similar to those from where the formula was derived;
- Select a formula according to its degree of accuracy, shown in Table 3.6;
- Based on the analyses of Yang and Huang (2001), select a formula that is most accurate under the given range of flow and sediment conditions.

Step 6: When none of the existing sediment transport formulas can give satisfactory results, use the existing data collected from a river station and plot sediment load or concentration against water discharge, velocity, slope, depth, shear stress, stream power, unit stream power or dimensionless unit stream power, and Velikanov's parameter. The least scattered curve without systematic deviation from a one-to-one correlation between dependent and independent variables should be selected as the sediment rating curve for the station.

3.9 Summary

This chapter comprehensively reviews and evaluates basic approaches and theories used in the determination of noncohesive sediment transport rate or concentration. The basic approaches used for the development of sediment transport functions or formulas are the regime, regression, probabilistic, and deterministic approaches. The concept that the rate of sediment transport should be directly related to the rate of energy dissipation rate in transporting sediment has gained increasing acceptance in recent years. Formulas derived from the power approach are those based on stream power (Bagnold, Engelund and Hansen, and Ackers and White), unit stream power (Yang), power balance (Pacheco-Ceballos), and gravitational power (Velikanov, Dou, and Zhang). Comparisons between measured results and computed results from different formulas indicate that, on the average, formulas derived from the power approach, especially the unit stream power approach, can more accurately predict sediment transport rate than formulas derived from other approaches.

Due to the complexity of flow and sediment conditions of natural rivers, recommendations are made for engineers to select appropriate formulas under different flow and sediment conditions. Sediment particle fall velocity and resistance to flow are two of the important parameters used in sediment transport and fluvial hydraulic computations. This chapter compares and evaluates different methods used for fall velocity computation and the estimation of resistance to flow or roughness coefficient for alluvial channels. This chapter also addresses the need to consider nonequilibrium sediment transport and the impact of wash load on sediment transport.

3.10 References

Ackers, P., and W.R. White (1973). "Sediment Transport: New Approach and Analysis," *Journal of the Hydraulics Division*, ASCE, vol. 99, no. HY11, pp. 2041-2060.

Alonso, C.V. (1980). "Selecting a Formula to Estimate Sediment Transport Capacity in Nonvegetated Channels," *CREAMS A Field Scale Model for Chemicals, Runoff, and Erosion from Agricultural Management System*, edited by W.G. Knisel, U.S.D.A. Conservation Research Report no. 26, Chapter 5, pp. 426-439.

Alonso, C.V., W.H. Neibling, and G.R. Foster (1982). "Estimating Sediment Transport Capacity in Watershed Modeling," *Transactions of the ASCE*, vol.24, no.5, pp. 1211-1220 and 1226.

American Society of Civil Engineers (ASCE), Task Committee on Relations Between Morphology of Small Stream and Sediment Yield (1982). "Relationships Between Morphology of Small Streams and Sediment Yield," *Journal of the Hydraulics Division*, ASCE, vol. 108, no. HY11, pp. 1328-2365.

Bagnold, R.A. (1966). *An Approach to the Sediment Transport Problem from General Physics*, U.S. Geological Survey Professional Paper 422-J.

Bishop, A.A., D.B. Simons, and E.V. Richardson (1965). "A Total Bed-Material Transport," *Journal of the Hydraulics Division*, ASCE, vol. 91, no. HY2, pp. 175-191.

Blench, T. (1969). Mobile-Bed Fluvialogy, A Regime Theory Treatment of Canals and Rivers for Engineers and Hydrologists, The University of Alberta Press, Edmonton, Alberta, Canada.

Brooks, N.H. (1955). "Mechanics of Streams with Moveable Beds of Fine Sand," *Proceedings of the ASCE*, vol. 81, no. 668, pp. 668-1 through 668-28.

Brownlie, W.R. (1981). "Prediction of Flow Depth and Sediment Discharge in Open Channels," W.M. Keck Laboratory of Hydraulics and Water Resources, Report no. KH-R-43A, California Institute of Technology, Pasadena, California.

Bureau of Reclamation (1987). *Design of Small Dams*, Denver, Colorado.

Cassie, H.J. (1935). *Über Geschiebebewegung*, Preuss. Versuchsanst für Wasserbau und Schifibau, Berlin, Mitt., vol. 19 (translation on file at the U.S. Soil Conservation Service, Washington, D.C.).

Chang, H.H. (1991). *Test and Calibration on Fluvial-12 Model Using Data from Stony Creek*, Report prepared for the California Department of Transportation.

Chang, H.H. (1994). "Selection of Gravel-Transport Formula for Stream Modeling," *Journal of Hydraulic Engineering*, ASCE, vol. 120, no. 5, pp. 646-651.

Colby, B.R. (1964). "Practical Computations of Bed-Material Discharge," *Journal of the Hydraulics Division*, ASCE, vol. 90, no. HY2.

Colby, B.R., and C.H. Hembree (1955). *Computation of Total Sediment Discharge, Niobrara River Near Cody, Nebraska*, U.S. Geological Survey Water Supply Paper 1357.

Dou, G. (1974). "Similarity Theory and Its Application to the Design of Total Sediment Transport Model," *Research Bulletin of Nanjing Hydraulic Research Institute*, Nanjing, China (in Chinese).

DuBoys, M.P. (1879). "LeRhône et les Rivères à lit affouillable," *Annales de Ponts et Chausées*, Sel 5, vol. 18, pp. 141-195.

Einstein, H.A. (1950). *The Bedload Function for Sediment Transport in Open Channel Flow*, U.S. Department of Agriculture Soil Conservation Technical Bulletin No. 1026.

Einstein, H.A., and N.L. Barbarossa (1952). "River Channel Roughness," *Transactions of the ASCE*, vol. 117, pp. 1121-1132.

Einstein, H.A., and N. Chien (1952). *Second Approximation to the Solution of Suspended Load Theory*, Institute of Engineering Research, University of California, Issue 2, Series 47.

Engelund, F., and E. Hansen (1972). *A Monograph on Sediment Transport in Alluvial Streams*, Teknisk Forlag, Copenhagen.

Fortier, S., and F.C. Scobey (1926). "Permissible Canal Velocities," *Transactions of the ASCE*, vol.89.

German Association for Water and Land Improvement (1990). *Sediment Transport in Open Channels—Calculation Procedures for the Engineering Practice*, Bulletin no. 17, Verlag Paul Parey, Hamburg and Berlin, Germany.

Gilbert, K.G. (1914). *The Transportation of Debris by Running Waters*, U.S. Geological Survey Professional Paper 86.

Govers, G. (1987). "Initiation of Motion in Overland Flow," *Sedimentology*, no. 34, pp. 1157-1164.

Guy, H.P., D.B. Simons, and E.V. Richardson (1966). *Summary of Alluvial Channel Data from Flume Experiments, 1956-1961*," U.S. Geological Survey Professional Paper 462-1.

Han, Q. (1980). "A Study on the Non-Equilibrium Transport of Suspended Load," *Proceedings of the International Symposium on River Sedimentation*, Beijing, China, pp. 793-802 (in Chinese).

Han, Q., and M. He (1990). "A Mathematical Model for Reservoir Sedimentation and Fluvial Processes," *International Journal of Sediment Research*, vol. 5, no. 2, pp. 43-84.

Hjulstrom, F. (1935). "The Morphological Activity of Rivers as Illustrated by River Fyris," *Bulletin of the Geological Institute*, Uppsala, vol. 25, ch. 3.

HR Wallingford (1990). Sediment Transport, the Ackers and White Theory Revised, Report SR237, England.

Julien, P.Y. (1995). *Erosion and Sedimentation*, Cambridge, University Press, Cambridge, UK.

Kalinske, A.A. (1947). "Movement of Sediment as Bedload in Rivers," *American Geophysical Union Transactions*, vol. 28, pp. 615-620.

Karim, M.F., and J.F. Kennedy (1990). "Means of Coupled Velocity and Sediment-Discharge Relationships for Rivers," *Journal of Hydraulic Engineering*, ASCE, vol. 116, no. 8, pp. 973-996.

Kennedy, R.G. (1895). "Prevention of Silting in Irrigation Canals," *Institute of Civil Engineers*, Proceeding Paper no. 2826.

Lacey, G. (1929). "Stable Channel in Alluvium," *Institute of Civil Engineers*, Proceeding Paper no. 4736.

Lane, E.W. (1953). "Progress Report on Studies on the Design of Stable Channels of the Bureau of Reclamation," *Proceedings of ASCE*, vol. 79.

Laursen, E.M. (1958). "The Total Sediment Load of Streams," *Journal of the Hydraulics Division*, ASCE, vol. 84, no. HY1, 1530-1 through 1530-36.

Leopold, L.B., and T. Maddock, Jr. (1953). *The Hydraulic Geometry of Stream Channels and Some Physiographic Implications*, U.S. Geological Survey Professional Paper 252.

Liu, H.K. (1958). "Closure to Mechanics of Sediment-Ripple Formation," *Journal of the Hydraulics Division*, ASCE, vol. 84, no. HY5, pp. 1832-10 through 1932-12.

Meyer-Peter, E., and R. Müller (1948). "Formulas for Bedload Transport," *Proceedings*, the Second Meeting of the International Association for Hydraulic Structures Research, Stockholm.

Molinas, A., and C.T. Yang (1986). Computer Program User's Manual for GSTARS (Generalized Stream Tube Model for Alluvial River Simulation), U.S. Bureau of Reclamation, Denver, Colorado.

Pacheco-Ceballos, P. (1989). "Transport of Sediments: Analytical Solution," *Journal of Hydraulic Research*, vol. 27, no. 4, pp. 501-518.

Parker, G. (1977). "Discussion of 'Minimum Unit Stream Power and Fluvial Hydraulics,' by C.T. Yang," *Journal of the Hydraulics Division*, ASCE vol. 103, no. HY7, pp. 811-816.

Parker, G. (1990). "Surface-Based Bedload Transport Relation for Gravel Rivers, *Journal of Hydraulics Research*, vol. 28, no. 4, pp. 501-518.

Rottner, J. (1959). "A Formula for Bedload Transportation," *LaHouille Blanche*, vol. 14, no. 3, pp. 285-307.

Rouse, H. (1939). *An Analysis of Sediment Transportation in the Light of Fluid Turbulence*, Soil Conservation Service Report no. SCS-TP-25, U.S. Department of Agriculture, Washington, D.C.

Rubey, W. (1933). "Settling Velocities of Gravel, Sand, and Silt Particles," *American Journal of Science*, vol. 25.

Schoklitsch, A. (1934). "Der Geschiebetrieb und die Geschiebefracht," *Wasserkraft und Wasserwirtschaft*, vol. 29, no. 4, pp. 37-43.

Schulits, S., and R.D. Hill, Jr. (1968). *Bedload Formulas*, Pennsylvania State University, College of Engineering, State College, Pennsylvania.

Shen, H.W., and C.S. Hung (1972). "An Engineering Approach to Total Bed-Material Load by Regression Analysis," *Proceedings of the Sedimentation Symposium*, ch. 14, pp. 14-1 through 14-17.

Shields, A. (1936). *Application of Similarity Principles and Turbulence Research to Bedload Movement*, California Institute of Technology, Pasadena (translated from German).

Shumm, S.A., and H.R. Khan (1972). "Experimental Study of Channel Patterns," *Geological Society of America*, Bulletin no. 83, p. 407.

Stein, R.A. (1965). "Laboratory Studies of Total Load and Apparent Bedload," *Journal of Geophysical Research*, vol. 70, no. 8, pp. 1831-1842.

Stevens, H.H., and C.T. Yang (1989). *Computer Programs for 13 Commonly Used Sediment Transport Formulas*, U.S. Geological Survey Water Resources Investigation Report 89-4026.

Talapatra, S.L., and S.N. Ghosh (1983). "Incipient Motion Criteria for Flow Over a Mobile Bed Sill," *Proceedings of the Second International Symposium on River Sedimentation*, Nanjing, China, pp. 459-471.

Toffaleti, F.B. (1969). "Definitive Computations of Sand Discharge in Rivers," *Journal of the Hydraulics Division*, ASCE, vol. 95, no. HY1, pp. 225-246.

U.S. Army Corps of Engineers (1977, 1993). *Generalized Computer Program, HEC-6, Scour and Deposition in Rivers and Reservoirs, Users' Manual*, the Hydrologic Engineering Center, Davis, California (March 1977; revised August 1993).

U.S. Army Corps of Engineers (1982). General Design Review Conference, Los Angeles District, Los Angeles, California.

U.S. Bureau of Reclamation (1987). *Design of Small Dams*. Denver, Colorado.

U.S. Interagency Committee on Water Resources, Subcommittee on Sedimentation (1957). *Some Fundamentals of Particle Size Analysis*, Report no. 12.

Vanoni, V.A., ed. (1975). *Sedimentation Engineering*, ASCE Task Committee for the Preparation of the Manual on Sedimentation of the Sedimentation Committee of the Hydraulics Division (reprinted 1977).

Vanoni, V.A. (1978). "Predicting Sediment Discharge in Alluvial Channels," *Water Supply and Management*, Pergamon Press, Oxford, pp. 399-417.

Velikanov, M.A. (1954). "Gravitational Theory of Sediment Transport," *Journal of Science of the Soviet Union Geophysics*, vol. 4 (in Russian).

Vetter, M. (1989). *Total Sediment Transport in Open Channels*, Report no. 26, Institute of Hydrology, University of the German Federal Army, Munich, Germany (translated from *Gesamttransport van Sedimenten in Offenen Gerinnen* into English by the U.S. Bureau of Reclamation, Denver, Colorado).

White, C.M. (1940). "The Equilibrium of Grains on the Bed of an Alluvial Channel," *Proceedings of the Royal Society of London, Series A*, vol. 174, pp. 332-338.

White, W.R., H. Milli, and A.D. Crabe (1975). "Sediment Transport Theories: A Review," *Proceedings of the Institute of Civil Engineers*, London, Part 2, no. 59, pp. 265-292.

Yalin, MS., and E. Karahan (1979). "Inception of Sediment Transport," *Journal of the Hydraulics Division*, ASCE, vol. 105, no. HY11, pp. 1433-1443.

Yang, C.T. (1971). "Potential Energy and River Morphology," *Water Resources Research,* vol. 7, no. 2, pp. 312-322.

Yang, C.T. (1972). "Unit Stream Power and Sediment Transport," *Journal of the Hydraulic Division*, ASCE, vol. 18, no. HY10, pp. 1805-1826.

Yang, C.T. (1973). "Incipient Motion and Sediment Transport," *Journal of the Hydraulic Division*, ASCE, vol. 99, no. HY10, pp. 1679-1704.

Yang, C.T. (1976). "Minimum Unit Stream Power and Fluvial Hydraulics," *Journal of the Hydraulics Division*, ASCE, vol. 102, no. HY7, pp. 919-934.

Yang, C.T. (1976). "Discussion of Sediment Transport Theories—A Review, by W.R. White, H. Milli, and A.D. Crabble," *Institute of Civil Engineering*, Part 2, vol. 61, pp. 803-810.

Yang, C.T. (1979). "The Movement of Sediment in Rivers," *Geophysical Survey 3*, D. Reidel, Dordrecht, pp. 39-68.

Yang, C.T. (1983). "Rate of Energy Dissipation and River Sedimentation," *Proceedings of the 2nd International Symposium on River Sedimentation*, Nanjing, China, pp. 575-585.

Yang, C.T. (1984). "Unit Stream Power Equation for Gravel," *Journal of Hydraulic Engineering, ASCE*, vol. 110, no. 12, pp. 1783-1797.

Yang, C.T. (1985). "Mechanics of Suspended Sediment Transport," *Proceedings of Euromech 192: Transport of Suspended Solids in Open Channels*, ed. W. Bechteler, Institute of Hydromechanics, University of the Armed Forces, Munich/Nuremberg, Germany, pp. 87-91.

Yang, C.T. (1996). *Sediment Transport Theory and Practice*, The McGraw-Hill Companies, Inc., New York (reprint by Krieger Publishing Company, Malabar, Florida, 2003).

Yang, C.T. (2002). "Sediment Transport and Stream Power," *International Journal of Sediment Research*, vol. 17, no. 1, Beijing, China, pp. 31-38.

Yang, C.T., and C. Huang (2001). "Applicability of Sediment Transport Formulas," *International Journal of Sediment Research*, vol. 16, no. 3, Beijing, China, pp. 335-343.

Yang, C.T., and A. Molinas (1982). "Sediment Transport and Unit Stream Power Function," *Journal of the Hydraulics Division*, ASCE, vol. 108, no. HY6, pp. 776-793.

Yang, C.T., and F.J.M. Simões (2000). *User's Manual for GSTARS 2.1 (Generalized Sediment Transport Model for Alluvial River Simulation Version 2.1)*, U.S. Bureau of Reclamation, Technical Service Center, Denver, Colorado.

Yang, C.T., and F.J.M. Simões (2002). *User's Manual for GSTARS3 (Generalized Sediment Transport Model for Alluvial River Simulation Version 3.0)*, U.S. Bureau of Reclamation, Technical Service Center, Denver, Colorado.

Yang, C.T., and F.J.M. Simões (2005). "Wash Load and Bed-Material Load Transport in the Yellow River," *Journal of Hydraulic Engineering*, ASCE, vol. 131, no. 5, pp. 413-418.

Yang, C.T., and C.C.S. Song (1979). "Theory of Minimum Rate of Energy Dissipation," *Journal of the Hydraulics Division*, ASCE, vol. 105, no. HY7, pp. 769-784.

Yang, C.T., and C.C.S. Song (1986). "Theory of Minimum Energy and Energy Dissipation Rate," *Encyclopedia of Fluid Mechanics*, Chapter 11, edited by N.P. Cheremisinoff, Gulf Publishing Company, Houston, Texas.

Yang, C.T., and S. Wan (1991). "Comparison of Selected Bed-Material Load Formulas," *Journal of Hydraulics Engineering*, ASCE, vol. 117, no. 8, pp. 973-989.

Yang, C.T., A. Molinas, and B. Wu (1996). "Sediment Transport in the Yellow River," *Journal of Hydraulic Engineering*, ASCE, vol. 122, no. 5, pp. 237-244.

Yang, C.T., C.C.S. Song, and M.J. Woldenberg (1981). "Hydraulic Geometry and Minimum Rate of Energy Dissipation," *Water Resources Research*, vol. 17, no. 4, pp. 1014-1018.

Yang, C.T., A. Molinas, and C.C.S. Song (1989). "GSTARS – Generalized Stream Tube Model for Alluvial River Simulation," *Twelve Selected Computer Stream Sedimentation Model Developed in the United States* (U.S. Interagency Subcommittee Report on Sedimentation), S.S. Fan (editor), Federal Energy Regulatory Commission, Washington DC, pp. 148-178.

Yang, C.T., M.A. Treviño, and F.J.M. Simões (1998). *User's Manual for GSTARS 2.0 (Generalized Stream Tube Model for Alluvial River Simulation Version 2.0)*, U.S. Bureau of Reclamation, Technical Service Center, Denver, Colorado.

Zhang, R. (1959). "A Study of the Sediment Transport Capacity of the Middle and Lower Yangtze River," *Journal of Sediment Research*, Beijing, China, vol. 4, no. 2 (in Chinese).

Chapter 4
Cohesive Sediment Transport

Page

4.1 Introduction 4-1
4.2 Cohesive Sediment Processes 4-1
4.2.1 Aggregation 4-1
4.2.2 Deposition 4-5
4.2.3 Consolidation 4-7
4.2.4 Toxicant Adsorption and Desorption 4-9
4.2.5 Erosion 4-10
4.2.5.1 Physical Factors Affecting Erodibility 4-11
4.2.5.2 Electrochemical Factors Affecting Erodibility 4-12
4.2.5.3 Biological Factors Affecting Erodibility 4-12
4.2.6 Experimental Methods to Determine Erosion Parameters 4-14
4.2.6.1 Rotating Cylinder 4-16
4.2.6.2 Straight Flume Studies 4-16
4.2.6.3 Annular Flume 4-18
4.2.6.4 In-Situ Methods 4-19
4.2.7 Critical Shear Stress and Erosion Rate Formulae 4-20
4.2.8 Discussion of Cohesive Soil Erosion Parameters Determined Through Experiment 4-22
4.2.9 Published Results of Erosion Parameters 4-23
4.3 Numerical Models of Cohesive Sediment Transport 4-31
4.3.1 One-Dimensional Models 4-31
4.3.2 Two-Dimensional Models 4-32
4.3.3 Three-Dimensional Models 4-33
4.3.4 Numerical Models of Contaminant Transport 4-34
4.4 Numerical Model GSTAR-1D 4-35
4.4.1 Conceptual Model 4-36
4.4.2 Active Layer Calculation 4-38
4.4.3 Consolidation 4-40
4.4.4 Bed Merge 4-41
4.4.5 Example Application 4-42
4.5 Summary 4-46
4.6 References 4-46

Chapter 4
Cohesive Sediment Transport

by
Jianchun Huang, Robert C. Hilldale, and Blair P. Greimann

4.1 Introduction

Cohesive sediments are composed primarily of clay-sized material, which have strong interparticle forces due to their surface ionic charges. As particle size decreases, its surface area per unit volume (i.e. specific surface area) increases, and the interparticle forces, not the gravitational force, dominate the behavior of sediment. There is no clear boundary between cohesive sediment and non-cohesive sediment. The definition is usually site-specific. In general, finer sized grains are more cohesive. Sediment sizes smaller than 2 μm (clay) are generally considered cohesive sediment. Sediment of size greater than 60 μm is coarse non-cohesive sediment. Silt ($2\mu\text{m} - 60\mu\text{m}$) is considered to be between cohesive and non-cohesive sediment. Indeed, the cohesive properties of silt are primarily due to the existence of clay. Thus in engineering practice, silt and clay are both considered to be cohesive sediment.

Cohesive sediments consist of inorganic minerals and organic material (Hayter, 1983). Inorganic minerals consist of clay minerals (e.g. silica, alumina, montmorillonite, illite, and kaolinite) and non-clay minerals (e.g. quartz, carbonates, feldspar, and mica, among others). The organic materials may exist as plant and animal detritus and bacteria.

Cohesive sediments are a concern in many waterways and are closely linked to water quality. Many pollutants, such as heavy metals, pesticides, and nutrients preferentially adsorb to cohesive sediments. In addition to the contaminants adsorbed to the sediments, the sediments themselves are sometimes a water quality concern. The turbidity caused by sediment particles can restrict the penetration of sunlight and decrease food availability, thus affecting aquatic life. Turbidity also increases water treatment costs.

4.2 Cohesive Sediment Processes

4.2.1 Aggregation

Cohesive sediments tend to bind together (aggregate) to form large, low-density units. This process is strongly dependent upon the type of sediment, the type and concentration of ions in the water, and the flow condition (Mehta et al. 1989). Metallic or organic coatings on the particles may also influence the interparticle attraction of fine sediments. Cohesive sediments are composed primarily of clay-sized material, which have strong interparticle forces due to their surface ionic charges. As particle size decreases, the interparticle forces dominate the gravitational force, and the settling velocity is no longer a function of only particle size. A predictive model of the aggregation process was formulated by McAnally and Mehta (2001) through a dimensional analysis of the significant parameters in collision, aggregation, and disaggregation.

In engineering models, aggregation is often indirectly considered by the change in settling velocity. The weight of an individual fine sediment particle is not sufficient to cause settling when the particle is suspended in water because any small disturbance, such as a turbulence fluctuation, will overcome the weight of the particle. The cohesion of the sediment causes the small particles to bind together to form larger flocs. The flocs may grow when they collide with other particles or other flocs. They may also be broken up by turbulent stress. The floc structure (size, density, and shape) determines the settling velocity. The effective settling rate is the fall velocity multiplied by a hindrance factor, which represents the velocity reduction due to other particles.

Krone (1962) performed flume studies and found that the settling velocity increases with sediment concentration and proposed the following formula:

$$\omega = KC_s^{4/3} \tag{4.1}$$

where ω = settling velocity (m/s),
C_s = suspended sediment concentration (g/l), and
K = an empirical constant equal to approximately 0.001 depending on the type of sediment.

A similar relationship was used by Cole and Miles (1983):

$$\omega = KC_s \tag{4.2}$$

where K = an empirical constant with a value of about 0.001 to 0.002.

Van Leussen (1994) proposed an empirical relationship between settling velocity, concentration and shear stress:

$$\omega = KC_s^n \frac{1+aG}{1+bG^2} \tag{4.3a}$$

where a, b = empirical constants, and
G = dissipation parameter defined as:

$$G = \sqrt{\frac{\varepsilon}{\nu}} \tag{4.3b}$$

where ε = turbulent energy dissipation rate, and
ν = molecular viscosity.

Nezu and Nakagawa (1993) provided the empirical relationship for ε in simple steady, uniform open channel flow as:

$$\varepsilon = \frac{u_*^3}{\kappa D}\frac{1-\zeta}{\zeta} \tag{4.3c}$$

where u_* = shear velocity,
κ = Von Kárman constant,
ζ = z/D, the relative elevation above the bed,
z = elevation above the bed, and
D = river depth.

The above relationships are not valid at high concentration. Thorn (1981) showed that the settling velocity increases with concentration at low concentration, then attains a maximum value and thereafter decreases due to hindered settling at intermediate concentrations and structural flocculation at large concentrations. Nicholson and O'Connor (1986) used the following relationship for settling velocity computation to incorporate these effects:

$$\begin{aligned} \omega &= A_1 C_s^{B_1}, && C_s \le C_H \\ \omega &= A_1 C_H^{B_1}[1.0 - A_2(C_s - C_H)]^{B_2}, && C_s > C_H \end{aligned} \tag{4.4}$$

where C_H = 25g/l refers to the onset concentration of hindered settling,
A_1 = 6.0×10^{-4} m^4/kg/s,
A_2 = 1.0×10^{-2} m^3/kg,
B_1 = 1.0, and
B_2 = 5.0.

All of the above coefficients are experimentally determined constants depending on the sediment type and salinity.

Burban et al. (1990) linked the settling velocity with the median floc diameter d_m from laboratory data:

$$\omega = a d_m^b \tag{4.5a}$$

with

$$a = B_1 (C_s \tau)^{-0.85} \tag{4.5b}$$

$$b = -[0.8 + 0.5\log(C_s\tau - B_2)] \tag{4.5c}$$

where C_s = cohesive sediment concentration (g/cm^3),
τ = fluid shear stress (dyne/cm^2),
d_m = median floc diameter (cm), and
B_1 and B_2 = 9.6×10^{-4} and 7.5×10^{-6}, respectively, both are experimentally determined constants.

Lick and Lick (1988) and Gailani et al. (1991) provided an experimentally based equation for determining the median floc diameter, written as:

$$d_m = \left(\frac{\alpha_o}{C_s \tau}\right)^{1/2} \tag{4.5d}$$

where α_0 = experimentally determined constant (= 10^{-8} g-m^2/cm^3-s^2 in freshwater).

Thorn (1981) showed that the settling velocity increases with concentration when concentrations are low, then attains a maximum value and thereafter decreases due to hindered settling at intermediate concentrations and structural flocculation at large concentrations. Van Rijn (1993) summarized the influence of sediment concentration on the settling velocity for sediments all over the world (Figure 4.1). Some of these sediments are implemented in GSTAR-1D.

The settling velocities, due to sediment flocculating, are usually site-specific and need to be determined by experiment. GSTAR-1D allows the user to provide a set of user specified data as shown in Figure 4.2.

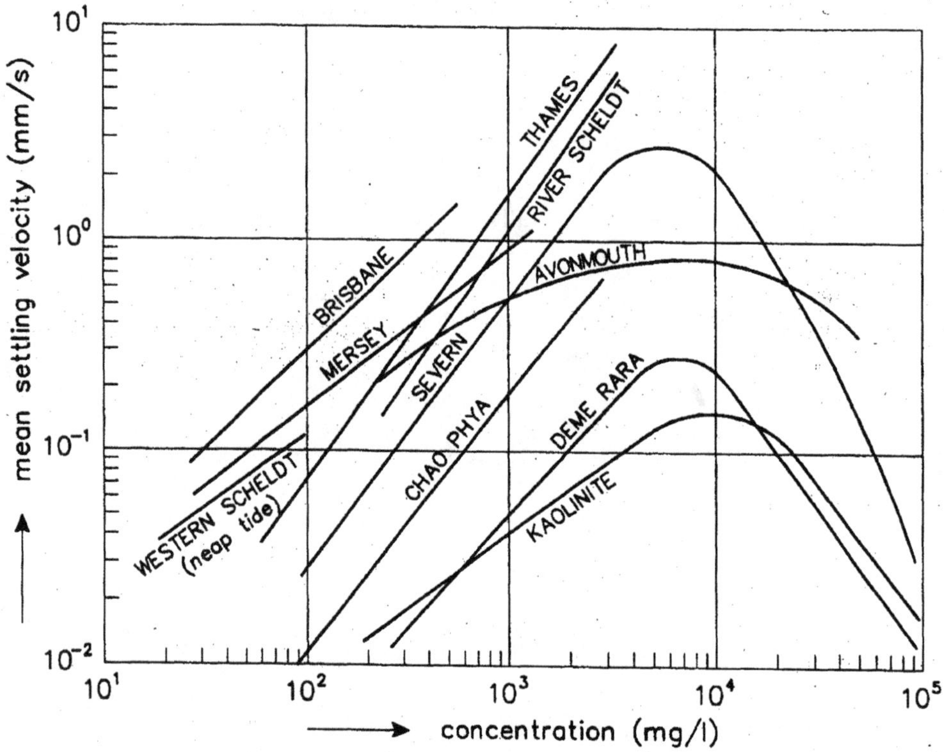

Figure 4.1. The influence of sediment concentration on the settling velocity (source: Van Rijn, 1993, Figure 11.4.2).

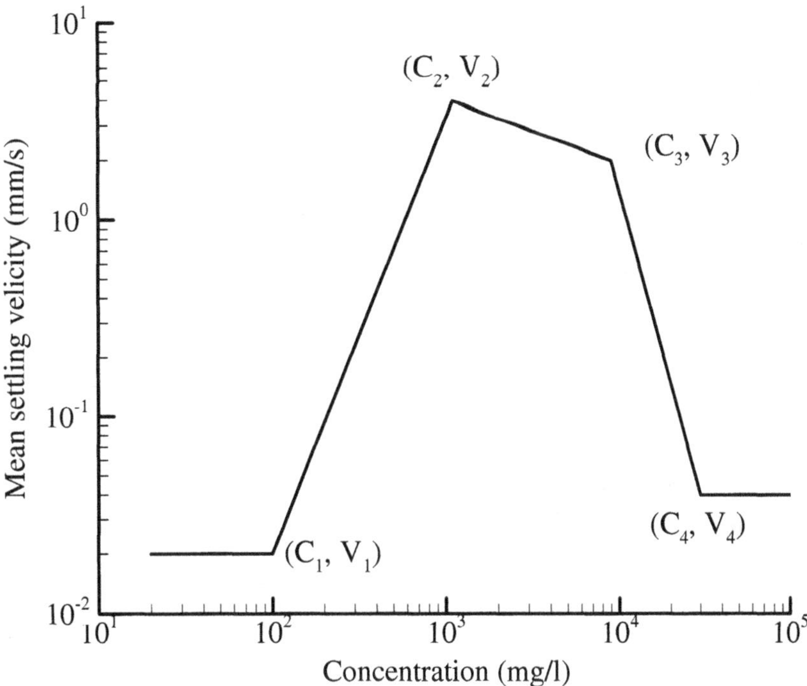

Figure 4.2. Input data illustration for settling velocity.

4.2.2 Deposition

Deposition occurs when the bottom shear stress is less than the critical shear stress. Only aggregates with sufficient shear strengths to withstand the highly disruptive shear stresses in the near bed region will deposit and adhere to the bed. Mehta and Partheniades (1973) performed laboratory studies on the depositional behavior of cohesive sediment and found that deposition is controlled by the bed shear stress, turbulence processes in the zone near the bed, settling velocity, type of sediment, depth of flow, suspension concentration and ionic constitution of the suspending fluid (also summarized in Hayter et al., 1999).

Two kinds of sediment deposition are included in GSTAR-1D, full and partial deposition. When the bed shear stress (τ) is smaller than the critical shear stress for full deposition ($\tau_{d,full}$), all sediment particles and flocs are deposited. Krone's (1962) deposition formulation is

$$Q_d = P_d \omega c \qquad \text{for } \tau \leq \tau_{d,full} \tag{4.6a}$$

where Q_d = deposition rate, and
P_d = the deposition probability.

The variable P_d is also the probability of particles sticking to the bed and not being re-entrained by the flow. A fraction of sediments settling to the near bed region cannot withstand the high shear stresses at the sediment-water interface and are broken up and re-suspended. The probability of deposition is given by

$$P_d = 1 - \tau / \tau_{d,full} \quad \text{for} \quad for \ \ \tau \leq \tau_{d,full} \tag{4.6b}$$

where τ = bottom shear stress, and
$\tau_{d,full}$ = critical shear stress for full deposition.

Many experiments were preformed to determine the values of critical shear stress for full deposition of cohesive sediments. They range between 0.06 and 1.1 N/m^2 depending upon the sediment type and concentration. Krone (1962) conducted a series of flume experiments to determine the critical shear stress for full deposition. For San Francisco Bay sediment, he found that $\tau_{d,full}$ = 0.06 N/m^2 when c < 0.3g/l; $\tau_{d,full}$ = 0.078 N/m^2 when 0.3 < c < 10g/l. Mehta and Partheniades (1975) found that $\tau_{d,full}$ = 0.15 N/m^2 for kaolinite in distilled water.

Partial deposition exists when the bed shear stress is greater than the critical shear stress for full deposition but smaller than the critical shear stress for partial deposition. At this range of bed shear stress, relatively strong flocs are deposited and relatively weak flocs remain in suspension. The partial deposition formulation is written as

$$Q_d = P_d \omega (c - c_{eq}) \quad \text{for} \quad \tau_{d,full} < \tau < \tau_{d,part} \tag{4.6c}$$

where c_{eq} = equilibrium sediment concentration.

The equilibrium sediment concentration is the concentration of relatively weak flocs that do not have sufficiently strong bonds and will be broken down before reaching the bed or will be eroded immediately after being deposited. The probability of deposition is given by

$$P_d = 1 - \tau / \tau_{d,part} \quad \text{for} \quad \tau_{d,full} < \tau < \tau_{d,part} \tag{4.6d}$$

There is no deposition when the bed shear stress is larger than the critical shear stress for partial deposition. The deposition rate is zero,

$$P_d = 0 \quad \text{for} \quad \tau \geq \tau_{d,part} \tag{4.6e}$$

At present, the behavior of critical shear stresses for full and partial depositions are not well understood, but the accuracy of the deposition model mainly depends on the use of their correct values. Thus, when the actual value of $\tau_{d,full}$ and $\tau_{d,partl}$ are uncertain, they become primary calibration parameters for determining the deposition rate.

4.2.3 Consolidation

Consolidation is another important phenomena in cohesive sediment transport. Two types of consolidation are usually considered: primary and secondary (Mehta et. al, 1989). Primary consolidation is caused by the self-weight of sediment, as well as the deposition of additional materials. Primary consolidation begins when the self-weight of the sediment exceeds the seepage force induced by the upward flow of pore water from the underlying sediment. During this stage, the self-weight of the particles expels the pore water and forces the particles closer together. The seepage force lessens as the bed continues to undergo self-weight consolidation. Primary consolidation ends when the seepage force has completely dissipated. Secondary consolidation is caused by the plastic deformation of the bed under a constant overburden. It begins during the primary consolidation and may last for weeks or months.

The consolidation process is incorporated into engineering models by representing the bed with a number of layers, each having a specific thickness, consolidation time, and critical shear stress. An idealized version of the consolidation process models the relationship by linking the layer bulk density, ρ, to the consolidation time, t (Nicholson and O'Connor, 1986):

$$\rho_b = \begin{cases} \rho_f, & t \le t_f \\ \rho_f + (\rho_\infty - \rho_f)\{1.0 - \exp[-A_2(t - t_f)]\}^{B_2}, & t_f < t < t_\infty \\ \rho_\infty, & t \ge t_\infty \end{cases} \tag{4.7}$$

where ρ_b = dry bulk density,
t = time, and
A_2 and B_2 = coefficients that account for the influence of mud type and salinity.

The subscripts f and ∞ represent the freshly deposited and the fully consolidated states, respectively.

Nicholson and O'Connor (1986) adopted the following values (Table 4.1) for the Grangemouth Harbor in Scotland, however the values of t_f, t_∞ and A_2 were not given.

Table 4.1. Model consolidation parameters of Nicholson and O'Connor

Consolidation parameter	Value
ρ_f	80 kg/m^3
ρ_∞	300 kg/m^3
B_2	0.45

Another similar consolidation model was applied by Teisson and Latteux (1986) in the Loire Estuary. The bed profile is considered to be composed of a number of layers of various thickness, each with a specified density, resistance to fluid shear, and duration of deposition. The bulk density for the Loire Estuary is given by:

$$\rho_b = \begin{cases} 136.2\log_{10}(t+5.42), & 0 < t \le 24h \quad t \text{ in hours} \\ 200 + 70\log_{10}(t), & t > 1 \text{ day} \quad t \text{ in days} \end{cases} \tag{4.8a}$$

Furthermore, they related the critical shear velocity for erosion to bed bulk density by,

$$u_{*e} = \begin{cases} 3.2\times10^{-5}\,\rho_b^{\;1.175}, & \rho_b < 240 g/l \\ 5.06\times10^{-8}\,\rho_b^{\;2.35}, & \rho_b > 240 g/l \end{cases} \tag{4.8b}$$

With these relationships, their consolidation model uses the values shown in Table 4.2.

Table 4.2. Consolidation model values

Layer	Density (g/l)	Duration	u_{*e} (cm/s)
1	100	2 hours	0.72
2	120	4 hours	0.88
3	145	8 hours	1.11
4	175	16 hours	1.38
5	205	1day 8 hours	1.66
6	230	2 days 16 hours	1.91
7	250	5 days 8 hours	2.18
8	270	10 days 16 hours	2.62
9	290	infinite	3.10

Another consolidation model was applied in the computer model SED2D WES (Letter et al., 2000) where the change in density with time is governed by the following equation:

$$\rho = \rho_f - (\rho_f - \rho_i)e^{-\beta t} \tag{4.9a}$$

where ρ = time-varying density,
ρ_i = initial density,
ρ_f = final ultimate density,
t = time, and
β = consolidation coefficient in 1/sec.

The consolidation coefficient β is calculated from user input for initial density and density at the reference time by solving Equation (4.9a):

$$\beta = \frac{1}{t_e}\ln\frac{\rho_f - \rho_i}{\rho_f - \rho_e} \tag{4.9b}$$

where ρ_e = density at reference time t_e.

4.2.4 Toxicant Adsorption and Desorption

Many pollutants, such as heavy metals, pesticides, and nutrients preferentially adsorb to cohesive sediments. Cohesive sediment serves as a carrier for toxicants, strongly relating water quality to cohesive sediment transport. Toxicants can be adsorbed onto sediment surfaces as adsorbed toxicants, and can be dissolved in the water as dissolved toxicants (adsorbed phase and dissolved phase). Adsorption is defined as the toxicants adhering to suspended sediments from the dissolved phase. Adsorbed toxicants can be removed from the water column after the cohesive sediment is deposited onto the river bed. Desorption is defined as the toxicant becoming dissolved in the water after being removed from the sediment surface. This usually happens when contaminated sediment is eroded from the river bed. Experiments (Thomann and DiToro, 1983) show that equilibrium between the dissolved and adsorbed phases is attained within minutes. Instantaneous local equilibrium is assumed for engineering problems.

Shrestha and Orlob (1996) presented a two-dimensional (2D) numerical model to simulate the fate and transport of cohesive sediments and associated nickel in southern San Francisco Bay. A partition coefficient, K_p, is used to define the distribution of the toxicant between the adsorbed and dissolved phases. The partition coefficient (m^3/kg) is defined as,

$$K_p = \frac{C_p}{C_d C_s} \tag{4.10}$$

where C_p = adsorbed toxicant concentration by weight (kg/m^3),
C_d = dissolved toxicant concentration by weight (kg/m^3), and
C_s = sediment concentration by weight (kg/m^3).

The partition coefficient for each toxicant is usually site-specific, and can be determined in the laboratory or in the field. The fraction of toxicant in the adsorbed phase, f_p is written as:

$$f_p = \frac{C_p}{C_t} = \frac{k_d C_s}{1 + k_d C_s} \tag{4.11}$$

where C_t = total toxicant concentration by weight (kg/m^3), and
k_d = equilibrium distribution coefficient (m^3/kg).

The sorbed mass of toxicant per unit mass of sediment is:

$$\frac{C_p}{C_s} = \frac{f_p C_t}{C_s} \text{ (kg/kg)} \tag{4.12}$$

If ΔM is the total mass of sediment deposited per unit area (kg/m^2) during a specific time period, then the toxicant mass associated with the deposited sediments is:

$$\Delta M_p = \frac{f_p C_t}{C_s} \Delta M \quad (\text{kg/m}^2) \tag{4.13}$$

On the other hand, the erosion of sediment brings toxicants back into the water column. If ΔM is the total mass of sediment eroded per unit area (kg/m^2) during a specific time period, then toxicant mass associated with the eroded sediment is:

$$\Delta M_p = \frac{f_p C_{t,b}}{C_{s,b}} \Delta M \quad (\text{kg/m}^2) \tag{4.14}$$

where $C_{s,b}$ = sediment concentration in bed (kg/m^3), and
$C_{t,b}$ = concentration of toxicant adsorbed to sediment in bed (kg/m^3).

4.2.5 Erosion

This section will address the major factors influencing a soil's ability to resist erosion, experimental methods to determine erodibility and a discussion on erosion parameters and formulae. Erosion of cohesive soils in this section refers to surface erosion (or fluvial or particle erosion), whereby individual particles or small aggregates are removed from the soil mass by hydrodynamic forces such as drag and lift (Millar and Quick, 1998). The ability of a cohesive soil to resist surface erosion is known as erosional strength (Zreik et al., 1998). Resistance to surface erosion differs from resistance to mass erosion. Mass erosion is determined by the soil's undrained strength, or yield strength (Millar and Quick, 1998). Mass erosion occurs when the yield strength is exceeded such as a slip failure of a streambank or when large flakes or chunks of soil are eroded from the streambed. Studies have shown (Kamphuis and Hall, 1983; Zreik et al., 1998; and Hilldale, 2001), that there is a strength difference of one to three orders of magnitude between erosional strength and yield strength.

Due to the complex and widely varied nature of particle bonds, much less is known about the properties influencing the bonding of cohesive soils. Unlike coarse sediments, cohesive sediments are not amenable to classification by grain size and distribution. These complex bonds greatly complicate modeling efforts of deposition and resuspension criteria for streams with a cohesive bed and/or banks. Table 4.3 shows the list of 28 parameters (Winterwerp, et al., 1990) that is used by Delft Hydraulics to characterize cohesive sediments. In addition to the physical and electro-chemical effects influencing cohesive soil behavior, recent research has indicated that biological effects also have considerable influence on cohesive sediment processes. Sometimes biological factors may be more important than the electrochemical effects when determining cohesive strength (Paterson, 1994). With this vast number of parameters affecting soil properties it is expected that there may be an interaction between the electro-chemical and biological effects. For example an electro-chemical process known to influence compaction of the bed may be disrupted by the presence of benthos.

Table 4.3. List of parameters used in the MAST-1 G6M project to characterize mud. This is a tentative list, resulting from the combination of the different lists used by the participants of the MAST G6M Cohesive Sediment Project. Some parameters are interdependent (after Winterwerp et al., 1990). Biological effects not included

Physico-chemical properties of the overflowing fluid	(19) specific surface
(1) chlorinity	(20) mineralogical composition
(2) temperature	(21) grain size distribution and sand content
(3) oxygen content	**Characteristics of bed structure**
(4) redox potential	(22) consolidation
(5) pH	a. consolidation curve and density profile
(6) Na-, K-, Mg-, Ca-, Fe-, Ai- ions	b. permeability
(7) sodium adsorption ratio (SAR)	c. pore pressure and effective stress
(8) suspended sediment concentration	(23) rheological parameters
Physico-chemical properties of the mud	a. upper and lower yield stress
(9) chlorinity	b. Bingham viscosity
(10) temperature	c. equilibrium slope of mud deposits
(11) oxygen content	(24) Atterberg limits (liquid and plastic limit)
(12) redox potential	**Water – bed exchange processes**
(13) pH	(25) settling velocity
(14) gas content	a. as a function of sediment concentration and floc density
(15) organic content	b. as a function of salinity
(16) Na-, K-, Mg-, Ca-, Fe-, Ai- ions	(26) critical shear stress for deposition
(17) cation exchange capacity (CEC)	(27) critical shear stress for erosion
(18) bulk density (density profile)	(28) erosion rate

4.2.5.1 Physical Factors Affecting Erodibility

Physical parameters affecting erodibility include: clay content, water content, clay type, temperature, bulk density (largely a reflection of the age of the deposit) and pore pressure, (Winterwerp et al., 1990). These properties vary greatly from one site to another. Rarely is a soil entirely made up of clay particles, rather fine sand and silt often make up a significant portion of the soil. For soils containing more than approximately 10% clay, the clay particles will assume control of the soil properties (Raudkivi, 1998). For a comprehensive discussion of the physical properties and minerology of the various types of clay particles, readers should refer to Loose Boundary Hydraulics, Chapter 10 (Raudkivi, 1998).

4.2.5.2 Electrochemical Factors Affecting Erodibility

The chemistry of the eroding fluid and the pore fluid influences the valence of clay particles and therefore plays a critical role in interparticle bonding. For example, increasing the electrolyte concentration or changing the cations in the pore fluid of a clay to one of a higher valence tends to cause flocculation of a clay–water suspension (Verwey and Overbeek, 1948). The Cation Exchange Capacity (CEC) is a measure of the type and amount of clay and is defined as the number of milliequivalents of exchangeable cations per 100 grams of dry soil. Increasing values of CEC have been shown (Ariathurai and Arulanandan, 1978) to decrease the erosion rate. In addition to the effects of cations and electrolytes, the pH of the pore fluid in the sediment matrix provides a significant contribution to the strength of the interparticle bond. Ravisanger et al. (2001) reported the following results on the effects that pH levels have on particle orientation and interparticle forces of kaolinite. The pore fluid pH influences the particle orientation during bonding by changing the surface or edge charges of the particles. These particle orientations are Edge-Edge (E-E), Edge-Face (E-F) and Face-Face (F-F). Briefly, low pH values ($pH \leq 5.5$) cause predominantly E-F associations, resulting in a stratified sediment bed exhibiting high erosion rates near the surface. Once the surficial sediments are eroded the erosion rate tends to become a relatively constant value. Intermediate pH conditions ($5 \leq pH \leq 7$) result in a reorientation of the bed structure from E-F to E-E particle associations. This results in a weaker bed structure due to the lack of surface contact, causing the bed to be more susceptible to erosion. At higher pH values ($pH > 7$) F-F particle orientations predominate and surface attraction forces of become significant because the area of surface contact has greatly increased, forming denser aggregates. This type of soil is fairly resistant to erosion.

Another significant factor in soil strength is salinity. The total content of dissolved salts in the pore fluid is a main factor in the susceptibility of cohesive soils to erosion (Sherard et al., 1972). These same authors also reported that the ratio of dissolved sodium ions to the other main basic cations (calcium and magnesium) in the pore water governs the susceptibility of cohesive soils to erosion. This is known as the Sodium Adsorption Ratio (SAR) and Arulanandan et al., (1975) showed that decreasing values of SAR decreased the erosion rate.

The presence of natural organic matter (NOM) in the eroding fluid is a result of decaying aquatic plant and animal matter (Schnitzer and Khan, 1972 and Thurman and Malcolm, 1983). Natural organic matter is measured as a percent of organic carbon adsorbed to the clay particles in the soil (Dennett et al., 1998). Adsorbed NOM can significantly affect the behavior of clay particles by influencing the interparticle bonding (Bennett, et al., 1991). Flume tests by Dennett et al. (1998) in which the level of NOM (% carbon) varied from 0.0% to 0.12% show an increased resistance to erosion with an increase in NOM. For the purpose of this discussion, the influence of NOM is not considered a biological effect due to the fact that living organisms are not directly involved the process.

4.2.5.3 Biological Factors Affecting Erodibility

The effects of organisms inhabiting stream or estuarial sediments may have three potential influences on the erodibility of the sediments; neutral (no effect), negative (decreasing soil stability) or positive (increasing soil stability). A positive effect on sediment stability is described as biogenic stabilization or biostabilization. This term has been defined as "a decrease in sediment erodibility caused directly or indirectly by biological action" (Paterson and Daborn, 1991). One of the most common negative effects of biological action is the reworking of the sediment by organisms known as bioturbation (a reworking or packaging of the sediment bed by organisms) (Paterson, 1997).

One way in which biogenic effects influence sediment stability is the growth of organisms on the bed surface (Paterson, 1997). The entrainment of particles is generally related to the hydrodynamic forces of drag and lift. Uneven surfaces with protrusions create a condition of hydraulic roughness. As organisms grow they have an effect of smoothing the bed by filling the inter-particle voids or creating a microbial mat (Manzenrider, 1983), reducing the hydraulic roughness. This decreases the stress in the near bed region having the effect of strengthening the bed by effectively increasing the velocity for which particles are entrained (Paterson, 1997).

It has been shown by Paterson (1997) that discrete particles (glass beads in this case) can become cohesive by the growth of bacteria and diatoms. The particles become covered with the developing growth of the bacteria and diatoms, causing cohesion to increase and roughness to decreases. Further binding is provided by bacterial secretion known as extracellular polymeric substances (EPS) forming cohesive networks between the diatoms. This biostabilization of originally non-cohesive particles points to the significant effect of biological processes.

Burrowing organisms have been shown to have both positive and negative effects on the stability of surficial soils. Laboratory research by Brekhovskikh et al. (1991) shows a possible 10-fold decrease in the critical stress for erosion when Oligochaeta (common burrowing worms in fresh water environments) are present. These effects were shown to largely depend on the population density and temperature. The same researchers found that Chironomids (common midges with burrowing larva in fresh water and sometimes saline environments) also had a negative effect on sediment stability; however, that influence decreased over time as the organisms excreted mucus and developed tube houses, cementing the bed and making it less erosive (the net effect was still negative because erosion was still less in the control flume).

Other research by Meadows et al. (Meadows and Tait, 1985 & 1989; Meadows et al., 1991; and by Meadows and Meadows, 1991) shows a strengthening of the bed by burrowing organisms. This work has demonstrated the potential of burrowing organisms to locally increase the critical shear stress of sediments. Results obtained by Black (1997) show the phenomenon of benthic adhesion within estuarine muds as a predominating influence over well documented electro-chemical particle-particle interactions. Black's (1997) research also indicated a positive effect of benthic diatoms by increasing inter-bond strength through the secretion of adhesive carbohydrate-rich mucus. It is likely that the effects of burrowing organisms and benthic diatoms are somewhat dependant on the soil type in addition to other site-specific conditions (e.g. salinity, pH, temperature).

A study (Littoral Investigation of Sediment Properties (LISP)-89, as reported in Daborn, 1993) performed by more that 30 scientists was carried out in a tidal flat to determine the threshold parameters for a numerical model. The critical shear stress was considered the essential parameter of this investigation. During this investigation it was found (Amos et al., 1988) that the sediment surface became significantly more resistant to erosion during the late summer. These results were obtained using an in-situ flume known as Sea Carousel. It was noted that several biological changes were occurring at the same time. During the investigation the foraging amphipod *Corophium volutator* was seen actively crawling at the soil surface (Boates and Smith, 1988). This foraging amphipod grazes on epipelic diatoms during ebbing tides. The diatoms have been shown by poisoning experiments and Sea Carousel flume measurements (Daborn, 1993) to increase the soil strength through secretions of soluble carbohydrates. The disappearance of *Corophium volutator* coincides with the arrival of migratory shore birds, which feed selectively on *Corophium* at rates estimated to be greater than 10,000 *Corophium* per bird per day (Hicklin and Smith, 1984). Daborn (1993) reports that more than 100,000 birds were in the estuary at their peak. With the arrival of the birds the numbers of the *Corophium* decreased and so did their behavior. The *Corophium* no longer grazed the surface but remained in their burrows to avoid predation (Boates and Smith, 1988). The conclusion reached by Daborn (1993) is that the decline of the *Corophium* activity produces greater amounts of chlorophyll, carbohydrates and organic matter in the soil due to the increased numbers of benthic diatoms. This has a secondary effect of increasing the soil's resistance to erosion.

The general conclusion of the studies mentioned above is that there is no comprehensive method to determine cohesive soil erodibility based solely on the known parameters of individual soils. Each site has specific physical, chemical and biological factors affecting the strength of the soil with no two being identical. Moreover, seasonal changes are likely to affect the presence of biota, since many of these species can not survive in low water temperatures. To successfully apply a numerical model of cohesive streambeds, it is necessary to test the individual soils for threshold conditions of erosion and deposition.

4.2.6 Experimental Methods to Determine Erosion Parameters

With the vast number of factors involved in the determination of the erodibility of cohesive soils, it becomes necessary to test cohesive soils for critical shear stress for erosion and deposition rather than using soil properties for predicting threshold values or using methods similar to those for coarse sediments. Some investigators create a sediment bed in the laboratory through settling of a premixed slurry or a placed bed using a rebuilt soil. This method generally simulates young soil deposits and also provides control over parameters such as bed sediment concentration (bulk density), soil type, and water quality (e.g. salinity, pH and electrolyte concentration). Since these types of experiments use a highly disturbed soil, critical parameters including chemistry and biology are often unaccounted for. Testing 'undisturbed' soil deposits will likely incorporate the chemical and biological effects. However this is a difficult task when one considers that any disturbance to the sample will potentially lead to biased results. Obtaining a completely undisturbed sample of a cohesive soil is unlikely; however, it is possible to minimize the disturbance to the sample. For example, transporting and storing cohesive soil samples in their natural water will not only prevent desiccation and minimize vibration during transport, but it also retains the chemical and biological properties that are instrumental in determining the soil's strength. Even if all the soil properties are retained for testing of an 'undisturbed' soil sample it is not likely that using the natural stream water is feasible, making it necessary to try to duplicate the chemical properties of the natural water (e.g. pH, temperature, salinity, electrolyte concentration, etc.). When one considers the uncertainties of extrapolating lab results to natural conditions, a strong case is made for in-situ testing of cohesive soils for threshold values. Recent investigators (e.g. Gust and Morris 1989, Black 1997, Ravens and Gschwend 1999) have used in-situ flumes to test soils in their natural environment. This seems to be the most promising method for determining the erodibility of cohesive soils.

Figure 4.3 illustrates an idealized graph of erosional and depositional characteristics determined from physical tests. In general, a physical test should provide erosion rates and critical shear stresses for deposition, surface erosion, and mass erosion. It should be mentioned that some test results are difficult to analyze due to wide scatter in data because of the inconsistent nature of cohesive soils.

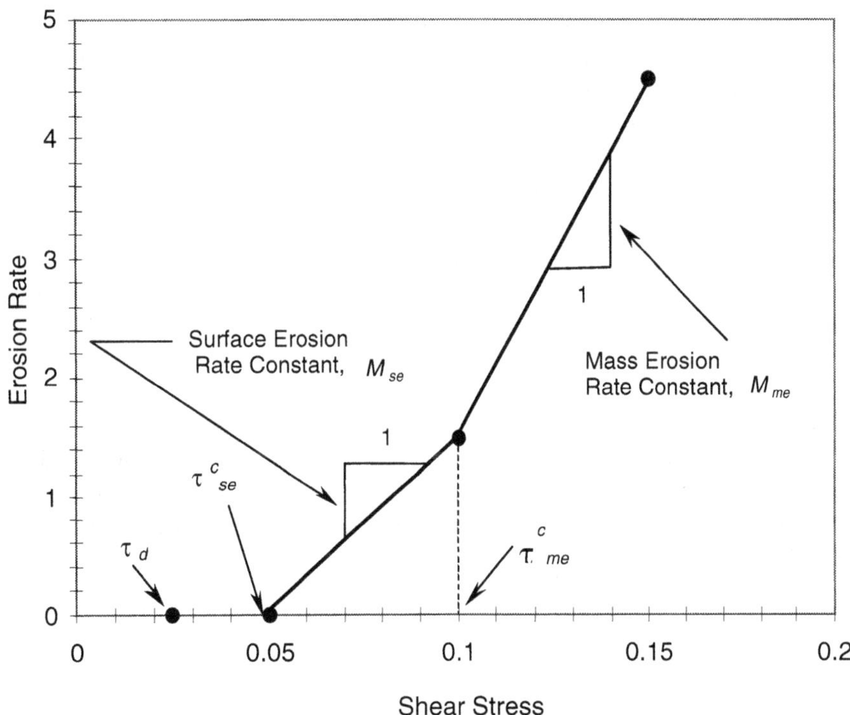

Figure 4.3. This idealized schematic illustrates the erosional characteristics that need to be determined from erosion tests and how they are plotted (after: Vermeyen, 1995). The values of critical shear stress for surface erosion (τ^c_{se}), critical shear stress for mass erosion (τ^c_{me}) and the subcritical condition of deposition (τ_d) is shown along with the mass and surface erosion rates. Units are intentionally omitted.

In the following sections (4.4.1 – 4.4.4), some common methods used to determine the erosion parameters of cohesive sediments are reviewed. These methods include the rotating cylinder, straight flume, annular flume, and in-situ testing methods.

4.2.6.1 Rotating Cylinder

A rotating cylinder used by Arulanandan et al. (1975) at the University of California at Davis is similar to the one developed by Masch et al. (1963). This device consists of two concentric cylinders separated by an annular space of 0.5 inches (1.27 cm). A soil sample 3.2 in. (8.13 cm) long and 3.0 in. (7.62 cm) in diameter is placed inside a Plexiglas cylinder that can be rotated at speeds up to 1500 rpm while the inner cylinder is stationary. The annular space is filled with an eroding fluid for the purpose of transmitting shear, creating a unidirectional current simulating stream flow. Stress on the soil is assumed to be nearly uniform due to a constant spacing between the soil and the outer cylinder and a uniform soil surface. This unidirectional flow around the soil cylinder created by this set-up can be described by the Navier-Stokes equations. The applied shear stress is calculated by the torque transmitted from the outer cylinder to the stationary inner cylinder. The erosion rate is determined by weighing the sample before and after each test to provide mass per unit time eroded from the sample for increasing values of shear stress. The erosion rate is plotted against the applied shear stress and the intercept on the applied shear stress axis defines the critical value. Results from Arulanandan et al., (1975) using the rotating cylinder are discussed later in this chapter.

A distinct advantage of this method is that the small volume of eroding fluid required allows for controlled conditions. Simulating, or actually using, natural stream water is possible. Altering the chemical composition of the eroding fluid to simulate the natural eroding fluid or varying a certain parameter is also possible since there is not a large volume of fluid involved. A similar device was used by Chapuis and Gatien (1986) to test intact samples without remolding. One drawback of an intact sample is the assumption of a smooth soil surface in determining a uniform shear stress. This assumption is valid for pure, remolded samples. However roots and/or larger discrete particles may protrude from the surface when a natural sample is used. This creates areas of locally increased shear stress, for which it is difficult to account. Chapuis and Gatien (1986) advise not cutting the sample with a thin-walled tube, stating that the surface of the sample becomes scaled due to surface remolding. These authors recommend a steel template and wire device to carve the sample.

4.2.6.2 Straight Flume Studies

Straight flumes have been used by many investigators (Vermeyen 1995; McNeil et al. 1996; Dennett et al., 1998 and Hilldale 2001) to determine the initiation of motion of cohesive sediments. Straight flumes allow the placement of 'undisturbed' or rebuilt soil samples in the bed of the flume to simulate stream flows over a soil bed. Straight flumes can present specific problems when testing cohesive sediments. Testing deposition is difficult and would ideally require a very long flume with a continuous soil bed and varied slopes to create erosional and depositional conditions. Recirculating pump systems can shear the suspended material due to the turbulence generated within the pumps, resulting in reduced floc sizes (Vermeyen, 1995) and making deposition experiments difficult to accurately duplicate. Maintaining conservation of mass for the soil is not likely in a recirculating system because soil may be deposited in pipe networks or the sump.

Hilldale (2001) used a recirculating flume with dimensions of 7.9 ft. long x 0.5 ft wide x 1.0 ft deep (2.4 m x 0.15 m x 0.31 m) to test the erosion criteria of cohesive bank soils. This system incorporated a tray containing an 'undisturbed' soil sample that was flush with the flume bed, which was built up with lead beads to create roughness upstream and downstream of the soil sample. This minimized fluid acceleration over the soil sample, which could skew results. To avoid edge effects, the soil sample did not extend the width of the flume. The soil sample was placed far enough downstream from the inlet so that the flow became fully developed before reaching the upstream edge of the soil sample. To test sediment entrainment, the fluid was sampled just downstream of the soil tray and compared to fluid samples taken upstream of the sample. When the downstream concentration was greater than the upstream (background) concentration, erosion was considered to have occurred. The uniform flow formula ($\tau = \rho gHS$) was used to calculate the applied shear stress over the sample. Dennett et al. (1998) found that determining shear stress using the uniform flow formula agreed with the slope of the logarithmic, near-bed velocity profile for similar test conditions. Several soil samples were tested varying the applied shear stress. The critical shear stress for erosion was determined from a plot of applied shear stress versus change in concentration. The applied shear stress at which the change in concentration increased from a zero value was determined to be the critical shear stress. No erosion rate was determined in this series of tests.

Vermeyen (1995) also tested an 'undisturbed' soil sample in a rectangular flume 12 ft. long x 0.96 ft. wide x 1.5 ft. deep (3.7 m x 0.29 m x 45.7 m). Two soil samples were placed end-to-end and were each 2.0 ft. (0.61 m) long and the full width of the flume. Erosion and deposition rates were quantified by measuring the sample's submerged weight. Velocity profiles were taken using an Acoustic Doppler Velocimeter (ADV) and shear stress was determined from the slope of the logarithmic velocity curve near the bed. The erosion rate and critical shear stress for erosion were determined following the method of Arulanandan et al. (1975). Deposition was also tested and quantified by weighing the sample before and after each test, which was run for a given amount of time.

Straight flumes incorporating similar processes have also been used by Westrich et al. (1997) and Kamphuis and Hall (1983). In the flume used by Westrich et al. (1997) sediment eroded by the fluid was monitored with a laser system placed just downstream of the sediment sample. Kamphuis and Hall (1983) used a flume tunnel capable of near bed flow velocities of 11.5 ft/s (3.5 m/s). Remolded soil samples were removed from the flume upon completion of the tests and evaluated.

Another straight flume method developed by McNeil et al. (1996) uses an innovative approach for exposing soil to the flow in order to examine the erodibility of natural soil deposits with depth. A straight, recirculating flume known as Sedflume (Sediment Erosion at Depth Flume) was designed and constructed to test undisturbed and reconstructed soil samples. This flume has a test section with an open bottom through which a rectangular coring tube 3.9 in. x 5.9 in. (10 cm x 15 cm) containing sediment can be attached. As the sediment erodes from the surface of the core, a piston inside the core can be raised to keep the sediment sample flush with the flume bed. The soil is sampled in rectangular cores taken from the river bottom, which are attached to the test section of the flume. The soil was sampled in soft deposits by pushing the core into the sediment bed. For deeper and/or more compact deposits a Vibracoring head was used to work the sample tube into the soil. Duplicate cores to those tested in the flume were tested for the sediment bulk properties, particle size distributions and total organic carbon. These 'undisturbed' samples of soil were tested for the erosion rate by measuring the depth of sediment eroded during the time of the test. A plot of the erosion rate versus depth was obtained. The critical shear stress for erosion was also evaluated by Sedflume on reconstructed soil samples, aged in the coring tube. This condition was identified visually, rather than measured quantitatively, which most likely resulted in a critical value for mass erosion.

4.2.6.3 Annular Flume

Annular flumes have long been used for testing cohesive soil properties (e.g. Partheniades et al., 1966; Mehta and Partheniades 1973; Fukuda and Lick 1980; Sheng 1988; Delo 1988; Maa 1989, and Krishnappan 1993). These flumes are of various diameters (3 to 22 ft, (1 – 7 m)) and have rectangular cross sections. A lid is placed in such a way so it just comes into contact with the water surface. Both lid (or ring) and flume are rotated in opposite directions to create a channel velocity. By rotating the ring and flume in opposite directions significant secondary currents are avoided in the corners of the channel (Krishnappan, 1993). These secondary currents have been reported to be as high as 15% of the tangential velocity, which exceeds that of the settling velocity of clay suspensions (Maa, 1989). These flumes are often equipped with Plexiglas windows for viewing and taps for extraction of fluid to test suspended sediments. Velocity distributions and applied shear stresses can be determined by using a laser and Preston tube to calculate velocity profiles and applied shear stresses, respectively, in clear water flows without sediment in the flume (Krishnappan, 1993). This data was obtained for various rotation speeds of the flume and lid. Suspended sediment is sampled from taps in the side of the flume and filtered to determine concentration. Zreik et al. (1998) used an annular flume and reported results in graphs of concentration versus time. These charts show rapid increases immediately after each incremental increase of shear stress. These same authors also report bed age, shear stress, test duration, erosion depth and temperature for each test performed.

The sediment bed of an annular flume can be prepared in a number of ways. For example Zreik et al. (1998) created the bed by allowing deposition of Boston Blue Clay from suspension in distilled water (prepared ahead of time). This procedure was followed by mixing the suspension by rotating the flume and followed by quiescent settling and aging of the bed for a specified amount of time, depending on the condition to be tested. Brekhovskikh et al. (1991) prepared the soil bed before adding water for the purpose of creating a soil bed containing benthos. Annular flumes provide a distinct advantage over straight flumes for testing deposition of cohesive soils. Fragile aggregates are not subject to break-up by pump propellers and do not settle in unwanted portions of the flume (e.g. headbox or pipes) as is the case in straight flumes. However annular flumes are not amenable to using 'undisturbed' soil samples obtained in the field although natural, disturbed sediments have been used.

4.2.6.4 In-Situ Methods

In light of the many factors affecting the erodibility of cohesive soils there is a strong case to be made for the use of in-situ testing methods. When soils are tested in-situ there is no error associated with extrapolating laboratory results to natural conditions. All physical, chemical and biological factors are as they exist in nature and there is no need for reconstructing soils for testing. A vast majority of in-situ testing methods reported in the literature are for estuarial applications (e.g. Gust and Morris 1989, Houwing and van Rijn 1995, and Ravens and Gschwend 1999). Researchers concerned with riverine and reservoir environments have not been as aggressive in developing this type of technology. Many of the devices mentioned below are developed for salt water applications that can be used or adapted for fresh water environments. A particular benefit would be the use of these devices in reservoirs that often consist of fluid mud that is much more difficult to duplicate in the lab.

The Seaflume system originally developed by Young and Southard (1978) has been modified by Gust and Morris (1989). This system is intended for salt water applications and can be deployed in depths up to 13,123 ft. (4000 m). The tripod structure is built from non-anodized aluminum having a base length of 11.5 ft. (3.5 m) and a height of 9.8 ft (3.0 m). Seaflume utilizes a straight flume 6.4 ft long by 1.4 ft wide by 0.64 ft high (1.95 m × 0.435 m × 0.195 m). The cross section is an inverted U-shape. The side walls of the flume cut approximately 2 in. (5 cm) into the sediment during deployment. Flow through the flume is controlled through a suction pump at the outlet. Velocity through the flume is measured by hot-wire sensors with a frequency response capable of resolving turbulent eddies. Shear stress is measured by epoxy-coated, flush mounted nickel grids. Differential signals from optical sensors positioned at the inlet and outlet are used to calculate instantaneous suspended sediment concentration. Data is stored on a digital cassette for evaluation upon recovery of the system. Output from Seaflume may consist of erosion rate versus time plots, erosion rate versus shear stress (log-log), inlet and outlet concentrations versus time, friction velocity versus time, and photographs of erosion within the flume.

A more portable device was built by Ravens and Gschwend (1999). This flume is 7.9 ft. (2.4 m) in length with a 3.3 ft. long (1.0 m) and consists of only the flume and the hose. The inlet section contains a sand coated, Plexiglas bottom. This inlet section allows the development of the boundary layer before the flow reaches the 3.9 ft (1.2 m) erosion section, which is open to the bed and has a laboratory tested uniform bottom stress. The outlet section of the flume has a 10.6 in. (27 cm) long floor protecting that portion of the bed from erosion due to the turbulence generated at the exit. The cross section is rectangular with a width of 4.7 in. (12 cm) and a height of 2.4 in. (6.0 cm). A mesh grid with 0.4 in. (1.0 cm) openings prevents large objects from entering the flume. A wide base running the length of the flume prevents the apparatus from sinking into the mud and two pieces of angle iron are mounted laterally to the bottom, which stabilizes the flume during operation. Water is pulled through the flume with a shipboard pump. Flow is monitored with a flow meter and a turbidimeter monitors the sediment suspension. Applied shear stress in the flume is based on a relationship between cross sectionally averaged velocity, shear stress and the Darcy-Weisbach friction factor. With a friction factor obtained from the Moody diagram (Moody, 1944), a power law fit was derived for the applied shear stress over a range of flow rates. Results obtained with this flume are discussed later in this chapter.

The development of a vertically deployed racetrack style flume is reported by Houwing and van Rijn (1995). The flume is 5.9 ft. (1.8 m) long and 2.3 ft. (0.7 m) in overall height. Flow is generated in the top section of the flume by a propeller, which is controlled by an adjustable oil pressure system. The lower portion of the flume is the test section where an open bottom exposes the flow to the bed. The test section has a rectangular cross section with a height of 3.9 in. (0.1 m) and a width of 7.9 in. (0.2 m). Flow is measured by means of an electro-magnetic flow meter and an optical sensor measures suspended sediment. Shear stress was determined from the velocity profile in the laboratory and an equation was derived so that velocity 1.0 in. (2.5 cm) above the bed is all that needs to be measured to obtain an applied shear stress.

As with any experiment, reproducibility is critical for dependable results. Spatial variations in bed topography, composition, and structure can be responsible for widely varied values of critical shear stress for erosion. This further mandates precise testing methods and devices so that variations in results can be attributed to bed variations and not the precision of the instrument.

Cornelisse et al. (1994) have published comparisons of several in-situ and lab erosion testing devices. These investigators concluded that reproducibility improves with an increased surface area.

4.2.7 Critical Shear Stress and Erosion Rate Formulae

This section covers erosion formulae commonly used for critical shear stress and erosion rates. Because there is no comprehensive theory regarding the erosion of cohesive soils the equations presented are empirical. The subsequent sections discuss the applicability of the following equations and how to determine the site specific parameters required for their use.

A formula for the surface erosion rate was presented by Ariathurai (1974) by fitting the experimental plots of erosion rate versus applied shear stress by Partheniades (1962):

$$Q_{se} = \begin{cases} M_{se} \dfrac{\tau - \tau_{se}^{c}}{\tau_{se}^{c}}, & \tau \geq \tau_{se}^{c} \\ 0, & \tau < \tau_{se}^{c} \end{cases} \tag{4.15}$$

where Q_{se} = surface erosion rate,

τ and τ_{se}^{c} = bed shear stress and critical surface erosion shear stress, respectively, and

M_{se} = surface erosion rate constant.

The excess bed shear stress, defined as $\tau - \tau_{se}^{c}$, is a measure of the erosion force. The critical erosion shear stress depends on a number of factors including sediment composition, bed structure, chemical compositions of the pore and eroding fluids, deposition history, and the organic matter and its state of oxidation (Ariathurai and Krone, 1976; Mehta et al., 1989). Usually, both M_{se} and τ_{se}^{c} change with the bed properties in depth and time. Field studies or laboratory measurements must be made to obtain the critical shear stress and the erosion rate.

Hwang and Mehta (1989) presented a relationship with the critical shear stress for surface erosion and the wet bulk density of the bed,

$$\tau_{se}^{c} = a_{se}(\rho_{wb} - \rho_l)^{b_{se}} + c_{se} \tag{4.16}$$

with a_{se}, b_{se}, c_{se}, and ρ_l having default values of 0.883, 0.2, 0.05 and 1.065, respectively for the stress in N/m^2 and the bed bulk density in g/cm^3. They are determined from field data. Hwang and Mehta (1989) also presented a relationship for the erosion constant as a function of the wet bulk density of the bed:

$$\log_{10} M_{se} = 0.23\exp\left(\frac{0.198}{\rho_{wb} - 1.0023}\right) \tag{4.17a}$$

where ρ_{wb} = wet bulk density of the deposit in g/cm^3, and
M_{se} = surface erosion rate constant in mg/cm^2/hr.

According to Teisson and Latteux (1986) and Cormault (1971) the experimental formulation of M_{se} for the Gironde Estuary is:

$$M_{se} = 0.55\left(\frac{\rho_b}{1000}\right)^3 \tag{4.17b}$$

where ρ_b = the dry bulk density of the deposit in g/l, and
M_{se} = surface erosion rate in kg/m^2/s.

In the numerical model of Nicholson and O'Connor (1986), the critical erosive stress is assumed to depend on the bed density:

$$\tau_{se}^{c} = \tau_{ef} + A(\rho_b - \rho_f)^B \tag{4.17c}$$

where τ_{ef} = critical shear stress of a freshly deposited bed,
ρ_b, ρ_f = dry bulk density of bed and freshly deposited bed, respectively, and
A and B = constants.

The parameters used in their model are shown in Table 4.4.

Table 4.4. Model erosion parameters

Erosion parameter	Value
τ_{ef}	0.8×10^{-1} N/m^2
ρ_f	80 kg/m^3
A	0.5×10^{-3} N m$^{5/2}$/kg$^{3/2}$
B	1.5

The rate of mass erosion over a time interval Δt is given by (Shrestha and Orlob, 1996)

$$Q_{me} = \frac{T}{\Delta t}\rho_s\left(\frac{\rho_b - \rho_w}{\rho_s - \rho_w}\right) \tag{4.18}$$

where T = thickness of the erodible layer, which is also a function of the excess bed shear stress,

Q_{me} = Mass erosion rate, and

ρ_s, ρ_b, ρ_w = density of sediment material, dry bulk density, and density of the suspending medium, respectively.

4.2.8 Discussion of Cohesive Soil Erosion Parameters Determined Through Experiment

Although direct measurement of cohesive soil properties provides the best possible results for determining erosional and depositional parameters, it is not always practical to do so. Monetary and time constraints often dictate available data. When testing of a specific soil is not possible it becomes necessary to make generalizations for modeling or analysis. This must be done with extreme caution and is not recommended for final results. This section presents some cautions regarding such generalizations and will provide a modeler or engineer with bounded values for erosion parameters, specifically the critical shear stress for particle erosion and ensuing erosion rates. These values vary by many orders of magnitude depending on the various soil parameters mentioned previously in this chapter, however in most cases some information is available about the soil and eroding fluid so that the values can be narrowed to arrive at a reasonable assumption.

Many studies have been performed linking mechanical soil properties to erosional properties, some of which are reviewed below. This has been done in an attempt to link erosion parameters of cohesive soils to some mechanical property. To date this has not been successfully accomplished due to the wide variation of parameters that determine the erodibility of cohesive soils. Smerdon and Beasley (1959) correlated critical shear stress for erosion with the plasticity index (PI), defined as the liquid limit (LL) minus the plastic limit (PL). These values are determined by the Atterberg limits test (Atterberg, 1911). Despite some scatter in the data (regression coefficient, R^2 = 0.77) this study showed an increase in resistance to erosion with increasing PI. Some investigators have since made similar correlations (Kamphuis and Hall, 1983 and Vermeyen, 1995) without being able to make generalizations about all cohesive soils.

Kamphuis and Hall (1983) made correlations of critical shear stress to unconfined compressive strength and vane shear strength. In this study critical shear stress was plotted with unconfined compressive strength and a regression with an R^2 value of 0.967 was obtained for 10 data points. This good fit of the data is likely due to critical shear stress being determined visually when pits began to appear in the soil sample. By definition this is mass erosion and this failure mechanism is more closely related to the soil's yield strength than is particle erosion. This observation was not reported in the literature. The investigators reported critical shear stress values in the range of 9 to 18 Pascals (Pa), much higher than those determined by other investigators for surface erosion. These values are another indication that critical shear stress for mass erosion was determined. Although a good correlation was obtained for critical shear stress and unconfined compressive strength these values can not be interpreted as applicable to any other soil types. Kamphuis and Hall (1983) arrived at this same conclusion.

Zreik et al. (1998) used a rotating annular flume to determine erosion properties for surface erosion using Boston blue clay. Critical shear stresses with depth were compared with undrained shear strength determined with a fall cone device. Their results show a one order of magnitude difference between the mechanical strength and the erosional strength (approximately 10 to 35 Pa and 0.1 to 1 Pa, respectively).

Although mechanical properties are not transferable from one soil type to another, they are consistent in showing that resistance to erosion is directly proportional to the soil's mechanical strength. It is worthwhile to mention that Partheniades (1965) determined that failure by shear stress applied by flowing water is different from failure by mechanical shear stress applied through the mass of the soil and that the Atterberg limits and mechanical composition can not be used as unique criteria for determining soil erodibility. The findings by Kamphuis and Hall (1983) and Zreik et al. (1998) support this.

4.2.9 Published Results of Erosion Parameters

Studies have been performed to relate erosion of cohesive soils with bulk density (Hwang and Mehta, 1989; Teisson and Latteux, 1986; van Rijn, 1993; Roberts et al., 1998; and others). Roberts et al. (1998) state that the bulk density alone could not be used to determine critical shear stress for erosion and erosion rates, however Hwang and Mehta (1989) were able to obtain expressions to determine the critical shear stress and erosion rate based on bulk density using data from a study performed with sediment from Lake Okeechobee. Because bulk properties are standard and easily determined it would be useful if erosion properties could be linked to bulk properties, however important factors such as sediment composition, salinity and other physico-chemical properties are not taken into account for this type of analysis. An interesting finding by Roberts et al. (1998) is that the erosion rates for sediment sizes greater than 0.125 mm (fine sand) is independent of bulk density, making bulk density a specific parameter for the erosion of cohesive soils. In spite of the shortcomings of determining erosion parameters with bulk density this may be the best generalization available. Mehta et al (1989) report that density provides an approximate indication of the erosional strength of the bed.

Hwang and Mehta (1989) performed erosion experiments to determine the critical shear stress and the erosion rate for surface and mass erosion using an annular flume with sediment and eroding fluid from Lake Okeechobee, Florida. By obtaining erosion rates and critical shear stress values for various wet bulk densities (ρ_{wb}) they developed Equations (4.16) and (4.17a) for surface erosion and the following relationships for mass erosion:

$$\tau_{me}^{c} = a_{me}\rho_{wb} + b_{me} \qquad (4.19)$$

$$M_{me} = const. \qquad (4.20)$$

where τ_{me}^{c} = the critical shear stress for mass erosion, in N/m^2,
M_{me} = the erosion rate constant for mass erosion, and
a_{me} and b_{me} = experimentally determined constants (9.808 and –9.934, respectively).

The constant value determined for the mass erosion rate is 224 mg/cm^2 hr. Using Equations (4.16), (4.17a), (4.19), and (4.20), Hwang and Mehta (1989) were able to apply a vertical sediment transport model for various bed depths with profiles of known bulk density. Changes in the concentration profile and bed elevation with time were the output for this model.

Van Rijn (1993) compiled data relating critical shear stress to dry bulk density and reports the following equation:

$$\tau_{se}^{c} = j(\rho_b)^k \qquad (4.21)$$

where ρ_b = dry bulk density of the soil in kg/m^3,
τ_{se}^{c} = critical surface erosion shear stress in units of N/m^2, and
j and k = coefficients determined by experiment.

The coefficient k was found to be in the range of 1 to 2.5 (van Rijn, 1993). Thorn and Parsons (1980) found $k = 2.3$ for mud from the Brisbane River, Australia, Grangemouth Estuary, Scotland and Belawan, Indonesia. Burt (1990) determined that $k = 1.5$ for mud from Cardiff Bay, England. Van Rijn makes no mention of any values for j in the literature. Re-evaluation of the data to obtain values for j would yield a large range of values due to the wide variation in k. Table 4.5 provides shear strength data from van Rijn (1993) and others related to the dry bulk density.

Table 4.6 provides erosion properties obtained by several investigators using various soils and testing methods. This table can be used as a guide to compare the results from various erosion experiments. Although there is a wide range of values in Table 4.6, erosion properties can be narrowed according to the soil type and eroding fluid. There is an exhaustive amount of literature providing erosion properties, however the data required to build such a table are not always presented in the literature. The data in Table 4.6 were re-analyzed to obtain the erosion rate parameters, with the exception of those shown for the study by Ravisangar et al. (2001), who published these values in the literature.

Table 4.5. Critical bed shear stress for surface erosion for different dry bulk densities (after van Rijn, 1993)

Soil Type	Sand	Organic	τ_{se}^{c} = critical shear stress for surface erosion [Pa]				
	[%]	[%]	ρ_b =100 [kg/m3]	ρ_b =150 [kg/m3]	ρ_b =200 [kg/m3]	ρ_b =250 [kg/m3]	ρ_b =300 [kg/m3]
Kaolinite (saline water)	0	0	-	0.05 – 0.10	0.30 – 0.40	-	-
Kaolinite (distilled water)	0	0	-	0.05 – 0.10	0.15 – 0.20	0.20 – 0.25	0.25 – 0.30
Hollands Diep 1 (lake)	9	10	0.15 – 0.25	0.30 – 0.40	0.40 – 0.50	0.60 – 0.80	-
Hollands Diep 2 (lake)	23	9	0.15 – 0.25	0.30 – 0.40	0.40 – 0.50	0.80 – 1.00	-
Ketelmeer (lake)	7	12	0.10 – 0.20	0.20 – 0.25	0.25 – 0.35	0.50 – 0.70	-
Biesbosch (lake)	8	8	0.20 – 0.25	0.25 – 0.30	0.30 – 0.35	0.50 – 0.70	-
Maas (river)	36	8	0.15 – 0.30	0.30 – 0.40	0.40 – 0.50	0.80 – 1.00	-
Breskens Harbour (estuary)	27	5	0.15 – 0.25	0.25 – 0.35	0.35 – 0.45	0.60 – 0.80	-
Delfzijl Harbour (estuary)	60	2	0.05 – 0.15	0.15 – 0.20	0.20 – 0.25	0.40 – 0.60	-
Loswal Noord (sea)	69	2	0.20 – 0.30	0.30 – 0.35	0.35 – 0.45	0.60 – 0.80	-
Brisbane, Grangemouth and Belawan	0	-	0.20 – 0.30	0.40 – 0.60	0.80 – 1.00	-	-
Loire	-	-	0.10 – 0.15	0.15 – 0.20	0.20 – 0.30	0.30 – 0.40	0.80 – 1.20
Cardiff Bay	-	-	0.20 – 0.30	0.40 – 0.50	0.60 – 0.70	0.70 – 0.90	-

Table 4.6. Erosion properties determined by several investigators using various methods and soil types

Investigator	Soil type	Flume type	Varied parameter	τ^c_{se} [Pa]	τ [Pa]	Q_{se} [kg/m^{2}s]	M_{se} [kg/m^{2}s]	$M_{se,ex}$ [kg/m^{2}s]	a [-]
(1)	(2)	(3)	(4)	(5)	(6)	(7)	(8)	(9)	(10)
Arulanandan et al. (1975)[a]	Yolo loam @ 0.1N salt concentration	Rotating cylinder	Sodium Adsorption Ratio SAR = (12.4)	0.5	0.7 - 1.9	1.8×10^{-3} - 14×10^{-3}	0.0054	-------	-------
			SAR = (9.4)	1.3	1.6 - 3.6	0.7×10^{-3} - 3.8×10^{-3}	0.0020	-------	-----
			SAR = (1.4)	3.8	5.6 - 6.7	0.9×10^{-3} - 2.5×10^{-3}	0.0055	-------	-----
		Rotating cylinder	SAR = (10.7)	0.05	0.4 - 1.2	5.0×10^{-3} - 15×10^{-3}	0.0009	-------	-----
			SAR = (3.0)	0.2	0.5 - 1.3	3.3×10^{-3} - 8.3×10^{-3}	0.0010	-------	-----
			SAR = (1.6)	0.5	1.0 - 2.1	1.2×10^{-3} - 3.5×10^{-3}	0.0015	-------	-----
Ariathurai and Arulanandan (1978)[a]	30% Illite (remaining material not provided in text)	Rotating cylinder	Temperature = (42°C)	1.2	1.3 - 2.4	0.4×10^{-3} - 4.0×10^{-3}	0.0040	-------	-----
			(23°C)	2.2	2.6 - 4.8	0.5×10^{-3} - 2.8×10^{-3}	0.0023	-------	-----
			(18°C)	2.4	2.6 - 6.0	0.2×10^{-3} - 2.8×10^{-3}	0.0019	-------	-----
			(9.5°C)	2.6	2.6 - 6.0	0.03×10^{-3} - 1.6×10^{-3}	0.0012	-------	-----
Partheniades (1965)[b]	S. F. Bay mud in water at ocean salinity	Straight flume	Placed bed	0.057	.011 - 0.93	0.15×10^{-6} - 5.0×10^{-6}	1.3×10^{-7}	-------	-----
			Flocculated bed	0.057	0.15 - 0.44	0.12×10^{-6} - 1.5×10^{-6}	2.7×10^{-7}	-------	-----
Nachtergale and Poesen (2002)[a,c]	Loess - undisturbed cylinder samples from crop land tested in fresh water	Straight flume	A Horizon	0.57	1.65 - 3.96	1.9×10^{-2} - 15×10^{-2}	0.0299	-------	-----
			B Horizon	0.66	1.65 - 3.96	3×10^{-2} - 4×10^{-2}	f	-------	-----
			C Horizon	1.65	1.65 - 3.96	1×10^{-2} - 41×10^{-2}	0.2827	-------	-----
Dennet et al. (1998)[a]	Kaolinite flocculated bed in fresh water	Straight flume	Natural Organic Matter (NOM) = 0.00%	0.77	0.96 - 1.82	3.7×10^{-3} - 22×10^{-3}	0.0155	-------	-------
			NOM = 0.002%	0.54	0.97 - 1.67	9.4×10^{-3} - 26×10^{-3}	0.0127	-------	-----
			NOM = 0.006%	0.39	0.97 - 1.67	18×10^{-3} - 40×10^{-3}	0.0121	-------	-----
			NOM = 0.012%	0.36	0.97 - 1.67	19×10^{-3} - 37×10^{-3}	0.0107	-------	-----

Table 4.6. Erosion properties determined by several investigators using various methods and soil types

Investigator	Soil type	Flume type	Varied parameter	τ^c_{se} [Pa]	τ [Pa]	Q_{se} [kg/m^{2}s]	M_{se} [kg/m^{2}s]	$M_{se,ex}$ [kg/m^{2}s]	a [-]
(1)	(2)	(3)	(4)	(5)	(6)	(7)	(8)	(9)	(10)
Ravisangar et al. (2001)[a]	Kaolinite flocculated bed in fresh water	Straight flume	pH of pore and eroding fluid = 3.5	0.91	0.97 - 1.67	$1.3x10^{-3}$- $12x10^{-3}$	0.0151	-------	-------
			pH = 4.0	0.91	0.97 - 1.67	$1.9x10^{-3}$- $16x10^{-3}$	0.0186	-------	-------
			pH = 5.0	0.88	0.97 - 1.67	$5.5x10^{-3}$- $30x10^{-3}$	0.0279	-------	-------
			pH = 6.0	0.12	0.97 - 1.67	$14x10^{-3}$- $25x10^{-3}$	0.0021	-------	-------
			pH = 7.0	-----	0.97 - 1.5	$6.0x10^{-3}$- $9.2x10^{-3}$	f	-------	-------
			pH = 8.0	-----	0.97 - 1.5	$2.8x10^{-3}$- $3.9x10^{-3}$	f	-------	-------
Westrich et al. (1997)	Undisturbed reservoir mud in fresh water	Straight flume	Upper-most horizon	0.74	0.77 - 1.76	$0.16x10^{-4}$- $4.0x10^{-4}$	0.00031	-------	-------
			Second horizon	0.66	0.69 - 1.68	$0.77x10^{-4}$- $14.4x10^{-4}$	0.00087	-------	-------
Johansen et al. (1997)[d,e]	Reconstructed bay (Wadden Sea) mud in water at ocean salinity	Annular flume	Consolidation time = 12 hrs, depth = 0-0.75 cm	0.055-0.20	0.088 - 0.25	$7.9x10^{-6}$- $64x10^{-6}$	-------	0.6776[g]	3.1891[g]
			Consolidation time = 24 hrs, depth = 0-0.81 cm	0.055 - 0.25	0.088 - 0.31	$4.6x10^{-6}$- $53x10^{-6}$	-------	0.3753[g]	3.1415[g]
			Consolidation time = 48 hrs, depth = 0-0.29 cm	0.084 - 0.37	0.11 - 0.44	$1.6x10^{-6}$- $44x10^{-6}$	-------	2.8612[g]	4.1504[g]
Ravens and Gschwend (1999)[a,g]	Boston Harbor mud, in-situ test, 12% clay, 51% silt, saline environment	In-situ flume	Applied shear stress	0.12	0.15-0.49	$0.13x10^{-3}$- $1.3x10^{-3}$	-------	0.0032	1

a. τ^c_{se} was determined with x-intercept of shear stress vs. E curve.

b. Both initial and final erosion rates were reported in the literature however only final erosion rates are reported in this table.

c. Five soil horizons were tested using soil sampled with cylinders from agricultural lands, only results from the top horizon are reported in this table.

d. Applied shear stresses are not reported and are calculated from a best-fit equation and averaged over the three runs.

e. Critical shear stress calculated for several depths.

f. Erosion rate constant, *M* is indeterminate because the slope is near or less than 0.

g. A dimensionless form was not possible, shear stress values must be in Pa (N/m^2).

Many investigations indicate that erosion rates are linear with increasing applied shear stress and can be described by Equation (4.15). Erosion rates for soils exhibiting this property can be determined with the knowledge of the critical shear stress $\left(\tau_{se}^{c}\right)$, applied shear stress (τ), and an erosion rate constant, (M_{se}) shown in columns five, six and eight of Table 4.6. Because the shear stress term is dimensionless the constant will have units associated with it, in this case the units are in [kg/m^{2}s]. This value is obtained by plotting the dimensionless parameter $\left(\frac{\tau}{\tau_{se}^{c}}-1\right)$ against the erosion rate, Q_{se} (Table 4.6, column seven) to find the slope of the line for the value M_{se}.

The erosion rate constant will have the same units as the plotted erosion rate. The applicability of the erosion rates shown in column seven is limited to surface (or particle) erosion. Van Rijn (1993) reports values for M_{se} in the range of 0.00001 to 0.0005 although Table 4.6 shows a much wider variation (0.00000013 to 0.2827), noting that the greatest value for M_{se} is obtained from loess taken from cropland. It is important to note that Equation (4.15) assumes a constant critical shear stress with depth, making it necessary to obtain multiple erosion rate constants for stratified soils. This has been demonstrated in Table 4.6 for experiments performed by Nachtergaele and Poesen (2002) and Westrich et al. (1997).

Soil stratification occurs in heterogeneous soils due to differential settling and depositional history. Deposited soils undergo self weight consolidation, whereby overburden stress causes the soil matrix to collapse, expelling the interstitial fluid and resulting in a decreased void space (Teisson et al., 1993). Decreasing the void space creates a stronger soil matrix due to increased surface contact of the particles. Consolidation will result in greater bed shear strength with depth and is generally related to age unless some mechanical method of compaction is used in laboratory situations. The strengthening of a soil mass with depth is finite and a maximum bed strength will develop that will remain mostly constant, assuming homogeneous soil composition.

The depth at which the bulk density and/or mechanical strength becomes relatively constant appears to occur several centimeters below the sediment surface (Figure 4.4) This gradient may be significant in flume studies or estuaries when cyclic stresses related to tides suspend and redeposit bottom sediment; however, the strength gradient near the soil surface is of little relevance in applications related to erosion in river channels or when deconstructing reservoirs when very little deposition occurs and erosion processes dominate. An assumption of a constant erosion rate with depth is generally acceptable in these situations for homogeneous soils.

Some analyses relate excess shear stress $\left(\tau-\tau_{se}^{c}\right)$ to the erosion rate, as shown in Equation (4.22), where the subscript *se,ex* indicates surface erosion rate determined with excess shear stress. The erosion rate for this type of analysis requires the erosion rate constant ($M_{se,ex}$) to have units of kg/m^{2}sPa (e.g., Ravens and Gschwend, 1999 in Table 4.6). This analysis is very similar to Equation (4.15), the difference being that Equation (4.22) does not have a dimensionless shear stress term. If a soil exhibits exponential surface erosion rates with increasing shear stress, the α

term will have a value other than one. For example, the study by Johansen et al. (1997) reports exponential erosion rates conforming to Equation (4.22) with an α value > 1. The constants for an exponential relationship are shown in columns nine and ten of Table 4.6, where $M_{se,ex}$ is an erosion rate constant associated with a dimensional excess shear stress and α is the exponent, both determined by a best fit equation of excess shear stress $(\tau - \tau_{se}^{c})$ plotted against the erosion rate (Q_{se}).

$$Q_{se} = M_{se,ex}\left(\tau - \tau_{se}^{c}\right)^{\alpha} \tag{4.22}$$

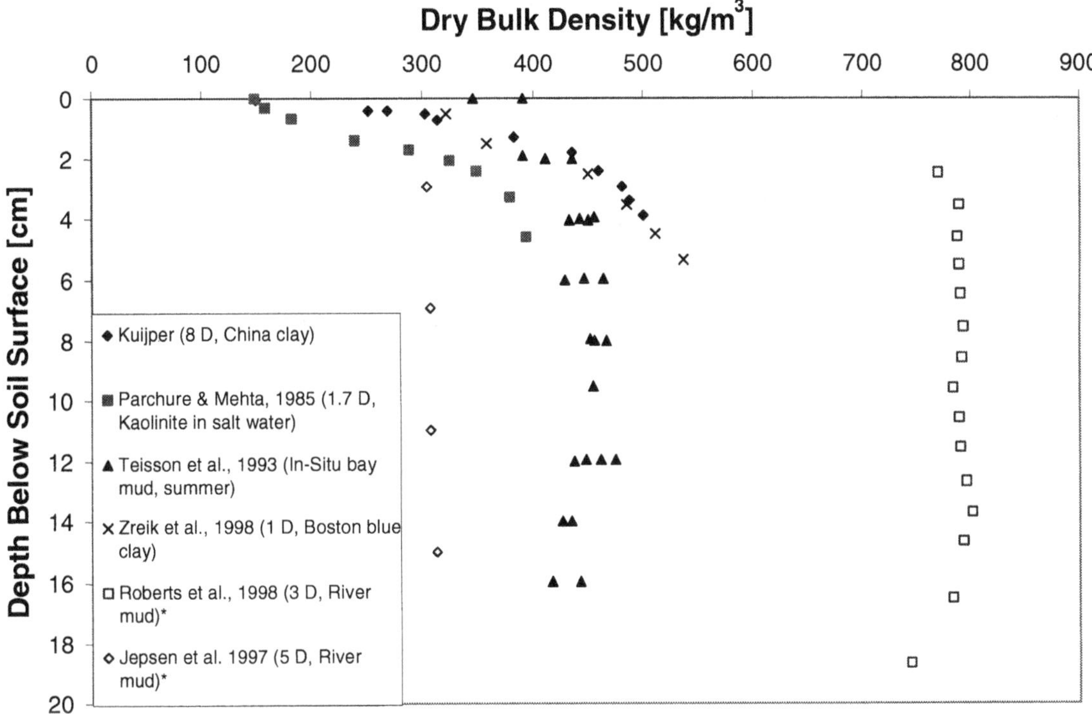

Figure 4.4. Chart comparing various bulk densities with depth. Ages of reconstructed soils range from 1.7 days (D) to 8 days. *Dry density obtained from reported wet bulk density by subtracting 1000 kg/m^3.

An equation taking the form of Equation (4.22) does not provide a dimensionless shear stress term, making it necessary to use units of Pa (N/m^2) for applied (τ) and critical (τ_{se}^{c}) shear stress if the erosion rate constant shown in column nine of Table 4.6 is to be used to obtain a rate in [kg/m^{2}s]. It is possible that an equation such as Equation (4.22) could represent both surface and mass erosion with mass erosion occurring soon after the break in slope on the graph, however, this has not been demonstrated in the literature.

For testing the upper portion of an estuarial sediment bed, Parchure and Mehta (1985) found that the methods described above were not applicable for the very soft, partially consolidated soils found in the upper few centimeters of the bed. These investigators incorporated the rate process theory to describe the time dependent particulate deformation of soft cohesive soils, which is applicable to soils exhibiting varying strength with depth (unlike Equation 4.15). This method uses the threshold energy concept to describe an erosion rate based on the excess shear stress $\left(\tau - \tau_{se}^{c}\right)$ and ln(Q_{se}) (Kelly and Gularte, 1981). The resulting equation is shown below (Parchure and Mehta, 1985):

$$\ln \frac{Q_{se}}{Q_f} = \beta\left(\tau - \tau_{se}^{c}\right)^{1/2} \tag{4.23}$$

where Q_f = Q_{se} when $\tau = \tau_{se}^{c}$, defined as the point at which no velocity dependent surface erosion occurs, and

β = a factor inversely proportional to temperature.

Use of this equation assumes the knowledge of the increase in erosional strength with depth. Floc erosion Q_f occurs in spite of the fact that $\tau = \tau_{se}^{c}$ because there is some entrainment due to the stochastic nature of these parameters. Determining Q_f experimentally can be very time intensive and may be alternately obtained when the ln(Q_{se}) is plotted against $\left(\tau - \tau_{se}^{c}\right)^{1/2}$, where Q_f is the intercept of the ln(Q_{se}) axis (Parchure and Mehta, 1985). Values of β and Q_f obtained by several authors are contained in Table 4.7.

Table 4.7. Erosion rate parameters used in Equation (4.23) (After Parchure and Mehta, 1985 with additional data from Kuijper et al., 1989 and Winterwerp et al., 1993)

Investigator	Soil Type	β (m/N$^{1/2}$)	Q_f x10^6 (kg/m^{2}s)
Parchure and Mehta (1985)	Kaolinite in tap water	18.4	0.83
	Kaolinite in salt water	17.2	2.3
	Lake mud	13.6	5.3
Partheniades (1965)	Bay mud	8.3	0.067
Lee (1979)	Lake mud	8.3	0.7
Thorn and Parsons (1977)	Estuarial mud	8.3	0.7
Thorn and Parsons (1979)	Estuarial mud	4.2	3.1
Dixit (1982)	Kaolinite	25.6	1.0
Kuijper et al. (1989)	China clay in annular flume	12 - 28	0.3 – 3.3
	China clay in tidal flume	27	0.2 – 1.1
Winterwerp et al. (1993)	Harbor mud	10	0.5

The laboratory use of Equation (4.23) appears to be limited to annular flumes, with soft deposited beds. This set-up best represents estuarial situations where cyclic conditions related to tides dominate the erosion process, where a soft fluid mud layer exists at the sediment-water interface.

For this type of application sediment is deposited, resuspended and deposited again, without ample time for complete consolidation. The top several centimeters of the soil are critical, as this is generally the extent of the erosion (over shorter time periods, e.g. days). For this situation an increase in bed shear strength with depth is significant, as opposed to riverine applications where a bed may be expected to undergo only erosion of a consolidated bed.

Most applications related to erosion of cohesive soils involve surface erosion, which has been the focus of the preceding discussion. When mass erosion occurs it is generally indicated on a graph by a break in slope, so that the erosion rate is shown to increase (see Figure 4.3). A separate critical shear stress (τ_{me}^{c}) and erosion rate constant (M_{me}) would be necessary for mass erosion and can be plotted on the same graph.

4.3 Numerical Models of Cohesive Sediment Transport

Numerical models are becoming a useful tool to predict cohesive sediment transport. Numerical models solve the mass transport equation for suspended sediment and the mass conservation equation for bed sediment after the hydraulic field is solved. In the following, numerical models are reviewed based on the number of dimensions of the model. Chapter 5 emphasizes non-cohesive sediment transport numerical models.

4.3.1 One-Dimensional Models

One-dimensional models solve the unsteady, cross-sectionally averaged equation for the mass balance of suspended sediment:

$$\frac{\partial(Ac_i)}{\partial t}+\frac{\partial(Qc_i)}{\partial x}=\frac{\partial}{\partial x}(AD_x\frac{\partial c_i}{\partial x})+S_i \tag{4.24}$$

where A = cross-sectional area (m^2),
Q = discharge (m^3/s),
c_i = cross-sectionally averaged sediment volume concentration (m^3/m^3) of constituent i,
t = time,
u = cross-sectionally averaged velocity components (m/s) in the streamwise direction x,
D_x = dispersion coefficient (m^2/s) in the streamwise direction x, and
S_i = source (erosion) and sink (deposition) terms (m^2/s) for constituent i.

Most 1D models are mainly designed for non-cohesive sediment transport with the capacities to simulate simple processes of cohesive sediment transport. These models include HEC-6 (U.S. Army Corps of Engineers, 1993), GSTARS 2.1, and GSTARS3 (Yang and Simões, 2000, 2002) and GSTAR-1D (Yang et al., 2004, 2005).

EFDC1D (Hamrick 2001) is a 1D sediment transport model that includes settling, deposition and resuspension of multiple size classes of cohesive and non-cohesive sediments. A bed consolidation model is implemented to predict time variations of bed depth, void ratio, bulk density and shear strength. Other contributions to 1D models of cohesive sediment transport were made by Odd and Owen (1972), Scarlatos (1981), and Li and Amos (1995).

4.3.2 Two-Dimensional Models

Two-dimensional models solve the depth-averaged or width-averaged convection-diffusion equation with appropriate boundary conditions. The 2D depth-averaged advection-dispersion equation for cohesive sediment transport is:

$$\frac{\partial(Dc_i)}{\partial t}+\frac{\partial(Duc_i)}{\partial x}+\frac{\partial(Dvc_i)}{\partial y}=\frac{\partial}{\partial x}(D_x D\frac{\partial c_i}{\partial x})+\frac{\partial}{\partial y}(D_y D\frac{\partial c_i}{\partial y})+S_i \tag{4.25}$$

where D = depth (m),
c_i = depth-averaged concentration (m^3/m^3) of constituent i,
t = time (s),
u, v = depth-averaged velocity components (m/s) in the streamwise and transverse or vertical directions, x and y, respectively,
D_x, D_y = dispersion coefficients (m^2/s) in the x and y directions, respectively, and
S_i = source (erosion) and sink (deposition) terms (m/s) for constituent i.

Usually, three types of boundary conditions are encountered in a 2D numerical model. At the inlet boundary, the sediment concentration is given. At the outlet boundary, there is no concentration gradient. At solid boundaries, there is no horizontal flux normal to the solid surface. Two-dimensional models have been applied to channel sedimentation and harbor sedimentation studies, where the variation of flow parameters with depth or width can be neglected.

SED2D WES (Letter et al., 2000) is a 2D finite element model for cohesive sediment transport developed by the US Army Waterways Experiment Station based on the original works of Ariathurai (1974) and Ariathurai and Krone (1976). Both cohesive and non-cohesive sediment can be simulated, but the model considers a single, effective grain size during each simulation.

Hayter (1983) developed an uncoupled, unsteady, 2D, finite element model for cohesive sediment transport. Further development of this model results in a well-documented numerical model HSCTM-2D (Hayter et al., 1999) with improved algorithm efficiency.

More 2D models were developed for applications to a variety of problems involving channel sedimentation, harbor sedimentation, and delta growth (Onishi, 1981; Cole and Miles, 1983; Teisson and Latteux, 1986 Ziegler and Lick, 1986; Gailani et al., 1991; and Ziegler and Nisbet, 1995; and Shrestha and Orlob, 1996).

4.3.3 Three-Dimensional Models

Three-dimensional (3D) models solve the full convection-diffusion equation, with appropriate boundary conditions. The general equation of transport of cohesive sediment is written:

$$\frac{\partial c_i}{\partial t}+\frac{\partial (uc_i)}{\partial x}+\frac{\partial (vc_i)}{\partial y}+\frac{\partial (wc_i)}{\partial z} = \frac{\partial}{\partial x}(D_x\frac{\partial c_i}{\partial x})+\frac{\partial}{\partial y}(D_y\frac{\partial c_i}{\partial y})+\frac{\partial}{\partial z}(D_z\frac{\partial c_i}{\partial z})+\frac{\partial}{\partial z}(\omega c_i) \tag{4.26}$$

where c_i = volume concentration (m^3/m^3) of sediment constituent i,
t = time (s),
u, v, and w = velocity components (m/s) in the directions, x, y, and z, respectively
D_x, D_y, D_z = dispersion coefficients (m^2/s) in the x, y, and z directions, respectively, and
ω = sediment fall velocity (m/s).

The source and sink terms, i.e., erosion and deposition fluxes, are incorporated in the bed boundary conditions, written as:

$$(D_z\frac{\partial c_i}{\partial z})_B = Q_e - Q_d \tag{4.27}$$

where Q_e and Q_d= volume eroded and deposited per unit area per unit time (m/s), respectively, and the subscript B denotes the boundary condition at the bed.

At the water surface, the upward diffusion rate is equal to the downward settling rate, which yields the water surface boundary condition,

$$\left(D_x\frac{\partial c_i}{\partial z}\right)_s = -\omega c_i \tag{4.28}$$

where subscript s = the boundary condition at water surface.

At solid boundaries, there is no diffusion normal to the solid surface, that is

$$D_m\frac{\partial c_i}{\partial n_m} = 0 \tag{4.29}$$

where n_m = the unit normal vector of the boundary.

Until recently, 3D numerical models were only used in academic areas. They may be the future of engineering models, but there are gaps in applying them for engineering purposes. The computing time required limits them to relatively short reaches of river for short time durations. Another factor that limits the use of a 3D model is the amount of field data needed for the calibration and verification of the 3D model.

The U.S. Army Corps of Engineers developed a 3D sediment transport model, CH3D-WES, that solves the complete 3D advection-dispersion equation using a finite-difference method (Sheng, 1983). Spasojevic and Holly (1996) generalized the 3D model (Sheng, 1986) to include cohesive-sediment capability. A new conceptual model was designed, in which the cohesive sediment can move in suspension, rest on the bed as settled mud, or occasionally form a layer of fluid mud. The suspended sediment can only be deposited onto the fluid mud, and the bed sediment can only be eroded from the settled mud. A thin active layer was used for erosion modeling. EFDC (Environmental Fluid Dynamics Code, Tetra Tech, Inc., 2000) is another 3D surface water model for hydrodynamic and sediment transport simulations in rivers, lakes and estuaries. The physics of the EFDC model and many aspects of the computational scheme are similar to the U. S. Army Corps of Engineers' 3D code CH3D-WES. Other 3D models were developed and applied by Nicholson and O'Connor, (1986), and Hayter and Pakala (1989).

4.3.4 Numerical Models of Contaminant Transport

Numerical models can also be used to predict the fate of contaminants. Numerical models solve the advection-diffusion equation for contaminants after the hydraulic field and sediment transport is solved. A 2D advection-diffusion equation for contaminates is given in the following as a reference.

Assuming instantaneous local equilibrium and constant partition coefficient between two phases of toxicant, the 2D equation for the total toxicant concentration is written as:

$$\frac{\partial(hc_t)}{\partial t}+\frac{\partial(huc_t)}{\partial x}+\frac{\partial(hvc_t)}{\partial y}=\frac{\partial}{\partial x}(D_x h\frac{\partial c_t}{\partial x})+\frac{\partial}{\partial y}(D_y h\frac{\partial c_t}{\partial y})+S \tag{4.30}$$

where
h = depth (m),
c_t = depth-averaged toxicant volume concentration (m^3/m^3),
t = time (s),
u, v = depth-averaged velocity components (m/s) in the horizontal streamwise and transverse directions, x and y, respectively,
D_x, D_y = dispersion coefficients (m^2/s) in the x and y directions, respectively, and
S = source terms (m/s).

Deposition of suspended sediment removes toxicants from the water column. The source term from deposition can be obtained from Equation (4.13) as

$$S = -\frac{f_p C_t \rho_s}{C_s \rho_c} Q_d = -\frac{f_p c_t}{c_s} Q_d \tag{4.31}$$

where Q_d = the sediment deposition rate (m/s),
C_t, C_s = total toxicant and sediment concentration by weight (kg/m^3),
c_t, c_s = total toxicant and sediment concentration by volume (m^3/m^3),
f_p = the fraction of toxicant in the adsorbed phase, and
ρ_c, ρ_s = dry density (kg/m^3) of toxicant and sediment, respectively.

Erosion of bed sediment brings toxicants into the water column. The source term from sediment erosion can be obtained from Equation (4.14) as

$$S = \frac{f_p C_{t,b} \rho_s}{C_{s,b} \rho_c} Q_e = \frac{f_p c_{t,b}}{c_{s,b}} Q_e \tag{4.32}$$

where Q_e = sediment erosion rate, surface or mass (m/s), and
$c_{t,b}$, $c_{s,b}$ = total toxicant and sediment volume concentration (m^3/m^3) in the bed.

The source term from chemical degradation or decay can be written as:

$$S_{de} = \lambda h c_t \tag{4.33}$$

where λ = chemical degradation rate or radionuclide decay rate (s^{-1}).

Shrestha and Orlob (1996) solved the 2D toxicant transport equation similar to Equation (4.30) to predict the fate of heavy metals in South San Francisco Bay.

There are some other numerical models to solve the toxicant transport. One example is HSCTM-2D (Hayter et al., 1999), which solves separate 2D transport equations for both dissolved and adsorbed toxicants. The equation is more complex, however it takes into account the adsorption of dissolved contaminants and the desorption of adsorbed contaminants onto and from suspended sediments, respectively.

4.4 Numerical Model GSTAR-1D

The first version of GSTARS was developed by Molinas and Yang (1985) to simulate the flow conditions in a semi-two-dimensional manner and the change of channel geometry in a semi-three-dimensional manner. Significant efforts were made to improve the first version and GSTARS 2.1 and GSTARS3 were released by Yang and Simões (2000, 2002).

U.S. EPA (Environmental Protection Agency) and the BOR (Bureau of Reclamation) are funding partners in the development of this numerical model for 1D river simulation. While basic concepts of GSTARS (Yang and Simões, 2000) have been retained, the program has been mostly rewritten. The numerical model, named GSTAR-1D, solves hydraulic and both cohesive and non-cohesive sediment transport, for steady and unsteady flow problems and for single and interconnected channels.

GSTAR-1D (Yang, et al., 2004, 2005) is a hydraulic and sedimentation numerical model developed to simulate flows in rivers and channels with or without movable boundaries. Some of its features are:

- Computation of water surface profiles in a single channel or multi-channel looped networks
- Steady and unsteady flows
- Subcritical flows in a steady hydraulic simulation
- Subcritical, supercritical, and transcritical flows in an unsteady hydraulic simulation
- Steady and unsteady sediment transport
- Transport of cohesive and non-cohesive sediments
- Cohesive sediment aggregation, deposition, erosion, and consolidation
- Sixteen different non-cohesive sediment transport equations that are applicable to a wide range of hydraulic and sediment conditions
- Cross stream variation in hydraulic roughness
- Exchange of water and sediment between main channel and floodplains
- Fractional sediment transport, bed sorting, and armoring
- Computation of width changes using theories of minimum stream power and other minimizations
- Point and non-point sources of flow and sediments
- Internal boundary conditions, such as time-stage tables, rating curves, weirs, bridges, and radial gates.

4.4.1 Conceptual Model

This section describes the conceptual model to simulate cohesive sediment transport in natural river systems. Figure 4.5 is a diagram of the proposed conceptual model, in which the bed is composed of an active layer and a maximum of N-1 inactive layers. In this figure, h = layer thickness, $P_{n,k}$ = volume fraction of k-th size class in layer n. The proposed model will be able to simulate both non-cohesive and cohesive sediment at the same time. A maximum of nf size fractions will be used to represent the sediment size distributions. The existing model will be modified to include cohesive sediment processes, such as settling, deposition, erosion, and consolidation. Consolidation of the sediment bed will be modeled by discretizing the bed into several layers.

The bed profile is considered to be composed of a number of layers of various thickness and bulk density. In each layer, bulk density of the cohesive sediment increases according to the empirically derived consolidation rate, and the bulk density of the non-cohesive sediment is constant. During consolidation, the bed thickness decreases and no mixing exists between each layer.

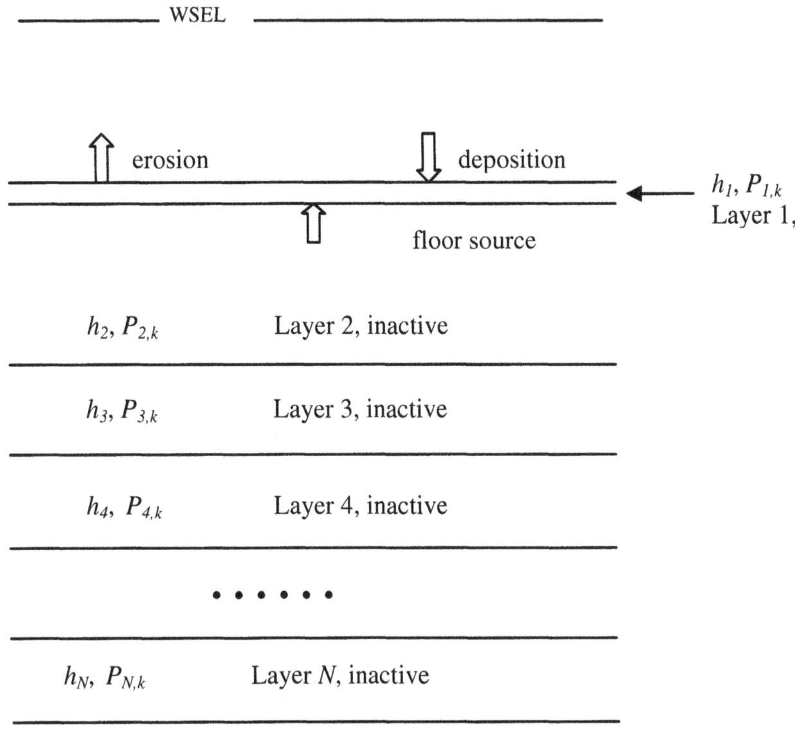

Figure 4.5. Diagram of the conceptual model.

The notion of an active layer provides an appropriate framework for erosion and deposition modeling. The active layer is defined as a thin upper layer of constant thickness. The active layer thickness is defined by the user as proportional to the geometric mean of the largest size class. Uniform size distribution and bulk density are assumed over the active layer depth. It is also assumed that all sediment particles of a given size class inside the active layer are equally exposed to the flow. Convincing experimental evidence shows that the presence of the fine cohesive sediment in the bed can cause the bed's strength to be greater than the shear stress required to entrain individual particles. The model assumes that the erosion rates of silts, sands and gravels are limited to the entrainment rate of the clay when more than 10% of the bed is composed of clay. The notion of an active layer allows the possibility of erosion of only fine sediment on the streambed surface. If the bed-shear stress is larger than the critical shear stress for the finer size classes, but smaller than that for coarser size classes, only the finer size classes are eroded. This will eventually armor the bed surface and prevent further erosion.

The active layer contains the bed material available for transport. During net erosion, the first inactive layer supplies material to the active layer. During net deposition, the additional material is moved to the first inactive layer. A range is set for the thickness of the first inactive layer. The lower limit is set to allow enough sediment to be supplied to the active layer during erosion. The upper limit is set to provide enough accuracy in simulating the bed layer. In the numerical model, the lower and upper limits are set to two and ten times the active layer thickness, respectively.

When the first inactive layer thickness becomes thinner than the lower limit, it merges with the second inactive layer. On the other hand, when it becomes thicker than the upper limit, it is divided into two inactive layers. All the others layer are shifted accordingly.

4.4.2 Active Layer Calculation

The use of an active layer provides a mechanism to model winnowing and armoring. Sediment can only be eroded from or deposited onto the active layer. The active layer thickness is defined by an auxiliary relation to represent the armoring thickness. One can define the thickness as proportional to the geometric mean of the largest size class containing at least 1% of the bed material at that location.

As the bed elevation descends or ascends during erosion or deposition, the active-layer floor changes its elevation to keep the active-layer thickness constant, as shown in Figure 4.6. The movement of the active-layer floor generates the active-layer floor source $S_{f,k}$ for the k-th size class. The kinematic condition of the k-th size fraction can be written as:

$$h_a \frac{dP_{a,k}}{dt} = -\frac{S_{e,k}}{\bar{\varepsilon}_k} + \frac{S_{f,k}}{\tilde{\varepsilon}_k} \tag{4.34}$$

where $P_{a,\,k}$ = active layer and size class, respectively

$S_{e,k}$ and $S_{f,k}$ = the active layer erosion source and floor source, respectively, and

$\bar{\varepsilon}_k$ and $\tilde{\varepsilon}_k$ = the volume of sediment in a unit bed layer volume (one minus porosity) of the erosion source and the floor source, respectively.

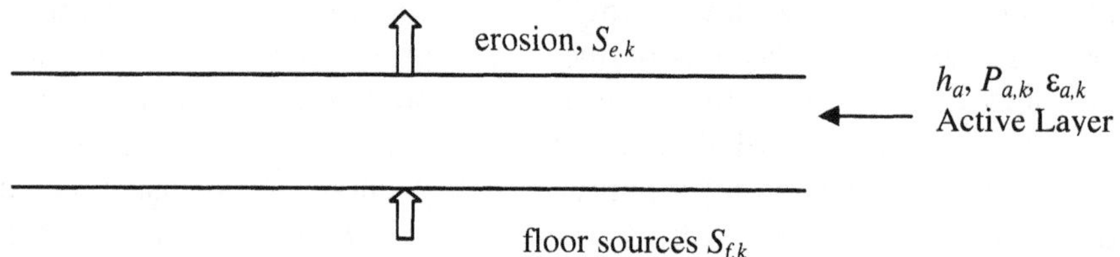

Figure 4.6 Diagram of the active layer model.

During net erosion, $\bar{\varepsilon}_k$ takes the value of the active layer ($\varepsilon_{a,k}$) and $\tilde{\varepsilon}_k$ takes the value of layer 2 ($\varepsilon_{2,k}$). On the other hand, during net deposition, $\bar{\varepsilon}_k$ takes the value of the freshly deposited sediment ($\varepsilon_{i,k}$) and $\tilde{\varepsilon}_k$ takes the value of the active layer ($\varepsilon_{a,k}$). Thus $\bar{\varepsilon}_k$ and $\tilde{\varepsilon}_k$ can be expressed as:

$$\bar{\varepsilon}_k = \begin{cases} \varepsilon_{a,k} & net\, erosion \\ \varepsilon_{i,k} & net\, deposition \end{cases} \tag{4.35}$$

$$\tilde{\varepsilon}_k = \begin{cases} \varepsilon_{2,k} & net\,erosion \\ \varepsilon_{a,k} & net\,deposition \end{cases} \tag{4.36}$$

where the subscript *i* represents freshly deposited sediment and:

$$\varepsilon_{n,k} = 1 - p_{n,k} \tag{4.37}$$

where $p_{n,k}$ = the porosity of the *k*-th size class in layer *n*.

By using the porosity of sediment, one can express the bulk density as:

$$\rho_{n,k} = \rho_s (1 - p_{n,k}) = \varepsilon_{n,k} \rho_s \tag{4.38}$$

where ρ_s = the density of sediment.

Summation of Equation (4.34) with the basic constraint of size fractions $\sum_k P_{n,k} = 1$ gives the global mass-conservation equation for the active layer:

$$\sum_k \frac{S_{e,k}}{\tilde{\varepsilon}_k} = \sum_k \frac{S_{f,k}}{\tilde{\varepsilon}_k} \tag{4.39}$$

that shows that the change of the bed elevation due to erosion (or deposition) is the same as the change in the active-layer floor and the active-layer thickness is kept constant.

According to the definition of the volume fraction of a size class, Equation (4.34) can be written as:

$$h_a \frac{dP_{a,k}}{dt} = -\frac{S_{e,k}}{\tilde{\varepsilon}_k} + \tilde{P}_k \sum_k \frac{S_{f,k}}{\tilde{\varepsilon}_k} = -\frac{S_{e,k}}{\tilde{\varepsilon}_k} + \tilde{P}_k \sum_k \frac{S_{e,k}}{\tilde{\varepsilon}_k} \tag{4.40}$$

where $\tilde{P}_k$ can be expressed as:

$$\tilde{P}_k = \begin{cases} P_{2,k} & net\,erosion \\ P_{a,k} & net\,deposition \end{cases} \tag{4.41}$$

where the subscripts *a* and *2* represent the active layer and layer 2, (the first inactive layer) respectively, and the subscript *k* represents the size class.

The mass-conservation equation for *k*-th size class in active layer is:

$$\frac{d}{dt}(\varepsilon_{a,k} P_{a,k} h_a) = -S_{e,k} + S_{f,k} \tag{4.42}$$

By substituting Equation (4.40) into Equation (4.42), one can express the change of $\varepsilon_{a,k}$ as:

$$\frac{d}{dt}\varepsilon_{a,k} = -(1 - \frac{\varepsilon_{a,k}}{\bar{\varepsilon}_k})\frac{S_{e,k}}{P_{a,k} h_a} + (1 - \frac{\varepsilon_{a,k}}{\tilde{\varepsilon}_k})\frac{\tilde{\varepsilon}_k \tilde{P}_k}{P_{a,k} h_a}\sum_k \frac{S_{e,k}}{\bar{\varepsilon}_k} \tag{4.43}$$

Equations (4.40) and (4.43) are essentially the equations for erosion and deposition, respectively, of the active layer.

4.4.3 Consolidation

Consolidation changes the thickness of the bed through changes in porosity. It should be noted that the consolidation process also affects the size-fraction distribution within the bed. Due to the slow rate at which consolidation occurs, the simulation of erosion and deposition are uncoupled from the bed consolidation process. Simulation of bed consolidation should be applied to both the active layer and inactive layers.

During consolidation, the mass of each size fraction remains constant. The mass-conservation equation for the sediment in each layer is:

$$\varepsilon_{n,k} P_{n,k} h_n = Const \tag{4.44}$$

where $P_{n,k}$ = volume fraction of *k*-th size class in layer *n*, and
h_n = thickness of layer *n*.

Equation (4.44) can also be written as:

$$P_{n,k}^{t+\Delta t} h_n^{t+\Delta t} = \varepsilon_{n,k}^t P_{n,k}^t h_n^t / \varepsilon_{n,k}^{t+\Delta t} \tag{4.45}$$

where t = the time step before consolidation, and
$t + \Delta t$ = the time step after consolidation.

Summation of Equation (4.45) with the basic constraint of size fractions $\sum_k P_{n,k} = 1$ gives the global mass conservation equation for sediment in layer *n*, i.e.:

$$h_n^{t+\Delta t} = \sum_k (\varepsilon_{n,k}^t P_{n,k}^t h_n^t / \varepsilon_{n,k}^{t+\Delta t}) \tag{4.46}$$

Equations (4.45) and (4.46) yield the expression for the size fraction change as:

$$P_{n,k}^{t+\Delta t} = \frac{\varepsilon_{n,k}^{t} P_{n,k}^{t} h_n^t / \varepsilon_{n,k}^{t+\Delta t}}{\sum_k (\varepsilon_{n,k}^{t} P_{n,k}^{t} h_n^t / \varepsilon_{n,k}^{t+\Delta t})} \tag{4.47}$$

Equations (4.46) and (4.47) are essentially equations for bed consolidation.

A similar relationship to the one used in SED2D WES (Letter, et al., 2000) is used here to calculate the change of volume concentration of sediment at bed $\varepsilon_{n,k}$.

$$\varepsilon = \varepsilon_f - (\varepsilon_f - \varepsilon_i)e^{-\beta t} \tag{4.48}$$

where β is the consolidation coefficient computed from user input for initial density as:

$$\beta = \log\left(\frac{\rho_f - \rho_i}{\rho_f - \rho_e}\right) \tag{4.49}$$

where ρ_i = fully consolidated density,
ρ_f = density, and
ρ_e = density at reference time t_e.

So the change of ε can be written as:

$$\frac{d\varepsilon}{dt} = \beta(\varepsilon_f - \varepsilon) \tag{4.50}$$

An explicit Euler scheme is used to calculate the sediment volume in a unit volume after consolidation and is shown here:

$$\varepsilon_{n,k}^{t+\Delta t} = \varepsilon_{n,k}^{t} + d\varepsilon = \varepsilon_{n,k}^{t} + \beta(\varepsilon_f - \varepsilon)dt \tag{4.51}$$

The bed thickness and sediment size fraction can be calculated with Equations (4.46) and (4.47).

4.4.4 Bed Merge

After each time step, the first inactive layer thickness is checked. Once the thickness decreases below the minimum limit, the content of the layer is merged with the underlying layer. The minimum limit of the first inactive layer thickness is set to allow enough sediment supply to the active layer during net erosion. It is obvious that the bed merge is not a physical process, but is a requirement of the discretized representation of the sediment bed.

During the merge of two layers, the new layer thickness is the sum of the two merged layers:

$$h = h_n + h_{n+1} \tag{4.52}$$

and the volume size fraction is:

$$P_k = \frac{P_{n,k}h_n + P_{n+1,k}h_{n+1}}{h_n + h_{n+1}} \tag{4.53}$$

The mass conservation equation is used to obtain ε_k:

$$\varepsilon_k = \frac{\varepsilon_{n,k}P_{n,k}h_n + \varepsilon_{n+1,k}P_{n+1,k}h_{n+1}}{P_k h} \tag{4.54}$$

4.4.5 Example Application

The GSTAR-1D was applied to the California Aqueduct near Arroyo Pasajero to study the influence of rainfall-runoff on sedimentation and water quality (Klumpp, et al., 2003). An unsteady flow and unsteady sediment model was used to simulated a duration of 800 hrs. GSTAR-1D is a numerical model for predicting cohesive and non-cohesive sediment transport, using steady or unsteady flow and can be applied to a simple channel as well as channel networks. The example presented here is not intended to show the readers how to set-up input data files and use GSTAR-1D, however it will provide a general idea of GSTAR-1D's capabilities related to cohesive sediment transport. Interested readers should refer to the GSTAR-1D User's Manual (Yang et al., 2004, 2005) for details on how to use GSTAR-1D.

The studied reach of the California Aqueduct, or San Luis Canal (SLC), extends 75 miles from Check Structure 15 to Check Structure 21. The SLC was designed and built to distribute water for both agricultural and municipal uses. It was built with drain inlet structures to capture floodwaters generated west of the SLC. Rainfall-runoff is admitted to the SLC when the capacity of ponding areas or bypass structures is exceeded. The runoff carries many tons of sediment into the aqueduct. The input data includes the cross-section geometry, the six check structures and their radial gate operations. Main channel flow through the aqueduct was assumed to be 2000 cfs. Six lateral inflows were modeled for the 100 year flood (Table 4.8).

These lateral inflows were modeled in terms of discharge hydrographs and sediment inflows. The bed material along the aqueduct is approximately 2% sand (non-cohesive sediment) and 98% silt and clay (cohesive sediment).

Table 4.8. Lateral inflows used in the GSTAR-1D example

Lateral flow	River mileage (mi)	Maximum discharge (cfs)
Cantua 4 barrel 6 X 4 Drain Inlets	133.67	1,002
Cantua Concrete Flume	134.81	850
Cantua Concrete Weir	134.87	2,743
Salt Creek Drain Inlets	136	112
Salt Creek Drain Inlet metal pipe	137.08	100
Arroyo Pasajero Drain	159	4,500

The present model used a modified version of Equation (4.15) for surface erosion

$$Q_{se} = \begin{cases} P_{se}\left(\dfrac{\tau - \tau_{se}^{c}}{\tau_{me}^{c} - \tau_{se}^{c}}\right) & \tau \geq \tau_{se}^{c} \\ 0 & \tau < \tau_{se}^{c} \end{cases} \tag{4.55}$$

where τ_{me}^{c} = critical mass erosion shear stress.

The modified relationship is more consistent with the mass erosion rate used below. The parameters τ_{se}^{c} and P_{se} are site-specific and have to be determined experimentally. Mass erosion is usually arbitrarily dependent on the model setup and its time scale used. The presented model takes the similar equation for mass erosion as the surface erosion.

$$Q_{me} = P_{me}\left(\frac{\tau - \tau_{me}^{c}}{\tau_{me}^{c}}\right) + P_{se} \qquad \tau \geq \tau_{me}^{c} \tag{4.56}$$

where Q_{me} = mass erosion rate,

τ and τ_{me}^{c} = bed shear stress and critical mass erosion shear stress, respectively, and

P_{me} = mass erosion constant.

Because physical experiments were not able to be performed, observation was used to determine the parameters for cohesive sediment transport. The critical shear stresses for full deposition, partial deposition, surface erosion, and mass erosion were determined from the observations in the channel during various discharges. These parameters are listed in Table 4.9.

Table 4.9. Cohesive sediment parameters for erosion and deposition used in the GSTAR-1D example

Process	Discharge (cfs)	Shear Stress (lb/ft^2)
Full deposition	2,000	0.003
Partial deposition	2,000	0.003
Surface erosion	8,000	0.005
Mass erosion	>>10,000	0.01

The surface erosion rate was calibrated and set to 0.3 $lb/ft^2/hr$.

Eqs. (4.6a-e) are used for cohesive sediment deposition. The settling velocities, due to sediment flocculating, are usually site-specific and need to be determined experimentally. GSTAR-1D allows the user to provide a set of user specified data as shown in Figure 4.7.

The parameters used in the example are listed in Table 4.10.

Table 4.10. Cohesive sediment parameters for fall velocity used in the GSTAR-1D example

Point	C (mg/l)	V (mm/s)
1	200	0.2
2	6,000	0.2
3	20,000	0.35
4	100,000	0.35

An equilibrium sediment concentration for partial deposition of 265 mg/l was observed at the downstream end of the channel, therefore an equilibrium concentration of 265 mg/l was used in the model.

Figure 4.8 shows the bed elevations before and after the flood. The fine sands allowed into the aqueduct are deposited just downstream of the inlet, raising the channel bed elevation. The fine sands are eroded after the flood, and the bed geometry returns to its initial form after the flood.

Figure 4.9 shows concentration changes with the sediment inflow from the Cantua Creek four barrel drain inlet. The peak sediment inflow concentration from Cantua Creek is about 7,500 mg/l. The peak concentration downstream of the inlets is about 2,000 mg/l. The baseline conditions prior to the lateral sediment inflow is 200 mg/l.

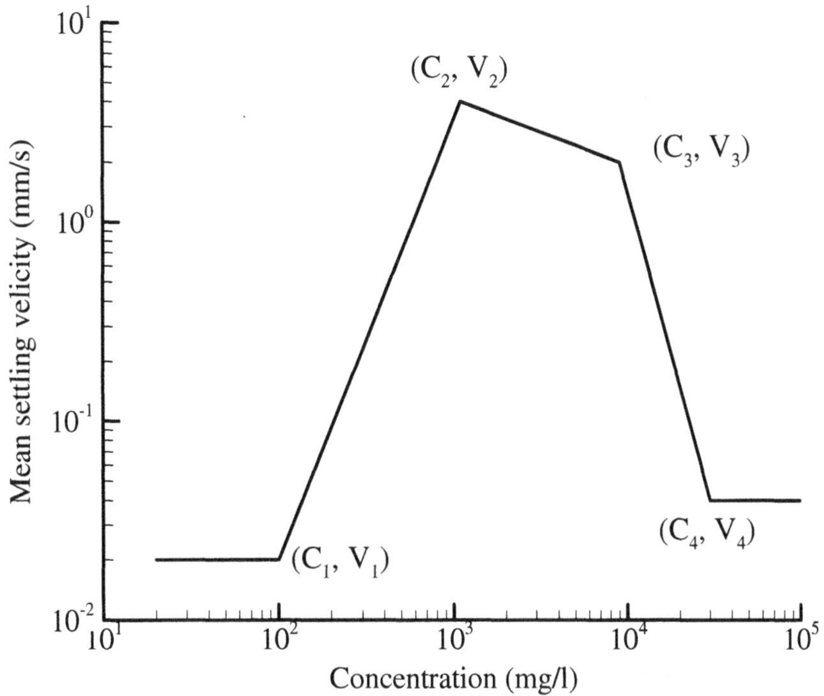

Figure 4.7. Input data illustration for settling velocity

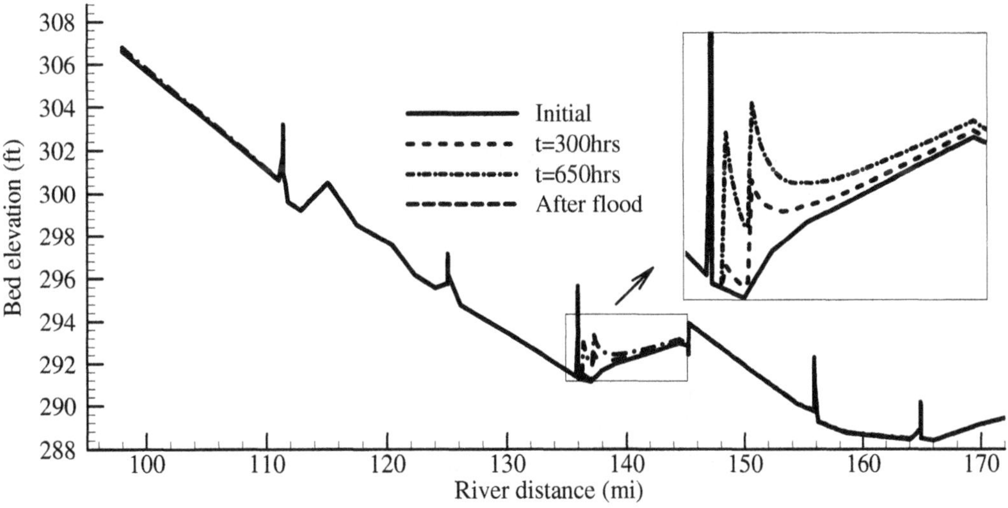

Figure 4.8. Bed elevation change of the SLC before and after a flood event.

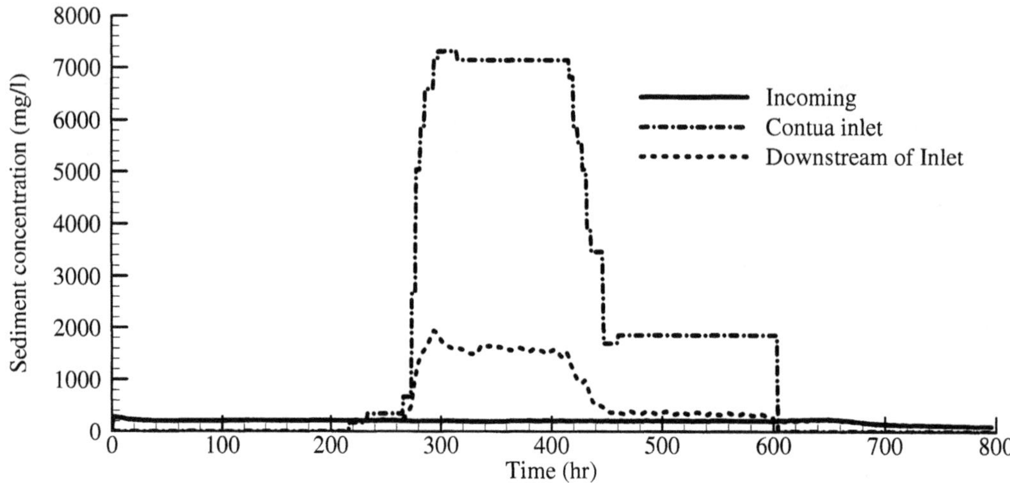

Figure 4.9. Sediment concentration changes with time at the four barrel inlet of the SLC.

4.5 Summary

Considerable advances in predicting cohesive sediment transport and related water quality have been made in the past decade. There are several numerical models that can be used to predict cohesive sediment transport movement, but they are subject to great uncertainty unless detailed measurements of important parameters can be made. In particular, it is necessary to measure the critical shear stresses for deposition and erosion. This chapter only summaries some basic processes involved in cohesive sediment transport. The methods summarized here are far from exhaustive.

It should also be noted that many traditional methods for non-cohesive sediment transport are also valid in cohesive sediment transport. For example, bathymetric analyses can be used to infer sediment movement, and geochronologic analyses can be used to age sediments. These types of field data collection are invaluable for validating numerical models.

4.6 References

Amos C.L., N. van Wagoner, and G.R. Daborn (1988). "The Influence of Subaerial Exposure on the Bulk Properties of Fine-Grained Intertidal Sediment From Minas Basin, Bay of Fundy," *Estuarine Coastal Shelf Sci.*, vol. 27, pp. 1-13.

Ariathurai, R. (1974). "A Finite Element Model for Sediment Transport in Estuaries," PhD dissertation, Univ. of California, Davis, California.

Ariathurai, R., and K. Arulanandan (1978). "Erosion Rates of Cohesive Soils," *Journal of the Hydraulics Division*, Proceedings of the ASCE, vol. 104, no. HY2, February, pp. 279-283.

Ariathurai, R., and Krone, R.B. (1976). "Finite element model for cohesive sediment transport," *Journal of Hydraulic Division*, ASCE, vol. 102, no 3, pp. 323-338.

Arulanandan, K., P. Loganathan, and R.B. Krone (1975). "Pore and Eroding Influences on Surface Erosion of Soil," *Journal. of the Geotechnical Engineering Division*, Proceedings of the ASCE, vol. 101 (GTI), pp. 51-66.

Atterberg, A. (1911). "Lerornas Főrhallande till Vatten, deras Plasticitetsgränser och Plasticitetsgrader," ("The Behavior of Clays with Water, Their Limits of Palsticity and Their Degrees of Plasticity,") *Kungliga Lantbruksakademiens Handlingar och Tidskrift,* vol. 50, no. 2, pp. 132-158: also in *Internationale Mitteilungen fűr Bodenkunde*, vol. 1, pp. 10-43 ("Uber die Physikalische Bodenuntersuchung und űber die Plastizität der Tone").

Bennett, R.H., N.R. O'Brien, and M.H. Hulbert (1991). "Determinants of Clay and Shale Microfabric Signatures: Processes and Mechanisms," in *Microstructure of Fine-grained Sediments, from Mud to Shale*, R. H. Bennett and M.R. Hulbert (eds.), Springer-Verlag, New York, New York.

Black, K.S. (1997). "Microbiological Factors Contributing to Erosion Resistance in Natural Cohesive Sediments," in *Cohesive Sediments*, N. Burt, R. Parker and J. Watts (eds.) John Wiley and Sons, pp. 231-244.

Boates, J.S., and P.C. Smith (1988). "Crawling Behaviour of the Amphipod *Corophium volutator* and Foraging by Semipalmated Sandpipers, *Calidris pusilla*," *Can. Journal of Zoology*, vol. 67, pp. 457-462.

Brekhovskikh, V.F., V.K. Debolsky, G.N. Vishnevskaya, and N.S. Zolotareva (1991). "Erosion of Cohesive Bottom Sediments: The Influence of the Benthos," *Journal of Hydraulic Research*, vol. 29, no. 2, pp. 149-160.

Burban, P.Y., Y.U. Xu, J. McNeil, and W. Lick (1990). "Settling Speeds of Flocs in Fresh Water and Seawater," *Journal of Geophysics Research*, vol. 95, no. 10, pp. 18213-18200.

Burt, T.N. (1990). "Cohesive Sediment and Physical Models," *Int. Conf. on Phys. Modeling of Transport*, MIT, Cambridge, Massachusetts.

Chapuis, R.P., and T. Gatien (1986). "An improved rotating cylinder technique for quantitative measurements of the scour resistance of clays," *Can. Geotechnical Journal*, vol. 23, pp. 83-87.

Cole, P., and G.V. Miles (1983). "Two-Dimensional Model of Mud Transport," *Journal of Hydraulic Engineering*, ASCE, vol. 109, no. 1, pp. 1-12.

Cormault, P. (1971). "Détermination Expérimentale du Débit Solide D'érosion Fins Cohésifs," 14e Congrés de l'A.I.R.H., Paris.

Cornelisse, J.M., H.P.J. Mulder, E.J. Houwing, H. Williamson, and G. Witte (1994). "On the Development of Instruments for In-Situ Erosion Measurements," in *Cohesive Sediments*, N. Burt, R. Parker, and J. Watts (eds.), John Wiley and Sons Ltd., pp. 175-186.

Daborn, G.R. (1993). "An Ecological Cascade Effect: Migratory Birds Affect Stability of Intertidal Sediments," *Limnology and Oceanography*, vol. 38, no. 1, pp. 225-231.

Delo, E.A. (1988). "The Behavior of Estuarine Muds During Tidal Cycles," *Rep. SR 138*, Hydraulic Research Station, Wallingford, UK.

Dennett, K.E., T.W. Sturm, A. Amirtharajah, and T. Mahmood (1998). "Effects of Adsorbed Natural Organic Matter on the Erosion of Kaolinite Sediments," *Water Environment Research*, vol. 70, no. 3, pp. 268-275.

Fukuda, M.K., and W. Lick (1980). "The Entrainment of Cohesive Sediments in Fresh Water," *Journal of Geophysical Research*, vol. 85, no. C5, pp. 2813-2824.

Gailani, J., C.K. Ziegler, and W. Lick (1991). "Transport of Suspended Solids in the Lower Fox River," *Journal of Great Lakes Research*, vol. 17, no. 4, pp. 479-494.

Gust, G., and M.J. Morris (1989). "Erosion Thresholds and Entrainment Rates of Undisturbed In-Situ Sediments," Journal of Coastal Research, Special Issue 5, Ft. Lauderdale, Florida.

Hamrick, J. (2001). "EFDC1D, A One Dimensional Hydrodynamic and Sediment Transport Model for River and Stream Networks - Model Theory and Users Guide," Tetra Tech, Inc., Fairfax, Virginia.

Hayter, E.J. (1983). "Prediction of Cohesive Sediment Movement in Estuarial Waters," Ph.D. Thesis, the University of Florida at Gainesville.

Hayter, E.J., and C.V. Pakala (1989). "Transport of Inorganic Contaminants in Estuarial Water," *Journal of Coastal Research*, Special Issue, no. 5, pp. 217-230.

Hayter, E.J., M.A. Bergs, R. Gu, S.C. McCutcheon, S.J. Smith, and H.J. Whiteley (1999). "HSCTM-2D, A Finite Element Model for Depth-Averaged Hydrodynamics, Sediment and Contaminant Transport," Report, National Exposure Research Laboratory, Office of Research and Development, U.S. EPA, Athens, Georgia.

Hicklin, P.W., and P.C. Smith (1984). "Selection of Foraging Sites and Invertebrate Prey by Migrant Palmated Sandpipers *Caldris pusilla* in Minas Basin, Bay of Fundy," *Can. Journal of Zoology,* vol. 62, pp. 2201-2210.

Hilldale, R.C. (2001). "Fluvial Erosion of Cohesive Sediments Considering Turbulence and Secondary Flow," M.S. Thesis, Dept. of Civil and Environmental Engineering, Washington State University, Pullman, Washington.

Houwing, E.J., and L.C. van Rijn (1995). "In-Situ Determination of the Critical Bed Shear Stress for Erosion of Cohesive Sediments," *Proceedings of the Coastal Engineering Conference*, vol. 2, Kobe, Japan, October 1994, pp. 2058-2069.

Hwang, K.N., and A.J. Mehta (1989). "Fine Sediment Erodibility in Lake Okeechobee," Coastal and Oceanographic Engineering Dept., Univ. of Florida, Report UFL/COEL-89/019, Gainesville, Florida.

Johansen, C., T. Larsen, and O. Petersen (1997). "Experiments on Erosion of Mud From the Danish Wadden Sea," in *Cohesive Sediments*, N. Burk, R. Parker, and J. Watts (eds.), John Wiley & Sons, pp. 305-314.

Kamphuis, J.W., and K.R. Hall (1983). "Cohesive Material Erosion by Unidirectional Current," *Journal of Hydraulic Engineering*, vol. 109, no. 1, January, pp. 49- 61.

Kelly, W.E., and R.C. Gularte (1981). "Erosion Resistance of Cohesive Soils," *Journal of the Hydraulics Division*, Proceedings of the ASCE, vol. 107, no. HY10, October, pp. 1211-1224.

Klumpp, C., J. Huang, and B.P. Greimann (2003). "Sediment Model of the Arroyo Pasajero and California Aqueduct," Bureau of Reclamation Report prepared for the CA Dept. of Water Resources, in preparation. (still in preparation)?

Kuijper, C., J.M. Cornelisse, and J.C. Winterwerp (1989). "Research on Erosive Properties of Cohesive Sediments," *Journal of Geophysical Research*, vol. 94, no. 10, October, pp. 14341 – 14350.

Krishnappan, B.G. (1993). "Rotating Cylinder Flume," *Journal Hydraulic Engineering*, vol. 119, no. 6, pp. 758-767.

Krone, R.B. (1962). "Flume Studies of the Transport of Sediment in Estuarial Shoaling Processes," Technical Report, Hydraulic Engineering Laboratory, University of California, Berkeley California.

Laursen, E.M. (1958). "The total sediment load of streams," *Journal of Hydraulic Division, ASCE*, vol. 84(1), pp. 1531-1536.

Letter, J.V., B.P. Donnell, W.H. McAnally, and W.A. Thomas (2000). "Users Guide to SED2D WES version 4.5," U.S. Army, Engineer Research And Development Center, Waterways Experiment Station, Coastal and Hydraulics Laboratory.

Li, M.Z., and C.L. Amos (1995). "SEDTRANS92: A Sediment Transport Model for Continental Shelves," *Computers and Geosciences*, vol. 21, no. 4, pp. 533-554.

Lick, W., and J. Lick (1988). "Aggregation and Disaggregation of Fine-Grained Lake Sediments," *Journal of Great Lakes Research*, vol. 14, no. 4, pp. 514-523.

Maa, J.P.Y. (1989). "The Bed Shear Stress of an Annular Sea-Bed Flume," *Proc., Water Quality Management*, Hamburg, Germany, pp. 271-276.

Manzenrider, H. (1983). Retardation of Initial Erosion Under Biological Effects in Sandy Tidal Flats. Leichtweiss, Inst. Tech. University Branschweig, 423 -435.

Masch, F.D., W.H. Espey Jr., and W.L. Moore (1963). "Measurements of the Shear Resistance of Cohesive Sediments," *Proceedings of the Federal Inter-Agency Sedimentation Conference*, Agricultural Research Service, Publication No. 970, Washington DC., pp. 151-155.

McAnally, W.H., and A.J. Mehta (2001). "Collisional Aggregation of fine Estuarial Sediment," Coastal and Estuarine Fine Sediment Processes, *Proceedings in Marine Science 3*.

McNeil, J., C. Taylor, and W. Lick (1996). "Measurements of Erosion of Undisturbed Bottom Sediments With Depth," *Journal of Hydraulic Engineering*, vol. 122, no. 6, pp. 316-324.

Meadows, P.S., and A. Meadows (1991). *The Environmental Impact of Burrows and Burrowing Animals,* Proceedings of a Symposium of the Zoological Society of London, Clarendon Press, Oxford.

Meadows, P.S., and J. Tait (1985). "Bioturbation, Geotechnics and Microbiology at the Sediment-Water Interface in Deep Sea Sediments," *Proceedings of the 19th European Marine Biology Symposium,* Gibbs, P.E. (ed.), Cambridge University Press, pp. 191-199.

Meadows, P.S., and J. Tait (1989). "Modification of Sediment Permeability and Shear Strength by Two Burrowing Invertebrates," *Marine Biology*, vol. 101, pp. 75-82.

Meadows, P.S., J. Tait,, and S.A. Hussain (1991). "Effects of Estuarine Infauna on Sediment Stability and Particle Sedimentation," *Hydrobiologia*, vol. 190, pp. 263-266.

Mehta, A.J., and E. Partheniades (1973). "Depositional Behavior of Cohesive Sediments," *Tech report No. 16*, Univ. of Florida, Gainesville, Florida.

Mehta, A.J., E.J. Hayter, W.R. Parker, R.B. Krone, and A.M. Teeter (1989). "Cohesive Sediment Transport. I: Process Description," *Journal of Hydraulic Engineering*, vol. 115, no. 8, pp. 1076-1093.

Millar, R.G., and M.C. Quick (1998). "Stable width and depth of gravel bed rivers with cohesive banks," *Journal of Hydraulic Engineering*, vol. 124(10), 1005-1013.

Molinas, A., and C.T. Yang (1985). "Generalized water surface profile computations," *Journal of the Hydraulic Diversion, ASCE*, vol. 111, no. HY3, pp. 381-397.

Moody, L.F. (1944). "Friction Factors for Pipe Flow," *Trans. ASME*, vol. 66.

Nachtergaele, J., and J. Poesen (2002). "Spatial and Temporal Variations in Resistance of Loess-Derived Soils to Ephemeral Gully Erosion," *European Journal of Soil Science*, vol. 53, September, pp. 449-463.

Nezu, I., and H. Nakagawa (1993). "*Turbulence in Open-Channel Flows,*" International Association for Hydraulic Research, Monograph Series, A.A. Balkema, Rotterdam, The Netherlands.

Nicholson, J., and B.A. O'Connor (1986). "Cohesive Sediment Transport Model," *Journal of Hydraulic Engineering*, vol. 112, no. 7, pp. 621-640.

Odd, N.V.M., and M.W. Owen (1972). "A Two-Layer Model for Mud Transport in the Thames Estuary," *Proc. of the Institution of Civil Engineers*, London, Supplement (ix), pp. 175-205.

Onishi, Y. (1981). "Sediment-Containment Transport Model," *Journal of Hydraulic Division*, ASCE, vol. 107, no. 9, pp. 1089-1107.

Parchure, T.M., and A.J. Mehta (1985). "Erosion of Soft Cohesive Sediment Deposits," *Journal of Hydraulic Engineering*, ASCE, vol. 111, no. 10, October, pp. 1308-1326.

Partheniades, E. (1962). *A Study of Erosion and Deposition of Cohesive Soils in Salt Water*, Ph.D. Thesis, University of California, Berkeley.

Partheniades, E. (1965). "Erosion and Deposition of Cohesive Soils," *Journal of the Hydraulics Division, Proceedings of the ASCE*, vol. 91, no. HY1, January, pp. 105-139.

Partheniades, E., J.F. Kennedy, R.J. Etter, and R.P. Hoyer (1966). "Investigations of the Depositional Behavior of Fine Cohesive Sediments in an Annular Rotating Channel.," *Hydrodynamics Lab Report No. 96*, MIT, Cambridge Massachusetts.

Paterson, D.M. (1994). "Microbiological Mediation of Sediment Structure and Behaviour," in *Microbial Mats* NATO ASI, vol. G 35, P Caumette and L.J. Stal (eds.), Springer-Verlag, pp. 97-109.

Paterson, D.M. (1997). "Biological Mediation of Sediment Erodibility: Ecology and Physical Dynamics," in *Cohesive Sediments,* N. Burt, R. Parker, and J. Watts (eds.), John Wiley and Sons, pp. 215-229.

Paterson, D.M., and G.R. Daborn (1991). "Sediment Stabilization by Biological Action: Significance for Coastal Engineering," in *Developments in Coastal Engineering,,* Peregrine, D. H. and J.H. Loveless (eds.), University of Bristol Press, pp. 111-119.

Raudkivi, A.J. (1998). *Loose Boundary Hydraulics*, A.A. Balkema, Rotterdam, The Netherlands, pp. 271-311.

Ravens, T.M., and P.M. Gschwend (1999). "Flume Measurements of Sediment Erodibility in Boston Harbor," *Journal of Hydraulic Engineering*, vol. 125, no. 10, pp. 998-1005.

Ravisanger, V., K.E. Dennett, T.W. Sturm, and A. Amirtharajah (2001). "Effect of Sediment pH on Resuspension of Kaolinite Sediments," *Journal of Environmental Engineering*, vol. 127, no. 6, June, pp. 531-538.

Roberts, J, R. Jepsen, D. Gotthard, and W. Lick (1998). "Effects of Particle Size and Bulk Density on Erosion of Quartz Particles," *Journal of Hydraulic Engineering*, vol. 124, no. 12, pp. 1261-1267.

Scarlatos, P.D. (1981). "On the Numerical Model of Cohesive Sediment Transport," *Journal of Hydraulic Research*, vol. 19, no. 1, pp. 61-68.

Schnitzer, M., and S.U. Khan (1972). *Humic Substances in the Environment,* Marcel Dekker, New York, New York.

Sheng, Y.P. (1983). Mathematical Modeling of Three-Dimensional Coastal Currents and Sediment Dispersion: Model Development and Application, technical report CERC-83-2, U.S. Army WES, Vicksburg, Mississippi.

Sheng, Y.P. (1988). "Consideration of Flow in Rotating Annuli for Sediment Erosion and Deposition Studies," *Journal of Coastal Research*, vol. S15, pp. 207-216.

Sheng, Y.P. (1986). "A Three-Dimensional Mathematical Model of Coastal, Estuarine and Lake Currents Using Boundary-Fitted Grid," Report No. 585, A.R.A.P. Group of Titan Systems, Princeton, New Jersey.

Sherard, J.L., R.S. Decker, and N.L. Ryka (1972). "Piping in Earth Dams of Dispersive Clay," presented at the ASCE Soil Mechanics and Foundation Conference, Perdue University, Lafayette, IN, June 12-14.

Shrestha, P.L., and G.T. Orlob (1996). "Multiphase Distribution of Cohesive Sediments and Heavy Metals in Estuarine Systems," *Journal of Environmental Engineering*, ASCE, vol. 122, no. 8, pp. 730-740.

Smerdon, E.T., and R.P. Beasley (1959). "The Tractive Force Theory Applied to Stability of Open Channels in Cohesive Soils," *Research Bulletin 715*, University of Missouri Agricultural Experiment Station, Columbia, Missouri.

Spasojevic, M., and F.M. Holly (1996). "Cohesive Sediment Capability in CH3D Phases II and III: Formulation and Implementation," IIHR Report no. 386, Iowa Institute of Hydraulic Research, The University of Iowa, Iowa City, Iowa.

Teisson, C., and B. Latteux (1986). "A Depth-Integrated Bidimensional Model of Suspended Sediment Transport," Proceedings., 3d International Symposium On River Sedimentation, S.Y. Wang, H.W. Shen, and L.Z. Ding (eds.), Jackson, MS, pp. 421-429.

Teisson, C., M. Ockenden, P. Le Hir, C. Kraneberg, and L. Hamm (1993). "Cohesive Sediment Transport Processes," *Coastal Engineering,* Elsevier Publishers, vol. 21, pp. 129-162.

Tetra Tech, Inc. (2000). "EFDC Technical Memorandum: Theoretical and Computational Aspects of Sediment Transport in the EFDC Model, 2nd Draft," Tetra Tech, Inc., Fairfax, Virginia.

Thomann, R.V., and D.M. DiToro (1983). "Physico-Chemical Model of Toxic Substances in the Great Lakes," CR805916 and CR 807853, U.S. Environmental Protection Agency (EPA), Large Lakes Res. Station, Grosse Isle, Michigan.

Thorn, M.F.C. (1981). "Physical Processes of Siltation in Tidal Channels," *Proceedings of the Conference on Hydraulic Modelling Applied to Maritime Engineering Problems*, Institution of Civil Engineers, London, England, 1981, pp. 47-55.

Thorn, M.F.C., and J.G. Parsons (1977). "Properties of Grangemouth Mud," Report No. EX 781, Hydraulics Research Station, Wallingford, England, July.

Thorn, M.F.C., and J.G. Parsons (1980). "Erosion of Cohesive Sediments in Estuaries," *Third Int. Symp. On Dredging Technology*, Bordeaux, France.

Thurman, E.M., and R.L. Malcolm (1983). "Structural Study of Humic Substances: New Approaches and Methods," in *Aquatic and Terrestrial Humic Substances,* R.F. Christman and E.T. Gjessing (eds.), Ann Arbor Science, Ann Arbor, Michigan.

U.S. Army Corps of Engineers (1993). The Hydraulic Engineering Center, *HEC-6 Scour and Deposition in Rivers and Reservoirs, User's Manual,* March 1977 (revised 1993).

van Leussen, W. (1994). "Estuarine Macroflocs and Their Role in Fine-Grained Sediment Transport," "Ph.D. thesis, Utrecht University (NL).

van Rijn, L. (1993). Principles of Sediment Transport in Rivers, Estuaries and Coastal Seas, Aqua Publications, The Netherlands.

Vermeyen, T. (1995). "Erosional and Depositional Characteristics of Cohesive Sediments Found in Elephant Butte Reservoir, New Mexico," Technical Report R-95-15, Water Resources Services, Technical Service Center, Bureau of Reclamation, Denver, CO.

Verwey, E.J.W., and J.T.G. Overbeek (1948). *Theory of the Stability of Hydrophobic Colloids*, Elsevier Publishing Co., New York, New York.

Westrich, B., R. Scharf, and V. Schürlein (1997). "Measurement of Cohesive Sediment Erodibility in a Laboratory Flume," *Environmental and Coastal Hydraulics: Protecting the Aquatic Habitat*, Proceedings of the 27th Congress of the IAHR, Part B-1, Aug. 10-15, San Francisco, pp. 209-215.

Winterwerp, J.C., J.M. Cornelisse, and C. Kuijper (1990). "Parameters to Characterize Natural Muds," in Abstract Volume, Int. Workshop on Cohesive Sediments, Brussels, KBIN, Brussels, 103-105.

Winterwerp, J.C., J.M. Cornelisse, and C. Kuijper (1993). "A Laboratory Study on the Behavior of Mud from the Western Scheldt Under Tidal Conditions," *Coastal and Estuarine Studies*, vol. 42, Nearshore and Estuarine Sed. Transport, AGU.

Yang, C.T., and F.J.M. Simões (2000). User's Manual for GSTARS 2.1 (Generalized Sediment Transport model for Alluvial River Simulation version 2.1), Bureau of Reclamation, Technical Service Center, Denver, Colorado.

Yang, C.T., and F.J.M. Simões (2002). *User's Manual for GSTARS3 (Generalized Sediment Transport Model for Alluvial River Simulation version 3.0),* Bureau of Reclamation, Technical Service Center, Denver, Colorado.

Yang, C.T., J. Huang, and B.P. Greimann (2004, 2005). *User's manual for GSTAR-1D (Generalized Sediment Transport model for Alluvial Rivers - One Dimension)*, version 1.0, U.S. Bureau of Reclamation, Technical Service Center, Denver, Colorado.

Young, R.A., and J.B. Southard (1978), "Erosion of Fine-Grained Sediments: Seafloor and Laboratory Experiments," *Bulletin of the American Geological Society*, vol. 89, pp. 663-672.

Ziegler, C.K., and W. Lick (1986). "A Numerical Model of the Resuspension, Deposition and Transport of Fine-Grained Sediments in Shallow Water," UCSB Rep. ME-86-3, University of California, Santa Barbara, California.

Ziegler, C.K., and B.S. Nisbet (1994). "Fine-Grained Sediment Transport in Pawtuxet River, Rhode Island," *Journal of Hydraulic Engineering*, ASCE, vol. 120, no. 5, pp. 561-576.

Ziegler, C.K., and B.S. Nisbet (1995). "Long-Term Simulation of Fine-Grained Sediment Transport in a Large Reservoir," *Journal of Hydraulic Engineering*, vol. 121, no. 11, pp. 773-781.

Zreik, D.A., B.G. Krishnappan, J.T. Germaine, O.S. Madsen, and C.C. Ladd (1998). "Erosional and Mechanical Strengths of Deposited Cohesive Sediments," *Journal of Hydraulic Engineering*, vol. 124, no. 11, pp. 1076 – 1085.

Chapter 5
Sediment Modeling for Rivers and Reservoirs

Page

5.1 Introduction ... 5-1
5.1.1 The Numerical Modeling Cycle ... 5-1
5.2 Mathematical Models ... 5-3
5.2.1 Three-Dimensional Models ... 5-3
5.2.2 Two-Dimensional Models ... 5-6
5.2.3 One-Dimensional Models ... 5-9
5.2.4 Bed Evolution ... 5-11
5.2.5 Auxiliary Equations ... 5-16
5.2.5.1 Flow Resistance ... 5-16
5.2.5.2 Sediment Transport ... 5-23
5.3 Numerical Solution Methods ... 5-26
5.3.1 Finite Difference Methods ... 5-27
5.3.2 Finite Element Methods ... 5-30
5.3.3 Finite Volume Methods ... 5-31
5.3.4 Other Discretization Methods ... 5-32
5.4 Modeling Morphologic Evolution ... 5-34
5.5 Reservoir Sedimentation Modeling ... 5-40
5.5.1 Reservoir Hydraulics ... 5-41
5.5.2 Sediment Transport in Reservoirs ... 5-44
5.5.3 Turbid Underflows ... 5-47
5.5.3.1 Plunge Point ... 5-48
5.5.3.2 Governing Equations ... 5-51
5.5.3.3 Additional Relationships ... 5-54
5.5.4 Difference Between Reservoirs and Other Bodies of Water ... 5-58
5.6 Data Requirements ... 5-59
5.7 One-Dimensional Model Comparison ... 5-62
5.8 Example: The GSTARS Models ... 5-62
5.8.1 Streamlines and Stream Tubes ... 5-64
5.8.2 Backwater Computations ... 5-65
5.8.3 Sediment Routing ... 5-67
5.8.4 Total Stream Power Minimization ... 5-73
5.8.5 Channel Side Slope Adjustments ... 5-74
5.8.6 Application Examples ... 5-75
5.9 Summary ... 5-82
5.10 References ... 5-83

Chapter 5
Sedimentation Modeling for Rivers and Reservoirs

by

Francisco J.M. Simões and Chih Ted Yang

5.1 Introduction

The study of natural river changes and the interference of man in natural water bodies is a difficult but important activity, as increasing and shifting populations place more demands on the natural sources of fresh water. Although the basic mechanical principles for these studies are well established, a complete analytical solution is not known but for the most basic cases. The complexity of the flow movement and its interaction with its boundaries, which are themselves deformable, have precluded the development of closed form solutions to the governing equations that describe the mechanical behavior of fluid and solid-fluid mixtures. As a result, alternative techniques have been developed to provide quantitative predictions of these phenomena as an aid to engineering projects and river restoration efforts. Modeling is one such technique.

There are two types of models: mathematical models and physical models (sometimes also called scale models). This chapter provides an overview of mathematical and numerical modeling, which is based on computation techniques, as opposed to physical modeling, which is based on traditional laboratory techniques and measurement.

Numerical modeling has become very popular in the past few decades, mainly due to the increasing availability of more powerful and affordable computing platforms. Much progress has been made, particularly in the fields of sediment transport, water quality, and multidimensional fluid flow and turbulence. Many computer models are now available for users to purchase. Some of the models are in public domain and can be obtained free of charge. Graphical user interfaces, automatic grid generators, geographic information systems, and improved data collection techniques (such as LiDAR, Light Distancing and Ranging) promise to further expedite the use of numerical models as a popular tool for solving river engineering problems.

5.1.1 The Numerical Modeling Cycle

In general, numerical models are used for the same reasons as physical models; i.e., the problem at hand cannot be solved directly for the prototype. The process from prototype data to the modeling and to final interpretation of the results (i.e., the modeling cycle) is complex and prone to many errors. Careful engineering judgment must be exercised at every step. The modeling cycle is schematically represented in Figure 5.1.

The prototype is the reality to be studied. It is defined by data and by knowledge. The data represents boundary conditions, such as bathymetry, water discharges, sediment particle size distributions, vegetation types, etc. The knowledge contains the physical processes that are known to determine the system's behavior, such as flow turbulence, sediment transport mechanisms, mixing processes, etc. Understanding the prototype and data collection constitute the first step of the cycle.

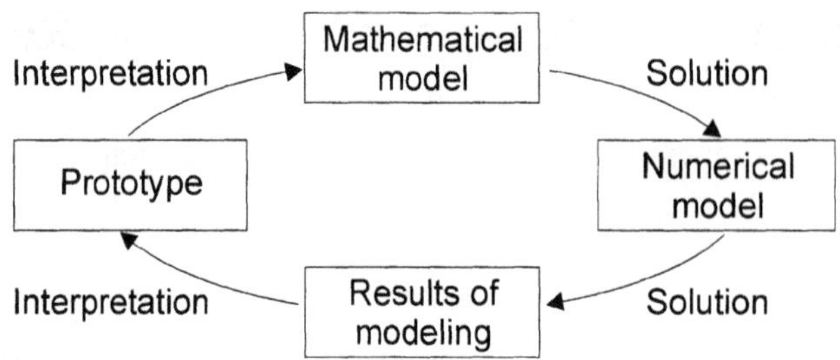

Figure 5.1. Computer modeling cycle from prototype to the modeling results. The cycle starts with the prototype being studied and ends with the interpretation of model results to withdraw conclusions about it.

In the first interpretation step, all the relevant physical processes that were identified in the prototype are translated into governing equations that are compiled into the mathematical model. A mathematical model, therefore, constitutes the first approximation to the problem. It is the prerequisite for a numerical model. At this time, many simplifying approximations are made, such as steady versus unsteady and one- versus two- versus three-dimensional formulations, simplifying descriptions of turbulence, etc. In water resources, one usually (but not always) arrives to the set-up of a boundary value problem whose governing equations contain partial differential equations and non-linear terms.

Next, a solution step is required to solve the mathematical model. The numerical model embodies the numerical techniques used to solve the set of governing equations that forms the mathematical model. In this step, one chooses, for example, finite difference versus finite element versus finite volume discretization techniques; selects the approach to deal with the non-linear terms; etc. Note that this is a further approximating step because the partial differential equations are transformed into algebraic equations, which are approximate but not equivalent to the former.

Another solution step involves the solution of the numerical model in a computer and provides the results of modeling. This step embodies further approximations and simplifications, such as those associated with unknown boundary conditions, imprecise bathymetry, unknown water and/or sediment discharges, friction factors, etc.

Finally, the data needs to be interpreted and placed in the appropriate prototype context. This last step closes the modeling cycle and ultimately provides the answer to the problem that drives the modeling efforts.

The choice of model for each specific problem should take into account the requirements of the problem, the knowledge about the system, and the data available. On one hand, the model must take into account all the significant phenomena that are known to occur in the system and that will influence the aspects that are being studied. On the other hand, model complexity is limited by the available data. At this time, there is no universal model that can be applied to every problem, and it may not even be desirable to have such a model. The specific requirements of each problem should be analyzed and the model chosen should reflect this analysis in its features and complexity. There is no lack of computer models for engineers to choose from. The success of a study depends, to a large degree, on the engineer's understanding of fluvial processes, associated theories, and the capabilities and limitations of computer models. In many cases, the selection of a modeler is more important than the selection of the computer model.

5.2 Mathematical Models

5.2.1 Three-Dimensional Models

The flow phenomena in natural rivers are three dimensional, especially those at or near a meander bend, local expansion and contraction, or a hydraulic structure. Turbulence is an essentially three-dimensional phenomenon, and three-dimensional models are particularly useful for the simulation of turbulent heat and mass transport. These models are usually based on the Reynolds-averaged form of the Navier-Stokes equations, using additional equations of varied degree of complexity for the turbulence closure.

The derivation of the governing equations can be found in many basic textbooks on fluid dynamics; therefore, they will only be presented here without further consideration. Interested readers are directed to textbooks such as the ones by White (1991) and by Warsi (1993). The Navier-Stokes equations represent the statement of Newton's second law for fluids (i.e., the conservation of momentum), and in the Cartesian coordinate system and for incompressible fluids, they can be written as

$$\frac{\partial u_i}{\partial t}+\frac{\partial u_i u_j}{\partial x_j}=\frac{F_i}{\rho}-\frac{1}{\rho}\frac{\partial p}{\partial x_i}+\frac{\partial}{\partial x_j}\left(\nu\frac{\partial u_i}{\partial x_j}-\overline{u_i' u_j'}\right) \tag{5.1}$$

where

i,j = Cartesian directions (= 1 for x, = 2 for y, and = 3 for z),
j = Cartesian directions perpendicular to i
u_i = Cartesian component of the velocity along the x_i direction (i = 1,2,3),
ρ = fluid density,
p = pressure,
F_i = component of the body forces per unit volume in the i-direction,
ν = kinematic molecular viscosity,
$-\rho\overline{u_i' u_j'}$ = turbulent stresses,

and the indexed summation convention is used (see Figure 5.2 for convention used). Equations (5.1) constitute a system of equations, one for each coordinate direction (i.e., for i = 1, 2, and 3).

The body forces include gravitational, buoyancy, and Coriolis forces, or any other body forces that may be present (such as magnetic forces in magnetohydrodynamic fluids). Additionally, in turbulent flows, the molecular viscosity term may be safely neglected in comparison with the turbulent stresses.

Conservation of mass is expressed by the continuity equation for incompressible fluids:

$$\frac{\partial u_i}{\partial x_i} = 0 \tag{5.2}$$

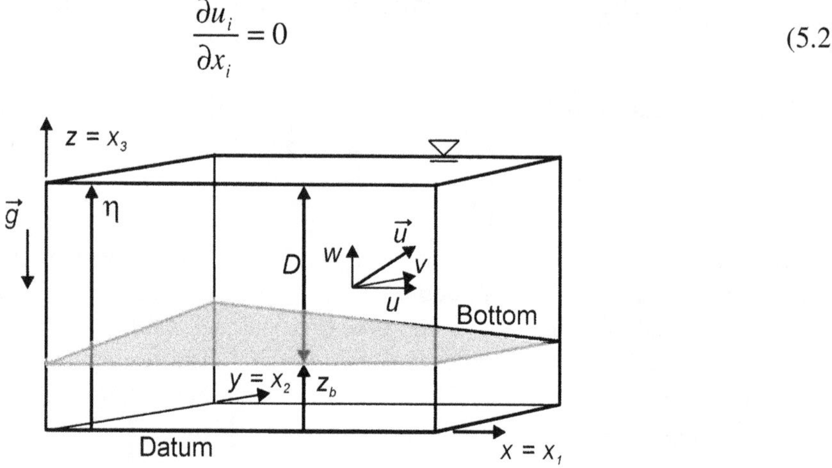

Figure 5.2. Sketch showing the coordinate system used and the definition of some of the variables. Note that $u = u_1$, $v = u_2$, *and* $w = u_3$.

The transport of constituents, such as dissolved and suspended solids, requires one equation per substance transported. This is a convection-diffusion type of equation that can be written in general as

$$\frac{\partial c}{\partial t} + u_i \frac{\partial c}{\partial x_i} = \frac{\partial}{\partial x_i}\left(\Gamma \frac{\partial c}{\partial x_i} - \overline{u_i' c'}\right) + S_c \tag{5.3}$$

where
c = scalar quantity per unit of mass,
Γ = molecular diffusivity coefficient,
$-\overline{u_i' c'}$ = turbulent diffusion of c, and
S_c = source/sink (i.e., creation/destruction) of c.

The turbulence terms ($-\rho\overline{u_i' u_j'}$ and $-\overline{u_i' c'}$) result from averaging the original Navier-Stokes equations using the Reynolds decomposition (Tennekes and Lumley, 1972) for a more detailed explanation about the technique) and require additional closure equations. One of the commonly used closure techniques is given by the k-ε model (Rastogi and Rodi, 1978), but there are many other alternative choices. The reader is directed to the turbulence modeling monograph by Rodi (1993) for further details about this subject.

In free surface flows, an additional equation is required to solve for the position of the free surface. A common technique is to use a rigid lid approximation in which the flow is solved in the same manner as pressurized flow by assuming a rigid frictionless boundary at the approximate position where the free surface is located. This eliminates the need to use an additional differential equation to compute the free surface position: the free surface location can be computed from the flow pressure by extrapolating (or interpolating) to the location where $p = p_a$, where p_a is the atmospheric pressure. Accuracy is lost when the free surface location differs significantly from the rigid lid location (say, by 10% or more of the flow depth), which may occur in bends and around obstacles.

The free surface elevation may also be computed by solving either the kinematic condition at the free surface,

$$\frac{\partial \eta}{\partial t} + u_s \frac{\partial \eta}{\partial x} + v_s \frac{\partial \eta}{\partial y} - w_s = 0 \tag{5.4}$$

or by using the depth-integrated continuity equation,

$$\frac{\partial D}{\partial t} + \frac{\partial (DU)}{\partial x} + \frac{\partial (DV)}{\partial y} = 0 \tag{5.5}$$

where η = free surface elevation,
u_s, v_s, w_s = components of the velocity vector at the free surface,
D = water depth, and
U, V = components of the depth-averaged velocity vector (to be defined in the next section).

The depth-averaged continuity equation offers the advantage of using the principle of mass conservation, therefore helping to enforce the incompressibility constraint. Note that the kinematic condition, Equation (5.4), is used in the derivation of Equation (5.5) (for details about the derivation of Equation (5.5) see Pinder and Gray, 1977). Furthermore, the use of a depth-averaged velocity in Equation (5.5), does not mean that the depth-averaged momentum equations (as described in the next section) need to be used, because U and V can be computed directly from the three-dimensional velocity field, as done by Simões (1995).

An important simplification to the system of Equations (5.1) is accomplished when the vertical acceleration terms can be neglected with respect to the pressure and body forces. In this case, the third momentum equation (z-momentum) reduces to

$$0 = -\frac{1}{\rho}\frac{\partial p}{\partial z} - g \tag{5.6}$$

where g = acceleration due to gravity (with a single component along the negative z- direction).

Equation (5.6) is the hydrostatic pressure approximation. This is a frequently used approximation in free surface flows, which is valid when the streamlines are only weakly curved in the vertical plane (i.e., they are nearly parallel to the bottom of the channel). Using the hydrostatic pressure approximation, the pressure gradient can be replaced by the free surface slope in neutrally stratified flows:

$$\frac{1}{\rho}\frac{\partial p}{\partial x}=g\frac{\partial \eta}{\partial x},\quad \frac{1}{\rho}\frac{\partial p}{\partial y}=g\frac{\partial \eta}{\partial y} \tag{5.7}$$

This allows the elimination of one unknown (the pressure p). The third momentum equation is not solved. Instead, the vertical component of the velocity at any vertical level z, w_z, is calculated directly from integrating Equation (5.2) along the vertical direction:

$$w_z=-\int_{z_b}^{z}\left(\frac{\partial u}{\partial x}+\frac{\partial v}{\partial y}\right)dz \tag{5.8}$$

Three-dimensional modeling is a very powerful tool in river engineering, but it also has high computational demands (i.e., faster computers with large memory space are needed). It also requires vast amounts of data for proper model setup, which takes time and is expensive to obtain. These requirements have, until recently, limited their use, but newer, faster, and more affordable computers, together with new data collection instrumentation, may overcome these limitations in the near future.

5.2.2 Two-Dimensional Models

Two-dimensional models for flow and sediment transport are becoming widely used due to the advent of fast personal computers and to the existence of a significant number of commercially available models.

Two-dimensional models can be classified into two-dimensional vertically averaged and two-dimensional horizontally averaged models. The former scheme is used where depth-averaged velocity or other hydraulic parameters can adequately describe the variation of hydraulic conditions across a channel. The latter scheme is used where width-averaged hydraulic parameters can adequately describe the variation of hydraulic conditions in the vertical direction. Most two-dimensional sediment transport models are depth-averaged models; hence, we focus on those in this section.

Two-dimensional, depth-averaged models result from vertically averaging the governing equations, Equations (5.1) and (5.2), after a few simplifying assumptions. First, integrating the continuity equation, making use of the kinematic condition at the free surface—Equation (5.4)—and the fact that the normal component of the velocity must vanish at the solid bed, one obtains

$$\frac{\partial D}{\partial t}+\frac{\partial (DU)}{\partial x}+\frac{\partial (DV)}{\partial y}=0 \tag{5.9}$$

where U and V = depth-averaged velocities defined as

$$U = \frac{1}{D}\int_{z_b}^{\eta} u dz \quad \text{and} \quad V = \frac{1}{D}\int_{z_b}^{\eta} v dz \tag{5.10}$$

where z_b = bed elevation (see Figure 5.2).

The momentum equation, Equation (5.1), can be averaged in the same way, but this time non-linear terms appear, such as

$$\int_{z_b}^{\eta} uv dz = DUV + \int_{z_b}^{\eta} (u-U)(v-V) dz \tag{5.11}$$

Including the Coriolis and pressure terms, whose integration is trivial, the depth-averaged momentum equations become

$$\frac{\partial(DU)}{\partial t} + \frac{\partial(DU^2)}{\partial x} + \frac{\partial(DUV)}{\partial y} = F_x + fDV - gD\frac{\partial \eta}{\partial x} - \frac{gD^2}{2\rho_0}\frac{\partial \rho}{\partial x} - \frac{\tau_{bx}}{\rho_0} + \frac{\partial(D\tau_{xx})}{\partial x} + \frac{\partial(D\tau_{xy})}{\partial y} \tag{5.12}$$

$$\frac{\partial(DV)}{\partial t} + \frac{\partial(DUV)}{\partial x} + \frac{\partial(DV^2)}{\partial y} = F_y - fDU - gD\frac{\partial \eta}{\partial y} - \frac{gD^2}{2\rho_0}\frac{\partial \rho}{\partial y} - \frac{\tau_{by}}{\rho_0} + \frac{\partial(D\tau_{xy})}{\partial x} + \frac{\partial(D\tau_{yy})}{\partial y} \tag{5.13}$$

where f = Coriolis parameter (= $2\Omega \sin\phi$),
Ω = angular rate of earth's revolution,
ϕ = geographic latitude,
F_i = driving forces ($i = x,y$),
ρ_0 = density of a reference state, and
τ_{bi} = bottom stresses ($i = x,y$).

The above equations are sometimes called the shallow-water equations or the depth-averaged Navier-Stokes equations. The cross-stresses τ_{ij} include viscous friction, turbulent friction, and the non-linear terms resulting from the vertical averaging process (e.g., Equation (5.11)), which are usually called the radiation stresses:

$$\tau_{ij} = \frac{1}{D}\int_{z_b}^{\eta}\left[\nu\left(\frac{\partial u_i}{\partial x_j} + \frac{\partial u_j}{\partial x_i}\right) - \overline{u_i' u_j'} + (u_i - U_i)(u_j - U_j)\right] dz \tag{5.14}$$

In most natural bodies of water, the molecular viscosity terms can be safely neglected in comparison with the turbulence terms. The radiation stresses are often neglected, but they represent important physical phenomena. For example, in bends they are at least partly responsible for shifting the high velocity part of the flow profile from the inner bank at the upstream region to the outer bank at the downstream region of the bend (Shimizu et al., 1991). In general, however, the terms of Equation (5.14) are collapsed in the form of diffusion coefficients and written as

$$\tau_{ij} = D_i\left(\frac{\partial U_i}{\partial x_j} + \frac{\partial U_j}{\partial x_i} - \frac{1}{2}\delta_{ij}\frac{\partial U_k}{\partial x_k}\right) \tag{5.15}$$

where δ_{ij} = Kronecker delta (= 1 if $i = k$, 0 otherwise), and
D_i = diffusion coefficient in the ith direction (in general, $D_1 = D_2 = D_H$).

In turbulent flow, the diffusion coefficients can be prescribed or computed from any of the many existing turbulence models (see Rodi (1993) for more details), and the bottom shear stresses are assumed to have the same direction of the depth-mean velocity and be proportional to the square of its magnitude:

$$\frac{\tau_{bx}}{\rho} = C_f u\sqrt{u^2+v^2} \quad \text{and} \quad \frac{\tau_{by}}{\rho} = C_f v\sqrt{u^2+v^2} \tag{5.16}$$

where C_f = standard friction coefficient $(C_f \approx 0.003)$.

Note that Equation (5.16) can also be written in terms of the Manning's roughness coefficient, n, or in terms of Chézy's roughness coefficient, C:

$$\frac{\tau_{bx}}{\rho} = \frac{gn^2}{D^{1/3}}u\sqrt{u^2+v^2} = \frac{g}{C^2}u\sqrt{u^2+v^2} \quad \text{and} \quad \frac{\tau_{by}}{\rho} = \frac{gn^2}{D^{1/3}}v\sqrt{u^2+v^2} = \frac{g}{C^2}v\sqrt{u^2+v^2} \tag{5.17}$$

The driving forces remaining in Equations (5.12) and (5.13) include such effects as atmospheric pressure gradients, wind stresses, density gradients, and tidal stresses.

Finally, the vertically-averaged form for the transport of a dissolved or suspended (very fine particles) constituent is

$$\frac{\partial(Dc)}{\partial t} + \frac{\partial(DUc)}{\partial x} + \frac{\partial(DVc)}{\partial y} = \frac{\partial}{\partial x}\left(DK_x\frac{\partial c}{\partial x}\right) + \frac{\partial}{\partial y}\left(DK_y\frac{\partial c}{\partial y}\right) + S_c \tag{5.18}$$

where K_x, K_y = diffusion coefficients in the x- and y-directions, respectively, and
c = depth-averaged concentration.

In general, $K_x = K_y = K_H$, and K_H is directly related to D_H.

The shallow-water equations can be written in many possible forms. Those forms may include different terms than the ones considered above (corresponding to other physical effects), or they may be written in terms of curvilinear coordinates, for example. Many other aspects that are of interest, but that are outside the scope of this chapter, are described with much greater detail by Vreugdenhil (1994).

5.2.3 One-Dimensional Models

Most of the sediment transport models used in river engineering are one dimensional, especially those used for long-term simulation of a long river reach. One-dimensional models generally require the least amount of field data for calibration and testing. The numerical solutions are more stable and require the least amount of computer time and capacity. One-dimensional models are not suitable, however, for simulating truly two- or three-dimensional local phenomena.

One-dimensional models are usually based on the same conservation principles as the multi-dimensional models described in the previous two sections; i.e., the conservation of mass and momentum. Conservation of mass (continuity equation) can be expressed as

$$\frac{\partial A}{\partial t} + \frac{\partial Q}{\partial x} = q_l \tag{5.19}$$

where A = cross-sectional area of the flow,
Q = water discharge, and
q_l = lateral inflow per unit length.

Conservation of momentum:

$$\frac{\partial Q}{\partial t} + \frac{\partial}{\partial x}\left(\beta \frac{Q^2}{A}\right) + gA\frac{\partial \eta}{\partial x} + gA\left(S_f - S_0\right) = 0 \tag{5.20}$$

where S_f = friction slope,
S_0 = bed slope, and
β = momentum correction coefficient $(\beta \approx 1)$.

Equations (5.19) and (5.20) are known as the de Saint Venant equations. They assume that all the main variables are uniform across the cross-section, that the bed slope is small, and that all curvature effects are neglected. For a full discussion about these equations, including alternative forms, and for a detailed derivation, see Montes (1998).

The friction slope is assumed to be a function of the flow, such that

$$S_f = \frac{Q|Q|}{K^2} \tag{5.21}$$

where K = conveyance.

The conveyance is calculated using a resistance function, such as Manning's or Chézy's.

Special internal boundary conditions need to be considered in cases where flow is not well represented by the one-dimensional flow equations. Such situations may be encountered in flow over weirs or through gates. Some examples are presented in Table 5.1.

Table 5.1. Governing equations for one-dimensional flow for a number of special type of internal boundary conditions (h_a, h_b, and h_c denote the hydraulic head at points *a*, *b*, and *c*, respectively)

Flow type	Boundary conditions
Flow over weirs 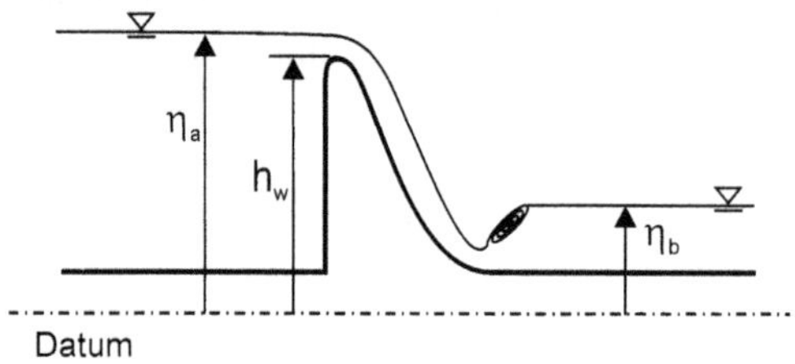	$Q_a = Q_b = f(\eta_a, h_w, \text{weir type and size})$
Flow through gates 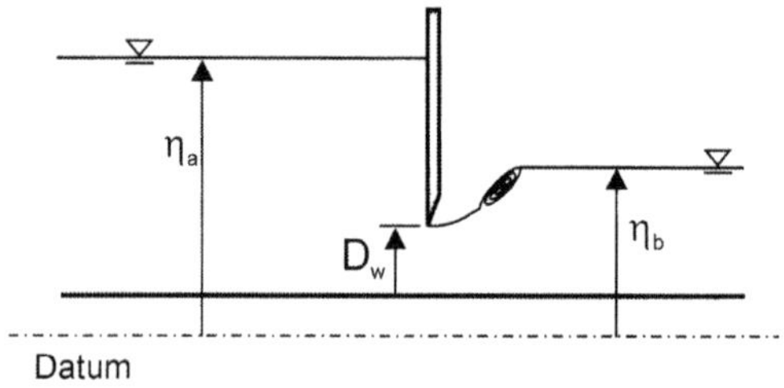	$Q_a = Q_b = f(\eta_a, \eta_b, D_w, \text{gate type, size, opening})$
Channel junctions 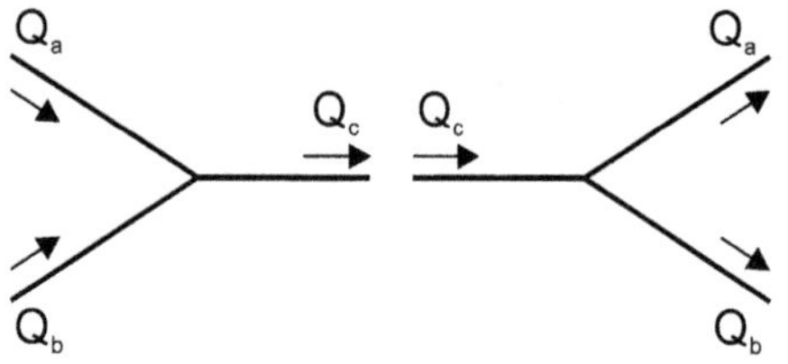	$Q_c = Q_a + Q_b$ $h_c + \frac{V_c^2}{2g} + h_l = h_a + \frac{V_a^2}{2g}$ $h_c + \frac{V_c^2}{2g} + h_l = h_b + \frac{V_b^2}{2g}$ where h_l is an energy loss coefficient
Channel contractions and expansions 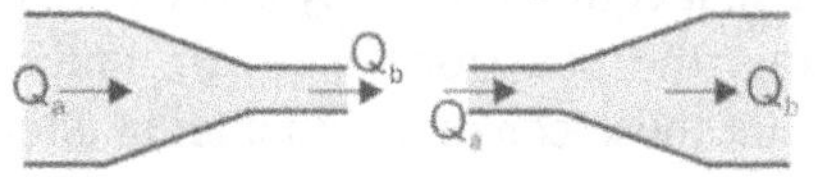	$Q_a = Q_b$ $h_b + \frac{V_b^2}{2g} + h_l = h_a + \frac{V_a^2}{2g}$

Usually ignored by most models is the fact that the physical coordinate x is not the same as the local coordinate that follows (it is tangent to) the streamline direction, s: x is a distance in an unchanging coordinate system, while s is the true distance traveled by the water. In fact, the equations above are correct only if $dx = ds$ (i.e., if the ratio of channel length (s) to the downstream distance (x) remains equal to 1). That is not the case for most riverflows, especially in the case of large increases in discharge over highly sinuous paths. As the flow rates increase and the stages rise, the main body of water tends to assume different paths, especially in channels with compound cross-sections. In those circumstances, DeLong (1989) has shown that the appropriate metric coefficient relating the true channel distance to the reference length, defined by

$$M_a = \frac{1}{A} \int_A \frac{dx}{ds} dA \tag{5.22}$$

needs to be added to Equation (5.19). This coefficient represents an area-weighted sinuosity. Similarly, a flow-weighted sinuosity coefficient, defined by

$$M_q = \frac{1}{Q} \int_Q \frac{dx}{ds} dQ \tag{5.23}$$

needs to be incorporated into Equation (5.20). In Equation (5.23), dQ represents the increment in discharge corresponding to the incremental area dA. The reader is referred to DeLong (1989) for more details. However, it should be pointed out that momentum is a vector quantity; therefore, (unlike mass) its conservation cannot be enforced by a single scalar quantity describing the motion along a sinuous streamline.

5.2.4 Bed Evolution

In systems with boundaries that are subject to deposition and/or scour, it is necessary to model the movement of the sediment particles with the flow. Sediment transport modeling is a complex topic and is subject to much uncertainty. Sediment transport science has been covered in Chapters 3 and 4. In this section, some of the issues related to modeling these processes are presented.

Just as for the fluid flow, the mathematical model for sediment transport is usually based on conservation laws; i.e., conservation of suspended sediment load, bedload, and bed-material for each size fraction class. A number of additional auxiliary equations are needed, such as for the bed-material sorting (the process of exchange of sediment particles between the water stream and some conceptual model of a layered bed), bed resistance, sediment transport capacity, etc. Without being exhaustive, but keeping a good level of generality, the set of basic one-dimensional differential equations governing the transport of sediments can be written in the following manner:

Suspended-load transport:

$$\frac{\partial}{\partial t}(AC_j)+\frac{\partial}{\partial x}(QC_j)=\frac{\partial}{\partial x}\left(D_L A\frac{\partial C_j}{\partial x}\right)+\Phi_{s,j} \tag{5.24}$$

Bedload transport:

$$\frac{\partial G_j}{\partial t}+u_{bj}\frac{\partial G_j}{\partial x}=u_{bj}\Phi_{b,j} \tag{5.25}$$

Conservation of bed-material (sediment continuity):

$$\frac{\partial(\bar{C}A)}{\partial t}+(1-p)\frac{\partial A_s}{\partial t}+\frac{\partial Q_s}{\partial x}+C_L q_L+\sum_j \Phi_{s,j}+\sum_j \Phi_{b,j}=0 \tag{5.26}$$

Bed-material sorting:

$$(1-p)\frac{\partial(\beta_j A_m)}{\partial t}+(1-p)\beta_j\Phi_0 \mathrm{H}\{\Phi_0\}+(1-p)\beta_{0j}\Phi_0\mathrm{H}\{-\Phi_0\}+\Phi_{s,j}+\Phi_{b,j}=0 \tag{5.27}$$

where A_m = cross-sectional area of the active layer,
A_s = bed-material area above some datum,
C_j = suspended-load concentration for size class j,
$\bar{C}$ = total suspended-load concentration,
C_L = sediment concentration of lateral flow,
D_L = dispersion coefficient in the longitudinal direction,
G_j = bedload transport rate of size fraction j,
p = porosity of bed sediments,
Q_s = sediment flow rate,
q_L = lateral flow rate per unit of length,
u_{bj} = average velocity of the bedload in size fraction j,
β_{0j} = fraction of the bed-material underlying the active layer belonging to size class j,
β_j = fraction of the bed-material in the active layer belonging to size class j,
$\Phi_{s,j}$ = net flux of suspended load from the active layer to the water stream (source/sink term), and
$\Phi_{b,j}$ = exchange of sediments in size class j between the active layer and the bedload transport layer (source/sink term).

Additionally, H{Φ} is the step function defined by

$$H\{\Phi\} = \begin{cases} 1 & \Phi \geq 0 \\ 0 & \Phi < 0 \end{cases} \tag{5.28}$$

The quantity Φ_0 is defined by

$$\Phi_0 = \frac{\partial A_s}{\partial t} - \frac{\partial A_m}{\partial t} \tag{5.29}$$

Note also that, in Equation (5.26),

$$\frac{\partial Q_s}{\partial x} = \sum_j \frac{\partial G_j}{\partial x} \tag{5.30}$$

Equivalent equations can be developed for multi-dimensional flows, which are similar to Equations (5.3) and (5.18), for three- and two-dimensional models, respectively.

The diffusion coefficient D_L in the suspended-load equation—Equation (5.24)—is the result of combining a laminar flow equation using Fick's law for the diffusion, with a turbulent flow term that uses the diffusion analogy to represent the turbulent correlations. Because D_L is heavily influenced by channel geometry and embodies the essentially three-dimensional nature of turbulence, it is very difficult to provide good theoretical estimates of its value. In practice, it may vary by orders of magnitude within the same river or watercourse. When no direct reliable measurements of this quantity exist, D_L is reduced to a numerical parameter and is determined by model calibration. A more complete treatment of this subject can be found in Fischer et al. (1979).

The source/sink term in the transport Equations (5.24) and (5.25) can be evaluated using non-equilibrium transport concepts. In rivers and streams, it is usually acceptable to assume that the bed-material load discharge is equal to the sediment transport capacity of the flow; i.e., the bed-material load is transported in an equilibrium mode. In other words, the exchange of sediment between the bed and the fractions in transport is instantaneous. However, there are circumstances in which the spatial-delay and/or time-delay effects are important. For example, reservoir sedimentation processes and the siltation of estuaries are essentially non-equilibrium processes. In the laboratory, it has been observed that it may take a significant distance for a clear water inflow to reach its saturation sediment concentration.

Residual transport capacity for the j-th size fraction is defined as being the difference between its transport capacity, C_j^*, and the actual transport rate, C_j. Armanini and Di Silvio (1988) assumed the source term in Equation (5.24) to be directly proportional to the residual transport capacity:

$$\Phi_{s,j} = \frac{Q}{\lambda_{s,j}} \left(\beta_j C_j^* - C_j \right) \tag{5.31}$$

where $\lambda_{s,j}$ = characteristic length for the suspended load.

Similarly, for the source/sink term of the bedload transport equation we can write

$$\Phi_{b,j} = \frac{Q}{\lambda_{b,j}} \left(\beta_j G_j^* - G_j \right) \tag{5.32}$$

where $\lambda_{b,j}$ = characteristic length for the bedload, and
G_j^*, G_j = the carrying capacity and the actual transport rate of the bedload, respectively.

More details are presented in Armanini and Di Silvio (1988) and references therein, including methods for computing the characteristic lengths $\lambda_{s,j}$ and $\lambda_{b,j}$.

Bell and Sutherland's (1983) work suggests that the bedload discharge G_j^* is related to the bedload capacity G_j by a loading law of the form

$$\frac{\partial G_j}{\partial x} = K(t)\left(G_j^* - G_j \right) + a_1 \frac{G_j}{G_j^*} \frac{\partial G_j^*}{\partial x} + a_2 \frac{G_j}{\beta_j} \frac{\partial \beta_j}{\partial x} \tag{5.33}$$

where $K(t)$ = a coefficient with units of [L^{-1}], and
a_1, a_2 = dimensionless coefficients.

There are other approaches—for example, based on particle fall velocities and near bed concentrations at reference levels—but regardless of the approach used, non-equilibrium transport should be considered and, if necessary, included in a general sediment transport model.

The velocity of the bedload in Equation (5.25) can be found using one of the expressions provided in the literature. See Bagnold (1963) or van Rijn (1984a), for example.

Equation (5.26) can be simplified without great loss of generality. Firstly, in most circumstances, the change in suspended sediment concentration in any given cross-section is much smaller than the change in river bed; i.e.,

$$\frac{\partial (\bar{C} A)}{\partial t} << (1-p) \frac{\partial A_s}{\partial t} \tag{5.34}$$

Secondly, if the parameters in the sediment transport function for a cross-section can be assumed to remain constant during a time step, we can write

$$\frac{\partial Q_s}{\partial t} = 0 \quad \text{or} \quad \frac{\partial Q_s}{\partial x} = \frac{dQ_s}{dx} \tag{5.35}$$

This assumption is valid only if there is little variation of the cross-sectional geometry; i.e., no significant erosion and/or deposition occurs in a time step. This assumption allows the decoupling of water and sediment routing computations. In practice, this condition can be met by using a small enough time step.

Then, ignoring for now the source/sink terms, introducing Equations (5.34) and (5.35) into Equation (5.26) yields

$$(1-p)\frac{\partial A_s}{\partial t} + \frac{dQ_s}{dx} + C_L q_L = 0 \tag{5.36}$$

which is a form of the bed continuity equation widely used in numerical models.

Distribution of bed sediments during the erosion/deposition process is straightforward in two- and three-dimensional models, in which case the sediments are distributed uniformly across the computational cell. In one-dimensional models, however, special techniques must be used to represent the non-uniform cross-sectional variation of the deposited sediments. For example, in reservoirs and slow deposition in many rivers, sediment deposits are formed by filling the lowest parts of the channel first and by lifting the channel bed horizontally across the section, as depicted in Figure 5.3 (a).

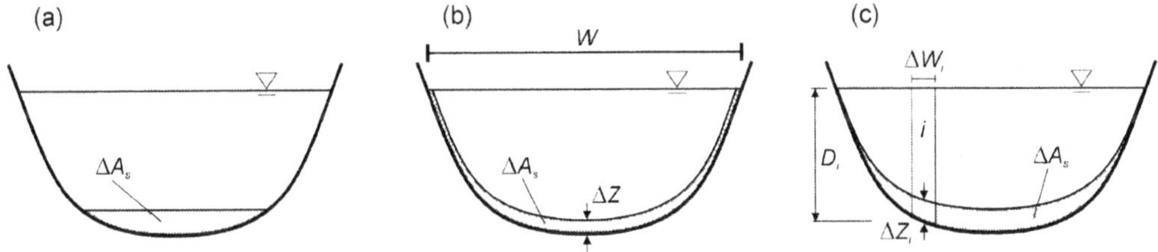

Figure 5.3. Sediment distribution methods of deposited and/or scoured sediments within a cross-section by a one-dimensional model. (a) horizontal distribution during deposition; (b) uniform distribution; and (c) cross-sectional distribution proportional to flow parameters.

The most common method used by one-dimensional models is by spreading the cross-sectional change, ΔA_s, with a constant thickness (measured along the vertical) across the wetted perimeter. The thickness ΔZ of the deposited/eroded materials is calculated from

$$\Delta Z = \frac{\Delta A_s}{W} \tag{5.37}$$

where W = channel top width.

This type of bed change is shown in Figure 5.3 (b). Other methods use selected flow parameters to compute the local bed variation. The cross-section is divided in slices of arbitrary width, ΔW_i, and the local bed variation ΔZ_i is computed for each of these slices. Common variables are the low depth D, the excess of bed shear stress $\tau - \tau_c$, and the conveyance K, yielding, respectively,

$$\Delta Z_i = \frac{D_i}{A} \Delta A_s \tag{5.38}$$

$$\Delta Z_i = \frac{(\tau_i - \tau_{ci})^m}{\sum_i (\tau_i - \tau_{ci})^m \Delta W_i} \Delta A_s \tag{5.39}$$

and

$$\Delta Z_i = \frac{K_i}{\sum_i K_i \Delta W_i} \Delta A_s \tag{5.40}$$

where m = an exponent,
τ_c = the Shields critical bed shear stress,

and the subscripts i refer to each of the slices used to subdivide the cross-section (see Figure 5.3 (c)).

5.2.5 Auxiliary Equations

5.2.5.1 Flow Resistance

The sets of differential equations presented in the previous sections require an additional set of relationships to define the boundary conditions. For two- and three-dimensional models, relationships are needed to represent the effects of solid boundaries on the flow field. These relationships are important because sediment transport processes are dominant in the near-bed region; therefore, it is important to have accurate predictions of the flow parameters in that region.

Near the bed, the equation of motion for steady, uniform turbulent flow is given by (see Figure 5.4)

$$\tau_z \Delta x \Delta y = \rho g (D - z) \Delta x \Delta y \sin\theta \tag{5.41}$$

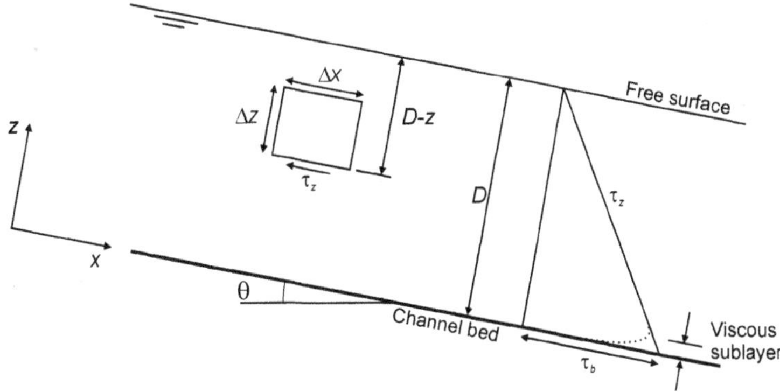

Figure 5.4. Vertical distribution of shear stress in uniform, turbulent, open channel flow.

Using $S_0 = \sin\theta$, at the bed (z = 0), Equation (5.41) becomes

$$\tau_b = \rho g D S_0 \tag{5.42}$$

where τ_b = bed shear stress.

Note that, by definition,

$$\tau_b = \rho U_*^2 \tag{5.43}$$

therefore,

$$U_* = \sqrt{gDS_0} \tag{5.44}$$

where U_* = the shear velocity.

The effects of the solid boundaries on the velocity distribution in turbulent flows are usually accounted for using an equivalent sand grain roughness, or Nikuradse roughness, k_s (Nikuradse, 1933). The bed roughness influences the near bed velocity profile due to the flow eddies generated by the roughness elements. These small eddies are quickly absorbed by the flow as they move away from the bed. A general form for the velocity distribution over the flow depth is given by

$$u = \frac{U_*}{\kappa} \log\left(\frac{z}{z_0}\right) \tag{5.45}$$

where κ = the von Kármán's constant (≈ 0.41 in clear free surface flows).

The zero-velocity level z_0 ($u = 0$ at $z = z_0$) was found to depend on the flow regime; i.e., if the solid boundaries are smooth or rough. In hydraulically smooth flow, the roughness elements are much smaller than the viscous sublayer—the viscous sublayer is a layer where the viscous stresses dominate over the turbulent stresses (see Figure 5.4)—while in hydraulically rough flow, the viscous sublayer does not exist; therefore, the velocity profile is not dependent on fluid viscosity. There is a transition range between the two flow regimes, called hydraulically transitional flow, where the velocity profile is affected by both viscosity and bottom roughness. The different flow regimes and their corresponding velocity profiles are summarized in Table 5.2.

Table 5.2. Zero-velocity level equations for the various flow regimes, depending on wall roughness. ν is the fluid viscosity

Flow regime	$R_{e*} \left(=\frac{U_* k_s}{\nu}\right)$	Zero-velocity level, z_0
Hydraulically smooth flow	$R_{e*} \le 5$	$0.11\frac{\nu}{U_*}$
Hydraulically rough flow	$R_{e*} \ge 70$	$0.033k_s$
Hydraulically transitional flow	$5 < R_{e*} < 70$	$0.11\frac{\nu}{U_*} + 0.033k_s$

Opposite to bed shear stress, wind stresses occur in the gas-liquid interface, or free surface, and are caused by the atmospheric circulation. A common semi-empirical relationship for the wind stresses is

$$\tau_s = \frac{C_f}{\rho_{air}} V^2 \tag{5.46}$$

where ρ_{air} = the density of air,
V = the wind velocity measured at the 10 m level, and
C_f = a drag coefficient and is of the order of 0.001.

The direction of the stresses is the same as of the wind.

In one-dimensional models, friction is used to compute the conveyance K, usually from an equation such as Manning's equation:

$$Q = KS_f^{1/2} = \frac{\delta}{n} AR^{2/3} S_f^{1/2} \tag{5.47}$$

where Q = flow discharge,
A = flow area,
R = hydraulic radius (= A/P),

P = wetted perimeter,
S_f = free surface slope,
n = Manning's roughness coefficient, and
δ = a parameter (= 1 in the metric system, = 1.49 in English units).

Note that Equation (5.47) is a steady state equation that, nonetheless, is used in unsteady hydraulic models. There are other formulae that use different friction factors, such as the Chézy roughness coefficient, C, and the Darcy-Weisbach coefficient f:

$$K = CAR^{1/2} = A\left(\frac{8gR}{f}\right)^{1/2} \tag{5.48}$$

Using Equations (5.47) and (5.48), it is easy to see that simple relationships exist between n, C, and f.

In one-dimensional flows, the roughness coefficient contains more than just skin friction losses. The overall resistance to the flow posed by channel meandering, vegetation types and density, cross-sectional changes in shape and size, irregularity of the cross-sections, etc., is included in the overall n for each channel reach. Estimating roughness is not a trivial task and requires considerable judgment. There are published flow resistance formulae that are more or less successful when applied to specific situations, but their lack of generality precludes their use in a numerical model for broad applications (see, for example, Klaassen et al. (1986) for more details). Some help exists in the form of tables, such as those that can be found in Chow (1959) and Henderson (1966). Barnes (1967) provides a photographic guide. The method by Cowan (1956) is summarized here. The basis of this method is the selection of a basic Manning's n value from a short set and to the application of modifiers according to the different characteristics of the channel. The method can be applied in steps, with the help of Table 5.3:

1. Select a basic n_0.
2. Add a modifier n_1 for roughness or degree of irregularity.
3. Add a modifier n_2 for variations in size and shape of the cross-section.
4. Add a modifier n_3 for obstructions (debris, stumps, exposed roots, logs,).
5. Add a modifier n_4 for vegetation.
6. Add a modifier n_5 for meandering.

The final value of the Manning's n is given by

$$n = n_0 + n_1 + n_2 + n_3 + n_4 + n_5 \tag{5.49}$$

The hydraulic resistance characteristics of vegetation depend on a number of parameters, such as plant flexibility or stiffness, density of vegetation, plant leaf characteristics (area, shape, and density), etc. For example, rigid vegetation increases flow resistance because the water has to expend more work to go around the obstacles posed by the individual plant stems and stalks. On the other hand, some types of highly flexible grass that bend easily with the flow have the effect of paving the bed, making it smoother and reducing its drag, and, therefore, decreasing flow resistance.

Table 5.3. Modifiers for basic Manning's n in the method by Cowan (1956), with modifications from Arcement and Schneider (1987)

Basic Manning's roughness values (n_0)			
Concrete	0.011-0.018	Gravel	0.028-0.035
Rock cut	0.025	Coarse gravel	0.026
Firm soil	0.020-0.032	Cobble	0.030-0.050
Coarse sand	0.026-0.035	Boulder	0.040-0.070
Fine gravel	0.024		
Modifier for degree of irregularity (n_1)			
Smooth	0.000	Moderate	0.006-0.010
Minor	0.001-0.005	Severe	0.011-0.020
Modifier for cross-sectional changes in size and shape (n_2)			
Gradual	0.000	Frequent	0.010-0.015
Occasional	0.005		
Modifier for the effect of obstructions (n_3)			
Negligible	0.000-0.004	Appreciable	0.020-0.030
Minor	0.005-0.019	Severe	0.060
Modifier for vegetation (n_4)			
Small	0.001-0.010	Very large	0.050-0.100
Medium	0.011-0.025	Extreme	0.100-0.200
Large	0.025-0.050		

Modifier for channel meandering (n_5)	
L_m/L_s	n_5
1.0-1.2 (minor)	0.0
1.2-1.5 (appreciable)	$0.15(n_0 + n_1 + n_2 + n_3 + n_4)$
> 1.5 (severe)	$0.30(n_0 + n_1 + n_2 + n_3 + n_4)$

L_m = meander length; L_s = length of straight reach

The effects caused by vegetation are complex, and there are no generally valid predictive models for their effects. Multi-dimensional modeling of turbulent dispersion using advanced turbulence models has been carried out—e.g., Shimizu and Tsujimoto (1994) and Naot et al. (1995) and (1996)—but its application to engineering models is difficult and requires significant computational effort. Simpler predictors for one-dimensional modeling have been developed by many authors, but they are usually based on limited amounts of data and are valid only for the region in which they were derived. Some of those expressions are presented in Table 5.4. Their use in any model should always be verified and supported with field measurements.

Table 5.4. Resistance relations for flow through vegetated channels

Author	Resistance equation	Notes
Gwinn and Ree (1980)	$n = \dfrac{1}{2.08 + 2.30\xi + 6\log(10.8VR)}$	R = hydraulic radius; ξ = a coefficient that depends on five retardance classes; the goodness-of-fit of the expression was not reported.
Kouwen and Li (1980)	$\dfrac{1}{\sqrt{f}} = a + \kappa \log\dfrac{R}{k_s}$; $k_s = 0.14h_v\left[\dfrac{\left(\dfrac{MEI}{\tau}\right)^{0.25}}{h_v}\right]^{1.59}$	a = dimensionless coefficient that is a function of the cross-sectional shape; h_v = local height of the vegetation; M = stem density; E = stem modulus of elasticity; I = stem area's second moment of inertia; Temple (1987) has correlated MEI with the vegetation height h_v for a range of dormant and growing grasses.
Pitlo (1986)	$n = \dfrac{0.0343}{(1 - D_v) + 0.0016}$	D_v = vegetation density; i.e., fraction of channel cross-section occupied by the submerged vegetation; the goodness-of-fit was not reported; based on measurements in one channel.
HR Wallingford (1992)	$n = 0.0337 + 0.0239\dfrac{D_v}{VR}$	Based on measurements taken in the Candover Brook, Hampshire, United Kingdom.
Bakry et al. (1992)	$n = a_1 D_h^{b_1}$; $n = a_2 + b_2 \log(VR)$; $n = a_3 + b_3 D_v$	D_h = hydraulic depth (= A/W); a_1 = coefficient ranging between 0.0087 and 0.0634; b_1 = coefficient ranging between -0.404 and 2.566; a_2 = coefficient ranging between -0.067 and 3.798; b_2 = coefficient ranging between -0.089 and 0.001; a_3 = coefficient ranging between 0.032 and 0.049; b_3 = coefficient ranging between 0.0072 and 0.12; units are metric.

Somewhat similar to the flow through vegetated channels is the case of mountain rivers, where flow resistance is dominated by grain roughness in gravel beds, rather than by the vegetation effects. Flow resistance is high at low stage and submergence and tends to decrease with increasing submergence (submergence is defined as the ratio between the water depth D and the size of the roughness elements, typically d_{84}). Many flow resistance relationships have been developed, but most suffer from a high content of empiricism and are site dependent. Errors on the order of 30% or higher are common. A summary of some of the most well-known relationships is presented in Table 5.5. These empirical formulae should be applied with care, within the range for which they were developed, and require careful verification and validation using field data.

Table 5.5. Resistance formulae for flow in mountain rivers

$$\sqrt{\frac{8}{f}} = A + B\log(X)$$

Author	A	B	X	Range of validity
Bray (1979)	0.701	6.68	$\frac{D}{d_{50}}$	$2.5 \le \frac{D}{d_{65}} \le 120$
Thompson and Campbel (1979)	5.66	$-0.566\frac{k_s}{R}$	$12\frac{R}{k_s}$	$\frac{D}{d_{84}} > 1.2$
Griffiths (1981)	2.15	5.60	$\frac{R}{d_{50}}$	$1 \le \frac{R}{d_{50}} \le 200$
Bathurst (1985)	4	5.62	$\frac{D}{d_{84}}$	$0.3 < \frac{R}{d_{84}} < 1$
Bray and Davar (1987)	3.1	5.7	$\frac{R}{d_{84}}$	$\frac{D}{d_{50}} > 1$

$$\sqrt{\frac{8}{f}} = AX^{B}$$

Author	A	B	X	Range of validity
Bray (1979)	5.03	0.268	$\frac{D}{d_{90}}$	$2.5 \le \frac{D}{d_{65}} \le 120$
Griffiths (1981)	3.54	0.287	$\frac{D}{d_{50}}$	$0.0085 < S_0 < 0.011$
Bray and Davar (1987)	5.4	0.25	$\frac{R}{d_{84}}$	$\frac{D}{d_{50}} > 1$
Bathurst (2002)	3.84 3.10	0.547 0.93	$\frac{D}{d_{84}}$	$\frac{D}{d_{84}} < 11;\quad 0.002 \le S_0 \le 0.040$

5.2.5.2 Sediment Transport

Sediment transport capacity—e.g., the bedload capacity G_j^* of Equation (5.32)—is computed using common sediment transport formulae, a subject that is covered in detail in Chapter 3 of this manual for non-cohesive sediments, and in Chapter 4 for cohesive sediments. Note, however, that most formulations are based on steady, uniform conditions. For unsteady transport under flood conditions, other approaches may have to be used—see Song and Graf (1995) and Bestawy (1997). Furthermore, in natural waterways, the presence of vegetation may require the use of different methods. Sediment transport in vegetated areas is an important topic due to increasing eco-hydraulics applications in water quality. For example, in grassed areas where the transported particles are usually very small and there is no bedload, the suspended-load transport equation, Equation (5.24), can be used, but with appropriately prescribed dispersion coefficients and sink terms. Deletic (2000) has developed expressions specifically for the trapping efficiency of grassed areas, from which the sink term can be readily calculated. Simões (2001) has successfully modeled the three-dimensional flow through sparse rigid vegetation, in which the dispersion coefficients were computed using rather simple turbulence closures.

Another factor contributing to the increased complexity of this subject is that there is no universal way to deal with sediment mixtures. There are almost as many approaches as there are authors. Furthermore, many of the methodologies are problem dependent and lack sufficient generality to be applicable (with reliability) to a wide range of problems. Here, some of the most common and useful methodologies are briefly presented, with the intent of underlining the importance of this subject.

By fractional transport capacity, we mean the technique to compute the transport rate of sediment mixtures with significant spread in particle sizes. It is important not only to compute the total transport capacity, but also the individual capacities for each of the particle size classes. There are essentially four ways to accomplish this task: (1) direct computation for each size fraction, (2) correction of bed shear stresses, (3) by fractioning the capacity of each size class, and (4) by using a distribution function.

The direct computation of each size fraction (e.g., Einstein, 1950) works by computing directly the sediment transport rate for each grain size present in the mixture, q_{sj} (the lower case denotes the quantity per unit width). Then, the total transport rate per unit width is computed from

$$q_s = \sum_j q_{sj} \tag{5.50}$$

Einstein (1950) was the first to recognize the effect of the presence of the larger particle sizes on the transport rate of the smaller sizes. He introduced a hiding factor to account for that effect, an approach that is now used by many, sometimes in modified and/or simplified manner.

The correction of bed shear stress approach works by introducing a correction factor to the computation of the shear stress acting upon the different particle sizes present in the bed. Thus, transport rate predictors for uniform bed-material are extended for sediment mixtures. If τ_j^* is the dimensionless shear stress acting upon the j-th size class particles, then the transport rate per unit width for that class is

$$q_{sj} = f\left(\xi_j \tau_j^*\right) \quad \text{or} \quad q_{sj} = f\left(\tau_j^* - \xi_j' \tau_{cj}^*\right) \tag{5.51}$$

where ξ_j, ξ_j' = correction factors accounting for particle sheltering and exposure effects, and

τ_{cj}^* = dimensionless critical (Shields) shear stress for particles in size class j.

The total transport rate is computed from Equation (5.50). The first form of Equation (5.51) was introduced by Egiazaroff (1965) and has been used by many researchers since then.

The fractioning of the capacity of each size class works in the following way: first, the potential transport capacity for each size fraction j, C_j, is computed from uniform sediment formulae as if the size fraction was the only sediment present in the bed. Then, it is reduced to match the availability of that particular size class, i.e.,

$$C_{tj} = p_j C_j \tag{5.52}$$

where p_j = percentage of material belonging to size class j present in the bed, and
C_{tj} = actual fractional transport potential for the j-th size class.

The total transport potential, C_t, is given by

$$C_t = \sum_j C_{tj} \tag{5.53}$$

Equation (5.53) represents the most widely used form of fractional transport used in numerical modeling. It has many shortcomings (see Hsu and Holly, 1992), among which is the fact that it predicts zero transport capacity for fractions that are not present in the bed—for example, it predicts zero transport of any sand entering a reach whose bed is composed only of gravel, irrespective of the hydraulic conditions. To overcome this problem, instead of p_j, many models use $\varepsilon p_j + (1-\varepsilon) p_j^*$, where p_j^* is the percentage of the j-th size class present in transport (i.e., entering the reach), and ε is a weighting factor $(0 \le \varepsilon \le 1)$. For many circumstances, it was found that $\varepsilon = 0.7$ provides a good approximation; however, this value is not universal.

Finally, the last method uses a distribution function to compute the transport capacity for each individual size class. This is accomplished by first computing the total transport capacity using a bed-material load equation and then distributing it into fractional transport capacities by using a distribution function:

$$C_{tj} = F_j C_t \quad \text{with} \quad \sum_j F_j = 1 \tag{5.54}$$

The advantage of this method is the fact that the distribution function F_j does not have to resemble the size distribution of the bed-material and that it can include the sheltering and exposure effects by relating F_j to both the sediment properties and the hydraulic conditions. This approach is less used than the three previously described methods. One well-known application of this concept is the work by Karim and Kennedy (1982).

Intimately connected to selective transport is the concept of bed sorting. As a result of computing sediment transport by size fraction, particles of different sizes are transported at different rates. Depending on the hydraulic parameters, the incoming sediment distribution, and the bed composition, some particle sizes may be eroded, while others may be deposited or may be immovable. Consequently, several different processes may take place. For example, all the finer particles may be eroded, leaving a layer of coarser particles for which there is no carrying capacity. No more erosion may occur for those hydraulic conditions, and the bed is said to be armored. This armor layer prevents the scour of the underlying materials, and the sediment available for transport becomes limited to the amount of sediment entering the reach. Future hydraulic events, however, such as an increase of flow velocity, may increase the flow carrying capacity, causing the armor layer to break and restart the erosion processes in the reach. Many different processes may occur simultaneously within the same channel reach. These depend not only on the composition of the supplied sediment (i.e., the sediment entering the reach), but also on bed composition within that reach.

It is important to track bed composition during flood events. During high flows, armor layers are often ruptured, exposing the underlying material, which is finer and more susceptible to erosion. This may be particularly important in reaches just downstream from dams, where reduced sediment supply usually results in base level lowering until a certain equilibrium condition is reached—that is, until the bed becomes armored for the prevailing (regulated) hydraulic conditions. If the armor layer is removed, a new degradation process is initiated, which may lead to further erosion long after the flood takes place. Furthermore, a certain type of sediment transport seems to happen only during flood situations, such as the transport characterized by dunes of finer sediment moving over gravel beds with a mobile armor layer (paved beds). This type of transport has the effect of destabilizing the armor layer, even if the flow is too weak to break that armor under the normal (non-flood) hydraulic regime, resulting in potentially additional degradation that normally would not occur (Klaassen, 1990).

There are many different approaches to bed sorting. Some of the most common are, perhaps, the ones by Bennett and Nordin (1977) and Borah et al. (1982), but many other approaches are also used. Most of the existing bed sorting algorithms are highly case dependent and lack sufficient generality. Therefore, they have to be selected carefully. For a more detailed treatment of the subject, see, for example, Armanini (1995).

5.3 Numerical Solution Methods

In most cases, there are no known analytical solutions for the governing equations described in the previous sections, except for the simplest of geometric configurations and flow regimes. However, there is a vast body of numerical mathematics that can be used to find the solutions for an alternative, but approximate, numerical (discrete) statement to the continuous problem expressed by the governing partial differential equations and their corresponding boundary conditions. The resulting numerical description of the mathematical model at hand consists of a set of algebraic equations that can be programmed and solved in a computer: the numerical model.

This section will offer a brief description of some of the methods most commonly used to solve the fluid flow equations. The techniques of interest are those based on some type of gridded discretization of the problem at hand, in which the continuous variables for which the solution is sought are solved only at specific discrete locations of the physical domain. The algebraic equations that form the numerical model are functions of those discrete quantities.

For the same problem (i.e., the same set of differential governing equations and boundary conditions), it is possible to obtain very distinct sets of algebraic numerical equations, depending on the technique used to discretize the equations. This chapter will be concerned with three such techniques: finite differences, finite elements, and finite volumes. In practice, there are many other techniques available to numerically solve the fluid flow equations. On the other hand, it is possible to state certain finite difference and finite volume techniques as a subset of the finite element technique. We will not deal with these fine points here, but the interested reader is directed to standard textbooks in computational fluid dynamics, such as those by Ferziger and Peric (2002) and by Wesseling (2001), for example.

As mentioned above, the translation from continuous to discrete replaces one problem (the continuous formulation) by another (the discrete formulation). The latter should provide a solution that should converge to the solution of the former. In this context, convergence is a term that denotes a relationship between the numerical and the analytical solutions. Convergence should be obtained as the grid spacing (Δx in one-dimensional problems) and time step (Δt in unsteady problems) are refined; i.e., as $\Delta x, \Delta t \rightarrow 0$. Figure 5.5 depicts several types of solution behavior for the discrete problem equations, including instability and convergence to the wrong solution.

Convergence is ensured by Lax's theorem—which has been proven for linear problems only, but which is nonetheless at the foundation of computational fluid dynamics—which states that consistence and stability are sufficient to ensure the convergence of a numerical scheme. Consistence is a term applied to the algebraic equations: a numerical scheme is said to be consistent with the partial differential equation if its truncation error disappears in the limit when $\Delta x, \Delta t \rightarrow 0$. Stability is a term that applies to the numerical solution itself: a solution is stable if it remains bounded at all times during the computations.

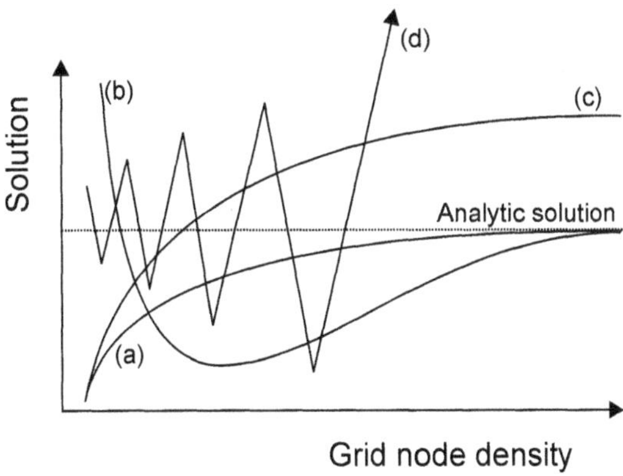

Figure 5.5. Different possible solution behaviors for numerical solution schemes: the solutions produced by schemes (a) and (b) converge to the analytical solution upon grid refinement; (c) converges to the wrong solution; and (d) does not converge at all.

5.3.1 Finite Difference Methods

Finite difference methods are probably the most simple and most common methods employed in fluid flow models, as well as in other disciplines requiring the numerical solution of partial differential equations. They are based on the approximation of the individual derivative terms in the equations by discrete differences, thus converting them into sets of simultaneous algebraic equations with the unknowns defined at discrete points over the entire domain of the problem. For example, the partial derivative of $u(x,y)$ at point (i,j) of the discretized domain in the x-direction can be written as

$$\frac{\partial u(x,y)}{\partial x} = \lim_{\Delta x \to 0} \frac{u(x+\Delta x, y) - u(x)}{\Delta x} \approx \frac{u_{i+1,j} - u_{ij}}{\Delta x_{ij}} \tag{5.55}$$

and similarly for the y-direction (see Figure 5.6).

Note that Equation (5.55) does not represent the only possible choice of discretization. For example, without loss of mathematical rigor, one could choose instead

$$\frac{\partial u}{\partial x} \approx \frac{u_{ij} - u_{i-1,j}}{\Delta x_{i-1,j}} \tag{5.56}$$

Equation (5.55) is called forward difference, and Equation (5.56) is called backward difference. One could define central differences, or even use multiple Δx to define differential operators that span many grid nodes. On the other hand, the grid can be quite complex. Curvilinear non-orthogonal grids are indeed used to describe complex flow domains, but the corresponding differential operators are also more complicated than the ones shown above. For a more detailed overview of finite difference methods in computational fluid dynamics, the reader is directed to Tannehill et al. (1997).

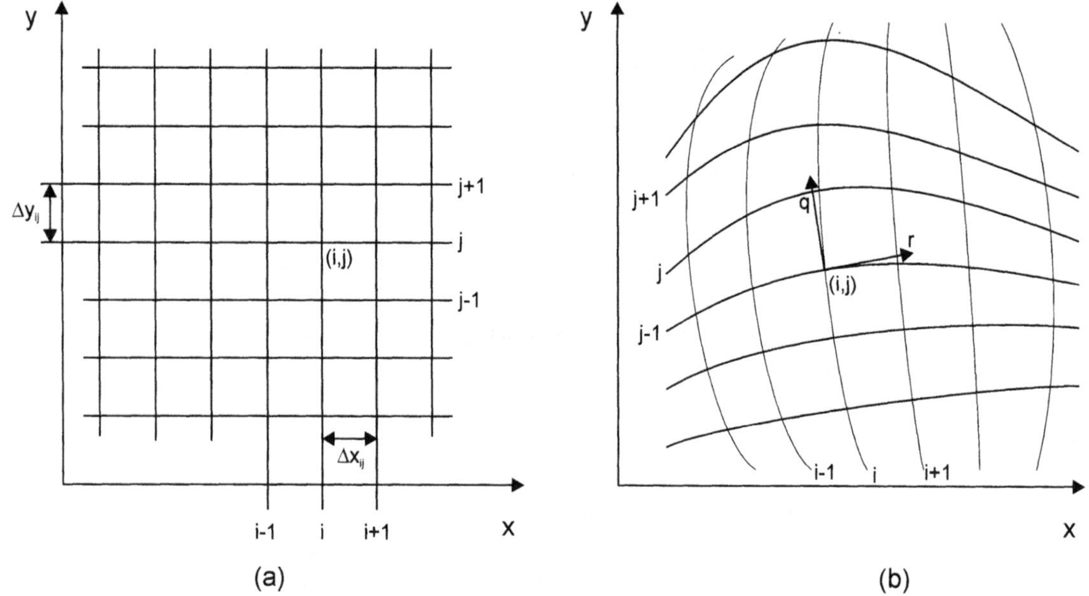

Figure 5.6. Typical mesh systems used in finite difference methods. (a) Cartesian orthogonal mesh; (b) curvilinear mesh (not necessarily orthogonal). The local coordinate system for point (*i,j*) in (b) is defined by the unit vectors *q* and *r*, which are tangent to the grid lines. The points where the variables are defined are located at the intersection of the grid lines.

In one-dimensional free surface flow, the most popular scheme to solve the de Saint Venant equations is the Preissman scheme (Preissman, 1961). For an extensive coverage of the method, the reader is directed to Cunge et al. (1980). Here, only a brief description is presented.

The Preissman scheme is a four-point scheme (also called box scheme), as shown schematically in Figure 5.7. If $f(x,t)$ is any of the flow variables of interest (e.g., water depth and discharge), then

$$\frac{\partial f}{\partial t} \approx (1-\psi)\frac{f_j^{n+1} - f_j^n}{\Delta t} + \psi\frac{f_{j+1}^{n+1} - f_{j+1}^n}{\Delta t} \tag{5.57}$$

$$\frac{\partial f}{\partial x} \approx (1-\theta)\frac{f_{j+1}^n - f_j^n}{\Delta x} + \theta\frac{f_{j+1}^{n+1} - f_j^{n+1}}{\Delta x} \tag{5.58}$$

and

$$f=(1-\psi)(1-\theta)f_j^n+\psi(1-\theta)f_{j+1}^n+\theta(1-\psi)f_j^{n+1}+\psi\theta f_{j+1}^{n+1} \tag{5.59}$$

where θ, ψ = weighting coefficients.

Direct application of the Preissman scheme to the de Saint Venant equations, Equations (5.19) and (5.20), results in a non-linear system of algebraic equations. To avoid having to deal with the numerical difficulties normally associated with non-linear systems, in practice the system is linearized by using a Taylor series expansion of the system coefficients and then by rewriting it in terms of the variations of the unknowns rather than solving for the unknowns themselves directly. In other words, if the original system is expressed in terms of the free surface η and the discharge Q as the dependent variables, after the Taylor series expansion the algebraic system of equations is rewritten in terms of $\Delta\eta_j$ and ΔQ_j, where $\Delta\eta_j$ and ΔQ_j represent the variation of η_j and Q_j in a time step Δt and for each discretization point j. The system can then be solved using traditional iterative (such as the Newton iteration method) or direct methods (such as the double-sweep method).

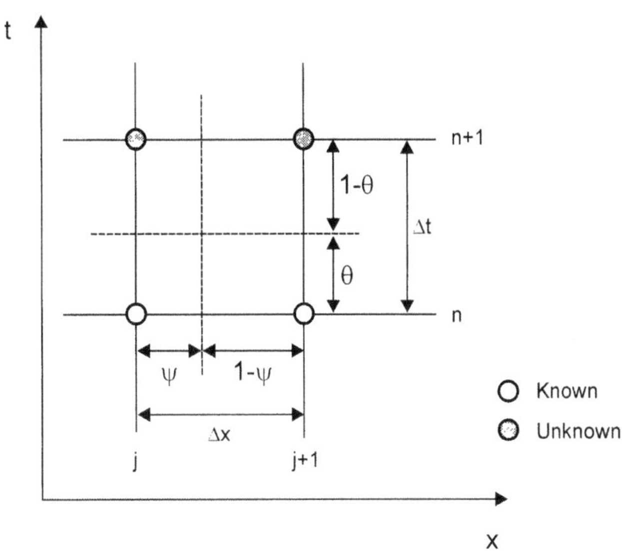

Figure 5.7. The four-point Preissman finite difference operator.

The weighting coefficients are used to control the numerical error and stability of the scheme. $\psi = \theta = ½$ produces a scheme whose truncation errors are of order Δt^2 and Δx^2. The scheme is stable and non-dissipative (i.e., it does not smear the solution). Increasing θ introduces truncation errors that cause dissipation. Using $\theta = 1$ yields the largest numerical dissipation, which produces poor results in unsteady problems but provides the fastest convergence (useful for steady-state problems). In practice, it has been found that $\theta = 0.67$ is a good value for most problems. If $\theta < ½$, the Preissman scheme is always unstable.

5.3.2 Finite Element Methods

Finite element methods have been used successfully for fluid flow problems since the 1960's. They are particularly useful to solve problems with complex geometries, as they do not require the structured grid system needed in finite difference techniques. In an unstructured grid, the computational nodes do not need to be defined in an ordered manner, as opposed to structured grids where each node is identified by an (i,j) pair (or (i,j,k) trio in three-dimensional models), such as those shown in Figure 5.6. An example of unstructured grid is shown in Figure 5.8.

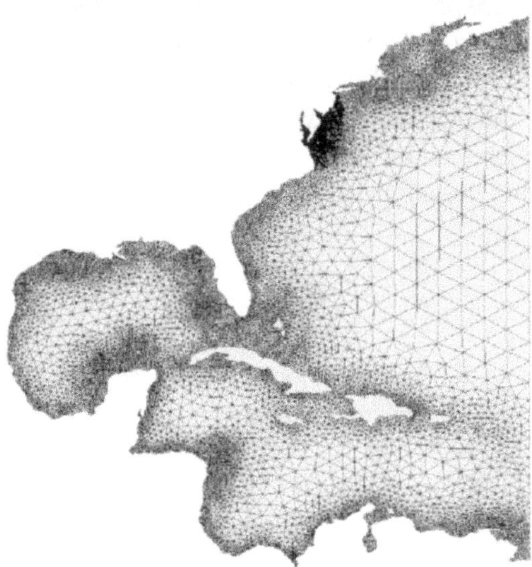

Figure 5.8. Finite element mesh for part of the American Atlantic coast, as generated by the two-dimensional automatic finite element grid generator software, CCALMR, Oregon Health and Science University. Note that the darker areas represent a very high density of triangles that cannot be resolved at the scale used in this figure.

There are two main approaches for the formulation of finite element methods: variational methods and weighted residual methods. In variational methods, the variational principle for the governing equation is minimized. In general fluid mechanics problems, exact forms of the variational principles for the governing non-linear equations are difficult to find (unlike in the linear equations encountered in solid mechanics); therefore, weighted residual methods are much more popular. Residual methods are based on minimizing some sort of error, or residual, of the governing equations. Let Ξ be the residual of the differential equation (for example, $\Xi = \Delta^2\psi$). Mathematically, minimization of Ξ to zero can be achieved by orthogonally projecting Ξ on a subspace of weighting functions, W_f; i.e., by taking the inner product of the residual and the weighting functions:

$$\left(\Xi, W_f\right) = \int_0^1 W_f \Xi dx = 0, \quad 0 < x < 1 \tag{5.60}$$

This process provides a mathematical framework to derive algebraic equations for any differential equation.

In finite element methods, the mathematical domain is divided into non-overlapping polyhedral subdomains (the elements) and Equation (5.60) is enforced in each subdomain, taking into consideration the boundary conditions. Within each element, the dependent variables are approximated by interpolating functions, Φ_a. The form assumed by Φ_a is determined by the type of element used. Some of the most commonly used elements for fluid mechanics applications are presented in Figure 5.9.

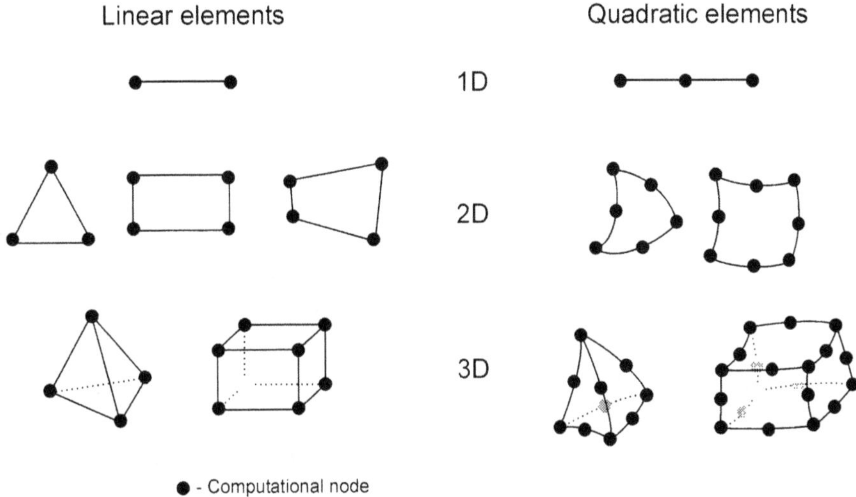

Figure 5.9. Some of the more common finite elements used in fluid flow modeling.

Note that the interpolating functions Φ_a can be used in lieu of the weighting functions W_f in Equation (5.60). In this case, the scheme is known as the Galerkin method. There are many other variations of the theme; i.e., in which Φ_a and W_f take different forms, and where Equation (5.60) also gets modified. Some of those methods commonly employed in computational fluid dynamics are the generalized Galerkin, the Taylor-Galerkin, and the Petrov-Galerkin methods. It is not in the scope of this text to produce detailed derivations of the methods. The interested reader should refer to Chung (2002) for a more comprehensive coverage of this subject.

5.3.3 Finite Volume Methods

Finite volume methods use conservation laws; i.e., the integral forms of the governing equations. The domain of computation is subdivided into an arbitrary number of control volumes, and the equations are discretized by accounting for the several fluxes crossing the control volume boundaries. There are two main types of techniques to define the shape and position of the control volumes with respect to the discrete grid points where the dependent variables are calculated: the node-centered scheme and the cell-centered scheme. Both schemes are schematically pictured in Figure 5.10. The node-centered scheme places the grid nodes at the centroids of the control volume, making the control volumes "identical" to the grid cells. In cell-centered schemes, the control volume is formed by connecting adjacent grid nodes.

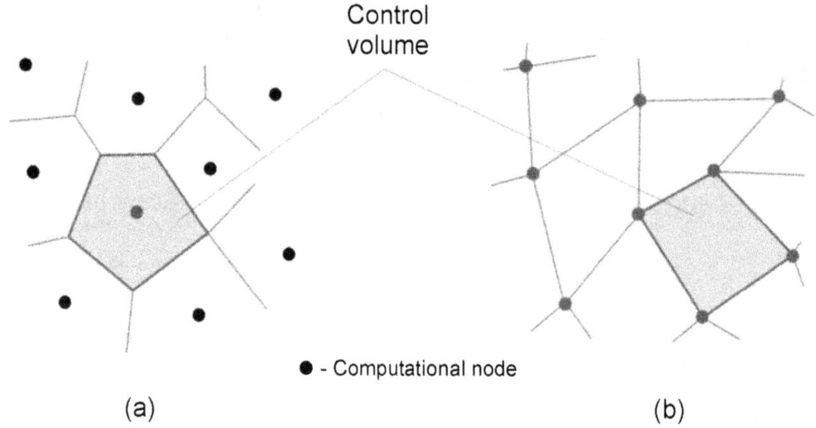

Figure 5.10. Representation of the control volumes formed by node-centered (a) and cell-centered (b) formulations used in finite volume discretizations.

The main advantage of finite volume methods is that the spatial discretization is done directly in the physical space, without the need to make any transformations between coordinate systems. It is a very flexible method that can be implemented in both structured and unstructured grid systems. Because the method is based directly on physical conservation principles, mass, momentum, and energy are automatically conserved by the numerical scheme.

Under certain conditions, the finite volume method is equivalent to the finite difference method or to particular forms of lower order finite element methods. The user is directed to Chung (2002) for more details on finite volume methods and their relationship with finite difference and finite element methods.

5.3.4 Other Discretization Methods

There are many other numerical discretization techniques that, for certain specific applications, offer significant advantages over the methods presented in the previous sections. It is outside of the scope of this chapter to cover them all, but a very brief overview will be presented of some selected methods. One such method is the spectral element method (SEM), used first by Patera (1984). The method is a particular type of the method of weighted residuals, sometimes also used with the Galerkin formulation, in which special "spectral" functions are used, usually Chebychev, Legendre, or Laguerre polynomials. These functions provide a physically more realistic description of flow phenomena than those used in conventional finite element methods, therefore leading to solutions with higher accuracy. In practice, however, their application is limited to simple geometries and simple boundary conditions.

The SEM tries to combine the advantages of the finite element method—especially its flexibility—with the greater accuracy of spectral schemes (e.g., Canuto et al., 1996). Its advantage lies in its non-diffusive approximation of the convection terms. Apart from its limited gamut of applications, its principal disadvantage is the much higher numerical effort required in comparison to the more traditional discretization methods.

Least squares methods have been used by many with the finite element formulation (e.g., Fix and Gunzburger, 1978). In this method, there is no need to do the integration by parts normally required in the standard Galerkin method. Instead, the inner products of the governing equations are minimized with respect to the nodal values of the variables. After the process is completed, higher order derivatives remain, requiring higher order trial functions than those used in the standard finite element methods.

A method based on boundary integral equations is the boundary element method (Brebbia, 1978). The solutions are obtained using the boundaries of a region, and interpolation functions coupled with the solutions of the governing equations are used to describe the interior of the domain. The equations are solved for nodes on the boundary alone. The values of the solution in the domain are calculated on the basis of the boundary information and the interpolation functions. The method has the advantage of requiring fewer equations to compute the solution, but the governing equations must be linear (or must first be linearized using a Kirchhoff transformation).

Some discretization methods use clusters of points for the spatial discretization in a gridless manner, rather than using points organized in connected grids in a conventional manner—see, for example, Batina (1993). In a gridless discretization, there are no coordinate transformations, nor is there the need to compute face areas or volumes. A least squares method is used to compute all the necessary gradients of the flow using a determined number of neighboring points surrounding each node. The points can be chosen along a certain direction to improve accuracy (e.g., in the characteristic directions). The differential form of the governing equations is used in a Cartesian coordinate system. The clusters of points may be denser in certain regions and sparser on others in order to better capture the flow gradients, in this respect having the flexibility of unstructured grid formulations. In spite of solving the conservation form of the flow equations, however, it is not yet clear that the gridless method can ensure conservation of mass, momentum, and energy.

Although the most common methods use Eulerian coordinates, there are instances in which a Lagrangian coordinate approach may be more appropriate. In Eulerian coordinate methods, the computational nodes, where the variables are calculated, are fixed in space, as opposed to Lagrangian coordinate methods, where the nodes are allowed to move with the fluid particles. Moreover, in many cases it is more advantageous to couple both Eulerian and Lagrangian methods, a method known as coupled Eulerian-Lagrangian (CEL)—Noh (1964). In CEL methods, the computational domain is separated in subsections, or subdomains, and the lines that define the boundary between each subdomain are approximated by time-dependent Lagrangian lines. In this framework, each subdomain is discretized by a time-independent Eulerian grid system which has its boundary prescribed by Lagrangian calculations. For each time step, first the Lagrangian computations are performed, then the Eulerian computations, then a further step that couples both computations. This last step determines which part of the Eulerian mesh is active and the pressures acting upon the Lagrangian boundaries. CEL methods have been applied in flows with moving boundaries, such as the interface that separates two distinct fluids in multiphase flows.

Many other methods that were not described in this section have been developed, such as the particle-in-cell method, Montecarlo methods, smooth particle hydrodynamics (used by astrophysicists in the analysis of dust clouds and exploding stars), and others whose application to computation fluid dynamics has been in fields other than those of river engineering. The interested reader can find descriptions of these methods in the relevant literature or in some of the textbooks in the references to this chapter.

5.4 Modeling Morphologic Evolution

In the category of morphologic evolution, we include models capable of computing not only bed changes, but also channel width changes. In the previous section, only fixed-width models were considered. Fixed-width models should be applied only to cases in which the prototype channel's width adjustments are not significant.

The causes behind river width adjustments are varied and involve many time scales and a wide range of fluvial processes and geotechnical mechanisms, making its modeling a challenge. In some instances, bank erosion is caused by large variations in discharge, especially by floods. In others, bank saturation and dewatering is the main mechanism of concern: as the river rises, the banks soften and get heavier due to saturation; when the river level falls, the supporting hydrostatic forces are removed, resulting in instability and collapse (this mechanism sometimes causes a wave of bank failure that proceeds rapidly upstream, a phenomenon known as explosive channel widening). Yet in other cases, bank retreat is less related to flow stage and intensity, but more to precipitation and ground-water events that generate erosion through sapping or piping. Non-fluvial processes that may cause bank erosion include freezing, precipitation, snowmelt, and vegetation. Human activity and trampling and grazing by livestock are also part of this latter category. Some of the processes and resulting failure mechanisms are represented in Figure 5.11. As a consequence of this gamut of different phenomena, it is important to identify the dominant erosion processes and failure mechanisms in the prototype and to include them in any conceptual or mathematical model of the same, which sometimes is a very difficult task to accomplish.

A variety of approaches are used in analyzing and predicting river width changes. One such approach involves the use of regime theory. Regime theory attempts to predict the form of equilibrium channels (e.g., width and depth) using basic hydraulic quantities (the flow discharge and sediment load)—Lacey (1920). In the past, such approaches were mostly empirical, resulting in equations that were not dimensionally homogeneous and whose range of validity was limited to the basins and data used in their derivation. Recently, Julien and Wargadalam (1995) attempted to provide a semi-theoretical basis to this approach by using the basic governing principles of open channel flow to derive a new set of relationships for equilibrium channels.

However, while regime equations are widely used by engineers, their use in modeling is very limited because they are unable to predict the rates of change of the main cross-sectional geometric parameters.

The most common approach used in models of bank erosion is based on mechanistic principles. The mechanistic approach uses geotechnical concepts for modeling bank mechanics. The bank retreat and advance processes are modeled as a result of fluvial erosion of the bank materials, as well as a result of near-bed degradation and/or increase in bank steepness and consequent geotechnical failure. The main controlling mechanism determining bank stability is related to the conditions at the base of the bank.

Mechanistic models can become very complex because each different failure mechanism (e.g., planar, rotational, and cantilever in Figure 5.11 (a)-(c)) requires a separate analysis. Furthermore, there are additional complexities due to the essentially three-dimensional nature of the flow near the toe of the banks, turbulence effects, roughness variability, variations in bed particle size, presence of cohesive sediment materials (there is a vast difference in the failure mechanics between cohesive and non-cohesive materials due to significant differences in their soil mechanics), vertical stratification and longitudinal variability of the bank materials, cross-sectional variation of the sediment transport rate, etc. Due to the limited scope of this monograph, a detailed exposition of these phenomena will not be included here. Instead, a summary of the most important failure types are presented next. The reader can find further details in ASCE (1998a) and (1998b) and in the references therein.

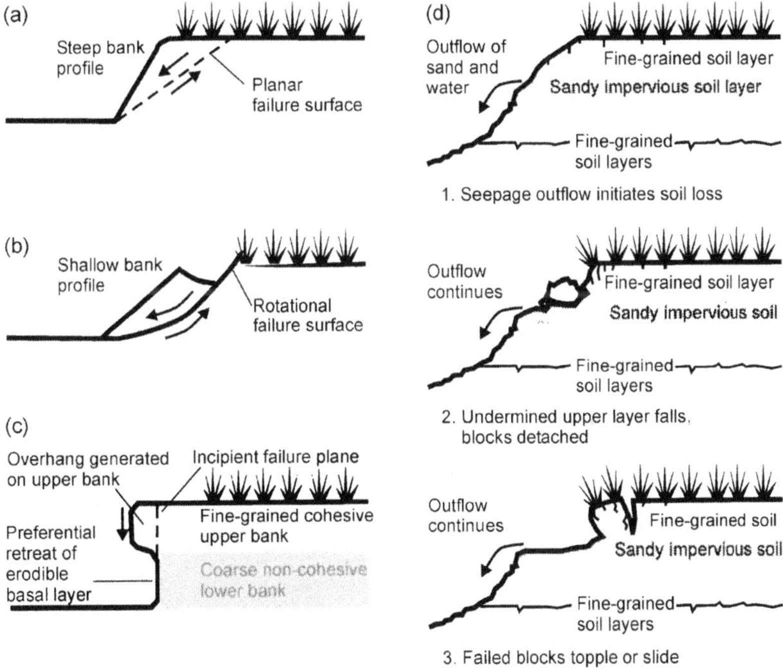

Figure 5.11. Dominant bank failure mechanisms due to geotechnical failure. (Adapted, with modifications, from Hagerty, 1991.)

The different types of bank failure are represented in Figure 5.12. The first type is the rotational slip, which may be defined as base, toe, or slope failure according to the point at which the failure arc intercepts the channel bed—see Figure 5.12 (a). This type of failure is easily analyzed using traditional geotechnical methods; e.g., Fredlund (1987).

Plane slip failure, Figure 5.12 (b), occurs mainly in very steep eroding riverbanks, often at the outer margin of meander bends and along severely incised channels. The approaches used to deal with this type of failure are based on the balance of forces acting normal to and along the failure (slip) plane of the potential failing block. The analysis leads to an equation that relates the critical height for mass failure, H_c, to a number of geotechnical parameters:

$$H_c = \frac{4c(\sin\theta\cos\phi)}{\gamma(1-\cos(\theta-\phi))} \tag{5.61}$$

where
c = cohesion,
γ = bank material unit weight,
ϕ = friction angle, and
θ = angle of bank slope, Figure 5.12(b).

Toppling failure results from tension cracks that develop downward from the ground surface, thus limiting the width of the potential failure block, as shown in Figure 5.12 (c). They develop parallel to the bank behind steep banks and are due to the tensile stress in the soil. The potential depth of tension cracking H_c is in the order of 10 ft.

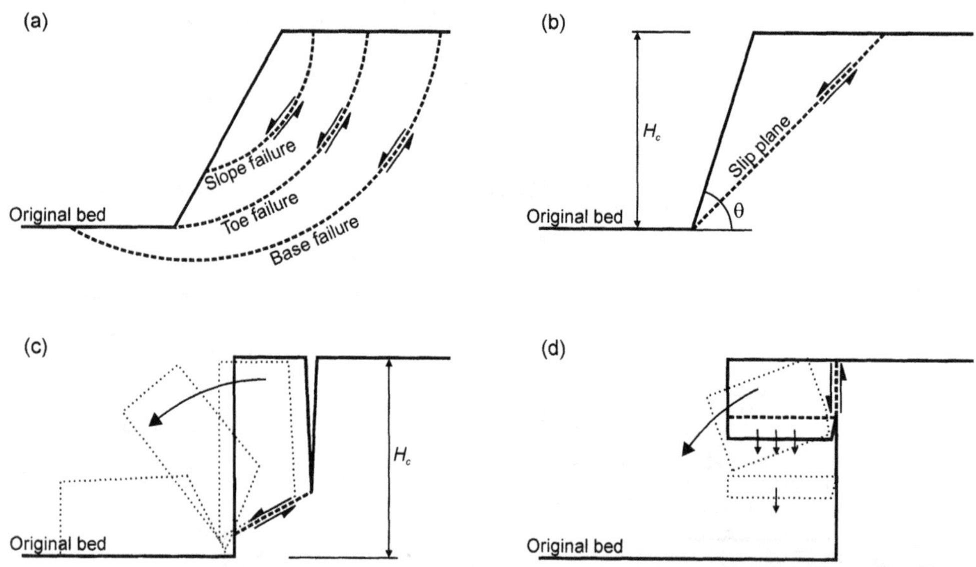

Figure 5.12. Types of bank failure usually employed in river widening models. (a) rotational slip failures, (b) plane slip failure, (c) toppling failure, and (d) cantilever failure.

Overhanging banks, Figure 5.12 (d), result from the erosion of an erodible layer below that of a layer more resistant to eroding, in a stratified bank. They may fail by tensile, beam, or shear (not shown) collapse, and they occur when the weight of the cantilever block exceeds the soil strength. This type of failure often occurs in areas with riparian and flood plain vegetation, where the plant roots contribute to reinforce the strength of the cantilever blocks.

As a result of the failure described above, mass is delivered to the toe of the bank slope, and its removal also needs to be properly modeled. This removal depends mostly on the rate of entrainment of this matter by the flow, as well as the capacity of the flow to transport the sediments downstream. Vastly different rates of removal may happen in practice, such as impeded removal (the bank failures supply debris at a higher rate than the flow can remove them), unimpeded removal (the debris delivery processes are balanced by ones removing them; therefore, no changes in basal slope or angle develop), and excess basal capacity (basal scour capability vastly exceeds the debris delivery rate). The basal matter removal rates have a vast importance in bank stability. For example, impeded removal results in basal accumulation, decreasing bank angle and height, and increasing stability, while excess basal capacity results in channel bed lowering, higher bank heights, and steeper angles, therefore decreasing stability. Basal debris behavior (whether the debris crumbles into non-cohesive individual particles, or whether cohesive effects dictate particle size, mass, or no erosion rates) and local hydraulics need to be accurately modeled in order to properly represent mass wasting removal from the toe.

The final approach to modeling width changes considered here is based on extremal hypothesis or theories. The fixed-width, one-dimensional models use the water depth D, the flow velocity V, and slope, S_f, as independent variables. The three independent equations that must be satisfied are the conservation of water,

$$Q = WDV \tag{5.62}$$

where W = the channel width;

a flow resistance equation (Chézy's equation is used for convenience),

$$V = C\left(\frac{WD}{P}S_f\right)^{1/2} \tag{5.63}$$

where C = Chézy's roughness coefficient, and
P = the wetted perimeter;

and a sediment transport equation,

$$Q_s = f\left(D, W, V, S_f, d, \text{etc.}\right) \tag{5.64}$$

where Q_s = sediment transport capacity, and
d = sediment particle size.

A fourth independent relationship must, however, be used if the channel width is to be considered as another independent variable. The fourth relationship is obtained by minimizing or maximizing some quantity, such as minimizing the rate of dissipation of energy.

The existing variable-width models in this category are based on the minimum energy dissipation rate theory developed by Song and Yang (1979a, 1979b, 1980) and Yang and Song (1979, 1986) and this general theory's special case, the minimum stream power theory, used by Chang and Hill (1976, 1977) and Chang (1979). The minimum energy dissipation rate theory states that when a closed and dissipative system reaches its state of dynamic equilibrium, its energy dissipation rate must be at its minimum value:

$$\varphi = \varphi_w + \varphi_s = \text{a minimum} \tag{5.65}$$

where φ = total rate of energy dissipation,
φ_w = rate of energy dissipation due to water movement, and
φ_s = rate of energy dissipation due to sediment movement.

The minimum value must be consistent with the constraints applied to the system. If the system is not at its dynamic equilibrium condition, its energy dissipation rate is not at its minimum value, but the system will adjust itself in a manner that will reduce its energy dissipation rate to a minimum value and regain equilibrium. Because of changing flow and sediment conditions, a natural river is seldom in its true equilibrium condition. However, a natural river will adjust its channel geometry, slope, pattern, roughness, etc., to minimize its energy dissipation rate subject to the water discharge and sediment load supplied from upstream.

For an alluvial channel or a river where the energy dissipation rate for transporting water is much higher than that required to transport sediment; i.e., $\varphi_w >> \varphi_s$, the theory of minimum energy dissipation rate can be replaced by a simplified theory of minimum stream power (Yang, 1992). For this case, a river will minimize its stream power, γQS, per unit of channel length subject to hydrologic, hydraulic, sediment, geometric, geologic, and manmade constraints (γ is the specific weight of water).

Note that modeling of morphologic changes does also include the modeling of bank deposition. In many systems, especially those involving the development of channel meandering, bank erosion and deposition are phenomena that are simultaneously present. For example, within the same cross-section, one bank may be retreating while the opposite bank may be advancing, as shown in Figure 5.13.

Just as in many other areas of application, the complexity and variety of geomorphic factors involved in river width adjustment phenomena require a careful and methodical approach to their modeling, with large support of field observations and prototype data. ASCE (1998b) proposed an eight-step approach to deal with these issues. In spite of being specific to bank evolution modeling, its basic principles are applicable to other areas of interest in hydraulic modeling; therefore, it is briefly presented here. A schematic view of the entire process is presented in Figure 5.14.

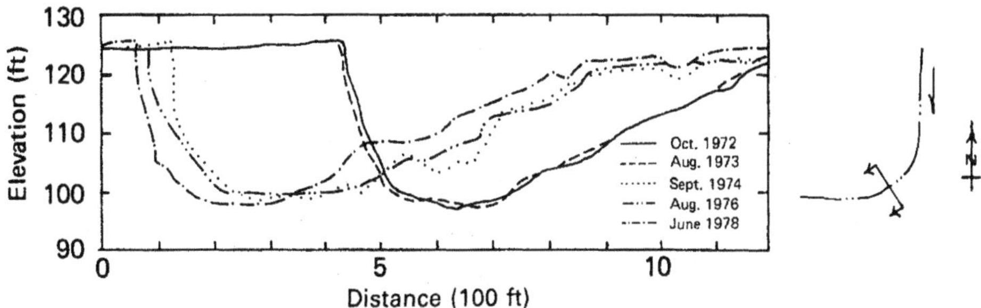

Figure 5.13. Development of mile 190.8 of the Sacramento River, 1972 through 1978, showing erosion in the left bank and deposition in the right bank, resulting in lateral bend migration (USACE, 1981).

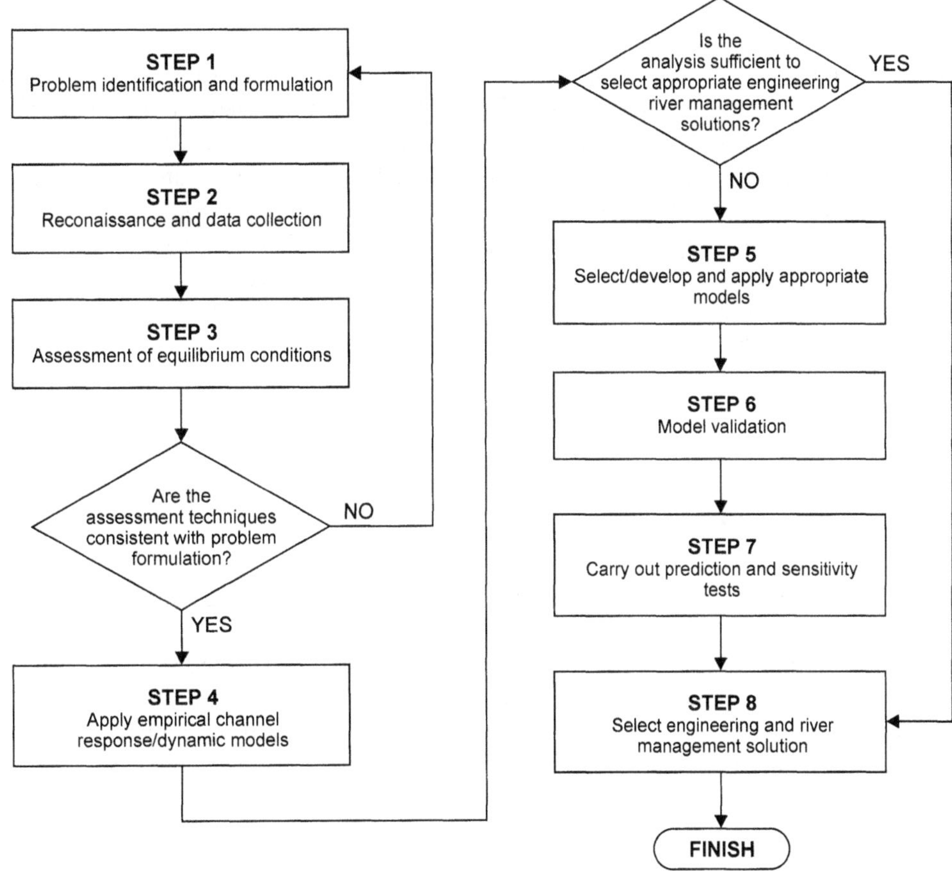

Figure 5.14. Procedure for the identification, analysis, and modeling of river width adjustment problems, after ASCE (1998b), with modifications.

Step 1, problem identification and formulation, involves determining the factors at play, which may be associated with river engineering factors, social activities, and may relate to existing conditions or future activities. It establishes the level of analysis required and defines the appropriate level of response. It is followed by step 2, field reconnaissance and data collection, in order to identify the nature and extent of the width adjustment problem. In this stage, channel characteristics are identified, as well as bank conditions and materials, hydrologic conditions, vegetation, any engineering works, and any other parameters that are considered relevant for the case at hand. During this stage, there is an assessment of the existing data available and, if necessary, it is the time to design and mount an adequate data gathering program.

At this stage, ASCE (1998b) recommends a simple assessment of the equilibrium conditions using the techniques of regime theory (step 3). The results of this analysis are compared with existing conditions to provide an indication of the present morphological status of the prototype. This makes it possible to determine the impact of the proposed engineering works.

In step 4, simple empirical channel response models are applied. This step may help to interpret existing and proposed conditions and to identify the dominant processes and trends at play. The information gathered at this stage forms a framework for the more detailed modeling work that follows (if appropriate). In step 5, more advanced models are developed and used, if necessary, to provide the more detailed information. These models should be validated with existing prototype data (step 6) and applied to current conditions, and also to assess the impacts of the proposed engineering works (step 7). At this stage, a sensitivity analysis involving all the pertinent parameters should be carried out, with particular emphasis on the parameters that are difficult to determine or that have significant spatial and/or temporal variability. In step 8, all the information gathered in the previous steps is used to formulate and implement the appropriate plan of action.

5.5 Reservoir Sedimentation Modeling

The basic governing equations involved in reservoir sedimentation processes are the same as in other bodies of water and are presented in the above sections. Other factors, however, may increase their complexity. Some of those factors are represented in Figure 5.15. Additionally, limnological variables may play a significant role. For example, there is a relationship between phytoplankton development and reservoir hydrodynamics: increased amounts of phytoplankton result in shallower thermoclines, warmer surface layers, and corresponding differences in hydrodynamics. The impact of limnological processes in reservoir circulation is a field that is poorly studied and that will not be covered in this chapter. Interested readers may refer to the survey paper by Bourget and Fortin (1995) and references therein.

The circulation in reservoirs is generally multi-dimensional, non-uniform, and unsteady. It is influenced by the hydrologic conditions of the reservoir and its watershed, by climate, by physiography, by the morphology of the reservoir, and by dam operation, among other parameters. Some of the water movements are periodical (e.g. seiche), and some are permanent (caused by the inflows of the main river and tributaries). Density stratification is usually present, due to the temperature, salinity, turbidity, and density of the reservoir waters. All these effects have a direct impact on the sedimentation processes in a reservoir, in its sediment trapping efficiency, and in the distribution of sediment deposits within the reservoir. As such, all the dominant physical processes must be included in a successful model. It is not the scope of this section to describe all of these complex processes with the detail they deserve. In this section, we will concentrate only on some of the most important aspects of reservoir hydraulics, sediment transport, and density currents.

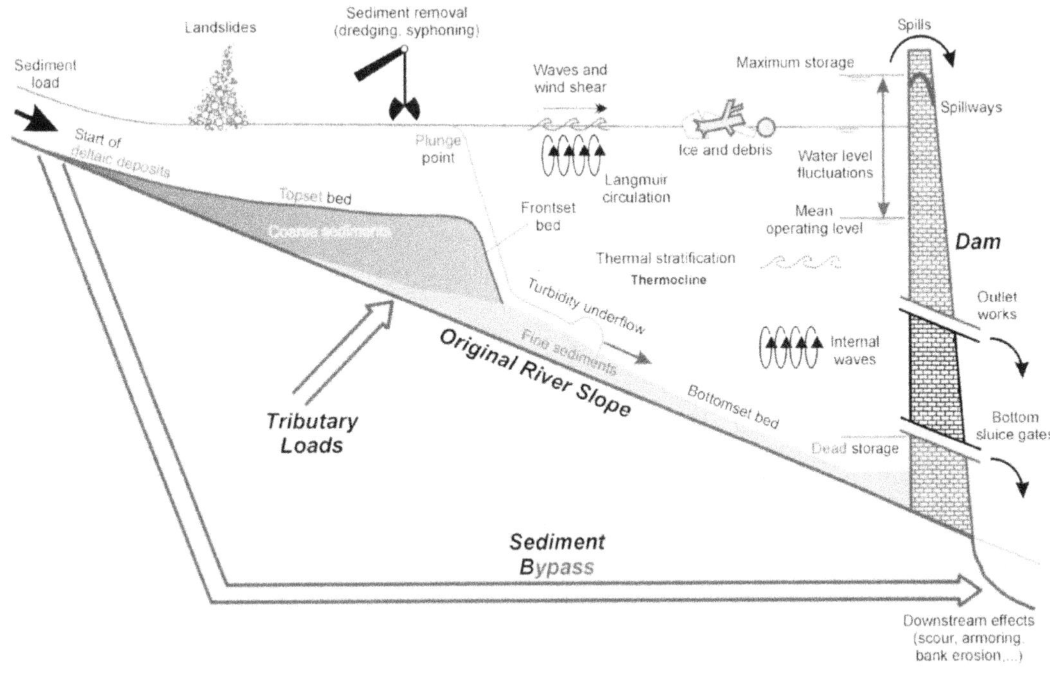

Figure 5.15. Schematic representation of the physical processes that may influence the sedimentation processes in a reservoir.

5.5.1 Reservoir Hydraulics

The equations governing fluid flow in reservoirs are the same as those governing the flow in rivers and other bodies of water, and they were presented in Section 5.2. Approximations to those equations are often made, depending on the dominant processes at play in the reservoir. Two- and one-dimensional models are widely used for engineering applications. One-dimensional models are appropriate in run-of-the-river reservoirs, where the flow is highly channelized and follows closely the thalweg, and where transverse mixing is well accomplished. On the other hand, when the reservoir pool is wide and without a single clear flow direction, multi-dimensional models must be used. Two different types of reservoirs where this might occur are shown in Figure 5.16. Tarbela Reservoir is a long and narrow reservoir, typical of mountain regions, where the flow behaves virtually like a one-dimensional river. In this case, the reservoir is defined by the area of the river where the backwater effects from the dam determine a level-pool free surface elevation. San Luis Reservoir has the configuration of a shallow, wide lake that is typical of low and flat regions, where flow circulation is essentially two-dimensional.

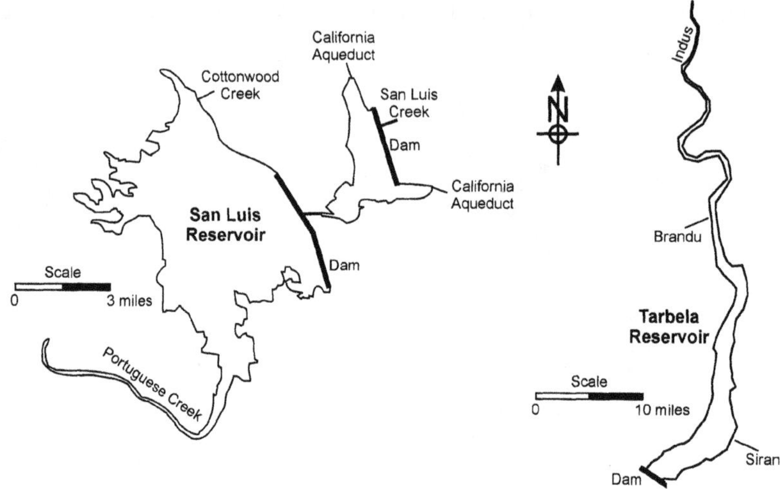

Figure 5.16. Pool contours for San Luis Reservoir, California, and for Tarbela Reservoir, Pakistan. Note the different scales in which the reservoirs are represented.

The type of mathematical model chosen for each particular application should reflect the physical characteristics of the reservoir. As an example, consider the de Saint Venant equation, Equation (5.20), rewritten using the velocity V and the depth of flow D as the dependent variables:

$$\underbrace{\overbrace{\frac{\partial(DV)}{\partial t}}^{\text{local}}+\overbrace{\frac{\partial(DV^2)}{\partial x}}^{\text{convective}}}_{\text{inertial force}}+\underbrace{\frac{\partial}{\partial x}\left(\frac{gD^2}{2}\right)}_{\text{pressure gradient}}+\underbrace{gDS_f}_{\text{friction}}-\underbrace{gDS_0}_{\text{weight}}=0 \tag{5.66}$$

The terms corresponding to the different forces acting on the control volume water column are shown in Equation (5.66). These forces are the dynamic mechanisms governing wave motion. Moreover, usually only some of these forces are significant and the others may be neglected. As a consequence, different types of wave modeling can be used (it may be helpful to follow this discussion using Table 5.6). Inertial or small gravity waves are waves dominated by inertial and pressure forces. Waves in which friction, gravity, and pressure forces dominate—where the inertial terms are negligible—are called diffusion waves. When only gravity and friction forces are present, the waves are called kinematic waves. When all forces in Equation (5.66) are important, then we talk of dynamic waves. Dynamic waves are the most general type of waves in open channel flow, and they are said to be steady when the local acceleration term—first term in Equation (5.66)—can be neglected.

Table 5.6. Shallow water wave types and the dominating forces that characterize them. The column tables represent, in order from left to right, the terms in Equation (5.66), and a cross mark (x) indicates that the corresponding term is important.

Wave type	Dominant forces				
	Local inertial	Convective inertial	Pressure	Friction	Gravity
Inertial waves	x	x	x		
Diffusion waves			x	x	x
Kinematic waves				x	x
Steady dynamic waves		x	x	x	x
Dynamic waves	x	x	x	x	x

For flood-wave modeling, Singh and Li (1993) developed a method to determine which type of flood-wave model is important in reservoirs. For the kinematic wave model to be important, the following inequality must be verified:

$$T \geq a \frac{D_0}{V_0 S_0} \tag{5.67}$$

where
a = a constant (= 138 if Manning's n is used, = 171 if Chézy's coefficient is used),
D_0 = mean flow depth of reservoir before arrival of the flood (in m),
V_0 = mean flow velocity in reservoir before arrival of the flood (m/s),
S_0 = bed slope, and
T = wave period (in s).

For the diffusion model, they indicate that

$$T \geq \frac{40}{S_0 \sqrt{g/D_0}} \tag{5.68}$$

and for inertial wave model,

$$T \leq \frac{0.01}{S_0 \sqrt{g/D_0}} \tag{5.69}$$

5.5.2 Sediment Transport in Reservoirs

Most studies of sedimentation in reservoirs are concerned with the silting processes; i.e., the amount of sediments trapped in the reservoir and their distribution within the reservoir. Regardless of whether the reservoir's purpose is flood protection or water supply, the rate of loss of its capacity is of interest because it determines the reservoir's useful life. The movement of the sediments within the reservoir is also important, due to the presence of contaminants that affect the quality of the reservoir's waters. There are several widely used empirical methods to determine the trapping efficiency and the distribution of sediments in reservoirs that will not be considered here. For detailed presentations of those methods, the reader is referred to Bureau of Reclamation (1987). Here, it suffices to point out that the movement of sediments in reservoirs is governed by the same flow and sediment transport equations presented in earlier sections. Next, we briefly present some specific aspects of interest in reservoir sedimentation, of which we exclude sediment density currents, to be presented in the following section.

The movement, deposition, and erosion of sediments in reservoirs in most models are treated similarly as the same processes in rivers or other quiescent bodies of water. There is little distinction between sedimentation formulae for river and channels, and for reservoirs. However, there are noteworthy differences between the two. For example, sediment deposits in reservoirs generally contain much finer materials (both in particle diameter and in relative percentages) than are found in their tributaries. Fine sediments that are usually considered wash load and, therefore, do not need to be modeled in river systems, find their way into reservoirs, where they are trapped. Consequently, silts and cohesive fines are often dominant in reservoirs. Cohesive sediments are considered in Chapter 4 of this manual and will be not pursued further here.

Most sediment transport equations were derived in very idealized conditions for uniform channel flow, and few were developed specifically for reservoirs. Because most equations are highly dependent on the data that were used to determine the values of their coefficients, it seems important to use equations that may have included reservoir data in their derivation or, better yet, that were derived specifically for reservoirs. Unfortunately, not many exist, and the authors know only of three, which are summarized next.

For reservoirs in South Africa, Rooseboom (1975) found that an equation based on unit stream power theory describes well the carrying capacity of flow through reservoirs. He proposed the following equation:

$$\log C_t = \alpha + \beta \log\left(V_m S_f\right) \tag{5.70}$$

where C_t = total sediment concentration, and
V_m = mean flow velocity.

The product $V_m S_f$, which is called the average unit stream power, can be expressed in terms of Chézy's equation, yielding

$$V_m S_f = \frac{PQ^3}{C^2 A^4} \tag{5.71}$$

where Q = flow discharge,
P = wetted perimeter,
A = flow area, and
C = Chézy roughness coefficient.

The values of the constants α and β were determined using field measurements in reservoirs and were found to be $\alpha = 5.30$ and $\beta = 0.283$. With these parameters, Equation (5.70) can be written as

$$C_t = 2\times10^5 \left(V_m S_f\right)^{0.283} \tag{5.72}$$

where $V_m S_f$ is calculated in m/s and C_t is in mg/l.

Westrich and Juraschek (1985) developed a sediment transport equation for silt-sized material in reservoirs:

$$C_v = \frac{0.0018\tau_b V}{(s-1)\rho g D \omega_s} \tag{5.73}$$

where C_v = sediment capacity concentration (by volume),
τ_b = bed shear stress,
s = specific gravity of silt,
ρ = fluid density,
g = acceleration due to gravity,
D = water depth, and
ω_s = settling velocity of the sediment particles.

Equation (5.73) was derived in the laboratory with particles having a settling velocity ranging from 0.6 to 9 mm/s. Note that the predicted transport capacities obtained from Equation (5.73) do not depend on bed-material composition, but only on the material in suspension. This equation also has the advantage of having been derived for silt sizes, while most sediment transport equations were derived for sand and gravel sizes (see Chapter 3). Atkinson (1992) expanded Equation (5.73) to include suspended sediment mixtures, including the effects that the presence of each size particle has on the fall velocity of the other size particles.

The Tsinghua University's equation (International Research and Training Center on Erosion and Sedimentation, 1985) is an empirical equation especially derived for calculating the transport capacity of flushing flows in reservoirs:

$$Q_s = \Omega \frac{Q^{1.6} S_0^{1.2}}{W^{0.6}} \tag{5.74}$$

where Q_s = sediment discharge (metric tons/s),
Q = water discharge (m^3/s),
S_0 = bed slope,
W = channel width (m), and
Ω = a factor that depends on sediment type.

The recommended values for Ω are presented in Table 5.7. The Tsinghua University's equation was derived from data of flushing reservoirs in China. The scatter of the data used is considerable, albeit not unusually high. Furthermore, the practice in China is to flush the reservoirs annually; therefore, little consolidation takes place between flushing events. In these conditions, the importance of reservoir operations is reduced. Extrapolation to reservoirs in other parts of the world should be done with caution.

Table 5.7. Values of the factor Ω in Tsinghua University's equation

Value of Ω	Type of sediments
1600	Loess sediments
650	Other sediments with median size finer than 0.1 mm
300	Sediments with median size larger than 0.1 mm
180	For flushing with a low discharge

During reservoir flushing operations, sediments are removed by hydraulic incision of the reservoir deposits. The thus formed channels have side slopes that vary across a wide range of values. Observed values are as low as 1.4° for poorly consolidated material, and as high as 90° in highly consolidated sediments. Because the effectiveness of the flushing operations depends on the size of the channels formed during the drawdown, it is important to have accurate predictions of the channel's side slopes.

In reservoirs, the side slopes of the channels formed during flushing depend on sediment properties, degree of consolidation, depth of deposits, and range of water fluctuations during flushing. Although the techniques described in Section 5.4 could be applied in this situation, specific techniques for reservoirs developed by Atkinson (1996) will be presented here.

Atkinson (1996) used observations of side slopes in several reservoirs from all over the world in his work. He also used theoretical concepts and laboratory observations of estuarine muds. He proposed two prediction methods. In the first method, he adopted earlier work by Migniot (1981) and used reservoir data to find an expression for the angle α at which the slope is just stable:

$$\tan \alpha = 6.30 \rho_d^{4.7} \tag{5.75}$$

where ρ_d = dry density of sediments, in metric ton/m^3.

A second method was developed using a numerical model, in which the results were presented in graphical format, as shown in Figure 5.17. Both methods were tested using observations in several reservoirs. As shown in Figure 5.18, there are discrepancies between computed and observed values, with Equation (5.75) overpredicting the measurements by an order of magnitude, while the method of Figure 5.17 seems to underpredict them by the same order of magnitude. Care should be exercised when using any of these methods in reservoir sedimentation models.

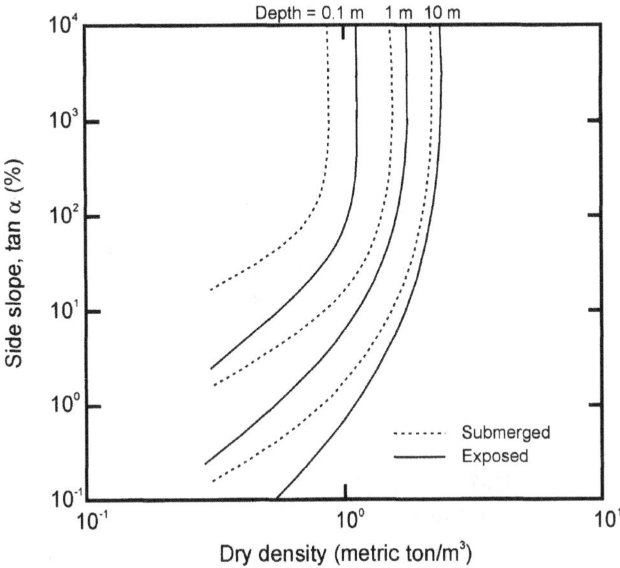

Figure 5.17. Side slope prediction at the limit of stability (Atkinson, 1996).

5.5.3 Turbid Underflows

Density currents are flows of fluids with different densities and occur normally in the stratified environments of lakes and reservoirs. They are gravity-driven flows caused primarily by the difference in the densities $\Delta\rho$ of the fluids involved. In a reservoir, they occur mainly because of the differences in density between the impounded and the inflowing waters. Density currents can be caused by differences in temperature ($\Delta\rho \approx 2$ kg/m^3), salinity ($\Delta\rho \approx 20$ kg/m^3), turbidity ($\Delta\rho \approx$ 20 to 200 kg/m^3), or a combination of any of these factors. Turbidity currents are density currents caused primarily by the presence of turbidity (they transport granular material), and they occur as underflows; i.e., they plunge under the lower density waters of the impoundment. Turbid density currents are important because they can significantly influence the distribution of sediments within a reservoir. If they reach the dam, they can be vented, allowing the removal of sediments without significant drawdown of the pool level. This section concerns the modeling of turbidity currents and their impacts on reservoir bathymetry. A more general treatment of density currents can be found in Turner (1973).

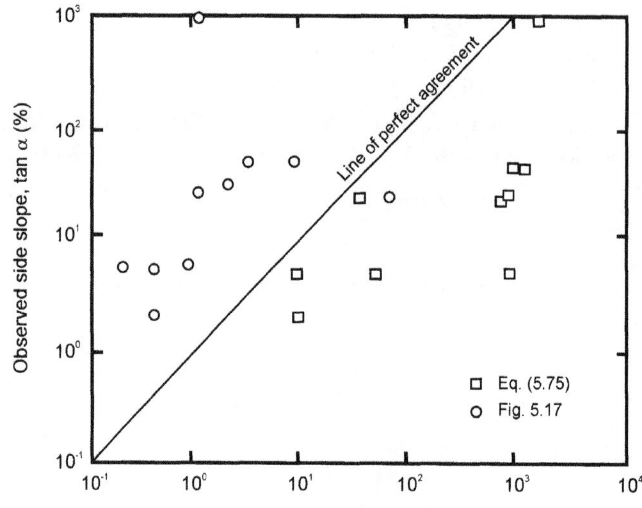

Figure 5.18. Comparison between the side slope prediction methods of Atkinson (1996).

5.5.3.1 Plunge Point

As the sediment transporting inflow waters enter the impounded waters, they plunge beneath the clear waters and travel downstream along the submerged thalweg. The point where the turbid inflowing water plunges beneath the ambient water is called the plunge point, or plunge line. The location of the plunge point is determined by the balance between stream momentum, the pressure gradient across the turbid-clear water interface, and the resisting shear forces. It is also influenced by morphologic factors, such as bed slope, bed roughness, and cross-sectional shape and area. In practice, the plunge point can be estimated from the densimetric Froude number F_p, defined as

$$F_p = \frac{V}{\sqrt{\varepsilon_i g D_p}} \tag{5.76}$$

where V = inflow velocity,
ε_i = relative density difference, $= (\rho_i - \rho_r)/\rho_r$,
ρ_i = density of the inflow water,
ρ_r = density of the receiving (ambient) water,
g = acceleration due to gravity, and
D_p = depth at the plunge point.

F_p has been observed to vary in the range between 0.2 and 0.8, and some of the values are presented in Table 5.8. Furthermore, Equation (5.76) is usually manipulated assuming a certain cross-sectional geometry, so that D_p can be found explicitly. A number of different authors assume different cross-sectional geometries, resulting in different expressions for the plunge depth equation. Some of the most common ones are presented in Table 5.9.

Table 5.8.—Values of the densimetric Froude number at the plunge point, after Morris and Fan (1998)

	Source of data	F_p
Laboratory data	Turbid water, 3-19 g/l	0.78
	Turbid water, 10-30 g/l	0.55-0.75
	Turbid water, 100-300 g/l	0.2-0.4
	Saline water	0.3-0.8
	Cold water	0.67
Field data	Liujiaxia Reservoir, Tao River	0.78
	Guanting Reservoir	0.5-0.78

In Table 5.9, the method of Akiyama and Stefan (1981) is the only one using the initial mixing coefficient, γ, which is defined as the ratio of entrained flow to the incoming flow:

$$\gamma = \frac{Q_{b0} - Q_i}{Q_i} \qquad (5.77)$$

where Q_{b0} = flow rate immediately downstream of the plunge point.

Note that if γ is known, Equation (5.77) can be used to compute Q_{b0}:

$$Q_{b0} = (1+\gamma)Q_i \qquad (5.78)$$

The empirical coefficient γ represents the initial mixing that occurs when the plunging turbidity current is subjected to the entrainment of the clear ambient water. The density of the inflow is correspondently decreased:

$$\rho_{b0} = \frac{\gamma\rho_r + \rho_i}{1+\gamma} \qquad (5.79)$$

where ρ_{b0} = density after entrainment at the plunge point.

Table 5.9. Formulations of the plunge depth D_p, according to several authors, with corresponding densimetric Froude numbers, F_p, where applicable

Authors	h_p	Notes
Savage and Brimberg (1975)	$D_p = \left(\frac{q_i^2}{g\varepsilon_i F_p^2}\right)^{1/3}$ $F_p^2 = \frac{2.05}{(1+\alpha)}\left(\frac{S_0}{f_b}\right)^{0.487}$	q_i is the inflow discharge per unit width, f_b is the bed friction coefficient, and α is the ratio of interfacial friction to bed friction.
Hebbert et al. (1979)	$D_p = \frac{1}{2\zeta}\left(\frac{2Q_i^2}{F_p^2 g\varepsilon_i (\tan\phi)^2}\right)^{1/5}$ $F_p^2 = \frac{\sin S_0 \tan\phi}{C_d}\left(1-0.85\sqrt{C_d}\sin S_0\right)$	Assumes triangular cross-section defined by the half-angle ϕ at the thalweg, for which $\zeta = 0.97$. Q_i is the inflow rate. C_d is a drag coefficient.
Jain (1981)	$D_p = 0.814\left(\frac{\alpha}{1+\alpha}\right)^{0.126}\left(\frac{f_t}{S_0}\right)^{0.325}\left(\frac{q_i^2}{g\varepsilon_i}\right)^{1/3}$	f_t is the total friction coefficient (i.e., interfacial plus bed friction).
Akiyama and Stefan (1981)	$D_p = 1.1(1+\gamma)\left(\frac{f_t q_i^2}{g\varepsilon_i S_0}\right)^{1/3}$	γ is the initial mixing coefficient.
Ford and Johnson (1983)	$D_p = \left(\frac{1}{F_p^2}\right)^{1/3}\left(\frac{Q_i^2}{W^2 g\varepsilon_i}\right)^{1/3}$	If F_p is known, D_p can be found iteratively by assuming a width W and calculating the depth.
Akiyama and Stefan (1985)	$D_p = \left(\frac{q_i^2}{g\varepsilon_i F_p^2}\right)^{1/3}$	Valid for channel slopes in the range $0.017 < S_0 < 0.124$.
Morris and Fan (1998)	$D_p = (g\varepsilon_i)^{-1/3}\left(\frac{Q_i}{F_p W}\right)^{2/3}$	Assumes a rectangular cross-section of width W.

From physical reasoning, it is expected that the mixing ratio will increase as the convective forces become dominant over the buoyant forces. Jirka and Watanabe (1980) represented this effect using an empirical relationship,

$$\gamma = 1.2F_p + 0.2 \tag{5.80}$$

which was derived for cooling ponds where $F_p \geq 0.167$. Unfortunately, the current knowledge about inflow mixing is incomplete. For example, there is the indication that for some reservoirs γ is independent of F_p. A more detailed discussion about this subject is given in Ford and Johnson (1983).

The stability of the density flows downstream from the plunge point is governed by the parameter θ (Rouse, 1950):

$$\theta = \frac{\nu g \varepsilon_i}{V^3} = \frac{1}{F_d^2}\frac{1}{R_e} \tag{5.81}$$

where F_d = the densimetric Froude number and R_e is the Reynolds number, defined as

$$R_e = \frac{VD}{\nu} \quad \text{and} \quad F_d = \frac{V}{\sqrt{g \varepsilon_i D}} \tag{5.82}$$

According to Rouse (1950), underflows are stable if $R_e F_d^{3/2} < 440$. Other criteria were given by Ippen and Harleman (1952) and by Keulegan (1949). The former applies to laminar flows and requires $\theta > 1/R_e$ for stability, while the latter indicates that stability requires $\theta < 0.127$ for laminar flows and $\theta < 0.178$ for turbulent flows.

5.5.3.2 Governing Equations

The governing equations for turbid underflows are based on the same conservation principles described in Section 5.2; namely, the conservation of mass, Equation (5.2), and conservation of momentum, Equations (5.1). To illustrate the use of those equations in the modeling of turbidity currents, this section will present the basic governing equations for the case of the steady-state flow in a vertical, two-dimensional plane under a stagnant fluid of density ρ_r. In this case $v = 0$; therefore, the continuity equation—Equation (5.2)—becomes

$$\frac{\partial u}{\partial x} + \frac{\partial w}{\partial z} = 0 \tag{5.83}$$

Integrating Equation (5.83) over the depth, from 0 to D, and taking into consideration that the horizontal velocity u is zero at the fluid's interface and the vertical velocity w is zero at the bed, one obtains

$$\frac{\partial}{\partial x}(DU) = -w_h \tag{5.84}$$

where U is the depth-averaged longitudinal velocity, defined by Equation (5.10), and w_h is the vertical velocity at the fluids interface. Figure 5.19 provides a sketch showing the symbols used for turbidity underflows. w_h is defined as the velocity of entrainment of the ambient fluid into the turbid current and is generally assumed to be proportional to the velocity U of the current:

$$-w_h = E_w U \tag{5.85}$$

where E_w = entrainment coefficient of the ambient fluid.

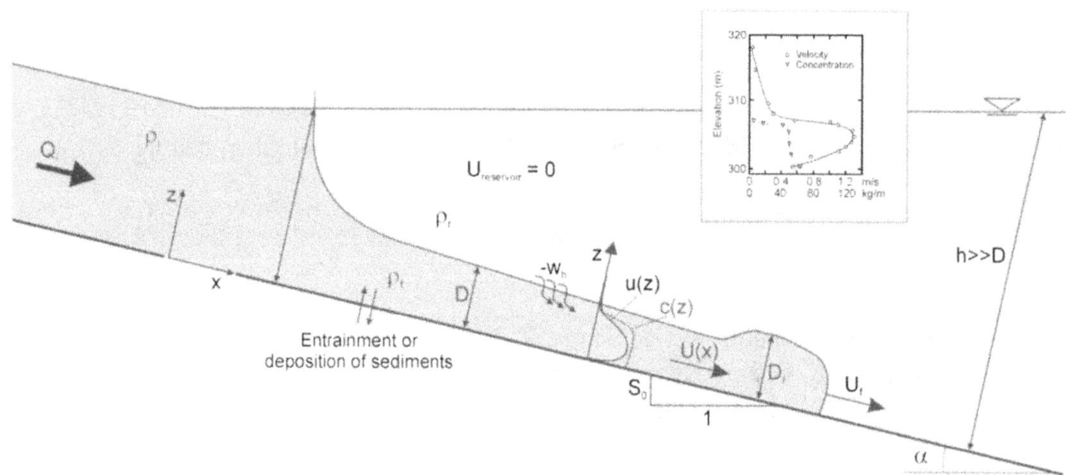

Figure 5.19. Sketch showing the symbols used to describe turbidity underflows. The inset shows typical velocity and concentration profiles in reservoirs, as measured by Fan (1991).

Conservation of longitudinal momentum (along the x axis) can be taken directly from Equation (5.1) for $i = 1$ and, with the assumptions discussed above, written for the turbid underflow as

$$\frac{\partial u^2}{\partial x}+\frac{\partial uw}{\partial z}=-\frac{1}{\rho_t}\frac{\partial}{\partial x}\left(p_t+\rho_t gz\cos\alpha\right)+\frac{1}{\rho_t}\frac{\partial \tau_{zx}}{\partial z} \tag{5.86}$$

The pressure due to the above clear water is assumed to be hydrostatic, and within the turbid flow itself, p_t is decomposed in part due to the ambient fluid, p_a, and in part due to the presence of sediments, p_s (= $p_t - p_a$). Using this approximation, and after integrating along the vertical, Equation (5.86) becomes

$$\frac{\partial\left(DU^2\right)}{\partial x}=-\frac{1}{2}g\varepsilon_t\cos\alpha\frac{\partial\left(C_sD^2\right)}{\partial x}+g\varepsilon_tC_sD\sin\alpha-U_*^2 \tag{5.87}$$

The full derivation can be found in more detail in Parker et al. (1987), for example. In Equation (5.87), ε_t is the relative density difference, = $(\rho_t - \rho_r)/\rho_r$, U_* is the shear velocity, and C_s is the depth-integrated value of the concentration in the turbid flow, defined as

$$C_s=\frac{1}{D}\int_0^D c(z)dz \tag{5.88}$$

Finally, the equation for the conservation of the solid particles can be obtained in a similar way from Equation (5.3) with the same type of simplifying assumptions. Expressing the turbulent Reynolds flux of sediments using a diffusion analogy; i.e., using

$$-\frac{\partial\left(\overline{c'w'}\right)}{\partial z} = D_z \frac{\partial^2 c}{\partial z^2} \tag{5.89}$$

we can write

$$\frac{\partial(uc)}{\partial x} + \frac{\partial(wc)}{\partial z} = \omega_s \cos\alpha \frac{\partial c}{\partial z} + D_z \frac{\partial^2 c}{\partial z^2} \tag{5.90}$$

where ω_s = sediments fall velocity.

Integrating Equation (5.90) over the depth yields

$$\frac{\partial(DUC)}{\partial x} = \omega_s (E_s - c_0) \tag{5.91}$$

In deriving Equation (5.91), the entrainment coefficient E_s proposed by Parker et al. (1987) was used to represent the erosion of bed sediments. The term $\omega_s c_0$ represents the deposition of sediments on the bed, where c_0 is the sediment concentration at the bed.

Equations (5.84), (5.87), and (5.91) represent conservation of fluid mass, momentum, and sediment mass, and constitute the set of governing equations for modeling steady, one-dimensional, turbid underflows. The partial derivatives ($\partial/\partial x$) in those equations can be replaced by total derivatives (d/dx).

Finally, it is noteworthy to point out that alternative forms of the governing equations are useful in certain circumstances. They express the profiles of the interface between the turbid water and the clear water above using the depth D and the Richardson number R_i, defined as

$$R_i = \frac{g \varepsilon_t D C_s \cos\alpha}{U^2} \tag{5.92}$$

They are:

$$\frac{dD}{dx} = \frac{1}{1-R_i}\left[\frac{1}{2}(4-R_i)E_w + \frac{1}{2}R_i \frac{\omega_s}{UC_s}(E_s - c_0) - R_i \tan\alpha + \left(\frac{U_*}{U}\right)^2\right] \tag{5.93}$$

$$\frac{dR_i}{dx} = \frac{3R_i}{D(1-R_i)}\left[\frac{1}{2}(2+R_i)\left(E_w + \frac{1}{3}\frac{\omega_s}{UC_s}(E_s - c_0)\right) - R_i \tan\alpha + \left(\frac{U_*}{U}\right)^2\right] \tag{5.94}$$

5.5.3.3 Additional Relationships

The governing equations presented in the previous section work with three unknowns: D, U, and C_s. However, relationships for the parameters E_s, E_w, ω_s, U_*, and c_0 are needed. Empirical relationships for these parameters have been developed by several authors. In this section, a brief presentation of some of the most commonly used parameters is made.

Determination of flow resistance is a topic that still needs research. In many cases, the traveling velocity of the turbidity currents is constant, and the currents can be well approximated by uniform flow. The shear velocity, which is defined by Equation (5.43), can be expressed in terms of a friction coefficient, f_t, using a Darcy-Weisbach type of equation:

$$f_t = \frac{\tau_b}{\rho U^2/8} = 8\left(\frac{U_*}{U}\right)^2 \tag{5.95}$$

The total friction coefficient f_t is the sum of the bed friction acting on the wetted perimeter, f, and the friction acting upon the clear-turbid liquid interface. Under the conditions of uniform, conservative underflow, the velocity can be estimated by a Chézy or Darcy-Weisbach type of equation:

$$U = C\sqrt{g\varepsilon_t D S_0} = \sqrt{\frac{8}{f_t}}\sqrt{g\varepsilon_t D S_0} \tag{5.96}$$

C ranges between 280 and 560 cm$^{1/2}$/s. A frequently used relationship for f_t is given by (Harleman, 1961)

$$f_t = (1+\alpha)f \tag{5.97}$$

where f = friction factor taken from the Moody diagram for pipe flow, and
α = factor representing the shear distribution at the interface as a function of bed shear (= 0.43).

Parker et al. (1987) wrote

$$U_*^2 = c_D U^2 \tag{5.98}$$

in which the value of the friction coefficient c_D ranges from 10^{-3} to 10^{-1} for Reynolds numbers—see Equation (5.82)—between 4×10^2 and 2×10^6. From Equation (5.95), it is clear that $c_D = f_t/8$.

Procedures to compute the values of the fall velocity ω_s can be found in any standard textbook about sediment transport, some of which are presented in Chapter 3. The value of the reference concentration c_0 is evaluated close to the bed, at a height of $0.05h$, and can be expressed as a function of the grain size as

$$c_0 = r_0 C_s \tag{5.99}$$

where the shape factor r_0 is a function of grain size. Parker et al. (1987) found $r_0 \approx 2$, remaining more or less constant in the range $1 < U_* / \omega_s < 50$ for uniform sediments. For non-uniform sediments

$$r_{0i} = 1.64 + 0.40 \delta_i^{1.64} \tag{5.100}$$

where δ_i = normalized grain size, = d_i/d_{50},

and Equation (5.99) becomes

$$c_{0i} = r_{0i} C_{si} \tag{5.101}$$

The entrainment coefficient for the bed sediments, E_s, can be computed after Garcia and Parker (1991):

$$E_s = \frac{a Z_u^5}{1 + \dfrac{a}{0.3} Z_u^5} \tag{5.102}$$

where a = a numerical constant (= 1.3×10^{-7}), and
Z_u = a similarity variable defined by

$$Z_u = \begin{cases} 0.586 \dfrac{U_*}{\omega_s} R_{ep}^{1.23} & \text{if} \quad 1.0 \le R_{ep} \le 3.5 \\ \dfrac{U_*}{\omega_s} R_{ep}^{0.60} & \text{if} \quad R_{ep} > 3.5 \end{cases} \tag{5.103}$$

with the particle Reynolds number R_{ep} defined by

$$R_{ep} = \frac{\sqrt{g \varepsilon_t d^3}}{\nu} \tag{5.104}$$

where d = mean diameter of the sediment particles.

The coefficient for the entrainment of the ambient fluid, E_w, can also be prescribed by an empirical relationship given by Parker et al. (1987):

$$E_w = \frac{0.075}{\sqrt{1+718R_i^{2.4}}} \tag{5.105}$$

where the bulk Richardson number R_i is defined by Equation (5.92). The scatter of the experimental data is rather high. Alternatively, a simpler relationship proposed by Egashira and Ashida (1980) can be used:

$$E_w = 0.0015R_i^{-1} \tag{5.106}$$

Many relationships have been proposed to determine the velocity of the advancement of the turbid underflow, U_f. An analysis based on the difference in pressure across the front of the advancing flow, assuming that $\rho_r \approx \rho_t$, yields (Turner, 1973)

$$U_f = \sqrt{2\varepsilon_t D} \tag{5.107}$$

A simple relationship for small slopes is (e.g. Turner, 1973)

$$U_f = 0.75\sqrt{\varepsilon_t D_f} \tag{5.108}$$

with D_f defined in Figure 5.19, an equation that has been well verified by others was developed by Britter and Linden (1980):

$$\frac{U_f}{\sqrt[3]{\varepsilon_t q_i}} = 1.5 \pm 0.2 \tag{5.109}$$

where q_i = inflow discharge.

To conclude this section, the venting of the turbidity currents that reach the dam will be considered. The venting of the sediment-laden flows is done by low level outlet gates, but it is necessary to have adequate information to avoid discharging clear water instead of the desired turbid fluid. This information includes not only the timing of arrival and concentration of the currents, but also the position of the interface between the clear reservoir water and the turbid underflow. The results presented next follow the discussion in Section 14.7.5 of Morris and Fan (1998).

The limiting height for the aspiration of a density current, based on experiments with saline water, is given by the following expressions:

$$\frac{\varepsilon_t g h_a^3}{q^2} = 0.43 \text{ for slots} \tag{5.110}$$

$$\frac{\varepsilon_t g h_a^5}{Q} = 0.154 \text{ for orifices} \tag{5.111}$$

where h_a = aspiration height (defined in Figure 5.20),
Q = total discharge through orifice, and
q = discharge per unit width for slotted gates.

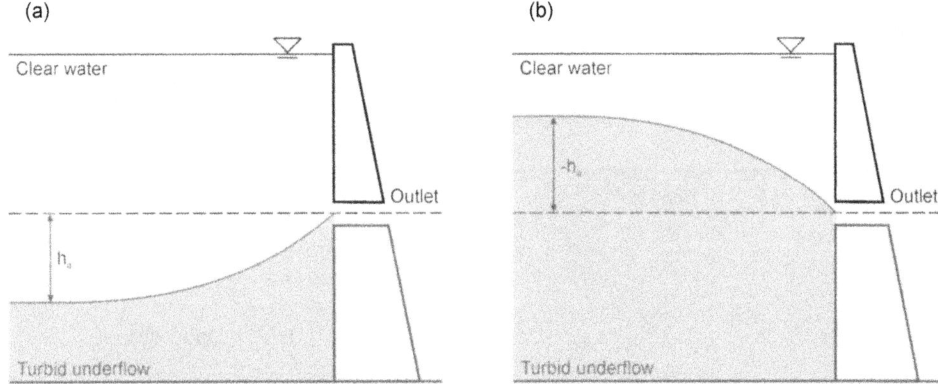

Figure 5.20. Venting of turbid underflows by the dam's bottom outlets showing (a) the limit height for aspiration of the turbidity current from below an outlet that is too high, and (b) the limit height for aspiration of clear water from above an outlet that is below the turbid-clear water interface. (Adapted from Morris and Fan (1998), with modifications.)

More recent research produced the relationships of Table 5.10. Observations from releases in the Sanmexia Reservoir during 1961 and 1962 produced the following orifice equation:

$$h_a = \Gamma \left(\frac{Q^2}{\varepsilon_t g} \right)^{1/5} \tag{5.112}$$

where Γ = 0.9 for the lower aspiration limit, Figure 5.20 (a), in 1961,
= 1.1 for the lower aspiration limit in 1962, and
= -0.4 for the clear water aspiration limit, Figure 5.20 (b), in both years.

Table 5.10. Expressions for the limiting heights for venting density underflows from bottom outlets

	Slot	Orifice
Limiting depth for aspiration of turbid water from below, Figure 5.20 (a)	$\left(\frac{\varepsilon_t g h_a^3}{q^2} \right)^{1/3} = 0.75$	$\left(\frac{\varepsilon_t g h_a^5}{Q^2} \right)^{1/5} = 0.8$
Limiting height for aspiration of clear water from above, Figure 5.20 (b)	$\left(\frac{\varepsilon_t g (h_a)^5}{q^2} \right)^{1/3} = -0.75$	$\left(\frac{\varepsilon_t g (h_a)^5}{Q^2} \right)^{1/5} = -1.2$

5.5.4 Difference Between Reservoirs and Other Bodies of Water

The previous subsections were devoted to presenting some of the most important processes directly involved with sedimentation modeling in reservoirs. However, these basic processes are virtually identical to those governing other bodies of pooled water, such as retention basins, ponds, pools, and lakes. Nonetheless, these are often distinguished as different types of water bodies. Due to the importance of lake circulation, which stems from ecological and water quality concerns, we will briefly address some of the most important differences between lakes and reservoirs as they may relate to modeling (see also Table 5.11).

Table 5.11. Principal differences between lakes and reservoirs (Marzolf, 2003)

	Lakes	Reservoirs
Ratio of surface to watershed area	Smaller, ~10	Larger, >500
Retention time	Longer	Shorter
Depth	Deeper	Shallower
Inflows	None to several	One dominant
Outflows	None to one	One
Transparency	Clear	Variable

The differences between lake and reservoir circulation and sedimentation result from the differences between specific combinations of the general forcing mechanisms that dominate the particular behavior of each water body. A major difference between the circulation of the two results from dam operation and the position of the dam outlets. Deep sluice gates remove water that is much different in nature than the water removed via the usually shallower lake outlets. In some dams, gates at different levels are operated, in turn, to selectively remove warmer or colder water from the reservoir—a requirement of downstream river management practices that has direct impacts on its ecology. Some gates are specifically designed to vent density currents. Some reservoirs are emptied periodically. As a result, the reservoir levels have much greater variation than those of lakes, as does the position of the thermocline, the latter having a major impact on reservoir limnology.

Recent studies indicate significant limnological differences between lakes and reservoirs. Straškraba (1998) shows a significant difference between the retention times of phosphorus for lakes and reservoirs (Figure 5.21). Retention of phosphorus results from the direct sedimentation of the phosphorus-carrying particles present in the inflowing river water, as well as from the sedimentation of particles formed within the water body itself—mostly by phytoplankton. This difference, which is due to sedimentation, indicates significant differences in the sedimentation processes between lakes and reservoirs, including their respective diffusion and resuspension processes. Due to the complex cycling of phosphorus, which is affected by chemical and biological processes by phytoplankton, zooplankton, and fish, other significant limnological differences may be expected.

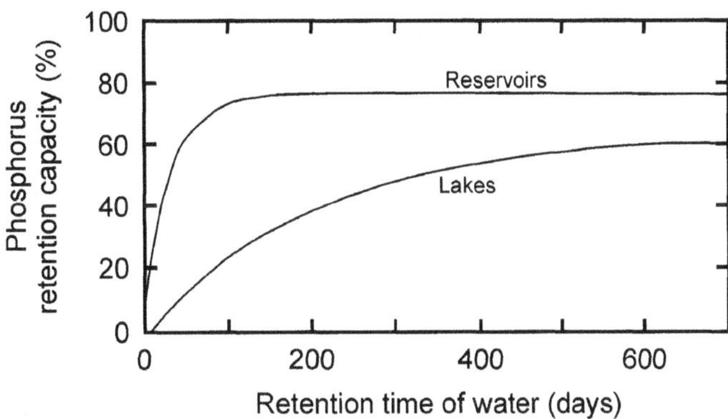

Figure 5.21. Phosphorus retention capacity in lakes and reservoirs, adopted with modification from Straškraba (1998). The retention capacity is given as a percentage of the lake/reservoir output as related to the loads, due to their respective inflows.

Another important feedback mechanism described by Straškraba relates phytoplankton development and hydrodynamics. Increase in phytoplankton results in decreased light penetration, decreased depth of the seasonal thermocline, and increased temperature of the uppermost water strata. This increase in temperature contributes to increased phytoplankton growth, therefore causing a positive feedback phenomenon. Of course, the depth of the thermocline and the temperature of the uppermost layer have a substantial and direct impact on the circulation of the water body. This feedback can be reproduced in a model only if physical and biological parameters are both included, which is not done in most hydrodynamic models. Other complex relations between limnology and hydrodynamics can be drawn. Here, it suffices to say that a model may have to include these effects if accurate computations of the circulation and sedimentation of reservoirs, lakes, and other bodies of water are sought. The particular type of interaction depends on the specific dominating processes at play, which may differ significantly between each water body. It is the responsibility of the modeler to identify and implement such processes, and multidisciplinary research in this topic is still vastly needed.

5.6 Data Requirements

In general, the basic data requirements for loose boundary hydraulic models can be grouped in three broad categories: geometric data, hydraulic data, and sediment data. These data establish the boundary conditions necessary to solve the governing equations and are an integral part of a model. The term "model" thus refers to the ensemble of the set of governing equations, their numeric solution technique, their implementation in a computer program, and the data that defines the prototype. In practice, data collection is more often the responsibility of the user of the computer program that implements the numerical part of the model, or of someone else that then conveys that data to the user. However, data collection and preparation are no less important than the computer model itself, and often play the dominant role in determining the accuracy and applicability of the final numerical solutions generated by the computer.

The geometry data defines the topography of the reach to be simulated; i.e., the channel bed, banks, and flood plains. In two- and three-dimensional models, the data is most often presented as a set of points given by its x, y, and z coordinates. The data is then interpolated to the locations of the grid nodes used in the discretization of the problem. Interpolation may be done using standard triangulation techniques, or with more sophisticated techniques, such as kriging. In river channels, special techniques are often used that, in the interpolation process, emphasize the contribution of the data points along the flow streamlines, as they have the potential of better capturing the longitudinal characteristics of the riverbed (i.e., the meandering of the thalweg).

In one-dimensional models, the geometry is usually defined by cross-sections. Each cross-section is a line representing a particular section of the modeled reach and is given by a set of points, each defined by a lateral distance and a bed elevation above a common datum. This line provides the information about the section shape and the locations of the subchannels, and should be taken between locations well above the highest stage levels. It should also be taken perpendicularly to the flow streamlines. Additionally, the distance between the cross-sections needs to be given, and this distance must be collected along the flow streamlines. Under bankfull conditions, this distance can be taken along the channel's thalweg, but this may vary under flooding conditions.

There are techniques that provide specific guidelines on how to collect cross-section data in the field. For example, some of the criteria recommended by Samuels (1990) are as follows:

- Select all sites of key interest;
- Select cross-sections adjacent to major structures and control points;
- Select cross-sections representative of the river geometry;
- As a first estimate, select cross-sections $20W$ apart, where W is the cross-section's top width;
- Select sections a maximum of $0.2D/S_w$ apart, where D is the water depth and S_w is the free surface slope;
- For unsteady flow modeling, select sections a maximum of $L/30$ apart, where L is the length scale of the physically important wave (flood or tide);
- Select sections a minimum of $10^{INT(\log Z)-\varepsilon}/(\delta S - S_w)$ apart, where ε is the machine precision, Z is the water surface elevation, $INT()$ is the function that represents the integer part of its argument, and δS is the relative error in the slope;
- The ratio of the areas between two adjacent cross-sections should lie between 2/3 and 3/2.

Hydraulic data encompasses the necessary upstream and downstream flow conditions, as well as friction factors and local head losses. Subcritical flows require the flow discharge at the upstream boundary and the stage at the downstream end, while supercritical flows require both the discharge and the stage at the upstream boundary (the analysis leading to this result can be found in any hydraulics textbook, such as Henderson, 1966). In two- and three-dimensional flows, this is equivalent to prescribing flow velocity field at the upstream boundary, and a stage at the downstream boundary. Because usually only the upstream discharge is known, a vector velocity field must be synthesized, which is usually done by distributing the discharge proportionally to the conveyance along each node of the computational boundary, and then fitting a logarithmic profile along the vertical direction. Thus, any number of hydrographs can be discretized on an arbitrarily complex computational domain.

The stage can be defined in a number of ways. Stage-discharge rating curves, an elevation hydrograph, or a water-surface-slope hydrograph are common ways to achieve that purpose. More complex conditions may be defined in the case in which the downstream boundary is a dam. In such cases, the reservoir operational scheme may be used. When the dam outlet works are used, relationships for the gates and spillways may have to be employed. These relationships are a function of the head at the dam, and more complex iterative schemes need to be used. In tidal regions, special relationships may be necessary.

In some models, especially if steady-state solutions are sought, special downstream boundary conditions are employed. These boundary conditions, generally known as non-reflective boundary conditions, prevent wave forms generated by spurious numerical solutions from being reflected back into the computational domain, as would happen in the case of a clamped down, free surface elevation. The use of these techniques allows the spurious waves to flow out of the computational domain and may significantly increase numerical convergence rates. A description of such techniques can be found in Keller and Givoli (1989).

Friction factors play an important role in determining stages and flow velocities, and they must be given in the form of numerical values associated with particular regions of the bed, or in the form of relationships that allow their representation as a function of other parameters—usually using hydraulic and/or sediment quantities. Local energy loss coefficients must also be given, such as those due to channel bends, natural or manmade obstacles to the flow, bridge piers and abutments, etc. Once again, these may be prescribed or may be calculated, and they commonly require iterative procedures.

Sediment data encompasses all the necessary information for sediment transport computations. The sediment inflow hydrograph must be given at the inflow boundaries. For two- and three-dimensional models, it is also necessary to specify the sediment concentration distribution along those boundaries, as well as the separation between bedload and suspended load. The sediment particle-size distributions are also needed. Bed size distributions need also to be determined for each computational grid node (or for each cross-section, in the case of one-dimensional models). Additionally, especially in the case of scour computations and bank widening, it is also necessary to provide the underlying bed-material size distribution. Water temperature variations must also be prescribed or modeled because they have an indirect impact on sediment transport via the sediment particles' fall velocities.

In practice, it is difficult to determine a priori some of the hydraulic parameters necessary for a successful simulation. For that reason, usually there is a model calibration stage in which stage and discharge observations along the study reach are used to adjust the values of those parameters, such as bed roughness, discharge coefficients, or other parameters particular to the model employed. Similarly, there should be a calibration stage for the sediment transport calculations. Observations of the sediment outflow quantities, of variations in channel width and bed elevation, and of changes in sediment particle-size distributions can be used to properly adjust model parameters.

5.7 One-Dimensional Model Comparison

Most sediment routing models for long-term simulation of long reaches of a river are one-dimensional models. Only one-dimensional models are considered in the following comparisons. There are many sediment transport models, and each model has its strengths and weaknesses. Comprehensive reviews of the capabilities and performance of these models are provided in reports by the National Research Council (1983), and Fan (1988), among others. Fifteen U.S. Federal agencies participated in a Federal Interagency Stream Restoration Working Group (1998) to produce a handbook on *Stream Corridor Restoration Principles, Processes, and Practices*. They selected the following eight models for comparison: CHARIMA (Holly et al., 1990), FLUVIAL-12 (Chang, 1990), HEC-6 (USACE, 1993), TAS-2 (McAnally and Thomas, 1985) MEANDER (Johannesson and Parker, 1985), USGS (Nelson and Smith, 1989), D-O-T (Darby and Thorne, 1996, and Osman and Thorne, 1988), and GSTARS (Molinas and Yang, 1986). Table 5.12 summarizes the comparisons of these eight models. Because the U.S. Bureau of Reclamation has now replaced GSTARS with GSTARS 2.1 and GSTARS3 (Yang and Simões, 2000 and 2002), the newer versions of GSTARS are included in Table 5.12. HEC-6, TABS-2, USGS, and GSTARS 2.1 and GSTARS3 are Federal models in the public domain; CHARIMA, FLUVIAL-12, MEANDER, and D-O-T are academic or privately owned models. GSTARS3 comprises reservoir sedimentation modeling, while GSTARS 2.1 emphasizes river sedimentation.

5.8 Example: The GSTARS Models

In this section, the basic principles and their implementation in the GSTARS 2.1 and GSTARS3 models are presented as an example of alluvial river models. GSTARS3 is a recent extension of GSTARS 2.1 to reservoir sedimentation. The following presentation will concentrate on presenting GSTARS3. Examples of application of both models will be given in Section 5.8.6 below. The Generalized Sediment Transport model for Alluvial River Simulation, version 3 (GSTARS3, Yang and Simões, 2002) is a publicly and freely available model developed by the U.S. Bureau of Reclamation. Unlike many of the other one-dimensional alluvial river modeling computer programs, GSTARS3 has the goal to simulate the flow conditions in a semi-two-dimensional manner and the change of channel geometry in a semi-three-dimensional manner. By using stream tubes within an essentially one-dimensional backwater model, this can be accomplished without the intensive data and computational requirements of the more sophisticated, truly two- and three-dimensional models.

Table 5.12. Comparison of sediment transport models (Y = yes; N = no)

<table>
<tr><th>Model</th><th>CHAR IMA</th><th>Fluvial-12</th><th>HEC-6</th><th>TABS-2</th><th>Mean-der</th><th>USGS</th><th>D-O-T</th><th>GSTARS 2.1/ GSTARS3</th></tr>
<tr><td colspan="9">Discretization and formulation:</td></tr>
<tr><td>Unsteady flow | stepped hydrograph</td><td>Y|Y</td><td>Y|Y</td><td>N|Y</td><td>Y|Y</td><td>N|Y</td><td>Y|Y</td><td>N|Y</td><td>N|Y</td></tr>
<tr><td>One-dimensional | quasi two-dimensional</td><td>Y|N</td><td>Y|Y</td><td>Y|N</td><td>N|N</td><td>N|N</td><td>N</td><td>Y|Y</td><td>Y|Y</td></tr>
<tr><td>Two-dimensional | depth-average flow</td><td>N</td><td>N</td><td>N</td><td>Y|Y</td><td>Y|Y</td><td>Y|Y</td><td>N</td><td>N|Y</td></tr>
<tr><td>Deformable bed | banks</td><td>Y|N</td><td>Y|Y</td><td>Y|N</td><td>Y|N</td><td>Y|N</td><td>Y|N</td><td>Y|Y</td><td>Y|Y</td></tr>
<tr><td>Graded sediment load</td><td>Y</td><td>Y</td><td>Y</td><td>Y</td><td>Y</td><td>N</td><td>Y</td><td>Y</td></tr>
<tr><td>Non-uniform grid</td><td>Y</td><td>Y</td><td>Y</td><td>Y</td><td>Y</td><td>Y</td><td>Y</td><td>Y</td></tr>
<tr><td>Variable time stepping</td><td>Y</td><td>N</td><td>Y</td><td>N</td><td>N</td><td>N</td><td>N</td><td>Y</td></tr>
<tr><td colspan="9">Numerical solution scheme:</td></tr>
<tr><td>Standard step method</td><td>N</td><td>Y</td><td>Y</td><td>N</td><td>N</td><td>N</td><td>Y</td><td>Y</td></tr>
<tr><td>Finite difference</td><td>Y</td><td>N</td><td>Y</td><td>N</td><td>Y</td><td>Y</td><td>Y</td><td>Y</td></tr>
<tr><td>Finite element</td><td>N</td><td>N</td><td>N</td><td>Y</td><td>N</td><td>N</td><td>N</td><td>N</td></tr>
<tr><td colspan="9">Modeling capabilities:</td></tr>
<tr><td>Upstream water and sediment hydrographs</td><td>Y</td><td>Y</td><td>Y</td><td>Y</td><td>Y</td><td>Y</td><td>Y</td><td>Y</td></tr>
<tr><td>Downstream stage specification</td><td>Y</td><td>Y</td><td>Y</td><td>Y</td><td>Y</td><td>N</td><td>Y</td><td>Y</td></tr>
<tr><td>Flood plain sedimentation</td><td>N</td><td>N</td><td>N</td><td>Y</td><td>N</td><td>N</td><td>N</td><td>N</td></tr>
<tr><td>Suspended | total sediment transport</td><td>Y|N</td><td>Y|N</td><td>N|Y</td><td>Y|N</td><td>N|N</td><td>N|Y</td><td>N|Y</td><td>N|Y</td></tr>
<tr><td>Bedload transport</td><td>Y</td><td>Y</td><td>Y</td><td>N</td><td>Y</td><td>N</td><td>N</td><td>Y</td></tr>
<tr><td>Cohesive sediments</td><td>N</td><td>N</td><td>Y</td><td>Y</td><td>N</td><td>Y</td><td>N</td><td>Y</td></tr>
<tr><td>Bed armoring</td><td>Y</td><td>Y</td><td>Y</td><td>N</td><td>N</td><td>N</td><td>Y</td><td>Y</td></tr>
<tr><td>Hydraulic sorting of substrate material</td><td>Y</td><td>Y</td><td>Y</td><td>N</td><td>N</td><td>N</td><td>Y</td><td>Y</td></tr>
<tr><td>Fluvial erosion of streambanks</td><td>N</td><td>Y</td><td>N</td><td>N</td><td>N</td><td>N</td><td>Y</td><td>Y</td></tr>
<tr><td>Bank mass failure under gravity</td><td>N</td><td>N</td><td>N</td><td>N</td><td>N</td><td>N</td><td>Y</td><td>N</td></tr>
<tr><td>Straight | irregular non-prismatic reaches</td><td>Y|N</td><td>Y|N</td><td>Y|N</td><td>Y|Y</td><td>N|N</td><td>N|N</td><td>Y|Y</td><td>Y|Y</td></tr>
<tr><td>Branched | looped channel network</td><td>Y|Y</td><td>Y|N</td><td>Y|N</td><td>Y|Y</td><td>N|N</td><td>N|N</td><td>N|N</td><td>N|N</td></tr>
<tr><td>Channel beds</td><td>N</td><td>Y</td><td>N</td><td>Y</td><td>Y</td><td>N</td><td>Y</td><td>Y</td></tr>
<tr><td>Meandering belts</td><td>N</td><td>N</td><td>N</td><td>N</td><td>N</td><td>Y</td><td>N</td><td>N</td></tr>
<tr><td>Rivers</td><td>Y</td><td>Y</td><td>Y</td><td>Y</td><td>Y</td><td>Y</td><td>Y</td><td>Y</td></tr>
<tr><td>Bridge crossings</td><td>N</td><td>N</td><td>N</td><td>Y</td><td>N</td><td>N</td><td>N</td><td>N</td></tr>
<tr><td>Reservoirs</td><td>N</td><td>Y</td><td>Y</td><td>N</td><td>N</td><td>N</td><td>N</td><td>Y</td></tr>
<tr><td colspan="9">User support:</td></tr>
<tr><td>Model documentation</td><td>Y</td><td>Y</td><td>Y</td><td>Y</td><td>Y</td><td>Y</td><td>Y</td><td>Y</td></tr>
<tr><td>User guide | hot-line support</td><td>N|N</td><td>Y|N</td><td>Y|Y</td><td>Y|N</td><td>N|N</td><td>Y|N</td><td>N|N</td><td>Y|N</td></tr>
</table>

GSTARS3 was developed due to the need for a generalized water and sediment-routing computer model that could be used to solve complex river engineering problems for which limited data and resources were available. In order to be successful, such a model should have a number of capabilities, such as being able to compute hydraulic parameters for open channels with fixed as well as movable boundaries; having the capability of computing water surface profiles in the subcritical, supercritical, and mixed flow regimes (i.e., in combinations of subcritical and supercritical flows and through hydraulic jumps without interruption); being able to simulate and predict the hydraulic and sediment variations, in both the longitudinal and transverse directions; being capable of simulating and predicting the change of alluvial channel profile and cross-sectional geometry, regardless of whether the channel width is variable or fixed; and it should incorporate site-specific conditions, such as channel side stability and erosion limits.

GSTARS3 consists of four major parts. The first part is the use of both the energy and the momentum equations for the backwater computations. This feature allows the program to compute the water surface profiles through combinations of subcritical and supercritical flows. In these computations, GSTARS3 can handle irregular cross-sections, regardless of whether single channel or multiple channels separated by small islands or sand bars.

The second part is the use of the stream tube concept, which is used in water and sediment routing computations. Hydraulic parameters and sediment routing are computed for each stream tube, thereby providing a transversal variation in the cross-section in a semi-two-dimensional manner. The scour or deposition computed in each stream tube gives the variation of channel geometry in the vertical (or lateral) direction. Bed sorting and armoring in each stream tube follow the method proposed by Bennett and Nordin (1977), and the rate of sediment transport can be computed using any of the several transport functions implemented in the code (see below for more details).

The third part is the use of the theory of minimum energy dissipation rate in its simplified version of minimum total stream power to compute channel width and depth adjustments (see Section 5.4). The use of this theory allows the channel width to be treated as an unknown variable. Treating the channel width as an unknown variable is the most unique capability of GSTARS3. Whether a channel width or depth is adjusted at a given cross-section and at a given time step depends on which condition results in less total stream power.

The fourth part is the inclusion of channel bank side stability criteria based on the angle of repose of bank materials and sediment continuity.

5.8.1 Streamlines and Stream Tubes

The basic concept and theory regarding streamlines, stream tubes, and stream functions can be found in most basic textbooks of fluid mechanics. In this section, only some of the basic concepts are given, as they are applicable to GSTARS3 computations.

By definition, a streamline is a conceptual line to which the velocity vector of the fluid is tangent at each and every point, at each instant in time. Stream tubes are conceptual tubes whose walls are defined by streamlines. The discharge of water is constant along a stream tube because no fluid can cross the stream tube boundaries. Therefore, the variation of the velocity along a stream tube is inversely proportional to the stream tube area. Figure 5.22 illustrates the basic concept of stream tubes used in GSTARS3.

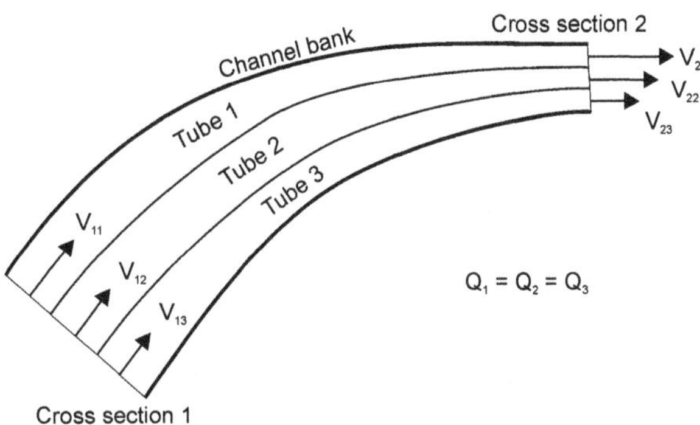

Figure 5.22. Top view of a hypothetical river reach illustrating the use of equal conveyance stream tubes in GSTARS3.

For steady and incompressible fluids, the total head, H_t, along a stream tube of an ideal fluid is constant:

$$\frac{F_p}{\gamma} + \frac{V^2}{2g} + D = H_t = \text{Constant} \tag{5.113}$$

where
F_p = pressure acting on the cross-section,
γ = unit weight of water,
g = acceleration due to gravity,
D = hydraulic head, and
V = flow velocity.

In GSTARS3, however, H_t is reduced along the direction of the flow, due to friction and other local losses.

5.8.2 Backwater Computations

The basic concepts and backwater computational procedures can be found in most open-channel hydraulics textbooks (e.g., Henderson, 1966). For quasi-steady flows, discharge hydrographs are approximated by bursts of constant discharge. During each constant discharge burst, steady-state equations are used for the backwater computations. GSTARS3 solves the energy equation based on the standard-step method:

$$z_{b1} + D_1 + \alpha_1 \frac{V_1^2}{2g} = z_{b2} + D_2 + \alpha_2 \frac{V_2^2}{2g} + H_t \tag{5.114}$$

where α = velocity distribution coefficient,
D = flow depth,
z_b = bed elevation above datum, and
H_t = total energy loss between sections 1 and 2.

The subscripts 1 and 2 denote two adjacent sections. The standard-step method for water surface profile computations is a trial-and-error iterative procedure to balance Equation (5.114). GSTARS3 uses the fixed-point iteration method described by Henderson (1966) for making estimated guesses to shorten the trial-and-error procedure. A more detailed description of this procedure can be found in Molinas and Yang (1985).

The energy equation is applied if there is no change of flow regime throughout the study reach. If there are changes in flow regime; i.e., if the flow changes from subcritical to supercritical or vice versa, the momentum equation is used:

$$\frac{Q\gamma}{g}(\beta_2 V_2 - \beta_1 V_1) = F_{p1} - F_{p2} + W \sin\phi - F_f \tag{5.115}$$

where β = momentum correction coefficient,
F_p = pressure acting on a given cross-section,
W = weight of water enclosed between sections 1 and 2,
ϕ = angle of inclination of the channel, and
F_f = total external friction force acting along the channel boundary.

The appropriate use of the two equations allows carrying backwater computations for subcritical, supercritical, or any combination of both flow conditions, even when hydraulic jumps are involved. Details of these computations were presented by Molinas and Yang (1985).

GSTARS3 uses the stream tube concept to accomplish a semi-two-dimensional approximation of the region being modeled. This allows the program to consider not only longitudinal, but also lateral variations of the hydraulics and sediment activity at each cross-section of the study. The water surface profiles are computed first. The channel is then divided into a selected number of stream tubes with the following characteristics: the total discharge carried by the channel is distributed equally among the stream tubes; stream tubes are bounded by channel boundaries and by imaginary vertical walls; the discharge along a stream tube is constant; and there is no exchange of water or sediments through stream tube boundaries (except due to stream curvature effects, discussed later in section 5.8.3.

Due to the nature of the backwater computations, the water surface elevation is assumed to be horizontal across each cross-section. The lateral locations of the stream tubes are computed at each time step from the channel conveyance; i.e., stream tube boundaries are set to provide equal conveyance stream tubes. Stream tube locations are computed for each time step; therefore, they are allowed to vary with time. Sediment routing is carried out independently for each stream tube and for each time step. Bed-material composition is computed for each tube at the beginning of the time step, and bed sorting and armoring computations are also carried out separately for each stream tube. In GSTARS3, lateral variations of bed-material composition are accounted for, and this variation is included in the computations of the bed-material composition and sorting for each stream tube.

GSTARS3 is not a truly two-dimensional program; therefore, it cannot simulate areas with recirculating flows or eddies. Other limitations include the inability to simulate secondary flows, reverse flows, water surface variations in the transverse direction, and hydrograph attenuation that result from the use of the simplified governing equations described in this chapter.

5.8.3 Sediment Routing

Sediment routing is done separately for each stream tube using Equation (5.36). The change in the volume of bed sediment due to deposition or scour, ΔA_s, is approximated as

$$\Delta A_s = \left(aW_{i-1} + bW_i + cW_{i+1}\right)\Delta z_{bi} \tag{5.116}$$

where W_i = top width of cross-section i for the stream tube at hand,
Δz_{bi} = change in bed elevation (positive for aggradation, negative for scour), and
a,b,c = constants that must satisfy $a+b+c=1$.

There are many possible choices for the values of a, b, and c. For example, $a=c=0$ and $b=1$ is a frequently used combination that is equivalent to assuming that the top width at station i represents the top width for the entire reach. If $b=c=0.5$ and $a=0$, emphasis is given to the downstream end of the reach. In practice, it is observed that giving emphasis to the downstream end of the reach may improve the stability of the calculations.

Using Equation (5.116), the partial derivative terms of Equation (5.36) are approximated as follows:

$$\frac{\partial A_s}{\partial t} \approx \frac{\left(aW_{i-1} + bW_i + cW_{i+1}\right)\Delta z_{bi}}{\Delta t} \tag{5.117}$$

$$\frac{dQ_s}{dt} \approx \frac{Q_{s,i} - Q_{s,i-1}}{\frac{1}{2}\left(\Delta x_i + \Delta x_{i-1}\right)} \tag{5.118}$$

$$\Delta z_{bi,j} = \frac{\Delta t}{(1-p_i)} \frac{q_L(\Delta x_i + \Delta x_{i-1}) + 2(Q_{s,i-1,j} - Q_{s,i,j})}{(aW_{i-1} + bW_i + cW_{i+1})(\Delta x_i + \Delta x_{i-1})} \tag{5.119}$$

where
j = size fraction index,
p_i = porosity of sediment in a unit bed layer at cross-section i,
q_L = lateral sediment discharge per unit channel length, and
$Q_{s,i,j}$ = computed volumetric sediment discharge for size j at cross-section i.

The total bed elevation change for the stream tube at cross-section i, Δz_{bi}, is computed from

$$\Delta z_{bi} = \sum_j \Delta z_{bi,j} \tag{5.120}$$

The new channel cross-section at station i, to be used at the next time iteration, is determined by adding the bed elevation change to the old bed elevation. Figure 5.23 provides a definition of variables for sediment routing in a stream tube, and Figure 5.24 provides a simplified flow chart of the overall process followed in computing the changes in bed elevation.

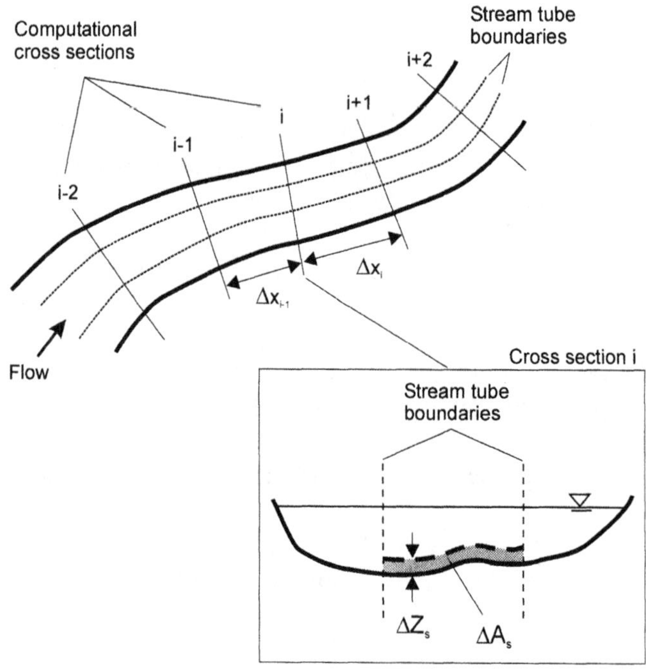

Figure 5.23. Definition of variables in the sediment routing equations, applied to one stream tube.

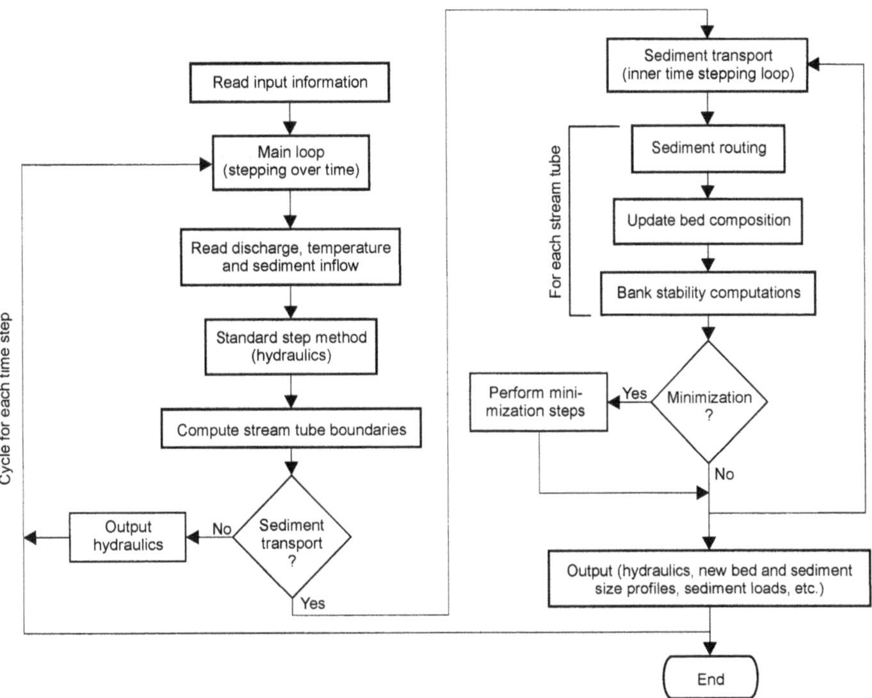

Figure 5.24. Simplified flow chart of program GSTARS3.

Accounting of bed composition is accomplished by the procedure proposed by Bennett and Nordin (1977). This method uses two or three conceptual layers (three layers for deposition and two layers for scour). The process is illustrated in Figure 5.25. The top layer, which contains the bed-material available for transport, is called the active layer. Beneath the active layer is the inactive layer, which is the layer used for storage. Below these two layers is the undisturbed bed, with the initial bed-material composition.

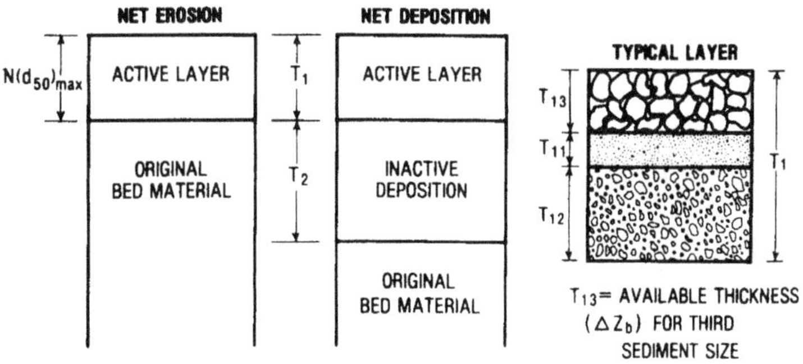

Figure 5.25. Bed composition accounting procedures (from Bennett and Nordin, 1977).

The active layer is the most important concept in this procedure. It contains all the sediment that is available for transport at each time step. The thickness of the active layer is defined by the user as proportional to the geometric mean of the largest size class containing at least 1 percent of the bed-material at that location. Active layer thickness is, therefore, closely related to the time step duration. Erosion of a particular size class of bed-material is limited by the amount of sediments of that size class present in the active layer. If the flow carrying capacity for a particular size class is greater than what is available for transport in the active layer, the term "availability limited" is used (Bennett and Nordin, 1977). On the other hand, if more material is available than that necessary to fulfill the carrying capacity computed by a particular sediment transport equation, the term "capacity limited" is used.

The inactive layer is used when net deposition occurs. The deposition thickness of each size fraction is added to the inactive layer, which, in turn, is added to the thickness of the active layer. The size composition and thickness of the inactive layer are computed first, after which a new active layer is recomputed and the channel bed elevation updated.

The procedures described above are carried out separately along each stream tube. Since the locations of stream tube boundaries change with changing flow conditions and channel geometry, those processes had to be adapted for use in GSTARS3. Bed-material is accounted for at the end of each time step for each stream tube. Bed-material composition is stored at each point used to describe the geometry for all the cross-sections. The values of the active and inactive layer thickness are also stored at those points. At the beginning of the next time step, after the new locations of the stream tube boundaries are determined, these values are used to compute the new layer thicknesses and bed composition for each stream tube.

Sediment transport capacities can be calculated from one of the 15 transport formulae programmed in GSTARS3, plus methods for the erosion and deposition of silt and clay, including high concentration of fines and flocculation and hindered settling. Sediment transport is computed by size fraction. The fractional transport method represented by Equations (5.52) and (5.53) is used. The hydraulic parameters used to compute the sediment carrying capacities in each reach (and stream tube) are computed as weighted averages from the hydraulic parameters from nearby stations. For each station i, the representative values of the area (A_{Ri}), depth (D_{Ri}), velocity (V_{Ri}), and friction slope (S_{Ri}) are computed as follows:

$$A_{Ri} = aA_{i-1} + bA_i + cA_{i+1} \tag{5.121}$$

$$D_{Ri} = aD_{i-1} + bD_i + cD_{i+1} \tag{5.122}$$

$$V_{Ri} = aV_{i-1} + bV_i + cV_{i+1} \tag{5.123}$$

$$S_{Ri} = aS_{i-1} + bS_i + cS_{i+1} \tag{5.124}$$

The weighting parameters *a*, *b*, and *c* can be chosen in any combination that satisfies $a+b+c=0$. For example, in rivers whose properties change more rapidly from section to section, a scheme incorporating information from the upstream and downstream reaches may be more appropriate. The values of $a=c=0.25$ and $b=0.5$ may be adopted in those circumstances. By changing *a*, *b*, and *c* appropriately, the user can use the parameters that favor stability or that favor sensitivity. Usually, more sensitive schemes are less stable, and vice-versa.

Non-equilibrium sediment transport is taken into account using the method developed by Han (1980). In this method, which is based on the analytical solution of the convection-diffusion equation, the non-equilibrium sediment transport rate for each particle size class *j* is computed from

$$C_i = C_i^* + \left(C_{i-1} - C_i^*\right)\exp\left\{-\frac{\chi w_s \Delta x}{q}\right\} + \left(C_{i-1}^* - C_i^*\right)\left(\frac{q}{\chi w_s \Delta x}\right)\left[1-\exp\left\{-\frac{\chi w_s \Delta x}{q}\right\}\right] \quad (5.125)$$

where C = actual sediment concentration,
C^* = sediment carrying capacity, computed from a standard formula,
q = flow discharge per unit width,
w_s = sediment fall velocity,
i = cross-section index (increasing from upstream to downstream), and
χ = dimensionless recovery factor.

In Equation (5.125), the particle size class index was dropped for convenience. Han and He (1990) recommend for χ a value of 0.25 for deposition and 1.0 for entrainment.

GSTARS3 computes the exchange of sediment across stream tube boundaries in certain specific circumstances. The movement of a sediment particle will have a direction which, in general, is neither the direction of the flow nor the direction of the bed shear stress. For example, in a bend of a channel with a sloping bed (such as the one in Figure 5.26), the larger particles will tend to roll down the slope (gravitational forces dominate), while the smaller particles may move up the slope (lift forces due to secondary currents dominate)—see, for example, Ikeda et al. (1987). A non-zero transverse flux results in exchange of sediments across stream tube boundaries. This exchange does not violate the theoretical assumptions behind the use of stream tubes because the trajectories of the sediment particles are not the same as the trajectories of the fluid elements (streamlines). Therefore, although there is no exchange of water between stream tubes, sediment may cross stream tube boundaries, and the use of stream tubes is still theoretically justified.

GSTARS3 includes the effects of stream curvature that contribute to the radial (transverse) flux of sediments, q_r, near the bed. The two effects considered are transverse bed slope and secondary flows. The effects due to secondary flows are modeled following Kikkawa et al. (1976), in which the angle that the bed shear stress vector makes with the downstream direction, β, is given by

$$\beta = \frac{DV}{U_* A_r R}\left(-4.167 + 2.640\frac{U_*}{\kappa V}\right) \tag{5.126}$$

where
V = average velocity along the channel's centerline,
U_* = shear velocity along the centerline,
D = water depth,
R = radius of curvature of the channel,
A_r = an empirical coefficient (for rough boundaries A_r = 8.5), and
κ = von Kàrmàn constant (= 0.41).

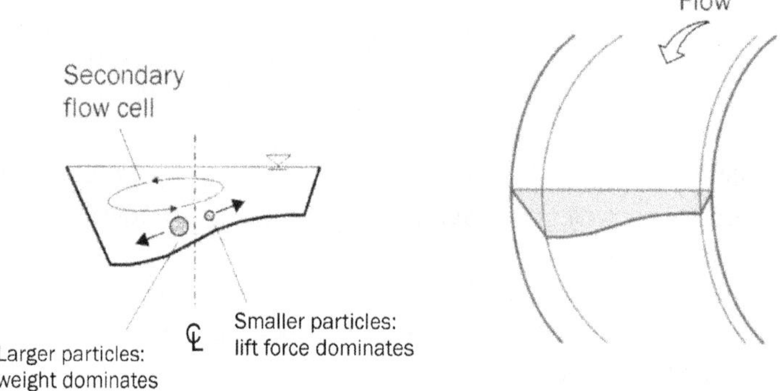

Figure 5.26. Bed sorting in bends, due to transverse bed slope and secondary currents.

In a bed with transverse slope, the gravity forces cause the direction of the sediment particles to be different from that of the water particles. Following Ikeda et al. (1987), the effects due to a transverse bed slope can be added to those due to curvature, such that

$$\frac{q_r}{q_s} = \tan\sigma = \tan\beta + \frac{1+\alpha\mu}{\lambda\mu}\sqrt{\frac{\tau_0^*}{\tau^*}}\tan\delta \tag{5.127}$$

where
q_s = unit sediment transport rate in the channel's longitudinal direction,
σ = angle between the direction of transport and the channel's downstream direction,
τ_0^*, τ^* = non-dimensional critical shear stress and bed shear stress, respectively,
δ = transverse bed slope,
α = rate of lift to drag coefficients on sediment particles (determined experimentally to be equal to 0.85),
λ = sheltering coefficient (= 0.59), and
μ = dynamic Coulomb friction factor (= 0.43).

The direction of sediment transport is calculated from Equation (5.127). The components of the sediment transport direction vector are given by

$$q_s = q_t \cos\sigma \tag{5.128}$$

$$q_r = q_t \sin\sigma \tag{5.129}$$

where q_t = sediment transport rate per unit width.

Equation (5.36) is then solved using $Q_s = q_s \Delta y$ and $C_L q_L = q_r$, where Δy = stream tube width. The above methods are applied only to sediment moving as bedload. GSTARS3 uses van Rijn's (1984b) method to determine if a particle of a given size is in suspension or moves as bedload.

5.8.4 Total Stream Power Minimization

The basic minimization procedures in GSTARS3 are based on the total stream power minimization theory—see Section 5.4. First, in order to apply the minimization procedure to channel reaches with gradually varied flows, γQS is integrated along the channel:

$$\varphi_T = \int \gamma QS dx \tag{5.130}$$

where φ_T is defined as the total stream power. This expression is discretized following Chang (1982):

$$\varphi_T = \sum_{i=1}^{N-1} \gamma \left(\frac{Q_i S_i + Q_{i+1} S_{i+1}}{2} \right) \Delta x_i \tag{5.131}$$

where N = number of cross-sections along the reach,
Δx_i = distance between cross-section i and $i + 1$,
Q_i = discharge and slope at station i, and
S_i = slope at station i.

The direction for channel adjustments is chosen by minimizing the integral represented by Equation (5.131) for total stream power at different stations. This process is repeated for each time step: if alteration of the channel width results in lower total stream power than raising or lowering of the channel's bed, then channel adjustments progress in the lateral direction; otherwise, the adjustments are made in the vertical direction.

Figure 5.27 is used to illustrate the process described above. When erosion takes place, channel adjustments can proceed either by deepening or by widening the cross-section. Both channel widening and deepening can reduce the total stream power for the reach, but GSTARS3 selects the adjustment that results in the minimum total stream power for the reach. If deposition is predicted by the sediment routing computations, then either the bed is raised or the cross-section is narrowed, but the choice must also result in a minimum of the total stream power for the reach. However, in each case, the amount of scour and/or deposition is limited by the predicted sediment load, and geological or manmade restrictions are also accommodated by the computational algorithms.

Quantitatively, the amount of channel adjustment during each time step is determined by the sediment continuity equation; i.e., Equation (5.119) for each stream tube. Channel widening or narrowing can take place only at the stream tubes adjacent to the banks. In this case, the hydraulic radius, R_h, replaces the top width, W, in Equation (5.119). For stream tubes that are not adjacent to the banks (i.e., interior tubes), bed adjustments can be made only in the vertical direction.

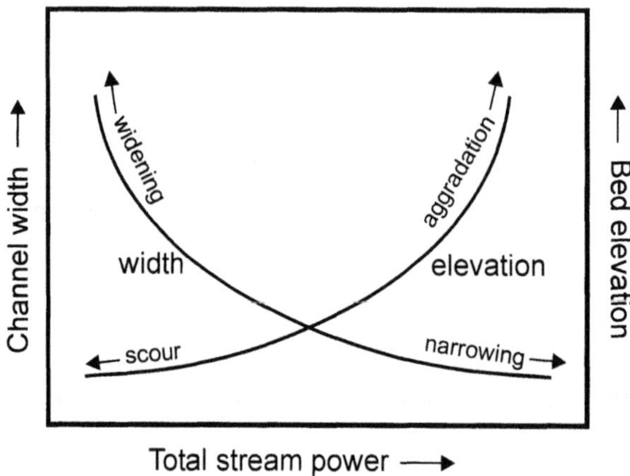

Figure 5.27. Total stream power variation as a function of changes in channel width and bed elevation, with constant discharge and downstream stage.

5.8.5 Channel Side Slope Adjustments

Channel geometry adjustment can take place in both lateral and vertical directions. For an interior stream tube, scour or deposition can take place only on the bed, and the computation of depth change shown in Equation (5.119) is straightforward. For an exterior stream tube, however, the change can take place on the bed or at the bank. As erosion progresses, the steepness of the bank slope tends to increase. The maximum allowable bank slope depends on the stability of bank materials. When erosion undermines the lower portion of the bank and the slope increases to a critical value, the bank may collapse to a stable slope. The bank slope should not be allowed to increase beyond a certain critical value. The critical angle may vary from case to case, depending on the type of soil and the existence of natural or artificial protection.

GSTARS3 checks the transverse bed slope for violation of a prescribed critical slope (i.e., the critical angle of repose of the bed-material). Each cross-section is scanned at the end of each time step to determine if any vertical or horizontal adjustments have caused the banks to become too steep. If any violations occur, the two points adjacent to the segment are adjusted vertically until the slope equals the user-provided critical slope. For the situation shown in Figure 5.28, the bank is adjusted from *abde* to *ab'd'e*, so that the calculated angle, θ, is equal to the critical angle, θ_c. The adjustments are governed by conservation of mass:

$$A_1 + A_2 = A_3 + A_4 \tag{5.132}$$

where A_1 = area of triangle *abb'a*,
A_2 = area of triangle *bcb'b*,
A_3 = area of triangle *cd'dc*, and
A_4 = area of triangle *d'edd'*.

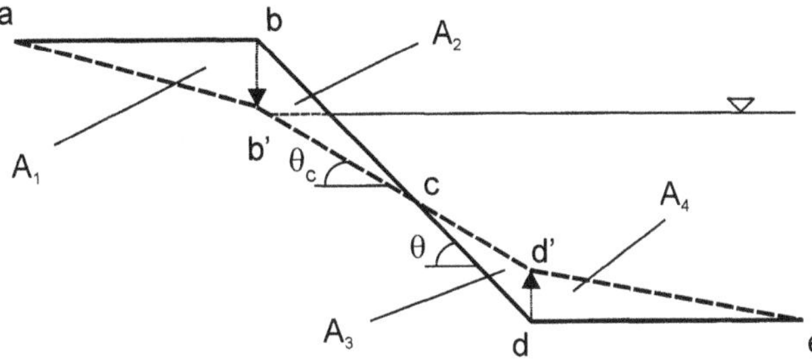

Figure 5.28. Representation of the bed adjustment to meet the angle of repose criteria.

5.8.6 Application Examples

To illustrate the application of the GSTARS models, three case studies applied to river and reservoir sedimentation problems are presented below. In the first application, the GSTARS 2.1 model is applied to predict river width changes. This case was taken from the GSTARS 2.1 User's Manual (Yang and Simões, 2000), where it appears as the third example problem in Appendix B. The data used is from a study conducted on Lake Mescalero Dam and Dike.

The Lake Mescalero Dam and Dike were constructed in 1974 and are located at the confluence of Ciewegita and Carrige Creeks on the Mescalero Apache Indian Reservation, about 2.5 miles southwest of Ruidoso, New Mexico. The data from the channel immediately downstream from the emergency spillway is used. The spillway is located in a bedrock cut in the sandstone and shale of the left abutment of the dam. The spillway consists of a 290-ft-long approach channel; a 103.8-ft-wide concrete weir; a 138-ft-long, concrete-lined discharge chute; a 15-ft-long flip bucket; and a concrete erosion cutoff wall located about 250 ft downstream from the flip bucket. The spillway crest rises 5.0 ft above the spillway approach channel floor and the discharge chute floor, to a crest elevation of 6905.0 ft.

The data for the flood of December 20, 1984 through December 31, 1984, was used. The water surface elevation of the reservoir was available for that period. Cross-sectional data were taken from a detailed topographic survey of the study area carried out on June 12, 1979. Figure 5.29 shows the layout of the channel below the spillway at that time. Also, cross-sectional channel geometry was available at two locations after the 1984 flood. These sections are used for comparison purposes.

The water surface elevation of the reservoir was available for the period of the flood. The discharge hydrograph was generated using the water levels and a standard weir equation, $Q = CLH^{3/2}$, with $C = 3.65$. The flood hydrograph is shown in Figure 5.29. The sediment data used was based on bed samples collected from several sites along the channel. Two typical size distributions are also shown in Figure 5.29. On the basis of the mean sediment size, a Manning's roughness coefficient of 0.06 was chosen for stage-discharge computations.

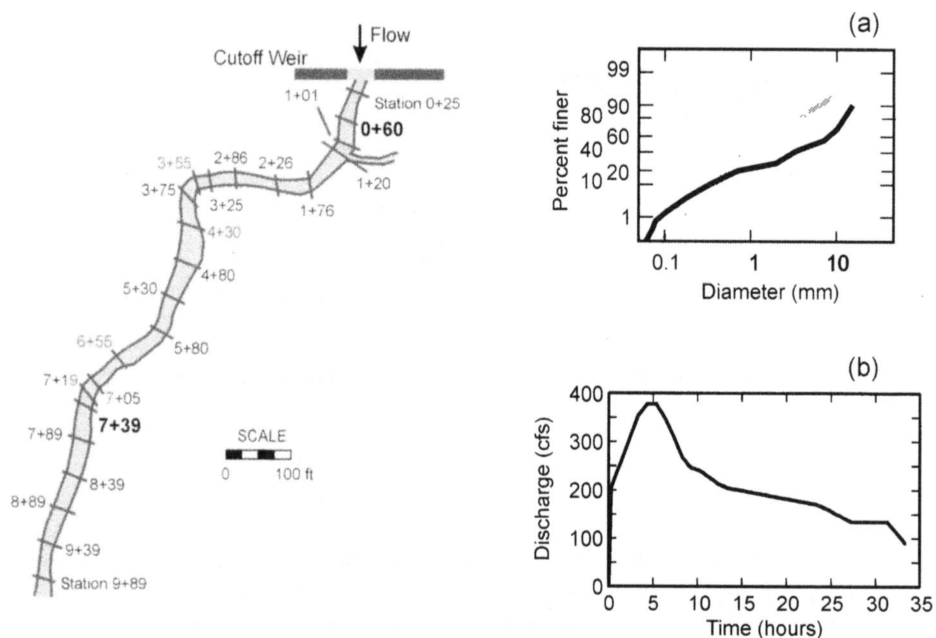

Figure 5.29. Layout of the cross-sections used in the simulation of scour downstream from Lake Mescalero's unprotected spillway. The cross-sections used to compare actual data with the simulations are highlighted. The figure also shows (a) typical particle size distributions found in the stream corridor and (b) the spillway hydrograph.

The cross-sections used in the comparisons are sections 0+60 and 7+39 (see Figure 5.29). Section 0+60 is located 60 ft downstream from the spillway, and section 7+30 is located 739 ft downstream from the spillway. Two runs were performed: one using stream power minimization to compute channel with changes and the other using only vertical scour (i.e., with fixed channel width). The comparison between the measured and the computed cross-sections is shown in Figure 5.30. Significant differences are observed when stream power calculations are activated, especially for section 0+60. Both the shape of the cross-sections and their thalwegs were more accurately predicted when the stream power minimization computations were activated.

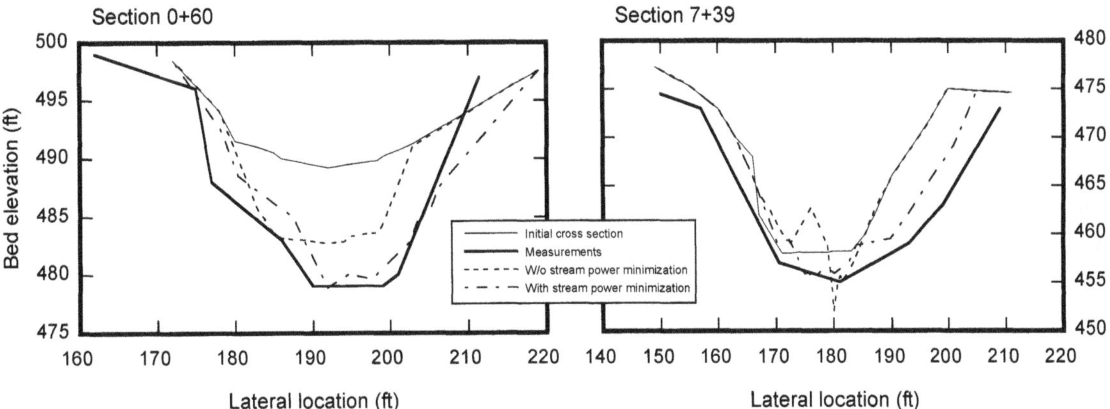

Figure 5.30. Comparison between measurements and simulation for two cross-sections of Lake Mescalero's spillway channel.

In the second example, GSTARS3 is used, and the experimental measurements of Swamee (1974) were chosen for the analysis. In Swamee's experiments, a small laboratory flume was used, to which a barrier was placed downstream to simulate a dam. Medium sand was used to study the deposition upstream of the barrier. The main characteristics of the experimental setup are as follows:

Channel width (cm)	20
Channel length (m)	32
Slope of bed	3.5×10^{-3}
Flow discharge (l/s)	2.0
Sediment d_{50} (mm)	0.4
Sediment discharge (l/s)	6.8×10^{-4}
Average temperature (°C)	12
Barrier height (cm)	9.0

In Swamee's experiments, an initial bed of sand (the same sand used for the input of sediment load) was used. Clear water at the prescribed discharge was fed to the channel for a sufficient amount of time to establish uniform flow conditions. The initial slope was determined after such an equilibrium was achieved. Then, a barrier was placed at the downstream end of the flume, thus raising the water surface at that end (profile of type M1). The barrier height was chosen to ensure that all the sediment in transport was trapped in its headwater reservoir. The sediment load was fed at the upstream end of the flume for a total run time of 70 hours.

An initial calibration was carried out to determine the main roughness characteristics of the channel. Note that Strickler's equation yields a Manning's n value of

$$n = \frac{d_{50}^{1/6}}{25.6} = 0.011 \tag{5.133}$$

where d_{50} is in ft.

If we use d_{90} in Equation (5.133), we obtain $n = 0.012$ (d_{90} = 0.70 mm). A sensitivity analysis was carried out using GSTARS3 and several n values. The results are shown in Figure 5.31. In practice, Swamee reported having observed bed forms (ripples and/or dunes) developing during the run. He was not specific about them; therefore, we could not use an adequate roughness estimator. However, he stated that the bed measurements were made by leveling the bed at the measuring stations, in segments with 0.5 m in length. Therefore, the measurements represented the average bed profile over that distance. That effect was well represented by the irregularity (waviness) shown by the bed and free surface elevation measurements. For the final runs, a n value of 0.020 was adopted.

Swamee measured the sediment load in transport and expressed it as a function of the bed shear stress:

$$q_{t*} = 100\tau_*^3 \tag{5.134}$$

where q_{t*} = non-dimensional unit-width sediment transport rate, and
τ_* = non-dimensional bed shear stress.

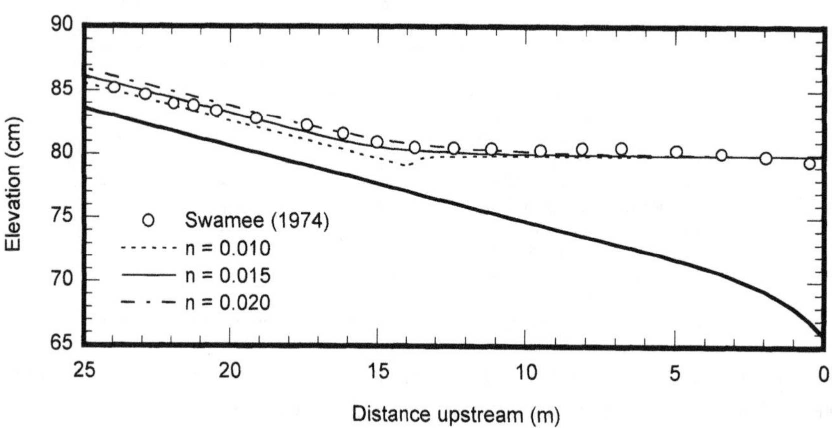

Figure 5.31. Sensitivity runs for the equilibrium condition of run No. 1 of Swamee (1974). The bed of the channel is represented by the lower thick line.

They are defined as

$$q_{t*} = \frac{q_t}{U_* d} \tag{5.135}$$

and

$$\tau_* = \frac{\tau - \tau_c}{(\gamma_s - \gamma)d} \tag{5.136}$$

where q_t = total bedload per unit width (units of L^2T^{-1}),
U_* = shear velocity,
d = size of the sediment particles,
τ = bed shear stress,
τ_c = critical (Shields) bed shear stress, and
γ, γ_s = unit weight of water and sediment, respectively.

The values of the coefficient and exponent were determined by data fitting. Equation (5.134) was implemented in GSTARS3 and used in the present runs.

The GSTARS3 results are presented in Figure 5.32. Agreement between measurement and simulation is close overall. The small oscillations observed in the experimental data can be attributed to a number of different factors. They are not observed in the numerical model because of the uniform boundary conditions used (constant water and sediment discharge, uniform sediment size, channel with constant width, and water surface level without kinks). This reflects the stability of the computations, because under such uniform conditions only numerical oscillations could cause disturbances in the predicted bed profiles. It is clear that in the numerical experiments presented here, GSTARS3 can produce accurate, stable, and oscillation-free solutions.

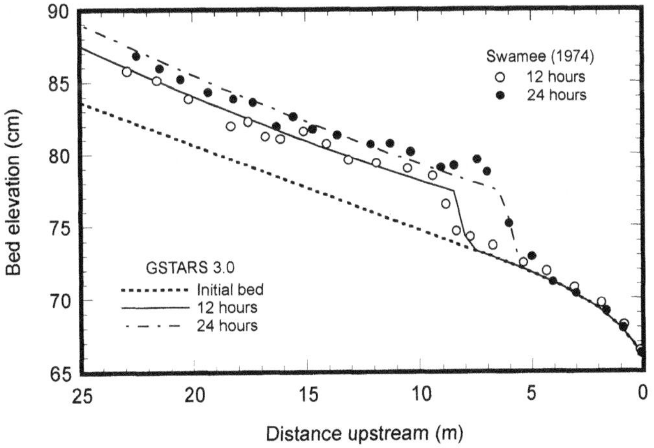

Figure 5.32. Comparison of experiments with simulations by GSTARS3 for two time instants: 12 and 24 hours after the start of the sediment loading.

Next, the GSTARS3 model is applied to Tarbela Dam. Tarbela Dam, in northern Pakistan along the Indus river, is the largest earth-filled dam in the world. The reservoir, with a capacity of 9.68×10^6 acre-feet, is a run-of-the-river type of reservoir with two major tributaries, the Siran and the Brandu. The reservoir's storage capacity has been continuously depleted since the dam was built in 1974, with an annual inflow rate of 240 million metric tons of sediment, mostly in the silt and clay range. This loss in capacity threatens irrigation water supply and hydropower production.

In this study, GSTARS3 is used to simulate 22 years of reservoir sedimentation (from 1974 through 1996) for a reach that spans 54.8 miles upstream from the dam. The hydrology of the system is given in Figure 5.33, together with the dam operation. The tributaries have a relatively small contribution when compared with the main stem discharge; therefore, they are not included in Figure 5.33 (but they were included in the computations).

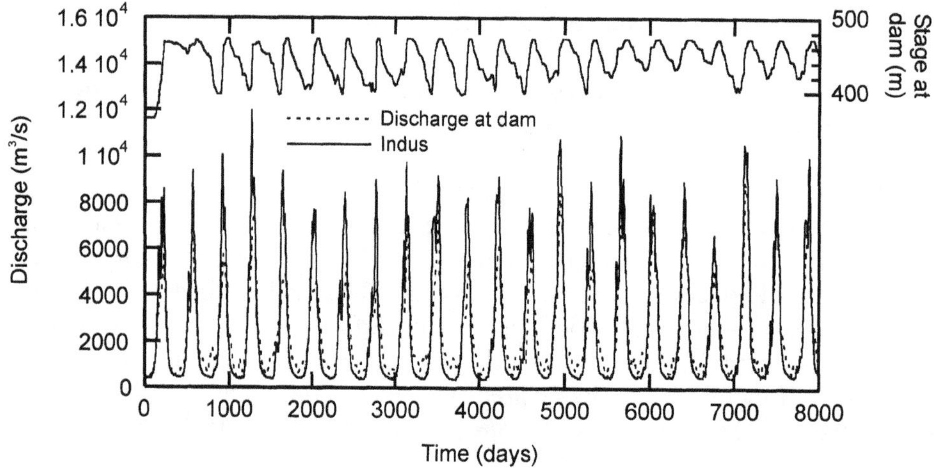

Figure 5.33. Hydrology and dam operation for Tarbela in the period of 1974 to 1996.

The Tarbela Reservoir bathymetry was discretized using existing surveyed cross-sections, which are marked in Figure 5.34. The GSTARS3 simulations were carried out using daily time steps for the hydraulic computations and 4.8 hours for sediment routing computations (8,040 time steps for hydraulics; 40,200 time steps for sediment). Yang's (1973) equation was extrapolated for the silt and clay range and was used in this simulation (the particle size distributions were in the range of 0.002 to 2.0 mm).

Computer runs on a PC-compatible desktop workstation running Microsoft Windows XP took less than 10 minutes. The simulation results for the thalweg are in good agreement with measurements, as shown in Figure 5.35. Two typical cross-sections are also shown in Figure 5.36. The quality of the simulation results produced by GSTARS3 shows that, at least for this case, it can be used for the long-term simulation of long rivers and river-like reservoirs.

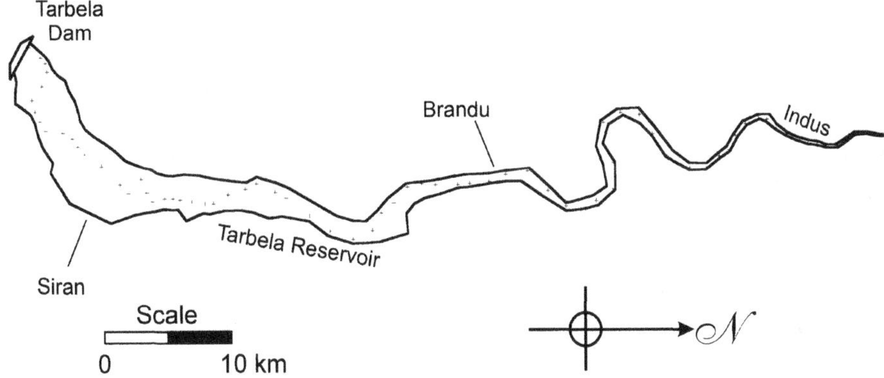

Figure 5.34. Tarbela Dam and Reservoir. The points (+) mark the locations of the cross-sections.

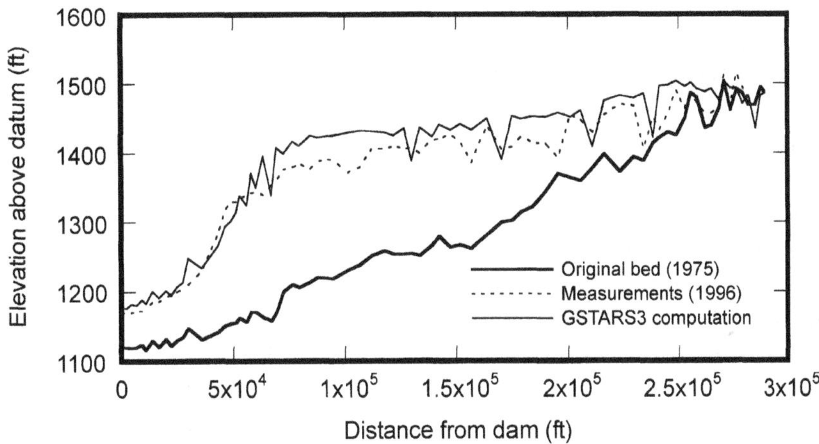

Figure 5.35. Results of the simulation of the Tarbela delta advancement over a period of 22 years.

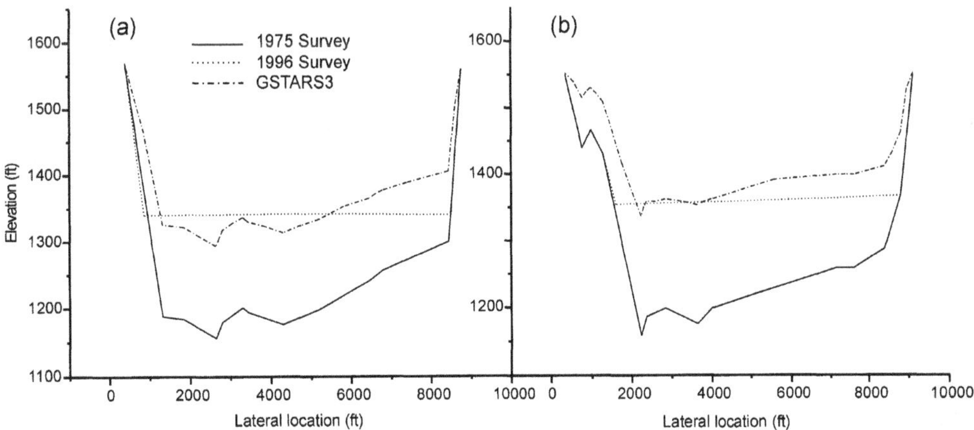

Figure 5.36. Cross-sections No. 24 and No. 30 in the simulation. Comparisons between the1975 and 1996 survey data and GSTARS3 simulation.

5.9 Summary

Sediment transport computer models have been increasingly used as tools for solving practical engineering problems. These models are also used to improve our understanding of river morphological processes. This chapter provides a brief review of basic theoretical concepts used in developing models and some of the practical approaches used for solving engineering problems.

Generally speaking, river hydraulics and sediment transport in natural rivers are three dimensional. Truly two- or three-dimensional models may be needed for solving localized problems, using detailed, site-specific field data for testing and calibration. One-dimensional models are more suitable for long-term simulation of a long river reach, where the lateral variations of hydraulic and sediment conditions can be ignored. From a practical point of view, a semi-two-dimensional model may be adequate for solving many river engineering problems.

Due to the changing hydrologic conditions of a river, hydraulic conditions in a river are unsteady from a theoretical point of view. With the possible exception of routing during a flood near its peak, however, most river hydraulic conditions can be approximated by a semi-steady hydrograph using constant-discharge bursts of short durations.

There are many well-established numerical schemes for solving flow and sediment transport model governing equations. The one-dimensional, finite difference uncoupled method is the one most commonly used in practice. With the advent of more powerful and affordable computing platforms, increasing number of commercially available models, and more efficient and accurate data collection techniques, there is an increasing demand for the use of multi-dimensional models in river engineering problems.

The simulated results from sediment-transport computer models are sensitive to the selection of sediment-transport formulas in the model. The user of a model should have a good understanding of sediment transport theories and the limits of application of different sediment transport formulas.

It is assumed in most sediment-transport models that channel width is a constant and cannot be adjusted. This unrealistic assumption can lead to erroneous results when applied to an alluvial river. The GSTARS3 model is based on the stream tube concept and the minimum stream power theory. Both the energy and momentum equations are used for water surface profile computation. GSTARS3 incorporates the concepts of sediment routing by size fraction, formation and destruction of an armor layer, channel width adjustments, and channel side stability, among other things, to simulate site-specific conditions. A model similar to GSTARS3 should be adequate for solving many semi-two-dimensional, quasi-steady river engineering problems with a minimum amount of field data required for calibration and testing.

Due to the limited scope of a chapter such as this, many important topics were only briefly touched upon or not covered at all. For example, it was not the purpose of this chapter to cover the subject of open-channel flow, which is described with much more detail than would ever be possible here in the classic works of Chow (1959) and Henderson (1966). Numerous textbooks cover a multitude of numerical solution methods, some of which are not even mentioned in this chapter but which play an important role in computational fluid dynamics. One such topic covers methods for the computation of flows with high gradients, such as those appearing in dam break flows; the interested reader can review this subject in Toro (2001). Mudflows and debris flows occur naturally and are important in many natural events, especially in regions with high gradient topography. Their rheology and modeling is complex and can be found in Coussot (1997) and Takahashi (1991), respectively. Flow turbulence is a field of active research and constant evolution. Turbulence in natural bodies of water is a topic covered in Nezu and Nakagawa (1993), and its modeling in hydraulics is covered in the monograph by Rodi (1993). The reader is advised to review the list of references at the end of this chapter for more in-depth coverage of many more topics, especially the textbooks, monographs, and survey papers, and the references therein.

5.10 References

Akiyama, J., and H. Stefan (1981). *Theory of Plunging Flow into a Reservoir*, St. Anthony Falls Hydraulic Laboratory Technical Memorandum I-97, University of Minnesota, Minneapolis, Minnesota.

Akiyama, J., and H. Stefan (1985). "Turbidity Current with Erosion and Deposition," *Journal of the Hydraulic Division of the ASCE*, vol. 111, no. HY12, pp. 1473-1496.

Arcement, G.J., and V.R. Schneider (1987). *Roughness Coefficients for Densely Vegetated Flood Plains*, U.S. Geological Survey Water–Resources Investigation Report 83–4247.

Armanini, A. (1995). "Non-uniform Sediment Transport: Dynamics of the Active Layer," *Journal of Hydraulic Research*, vol. 33, no. 5, pp. 611-622.

Armanini, A,. and G. Di Silvio (1988). "A One-dimensional Model for the Transport of a Sediment Mixture in Non-equilibrium Conditions," *Journal of Hydraulic Research*, vol. 26, no. 3, pp. 275-292.

ASCE (1998a). "River Width Adjustment. I: Processes and Mechanisms," *Journal of Hydraulic Engineering*, vol. 124, no. 9, pp. 881-902.

ASCE (1998b). "River Width Adjustment. II: Modeling," *Journal of Hydraulic Engineering*, vol. 124, no. 9, pp. 903-917.

Atkinson, E. (1992). *The Design of Sluices Settling Basins, a Numerical Modeling Approach*, Report OD 124, HR Wallingford Ltd., Wallingford, United Kingdom.

Atkinson, E. (1996). *The Feasibility of Flushing Sediment from Reservoirs*, Report OD 137, HR Wallingford Ltd., Wallingford, United Kingdom.

Bagnold, R.A. (1963). *The Nature of Saltation and of Bedload Transport in Water*, Proceedings of The Royal Society of London A332.

Bakri, M.F., T.K. Gates, and A.F. Khattab (1992). "Field-measured Hydraulic Resistance Characteristics in Vegetation-infested Canals," *Journal of Irrigation and Drainage Engineering*, vol. 118, no. 2, pp. 256-274.

Barnes, H.H. (1967). *Roughness Characteristics of Natural Channels*, U.S. Geological Survey Water-Supply Paper 1849.

Bathurst, J.C. (1985). "Flow Resistance Estimation in Mountain Rivers," *Journal of Hydraulic Engineering*, ASCE, vol. 111, no. 4, pp. 625-643.

Bathurst, J.C. (2002). "At-a-site Variation and Minimum Flow Resistance for Mountain Rivers," *Journal of Hydrology*, vol. 269, pp. 11-26.

Batina, J.T. (1993). "A Gridless Euler/Navier-Stokes Solution Algorithm for Complex Aircraft Configurations," AIAA Paper 93-0333.

Bell, R., and A. Sutherland (1983). "Nonequilibrium Bedload Transport by Steady Flows," *Journal of Hydraulic Engineering*, vol. 109, no. 3, pp. 351-367.

Bennett, J., and C. Nordin (1977). "Simulation of Sediment Transport and Armouring," *Hydrological Sciences Bulletin,* vol. 22, pp. 555-569.

Bestawy, A. (1997). *Bedload Transport and Bed Forms in Steady and Unsteady Flows*, Ph.D. dissertation, Catholic University of Leuven, Belgium.

Borah, D., C. Alonso, and S. Prasad (1982). "Routing Graded Sediments in Streams: Formulations," *Journal of the Hydraulic Division of the ASCE*, vol. 108, no. HY12, pp. 1486-1503.

Bourget, E., and M.J. Fortin (1995). "A Commentary on Current Approaches in the Aquatic Sciences," *Hydrobiologia*, vol. 300/301, pp. 1-16.

Bray, D.I. (1979). "Estimating Average Velocity in Gravel-bed Rivers," *Journal of the Hydraulic Division of the ASCE*, vol. 105, no. HY9, pp. 1103-1122.

Bray, D.I., and K.S. Davar (1987). "Resistance to Flow in Gravel-bed Rivers," *Canadian Journal of Civil Eng*ineering, vol. 14, no. 1, pp. 77-86.

Brebbia, C.A. (1978). *The Boundary Element Method for Engineers*, Pentech Press, London.

Britter, R.E., and P.F. Linden. (1980). "The Motion of the Front of a Gravity Current Traveling Down on an Incline," *Journal of Fluid Mechanics*, vol. 99, pp. 531-543.

Bureau of Reclamation (1987). *Design of Small Dams, 3rd edition,* Department of the Interior, Bureau of Reclamation, Denver, Colorado.

Canuto, C.M., A. Hussaini, and T. Zang. (1996). *Spectral Methods in Fluid Mechanics*, Springer Series in Computational Physics, Springer-Verlag, Berlin.

Chang, H.H. (1979). "Minimum Stream Power and River Channel Patterns," *Journal of Hydrology*, vol. 41, pp. 303-327.

Chang, H.H. (1982). "Mathematical Model for Erodible Channels," *Journal of the Hydraulic Division of the ASCE*, vol. 108, no. HY5, pp. 678-689.

Chang, H.H. (1990). Generalized Computer Program FLUVIAL-12, Mathematical Model for Erodible Channels, User's Manual.

Chang, H.H., and J. Hill (1976). "Computer Modeling of Erodible Flood Channels and Deltas," *Journal of the Hydraulic Division of the ASCE*, vol. 102, no. HY10, pp. 1461-1477.

Chang, H.H., and J. Hill (1977). "Minimum Stream Power for Rivers and Deltas," *Journal of the Hydraulic Division of the ASCE*, vol. 103, no. HY12, pp. 1375-1389.

Chow, V.T. (1959). *Open Channel Flow*. McGraw-Hill Publishing Company, New York.

Chung, T.J. (2002). *Computational Fluid Dynamics*, Cambridge University Press.

Coussot, P. (1997). *Mudflow Rheology and Dynamics*, IAHR Monograph, A.A. Balkema, Roterdam.

Cowan, W.L. (1956). "Estimating Hydraulic Roughness Coefficients," *Agricultural Engineering*, vol. 37 (July), pp. 473–475.

Cunge, J., F. Holly, and A. Verwey (1980). *Practical Aspects of Computational River Hydraulics*, Pitman, Marshfield, MA.

Darby, S.E., and C.R. Thorne (1996). "Development and Testing of Riverbank Stability," *Journal of Hydraulic Engineering*, vol. 122, no. 8, pp. 443-454.

Deletic, A. (2000). *Sediment Behaviour in Overland Flow Over Grassed Areas*, Ph.D. Thesis, University of Aberdeen, UK.

DeLong, L. (1989). "Mass Conservation: 1D Open Channel Flow Equations," *Journal of Hydraulic Engineering*, vol. 115, no. 2, pp. 263-269.

Egashira, S., and K. Ashida (1980). "Studies on the Structure of Density Stratified Flows," *Bulletin of the Disaster Prevention Research Institute*, vol. 29, Kyoto, Japan.

Egiazaroff, I. (1965). "Calculation of Nonuniform Sediment Concentrations," *Journal of the Hydraulic Division of the ASCE*, vol. 91, no. HY4, pp. 225-247.

Einstein, H. (1950). *The Bedload Function in Open Channel Flows*, Technical Bulletin no. 1026, U.S. Department of Agriculture, Soil Conservation Service.

Fan, J. (1991). *Density Currents in Reservoirs*, Workshop on Management of Reservoir Sedimentation, New Delhi, India.

Fan, S.S. (ed.) (1988). "Twelve Selected Computer Stream Sedimentation Models Developed in the United States," *Proceedings of the Interagency Symposium on Computer Stream Sedimentation Model, Denver, Colorado.* Published by the Federal Energy Regulatory Commission, Washington, DC.

Federal Interagency Stream Restoration Working Group (1998). *Stream Corridor Restoration Principles, Processes, and Practices.* (http://www.usda.gov/stream_restoration/)

Ferziger, J., and M. Peric (2002). *Computational Methods for Fluid Dynamics*, 2nd ed., Springer-Verlag.

Fischer, H., J. List, R. Koh, J. Imberger, and N. Brooks (1979). *Mixing in Inland and Coastal Waters*, Academic Press.

Fix, G., and M. Gunzburger (1978). "On the Least Squares Approximations to Indefinite Problems of the Mixed Type," *International Journal of Numerical Methods in Engineering*, vol. 12, pp. 453-469.

Ford, D., and M. Johnson (1983). *An Assessment of Reservoir Density Currents and Inflow Processes*, Technical Report E-83-7, U.S. Army Engineering Waterways Experiment Station, Vicksburg, Mississippi.

Fredlund, D.G. (1987). *Slope Stability Analysis Using PC-slope.* Geo-Slope Programming Ltd., Calgary, Alberta, Canada.

Garcia, M.H., and G. Parker (1991). "Entrainment of Bed Sediment into Suspension," *Journal of Hydraulic Engineering*, vol. 117, no. 4, pp. 414-435.

Griffiths, G.A. (1981). "Flow Resistance in Coarse Gravel Bed Rivers," *Journal of the Hydraulic Division of the ASCE*, vol. 107, no. HY7, pp. 899-918.

Gwinn, W.R., and W.O. Ree (1980). "Maintenance Effects on the Hydraulic Properties of a Vegetated-lined Channel," *Transactions of the ASAE*, vol. 23, no. 3, pp. 636-642.

Hagerty, D. (1991). "Piping/Sapping Erosion 1: Basic Considerations," *Journal of Hydraulic Engineering*, vol. 117, no. 8, pp. 997-998.

Han, Q. (1980). "A Study on the Nonequilibrium Transportation of Suspended Load." *Proceedings of the International Symposium on River Sedimentation*, Beijing, China, pp. 793-802. (In Chinese.)

Han, Q., and M. He (1990). "A Mathematical Model for Reservoir Sedimentation and Fluvial Processes," *International Journal of Sediment Research*, vol. 5, no. 2, International Research and Training Center on Erosion and Sedimentation, pp. 43-84.

Harleman, D. (1961). Stratified Flow. In *Handbook of Fluid Dynamics*. V. Streeter (ed.), McGraw-Hill, New York.

Hebbert, B., J. Imberger, I. Loh, and J. Petterson (1979). "Collie River Underflow into the Wellington Reservoir," *Journal of the Hydraulic Division of the ASCE*, vol. 105, no. HY5, pp. 533-545.

Henderson, F. (1966). *Open Channel Flow*, MacMillan Book Company, New York, NY.

Holly, M.F., J.C. Yang, P. Schwarz, J. Schaefer, S.H. Su, and R. Einhelling (1990). *CHARIMA - Numerical Simulation of Unsteady Water and Sediment Movement in Multiply Connected Networks of Mobile-bed Channels*, IIHR Research Report no. 343, Iowa Institute of Hydraulic Research, the University of Iowa, Iowa City, Iowa.

HR Wallingford (1992). *The Hydraulic Roughness of Vegetated Channels*, Report SR 305, HR Wallingford Ltd., Wallingford, United Kingdom.

Hsu, S., and F. Holly (1992). "Conceptual Bedload Transport Model and Verification for Sediment Mixtures," *Journal of Hydraulic Engineering*, vol. 118, no. 8, pp. 1135-1152.

Ikeda, S., M. Yamasaka, and M. Chiyoda (1987). "Bed Topography and Sorting in Bends," *Journal of Hydraulic Engineering*, vol. 113, no. 2, pp.190-206.

International Research and Training Center on Erosion and Sedimentation (1985). *Lecture Notes of the Training Course on Reservoir Sedimentation,* Sediment Research Laboratory of Tsinghua University, Beijing, China.

Ippen, A.T., and D.R.F. Harleman (1952). *Steady-state Characteristics of Sub-surface Flow*, Circular no. 521, U.S. National Bureau of Standards, pp. 79-93.

Jain, S.C. (1981). "Plunging Phenomena in Reservoirs," *Proceedings of the Symposium on Surface Water Impoundments*, H. Stefan (ed.), ASCE, pp. 1249-1257.

Jirka, G., and M. Watanabe (1980). "Thermal Structure of Cooling Ponds," *Journal of the Hydraulic Division of the ASCE*, vol. 106, no. HY5, pp. 701-715.

Johannesson, H., and G. Parker (1985). *Computer Simulated Migration of Meandering Rivers in Minnesota*, Project Report no. 242, St. Anthony Falls Hydraulic Laboratory, University of Minnesota.

Julien, P.Y., and J. Wargadalam (1995). "Alluvial Channel Geometry, Theory and Applications," *Journal of Hydraulic Engineering*, vol. 121, no. 4, pp. 312-325.

Karim, M., and J. Kennedy (1982). *Computer-based Predictors for Sediment Discharge and Friction Factor of Alluvial Streams*. IIHR Report no. 242, Iowa Institute of Hydraulic Research, University of Iowa, Iowa City, Iowa.

Keller, J.B., and D. Givoli (1989). "Exact Non-reflecting Boundary Conditions," *Journal of Computational Physics*, vol. 82, pp. 172-192.

Keulegan, G.H. (1949). "Interfacial Instability and Mixing in Stratified Fluids," *Journal of Research, NBS*, vol. 43, RP 2040.

Kikkawa, M., S. Ikeda, and A. Kitagawa (1976). "Flow and Bed Topography in Curved Open Channels," *Journal of the Hydraulic Division of the ASCE*, vol. 102, no. HY9, pp. 1327-1342.

Klaassen, G. (1990). *Sediment Transport in Armoured Rivers During Floods*, Report prepared for the Rijkswaterstaat, Delft Hydraulics, Holland.

Klaassen, G., H.J. Ogink, L.C. van Rijn (1986). *DHL-research on Bedforms, Resistance to Flow and Sediment Transport*, Delft Hydraulics Communication no. 362, July 1986.

Kouwen, N., and R.M. Li (1980). "Biomechanics of Vegetative Channel Linings," *Journal of the Hydraulic Division of the ASCE*, vol. 106, no. 6, pp. 713-728.

Lacey, G. (1920). "Stable Channels in Alluvium," *Proceedings of the Institute of Civil Engineers*, vol. 229, pp. 259-292.

Marzolf, R. (2003). Personal communication.

McAnally, W.H., and W.A. Thomas (1985). *User's Manual for the Generalized Computer Program System, Open-channel Flow and Sedimentation, TABS-2, Main Text*. U.S. Army Corps of Engineers, Waterways Experiment Station, Vicksburg, Mississippi.

Migniot, C. (1981). *Erosion and Sedimentation in Sea and River*. La Pratique des Sols et des Foundations, editions Le Moniteur, France.

Molinas, A., and C.T. Yang (1985). "Generalized Water Surface Profile Computations," *Journal of the Hydraulic Division of the ASCE*, vol. 111, no. HY3, pp.381-397.

Molinas, A., and C.T. Yang (1986). *Computer Program User's Manual for GSTARS (Generalized Stream Tube model for Alluvial River Simulation),* U.S. Bureau of Reclamation, Engineering and Research Center, Denver, Colorado, USA.

Montes, S. (1998). *Hydraulics of Open Channel Flow*, ASCE Press.

Morris, G.L., and J. Fan (1998). *Reservoir Sedimentation Handbook.* McGraw-Hill Publishing Company, New York.

Naot, D., I. Nezu, and H. Nakagawa (1995). "Periodic Behaviour of Partly Vegetated Open Channel," *HYDRA 2000*, vol. 1, Thomas Telford, London, pp.141-146.

Naot, D., I. Nezu, and H. Nakagawa (1996). "Hydrodynamic Behavior of Partly Vegetated Open Channels," *Journal of Hydraulic Engineering*, vol. 122, no. 11, pp.625-633.

National Research Council. (1983). *An Evaluation of Flood-level Prediction Using Alluvial River Models,* National Academy Press, Washington, DC.

Nelson, J.M., and J.D. Smith (1989). "Evolution and Stability of Erodible Channel Beds," in *River Meandering*, S. Ikeda and G. Parker (eds.), Water Resources Monograph 12, American Geophysical Union, Washington, DC.

Nezu, I., and H. Nakagawa (1993). *Turbulence in Open Channel Flows*, IAHR Monograph, A.A. Balkema, Roterdam.

Nikuradse, J. (1933). *Strömungsgesetze in Rauhen Rohren,* Forschung Arb. Ing. Wesen no. 361, Germany. (In German.)

Noh, W.F. (1964). "CEL: a Time-dependent, Two-space-dimensional, Coupled Eulerian-Lagrangian Code," In *Methods in Computational Physic,* F.H. Harlow (ed.), Academic Press, New York.

Osman, A.M., and C.R. Thorne (1988). "Riverbank Stability Analysis; I : Theory," *Journal of Hydraulic Engineering*, vol. 114, no. 2, pp.134-150.

Parker, G., M. Garcia, Y. Fukushima, and W. Yu (1987). "Experiments on Turbidity Currents Over an Erodible Bed," *Journal of Hydraulic Research*, vol. 25, no. 1, pp.123-147.

Patera, A. (1984). "A Spectral Method for Fluid Dynamics, Laminar Flow in a Channel Expansion," *Journal of Computational Physics*, vol. 54, pp.468-488.

Pinder, G., and W. Gray (1977). Finite Element Simulation in Surface and Subsurface Hydrology, Academic Press.

Pitlo, R.H. (1986). "Towards a Larger Capacity of Vegetated Channels," *Proceedings of the EWRS/AAB 7th Symposium On Aquatic Weeds*, pp.245-250.

Preissman, A. (1961). "Propagation des Intumescences Dans les Cannaux et Rivièrs," *1st Congress of the French Association for Computation*, Grenoble, France, pp.433-442.

Rastogi, A., and W. Rodi (1978). "Calculation of General Three-dimensional Turbulent Boundary Layers," *AIAA Journal*, vol. 16, pp.151-159.

Rodi, W. (1993). *Turbulence Models and Their Application in Hydraulics*, IAHR Monograph, A.A. Balkema, Roterdam.

Rooseboom, A. (1975). *A Sediment-production Map for South Africa*, Report 61, Department of Water Affairs, Pretoria, South Africa.

Rouse, H. (1950). *Elementary Mechanics of Fluids*. John Wiley and Sons, Inc., New York.

Samuels, P.G. (1990). "Cross-section Location in 1-D Models," *International Conference on River Hydraulics*, W. White (ed.), John Wiley & Sons, pp.339–350.

Savage, S., and J. Brimberg (1975). "Analysis of Plunging Phenomena in Water Reservoirs," *Journal of Hydraulic Research*, vol. 13, no. 2, pp.187-204.

Shimizu, Y., H. Yamaguchi, and T. Itakura (1991). "Three-dimensional Computation of Flow and Bed Deformation," *Journal of Hydraulic Engineering*, vol. 116, no. 9, pp.1090-1108.

Shimizu, Y., and T. Tsujimoto (1994). "Numerical Analysis of Turbulent Open-channel Flow Over a Vegetation Layer Using a k-ε Turbulence Model," *Journal of Hydroscience and Hydraulic Engineering*, vol. 11, no. 2, pp.57-67.

Simões, F.J.M. (1995). *Modeling Turbulent Flow and Mixing in Compound Channels*, Ph.D. dissertation, the University of Mississippi, University, Mississippi.

Simões, F.J.M. (2001). "Three-dimensional Modeling of Flow Through Sparse Vegetation," *Proceedings of the 7th Federal Interagency Sedimentation Conference*, Reno, Nevada, March 25-29, pp.2001.

Singh, V.P., and J. Li (1993). "Identification of Reservoir Flood-wave Models," *Journal of Hydraulic Research*, vol. 31, no. 6, pp.811-824.

Song, T., and W. Graf (1995). "Bedload Transport in Unsteady Open-channel Flow," in *Rapport Annuel*, Laboratoire de Recherches Hydrauliques, Ecole Polytechnique Federal, Lausanne, Switzerland.

Song, C.C.S., and C.T. Yang (1979a). "Theory of Minimum Rate of Energy Dissipation," *Journal of the Hydraulic Division of the ASCE*, vol. 105, no. HY7, pp.769-784.

Song, C.C.S., and C.T. Yang (1979b). "Velocity Profiles and Minimum Stream Power," *Journal of the Hydraulic Division of the ASCE*, vol. 105, no. HY8, pp.981-998.

Song, C.C.S., and C.T. Yang (1980). "Minimum Stream Power: Theory," *Journal of the Hydraulic Division of the ASCE*, vol. 106, no. HY9, pp.1477-1487.

Straškraba, M. (1998). "Coupling of Hydrobiology and Hydrodynamics: Lakes and Reservoirs," in *Physical Processes in Lakes and Oceans*, J. Imberger (ed.), American Geophysical Union, pp.623-644.

Swamee, P. (1974). *Analytic and Experimental Investigation of Streambed Variation Upstream of a Dam*, Ph.D. Thesis, Dept. of Civil Engineering, University of Roorkee, India.

Takahashi, T. (1991). *Debris Flow*, IAHR Monograph, A.A. Balkema, Roterdam.

Tannehill, J., D. Anderson, and R. Pletcher (1997). *Computational Fluid Mechanics and Heat Transfer*, 2nd ed., Taylor and Francis.

Temple, D.M. (1987). "Closure of 'Velocity Distribution Coefficients for Grass-lined Channels,'" *Journal of Hydraulic Engineering*, vol. 113, no. 9, pp.1224-1226.

Tennekes, H., and J. Lumley (1972). *A First Course in Turbulence*, MIT Press, Cambridge, Massachusetts.

Thompson, S.M., and P.L. Campbel (1979). "Hydraulics of a Large Channel Paved With Boulders," *Journal of Hydraulic Research*, vol. 17, no. 4, pp.341-354.

Toro, E.F. (2001). *Shock-capturing Methods for Free-surface Shallow Flows*, John Wiley and Sons, Ltd., New York.

Turner, J.S. (1973). *Buoyancy Effects in Fluids*, Cambridge University Press, Cambridge, United Kingdom.

USACE (1981). Sacramento River and Tributaries Bank Protection and Erosion Control Investigation, California, U.S. Army Corps of Engineers, Sacramento District.

USACE (1993). *HEC-6 Scour and Deposition in Rivers and Reservoirs User's Manual*, U.S. Army Corps of Engineers, Hydrologic Engineering Center, Davis, California.

van Rijn, L.C. (1984a). "Sediment Transport, Part I: Bedload Transport," *Journal of Hydraulic Engineering*, vol. 110, no. 10, pp.1431-1456.

van Rijn, L.C. (1984b). "Sediment Transport, Part III: Bed Forms and Alluvial Roughness," *Journal of Hydraulic Engineering, ASCE*, vol. 110, no. 12, pp.1733-1754.

Vreugdenhil, C. (1994). *Numerical Methods for Shallow-water Flow.* Kluwer Academic Publishers.

Warsi, Z. (1993). Fluid *Dynamics: Theoretical and Computational Approaches*, CRC Press, Boca Raton, Florida.

Wesseling, P. (2001). Principles of Computational Fluid Dynamics, Springer-Verlag.

Westrich, B., and M. Juraschek (1985). "Flow Transport Capacity for Suspended Sediment," *21st Congress of the IAHR*, vol. 3, Melbourne, Australia, 19-23 Aug. 1985, pp.590-594.

White, F. (1991). *Viscous Fluid Flow*, 2nd ed., McGraw-Hill Publishing Company, New York.

Yang, C.T. (1973). "Incipient Motion and Sediment Transport," *Journal of the Hydraulics Division of the ASCE*, vol. 99, no. HY10, pp.1679-1704.

Yang, C.T. (1992). "Force, Energy, Entropy, and Energy Dissipation Rate," in *Entropy and Energy Dissipation in Water Resources*, V.P. Sing and M. Fiorentino (eds.), Kluwer Academic Publisher, Netherlands, pp.63-89.

Yang, C.T., and C.C.S. Song (1979). "Theory of Minimum Rate of Energy Dissipation," *Journal of the Hydraulic Division of the ASCE*, vol. 105, no. HY7, pp.769-784.

Yang, C.T., and C.C.S. Song (1986). *Theory of Minimum Energy and Energy Dissipation Rate.* Encyclopedia of Fluid Mechanics, vol. 1, Chap. 11, Gulf Publishing Company, N. Cheremisinoff (ed.), pp.353-399.

Yang, C.T., and F.J.M. Simões (2000). *User's Manual for GSTARS 2.1 (Generalized Sediment Transport model for Alluvial River Simulation version 2.1)*, Technical Service Center, U.S. Bureau of Reclamation, Denver, Colorado.

Yang, C.T., and F.J.M Simões (2002). *User's Manual for GSTARS3 (Generalized Sediment Transport model for Alluvial River Simulation version 3.0)*, Technical Service Center, U.S. Bureau of Reclamation, Denver, Colorado.

Chapter 6
Sustainable Development and Use of Reservoirs

Page

6.1 Introduction 6-1
6.2 Sustainable Development and Use of Water Resources 6-1
6.3 Dynamic Adjustment of a River System 6-4
6.4 Planning 6-10
6.4.1 Background Information and Field Investigation 6-10
6.4.2 Basic Consideration 6-10
6.4.3 Sediment Control Measures 6-10
6.5 Design of Intakes 6-13
6.5.1 Location of Intakes 6-13
6.5.2 Types of Intakes for Sediment Control 6-14
6.6 Sediment Management for Large Reservoirs 6-16
6.7 Sediment Management for Small Reservoirs 6-17
6.7.1 Soil Conservation 6-17
6.7.2 Bypass of Incoming Sediment 6-18
6.7.3 Warping 6-18
6.7.4 Joint Operation of Reservoirs 6-18
6.7.5 Drawdown Flushing 6-19
6.7.6 Reservoir Emptying 6-19
6.7.7 Lateral Erosion 6-19
6.7.8 Siphoning Dredging 6-19
6.7.9 Dredging by Dredgers 6-20
6.7.10 Venting Density Current 6-20
6.7.11 Evaluation of Different Sediment Management Measures 6-20
6.8 Effective Management of Reservoir Sedimentation 6-21
6.9 Operational Rules 6-22
6.10 Cost of Sedimentation Prevention and Remediation 6-23
6.11 Reservoir Sustainability Criteria 6-24
6.12 Technical Tools 6-25
6.12.1 GSTARS 2.0/2.1 Models 6-25
6.12.2 GSTARS3 Model 6-27
6.12.3 GSTAR-1D Model 6-28
6.12.4 GSTAR-W Model 6-28
6.12.5 Economic Model 6-28
6.13 Summary 6-29
6.14 References 6-30

Chapter 6
Sustainable Development and Use of Reservoirs

by
Chih Ted Yang

6.1 Introduction

The construction of reservoirs is one of the most important practices for the development and management of water resources. Reservoir construction serves agricultural, flood control, hydropower generation, water supply, navigational, recreational, and environmental purposes.

Many large reservoirs were built without a thorough or systematic evaluation of the long-term environmental, social, and economic interactions of different alternatives. The "dead storage" concept has been used in the planning and design of reservoirs to store sediment for a predetermined useful life, say 50 to 100 years. However, many existing reservoirs have reached, or may soon reach, their designed useful life. In addition, environmental, social, economic, and political considerations, and the fact that suitable damsites are scarce now, necessitate new and innovative approaches for water resources development and management.

To cope with population increases and the increasing demands of higher standards of living, more reservoirs will be built, especially in developing countries. How to plan, design, construct, and operate reservoirs for sustainable use for generations to come is a challenge to engineers and policymakers.

This chapter explains the concept of sustainable development and use of water resources and the dynamic adjustments of river systems. Methods for planning, designing, operating, and maintaining a reservoir; reducing sediment inflow; and removing sediment in a reservoir are described. Yang (1997) provides a comprehensive review of these methods. New technical tools are introduced to enhance our ability to sustain the development and use of reservoirs where erosion and sedimentation are concerns.

6.2 Sustainable Development and Use of Water Resources

The description of "sustainable development" given by the World Commission on Environment and Development (WCED, 1987) is:

> Humanity has the ability to make development sustainable to ensure that it meets the needs of the present without compromising the ability of future generations to meet their own needs.

Concerning the "limits of development," the WCED (1987) states:

> The concept of sustainability implies limits: not absolute limits, but limitations imposed by the present state of technology and social organization on environmental resources and by the ability of the biosphere to absorb the effect of human activities.

Because the environment has an absorption limit, its overuse over a considerable time span can bring about irreversible changes. If this overuse continues, the irreversible changes can make the earth environmentally unsuitable for human habitation. It should be noted that the limitations are not fixed; they depend on the state of technology and the social organization which manages environmental resources (Takeuchi and Kundzewicz, 1998). In the use of "sustainable development,"

> Development does not necessarily mean growth in the quantitative sense. It is rather a qualitative process which does not have any limit. Development is a continuous process of changing state from one form to another seeking to meet ever changing objectives. Sustainable human existence, if ever possible, must be an endless process of change in social, cultural, and industrial states within a certain limit of the sustainable use of energy and resources. (Takeuchi and Kundzewicz, 1998)

The idea of sustainability was advanced in the Rio Declaration on Environment of Development issued at the United Nations Conference on Environment and Development (UNCED, 1992). Of the 27 principles declared, the following 5 principles address the concept of sustainable development:

> *Principle 1*: Human beings are at the centre of concerns for sustainable development. They are entitled to a healthy and productive life in harmony with nature.
>
> *Principle 3*: The right to development must be fulfilled so as to equitably meet developmental and environmental needs of present and future generations.
>
> *Principle 4*: In order to achieve sustainable development, environmental protection shall constitute an integral part of the development process and cannot be considered in isolation from it.
>
> *Principle 5*: All states and all people shall cooperate in the essential task of eradicating poverty as an indispensable requirement for sustainable development, in order to decrease the disparities in standards of living and better meet the needs of the majority of the people of the world.
>
> *Principle 8*: To achieve sustainable development and a higher quality of life for all people, states should reduce and eliminate unsustainable patterns of production and consumption and promote appropriate demographic policies.

The above five principles state that human beings have the right to develop a higher quality of life. In the process of developing a higher quality of life, environmental protection should be treated as an important and integrated part of development. Chapter 18 of the UNCED documentation, entitled "Protection of the quality and supply of freshwater resources: Application of integrated approaches to development, management and use of water resources," states that the following four principles should be pursued:

> (a) to promote a dynamic, interactive, iterative and multisectoral approach to water resources management, including the identification and protection of potential sources of freshwater supply, that integrates technological, socio-economic, environmental and human health considerations;
>
> (b) to plan for the sustainable and rational water utilization, protection, conservation and management of water resources based on community needs and priorities within the framework of national economic development policy;

(c) to design, implement and evaluate projects and programmes that are both economically efficient and socially appropriate within clearly defined strategies, based on an approach of full public participation, including that of women, young people, indigenous people and local communities in water management policy-making and decision-making;

(d) to identify and strengthen or develop, as required, in particular in developing countries, the appropriate institutional, legal and financial mechanisms to ensure that water policy and its implementation are a catalyst for sustainable social progress and economic growth.

The UNCED (1992) documentation also suggested the following three holistic approaches for sustainable water resource management:

"(a) consideration of various alternative means and components of water resource management in every stage of planning, design, construction, and operation;

(b) multidisciplinary approach to include engineering, biology, economics, sociology, health, laws, public administration, etc.;

(c) multisectoral exercise to include all levels and sectors of legislative and governmental units, cultural and interested groups, indigenous people, women and young people, non-governmental organizations and similar groups."

The worst enemy of sustainable use of reservoirs is sedimentation. Reservoir sedimentation replaces and depletes useful reservoir volume for flood control, hydropower generation, irrigation, water supply, recreation, and environmental purposes. The conventional concept of allocating "dead storage" for a predetermined useful life of 50 to 100 years is no longer acceptable. In the planning and design of new reservoirs, engineers must incorporate the concept of sustainable use and operation. For existing reservoirs, engineers should take appropriate remedial measures to prolong their useful functions within economic, social, political, and environmental constraints. If a reservoir has stopped serving its useful purposes, decommissioning should be considered to restore a river's natural pre-dam condition to the extent possible.

The following sections will discuss the concept of dynamic adjustment of a river system, methods used to reduce sediment inflow to a reservoir, reservoir sediment removal, and technical tools available for engineers.

In summary, sustainable development and use of reservoirs means that:

- We should develop and use reservoirs for the benefits of present and future generations in a socially, environmentally, and economically acceptable manner.
- Development can be growth in a quantitative sense, a qualitative sense, or both.
- Sustainable development and use of reservoirs implies that there are limitations imposed by the present state of technology. These limitations can be changed due to changing social, environmental, and economic considerations.

Basic principles in sustainable development and use of reservoirs are:

- Human beings are at the center of concerns for sustainable development and use. Humans are entitled to a healthy and productive life in harmony with nature.

- Along with the right to develop and use reservoirs comes the responsibility to meet the needs of present and future generations.

- To achieve sustainable development and use of reservoirs and a higher quality of life for all people, we should gradually reduce and eliminate unsustainable patterns of development and use subject to social, environmental, and economic considerations.

- Reservoir sedimentation shortens the useful life of reservoirs. Systematic and thorough consideration of technical, social, environmental, and economic factors should be made to prolong the useful life of reservoirs.

6.3 Dynamic Adjustment of a River System

A river is a dynamic system. Its channel roughness, geometry, and longitudinal profile continuously adjust in response to natural and human caused changes. Lane (1955) pioneered the concept of dynamic adjustment with the following qualitative relationship:

$$Q_s d \propto QS \tag{6.1}$$

where Q, Q_s = water and sediment discharges, respectively,
d = sediment particle diameter, and
S = channel slope.

Equation (6.1) states that $Q_s d$ is proportional to QS. "Because Q_s released from a reservoir is less than the amount of sediment a river is capable of transporting without the reservoir, downstream channel slope will be reduced through degradation" (Yang, 1996). It should be noted that Equation (6.1) provides a qualitative relationship which cannot be used by engineers for quantitative computations of dynamic adjustments of a river caused by the construction and operation of a reservoir.

Yang (1986) introduced the following quantitative equation for the prediction of dynamic adjustments of a river based on his unit stream power equations (Yang, 1973, 1979):

$$C_t = I\left(\frac{VS}{w}\right)^J \tag{6.2}$$

or

$$\frac{Q_t}{Q} = I\left(\frac{QS}{WDw}\right)^J \tag{6.3}$$

where V = average flow velocity,
S = water surface or energy slope,
w = channel width,
D = channel depth,
W = sediment fall velocity,
C_t = total bed-material concentration,
Q = water discharge,
Q_t = total bed-material load, and
I, J = coefficients.

Because sediment fall velocity is directly proportional to the square root of sediment diameter, Equation (6.3) can be rewritten as:

$$\frac{Q_t d^{J/2}}{K} = \frac{Q^{J+1} S^J}{W^J D^J} \tag{6.4}$$

where K = a site-specific parameter.

Most natural rivers have a J value between 0.8 and 1.5. If an average value of 1.0 is used, Equation (6.4) becomes:

$$\frac{Q_t d^{0.5}}{K} = \frac{Q^2 S}{WD} = \frac{Q^2 S}{A} \tag{6.5}$$

where A = channel cross-sectional area, and
d = median sediment particle diameter.

The following example illustrates how to use Equation (6.5) to predict river morphologic adjustments due to the proposed construction of Chi-Ban Hydropower Project in Taiwan. The example was based on the study done by Yang and Yeh (1992), which was later summarized by Yang (1996).

Example 6.1 Li-Wu Creek in eastern Taiwan, as shown in Figure 6.1, has a total length of 58 kilometers. The combined effects of tectonic movement and erosion caused by rapid flow created the narrow Taroko Gorge, more than 1,000 meters deep, between Chiu-Chu-Tung and Taroko. The proposed Chi-Pan Hydropower Project has a capacity of 160,000 kW. The project consists of a main dam at Ku-Yuan and three diversion dams at Pai-Yang, Ho-Shou, and Hsiu-Te. A regulation dam is also proposed on Li-Wu Creek just upstream of Tien-Hsiang. The construction of diversion dams can increase inflow to the Chi-Pan Powerplant. The regulation dam can regulate and maintain a minimum daytime flow of 5.55 m^3/s and nighttime flow of 4.5 m^3/s through the Taroko Gorge to meet minimum flow requirements for environmental reasons.

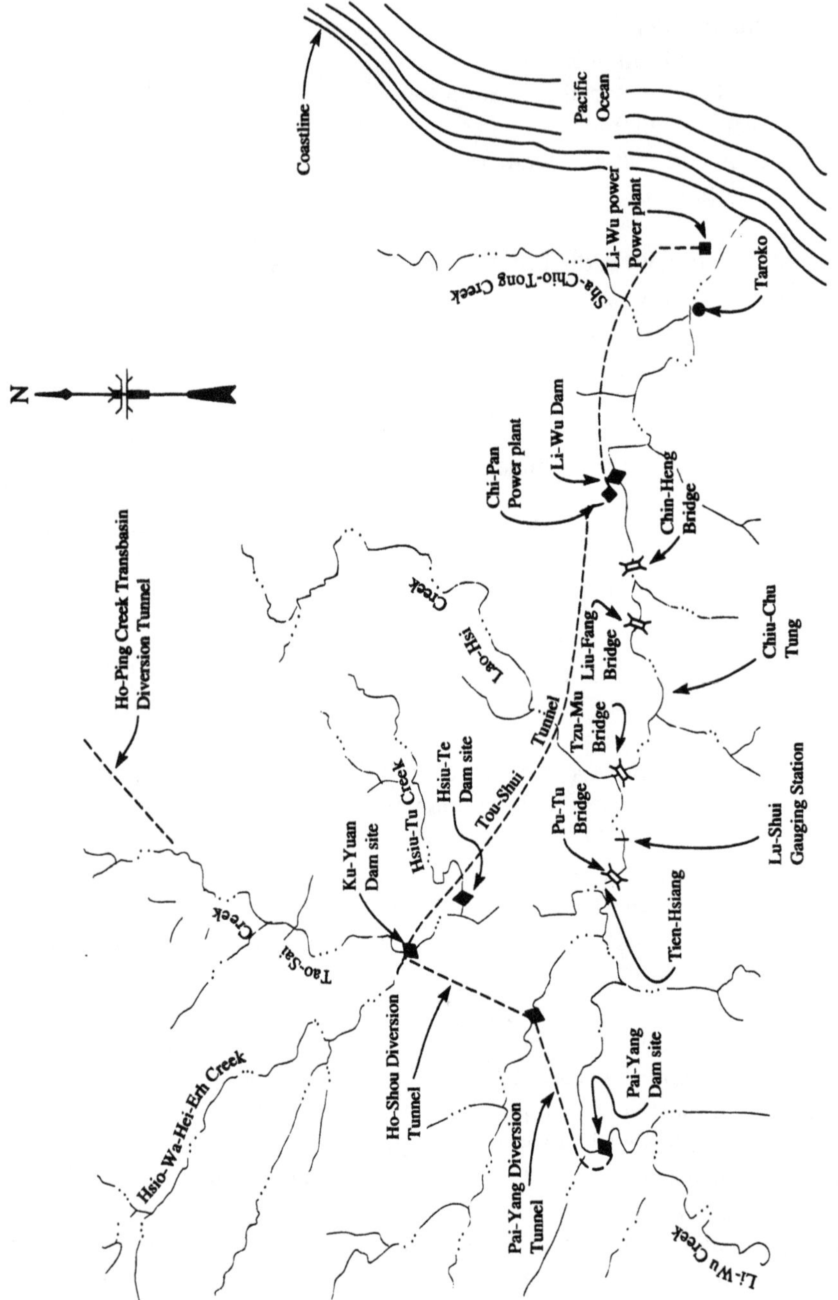

Figure 6.1. Chi-Pan Hydropower Project plan.

A transbasin diversion of 7 m^3/s from Ho-Ping Creek was also considered in order to increase the flow and sediment transport capacity in Li-Wu Creek. Environmentalists raised the concern that the proposed project might reduce Li-Wu Creek's sediment transport capacity to the extent that sediment might be deposited in the gorge, and the spectacular gorge would eventually disappear. Figure 6.2 shows the flow duration curves at Lu-Shui gauging station under different alternatives.

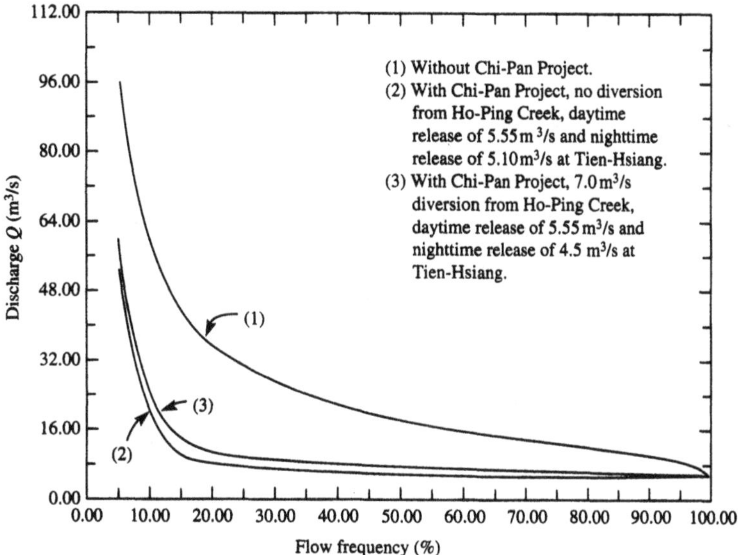

Figure 6.2. Flow duration curves at Lu-Shui gauging station.

Manning's roughness coefficient has a value of 0.035. Channel slopes at different stations can be obtained from 1:1000 topographic maps. Figure 6.3 shows bed-material size distributions at Lu-Shui station. Table 6.1 shows measured and computed sediment loads at Lu-Shui. Assess the sedimentation impact along the Taroko Gorge due to the proposed project.

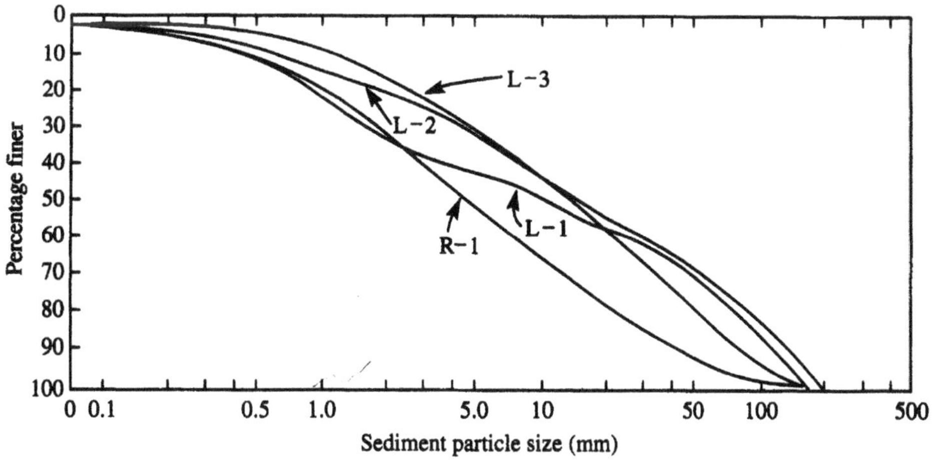

Figure 6.3. Bed-material size distribution at Lu-Shui gauging station.

Table 6.1. Computation of sediment transport rate at Lu-Shui gauging station

Flow duration (%)	Time interval (%)	Median (%)	Water discharge (m^3/s)	Einstein method (10^6 t/yr)		Schoklitsch method (10^6 t/yr)			Measured suspended load (10^6 t/yr)
				Bedload	Suspended load	d_{40}	d_{50}	d_{60}	
0-0.08	0.08	0.04	1,323.9	0.07*	0.05	0.12	0.08	0.06	0.96
0.08-0.02	0.12	0.14	802.9	0.06	0.04	0.10	0.08	0.05	0.79
0.2-0.5	0.3	0.35	444.4	0.09	0.04	0.14	0.10	0.08	0.77
0.5-1.5	1.0	1.0	247.8	0.16	0.06	0.27	0.19	0.14	1.17
1.5-5.0	3.5	3.25	132.2	0.35	0.12	0.50	0.36	0.25	1.12
5.0-15	10	10	59.95	0.61	0.17	0.64	0.45	0.32	0.56
15-25	10	20	35.38	0.45	0.11	0.37	0.26	0.18	0.12
25-35	10	30	26.98	0.36	0.08	0.28	0.20	0.13	0.06
35-45	10	40	21.48	0.29	0.07	0.22	0.16	0.10	0.03
45-55	10	50	18.09	0.25	0.05	0.19	0.13	0.09	0.02
55-65	10	60	15.34	0.20	0.04	0.16	0.11	0.07	0.01
65-75	10	70	13.19	0.17	0.04	0.14	0.09	0.06	0.01
75-85	10	80	11.32	0.14	0.03	0.12	0.08	0.05	0.01
85-95	10	90	9.59	0.11	0.02	0.10	0.07	0.04	—
95-100	5	97.5	6.99	0.03	0.01	0.04	0.02	0.01	—
Total				3.34	0.93	3.39	2.28	1.63	5.63

* For Q = 1323.0 m^3/s, Q_b = 2.643 t/s or Q_b = 2.643 x 365 x 86,400 x 0.08% = 0.07 x 10^6 t/yr.

Solution: Based on measured relationships among Q, A, d, a slope of 0.001, and computed Q_t from the Einstein method shown in Table 6.1; Table 6.2 shows the K values in Equation (6.5) at different discharges. The following regression equation can express the K values in Table 6.2:

$$\log K = 3.6063 - 0.3115 \log Q \qquad (6.6)$$

Table 6.2. Lu-Shui gauging station K values at different discharges

Q (m^3/s)	1,323.9	802.9	444.4	247.8	132.2
K	461	512	561	642	816
Q (m^3/s)	59.95	35.38	26.98	21.48	18.09
K	1,171	1512	1647	1712	1744
Q (m^3/s)	15.34	13.19	11.32	9.59	6.99
K	1756	1753	1736	1697	1504

Equation (6.6) has a correlation coefficient of 0.98. From Equations (6.5) and (6.6) and curve (1) in Figure 6.2, the average annual total bed-material load at Lu-Shui station is 4.2 x 10^6 metric tons, which is very close to the 4.27 x 10^6 metric tons computed from the Einstein method, as shown in Table 6.1. Assume that sediment size distribution, K values, and computed total bed-material loads at Lu-Shui station can be applied to other stations through the study reach. Channel cross-sectional areas under different discharges can be computed from Equation (6.5). Table 6.3 summarizes the mean annual discharge and bed-material transport capacity under different alternative plans obtained from Equations (6.5) and (6.6) and the *K* values in Table 6.2.

Table 6.3. Water discharge and sediment transport capacity at different stations along Li-Wu Creek

Station	Mean annual discharge (m^3/s)			Annual sediment transport capacity (10^6 t/yr)		
		Postproject			Postproject	
	Preproject	Without diversion	With diversion	Preproject	Without diversion	With diversion
Pu-Tu Bridge	32.41	16.04	17.92	10.97	5.29	5.96
Lu-Shui station	32.71	16.35	18.23	4.20	2.02	2.29
Tzu-Mu Bridge	35.24	18.88	20.72	2.91	1.56	1.71
Chiu-Chu-Tung	36.74	20.37	22.25	15.34	8.40	9.22
Liu-Fang Bridge	36.88	20.52	22.40	14.82	8.10	8.86
Chin-Heng Bridge	37.48	21.12	23.00	17.48	9.55	10.46
Li-Wu Dam	38.39	22.03	23.91	7.49	4.21	4.57

Table 6.3 shows that the reach between Lu-Shui station and Tzu-Mu Bridge has the lowest sediment transport capacity, while the gorge reach between Chiu-Chu-Tung and Chin-Heng Bridge has the highest sediment transport capacity. Field surveys and observations confirmed bedrock exposure with no deposition between Lu-Shui station and Tzu-Mu Bridge. This indicates that the total bed-material entering the gorge reach should be less than 2.91 metric tons per year. The results in Table 6.3 indicate that the Li-Wu Creek sediment transport capacity through the gorge reach far exceeds the bed-material supply from upstream, with or without the transbasin diversion. Consequently, the current downcutting trend of the gorge reach should continue with the proposed project. Yang and Yeh (1992) have published a more detailed description of the project.

Yang (1971) introduced two basic laws in river morphology. They are the law of average stream fall and the law of least rate of energy dissipation. Yang derived these two laws from the concept of entropy and from thermodynamic laws. The law of average stream fall states that under the dynamic equilibrium condition, the ratio of the average fall between any two different order streams in the same river basin is unity. The law of least rate of energy dissipation states that during the evolution toward its equilibrium condition, a natural stream chooses its course of flow in such a manner that the rate of potential energy expenditure per unit mass of water along this course of flow is a minimum. This minimum value depends on the external constraints applied to the stream (Yang, 1971). The first law can be used to predict whether a river's longitudinal profile will aggrade or degrade in the future. The second law can be used to explain and predict a river's morphologic changes.

The theory of minimum energy dissipation rate and its simplified version of minimum unit stream power were first introduced by Yang (1976). This theory was later expanded (Yang and Song,

1979, 1986) as a basic theory for all closed and dissipative systems, including river systems. The theory states that for a closed and dissipative system at its dynamic equilibrium condition, its energy dissipation rate is at its minimum value. The minimum value depends on the constraints applied to the system. If the system is not at its dynamic equilibrium condition, its energy dissipation rate is not at its minimum value. However, the system will adjust itself in such a manner that the energy dissipation rate can be reduced and regain equilibrium. The theories of minimum unit stream power and minimum stream power are simplified and special versions of the general theory of minimum energy dissipation rate. This theory is consistent with, but more general than, the law of least rate of energy dissipation. This theory or its simplified theories of minimum stream power and minimum unit stream power can be applied to solve a wide range of fluid mechanics and river morphology problems. The theory of minimum stream power also provides the theoretical basis for the GSTARS 2.0 (Yang et al., 1998), GSTARS 2.1 (Yang and Simões, 2000), and GSTARS3 (Yang and Simões, 2002) computer models, which are discussed in Chapter 5.

6.4 Planning

6.4.1 Background Information and Field Investigation

Planning starts with collecting background information for field investigations. Typical background information is identified in Table 6.4. Table 6.5 provides a checklist of tasks to accomplish during a reconnaissance inspection, and Table 6.6 provides a list of related observations (Dorough and Yang, 1996).

6.4.2 Basic Consideration

It is important to minimize sediment inflow to a reservoir. This process starts with selection of a reservoir site. Areas with high rates of erosion should be avoided. If this is not possible, a reservoir can be located offstream, and water can be diverted to it from the main river. In most cases, out of necessity, reservoirs are built on streams, and sediment control measures must be implemented to reduce the capacity loss of the reservoir.

In locations where the reservoir traps large volumes of sediment, the accumulated sediment can impair the effectiveness of the reservoir. In those cases, whether to remove sediment from the reservoir becomes an important consideration.

6.4.3 Sediment Control Measures

Sediment control measures include soil conservation practices, control of distribution of deposits, construction of debris basins, control of turbidity, and design of a proper intake structure to reduce sediment inflow. The following is a brief description of some issues related to sediment control measures (Wu et al., 1996):

Table 6.4. Background information for field investigations (Dorough and Yang, 1996)

Reservoir survey data	Provide point measurements of sediment yield. This serves as an indicator of sediment production from a localized contributing drainage area.
Topographic maps (suggest 1:24,000 scale)	Illustrate topographic relief and indications of land use and road and highway access. When combined with aerial photographs, this may help to identify non-sediment-contributing areas (i.e., lakes, ponds, and reservoirs).
Road maps	Provide road and highway access routes. Identify stream locations and locations of roads, highways, railroads, and other constraints to sediment movement within the basin.
Aerial photographs	Provide broad aerial view of the study area. Depict land usage and vegetation cover. May illustrate locations of gullying and channel head cutting and provide opportunity to assess changes with time. Show roads and other access routes.
Sediment sampling data	Additional indicator of sediment yield from the contributing drainage area.
Hand-held photographs	Provide historical view of field characteristics. Useful onsite for comparison against existing gully and channel head-cutting to assess time-related changes.
Soil conservation maps	Depict locations and types of soil conservation structures. May help in identifying locations of non-sediment-contributing areas or areas with significant reductions in sediment yield. Permit sedimentation specialist to conduct a site inspection of structure effectiveness.
Soil conservation practices	Provide information on soil conservation practices, locations, and projections on reductions in soil loss. May identify future plans.
Land-use maps	Delineate types and perimeters of different land usages. Permit field engineer to compare erosion potential from area to area.
Urbanization projections	Permit the sedimentation specialist to review areas that are undergoing, or may undergo, urbanization during the project life.
Regional sediment reports	Provide preliminary information on basin-wide sediment yield rates; may be especially useful if reports contain maps delineating isogram sediment yield lines.
Reservoirs, lakes, ponds, and strip mine sites	Reservoirs, lakes, ponds, and strip mines (other than spoil pile sites) serve as non-contributing sediment yield areas and should be noted on topographic maps and aerial photographs for inspection in the field.

Table 6.5. Field reconnaissance tasks (USACE, 1989)

Verify topographic maps
Note boundary conditions
Note bed and bank material slope
Note slope of stream in general and any break points
Obtain representative samples of bed-material
Note condition of banks, whether stable or caving, and the type of material found in the streambed and banks
Record conditions by location
Record drift accumulations, debris
Estimate percent of bed-material that is naturally armored
Note problem areas and attempt to ascertain the cause
Note changes in bed gradation and take representative samples for sediment study

Table 6.6. Field reconnaissance observations (USACE, 1989)

Note channel mining observations
Note tributary entry points, amounts of flow, turbidity of flow, condition of the tributary
Note diversion points
Note natural grade controls such as rock outcrops
Note presence of protection measures, their size, why they were placed
Note gauge locations, type of gauge
Note structural feature locations and observe bank and bed conditions in the vicinity of the structures
Note existing similar projects on same or adjacent streams—how they are performing
Note overbank conditions—areas of scour or deposition; if deposition exists, obtain samples and measure depth, and note extent on map
Take velocity measurements at several locations using surface floats, pacing, and stop watch
Talk with locals to identify problem areas; get an estimate of time or problem. Also, inquire as to local land use history—when urbanized, cleared, etc.

- *Soil conservation practices.*—Soil conservation practices greatly reduce erosion of the land surface, channel bank cutting, and head-cutting. Where a reservoir controls a small drainage of only 1 or 2 square miles, the sediment contribution to the reservoir can be reduced by as much as 95 percent by intensive conservation measures. For large reservoirs with large drainage areas, however, it should be recognized that it may not be economically feasible to control or greatly reduce the sediment inflow.

- *Control of distribution of deposits.*—When a stream enters a wide reservoir, the location of the deposits can be controlled to some extent by judicious breaching of the channel formed in the delta. In this manner, the major part of the sediment, including silts and clays, can be retained in the upstream portion of the pool.

- *Construction of debris basins.*—The U.S. Army Corps of Engineers, the U.S. Bureau of Reclamation, and the U.S. Natural Resources Conservation Service, among others, have designed and installed debris basins in mountainous areas. These are essentially reservoirs, generally located in canyons, designed to catch the coarse sediments and prevent their being carried downstream onto residential or other critical areas. In some cases where the topography permits adequate storage, the debris basins are designed to hold a 5-year or more sediment inflow, but the basins are more often designed to hold the sediment from a 100-year frequency storm plus the sediment yield of 2 or 3 average years. The basins are maintained by periodically removing the accumulated sediment by mechanical means. Where these structures are above reservoirs, they can be an effective means of reducing the amount of sediment transported to the reservoirs.

- *Trap efficiency.*—The actual accumulation of sediment in a reservoir depends on the proportion of the inflowing sediment that will be retained in the reservoir (i.e., trap efficiency). Two empirical relationships commonly used by engineers for estimating trap efficiency are those proposed by Brune (1953) and Churchill (1948).

- *Sediment storage planning.*—The major steps of reservoir sediment storage planning are determining:
  - watershed sediment yield
  - sediment inflow
  - amount of sediment trapped
  - location of reservoir sediments
  - distribution of reservoir sediments
  - sediment control measures needed

- *Estimation of reservoir life.*—Most of the prediction methods are based on measurements or computer model simulations. However, existing records can give us a general indication of sedimentation rate and useful life of reservoirs.

A study of several large reservoirs with large trap efficiencies (Table 6.7) indicates that the ratio of storage capacity to average annual water inflow is a good indicator of the reservoir's storage life. When the ratio of acre-feet of storage capacity to acre-feet per year of inflow is less than 1, the storage life of the reservoir is measured in tens of years. When the ratio is larger than 1.5, the storage life is measured in hundreds of years (Richardson, 1996).

Table 6.7. Statistics of listed reservoirs (Richardson, 1996)

River	Lake or dam	Millions			Ratio capacity/ inflow	Trap efficiency estimated	Reservoir life years
		Storage capacity (ac-ft)	Water inflow (ac-ft/yr)	Sediment inflow (t/yr)			
Nile	Nasser	133.1	68.1	140	1.95	99%	1,000+
Nile	Old Aswan	5.0	67.8		0.07	0.01%	Forever
Indus	Tarbela	11.1	63.0	397.4	0.18	99%	50 to 100
Colorado	Mead	32.5	13.4	142.9	2.43	99%	1,000+

6.5 Design of Intakes

6.5.1 Location of Intakes

Tan (1996) summarized the basic criteria for selection of intake structure locations as follows:

- An intake should be located on a stable reach without frequent wandering of a flow path.

- The concave bank of a bend is an ideal location for an intake because it can divert relatively clear water with low sediment concentration.

- An intake on a straight reach is usually not favorable for preventing sediment from entering it.

- An intake should be as far as possible from the confluence of a tributary to avoid disturbing and complicating the flow pattern in front of the intake.

6.5.2 Types of Intakes for Sediment Control

Five types of intakes can be used for sediment control. They are intakes with sluice gates, lateral intakes, tiered intakes, bend-type intakes, bottom-grate-type intakes, and combined intakes (Tan, 1996). Each type of intake is adapted to certain conditions.

Intakes with sluice gates.—This type of intake is widely used. The diversion works are generally composed of flood sluice gates, desilting sluice gates, intakes, and other structures. The intake is usually located near desilting sluices and flood sluices to keep the main flow close to the intake. The axis of intake may be parallel, oblique, or perpendicular to the riverflow. Generally, a smaller angle is favorable for decreasing sediment entering an intake.

Lateral intakes.—This type of intake is still widely used in various diversion works because of the simplicity in the layout. An effective measure to improve the efficiency of lateral intakes for sediment control is to enhance the sediment-carrying capacity of flow in front of intakes by means of facilities such as circulation sluicing flumes, sediment intercepting devices, and releasing galleries. The approaching flow may also be regulated and directed toward the intakes by river training works such as guide walls, spurs, and vanes.

Tiered intakes.—The tiered intake has a double set of stairs of gate openings separated by a horizontal diaphragm. The upper opening is for a water diversion, and the lower opening is for sediment sluicing. This type of intake has been used in some mountain rivers in Middle Asia which are characterized by steep slopes (1/60 to 1/120), coarse sediment, and large seasonal variation of discharge. Tiered intakes also were constructed at Milburn Diversion Dam and Arcadia Diversion Dam on the Middle Loup River in Nebraska in the United States, where they were called short undersluice tunnels.

Generally, the tiered intake is appropriate for the conditions of a small watershed and a large river discharge (much more than diversion discharge). However, tiered intake structures are relatively complicated, water consumption for releasing sediment is large, and preventing floaters (timber, ice, garbage, etc.) from entering the intake is difficult. In practice, these factors limit, to a certain extent, the wide application of tiered intakes.

Bend-type intakes.—Bend-type intake refers to a combination of intake, sluice, and a natural, trained, or artificial bend in which the bend plays a key role for sediment control. The intake is commonly built on the concave bank at the end of a bend to divert water by meeting the main current and to release sediment toward the side under the effect of secondary flow in the bend.

A typical, artificial, bend-type intake consists of an upper flood escape, a diversion bend, and a water intake and sediment sluice at the end of the diversion bend. In fact, the effect of secondary flow in a bend, and the relevant principle of diverting water by meeting the main current and releasing sediment toward the side, are applied here twice. First, in inlet flood escape, most of the incoming sediment is flushed downstream through the flood escape. Second, in the combination of the intake and the sediment sluice, most of the sediment that enters the bend is released through the sediment sluice. Consequently, little sediment can enter the intake.

The bend-type intake is commonly suitable for mountain rivers carrying bedloads of gravel, cobbles, and coarse sand. Bend-type intakes have been widely built in Middle Asia and Western China.

Bottom grate-type intakes.—A bottom grate-type intake is usually built on medium and small mountain creeks with steep slopes and torrential currents carrying a great amount of large-sized bedload, such as gravels, pebbles, and cobbles.

This type of intake consists mainly of a weir with a bottom grate intake, diversion gallery, or canal, sluice, and spillway. When the current carrying sediment overflows the bottom grate on top of the weir, sediment coarser than the spacing between grate bars is excluded by the bars and carried downstream by the flow; the water passes through the grate and into a diversion gallery. A bottom grate intake commonly has a small diversion discharge and is characterized by a simple structure, easy construction, low cost, convenience of management, and high efficiency of coarse sand prevention. Therefore, the bottom grate intake is widely used in some rivers to keep the upstream channel straight. Some blocks or guide vanes can be provided for improving the flow condition.

Combined intakes.—A combination intake composed of a vortex tube, curved desilting channel, and settling basin can be used to control the amount of sediment inflow (Tan, 1996).

The vortex tube is a tube with a top slot that is set in the bottom of a diversion canal for releasing bedload. Water entering the tube forms a strong spiral flow that can raise coarse particles and cause them to be suspended and carried downstream. A major problem of the vortex tube is the inlet dimension and possible blockage by small trash. In a stream with floating and submerged trash and twigs, some provisions for clearing are necessary.

A curved desilting channel is a section of a diversion channel that has a broadened cross-section and curved route. This type of channel is used mainly to allow coarse sand and gravel to settle out and keep from entering the diversion canal. Because velocity decreases in the broadened section, the coarse particles descend to the channel bed and move toward a convex bank under the effect of circulation flow developed at the curved channel. Then, the coarse particles are flushed out through sediment sluicing galleries placed at the bottom of the convex bank.

A settling basin is a wider and deeper section of a diversion canal. It is mainly used to precipitate suspended load by providing a greater cross-section with less velocity and to exclude the suspended load through sediment-releasing facilities. A settling basin commonly consists of a diffuser, settling chamber, contractor, and sediment-releasing works.

6.6 Sediment Management for Large Reservoirs

The basic sediment management principle for large reservoirs is to alleviate reservoir sedimentation by sluicing density current and hyperconcentrated density current, and storing clear and flushing turbid water (Wan, 1996). Density current is a sediment-laden flow of greater density than that of clear water entering a reservoir.

A density current may carry a large amount of sediment and pass a long distance along a reservoir bed without mixing with surrounding clear water. Whether a density current can keep moving depends on two conditions. First, the incoming sediment-laden flow must be maintained. Once the incoming flow recedes and a turbid water supply is no longer available at the inlet of the reservoir, the density current will stop moving. Second, the density current must be strong enough to overcome the resistance along the path through the reservoir.

A density current reaching the damsite can be released only if some bottom sluices at a low enough elevation are open. Depending on the transition of its kinetic energy into potential energy, a density current can climb a certain height, but the height is not great.

The continuous motion of a density current depends on the sediment concentration of the incoming flow and the content of fine particles in it; the duration of the incoming flow; the magnitude of flow discharge; and the topography along its course, such as bends, abrupt expansions, and the slope of the reservoir bottom.

In reservoir sediment management, proper timing of the operation of sluices is important once a density current is formed at the inlet. If bottom sluices are opened much before the density current reaches the damsite, clear water will be released and wasted. If bottom sluices are closed when a density current reaches the damsite, sediment particles in the density current will deposit and cause siltation.

Due to the large density difference between the current and the clear water, a hyperconcentrated density current moves at a much higher velocity than a common density current. Because hyper-concentrated fluid possesses a higher yield stress and high viscosity, the interfacial mixing between the upper clear water and the lower hyperconcentrated density current is weak, and sediment particles in the hyperconcentrated water either do not settle or settle very slowly. Consequently, little deposition occurs, no obvious sorting phenomenon happens, and high concentration can be maintained while a hyperconcentrated density current passes through a reservoir.

For most rivers, runoff and sediment load are unevenly distributed among seasons. Most incoming flow and sediment are concentrated in flood season, with sediment load being the most uneven. For instance, at Sanmenxia on the Yellow River, 60 percent of the runoff and 85 percent of the sediment come in the flood season. The reservoir pool is lowered in the flood season so that the slope and velocities are high, increasing the sediment-carrying capacity of the flow. All of the incoming sediment in the flood season, and sometimes even the deposition in the preceding non-flood season, can be flushed out of the reservoir. In this way, sedimentation in the reservoir can be avoided or mitigated, and the storage capacity of the reservoir can be preserved for long-term usage. Chinese engineers refer to such a strategy as "storing clear water in the non-flood season and flushing turbid water in the flood season."

The uneven distribution of incoming runoff and sediment load and some amounts of available incoming runoff in non-flood season are the necessary hydrological conditions for adopting such a strategy. To lower the pool during floods, the dam must be equipped with bottom sluices of large enough sizes. Even if the pool is lowered, only the sediment in the channel within a certain width can be flushed. Flood plain deposits cannot be eroded. Therefore, a gorge-shaped reservoir (i.e., a reservoir with narrow cross-section) favors the adoption of such a strategy (Wan, 1996).

Associated with the storage of water in a reservoir, flow velocity decreases in backwater regions and sediment carried by the flow is deposited and creates a delta. The water level over the delta region rises simultaneously with the downstream development of the delta. This rise further decreases the flow velocity in this region and causes more deposition. Such a feedback phenomenon is called retrogressive deposition (Wan, 1996).

Retrogressive deposition may slow down incoming flow and attract more sediment more rapidly. This could create favorable conditions for dense aquatic growth and marsh weeds in the backwater reach, and the conditions would then spread upstream gradually. As a result, the delta can develop rapidly. The impact of retrogressive deposition on the environment should be studied.

6.7 Sediment Management for Small Reservoirs

Although the general strategies for reservoir sediment management are the same for large reservoirs and small reservoirs, sediment management practices for those two categories of reservoirs are often as different as their magnitudes. Some of the measures commonly used to reduce reservoir sedimentation, mainly for small reservoirs, are summarized in the following sections.

6.7.1 Soil Conservation

In the upstream watershed of a reservoir, three basic patterns of soil conservation measures are commonly taken to reduce sediment load entering the reservoir: structural measures, vegetative measures, and tillage practice. Structural measures include terraced farmlands, flood interception and diversion works, gully head protection works, bank protection works, check dams, and silt-trapping dams. Vegetative measures include growing soil and water conservation forests, closing off hillsides, and reforestation. Tillage practice includes contour farming, ridge and furrow farming, pit planting, rotation cropping of grain and grass, deep ploughing, intercropping and interplanting, and no-tillage farming.

The effectiveness of soil conservation measures in reducing sediment inflow to a reservoir is different for different watershed sizes. For a large watershed with poor natural conditions, soil conservation can hardly be effective in the short term. Nevertheless, if the watershed is not large, the effect of soil conservation can be felt in a short period.

6.7.2 Bypass of Incoming Sediment

Rivers, especially sediment-laden rivers, carry most of the annual sediment load during the flood season. Bypassing heavily sediment-laden flows through a channel or tunnel may avoid serious reservoir sedimentation. The bypassed flows may be used for warping, where possible. Such a combination may bring about high efficiency in sediment management.

When heavily sediment-laden flows are bypassed through a tunnel or channel, reservoir sedimentation may be alleviated to some extent. In most cases, however, the construction cost of such a facility is high. Where a unique topography is available, the cost of construction may be reduced and bypassing facilities may be practical.

6.7.3 Warping

Warping has been used around the world. It has a history of more than 1,000 years in North and Northwest China as a means of filling low land and improving the quality of salinized land. Now, this practice may have a dual role, not only improving the land but also reducing sediment load entering reservoirs. Warping is commonly carried out in flood seasons, when the sediment load is mainly concentrated, especially in sediment-laden rivers. Warping can also be used downstream from dams when hyperconcentrated flow is flushed out of reservoirs.

6.7.4 Joint Operation of Reservoirs

Joint operation of reservoirs is a rational scheme to fully use the water resources of a river with cascade development. For sedimentation management of reservoirs built on sediment-laden rivers, such an operation may also be beneficial to mitigate reservoir sedimentation and to fully use the water and sediment resources, provided a reasonable sequence of cascade development is made. There are various patterns of joint operation of reservoirs built in semi-arid and arid areas. The idea is to use the upper reservoir to impound floods and trap sediment and to use the lower reservoir to impound clear water for water supply.

Another idea is to use the upper reservoir for flood detention and the lower reservoir for flood impoundment. Irrigation water in the lower reservoir is used first; when it is exhausted, the water in the upper reservoir is used. The released water from the upper reservoir may not only erode the deposits in the lower reservoir, but also cause warping by the sediment-saturated water.

6.7.5 Drawdown Flushing

Drawdown flushing is a commonly used method of recovering lost storage of reservoirs. It may be adopted in both large and small reservoirs. The efficiency of drawdown flushing depends on the configuration of the reservoir, the characteristics of the outlet, the incoming and outgoing discharges, sediment concentrations, and other factors.

6.7.6 Reservoir Emptying

Reservoir emptying operations may be used for small or medium-size reservoirs to recover a part of the storage capacity if temporary loss of water supply is acceptable. In the process of reservoir emptying, three types of sediment flushing occur: retrogressive erosion and longitudinal erosion, sediment flushing during detention by the base flow, and density current venting. Data on the three types of sediment flushing are shown for Dongxia Reservoir in Table 6.8.

Table 6.8. Sediment flushing from 1978 to 1981, Dongxia Reservoir (Zhou, 1996)

Types of sediment flushing	Retrogressive erosion	Detention	Density current
Duration (hr)	93	3,552	814
Average sediment concentration (kg/m^3)	172	125	
Average discharge (m^3/s)	14.5	2.8	
Water used (million tons)	4.87	35.1	24.7
Sediment flushed (million tons)	0.83	4.4	2.44
Unit water consumption (m^3/t)	5.9	8	10

6.7.7 Lateral Erosion

The technique of lateral erosion is to break the flood plain deposits and flush them out by the combined actions of scouring and gravitational erosion caused by the great transverse gradient of the flood plains. In so doing, it is necessary to build a low dam at the upstream end of a reservoir to divert water into diversion canals along the perimeter of the reservoir. The flow is collected in trenches on the flood plains.

During lateral erosion, because the surface slope of the flood plain is steep and the flow has a high undercutting capability, intensive caving-in occurs at both sides of the collecting trench. The sediment concentration of the flow may be as high as 250 kg/m^3. This technique has the advantage of high efficiency and low cost, and no machines or power are required.

6.7.8 Siphoning Dredging

Siphoning dredging makes use of the head difference between the upstream and downstream levels of the dam as the source of power for the suction of deposits from the reservoir to the downstream side of the dam.

Siphoning dredging has a wide range of applications in small and medium-size reservoirs. In China, it has been applied in semi-arid areas where the flushed water and sediment mixture has been diverted into farmland for warping. Such an application is valuable to solve reservoir sedimentation and to fulfill the demand of irrigation if the head difference is adequate and the distance between upstream and downstream ends of the siphon is not too great.

6.7.9 Dredging by Dredgers

Dredging is used to remove reservoir deposits when other measures are not suitable for various reasons. In general, dredging is an expensive measure. However, when the dredged material may be used as construction material, it may be cost effective.

6.7.10 Venting Density Current

Density currents have been observed in many reservoirs around the world. The conditions necessary to form a density current, and allow it to reach the dam and be vented out if the outlet is opened in time, have been studied extensively, both from the data of field measurements and laboratory tests. Venting of density currents is one of the key measures for discharging sediment from several reservoirs in China, especially from impounding reservoirs. Density current venting may be carried out under the condition of impoundment, thus maintaining the high benefit of the reservoirs.

6.7.11 Evaluation of Different Sediment Management Measures

An in-depth evaluation of different sediment management measures by Zhou (1996) lead to the following conclusions:

- Sediment management strategies for large reservoirs may be applied to small reservoirs, but not vice versa.

- Soil conservation measures are the most commonly adopted measures of reducing sediment inflow to reservoirs. They are effective in a relatively short period in small watersheds. They are also effective for a mid-ranged period in large watersheds.

- Bypassing the inflow of heavily sediment-laden floods to reservoirs, warping, or joint operation of reservoirs must be done under specific conditions before they can be adopted for reducing sediment inflow to reservoirs.

- Lost storage capacity can be partially recovered by drawdown flushing, reservoir emptying, lateral erosion, siphoning dredging, dredging, venting of density currents, and other methods. The method most suitable for a specific reservoir depends on many factors. In the planning and design stage, this issue should be carefully studied and an optimum choice made. Due to the complexity of sediment problems, however, the issue of recovering lost storage capacity should be re-examined during the operation of the project.

6.8 Effective Management of Reservoir Sedimentation

Effective management of reservoir sedimentation should consider the following factors (Tomasi, 1996):

- *Legislation.*—In most countries, laws and regulations prescribe quality standards for the material removed from a reservoir and regulate the performance of the maintenance operations should be established. Examples include the maximum values of sediment concentrations in a flushing operation and the maximum sediment and water quality parameters downstream.

- *Territorial constraints.*—Reservoir topographic features, riverbed characteristics prior to dam construction, and local morphology and land use may drastically limit the methodology to be adopted. For example, the absence of a place to store material removed from the reservoir, difficulties in accessing the reservoir, and other factors may preclude excavation or dredging.

- *Human activities.*—Socioeconomic development of the area may strongly influence the quality of the deposited material. In industrialized, overcrowded areas, water and sediment pollution problems will certainly be more frequent than in less populated areas. Environmental protection is often a high-priority constraint in several countries.

- *Economic aspects.*—Besides the economic value of the recovered storage, sediment may also be valuable for industrial purposes, such as construction activities and landscape improvement. The overall economic (cost-benefit) balance must also be carefully evaluated. In particular cases, such as drinking and irrigation water reservoirs in semi-arid areas, preserving reservoir capacity could be the dominant objective.

- *Safety.*—Human lives must be protected by performing periodic maintenance on the submerged structures and testing the operation of such systems. This implies the removal of sediments from the reservoirs by any means.

- *Type, size, and elevation of outlet structures.*—Many dams do not have bottom outlets or structures located low enough to enable drawdown flushing. The installation of bottom outlets after dam construction, solely for the purpose of sluicing and flushing, is believed to be uneconomical and, in many cases, from a technical point of view, infeasible.

- *Incoming flow and sediment*.—Inflow of water, sediment yield, and sediment aggradation are the basic factors to be considered in effective reservoir management. Fine sediment load may be generally reduced by soil conservation work in the watershed. Coarse sediment transported by the river can be temporarily intercepted (and subsequently removed) by trenches, check dams, debris basins, or other hydraulic structures upstream from the dam.

The selection of one or more sediment control strategies must be made by a multidisciplinary team with a multicriterion approach that should be flexible. Indeed, certain water uses, such as environmental purposes, tourism, and recreation, and certain constraints have acquired a higher priority in the last decades. The selected methodology or combination of methodologies should represent a compromise among the parties that use the water. In this compromise, the environment, social needs, and other concerns must be evaluated carefully.

6.9 Operational Rules

The operational rules of reservoirs have significant influence on reservoir sedimentation. Three basic types of operating rules have been adopted around the world: impoundment, impounding the clear and discharging the muddy, and flood detention. The first two types are more often adopted in China. In Table 6.9, some basic characteristics of reservoir operating rules are listed.

Table 6.9. Operating rules of reservoirs (Zhou, 1996)

Operating rule	Regulation of sediment	Method of sediment sluicing	Period of sediment sluicing
Sediment totally impoundment trapped	None	None or dredging	None
Sediment partly trapped	None	Density current venting, sluicing	Beginning of flood seasons
Impounding clear and discharging turbid	Yearly or seasonally	Discharging sediment during detention, density current, venting, etc.	Flood seasons
Detention	None	Discharging sediment during detention, reservoir emptying	Flood seasons

Unfortunately, in the United States, the basic operating rules for reservoirs have been neglected for almost half a century. The above rules have been seriously considered as measures to alleviate reservoir sedimentation only after sedimentation problems in reservoirs built on heavily sediment-laden rivers have become acute.

If the operating rule of impoundment is adopted for a reservoir suitable for impoundment and detention, the benefit provided by the reservoir may be high in the short term, but reduced in the long term by the loss of reservoir storage. For a reservoir adopting the impoundment and detention operating rule, there are some conflicts between various function of the reservoir, as follows:

- *Conflicts with power generation.*—To fulfill the demands of sediment discharge, the pool level fluctuates significantly, resulting in the following problems:
  - The annual energy output of the project will be reduced somewhat, although the total energy output is larger in the long term.
  - The water level in the flood season is reduced, so it is necessary to design a turbine with a large discharge capacity to compensate for the reduction of head.
  - The variation of the water level in the reservoir is quite large, so the design of the turbine is more difficult than under typical conditions.

Because the flow with high sediment concentration may pass through the turbine, abrasion of the runner and other turbine parts may be serious, resulting in reduced efficiency and lifespan of the turbine. Cooling water systems may be choked by sediment particles.

- *Conflicts with irrigation.*—Because discharging sediment requires a large amount of water, conflicts between sediment discharge needs and irrigation needs may develop.
- *Changes in the downstream channel.*—Reservoirs under the operating rule of impoundment and detention alter the relationship between waterflow and sediment loads released from the reservoir, but the change is not as large as that under the operating rule of impoundment. Nearly clear water will be released from impounding reservoirs, resulting in downstream channel degradation. If a reservoir is operated in two stages, impoundment in the first stage followed by impoundment and detention in the second stage after the deposition of considerable volumes of sediment in the reservoir, the discharged sediment load will induce a new problem in the downstream river channel, which would have adjusted to the conditions of clear water releases.

6.10 Cost of Sedimentation Prevention and Remediation

Cost and other economic considerations are driving factors in determining the type, amount, and timing of remediation of reservoir sedimentation. Reservoir dredging and other remediation measures can be very costly, sometimes even exceeding the initial cost of constructing the reservoir. Capital, like many other commodities, is a limited resource and is best expended where there is the expectation of greatest benefit. Unfortunately, the benefits of remediation measures are not always obvious to those who manage the purse strings, and funding for remedial work is, therefore, difficult to obtain, even in wealthy countries. It is up to the sedimentation engineer and reservoir operators to justify the need for remediation and to evaluate the costs and economic benefits of remediation to obtain funding approvals, regardless of whether the reservoir is privately or publicly owned (Harrison, 1996).

There can be substantial indirect economic losses associated with lost water supplies or lost hydrogeneration due to sedimentation. In underdeveloped regions of the world, there may not be alternative power or alternative water supplies when normal reservoir operations are shut down, resulting in interruption of service to customers for extended periods of time. The societal costs of such interruptions can range from personal hardship, to the shutting down of businesses and factories, to loss of life. Unreliable water and electric power supplies, for whatever reason, also may constrain the economic development of regions. In developed regions, other reservoirs and alternative power supplies, such as nuclear or fossil-fueled plants, would likely make up for sediment-related losses of supply. However, such alternatives might come at a substantially greater cost.

Costs for remedial actions vary widely. Dredging may range in cost from less than $2 to more than $50 per cubic meter, depending on site conditions and quantities. Flushing and sluicing options may, or may not, be less costly than dredging, depending on the value of the lost water, lost power generation, and environmental impacts. Also, flushing and sluicing may require substantial expenditures for constructing new, low-level sluice openings at a dam, if they were not provided in the original design. Watershed restoration programs may be relatively inexpensive for the reservoir operator if other beneficiaries can be enlisted to share in the costs. However, watershed management may take many years to become significantly effective, requiring interim dredging or other measures to continue operation of the reservoir. Strategies combining two or more types of remediation may be undertaken to achieve the most cost-effective, long-term solution, such as watershed management and sediment passthrough. The best and most cost-effective remediation plans for such reservoirs must be determined by identifying and evaluating all the alternatives and the resources available for implementation. However, even the best plans are of little value if no funds are available for implementation. In that case, compromises may be necessary (Harrison, 1996).

6.11 Reservoir Sustainability Criteria

Criteria to ensure the sustainability of reservoirs can be divided into three categories: safety, reduction of sediment inflow, and environmental compatibility. Safety standards must be met for dams, rock slopes, earthquakes, floodflows, structural behavior, aging, and rehabilitation. Safety concerns should be addressed in the engineering design, construction, and operation of dam projects (Veltrop, 1996).

A variety of methods have been applied over the years to reduce sediment inflow. These include erosion control through soil conservation, contour building and terracing of steep slopes in catchment areas, trapping sediments behind check or debris dams, reforestation or revegetation of denuded areas, and providing vegetation screens of shrubs or weeds. Because control of large drainage areas is difficult, prevention of sediment inflow is often impractical. Therefore, efforts to remove sediments from reservoirs abound.

Sustainability also requires: (1) studies and implementation of mitigatory measures to counter the impacts of dams and reservoirs on the natural environment (maintenance of ecosystems, water quality, temperature, etc.), (2) social acceptance, (3) quality of life, and (4) economic justification and financial support. Public acceptance of dams and reservoirs requires that these engineering works be in tune with the current ideas and values of society.

Sustainable use of reservoirs also requires:

- Greater use of hydrometeorological forecasts to improve the efficiency of reservoir operation
- Optimum operation of reservoirs to resolve conflicting release requirements, including establishing seasonal minimum instream releases
- Ecologically friendly operating rules
- Operational control of water levels to influence the habitat of disease carriers
- Reduced water losses through evaporation control
- Documented socioeconomic impacts of reservoir sedimentation, including the cost of storage loss

6.12 Technical Tools

Computer models are useful tools to simulate and predict the effects of erosion and sedimentation on the sustainable development and use of reservoirs. They are technical tools that enable planners, engineers, and policymakers to select an optimum solution among different alternatives. There are two types of computer models related to erosion and sedimentation: (1) those related to the erosion and sedimentation processes in rivers and reservoirs, and (2) those related to economic analysis of sustainable development and use of reservoirs. A brief introduction of some of the models, especially those developed by the U.S. Bureau of Reclamation, is given in the following sections.

6.12.1 GSTARS 2.0/2.1 Models

A generalized erosion and sedimentation model for alluvial rivers should be able to:

- Compute hydraulic parameters for open channels with fixed, as well as movable, boundaries
- Compute water surface profiles in the subcritical, supercritical, and mixed flow regimes (i.e., in combinations of subcritical and supercritical flows without interruption)
- Simulate and predict the hydraulic and sediment variations, both in the longitudinal and the transverse directions

- Simulate and predict the change of alluvial channel profile and cross-sectional geometry, regardless of whether the channel width is variable or fixed

- Incorporate site-specific conditions such as channel side stability and erosion limits

- Simulate bed sorting and armoring

The Generalized Stream Tube model for Alluvial River Simulation Version 2.0 (GSTARS 2.0), released by Reclamation (Yang et al., 1998), and its improved version GSTARS 2.1 (Yang and Simões, 2000) have the above capabilities. The stream tube concept was used in GSTARS 2.0/2.1 for water and sediment routing in a semi-two-dimensional manner. The adjustment of bed provides the variation in the vertical direction. Conjunctive use of energy and momentum equations enable computation of water surface profiles through subcritical, critical, and supercritical flows without interruption. The use of minimum total stream power theory (Yang and Song, 1986) provides the theoretical basis for the determination of channel width and depth adjustments. GSTARS 2.0/2.1 contains the following sediment transport formulas from which a user may choose:

- Ackers and White's 1973 method
- Engelund and Hansen's 1972 method
- Krone's 1962 and Ariathurai and Krone's 1976 methods for cohesive sediment transport
- Laursen's 1958 formula
- Meyer-Peter and Müller's 1948 formula
- Parker's 1990 method
- Revised Ackers and White's 1990 method
- Toffaleti's 1969 method
- Yang's 1973 sand and 1984 gravel transport formulas
- Yang's 1979 sand and 1984 gravel transport formulas
- Yang's 1996 modified formula for sediment-laden flows

Non-equilibrium sediment transport can also be simulated based on the theory introduced by Han (1980).

Some of the possible applications and features of GSTARS 2.0/2.1 are:

- The model can be used for water surface profile computations with or without sediment transport.

- It can compute water surface profiles through subcritical and supercritical flow conditions, including hydraulic jumps, without interruption.

- It can compute the longitudinal and transversal variations of flow and sediment conditions in a semi-two-dimensional manner, based on the stream tube concept. If only one stream tube is selected, the model becomes one dimensional. If multiple stream tubes are selected, both the lateral and vertical bed elevation changes can be simulated.

- The bed sorting and armoring computation based on sediment size fractions can provide a realistic simulation of the bed-armoring process.
- The model can simulate channel geometry changes in width and depth simultaneously, based on minimum total stream power.
- The channel side stability option allows simulation of channel geometry change, based on the angle of repose of bank materials and sediment continuity.

GSTARS 2.0/2.1 is a general numerical model developed for a personal computer to simulate and predict river morphologic changes caused by natural and engineering events. Although GSTARS 2.0/2.1 is intended to be used as a general engineering tool for solving fluvial hydraulic problems, it does have the following limitations from a theoretical point of view:

- GSTARS 2.0/2.1 is a quasi-steady flow model. Water discharge hydrographs are approximated by bursts of constant discharges. Consequently, GSTARS 2.0/2.1 should not be applied to rapid, varied, unsteady flow conditions.
- GSTARS 2.0/2.1 is a semi-two-dimensional model for flow simulation and a semi-three-dimensional model for simulation of channel geometry change. It should not be applied to situations where a truly two-dimensional or truly three-dimensional model is needed for detailed simulation of local conditions. However, GSTARS 2.0/2.1 should be adequate for solving most river engineering problems.
- GSTARS 2.0/2.1 is based on the stream tube concept. The phenomena of secondary current, diffusion, and superelevation are ignored.

GSTARS 2.0 is written for DOS PC operation. An improved and revised version, GSTARS 2.1, for Windows operation with graphic interface, was released by Reclamation in the year 2000 (Yang and Simões, 2000) to replace GSTARS 2.0.

6.12.2 GSTARS3 Model

The Generalized Sediment Transport model for Alluvial River Simulation Version 3.0 (GSTARS3) was developed by Reclamation to simulate the process of river and reservoir sedimentation (Yang and Simões, 2002). In addition to the capabilities of GSTARS 2.1 and GSTARS 2.0, GSTARS3 has the ability to simulate the formation of deltas, non-equilibrium sediment transport in a reservoir, the exchange of sediment across stream tubes, more extensive cohesive material transport capabilities, and fractional transport of sediment mixtures of vastly different sizes. GSTARS3 has the ability to simulate different reservoir operation schemes, including sluicing to remove sediment from a reservoir. Yang and Simões (1999) provided a summary report on the development of GSTARS 2.0, 2.1, and 3. Chapter 5 gives the basic principles and approaches used in sediment transport computer models, including those used in GSTARS 2.0/2.1/3.

6.12.3 GSTAR-1D Model

A new model developed by Reclamation is GSTAR-1D. GSTAR-1D was developed in cooperation with the U.S. Environmental Protection Agency for use in Total Maximum Daily Load (TMDL) studies. Its development was driven by the need to incorporate more physical processes associated with cohesive sediment transport and the ability to model unsteady flow. The user's manual and program have been released by the Bureau of Reclamation (Yang et al., 2004, 2005). In addition to most of the capabilities of GSTARS 2.1, GSTAR-1D has the following features:

- Computation of water surface profiles in single channels, simple channel networks, and complex networks
- Steady and unsteady flow models
- Cohesive sediment aggregation, deposition, erosion, and consolidation
- Multiple bed layers
- Flood plain simulation
- Exchange of water and sediment between main channel and flood plains
- Ability to limit erosion and deposition at cross-sections
- Point and non-point sources of flow and sediments
- Internal boundary conditions, such as time-stage tables, rating curves, weirs, culverts, bridges, dams, and radial gates

6.12.4 GSTAR-W Model

Another model being developed by Reclamation is GSTAR-W (Yang, 2002). This is an erosion and sedimentation model for river systems in a watershed. GSTAR-W will integrate capabilities of GSTARS 2.1, GSTARS3, and GSTAR-1D for river and reservoir systems in a watershed for the determination of TMDL under different hydrologic, hydraulic, topographic, and geologic conditions at any location in a watershed. The concepts, theoretical basis, and approaches used in the GSTAR-W model are summarized by Yang et al. (2003).

6.12.5 Economic Model

To determine the sustainability of a reservoir requires an economic model, in addition to hydraulic and sedimentation models. An economic model should select the size and other controllable characteristics of a reservoir, sediment management strategy, and the useful life of the reservoir to maximize the net benefits. Palmieri et al. (1998) express this concept by the following equation:

$$\text{maximize} \int_0^T NB(P, CD, CH, CP_0, CP_t, X_t) e^{-rt} dt \\ - C_0(P, CD, CH, CP_0) + SV(P, CD, CH, CP_0) e^{-rT} \tag{6.7}$$

subject to

$$dCP/dt = -M_t + X_t \\ CP_t \geq 0 \tag{6.8}$$

where
NB = net benefits (i.e., benefits - costs of operating and maintaining the dam),
P = vector of prices,
CH = vector of site characteristics,
CD = vector of dam characteristics and functions,
CP_0, CP_t = initial reservoir capacity and remaining capacity at time t, respectively,
M_t = mean sediment yield at time t,
X_t = sediment removal at time t,
C_0 = initial setup costs,
r = interest rate,
SV = salvage value,
T = operating lifetime, and
t = time.

The economic model can be used to select the optimum initial design reservoir capacity CP_0, dam characteristics and functions CD, sediment removal X_t, and a reservoir economic operating lifetime T.

6.13 Summary

This chapter provides a brief review and summary of technology available in the sustainable development and use of reservoirs where erosion and sedimentation should be considered. This study reached the following conclusions:

- The concept of "dead storage" for reservoir sedimentation of limited useful life is not consistent with the concept of sustainable use of reservoirs.

- The concept of sustainability should be used for future reservoir planning, design, and operation.

- Thorough and systematic approaches should be used in evaluating the impacts of erosion and sedimentation on sustainable development and use of reservoirs.

- Different engineering methods are available to remove or reduce the amount of sediment in a reservoir.

- River hydraulics and sediment transport computer models are available, or are being developed, to make detailed analyses and predictions of the impacts of river morphology and sedimentation processes on sustainable use and operation of a reservoir. The minimum energy dissipation rate theory, or the simplified versions of minimum unit stream power and minimum total stream power, can be used as a theoretical basis for developing these models.

- An economic model to maximize the net benefit should be used in the selection of reservoir size and useful life, subject to controllable reservoir characteristics and management strategy.

6.14 References

Bruce, G.N. (1953). "Trap Efficiency of Reservoirs," *Transactions of the American Geophysical Union*, vol. 34, no. 3.

Churchill, M.A. (1948). Discussion of "Analysis and Use of Reservoir Sedimentation Data," by L.D. Gottschalk, in *Proceedings of the Federal Interagency Sedimentation Conference*, Denver, Colorado.

Dorough, W.C., and C.T. Yang (1996). "Field Investigation for Sediment Planning of Reservoir," in *Proceedings of the International Conference on Reservoir Sedimentation*, vol. 2, sec. V, Chapter 5.3, Fort Collins, Colorado.

Fread, D.L. (1977). "The Development and Testing of a Dam-Break Flood Forecasting Model," in *Proceedings of the Dam-Break Flood Routing Model Workshop*, Bethesda, Maryland, pp. 164-197.

Fread, D.L., and J.M. Lewis (1998). *NWS FLDWAV Model*, National Weather Service, Office of Hydrology, Silver Spring, Maryland.

Han, Q. (1980). "A Study on the Non-Equilibrium Transport of Suspended Load," *Proceedings of the International Symposium on River Sedimentation*, Beijing, China, pp. 793-802 (in Chinese).

Harrison, L.L. (1996). "Overview of Impacts and Benefits of Preventive and Remedial Measures," in *Proceedings of the International Conference on Reservoir Sedimentation*, vol. 2, sec. VII, Fort Collins, Colorado.

Lane, E.W. (1955). "Design of Stable Channel," *Transactions of the ASCE*, vol. 120, pp. 1234-1279.

Palmieri, A., F. Shak, and A. Dinar (1998). *Sustainability vs. Non-Sustainability: An Economic Evaluation of Alternative Sediment Management Strategies*, paper presented at a seminar at the World Bank, Washington DC.

Richardson, E.V. (1996). "Estimating Reservoir Life," in *Proceedings of the International Conference on Reservoir Sedimentation,* vol. 2, sec. V, Fort Collins, Colorado, pp. 777-781.

Takeuchi, K., and Z.W. Kundzewicz (1998). "Sustainability and Reservoirs," *Sustainable Reservoir Development and Management,* K. Takeenchi, M. Hamlin, Z.W. Kundzewicz, D. Rosbjerg, and S.P. Simonovic (eds.), International Association of Hydrological Sciences Publication no. 251, IAHS Press, Institute of Hydrology, Wallingford, Oxfordshire OX108BB, United Kingdom.

Tan, Y. (1996). "Design of Silt Related Hydraulic Structures," in *Proceedings of the International Conference on Reservoir Sedimentation,* vol. 1, sec. I, Fort Collins, Colorado, pp.675-731.

Tomasi, L. (1996). "Operation and Maintenance Problems Due to Sedimentation in Reservoirs," in *Proceedings of the International Conference on Reservoir Sedimentation*, vol. I, sec. I, Fort Collins, Colorado, pp.15-28.

United Nations Conference on Environment and Development (UNCED) Agenda 21 (1992). *Programme of Action for Sustainable Development*, Rio de Janeiro, Brazil, June 3-14.

U.S. Army Corps of Engineers (1989). *Sedimentation Investigation of Rivers and Reservoirs,* Engineer Manual, EM 1110-2-4000.

Veltrop, J.A. (1996). "Sediment Control is Essential for Sustainability of Reservoirs," in *Proceedings of the International Conference on Reservoir Sedimentation,* vol. 2, sec. VII, Fort Collins, Colorado, pp. 1141-1156.

Wan, Z. (1996). "Reservoir Sediment Management Strategies for Large Dams," in *Proceedings of the International Conference on Reservoir Sedimentation*, vol. 2, sec. VII, Fort Collins, Colorado, pp. 829-849.

World Commission on Environment and Development (WCED) (1987). *Our Common Future,* Oxford University Press, United Kingdom.

Wu, C.M., M.A. Samad, J.C. Yang, W.S. Liang, and D.C. Baird (1996). "Planning and Design," in *Proceedings of the International Conference on Reservoir Sedimentation*, vol. 2, sec. V, Chapter 5.1, Fort Collins, Colorado, pp. 571-611.

Yang, C.T. (1971). "Potential Energy and River Morphology," *Water Resources Research,* vol. 7, no. 2, pp. 311-322.

Yang, C.T. (1973). "Incipient Motion and Sediment Transport," *Journal of the Hydraulics Division,* ASCE vol. 99, no. HY10, pp. 1679-1704.

Yang, C.T. (1976). "Minimum Unit Stream Power and Fluvial Hydraulics," *Journal of the Hydraulics Division,* ASCE vol. 102, no. HY7, pp. 919-934.

Yang, C.T. (1979). "Unit Stream Power Equations for Total Load," *Journal of Hydrology,* vol. 40, pp. 128-138.

Yang, C.T. (1986). "Dynamic Adjustment of Rivers," in *Proceedings of the 3rd International Symposium on River Sedimentation,* Jackson, Mississippi, pp. 118-132.

Yang, C.T. (1996). *Sediment Transport Theory and Practice,* The McGraw-Hill Companies, Inc., New York, 396 p. (reprint by Krieger Publishing Company, Malabar, Florida, 2003).

Yang, C.T. (1997). "Literature Review of Reservoir Sedimentation Process, Planning, Design, Operation, Maintenance, and Remedial Measures," *Technical Assistance and Cooperative Reservoir Sediment Study Progress Report,* vol. 1, U.S. Bureau of Reclamation, Technical Service Center, Denver, Colorado.

Yang, C.T. (2002). Total Maximum Daily Load of Sediment, International Workshop on Ecological, Sociological and Economic Implications of Sediment Management in Reservoirs, Paestum, Italy.

Yang, C.T., and C.C.S. Song (1979). "Theory of Minimum Rate of Energy Dissipation," *Journal of the Hydraulics Division,* ASCE vol. 105, no. HY7, pp. 769-784.

Yang, C.T., and C.C.S. Song (1986). "Theory of Minimum Energy and Energy Dissipation Rate," *Encyclopedia of Fluid Mechanics,* Chapter 11, Gulf Publishing Company, Houston, Texas, pp. 353-399.

Yang, C.T., and K.C. Yeh (1992). "Hydropower Development and River Morphology," *Proceedings of the 5th International Symposium on River Sedimentation,* Karlsruhe, Germany, vol. 1, pp. 221-227.

Yang, C.T., and F.J.M. Simões (1999). *Progress Report on the Development of GSTARS3,* U.S. Bureau of Reclamation, Technical Service Center, Denver, Colorado.

Yang, C.T., and F.J.M. Simões (2000). *User's Manual for GSTARS 2.1 (Generalized Stream Tube model for Alluvial River Simulation Version 2.1),* U.S. Bureau of Reclamation, Technical Service Center, Denver, Colorado.

Yang, C.T., and F.J.M. Simões (2002). *User's Manual for GSTARS3 (Generalized Sediment Transport model for Alluvial River Simulation Version 3.0),* U.S. Bureau of Reclamation, Technical Service Center, Denver, Colorado.

Yang, C.T., M.A. Treviño, and F.J.M. Simões (1998). *User's Manual for GSTARS 2.0 (Generalized Stream Tube model for Alluvial River Simulation Version 2.0)*, U.S. Bureau of Reclamation, Technical Service Center, Denver, Colorado.

Yang, C.T., J. Huang, and B. Greimann (2002). *GSTARS-2C Progress Report*, submitted to the U.S. Environmental Protection Agency, Athens, Georgia, January 2002.

Yang, C.T., J. Huang, and B.P. Greimann (2004, 2005). *User's Manual for GTAR-1D (Generalized Sediment Transport Model for Alluvial Rivers - One Dimensional, version 1.0)*, U.S. Bureau of Reclamation, Technical Service Center, Denver, Colorado.

Yang, C.T., Y.G. Lai, T.J. Randle, and J.A. Daraio (2003). "Development of a Numerical Model to Predict Sediment Delivery to River and Reservoir Systems," Project Progress Report no. 1, "Review and Evaluation of Erosion Models and Description of GSTAR-W Approach," U.S. Bureau of Reclamation, Technical Service Center, Denver, Colorado.

Zhou, Z. (1996). "Overview of Preventative and Remedial Measures," in *Proceedings of the International Conference on Reservoir Sedimentation*, vol. 2, sec. VII, Fort Collins, Colorado, pp. 903-919.

Chapter 7
River Processes and Restoration

Page

7.1 Introduction 7-1
7.2 Conceptual Model 7-2
7.3 Data Collection, Analytical, and Numerical Modeling Tools 7-3
7.3.1 Data Collection Activities 7-3
7.3.2 Geomorphic Processes 7-6
7.3.2.1 Geology 7-10
7.3.2.2 Climate 7-10
7.3.2.3 Topography 7-11
7.3.2.4 Soils 7-11
7.3.2.5 Vegetation 7-11
7.3.2.6 Channel Morphology 7-12
7.3.2.7 Geomorphic Mapping 7-13
7.3.2.8 Channel Geometry Analysis 7-15
7.3.2.9 Stream Classification 7-15
7.3.2.10 Channel Adjustments and Equilibrium 7-16
7.3.2.11 Geomorphic Summary 7-18
7.3.3 Disturbances Affecting the River Corridor 7-19
7.3.3.1 Dams 7-20
7.3.3.2 Diversions 7-21
7.3.3.3 Levees 7-22
7.3.3.4 Roads in the River Corridor 7-23
7.3.3.5 Bridges 7-23
7.3.3.6 Bank Protection 7-24
7.3.3.7 Removal of Vegetation and Woody Debris 7-25
7.3.3.8 Forestry Practices 7-25
7.3.3.9 Grazing (bank erosion) 7-26
7.3.3.10 Gravel Mining 7-26
7.3.3.11 Urbanization 7-26
7.3.3.12 Recreation 7-27
7.3.4 Hydrologic Analysis 7-27
7.3.4.1 Historical Discharge Data 7-27
7.3.4.2 Flood Frequency Analysis 7-28
7.3.4.3 Flow Duration Analysis 7-28
7.3.4.4 Ground Water Interaction 7-28
7.3.4.5 Channel Forming Discharge 7-29

Page

7.3.5 Hydraulic Analysis and Modeling 7-30
7.3.5.1 Topographic Data Needed 7-30
7.3.5.2 Longitudinal Slope and Geometry Data 7-31
7.3.5.3 Physical and Numerical Models 7-31
7.3.6 Sediment Transport Analysis and Modeling 7-32
7.3.6.1 Sources of Upstream Sediment Supply 7-32
7.3.6.2 Total Stream Power 7-33
7.3.6.3 Incipient Motion 7-33
7.3.6.4 Sediment Particle Size Analysis 7-33
7.3.6.5 Sediment-Discharge Rating Curves 7-36
7.3.6.6 Reservoir Sediment Outflows 7-38
7.3.6.7 Scour and Degradation 7-38
7.3.6.8 Sediment Transport Equations 7-39
7.3.6.9 Sediment Considerations for Stable Channel Design 7-39
7.3.6.10 Evaluation of Potential Contaminants 7-41
7.3.7 Biologic Function and Habitat 7-41
7.4 Sediment Restoration Options 7-42
7.4.1 Goals and Objectives 7-42
7.4.2 Fully Assess the Range of Options 7-43
7.4.2.1 Sediment and Flow 7-43
7.4.2.2 Local Versus System-wide 7-43
7.4.2.3 Natural Versus Restrained Systems 7-44
7.4.2.4 Monitoring Versus Modification 7-45
7.4.3 Restoration Treatments 7-45
7.4.3.1 Restoration of the Historic Channel Migration Zone 7-45
7.4.3.2 Levee Setback and Removal 7-46
7.4.3.3 Roadway Setback 7-47
7.4.3.4 Lengthening Bridge Spans 7-47
7.4.3.5 Side Channel, Vegetation, and Woody Debris Recovery 7-48
7.4.3.6 Changes to Channel Cross Section or Sizing 7-48
7.4.3.7 Changes to Channel and Flood Plain Roughness 7-49
7.4.3.8 Bank Stabilization Concepts 7-49
7.4.3.9 Grade Control Structures 7-50
7.4.3.10 New Channel Design and Relocations 7-51
7.4.3.11 Special Flow Releases From Dams 7-52
7.4.4 Biologic Function and Habitat 7-53
7.4.4.1 Channel and Cross-Section Shape 7-53
7.4.4.2 Channel Banks 7-54
7.4.4.3 Channel Planform Characteristics 7-54
7.4.4.4 Changes in Channel Grade 7-54

Page

7.4.4.5 Flow and Sediment Designs 7-55
7.4.5 Watershed Level Restoration 7-56
7.4.6 Uncertainty and Adaptive Management 7-56
7.5 Summary 7-57
7.6 References 7-58

Chapter 7
River Processes and Restoration

by
Timothy J. Randle, Jennifer A. Bountry, Paula W. Makar, Lisa M. Fotherby, Travis R. Bauer, and Peter J. Murphy

7.1 Introduction

River channels and flood plains convey water, transport sediment, and often support complex ecosystems that include aquatic and terrestrial plants, fish, and wildlife. Rivers are important resources that provide environmental, cultural, and economic benefits including municipal water use, irrigation, hydropower, navigation, fishing, and recreation. In achieving these benefits, the natural processes of the river and the ability of natural ecosystems to function over the long term are often impacted. This can result in unanticipated consequences, such as the loss of aquatic and terrestrial habitat for fish and wildlife.

River channels and flood plains convey runoff from rainfall and snowmelt. They can temporarily delay floodwaters, thereby reducing flood peaks. The processes of sediment erosion, transport, and deposition may cause local and system-wide adjustments in the bed and banks of the river channel, including the migration of a river channel across the flood plain. In a natural setting, undisturbed by anthropomorphic impacts or cataclysmic events, the river is dynamic and changes occur over time as a result of these sediment processes. However, when viewed over the long term, these changes tend to fluctuate about an equilibrium condition, known as dynamic equilibrium. Disturbances to the river corridor often affect these sediment processes, and the channel may become unstable (i.e. no longer in a state of dynamic equilibrium). With time, the disturbed channel may achieve a new state of dynamic equilibrium, but viable water management activities or appropriate aquatic habitat for fish and wildlife may not be sustainable during this period of transition. Disturbances to rivers and streams can be caused by dams, water diversions, levees, roads, bridges, bank protection, removal of vegetation and woody debris, logging, grazing, gravel mining, urbanization, and recreation.

Although river restoration can mean different things to different people, in this manual, restoration is defined as the full or partial restoration of natural processes and dynamic equilibrium. Natural processes in a river channel below a dam might be restored by modifying dam releases to replicate portions of the natural hydrologic cycle that are in proportion with the sediment supplies to the downstream river channel. In the case of a river with flood-control levees, the levees might be set back or removed to restore the connection between the main channel and flood plain. The challenge for restoration planning and design is to ensure that the river channel and flood plains are capable of conveying water, transporting sediment and large woody debris (if present), and supporting aquatic and terrestrial ecosystems in a long-term dynamic equilibrium.

The recent publication, *Stream Corridor Restoration: Principles, Processes, and Practices* (Federal Interagency Stream Restoration Working Group, 1998), provides a comprehensive discussion of the physical, chemical, and biological processes related to stream corridors and describes the aspects of restoration planning and design. Another recent publication, *Channel Restoration Design for Meandering Rivers* (Soar and Thorne, 2001), describes the analysis and restoration design procedures for meandering rivers.

This chapter of the *Erosion and Sedimentation Manual* focuses on physical river processes and sediment management aspects of river restoration. Accordingly, this chapter avoids the discussion of the legal and institutional policy issues. In Section 7.2, the technique of using a conceptual model to formulate hypotheses, determine analysis methods, and design data collection programs is described. Section 7.3 discusses potential data collection activities and the application of various analytical and numerical modeling tools, along with their applicability to various sediment management questions. Section 7.4 provides some management considerations for choosing a sediment restoration option, various techniques that have been implemented and their applicability to sediment issues, and a discussion of how adaptive management techniques can be incorporated into restoration projects.

7.2 Conceptual Model

Restoration projects begin with the resource management goals and objectives, such as improving riparian vegetation, fish habitat, wildlife habitat, water quality, or aesthetics. For some projects, the goals and objectives may be to limit or stop bank erosion for a reach where human impacts have accelerated the natural rates of channel migration. The goals and objectives help to determine the reasonable range of restoration options. The range of options, in turn, helps to determine the required analysis methods and the data collection needs.

Defining the restoration goals and objectives can be a difficult policy process. There are often many resource management agencies, landowners, and other special interest groups involved, and each party has different management objectives. There may also be conflicting interpretations of the processes affecting the river and the range of feasible restoration options that can be applied.

Understanding the key processes that have affected, and continue to affect, the river corridor is a prerequisite to river restoration projects. The formulation of a conceptual model can help the investigators, resource management agencies, landowners, and other interested groups begin to think about and understand the linked processes that affect the river corridor. This understanding can help refine the resource management goals and objectives and help determine the range of feasible restoration options.

The formulation of a conceptual model is a continuing and dynamic process that occurs throughout the study. The conceptual model can continue to be refined after project implementation as part of an adaptive management process (see Section 7.4.6). Initially, the conceptual model may be nothing more than a linked set of hypotheses that need to be tested, but these hypotheses will help determine the data collection and analysis needs to get the project started.

The conceptual model may incorporate different temporal and spatial scales, depending on the processes affecting the study reach. River processes can be driven by recent geologic processes that have evolved over thousands or tens of thousands of years. The understanding of geologic processes, such as landslides, debris flows, and uplift may require a temporal scale of centuries and a spatial scale that includes hundreds of square miles. The conceptual model should also consider the temporal scale of the disturbance to the river reach with a corresponding spatial scale that includes the river reach of interest and the distance to the source of possible disturbance. For hydrologic processes, a spatial area of the upstream watershed should be considered in combination with a temporal scale of several decades. However, the processes associated with individual floods, such as the downstream translation of discharge waves, river channel erosion, migration, and sediment deposition, would require a time scale of minutes, hours, or days, and a spatial scale measured in smaller increments of the watershed, possibly feet or miles. Unlike numerical models, a conceptual model can easily accommodate different time and space scales for different processes.

The first steps in developing a conceptual model include a literature review of technical studies in the area; an initial review of available maps, aerial photographs, and stream gaging data, and a field reconnaissance trip to the study area. The initial conceptual model will have some uncertainty, but it provides a tool to formulate an appropriate study plan including data collection and analysis activities, to help reduce uncertainty and answer study questions. A variety of data collection and analyses tools that can be used to test and improve the conceptual model are discussed in the next section.

7.3 Data Collection, Analytical, and Numerical Modeling Tools

The choice of data collection and analysis methods needs to be customized to the requirements of the project, since every project is unique. The following discussion provides a range of data collection activities and analytical and numerical tools that can be used in a restoration project, and the potential benefit of each in understanding the role of sediment processes in the river system being studied. Depending on the size of the project and resources available, analyses can range from simple computations to a multi-dimensional, customized model to analyze various restoration options. When identifying tasks for a restoration project, the objective should be to choose a suite of integrated, multi-discipline analyses that address management questions and research hypotheses posed in the conceptual model.

7.3.1 Data Collection Activities

The types of data collection activities should be driven by the analytical and numerical analyses that will be accomplished as part of the restoration project and by future monitoring needs. Some of the more typical examples of data collection activities that can be utilized to support several types of analyses are listed below:

- The collection of historic maps, aerial photographs, and ground photographs is essential for understanding the history of channel migration and land use over the last several decades. Future trends in channel migration are often predicted by examination of historic trends.

- Aerial field reconnaissance can be a very helpful way of viewing the watershed, especially in areas of dense forest or rugged terrain.

- Topographic surveys of the river channel, flood plains, terraces, and high water marks describe the relationship between riverflow and water surface elevation (stage-discharge rating curves). This relationship helps determine the bankfull discharge and the discharge required to inundate terraces at the flood plain boundaries. Topographic survey data are also essential for the numerical modeling of hydraulic and sediment transport processes.

    - Ground survey methods include using total stations, real-time kinematic global positioning system (RTK GPS) instruments, and depth sounders from boats. Aerial survey methods include photogrammetry and light detection and ranging (LIDAR) systems. Depth sounders from boats work best for river channels that are too deep to wade. RTK GPS can be used to track the horizontal and vertical coordinates of a moving boat. In 1 week, a raft equipped with a depth sounder and RTK GPS was used by the authors to survey 30 miles of the Hoh River near Forks, Washington (Piety et al., 2003). In river reaches where satellite coverage may be limited by mountains, canyon walls, or riparian forests, robotic total stations can be used to track the position of a moving boat over a distance of approximately 1,000 feet. This method was used in a 1-mile canyon reach of the Rouge River below Savage Rapids Dam, near Grants Pass, Oregon, and also within a 5-mile reach of the Teton River Canyon near Driggs, Idaho (Bountry and Randle, 2001; Randle et al., 2000).

    - The ortho-rectification of aerial photographs, using stereo pairs, can produce digital elevation models (grids or surfaces) and contour maps (photogrammetry) of the river corridor and also allow the rectification of historic aerial photographs. The grid size or contour interval that can be produced from photogrammetry depends on the ground survey control and the altitude at which the photographs are taken (see "Geospatial Positioning Accuracy Standards," *Part 3: National Standard for Spatial Data Accuracy,* by Federal Geographic Data Committee, 1998 and "Using the National Standard for Spatial Data Accuracy to measure and report geographic data quality," in *Positional Accuracy Handbook* by Land Management Information Center Minnesota Planning, 1999). The more precise the survey control and the lower the photograph altitude, the smaller the contour interval that can be achieved. Proper survey control depends on the use of photograph control panels or painted lines on paved surfaces. Where vegetation is present, photogrammetry is most effective when the base aerial photography is taken during the winter when deciduous trees lose their leaves. Photogrammetry cannot produce accurate ground elevations in densely forested areas, especially in conifer forests. Ground surveys using total stations along cleared lines would be necessary for areas of dense vegetation. Aerial photography taken at low flows is best for developing topography of exposed riverbed topography, however, aerial photographs taken during floods can also be useful because it directly measures the areas of inundation. Although photogrammetry cannot produce reliable elevations below water, the data above water can be combined with the channel bottom data surveyed by boat. The channel bottom survey should preferably be measured at a higher flow level than when aerial photographs are taken, so the data overlaps. Reclamation surveyed a 12-mile reach of the Snake River near Fort Hall, Idaho, during April 2000 using a combination of photogrammetry and a channel bottom survey by boat (Bureau of Reclamation, 2001). Photogrammetry was used for the gravel bars exposed above water, the flood plains, and terraces. A raft equipped with a depth sounder and RTK GPS was used to survey the wetted river channel. The data was then combined to make a continuous digital elevation model of the study area.

    - LIDAR surveys can provide useful topography data in forested areas. This method transmits light from an aircraft to the ground surface to measure elevations. The method requires that at least some of the light beams reach the forest floor and reflect back to the aircraft. Therefore, a LIDAR survey works best during the winter months when deciduous trees lose their leaves. Ortho-rectified aerial photographs collected simultaneously with the LIDAR data are very useful for processing the LIDAR data and for interpreting the topographic data produced from LIDAR surveys. Reclamation has used LIDAR to survey the forested flood plains and exposed gravel bars along the Elwha and Quinault Rivers of the Olympic Peninsula in Washington. The LIDAR survey of the Elwha River revealed the existence of a channel through a forested terrace that had not been identified before. A subsequent field inspection later verified the existence of this channel. LIDAR data on the Quinault River allowed a fairly quick identification of terrace surfaces and historical channels throughout the valley floor that would not have been detectable from aerial photography alone. Although the cost of LIDAR surveys may appear high, the survey can provide ground elevations in large forested areas that would be time consuming and expensive to survey through ground-based methods.

- Particle size gradation measurements of the bed and bank materials are necessary for computations of sediment transport capacity and the assessment of channel stability. Also, the comparison of bed-material particle size with longitudinal distance along the river can help determine where the sediment transport capacity of the river changes or identify additional sources of sediment from tributaries. Pebble counts (Bunte and Abt, 2001) can be used to measure the particle size distribution of gravels and coarser material exposed on the surface of the channel bed. The particle size distribution of surface and subsurface layers within a unit area (e.g., 1 m^2, 1 yd^2) can also be measured for cobbles, gravels, and sands by sieving and weighing in the field. Ground-based photography methods can also be used to quickly record the surface particles at many different locations. If a scale is placed in the photograph, the particle size distributions can then be measured from the photographs using commercially available software.

- Measurements along river and terrace banks, such as height, slope, material composition, and vegetation root density can be used to assess bank stability.

- Soil profiles and radiocarbon analysis of charcoal material in terrace deposits can be used to determine the minimum age of the deposits. This information is useful for defining the flood plain and channel migration zone boundaries, the frequency of large and rare floods, and the minimum age that the terrace was abandoned by the active river channel.

- Measurements of large woody debris (location, size, and type of wood) that are present along the active channel and flood plains are useful for assessing the rates and locations of lateral channel migration, the abundance of aquatic habitat, and the amount of additional roughness present in the channel (Abbe, 2000). Large and stable logjams can create fish habitat, protect riverbanks and islands from erosion, and slow the rates of lateral channel migration.

- Investigations of sediment sources to the study reach will help determine the primary processes and timing of sediment delivery. For example, sediment may be primarily supplied from certain portions of the upstream watershed, from landslides, from tributary debris flows, and from channel bank and bed erosion.

- Stream gaging of river stage and water discharge, suspended sediment load, and bedload is needed to quantify streamflow and sediment transport rates and to calibrate numerical models. Historical gaging measurements often also document channel geometry and velocity during a range of flows. Measurement of water temperature and turbidity levels may also be useful where aquatic habitat is of primary concern.

- Measurement of flow velocity (direction and magnitude) in the river channel can help provide information to calibrate numerical or physical models, provide an indication of where depositional areas exist, and show the presence of secondary currents that may initiate bank erosion.

7.3.2 Geomorphic Processes

A geomorphic analysis provides the context to help understand the river channel planform, historic channel paths and rates of migration, interactions with flood plains and terraces, and sediment sources and sinks. The analysis helps to identify upstream and downstream influences, geologic controls along the study reach, and human actions that have affected the natural processes. The analysis assists in identifying the cause(s) and magnitude of the disturbance to be restored, potentially useful methods for restoration, and likely channel response to restoration activities.

An example using the Hoh River near Forks, Washington, illustrates many of the processes occurring in and around a river. Figure 7.1 shows a composite cross section of the Hoh River valley, including the river channel, flood plain, and terraces. Because the relative amounts of water, sediment, and woody debris are continually changing, the channels and flood plain are continually adjusting to the variable supplies of these three components. Response to the changes may vary with time and location along the river, depending upon local conditions and the

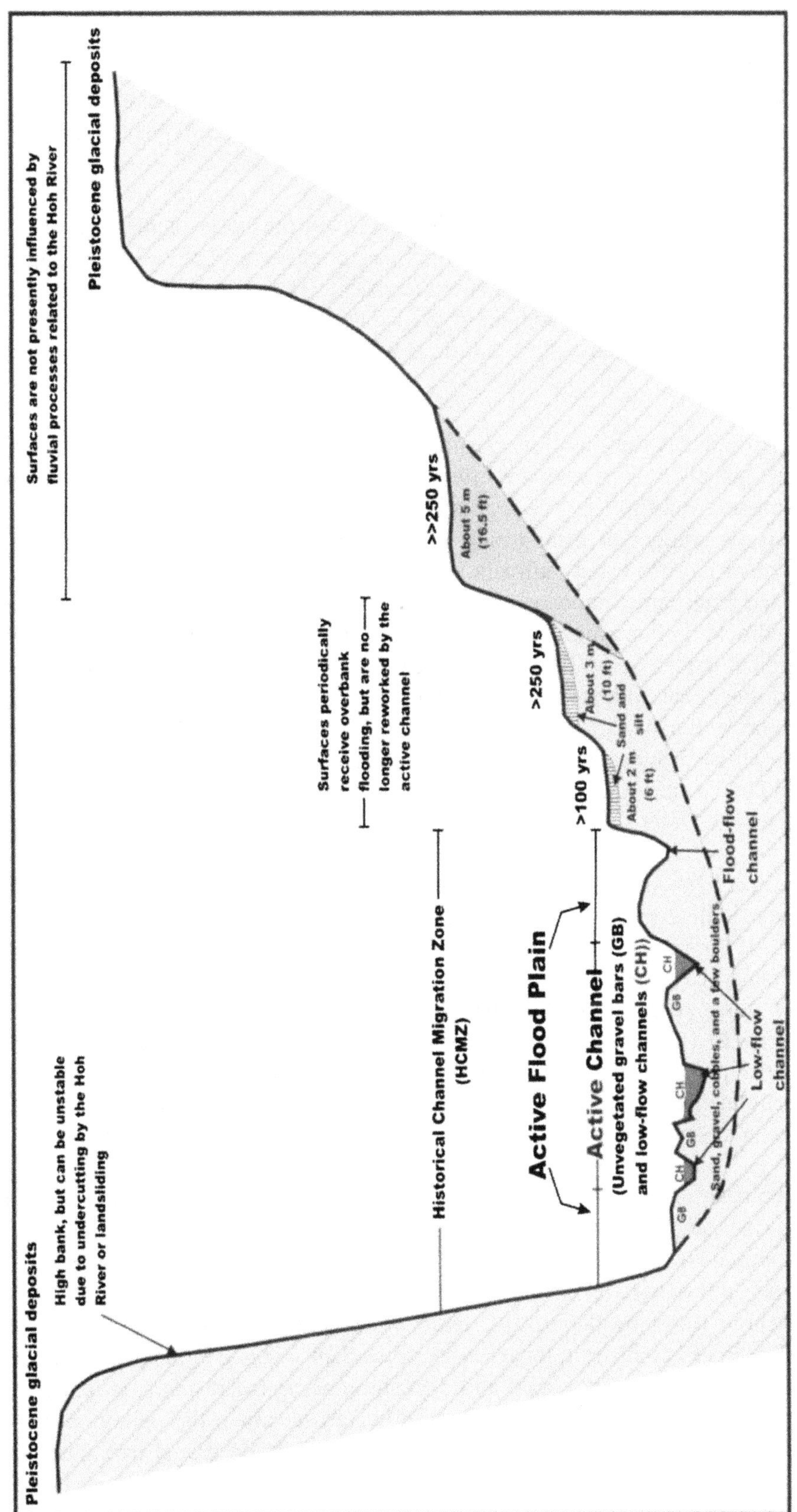

Figure 7.1. Generalized schematic cross section across the Hoh River valley, showing the relationships between the active channel, flood levels, and terraces

frequency of flooding. Water, sediment, and woody debris may move longitudinally down the valley, but alteration in these components may also result in vertical downcutting or aggradation, or in lateral erosion or deposition. In this way, natural river channels are dynamic and are subject to change over time, especially as bankfull and higher flows pass through the channel and rework sediment and woody debris.

The terraces in Figure 7.1 are numbered from T2 to T5 on the basis of their relative age and positions above the active river channel. This sketch is a composite of many sites along the river, so that a single locality may have fewer or more terraces. The estimated terrace ages are from Fonda (1974) and Swanson and Lienkaemper (1982). The ages are based primarily on the ages of trees growing on the terraces and reflect the time when the terraces became stable enough to support trees.

The active channel includes the low-flow channels and intervening and adjacent sand or gravel bars as shown in Figure 7.1. The bars are often unvegetated, but they may be sparsely covered with seedlings, grasses, and shrubs. The higher bars are often covered with finer sediment. This main channel may be straight or meandering in planform. The channel alignment and width depends on the rate of riverflow and sediment supplied from upstream. In an alluvial river reach, the channel bed and banks are composed of sand, gravel, and cobble-sized sediments and are not constrained by bedrock. In these alluvial river reaches, the channel width, depth, and longitudinal alignment will adjust over time so that the upstream supply rate of water and sediment can be conveyed through the reach without long-term erosion or deposition. The hydraulic capacity of the river to transport sediment increases primarily with the longitudinal slope of the river and the rate of riverflow, which includes an increase in flow velocity and depth. The longitudinal slope of the river is limited by the valley slope (straight channel) but can be less if the river meanders back and forth across the river valley.

If the upstream sediment supply rate is greater than the hydraulic transport capacity, then sediment can deposit in the river channel and the resulting alignment will tend to follow a straighter pattern. If the river channel alignment is already straight, additional sediment deposits will result in a braided river with multiple channels. If the upstream sediment supply rate is less than the sediment transport capacity, then the channel bed and banks can erode and the resulting alignment will tend to follow a more meandering pattern. If coarse particles remain on the riverbed while finer particles erode, then eventually a coarse layer of particles will armor the riverbed and limit channel bed incision. However, a subsequent flood can later erode the armor layer if there is enough sediment transport capacity to move the larger-sized particles.

When straight, the river channel may have the potential to locally undercut and erode banks on either side of the river, and it can locally deposit sediment during high flows in the form of mid-channel or longitudinal bars that run parallel with the river channel. When meandering, the river will typically erode the outside bank of the meander bend while maintaining enough channel width to convey high flows by depositing sediment along the inside of the bend. Sediment deposits on the inside of bends are referred to as point bars. This continual process of erosion along the outside bank of a meander bend and deposition along the inside bank allows the river channel to migrate laterally across the valley and downstream along the valley, while building and reworking the flood plain. The deposition of woody debris on gravel bars and subsequent formation of logjams also play an integral role in the occurrence of channel changes by creating multiple channel paths and either causing or mediating avulsions (Collins and Montgomery, 2002). Large woody debris also increases the channel roughness. If logjams are large enough to span the river channel, they can locally increase the river stage, which creates waterfalls over the logjams, short scour pools immediately downstream, and longer pools upstream. The upstream pools and the increased roughness can reduce the sediment transport capacity of the river channel.

In most wide valleys with a flat floor, one can expect that the river has been in many positions across the valley floor at some point in the past (Leopold et al., 1964 and Leopold, 1994). Terraces within the valley are remnants of old flood plains and consist of old channel deposits of gravel and sand. A change in the relative proportion of water and sediment resulted in incision into these channel deposits, which left them exposed above the new active channel and flood plain (Figure 7.1). The old channel bed and flood plain, now a terrace, may still receive some water during higher flows, and the old channel deposits are often covered by fine sand and silt. If incision continues, the terrace may eventually be high enough above the level of the channel so that frequent floodflows no longer reach its surface. In this way, the terrace surfaces slowly stabilize, vegetation establishes, and soils begin to form. Higher terraces are often progressively older.

The historic channel migration zone boundary is defined by the area of the existing active channels and flood plains. The active river channel represents the low-flow river channel and sand/gravel bars that are frequently reworked during floods. The active flood plain includes areas adjacent to the active channel that contain side channels, secondary or flood plain channels, and low-elevation, vegetated surfaces that are frequently inundated by floods with recurrence intervals in the range of 1.5 to 5 years. The area outside of the historic channel migration zone boundary can still be inundated, but larger floods are required to overtop the terrace banks that form this boundary. The area within the historic channel migration zone is dynamic and continually being reworked and changing during floods. It represents the area where the majority of coarse sediment (sand, gravel, and cobble) and woody debris is either currently being transported (active river channel) or has been deposited.

The historic channel migration zone can be bounded by terraces, glacial deposits (till, outwash, lacustrine sediments), alluvial fans, bedrock, or bank protection (riprap, engineered logjams, bridge abutments, levees, road embankments). In some locations, future channel changes will expand the historic channel migration zone boundary where the boundary consists of erodible material, but this process is slow for stable systems. Significant long-term changes in river discharge and sediment supply (either natural or human caused) could cause major aggradation or incision and redefine the channel migration zone.

Information from several sources such as geology, climate, topography, soils, vegetation, channel morphology, and a chronology of disturbances both natural and anthropogenic is combined in a geomorphic analysis (Thorne, 2002). Processes and channel changes can be investigated at the local, reach, and watershed scales. Aerial photography and reconnaissance are excellent tools to view the larger scale attributes of the river system. Historical data including previous studies, maps, agricultural reports, and photographs can provide invaluable information about the river system over time. Time scales may also vary in an investigation, from a single year to decades or centuries. Geomorphic changes can occur over years to millions of years, but a stream restoration project generally considers the results of such changes in terms of years to decades (Federal Interagency Stream Restoration Working Group, 1998).

7.3.2.1 Geology

With respect to a river, geology can be thought of as providing the general setting for the movement of water and sediment. The characteristics of the geology are dependent on the composition and structure of the earth's crust. In particular, the lateral and vertical extent of bedrock potentially limits the channel position over time. A geomorphic analysis should include a discussion of the general lithology and structure of the study site, particularly with regard to its effect on the river. Structure is the shape and position of rock units and their relationships to each other. It includes faults, joints, fold patterns, layering, etc. Tectonic plate movement is a prime force in determining structure. Uplift of the underlying bedrock is often a cause of channel incision. As the general elevation of the land increases, the river may erode the bed to maintain its slope (Schumm, 2000). An extreme example is the Grand Canyon of Arizona.

Lithology is the chemical and mineralogical composition, texture, and internal strength of the rock. Texture can be described as unconsolidated, indurated (hard), fractured, or solid. Solid or indurated rock often provides a control point for bed elevation and slope. The amount of time it acts as a control will be influenced by the lithology. For example, limestone may be eroded faster than basalt in a wet climate. Erodibility, porosity, and permeability are characteristics defined by both structure and lithology. Spatial variability of these characteristics should be noted. Textural distribution of unconsolidated sediment can be very significant. Larger-sized sediment, such as cobbles or boulders, may armor the channel at the mouth of a tributary and act as a control point similar to solid rock.

7.3.2.2 Climate

The atmospheric aspects of the basin include such things as precipitation intensity, duration and frequency, temperature, and year-to-year and seasonal variability. Climate can be thought of as the meteorological aspects that influence the hydrology. Long-term climate changes can cause glaciers to advance and retreat and sea level to fall and rise. These changes can have profound effects on river systems that can include degradation and aggradation.

Short-term climate fluctuations of 3 to 10 years can be detected through analysis of instrument records of precipitation and flow, and through extensions of these records based on paleo techniques. The science of tree-ring analysis has considerably extended the record of climate based on instrument measurements. Instrumented data are available back to approximately 1895; however, the paleo technique of dating tree rings provides information back to the 1600's or 1700's in the Western U.S., while other paleo techniques can extend the record back a thousand years or more. Tree-ring data have been used to extend the period of record for temperature measurements, the Palmer Drought Severity Indices (PDSI), and for some hydrological records of streamflow. The PDSI is a drought model derived by Palmer (1965) that is computed from temperature and precipitation data to provide a measure of climatic stress on crops and water supplies

Dates and duration of climactic fluctuations (for example documenting a drought or wet period) are often used to identify the cause of changes in river behavior. The middle Rio Grande in central New Mexico has shown direct correlation between periods of channel narrowing and drought (Bauer and Makar, 2003). The channel narrowed through vegetation colonization and the attachment of bars and islands to the channel banks that were formed during the drier periods between the mid-1940's and the mid-1970's. Peak flows were not high enough to scour new vegetation, which then became established and resistant to removal. Later increases in flows tended to erode the channel bed rather than the banks in many locations. Current research topics that relate to river channels include the influence of the El Nino Southern Oscillation on long-term weather patterns.

7.3.2.3 Topography

Topography is primarily the relief, aspect, and elevation of a river basin. Relief is the amount of change in elevation, which has an influence on channel slope or gradient, runoff, sediment storage, and delivery to the river. Aspect and general site elevation, which may be less influential, affect vegetation type and growth, sediment deposition and storage, soil development, and microclimates.

7.3.2.4 Soils

The development of soils is influenced by several factors, including geology, climate and topography. Soils strongly influence runoff and the type of vegetation. Soils may control other aspects of the fluvial system such as water quality, slope and bank stability, and sediment supply. Soil formation is also commonly used to estimate the age of particular landforms (Birkeland, 1999). Soil profiles may therefore be useful in determining the age of terraces, rates of channel migration, and landscape stability or instability, such as periods of deposition and erosion.

7.3.2.5 Vegetation

Types of vegetation are largely determined by the four previously listed factors and may, in turn, influence the state of those factors. The age and successional stage of vegetation can be used to establish the age of fluvial deposits and landforms. The stability of banks can be strongly influenced by root structure and extent. Lateral migration of the river channel across the flood plain is often restricted by vegetation. Large woody debris (snags and logjams) can also be a controlling factor of the channel morphology.

7.3.2.6 Channel Morphology

Channel morphology is described by the shape, slope, and pattern of the channel and includes the flood plain and terraces. The shape of a natural channel is irregular, but can be represented by several variables, including the channel width at bankfull discharge, average and maximum depth, channel side or bank slope, cross-sectional area, and wetted perimeter. Channel bank slope is frequently different for each riverbank location and can be estimated by the slope of a line from the point defining channel top width in a bankfull stage to the point defining channel bottom width at a low-flow stage. The steepest section of bank slope can also be measured to assess bank stability. The hydraulic radius (cross-sectional channel area divided by the wetted perimeter) and the width-to-depth ratio are two other commonly used variables that are derived from the measured variables. Channel shape may also include bed forms such as ripples, dunes, pebble clusters, and point bars (Knighton, 1998). Flood plain width and terrace height are useful measures to help identify the channel migration zone. Dimensionless ratios between these parameters are often used as indicators to compare channel properties between reaches and to reference reaches.

Slope is the longitudinal gradient of the river. Within the constraints of the longitudinal valley slope and bedrock outcrops along the valley walls, the river channel slope is a function of the riverflow and sediment load. The river channel slope, or longitudinal profile, can be defined by measuring elevations along the channel thalweg (lowest point in the river cross section), the average riverbed elevation, or the water surface elevation at some reference discharge (e.g., bankfull discharge). Although the river slope can be locally steep at riffles and rapids, the average slope cannot be steeper than the valley slope. Reach-scale changes in riverbed material, sediment load, or discharge can be the major causes of reach-scale variations in profile, while factors such as geology, hydrology, and watershed size influence the shape of the generally concave bed profile on a basin scale (decreasing slope with distance downstream).

The sequence of riffles and pools is another aspect of channel form. Pool and riffle sequences are found in most river patterns and result from a combination of scour and deposition. The longitudinal spacing of pools and riffles is usually fairly regular. The channel thalweg moves from side to side across the channel. The deepest pool is usually found at the apex of the curve and along the outside of the meander bend. The shallow, coarser-grained riffle is usually found at the crossing point, which is the tangent between two meander bends. The sequence of pools and riffles may be unsymmetrical or skewed due to local factors. Leopold (1994) found that, for channels of all sizes, there is a remarkable relationship among the meander wavelength, channel width, and the radius of curvature. The meander wavelength is nearly always between 10 and 14 times the channel width, with an average of 11 times the channel width. The radius of curvature for the central portion of the meander curve averages about one-fifth of the meander wavelength. This means that the radius of curvature is about 2.3 times the channel width.

River planform has traditionally been divided into three classes: straight, meandering, and braided (Leopold and Wolman, 1957), with meandering being the most common planform type (Leopold, 1994). Schumm (1985) defined 14 patterns based on the type and amount of sediment load moving through the river, single or multi-channel character, and sinuosity. Schumm also assigned a relative stability to each planform type.

Multi-channel patterns may be braided or anastomosed. Anastomosed channels are multiple channels separated by stable bars and islands. They are sometimes considered a subset of anabranched channels, where a branch of the river diverges and then reenters the main stem downstream. Braided channels also have multiple flow paths. Mid-channel bars separate the flow at lower discharges but are submerged at high discharges, so there is a single channel at the surface. Braided channels are characterized by high sediment load and high stream power and they usually have erodible banks. The degree of braiding can be quantified using a variety of indices, which are generally of two types. One type calculates the total length of all the individual channels that extend the whole length of a given reach, divided by the mid-channel length of the main-stream channel length in the reach. The second type counts the number of active channels or braid bars per channel transect. Bridge (1993) states that the second type is preferable because it is not a function of individual channel sinuosity.

The degree of channel meandering can be characterized by sinuosity. Sinuosity for a given river reach is defined as channel length (or channel slope) divided by valley length (or valley slope). The straighter the river channel alignment, the lower the sinuosity and the steeper the slope. If the sinuosity is greater than 1.3, the channel is considered meandering (Federal Interagency Stream Restoration Working Group, 1998).

7.3.2.7 Geomorphic Mapping

The objective of geomorphic mapping is to identify and locate geomorphically significant features on the study site landscape that impact river processes. These features may include bedrock outcrops, landslides, debris fans, alluvial fans, bars, tributaries, distributaries, flood plain and terrace boundaries, and side channels. Typically, the map distinguishes categories or units of similar characteristics. These characteristics include age, origin or process, landform and material or structure. It is assumed that areas of the same unit type will behave in a similar manner. The map becomes an effective source of data that enhances understanding of the landscape and prediction of responses to changes in the system. It is useful in restoration projects to help identify factors that may control the system and to evaluate potentially self-sustaining modifications to the system.

Watershed topographic parameters such as slope angle, aspect, and relief; vegetation type; rock type, age, and structure; soils; land use; and surface drainage are commonly used in classifying watersheds. Dominant processes such as fluvial, eolian, weathering, glacial, ground water, tectonic, coastal, and anthropogenic activities may also be used to differentiate among watershed types. Vegetation can provide clues to the underlying material. For example, changes in species or successional stage often are a result of differing soil or geology. Fluvial features such as tributaries, terraces, flood plains, current and past channel patterns, sediment type and size, bed control, channel geometry, and slope can be used to classify units associated with the channel.

Soil profiles can be used to describe the soil stratigraphy and to assess the ages and continuity of successive deposits, including terraces. Stratigraphic data can be helpful in assessing bank erodibility. Soil profiles can be described from the examination of river and terrace banks, escarpments, pits, and drill logs. A particularly useful source of soil data is soil surveys. Soil surveys are commonly produced in the United States for agricultural areas. Soil types and ages can help identify landforms that have undergone similar processes. Soil types can be identified by sediment source, erodibility, land use, vegetation, mapping for wetland, upland, and chemical composition. Frequently, archaeological assessments use the stratigraphic context to locate potential sites. Sedimentation and erosion rates can often be estimated from radiocarbon dating of terraces or other fluvial features. Other useful data are geologic events, processes which may identify post-depositional modification, and climatic trends (Birkeland, 1999).

The level of detail and scale of the map is dependent on the intended use of the final product (Compton, 1962; Bureau of Reclamation, 1998). Drainage basin and valley parameters may be adequately represented at a scale of 1:24,000, but for channel features 1:10,000 or smaller is an appropriate scale. Detailed mapping of specific sites may require finer scales to portray features of interest. Project goals and objectives will help determine the appropriate scale.

Information sources for geomorphic mapping include historic maps; aerial and ground photographs; bathymetric, ground, and aerial surveys of the river channel, flood plains and terraces; land use records; and field inspections. Government agencies such as Natural Resources Conservation Service, Bureau of Reclamation, U.S. Army Corps of Engineers, U.S. Forest Service, state land-use agencies, local conservation districts, and local planning and zoning agencies are excellent resources. U.S. Geological Survey (USGS) has both topographic and geologic maps available, typically at large scales. Field surveys will provide more precise topography data than typical USGS maps, especially for areas along the active river channel and flood plains. Field verification of mapping results should be accomplished for all remote sensing data. The reliability of historical data must be assessed and cross-correlated to determine the significance of the information.

A Geographic Information System (GIS) is a tool for storing and analyzing spatial data sets. The Federal Geographic Data Committee defines GIS as "A computer system for the input, editing, storage, retrieval, analysis, synthesis, and output of location-based information. GIS may refer to hardware and software, or include data" (Office of Management and Budget, 2002). Maps, aerial photos, land use records, etc., can be digitized into separate "layers," describing a single type of information. Separate layers might include topography, geology, soils, geomorphology, and land use. The layers are geo-referenced and can be displayed together at the same scale. Relationships between disparate information sources can often be identified with this tool; for example, how certain channel features are dependent on geologic features. Several useful parameters including sinuosity, channel width, and planform classification can be measured using GIS mapping tools.

If the study reach is not in equilibrium, a reference reach with similar fluvial, geomorphic, and sedimentary processes may provide stable channel form information. Ideally, the reference reach is in an undisturbed condition and within the same watershed. Classification systems (see Section 7.3.2.9) are useful to identify reaches that are similar. Care must be taken in scaling measurements from reference reaches. Osterkamp et al. (1983) and Dodds and Rothman (2000) provide discussions of scaling relations.

7.3.2.8 Channel Geometry Analysis

Many methods have been proposed over the years to predict stable channel geometry. Most calculate width and depth and use discharge as the independent variable. Many equations neglect sediment entirely, others use the bed-material sediment size, and most assume a constant sediment-discharge relationship.

The hydraulic geometry approach (Leopold and Maddock, 1953), uses power equations to define relationships among channel variables based on bankfull discharge. Williams (1978) presents values of the coefficients of those power equations for 165 channels. The regression coefficients do vary between locations and flow regimes and must be checked for applicability to a specific site.

Several regime equations describing channel geometry have been developed based on measured data of many rivers. Lacey (1929), Simons and Albertson (1963), and Blench (1957) are commonly used for sand bed streams. Gravel bed equations include Kellerhals (1967), Bray (1982), and Hey and Thorne (1986). Julien and Wargadalam (1995) use a semi-theoretical approach based on the principals of open channel flow. Again, these equations are applicable only to the specific river conditions that the equations were based on and must be selected carefully.

Soar and Thorne (2001) present regime equations for both natural sand and gravel-bed rivers considered to be in dynamic equilibrium. Different equations are presented for riverbanks that are resistant to erosion because of riparian trees and for riverbanks that are more erosive because of a lack of riparian trees. For 58 sand bed rivers in the United States, these regime equations explained 85 of the variance in bankfull width as a function of bankfull discharge (see Table 7.1).

Other approaches use minimum stream power, sediment concentration, or energy dissipation to predict channel shape. Yang (1986, 1996), Chang (1988), and Bettes and White (1987) have used minimum stream power to provide a third relationship beyond sediment transport and alluvial resistance relationships to calculate channel geometry. Care must be taken to validate an approach for a particular location and time period.

7.3.2.9 Stream Classification

River channels can be classified based on such parameters as planform, slope, width-to-depth ratio, and bed material grain size. It is assumed that channels in the same class will act in a similar manner. Stream classification systems such as those by Montgomery and Buffington (1998) or Rosgen (1996) can help to define the stable channel dimensions and appropriate management actions. These systems have been used most frequently in gravel and cobble bed channels. A concise discussion of the two systems can be found in Bunte and Abt (2001). Rosgen in particular emphasizes the need for training in the stream classification method for appropriate application.

Table 7.1. Regime equations by Soar and Thorne (2001), which include the 95-percent confidence limits on the mean response

River type	Regime equations in metric units[1]	Regime equations in english units[2]	Number of rivers	Coefficient of determination (R^2)
Sandbed rivers in the United States with less than 50 percent tree-lined banks[3]	$W_b = 5.32\ Q_b^{0.5}\ e^{\pm 0.082}$	$W_b = 2.94\ Q_b^{0.5}\ e^{\pm 0.082}$	32	0.87
Sandbed rivers in the United States with more than 50 percent tree-lined banks. [3]	$W_b = 3.38\ Q_b^{0.5}\ e^{\pm 0.085}$	$W_b = 1.87\ Q_b^{0.5}\ e^{\pm 0.085}$	26	0.85
Gravelbed rivers in the United States with thin bank vegetation[3]	$W_b = 4.17\ Q_b^{0.5}\ e^{\pm 0.087}$	$W_b = 2.30\ Q_b^{0.5}\ e^{\pm 0.087}$	9	0.95
Gravelbed rivers in the United States with thick bank vegetation[3]	$W_b = 3.67\ Q_b^{0.5}\ e^{\pm 0.065}$	$W_b = 2.03\ Q_b^{0.5}\ e^{\pm 0.065}$	14	0.96
Gravelbed rivers in the United Kingdom with less than 5 percent tree and shrub vegetation along the banks[3]	$W_b = 3.75\ Q_b^{0.5}\ e^{\pm 0.064}$	$W_b = 2.07\ Q_b^{0.5}\ e^{\pm 0.064}$	29	0.92
Gravelbed rivers in the United Kingdom with at least 5 percent tree and shrub vegetation along the banks[3]	$W_b = 2.48\ Q_b^{0.5}\ e^{\pm 0.051}$	$W_b = 1.37\ Q_b^{0.5}\ e^{\pm 0.051}$	33	0.93

[1] The bankfull discharge (Q_b) is in m^3/s and the bankfull width (W_b) is in meters.
[2] The bankfull discharge (Q_b) is in ft^3/s and the bankfull width (W_b) is in feet.
Note: The term, $e^{\pm 0.0xx}$ in these equations, accounts for the 95-percent confidence limits on the mean response.

7.3.2.10 Channel Adjustments and Equilibrium

An alluvial river channel is continually adjusting to achieve dynamic equilibrium between discharge, sediment supply, channel geometry, and slope. A disturbance to the river system, such as the construction of a storage dam, occurrence of landslides, or removal of riparian vegetation can disrupt this equilibrium. For example, when a reservoir traps sediment, water released from the reservoir will commonly scour the riverbed below the dam, causing channel degradation. Degradation continues until a new dynamic equilibrium is established. Dynamic equilibrium for a river channel can be defined as the condition where, over the long term, the river's sediment transport capacity is in balance with the upstream sediment supply. The river will continue to adjust its bed and banks in response to changing hydrologic and sediment supply conditions. Lane (1955) developed a qualitative relationship between sediment load and size, and river slope and water discharge:

$$Q_s d \propto Q_w S \tag{7.1}$$

where Q_s = sediment load (of sizes represented in the riverbed),
d = sediment particle diameter of the riverbed,
Q_w = water discharge, and
S = river channel slope.

In Lane's relationship, when the sediment load is reduced, the average sediment particle diameter of the riverbed will tend to increase and the channel slope will tend to decrease. If water discharge is decreased, the river slope will tend to increase and the sediment particle diameter will tend to decrease.

Yang (1986, 1996) theoretically derived a quantitative equation for the prediction of dynamic adjustment of a river channel based on his unit stream power equations (1973, 1979).

$$\frac{Q_s d^{J/2}}{K} = \frac{Q_w^{J+1} S^J}{A^J} \tag{7.2}$$

where K = site-specific coefficient,
J = site-specific exponent,
A = channel area ($A = WD$),
W = channel width, and
D = hydraulic depth.

For most natural river channels, the exponent J has a value between 0.8 and 1.5. If an average J value of 1 is used, the above equation can be simplified to

$$\frac{Q_s d^{0.5}}{K} = \frac{Q_w^2 S}{WD} \tag{7.3}$$

Yang's equation is similar to Lane's relationship, but Yang's equation can be used directly to predict the dynamic adjustments of a river channel due to natural or anthropomorphic events, after the coefficient K has been determined for the river. Water discharge is raised to the second power, demonstrating that channel adjustments are most sensitive to changes in water discharge.

River channels can experience change in planform, slope, cross section, and bed topography. Planform changes can be a shift in classification (i.e., straight to meandering), where the rate of change might be the distance per time progression of the shift upstream or downstream. Within a classification, variables that can be measured include channel location, meander wavelength, bend radius of curvature, channel width, sinuosity, shape and size of main bars or islands, braiding intensity, and degree and character of anabranching.

Slope changes over time should compare elevation of the same feature type (i.e., thalweg to thalweg not thalweg to average bed) at the same location. Bed control locations may be identifiable in this analysis. Reach averages of slope may be used when channel locations change. Channel shape changes over time are evaluated through the cross section geometry and bed topography. A plot with serial cross section surveys is a very effective method to visualize change. Parameters such as width, depth, flow area, and bank erosion are easily calculated from the survey data. Bed topography is frequently assessed by reach (e.g., island acreage or number and spacing of pool and riffle sequences).

The amount and rate of change can be derived from comparison over time of the parameters discussed above, assuming enough data are available. Comparison of channel characteristics (such as channel capacity, stage-discharge rating curves, and bed material grain size) with a known stable reach (reference reach) can be very insightful. Reach averages may be used for comparison of all variables discussed above. Other data from a chronology of events and disturbances can be used in the analysis of channel change. The chronology may contain information from maps, historic construction documents, climate and hydrology data, and anecdotal evidence. The established sequence of events may provide needed information to correlate or explain the changes shown in historical data analysis.

Rates of change as affected by process are used to predict channel responses to disturbances. Trends of channel evolution can be unsteady, nonuniform, and complex (Simon and Thorne, 1996), so it may not be appropriate to predict only by historical trends. Another concern is a possible change in measurement techniques and equipment over time because the characteristic being compared may not be the same.

River channel stability may be assessed through the following indicators:

- Amounts and rates of historical changes of bed elevation and horizontal position,
- Sediment transport capacity,
- Channel bank stability,
- Planform characteristics such as sinuosity or meander wavelength,
- Comparison of channel planform with a known stable reference reach,
- Channel hydraulic capacity (bankfull discharge),
- Stage-discharge rating curves, and
- Bed-material grain size distributions to cite several examples.

More than one approach may be used to check various processes in the stability assessment.

7.3.2.11 Geomorphic Summary

Geomorphology can help identify critical reaches and how the project fits into the context of the entire system. Several steps are involved in the geomorphic analysis, including defining channel form and processes, mapping, assessing river channel stability, and identifying disturbances affecting the river corridor. Understanding how and why a system changed will assist in assessing the stability of the proposed restoration. Two publications with useful information on performing a geomorphic analysis are *Stream Reconnaissance Handbook* by Thorne (1998) and *Hydraulic Design of Stream Restoration Projects* by Copeland et al. (2001).

7.3.3 Disturbances Affecting the River Corridor

To have a successful river restoration project, it is essential to understand what types of disturbances have played a role in forming the current channel configuration, which disturbances are presently continuing, and which are likely to occur again in the future. For this section, a disturbance is defined as a detectable change to the natural sedimentation processes of erosion, transport, and deposition that alters the physical characteristics of the stream channel, flood plain, and riparian zone. A natural fluvial process such as landslides, fire, floods or human-induced impacts such as levees, bridges, and dams may cause the disturbance to occur. In general, natural disturbances may be less frequent than human disturbances, which often affect the river on a continual basis. The magnitude and duration of the disturbance directly depend on the location and characteristics of the process that triggers the change. For instance, if a landslide in the upstream watershed occurs during a storm event, a large pulse of sediment would be delivered to the river channel and turbidity would likely increase. However, the delivery of suspended sediment to the river would be limited to the duration of the storm event. On the other hand, a permanent levee along the river corridor may constrict the channel width and have long-term impacts on river channel bed-material size and sediment transport capacity. It is important to identify not only the existing disturbances in the river and watershed, but also the historical disturbances. Recognizing the historical disturbances can help identify natural disturbances in the watershed that are likely to occur again and human disturbances that have played a role in forming the current river channel configuration.

Natural disturbances such as floods, droughts, wind storms, fire, landslides, debris dam failures, volcanic eruptions, and earthquakes can all result in a change in sediment processes in a river system. Many natural disturbances like volcanic eruptions and earthquakes occur on an infrequent basis. These types of disturbances are hard to predict, but they can play an important role in understanding how the existing river corridor came to its present state. Other natural disturbances such as fires, floods, and landslides occur on a more frequent basis and are easier to predict the potential for repeat occurrence in the future.

A great source for locating available documentation, research, and analysis on natural disturbances is local, state, and federal government agencies. It is important to look for information not only at the site of interest, but also in areas further upstream that may contain disturbances that affect the area being considered for restoration. Another option for documenting historical disturbances is to use historical aerial photographs and maps. In many areas, digital ortho-rectified photographs are available that can be used to create a data base of existing and historical aerial photographs and maps in a geographic information system. With this modern technology, impacts from floods, occurrence of landslides, and other natural disturbances can be quickly analyzed. Another source of valuable information is interviews with local landowners. Although mostly undocumented information, landowners often experience natural disturbances to the river corridor firsthand and can be an excellent source for better understanding the impact and timing of historical events not otherwise documented.

Activities that can cause disturbance to the river corridor include the construction of dams, diversions, levees, roads, and bridges; bank protection;, removal of large woody debris; logging;

grazing;, gravel mining; urbanization;, and recreation. The most common form of human-induced disturbance of sediment processes is the construction of features that alter the hydraulics and geomorphic characteristics of the river channel and flood plain (e.g. channel width, depth, slope, roughness, and alignment). These features may cut off flood plain area, alter channel planform, reduce incoming sediment, induce channel and bank erosion, or change the capacity to transport and store sediment. The following topics briefly discuss some of the more typical human-induced disturbances and how they impact sediment processes on rivers. A more detailed discussion on these topics can be found in the *Stream Corridor Restoration Manual: Principles, Processes, and Practices* (1998).

7.3.3.1 Dams

According to the *National Inventory of Dams* (U.S. Army Corps of Engineers, 1999), there are currently over 76,000 dams in the United States. Many of these dams are built across rivers and alter sediment transport delivery rates and volumes. Dams create impoundments (reservoir pools) and, if these impoundments are large enough, they can trap sediment delivered from the upstream watershed and result in clear water releases downstream.

Large dams create significant impoundments that may take centuries to completely fill with sediment. If the clear water releases are sustained for long periods of time, net erosion can result as the existing sediments along the bed are gradually transported downstream. A reduction in the sediment supply to the downstream river channel can alter the sediment sizes present along the channel bed. While smaller dams can locally impact the sediment regime near the dam, they have limited impact on reducing sediment delivery downstream. They may initially trap sediment in the impoundment, but once the available storage is filled, sediments can once again be transported over the top of the dam during high flows.

In addition to reducing sediment delivery to the downstream river, dams can also alter natural flow regimes. The majority of sediment transport occurs at high flows when a significant portion of the channel is inundated. If dam operations reduce the frequency or magnitude of peak flows, this will reduce the sediment transport capacity of the downstream river. When significant sources of sediment are available in the downstream channel from downstream tributaries or bank erosion, this reduction in peak flows and sediment transport capacity can result in aggradation. All of these possibilities should be taken into consideration when assessing a river with upstream dams. The USGS has published two studies on the impacts of dams, which are *Downstream Effects of Dams on Alluvial Rivers* by Williams and Wolman (1984) and *Dams and Rivers Primer on the Downstream Effects of Dams* by Collier et al (1996). These two publications describe many of the different types of impacts on the downstream river channel that can occur as a result of dams.

Because there are natural processes that reduce the useful storage in a reservoir, these losses, over time, decrease the reservoir sediment trap efficiency and increase the portion of sediment released to the downstream river channel. Reservoir storage losses due to sedimentation may reduce the reservoir's detention of floodflows and change the frequency distribution of controlled and uncontrolled flood outflows. Thus, the hydrology of a river downstream from a reservoir changes, usually slowly, throughout the life of the reservoir. Some reservoirs have very small sediment inflow rates and very long lives. Those with little or no sediment inflow are usually off channel storage reservoirs.

The operating rules that govern the various outflows of water from the reservoir and its canals determine the hour-by-hour and day-by-day controlled flows that enter the downstream river. The controlled flows may range from a minimum of no outflow at all to large spillway outflows controlled by gates on the spillway crest. The controlled flows are frequently limited to a range, or ranges, of flows by the structures that release the water from the reservoir. The operating rules may release water in more limited ranges, and those ranges may depend on the season of the year, the water level in the reservoir, and downstream water needs.

Dams may also trap contaminated sediments that originate from historical mining sites or other unregulated nonpoint pollution sources upstream. When considering releases of sediments from reservoirs through either dam removal or low-level release operations, the potential for releasing contaminants should be assessed. An additional discussion related to impacts from decommissioning dams is discussed in Chapter 8.

7.3.3.2 Diversions

Water diversion structures remove water from a river to meet the needs of agricultural, industrial, or municipal water users. If a significant percentage of the total volume of flow is removed from the river, hydraulic and sediment processes will be affected. Removal of a significant portion of the total flow can reduce wetted width, depth, and velocities, which can, in turn, increase water temperatures and reduce the area inundated along the channel. If the reduction in flow occurs over a long enough period of time, these impacts will alter the aquatic habitat and, possibly, riparian vegetation along the river channel. As the natural streamflow decreases (especially during low-flow summer months), the amount of flow diversion becomes more noticeable. For instance, during a flood, the flow diversion would likely be a small proportion of the total flow. However, during a river's low-flow period, the flow diversion may be a large portion of the total flow. For this reason, flow diversions may have a significant impact on the aquatic habitat. If flow diversions significantly reduce the mean annual riverflow, then likely they would also significantly reduce the sediment transport capacity of the river channel, which could lead to aggradation of the riverbed.

Unlike storage reservoirs, which often release clear water with no sediment, diversion structures divert water but try not to divert sediment. Because the water volume is reduced, the transport capacity is reduced and additional deposition in the downstream channel may result. If this happens, increased deposition can result in higher levels of flooding and increased lateral channel migration and bank erosion. Restoration strategies on rivers with diversions should consider the timing, magnitude, and duration of flow diverted, as well as how this may impact the flow regime and the annual sediment transport capacity. In many areas where flow reduction from diversions has significantly altered habitat, restoration work has focused on defining a minimum riverflow in order to provide a particular amount of habitat area. Periodic high flows may be needed to increase the annual sediment transport capacity of the downstream river channel.

7.3.3.3 Levees

Levees are an artificial embankment (usually earth and rock) built along a river to protect an area from flooding or to confine water to a particular channel path. Levees built in the active channel or flood plains have a greater impact on river hydraulics and sediment transport than levees built along a terrace and outside of the historic channel migration zone. The larger the portion of channel and flood plain that a levee cuts off, the greater impact it will have on hydraulics, sediment transport, and geomorphic processes.

Flood plain sediments are naturally deposited and reworked by the river as the channel migrates across the flood plain. Flood plains also tend to temporarily store floodwaters and attenuate and slow flood peaks that pass to downstream reaches. When the flood plains are cutoff by levees, channel migration is restricted and flood peaks will pass more quickly to downstream river reaches and with less attenuation. As levees force more water to flow in the main channel, they result in higher river stage, depth, velocity, and sediment transport capacity. If levees force the river to flow in a straighter alignment, then slope is also increased, which further increases flow velocity and sediment transport capacity. These increases often result in channel incision or degradation, coarser bed material, and even armoring of the riverbed. The increased depth and velocity also increase the capacity to transport large woody debris, which reduces the number of logjams present in the constricted reach.

In some cases, levees are not built a consistent distance away from the river channel and can locally create channel constrictions. Where the levees form a constriction, they result in a local increase in the velocity and sediment transport capacity at the constriction, and cause backwater upstream. Upstream from the constriction, sediment can deposit in the backwater areas and aggrade the channel bed. In this manner, the levee constriction can act in a similar way to a bridge constriction. This fluctuation in distance from the levee to the riverbank can result in a variation in sediment transport capacity with alternating reaches of erosion (in constricted areas) and deposition (in backwater areas upstream of constriction). For example, levee constrictions along the Dungeness River near Sequim, Washington, resulted in 10 feet of local sediment deposition upstream from the constriction points (Bountry et al., 2002).

In the natural setting, a portion of the suspended sediment load (sand, silt, and clay) will deposit on flood plain surfaces during high flows as water overtops the riverbanks. When levees cut off the flood plain, all of the water and suspended sediment remain in the river channel contained between the levees. This can increase the turbidity of the river and reduce the ground water-river interaction processes. In addition to cutting off flood plains, levees can also cut off secondary channels that are important for fish and other aquatic animals and plants.

Restoration strategies typically involve the removal or setting back of the levees to allow flood flows to have full or partial access to the flood plain. Properties or easements within the affected flood plain may have to be purchased from private landowners. The degree to which a levee is set back depends on the management objectives and cost. The examination of historical photographs and maps will help determine the boundaries of the natural flood plain and the historic channel migration zone.

7.3.3.4 Roads in the River Corridor

Roads are designed and constructed to provide a safe access route for commercial and private transportation. However, the protection of roads within the river corridor may have an impact on river processes. The degree to which a road may impact a river depends on the elevation of the road relative to the natural topography and the alignment of the road relative to the river channel and flood plain.

The elevation of a road embankment crossing through a flood plain is typically higher than the natural flood plain topography. Therefore, the road embankment can act as a levee and cut off a portion of the flood plain during high flows and limit future channel migration. This can cause water to pond up behind the road embankment as it drains off the terraces or valley walls. If a significant portion of the flood plain is cut off, more water is forced to remain in the river channel, causing increased stage and velocity. However, if the road is positioned at the edge of the flood plain, so the percentage of flood plain area cut off is small, the impact is greatly minimized.

Roads constructed not only in the flood plain, but also parallel to the river channel, can create "hard points" where the road embankment comes into contact with the river. If the natural riverbank is made of a material resistant to erosion, such as bedrock, the impact on river processes is minimal because bank erosion and channel migration are already limited. When the natural bank is made of more erodible materials, such as alluvium or glacial deposits, the river can erode this bank over time as the channel migrates across the flood plain. When the river begins to erode the road embankment, rock riprap is typically used to stabilize the bank and prevent any further erosion from taking place. If designed properly, rock riprap can be an effective embankment protection strategy, but it can result in increased velocities and a deep thalweg adjacent to the embankment toe. The protected bank limits the amount of natural channel migration that can occur. If riprap prevented the natural growth of vegetation along a riverbank, there would be a lack of cover (shade) and recruitment of woody debris. Other options are to construct engineered logjams along the bank to deflect high-velocity flow away from the bank and create downstream eddies that promote sand and gravel bar deposition. Planting trees along the bank can also increase stability although some rock may still be required to protect the toe of the bank from erosion. The installation of bend-way weirs can also be used to deflect high-velocity flows from the riverbank.

Restoration strategies involving roads should consider what the natural bank or topography would be like if the road were not in place, how the road impacts river channel and flood plain processes, what options exist for protecting the road embankment from river erosion, and what options exist for relocating or setting back the road.

7.3.3.5 Bridges

The construction of bridges along a river channel can have a varying level of impact on river channel hydraulics and sediment processes. The extent of the impact is typically limited to several channel widths upstream and downstream of the bridge. The steeper the river, the less distance (both upstream and downstream) from the bridge the effect will extend. Even though a bridge may not constrict the active channel width, a constriction of the channel migration zone, due to an embankment across the flood plain, would prevent future channel migration. For example, when a meander bend migrates downstream and encounters the bridge approach embankment, the river is forced into a sharp bend and then must flow parallel to the embankment before flowing under the bridge. The combination of the sharp river bend and the continuing lateral channel migration often leads to extensive bank erosion immediately upstream from the bridge embankment.

When bridge piers and embankments constrict the active channel and flood plain, they alter the local capacity to transport water and sediment. The constriction causes higher velocities through the bridge, which increase transport capacity. However, a low-velocity backwater area forms upstream of the constriction, which reduces transport capacity and results in deposition of sediment and woody debris, if present in the system. Bridge spans can be lengthened to reduce or eliminate the impact on natural river processes. The amount that a bridge span is lengthened depends on the resource management objectives and costs. However, in many cases, bridges are built on natural geologic constrictions and do not create unnatural constrictions on the river. Restoration strategies at bridge locations should consider whether the bridge is formed at a natural constriction, and what level of impact, if any, the bridge exerts on the river hydraulics and sediment transport capacity.

7.3.3.6 Bank Protection

Bank protection is placed along a riverbank to prevent potential erosion of the bank during floods. However, bank protection typically does not raise the bank elevation like a levee and, therefore, does not prevent floodflows from overtopping the bank. However, bank protection does act to prevent lateral channel migration and erosion. Further, bank protection may cut off access to side channels if their entrance becomes blocked by the bank protection. An important factor in understanding the degree of impact from bank protection on natural river processes is to determine whether the bank is at the edge of a terrace or within the natural flood plain and channel migration zone. When bank protection is placed along a valley wall or along a terrace bank at the edge of the natural flood plain, the impact on channel processes is often negligible. This is because the natural rate of bank erosion and lateral migration along the valley wall is limited, due to geologic controls. When a terrace can be dated as being hundreds or thousands of years old, protecting the edge of the terrace and the historic channel migration zone, is often mitigation for some other impact, rather than a direct effect on natural river processes. However, when bank protection is placed along a younger surface within the flood plain, the impact on bank erosion and lateral migration processes may be significant. The degree of impact depends on the length of bank protection placed on the channel, the amount of active channel or flood plain that is cut off, and the combined effect from any bank protection structures on the opposite bank.

7.3.3.7 Removal of Vegetation and Woody Debris

In a natural setting, trees typically grow along alluvial riverbanks and provide a mechanism for protecting the banks and limiting the amount and rate of bank erosion during floods when high velocity riverflows are adjacent to the bank. Although bank erosion does occur in natural settings, the rate and extent of the erosion is usually limited by vegetation. The root structure of the vegetation increases the bank roughness and provides more resistance to erosion. As small amounts of the bank erode, large trees can fall into the river channel and line the bank, which tends to naturally protect the bank by deflecting high-velocity flows away from the bank.

Development along rivers often involves removal of riparian forests from the riverbanks and flood plain. When vegetation is removed, this important natural bank protection feature is lost. Subsequent flooding in the river can result in large amounts of bank erosion where vegetation has been cleared, especially along the outside curve of meander bends. Historical aerial photographs can be used to document when vegetation was cleared in a particular area, and the resulting bank erosion that has occurred from subsequent floods. Field investigations of the materials that comprise the bank are also important to assess the potential for bank erosion in areas where vegetation has been cleared.

In many river systems, woody debris was historically removed from the river channel to improve conveyance of water or for navigation purposes. The presence of large woody debris in the river channel tends to increases channel roughness. If large woody debris were removed from the river channel, the river would be left with excess energy. One possible response would be an increase in sinuosity and a corresponding reduction in river slope to reduce the excess energy. Such an increase in sinuosity could lead to an increase in bank erosion and, possibly, an expansion of the historic channel migration zone.

7.3.3.8 Forestry Practices

Forest harvest, road building, road maintenance, and other management activities in the upstream watershed can result in a net increase in water and sediment delivery to the mainstem river channel and tributaries within the watershed. Another potential impact from forestry activities is a loss of slope stability resulting in landslides or debris flows.

Increases in the coarse sediment supply from tributaries and the upstream river channel have the greatest potential to affect the physical characteristics of the channel planform and geometry, while increases in the suspended sediment load can result in increased turbidity, poor water quality, and degraded aquatic habitat. Many investigators have suggested that increases in coarse sediment from the watershed can be detected by increases in channel width, but that increases in peak flows from forestry practices can have the same result (MacDonald, 1991). Channel changes can often have multiple causes and can include both natural and human factors. Therefore, the impact from forestry practices must be carefully assessed through quantitative measures, where possible, and qualitative measures at a minimum. Useful information for this assessment includes the proximity of the forestry practice to the affected site, the relative percentage of land affected in the upstream watershed, the impact on the stream channel and landscape as a result of the forestry practice, and the existing land use following forest harvest.

7.3.3.9 Grazing (bank erosion)

Grazing or trampling of the vegetation along a river corridor can accelerate bank erosion. Grazing practices that allow free animal access to the river may also result in bank erosion from the physical trampling of the banks by animals accessing the river.

7.3.3.10 Gravel Mining

Gravel mining is the physical removal of sediment from a river channel and is often accomplished by scraping gravel bars or excavating material from areas adjacent to the active river channel with mechanical equipment. The impact of gravel mining can be determined by assessing the rate, duration, and volume of sediment removed relative to the incoming sediment load of the river. In addition, gravel mining in the active river channel results in a lowering of the channel bottom, which will have local impacts on hydraulics and sediment transport processes. For example, gravel would tend to deposit and refill the excavated areas. If a gravel pit in the channel is long enough, the hydraulic slope entering the gravel pit will become steep and headcut erosion of the river channel will tend to migrate upstream.

Records of gravel mining quantities and frequency are often difficult to find for historic operations, but, more recently, local governments have required permits to mine gravel and information is more readily available. The volume of gravel mined from the river can be evaluated to determine the impact on the sediment budget for a particular river reach.

Historically, mining occurred in and adjacent to a large number of rivers, and piles of sediment were often overturned and left behind in search of minerals and precious metals. The physical movement of river and flood plain materials can have a significant impact on hydraulics and sediment transport. In addition, sediments left behind in mining waste piles may contain contaminants that must be addressed, if present. Otherwise, these contaminants can enter the streamflow and affect water quality as sediments are mobilized during high flows.

Gravel mining or bar scalping has been used as a management tool to help control aggradation. Long-term aggradation should be documented through repeat cross-section measurements and, if present, the cause of the aggradation should be determined before employing gravel mining as a management strategy.

7.3.3.11 Urbanization

When areas along a river corridor are urbanized, the surfaces are often cleared of vegetation and significant flood plain areas become paved with asphalt or concrete. Creating impervious surfaces can increase the volume of water conveyed to the river channel during storm events and flooding. In a natural setting, water overtops riverbanks during flooding and can be temporarily stored along the river's flood plain. A large portion of the water that enters the flood plain will pond up and gradually seep through the soil into the ground water thus limiting the amount of floodwater that re-enters the channel during the peak of the flood. When flood plain surfaces become impervious, water will quickly run off that surface and flow back into the river channel, which can increase the volume of water delivered to the river. Determining the amount of area urbanized and proximity to the river channel can help assess the degree of impact of impervious surfaces. In addition, if contaminants from nonpoint surface runoff are present, they also need to be addressed in restoration plans.

7.3.3.12 Recreation

Most recreational river users are very respectful of the environment and work to preserve natural river conditions. However, certain activities by recreational river users can begin to impact natural river processes if sustained over long periods of time. One example might be wakes created by jet boats along a riverbank that lead to accelerated bank erosion. Another example is the compaction of soils and erosion from human and vehicle use on river access roads, trails, and camping sites. Restoration plans should take into account the amount of existing and planned future recreational use and access at a given site.

7.3.4 Hydrologic Analysis

A hydrologic analysis provides information that describes the magnitude, duration, and frequency of discharge on a river. It also provides an understanding of the seasonal variations of flow and the time at which channel forming flows are likely to occur. For example, floods may be short-duration events that occur from winter rains, long-duration events that occur during the spring snowmelt, or rain on snow events, or combinations of all three. Understanding the watershed hydrology is an important step in understanding the sediment processes in the river channel. The hydrologic analysis may also need to investigate the relationship between flow in tributaries and the main river channel, and the interactions with groundwater, to determine if river reaches tend to lose or gain water.

7.3.4.1 Historical Discharge Data

Hydrologic analyses are usually based on continuous discharge records from river gaging stations at one or more locations on a river. In addition to discharge data, gaging stations also often contain water surface elevation measurements, velocity data, and cross-section geometries used for measuring discharge. Discharge information can be downloaded from the USGS web page (www.usgs.gov). Other sources include the River Forecast Centers of the National Oceanographic and Atmospheric Administration, agencies or companies that operate dams, water treatment facilities, ferry boats, drawbridges, etc. When evaluating historical discharge records, it may also be of interest to look for trends in riverflow. River hydrology variations can be over short or long-term periods and can be caused from a number of influences such as climatic fluctuations, land use changes, river diversions, upstream reservoirs, and consumptive use.

When a stream gage does not exist, it may be possible to extrapolate the data from another location in the watershed or a nearby drainage area with similar characteristics. When streamflows have been significantly altered, for example, by storage reservoirs or flow diversions, it may be desirable to determine the natural historical flows without the presence of these impacts. A natural flow reconstruction would then help determine the effect that reservoirs and flow diversions have had on sediment transport processes.

7.3.4.2 Flood Frequency Analysis

Identifying the magnitude and frequency of potential future floods is an essential component of any river restoration project. Flood frequency analysis provides an estimate of the probability of future flood events. Usually, annual instantaneous peak flows are used for this analysis because they indicate the maximum discharge a river may be subjected to. A partial duration analysis uses all flood peaks above a certain base magnitude and is useful when multiple flood events within a single year are important. Flood frequency analyses are used to provide a range of riverflows for hydraulic and sediment transport analyses that may be used to compute the extent of flooding or hydraulic and sediment parameters. They are also used to identify potential design floods for bank protection and river restoration projects. Typical frequencies are the 1.5-year, 2-year, 5-year, 10-year, 25-year, 50-year, 100-year, and 200-year floods, depending on the extent of historical discharge data available.

7.3.4.3 Flow Duration Analysis

A flow duration analysis is used to show the percentage of time a given flow is equaled or exceeded within a given timeframe. A flow duration analysis is typically based on mean-daily flow data, rather than peak flood values as in the flood frequency analysis. The most commonly referenced values from this analysis are the flows that are exceeded 90, 75, 50, 25, and 10 percent of the time. Flow duration analyses can be performed for all flows within a given number of years. The flow duration analysis can also be performed on data that are segregated by season or month within a given number of years. A comparison of the flow-duration values over different time periods can be used to assess flow changes due to climate change or the influence of water resource development.

7.3.4.4 Ground Water Interaction

Evaluation of the hydrologic interaction between ground water and surface water in a river channel may also be of interest for a restoration project. Migration of the river channel and significant aggradation or incision of the channel bed can cause changes in the ground water and surface water relationship. For instance, side channels through the flood plain may provide viable aquatic habitat when there is a surface water connection with the main channel. However, the side channels can become disconnected from the main channel if the riverflow decreases too much or if the river migrates away from the side channel entrance. In some cases, the groundwater table may be high enough to maintain water in the side channel and maintain the aquatic habitat. The basic linkages between flow in the main river channel and side channels can be determined from simultaneous discharge measurements (including flow depth, width, and velocity) in the main river channel and in side channels. A study of well data and the direction of ground water flow can help identify areas where there may be an upwelling or downwelling of ground water into the side channels or main river channel. Such areas may provide import habitat for fish.

7.3.4.5 Channel Forming Discharge

The channel forming discharge is an important concept for the evaluation of river channels and is often used for design purposes. River channels continue to evolve with time. The river channel cross section and planform are actually formed by a range of discharge values. Large and infrequent floods have the capacity to temporarily shape and modify the channel, but they occur too infrequently over the long term to sustain the channel shape. Low flows occur frequently, but they have too little stream power to shape the channel. Therefore, a range of high flows that occur frequently enough are responsible for shaping the channel over the long term. The channel forming discharge is an index to this range of channel forming flows. It is also called the dominant or formative discharge and often equated to bankfull discharge, effective discharge, or specified by a flow recurrence interval commonly in the range of a 1- to 2.5-year event (Leopold, 1994). The duration of the channel forming flow and the quantity of sediment transported has a role in determining the channel forming flow that is as great as the role of the recurrence interval of the flow.

Ideally, the channel forming discharge is the flow that just fills the active channel to the top of its banks. The channel forming discharge will equal the average bankfull discharge on a river in a stable or equilibrium condition. The difficulty lies in finding a river in a stable or "undisturbed" condition. If there are recent changes (e.g., the historic flows of the river have changed, the sediment transport has been altered, or there have been physical changes to the geometry of the channel including slope), then the channel is no longer in a stable or equilibrium condition and the bankfull discharge represents a previous condition. Incised channels provide a standard example where a bankfull discharge will not represent the channel forming discharge. Even with a stable system, irregularities in a natural channel can make it difficult to pick out a bankfull discharge. Vegetation found along the channel banks can help define the bankfull discharge. Depending on the environmental conditions, the indicator could be the beginning of significant vegetation growth or the line of significant change in vegetation type. This is surprisingly apparent, even in an urban concrete lined channel, where the crack between sloped concrete panels can produce vegetation above a recurring flow line. In some cases, a range of bankfull values may be more appropriate than selecting a specific value.

The channel forming discharge has also been based on a recurrence interval. Leopold (1994) stated that most investigations have concluded that the bankfull discharge recurrence intervals range from 1.0 to 2.5 years, and the bankfull discharge recurrence interval is often assumed to be 1.5 years. For 58 sand bed rivers in the United States that were determined to be in a stable and natural condition, 83 percent had a bankfull discharge with a recurrence interval of between 1 and 2 years (Soar and Thorne, 2001).

Effective discharge is the discharge that transports the greatest amount of sediment over a period of many years or a few decades and has been used to estimate the channel forming discharge. This approach is based on the assumption that channel forming processes are primarily a function of sediment transport. The discharge and sediment transport history are combined to determine the flow range that transports the most sediment over the long term. Papers by Andrews (1980) and Biedenharn et al. (2000) describe the process for calculating effective discharge using equal discharge increments. Cohn (1995) discusses statistical concerns associated with effective discharge computations. Soar and Thorne (2001) found that the effective discharge, computed using equal discharge increments, underestimated the bankfull discharge at 86 percent of the 81 sand bed rivers they investigated. If much more than 50 percent of the long-term sediment load is transported by discharges greater than the computed effective discharge, the computed value is too low and requires adjustment (Soar and Thorne, 2001). The effective discharge can also be computed using probability intervals, which may be more representative of the bankfull discharge. In addition, the median sediment transporting discharge can be computed where half of the long-term sediment load is transported by discharges that are greater and the other half by discharges that are lower (see Chapter 9).

7.3.5 Hydraulic Analysis and Modeling

Restoration strategies often involve modifications to the existing river channel geometry and flood plain, either directly through construction at a particular site or indirectly by altering the flow and sediment regime of the river. A hydraulic analysis gives information regarding the slope, hydraulic geometry, unit stream power, shear stress, and overall energy of the river at a variety of discharges for the existing system, while also providing predictions of how the river will respond to any proposed modifications. A hydraulic analysis can be as simple as performing normal depth calculations of depth and velocity at a given location or as complicated as running a three-dimensional computer model or constructing a physical laboratory model to simulate the river channel hydraulics. When selecting the type of hydraulic analysis for restoration efforts, the key is to choose the level of analysis that will provide the necessary information for restoration planning, alternative selection, design, implementation, and monitoring efforts, while working within the project budget and timeframe.

7.3.5.1 Topographic Data Needed

The basic input data needed for any level of hydraulic analysis are the geometry and longitudinal slope of the river channel and flood plain. These data can be in the form of surveyed cross sections or continuous topography (methods of collecting this data are discussed in Chapter 9). No matter what method is used, a critical criterion for collecting data in natural channels is to ensure that the level of detail in the collected data matches the needs of the specified analysis.

For instance, if it was desired to use a numerical hydraulic model to simulate a flood over several miles of river, the spacing and detail of cross-section geometry could be much greater to represent average conditions than when modeling a short reach during low flows that would be much more sensitive to changes in channel geometry. Another source of data that can be useful is discharge measurement records at gaging stations. Gaging stations are usually located at stable sections. Available data often include monitoring of the channel geometry over time, which can help modelers better understand the range of natural fluctuation (erosion and deposition) along the channel bed as a result of floods, along with velocity measurements, which can be used to calibrate models.

7.3.5.2 Longitudinal Slope and Geometry Data

The first step in a hydraulic analysis is to evaluate the measured longitudinal slope and geometry data along the river. Are there areas of significant change in slope that may indicate a change in hydraulic and sediment transport capacity? Where are the expected areas of highest unit stream power or shear stress in the channel? What depositional features exist in the channel? What are the "wetted width to depth" ratios? Is the channel deep and narrow or wide and shallow? Are there multiple flow paths or a single channel? Do the active channel and flood plain have a consistent width or change as a result of manmade or natural geologic features? Combining these types of questions with the geomorphic analyses will help provide an assessment of the existing system and its capability to move water, sediment, and woody debris.

7.3.5.3 Physical and Numerical Models

To enable prediction of hydraulic parameters at a variety of discharges or to predict changes as a result of modifications to the river, a numerical or physical hydraulic model can be used. There are numerous models available, and research is always ongoing to improve capabilities in this field.

A physical model can also be a great tool for restoration projects involving modifications to existing water management facilities, or in the design of new facilities. Physical models re-create the actual facility at a smaller scale and can be used to look at both hydraulic and sediment impacts at a small, detailed level as design options are varied. Physical models have been used to evaluate the impact of design options on structures that affect fish protection and passage, aquatic habitat restoration, water quality, river sediment flushing, reservoir and river sedimentation related to hydraulic structures, and erosion control.

Numerical models can be used to compute hydraulic parameters that describe existing river conditions at a range of flows and to predict future conditions that may exist as a result of implementing various restoration options. In either case, the results from a numerical model should be checked with the conceptual model to see if they make logical sense. A simple, common approach for numerical hydraulic modeling is to use an "off-the-shelf software" model to evaluate parameters such as depth, velocity, wetted width, flood inundation, unit stream power, shear stress, Froude number, etc. Examples of one-dimensional models include the U.S. Army Corps of Engineers' HEC-RAS (2002); the Bureau of Reclamation's GSTAR-1D (Yang et al, 2004, 2005), GSTARS 2.1 (Yang and Simões 2000), and GSTARS3 (Yang and Simões 2002); and the Danish Hydraulic Institute's Mike 11 (2003). Numerical models can vary in complexity, ranging from steady to unsteady flow, one-dimensional to three-dimensional flow, up to models that integrate groundwater interaction, surface runoff, tributary inflows, diversions, structures, and sediment transport. In other more complex restoration cases, numerical models must be custom developed to properly address restoration project objectives. For instance, the Bureau of Reclamation developed a numerical model for the Platte River in Nebraska (Murphy and Randle, 2003) that integrates hydraulics, sediment transport, deposition, erosion, and the growth and removal of vegetation to address geomorphic changes in the active river channel and flood plain.

7.3.6 Sediment Transport Analysis and Modeling

When the river is in a dynamic equilibrium, the upstream sediment supply can be transported by the flow, and the net change in sediment deposition or scour along the riverbed is insignificant over the long term. When a restoration project is being considered, the natural dynamic equilibrium of the river may have been disturbed by human or natural influences. For example, a river channel might be degrading downstream from a dam or through a reach constricted by levees. The channel might have already reached a new equilibrium for the disturbed condition. Therefore, the existing stability of the river channel and the future stability under the restored condition should be assessed as part of the restoration design. In addition to field inspection and comparison with historic data, the channel stability can be assessed by comparing the upstream sediment supply with the river channel's hydraulic capacity to transport sediment and comparing how this capacity may change along the study reach. Channel degradation would be expected if the sediment transport capacity exceeds the supply, and aggradation would be expected if the supply exceeds the transport capacity.

7.3.6.1 Sources of Upstream Sediment Supply

The first step in understanding the role of sediment in a river is to identify the sources of sediment in the river system. Major sources of sediment often include tributary inflows, landslides and debris flows, hillslope erosion, erosion of banks, and sediment stored along the riverbed and flood plain. Land use changes in the flood plain and upstream watershed can result in changes to the amounts of runoff and sediment supply. For instance, upstream reservoirs, levees, and bank protection may limit additional sediment input to the system by trapping incoming sediment and preventing erosion, but logging roads and development may cause increases in runoff and sediment delivery. In restoration projects, it is important to understand how water management projects and land use changes may be impacting the sediment load of a river, both currently and in the future.

7.3.6.2 Total Stream Power

There are numerous techniques for evaluating the sediment transport capacity of the river (a detailed description of sediment transport capacity is discussed in Chapter 3). When sediment data are limited and the study reach is large, total stream power (γQS) can be a simple technique to show reaches that may have higher or lower sediment transport capacity. Total stream power is based on the product of discharge, the slope of the river, and the unit weight of water (Yang, 1996). In the upper reaches of the watershed, the longitudinal channel slope tends to be steep, but the riverflow tends to be low. In lower reaches of the watershed, the riverflows tend to increase, but the longitudinal slope of the river channel tends to decrease. Thus, total stream power could increase, decrease, or remain about the same with distance along the river system, depending on local geologic and human-built controls. Areas with relatively high total stream power would be expected to have a greater capacity to transport sediment than areas with low total stream power.

7.3.6.3 Incipient Motion

Incipient motion is important for the study of sediment transport, channel degradation, and stable channel design. Incipient motion defines the hydraulic conditions that begin to transport a sediment particle of a particular size along the riverbed. The incipient motion criteria are often used in sediment transport equations and in the design of bank stabilization to determine if the gradation of rock being used is adequate for the design flood. Depending on the application, incipient motion can represent the time at which a single particle begins to move, or the time at which the entire moveable bed is entrained and being transported. Discussions on the methodology for computing incipient motion can be found in Chapter 3 and in several references, such as *Sediment Transport: Theory and Practice* by Yang (1996) and *Computing Degradation and Local Scour, a Technical Guideline* by Pemberton and Lara (1984).

7.3.6.4 Sediment Particle Size Analysis

Sediment data can be collected to define the particle size distribution along the riverbed and the relationship of sediment transport to discharge. Particle size distribution data can provide information on size, shape, specific gravity, and fall velocity of the sediment particles present in the riverbed for a given reach or an entire river system. For instance, it may be of interest to determine if the median size of sediment along the riverbed is decreasing or increasing over a long reach where river slope and bankfull discharge may be changing with distance downstream. Bed-material particle size measurements can also be utilized to help define the geomorphic characteristics of river channel features; such as pools, riffles, and point bars; based on the horizontal and vertical structure of the particle deposits. New techniques are being researched to use bed-material sampling to assess cumulative watershed effects on the river system (Bevenger, et. al, 1995).

Meandering rivers have sinuous planforms and nonuniform bed slopes, with straight riffles connecting curved pools, and have complex spatial patterns of bed-material grain sizes (Leopold and Wolman, 1957; Bunte and Abt, 2001). While the transverse size distributions across riffles tend to be fairly uniform, the transverse size distributions across the curved meander bends tend to have finer bed material on the point bars that are found at the inside bank of each bend and coarser bed material towards the deep, outside part of the bend (Ikeda and Parker, 1989). The size of sediment deposited on a point bar typically decreases in the downstream direction. The sediment particle sizes found in the riffle are typically coarser than those found on the point bar. Particle sizes found in the pool, associated with a meander bend, are typically coarser than the sizes found in the riffle. A frequently found pattern is sand at the point bar, gravel in the riffles, and cobbles in the meander bend pools. In addition, the bed-material grain sizes are associated with the average slope of the river, averaged over many meander bends. The higher the average river slope is, the coarser the bed. Size distributions of bed material will also be affected by the time of sample collection. Finer sediment is often scoured and removed during the rising limb of a flood hydrograph and usually redeposited on the falling limb of the hydrograph, especially in the slower velocity pools. If samples are collected during or after the falling limb of the hydrograph, the bed material samples may not include coarser deposits that may be present underneath the finer sediments, especially in bend pools.

Braided rivers are frequently classified as straight and have uniform bed slopes, but these features depend on the form of the valley along which the riverflows (Leopold and Wolman, 1957; Best and Bristow, 1993). In general, the sinuosity and slope of a braided river follow the pattern of its valley. In any case, the spatial patterns of bed-material grain sizes are not as complex as those of a meandering river. The transverse distributions of bed material tend to be uniform, and the sinuous, riffle-pool sequence is generally absent from the braided river planform and profile. Bed material grain sizes tend to be uniform within reaches with similar channel properties such as slope and width.

An example of grain size distributions across a meandering river channel is given in Figure 7.2, and the variation in median grain size with distance along the channel is shown in Figure 7.3. An example of grain size distributions across a braided river channel is given in Figure 7.4, and the variation in median grain size with distance along the river channel is shown in Figure 7.5.

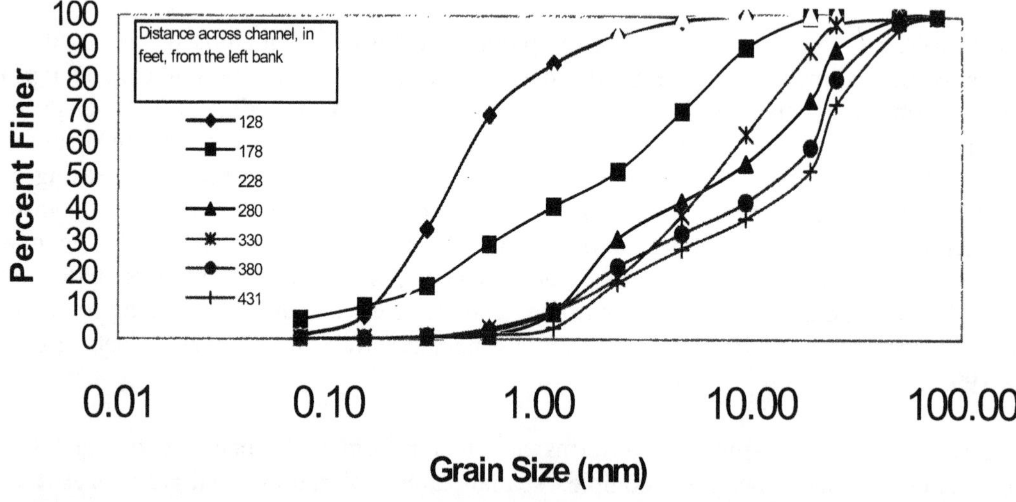

Figure 7.2. Sacramento River, California, grain size distributions across this meandering river channel.

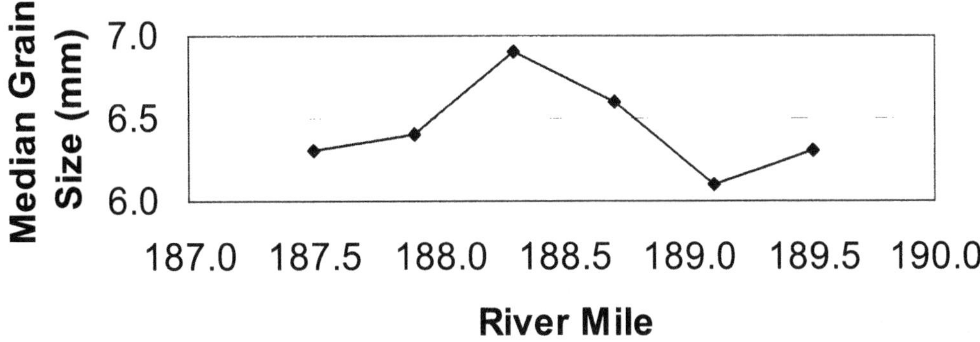

Figure 7.3. Sacramento River, California, median grain size along a longitudinal segment of this meandering river.

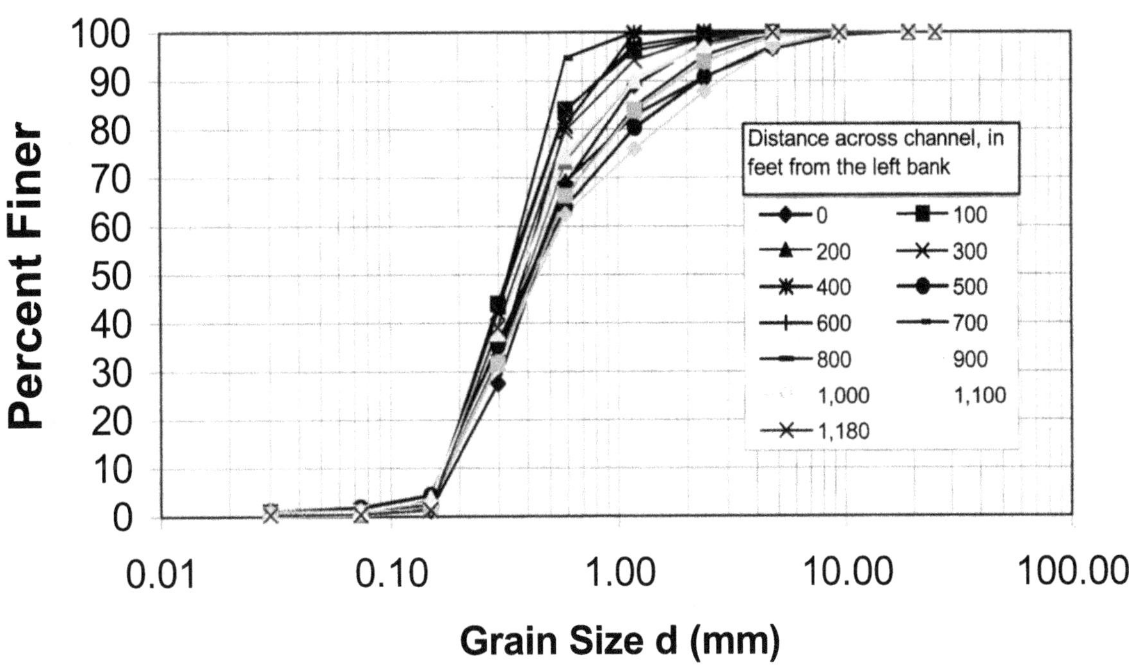

Figure 7.4. Platte River at Kearney, Nebraska, grain size distributions across this braided channel.

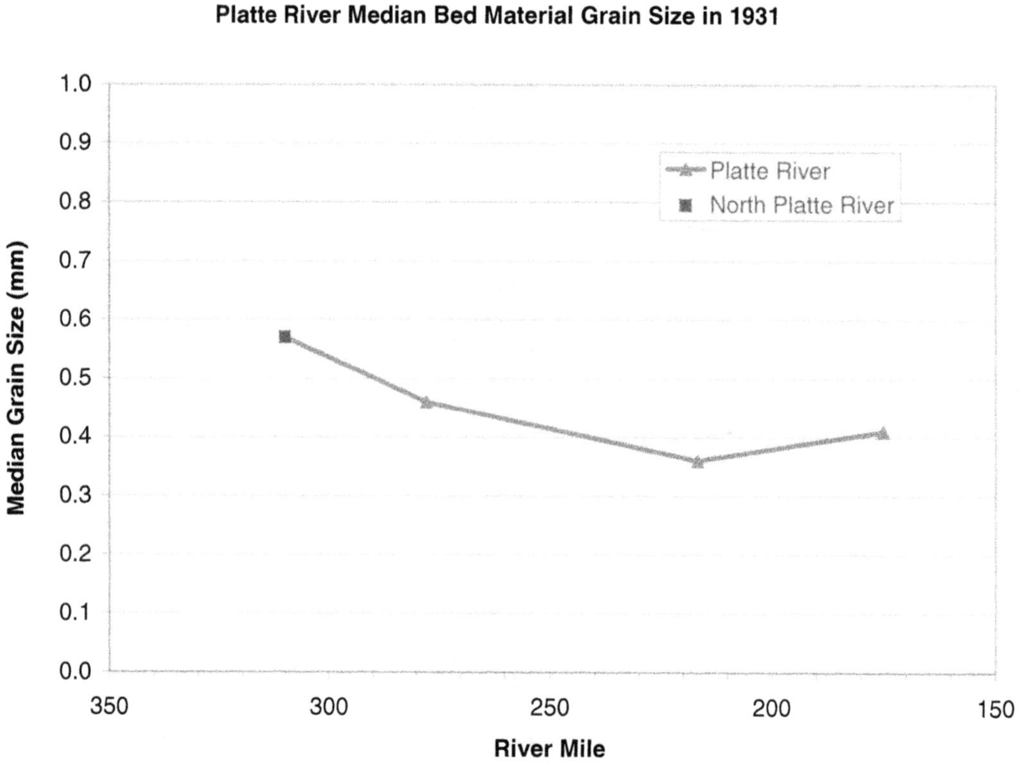

Platte River Median Bed Material Grain Size in 1931

River Mile	Median Grain Size (mm)

Figure 7.5. Platte River, Nebraska, median grain size along this braided river channel in 1931.

7.3.6.5 Sediment-Discharge Rating Curves

Sediment-discharge rating curves connect the river hydrology to sediment transport. Sediment rating curves are usually developed by measuring the depth-averaged, suspended sediment concentration and flow rate for a large number of vertical lines across a river channel at a stream gaging station (Edwards and Glysson, 1988). The suspended sediment load for the cross section is calculated by multiplying each average concentration by the flow rate corresponding to its vertical and summing over the cross section. Infrequently, bedload is measured across the channel. However, if the bedload is not sampled, it is sometimes computed using the modified Einstein Procedure (Colby and Hambree, 1955) or estimated at 2 to 15 percent of the suspended load (Strand and Pemberton, 1982). In either case, the bedload is added to the suspended load to calculate the total sediment transport at the cross section. The flow rate measured for the cross section is the sum of the flow rates for each vertical. The sediment transport rate and the flow rate define one point on the sediment-discharge rating curve. The sediment rating curve is developed by repeating this procedure many times to achieve the widest range of flows possible.

Typically, there is considerable scatter among the points on a sediment-discharge rating curve. Figure 7.6 shows measured discharge and sediment concentration and a best-fit, log-log sediment rating curve through the data. This best-fit curve can be used directly with the flow-duration curve to produce a sediment-transport duration curve, but that application is biased (Gilroy et al., 1990). The bias arises when the sediment load in real units is calculated from the log-log curve, and, therefore, a retransformation bias correction is often applied (Cohn, 1995).

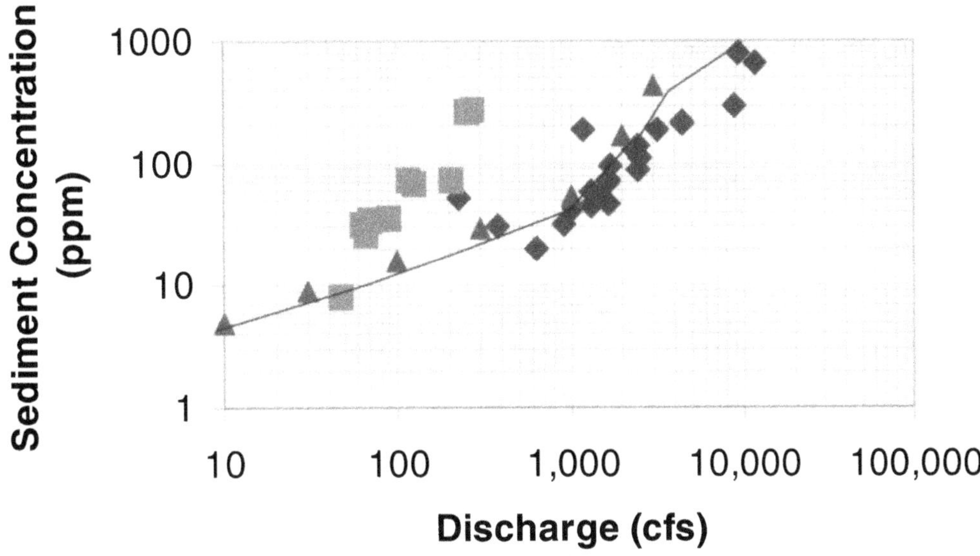

Figure 7.6 Suspended sand rating curve of the Platte River at Overton-Grand Island, Nebraska.

Seasonal differences associated with the annual cycle of runoff may require separate sediment rating curves for each season. Because of its slow nature, snowmelt typically erodes less sediment from the watershed than rainfall, especially thunderstorms and tropical cyclones. Different rating curves are often used for the snowmelt and thunderstorm seasons (Miller, 1951).

Additional variation in a sediment rating curve can occur if the supply of sediment to the riverbed is limited by a more irregular cause than a simple seasonal change. The erosional patterns of a watershed can vary greatly with landslides, fires, or other land use changes (such as grazing), which cause changes in the sediment supply to the downstream river channel.

When the suspended sediment concentrations and bedload samples are measured, the grain-size distributions of each sample can be developed to see if the samples have uniform grain sizes with standard deviation of less than 1.3. If the moving sediments are not uniform, then sediment-rating curves for each grain size can be developed. Because this level of detail is not frequently needed, the typical sediment rating curves do not distinguish among grain sizes.

Many times sediment rating curves are needed but data are not available. In those cases, an alternative is to use a predictive sediment-transport formula to estimate the sediment load for a given discharge and then develop the sediment rating curve.

7.3.6.6 Reservoir Sediment Outflows

Sediment transport downstream from dams is affected by reservoir sedimentation and hydrologic change caused by the dam and reservoir. The amount of sediment passing though a reservoir is highly dependant on the reservoir's life stage. When a large storage reservoir is first built, virtually no sediment is transported through the reservoir. All but the finest material will be deposited in the reservoir. Under these circumstances, water released from the dam can be assumed sediment free (Simons and Sentürk, 1992). As reservoir storage capacity is lost through reservoir sedimentation, and sediment is deposited near the outlet works, sediment outflow from the reservoir will increase. A typical response to dam construction and the rapid decrease in sediment supply is for the channel to degrade and armor if enough coarse material is present in the bed. Channel bed degradation allows the river to decrease its slope and sediment transport capacity in response to a decrease in sediment load. When the reservoir fills and sediment supply to the downstream channel increases, there may or may not be a noticeable channel response. Channel changes are often abrupt and difficult to predict. A possible response would be for the river to aggrade until the slope is steep enough to transport the additional sediment.

Sediment transport is also affected by the change in hydrology caused by the dam and reservoir operations. Peak flows are often reduced after dam construction (Julien, 2002) and sometimes there is even a reduction in the volume of water released into the channel due to diversions. In these cases, there is lower sediment transport capacity to transport the sediment in the channel, whether it is the beginning or end of the reservoir life cycle. In any case, the amount of sediment delivered to the channel downstream from a dam depends on the reservoir sediment trap efficiency and the configuration of the outlet structures in the dam. The ability of the channel to move that sediment, or sediment supplied from downstream tributaries, will depend on the amount and timing of releases from the dam.

7.3.6.7 Scour and Degradation

Rivers often experience changes in bed elevation. These changes can be caused by man or occur naturally. Whatever the cause, there are three generally accepted forms of bed elevation change: aggradation or degradation, general scour, and local scour (Simons and Sentürk, 1992). Aggradation and degradation are long-term processes that take place over long reaches. Bed level changes associated with aggradation or degradation are often related to changes in sediment load. General scour or contraction scour is often caused by a narrowing of the river that locally increases the velocity in a short reach. This type of scour can also occur during a flood where the bed scours on the rising limb and fills during the falling limb. Local scour is much more complicated than other forms of bed elevation change. Local scour is the removal of material near or around structures or channel obstructions (Simons and Sentürk, 1992). Another distinction is that local scour is caused by secondary flow currents that are directed toward the bed. This type of flow is difficult to model and requires complex three-dimensional models to describe the flow phenomena. Local scour is most easily estimated by using an equation developed specifically for a given structure (pier or abutment) and configuration. Bend scour exhibits characteristics of both general and local scour. There is often a decrease in width and an increase in velocity through a bend but the bend curvature creates secondary currents that are directed toward the bed and banks. Two good sources for additional information on bed level changes are *Computing Degradation and Local Scour* (Pemberton and Lara, 1984) and *Sediment Transport Technology Water and Sediment Dynamics* (Simons and Sentürk, 1992).

7.3.6.8 Sediment Transport Equations

When sufficient sediment and hydraulic data are available, sediment transport equations (see Chapters 3 and 4) and modeling can be very effective tools to not only better understand the existing transport capacity, but, more importantly, to predict how the existing system would respond to restoration options that alter the balance of sediment and flow, or the geometry of the channel. A variety of sediment equations and models are available, and research is always ongoing to improve the capabilities in this discipline. A discussion of available tools and data collection techniques is provided in Chapter 9 of this document.

7.3.6.9 Sediment Considerations for Stable Channel Design

Stable channel design parameters include the discharge capacity at low flow, bankfull flow, and floodflow; the channel width, depth, and slope; and the long-term sediment load. Discharge considerations alone are not sufficient. To be stable over the long term, the design must also take into account the river's sediment load. If the channel's sediment transport capacity is less than the incoming sediment load, channel aggradation and, possibly, bank erosion will occur. If the channel's sediment transport capacity exceeds the load of incoming sediment, channel degradation (erosion of the bed and banks) will occur. Increased sinuosity and river meandering may also result under the condition of excessive transport capacity.

A unique channel design can be determined by using the minimum rate of energy dissipation theory (Yang, 1976;, Yang and Song, 1979, 1984; and Yang 1996). This theory states that when a dynamic system reaches its equilibrium condition, its rate of energy dissipation is at a minimum. The minimum depends on the constraints applied to the system; in this case, water discharge, sediment load, channel roughness, and, possibly, the valley slope. For a uniform flow of a given channel width where the rate of energy dissipation due to sediment transport can be ignored, the rate of energy dissipation per unit weight of water is:

$$\frac{dY}{dt} = \frac{dx}{dt}\frac{dY}{dx} = VS = \textit{unit stream power} \tag{7.4}$$

where
Y = potential energy per unit weight of water,
V = mean channel velocity, and
S = channel slope.

Thus, the theory of minimum unit stream power requires that VS = minimum. The integration of VS across the channel will yield $Q_w S$, which is also equal to a minimum. The following four equations can be utilized for stable channel design:

1. Continuity equation

$$Q_w = VA \tag{7.5}$$

where Q_w = water discharge in the channel,
V = channel velocity, and
A = cross-sectional area of flow.

2. Manning's equation

$$V = \frac{1.486}{n}\left(\frac{A}{P}\right)^{2/3} S^{1/2} \tag{7.6}$$

3. A sediment transport function

$$Q_s = f(V, S, A, P, D, d, v, \omega) \tag{7.7}$$

where Q_s = sediment load,
P = wetted perimeter,
D = water depth,
d = sediment particle diameter,
v = fluid viscosity (primarily a function of water temperature), and
ω = particle fall velocity (primarily a function of the diameter, shape, and density of the sediment particle and of the fluid viscosity).

The chosen sediment transport equation needs to be verified or calibrated with measured hydraulic and sediment load data.

4. Minimum unit stream power

$$VS = \text{minimum} \tag{7.8}$$

The known terms are Q_w, Q_s, n, d, v, ω. The unknown terms are V, S, A, and P and can be solved by solution Equations (7.5) through (7.8). The bankfull channel width, computed from these four equations, should be checked for reasonableness with one of the regime equations presented in Table 7.1. If the computed slope is steeper than the valley slope, then the solution must be constrained by the valley slope. If the computed slope is less than the valley slope, then the sinuosity and the degree of meandering can be computed by comparison with the valley slope. The meander wavelength can best be determined from an empirical equation as a function of the bankfull channel width such as Equation (7.9) (Leopold, 1995):

$$\lambda = 10\, W^{1.03} \tag{7.9}$$

where λ = meander wavelength, in feet, and
W = bankfull channel width, in feet.

7.3.6.10 Evaluation of Potential Contaminants

In certain instances, evaluation of potential contaminants attached to river sediments may also be required for a restoration project. This analysis is often done where mining or other historical human activities may result in uncontrolled release of contaminants into a river system. Evaluation of contaminants is of particular interest in dealing with restoration projects where old mine tailings or sediments trapped in a reservoir may be released as a result of the project, but would otherwise remain in place. Typically, sediments are tested for levels of contaminants above the natural background levels present in the river system. Samples can be collected for testing from the reservoir using a variety of methods, including drilling from barges, using divers, or performing a reservoir drawdown to access exposed sediments on the margin of the reservoir.

7.3.7 Biologic Function and Habitat

In addition to the focus on the physical characteristics of a river system, discussed in the previous sections, the impact of physical characteristics on biologic function should also be addressed in the plan or design. The biologic community is directly and indirectly dependent on the physical characteristics of a channel, and alterations to the existing condition can result in biologic consequences. This is not to imply that physical change to the river brings only negative impacts. For example, natural disturbances such as flooding are sometimes required to renew biological function and process. Physical change to the river system by man, if thoroughly considered and in accord with the natural sciences, can also promote biologic function and benefits.

Since the design or management of physical features in a river can impact vegetation, macro-invertebrates, fish, and wildlife, biologic impacts should always be considered. The statement "if you build it, they will come" can be applied to channel restoration efforts. The physical conditions constructed are the invitation to colonization by the biologic communities associated with that niche. If the desired conditions are not correctly diagnosed and replicated, biological communities can also disappear at a rapid rate. In addition to analyzing the geomorphology, hydrology, hydraulics, and sediment transport of a river system, consider the target communities or species in the project area and determine how physical features can be optimized to promote desirable habitat. Consideration should also be given to the competition among species. For example, backwater habitats may be of more benefit to some fish species than others, and the establishments of backwater habitats may help some species to the ultimate detriment of others.

Depending on the project, variables such as slope, bankfull discharge, or channel width may be fixed, but these values can often be adjusted to promote environmental qualities, since geomorphic variables usually function effectively within a range of values. When the variable is open to determination based on geomorphology, hydraulics, and sediment transport, an analysis of biologic benefit should also aid the determination. For this reason, the river restoration study team should be interdisciplinary, with biologists as key members of the team.

7.4 Sediment Restoration Options

Choosing a restoration option involves not only an understanding of existing river processes, but also a determination of what the desired outcome of the project will be. The following discussions present some ideas for determining which restoration options will be considered for a project based on the goals and objectives of the project, the range of feasible options available for implementation, and some examples of options that have been utilized on other rivers.

7.4.1 Goals and Objectives

Restoration goals and objectives are most useful when they are expressed in terms of the final desired outcome, rather than in terms of a possible solution to achieve that outcome. For example, a restoration goal of significantly improving spawning habitat for a certain species of fish is more helpful than a goal of achieving a certain minimum flow. The minimum flow might help achieve the larger goal, but it may not be the right flow or make enough of a difference by itself. A restoration goal of allowing a meandering river channel to migrate across its flood plain, at some natural rate, is more helpful than prescribing a single meander amplitude for all locations.

Restoration projects should be consistent with natural processes in order to be sustainable. For example, streambank stabilization could be considered a restoration project if bank erosion were caused by some human disturbance, such as the redirection of flow velocities from an eroding bank by some upstream structure. However, river channel migration is a natural process and streambank erosion is common on the outside of meander bends. In some cases, the rates of channel migration have been accelerated by human disturbance, such as the clearing of riparian vegetation. Bank stabilization in this case would be consistent with natural processes if the stabilization project were designed to slow the rates of channel migration back to more natural conditions. The complete prevention of channel migration across the flood plain is not consistent with natural processes and would likely incur long-term maintenance costs.

Restoration projects often work best when they can treat the disturbance at the source. In some cases, the source can be fairly local, but in other cases the source of the disturbance may be far upstream. If the disturbance cannot be treated at its source due to legal, institutional, or economic reasons, then a restoration goal might be to help the river achieve a new equilibrium.

Technical understanding of the causes of the disturbance and the possible range of solutions can significantly alter the restoration goals and objectives. During the early 1980's, the general understanding amongst resource managers at Grand Canyon National Park was that sandbars along the Colorado River were destined to erode over time. Therefore, the management goal was to focus on slowing the rate of eventual erosion. After additional study (U.S. Department of the Interior, 1995), it was determined that the objectives of the restoration plan could be expanded from slowing the rate of erosion, to the sustainable maintenance of the sandbars along the river through altered flow management. A similar expansion of the objective, based on technical investigations, occurred in conjunction with the Platte River in Nebraska. Resource managers from the U.S. Fish and Wildlife Service were concerned about narrowing of the Platte River channel, and the initial objectives focused on restoration programs for slowing the rate of narrowing. Further studies (Murphy and Randle, 2003) indicated that an objective of some channel widening was possible with a concerted restoration plan for flow and land management actions.

7.4.2 Fully Assess the Range of Options

The following topics are proposed for consideration when formulating conceptual alternatives at the onset of a restoration project. Later in the restoration project analysis, these topics can help to produce a wider range of viable options or aid in identifying nonfeasible solutions.

7.4.2.1 Sediment and Flow

Both sediment and flow should be considered when searching for feasible solutions. There is a greater general awareness of the direct impacts that result to the channel, both physically and environmentally, from the manipulation of flows. What are less frequently taken into consideration are channel changes resulting from disturbances to the sediment transport regimen. A good restoration alternative should address both concerns. Stable channels can be planned using the procedures presented under the section 7.3.6.9.

7.4.2.2 Local Versus System-wide

An important step in the preliminary stages of a restoration project is to identify whether the problem is local or system wide, and match the solution to the scope of the problem. River processes, including sediment transport, are in general system-wide. However, many problems are treated locally, and the treatment may or may not be effective in fixing the problem.

For example, a local problem may be a short reach of riverbank where erosion has been induced as a result of a bridge located just downstream of the riverbank. It may be desired to limit additional bank erosion in order to control the alignment of the river as it approaches the bridge. In order to treat this problem, the first step would be to verify that the erosion is indeed localized and due to the bridge, rather than a reaction to a system-wide impact. Then, an effective solution can be generated to limit additional erosion of the bank, as long as potential impacts from implementing the solution are understood. For instance, the solution to the local bank erosion problem should not induce bank erosion on the opposite riverbank.

On the other hand, a bank erosion problem in a river tributary, resulting from bed degradation, could be a system-wide problem if the degradation resulted from a migrating head cut triggered by base level lowering in the main channel. A grade control structure in tandem with bank protection, both local solutions, may be sufficient for this system-wide problem. However, a full assessment would be needed to determine the causative factors for the head cut, anticipate the measure of future degradation, and incorporate these conclusions into the design of the grade control structure.

Streambed and bank erosion resulting from a reduction in sediment supply, or an increase in clear water discharge, from the upstream watershed is also a system-wide problem. In this case, a localized solution, the installation of bank protection, would be ineffectual and probably require costly maintenance. One preferred option might be an action plan that re-establishes the transport of sediment in the channel through changes in watershed management.

Increased sediment loads, causing system-wide problems, can result from sources such as bank erosion, a rejuvenated headcut on a tributary stream, or land clearing and development. Sedimentation problems can also be caused by the diversion of water from the stream without the corresponding diversion of sediment. Solutions can be local or system wide. The eroding high bank or rejuvenated tributary headcut could be stabilized locally. Depending on the problem, a system-wide solution might call for vegetated buffer areas along streambanks in a watershed to reduce the sediment supply to downstream reaches. When possible, annual discharges of high flows for a short duration, which do not produce negative flood impacts, can help move sediment from an aggrading reach. Potential solutions in urban areas might call for legislation requiring sediment traps on storm water systems or revised maintenance programs for winter road sanding operations.

7.4.2.3 Natural Versus Restrained Systems

Although the goal may be to fully or partially restore natural processes, restoration projects are often faced with the question of whether to incorporate features that limit channel migration and bank erosion to protect infrastructure and property. A restrained system is defined as a river with bank segments that are fixed to an existing position through the use of erosion-resistant materials, such as riprap. In cases where bank erosion is occurring, this is often perceived initially as the favored option to limit further bank erosion. However, when reviewed over the long term, hardening banks may be the less desirable option because it can generate negative consequences on river processes. For example, when banks are hardened on low elevation surfaces within the channel migration zone; this can limit natural channel migration and development of side channels.

When implementing restoration projects, it is important to assess whether long-term interests are best served by solutions that restrict the channel and flood plain, or by solutions that allow for a river to function in a more natural dynamic equilibrium. When viewed through a short window of time, restraining the flood plain or hardening the riverbanks may appear to be a good option. However, this option can be costly when maintenance and environmental costs are projected over the long term. Increases in the percentage of restrained bank tend to lead to a larger and greater complexity of problems for the river system over the long term. Such problems can include the failure of bank protection, channel instability on the opposite riverbank or the downstream river channel, and the limitation of natural channel migration.

Before accepting a restraining solution, determine if there is a feasible and cost-effective solution that allows the channel to migrate within the natural channel migration zone. The range of alternative solutions is as broad as the range of problems and could include levee setbacks, new flood plain zoning, watershed-wide management approaches, road or structure relocations, natural channel relocations, flow-diversion structures, property acquisition, or construction of longer bridge spans across the river channel.

7.4.2.4 Monitoring Versus Modification

Not all problems require immediate action. In some situations, a monitoring program may be a feasible and effective alternative to action-based solutions. Geomorphic assessments, risk analysis, probability studies, and cost-benefit ratios may indicate that a monitoring program, combined with simple actions only required during or after high flows, could be the preferred alternative. Adaptive management is a more sophisticated version of the monitoring alternative and is addressed in section 7.4.6.

Some general rules for implementing an effective monitoring program are listed below:

- Design the program with specific goals and objectives in mind.
- Make predictions about the parameters that are to be monitored.
- Limit the monitoring to repeatable procedures.
- Use quantitative measures of such things as bank erosion, channel width, bank height, and bed-material grain size.
- Take photographs and document the locations where the photographs are taken.
- Be realistic about the windows of time when field visits are practical and productive. For example, manual data collection during peak flow events is difficult to do safely.
- Design and incorporate adequate analysis time and labor costs with a framework for triggering decisions and needed actions.

7.4.3 Restoration Treatments

The previously discussed conceptual model and analysis tools provide a basis for making decisions on how to restore natural channel processes as part of accomplishing restoration project objectives and goals. Some common strategies that can be implemented and their relationship with restoration of sediment processes are discussed below.

7.4.3.1 Restoration of the Historic Channel Migration Zone

As discussed previously, the historic channel migration zone boundary has the potential to expand in the future where the boundary is composed of erodible material and is subjected to lateral erosion from the river during floods. In undeveloped areas, the rate of this expansion can be limited by old growth trees on terrace surfaces that protect the bank when the roots reach the interface between the bank and the water surface. In areas impacted by human development, clearing of the vegetation adjacent to the historic channel migration zone can accelerate the rate at which the historic channel migration zone boundary expands if the boundary is composed of erodible material such as alluvium or glacial outwash. Also, logging of the riparian flood plain within the historic channel migration zone can accelerate the rate of channel migration and erosion. Vegetation on bars normally dissipates energy (and water velocities) between the main channel and the boundaries of the historic channel migration zone. When it is removed, the rates of channel migration can be accelerated. Potential restoration strategies to minimize future erosion where infrastructure or property needs to be protected could include deflecting the river away from the bank, setting back the infrastructure farther away from the historic channel migration zone boundary, placing engineered bank protection, or a combination of these options. Because logging and development often result in the highest potential future rates of lateral erosion, strategies to limit erosion may also include long-term land use alternatives, such as revegetating the terrace surface and preventing development within or adjacent to the historic channel migration zone.

7.4.3.2 Levee Setback and Removal

The setback or removal of levees is often one of the most effective ways to restore flood plain access in rivers. Flood protection levees cut off access to the flood plain resulting in a higher peak flow, river stage, and velocities, and increased sediment and woody debris transport rates in the main channel. Rivers confined by levees often experience erosion along the channel bottom and a coarsening of sediment sizes present on the channel bed. If there is no property or infrastructure at risk, resource managers may decide to completely remove a levee to restore the flood plain. If some level of flood control must be maintained, the levee may be set back a certain distance from the active channel. When considering a levee setback or removal, the existing sediment processes in the active channel must also be assessed to determine if any modifications need to be incorporated.

For example, on the Dungeness River in Washington, levees exist along both sides of the river for several miles near the mouth. With the higher velocities and river stage created by the levees, one would expect erosion and armoring of the river channel. However, in some places the levees cut off or constrict nearly the entire flood plain, while in other places the levees were set back in the flood plain a distance of two to three times the active channel width. Similar to a bridge constriction, where the levees create a constriction, they locally increase the flow velocities and create backwater areas upstream where sediment deposits during floods. When the elevations of the existing channel bed were compared to the 1930's conditions, up to 10 feet of sediment deposition was documented in the backwater areas upstream from the levee constrictions (Bountry et al., 2002). The sediment deposition was so great that the channel bottom is now higher than the surrounding flood plains. If these levees were set back, the sediment deposition upstream from the levee constrictions must be excavated to maintain the active channel in its historic location. Otherwise, the river channel would immediately avulse (abruptly change course) across the flood plain and onto surfaces where the active river has not flowed for thousands of years.

7.4.3.3 Roadway Setback

When roads exist along river channels in the flood plain zone, the roads can be at risk for erosion from flooding due to the natural tendencies of the river to laterally migrate. In response to this action, roadways are often protected from erosion by using bank protection, including rock, woody debris, concrete, or other manmade devices, and by building the road at a higher elevation than the surrounding ground surface. An alternative in areas where the river is known to run alongside the road is to set back the road and restore the natural riverbank. Ideally, a road should be set back as far away from the river and flood plain boundaries as possible, preferably on a higher, older surface that is not likely be subjected to erosion in the near future. However, placing the road onto a steep hillside may actually initiate slope failures and cause additional sediment erosion issues. The benefit of road setback versus long-term bank protection maintenance costs and impacts on river processes at the existing site must be evaluated on a site-by-site basis.

7.4.3.4 Lengthening Bridge Spans

A bridge may be located at a natural geologic constriction and have a very limited impact on channel processes. However, in many cases, bridge spans constrict the natural channel width, causing a local increase in velocity, river stage, and sediment transport through the bridge and a backwater effect and depositional area upstream of the bridge. The impacts from a particular bridge typically do not expand upstream and downstream more than several channel widths. Additional bank protection in the vicinity of the bridge or other structures, such as levees, will increase the extent of effects on channel processes. When bridges are constricting the river channel and causing localized undesired impacts, a restoration option is to lengthen the bridge span and reduce the number and width of piers in the wetted channel. An analysis should be done to determine how bridge lengthening will meet project goals and what the effects on hydraulic and sediment processes will be.

Questions that should be addressed regarding modifications to bridges include:

- Is the existing bridge located at a natural geologic constriction or does the existing bridge cut off historical channel paths or flood plain?
- What will be the new slope of the river channel through the reach?
- How will sediment transport through the reach change?
- Does the bridge deck need to be raised because of expected deposition or will sediment transport through the reach remain high?
- If flooding is currently an issue, will the bridge lengthening alone be sufficient enough to reduce upstream river stage to an acceptable level during high flows?
- Does there need to be any channel modification or design work done in coordination with a bridge expansion project?

7.4.3.5 Side Channel, Vegetation, and Woody Debris Recovery

Side channels are an important resource in a natural river system for providing aquatic habitat. It may also be desired to restore side channel habitat as part of the restoration of flood plain or historic channel migration zone areas. For instance, consider a levee or road embankment that has been constructed such that it cuts off historical side channel connections with the main channel. If the restoration project is to set back the levee or road, side channels could be constructed in the setback area to speed-up the recovery of natural processes. As part of the recovery process, woody debris could also be incorporated if it was traditionally part of the natural system but was currently absent due to higher transport capacities from the constricted levee reach or historical removal.

Side channels are old channel paths of the main river channel and may one day contain the main channel again. Therefore, restoration designs of side channels are more likely to be successful if they can follow the historic paths of the main channel. Historic aerial photography, maps, and surveys of the river corridor, prior to disturbance, can provide an excellent guide for side channel design. Also, geomorphic field mapping of these historic channel paths likely will be necessary. Large woody debris may need to be placed at the upstream entrance to side channels to prevent the main channel from easily overtaking the side channel.

Recent research has shown that vegetation and woody debris can be a critical component in maintaining a natural rate of channel migration by stabilizing banks within and on the boundaries of the channel migration zone (Beeson and Doyle, 1995; Collins and Montgomery, 2002). When vegetation has been cleared on terrace surfaces adjacent to the channel migration zone, the terrace banks often experience high rates of erosion during floods, putting infrastructure and property at risk. When vegetation and woody debris have been cleared within the channel migration zone, the channel may become unstable and migration rates accelerated. Vegetation recovery efforts can restore the natural roughness in the channel and flood plain. This re-establishes slow velocity and energy dissipation in areas that would otherwise be subject to high velocity and larger volumes of runoff during floods. Recovery of native vegetation is often more aesthetically pleasing and considered a more natural solution to erosion and instability. While vegetation recovery can take longer than many other restoration strategies to be fully implemented, it can be combined with other restoration options to maintain more stability over the long term.

7.4.3.6 Changes to Channel Cross Section or Sizing

Deep, narrow channels provide efficient flow conveyance systems, but they are subject to higher erosive forces and often require hard channel linings. In these cases, sediment transport capacity may exceed the supply of sediment. In populated areas, channels are often confined and made deeper to enhance floodflow conveyance. To reverse that process in a restoration project, the hard lining of channel banks can be removed to allow the channel to adjust to new conditions over time, or a new channel cross section could be constructed to speed up the process.

A common natural channel configuration is a meandering thalweg path (deepest point in the channel) within the channel banks. The channel banks are not necessarily stable, but often migrate slowly over time within the flood plain. This configuration can be described as a staged design, including the low-flow channel (smallest) which is contained within the larger bankfull flow channel, which is, in turn, contained within the larger flood plain. As flows increase the wetted width increases and water overtops banks from one channel into the next larger channel. This staged configuration allows flows to spread out during floods and keeps flow depths and erosive forces in the main (low flow) channel at minimal levels. This also keeps smaller flows concentrated within the low-flow channel to maintain efficient sediment transport and biologic benefit. The natural channel will evolve to this configuration in the absence of geologic or anthropomorphic constraints until natural vegetation in the overbank areas is sufficient to withstand the erosive forces and protect the overbank from erosion.

If the channel is constructed with a simple, over-wide trapezoidal geometry, sediment transport capacity may be less than the incoming supply of sediment. Under this condition, sediment will settle out in the areas of lowest flow velocities, and, over time, will create the complex geometry of a low-flow channel, effective flow channel, and overbank terraces.

The complex channel configuration does not require costly bank protection and, for biologic benefit, offers a wider range of habitat niches over a greater spatial area. It also helps to promote the balance of sediment supply and transport. A drawback to this treatment is that a large amount of land is required to develop the active stream corridor. This land may serve dual uses, but the uses must be flood tolerant and adaptable to occasional channel migration. For instance, using the land for agricultural purposes may have a low risk of flood damage, and the land could actually benefit from periodic inundation. On the other hand, it would be a high risk to build residences in this area that could be damaged or put people in danger during floods.

7.4.3.7 Changes to Channel and Flood Plain Roughness

The roughness of the channel or flood plain has a strong influence on flow conveyance capacity and sediment transport. The management of roughness characteristics within the flow corridor can serve as effective restoration techniques. Grasses and other vegetation, which bend over easily during high flows, offer lower flow resistance, provide greater flow conveyance, and may be subject to higher erosive forces. Structures and substantial growths of vegetation on the flood plain, including trees and stiffer shrubs, increase roughness, decrease flow conveyance, and reduce the average erosive force acting on the flood plain. Vegetation on channel banks increases roughness, reduces velocity and shear stresses along the bank, and can be used at some locations to encourage sediment deposition for bank stabilization. Native riparian species of vegetation may be planted in the flood plain to restore the natural roughness. This vegetation may require some protection from people, floods, and grazing animals until it becomes established.

7.4.3.8 Bank Stabilization Concepts

Bank stabilization has historically been constructed on a reactionary basis in response to lateral erosion of banks from floods. There are several manuals that discuss streambank stabilization methods, such as the *WES Stream Investigation and Streambank Stabilization Handbook* (Biedenharn, 1997). An important point brought up in the manual is that "too often river engineers and scientists may be pressured by circumstances beyond their control to plan and construct riverbank stabilization works too quickly, without adequate time or resources for a conceptual evaluation of the problem." (Biedenharn, 1997). By following the steps outlined in this chapter, it is hoped that a better understanding can be gained of how proposed bank stabilization may or may not work with natural processes prior to installation.

The impact of bank stabilization on natural processes depends on the location of the bank relative to the channel migration zone boundaries, the rate of natural channel migration, and the material that composes the bank stabilization. Traditional bank stabilization uses hard, angular rock that protects against the high velocities and shear stress against the bank.

Stabilizing banks within the channel migration zone (using any type of material) has the biggest impact on altering natural channel dynamics and, where possible, should be avoided in restoration projects. Stabilizing terrace banks along the boundaries of the historic channel migration zone would have minimal impact on channel dynamics. Softer types of bank protection may be considered to better mimic the natural rates of channel migration. Softer bank protection can incorporate revegetation, root wads, geotextiles, and a number of other combinations that create a deformable rough surface and dissipate energy.

Another solution to bank stabilization may be to design structures that are tied into the bank and extend into the river channel to deflect the thalweg and high velocities away from the bank. Bend-way weirs and engineered logjams are two examples that are becoming more popular (Derrick, 1998; Abbe, 2000). These types of structures not only provide protection to banks, but structures using native materials are also aesthetically pleasing and develop aquatic habitat by forming cover and scour pools utilized by fish.

7.4.3.9 Grade Control Structures

Grade control features can be placed across the river channel to prevent head-cutting in degrading channels or to build up the bed of an incised stream to match an upstream riverbed elevation (Federal Interagency Stream Restoration Working Group, 1998). They work by providing a "hard point" in the riverbed that resists erosion forces and reduces the upstream energy slope. These types of structures work well in areas where the structure can be keyed into the banks and the channel cannot outflank the structure. Originally, grade control structures were built in a straight line perpendicular to the riverflow. Recent variations have incorporated "V" and "W" shaped structures to not only control the grade, but to direct the thalweg of the riverbed toward the center of the stream. This variation helps redirect erosive stream forces away from the banks of the river. The structures are typically composed of rock, steel sheetpile, wood, or a combination of these materials. Fish passage and aquatic habitat can be integrated into the structure if designed and sited properly. If improperly designed, the structure can become a fish passage barrier, can result in significant downstream scouring, or can cause upstream meandering that will result in the river outflanking the structure.

7.4.3.10 New Channel Design and Relocations

In the last few decades, there has been a growing awareness of the impacts from historic approaches to river-engineering problems undertaken in the early to mid-1900's. Historically, management goals were focused on increasing navigation and flood conveyance, providing access across rivers, and providing protection to adjacent development and infrastructure from flooding. Potential long-term impacts to river processes from these activities were not generally considered or understood. In many areas, these activities resulted in extensive environmental impacts. In recent decades, society began to place more value on the natural environment. Because of past impacts and society's changing values, restoration approaches that work with natural river processes and enhance or restore habitat have become part of the management objective. In highly disturbed rivers, these approaches may still require substantial construction work in the river corridor, but the outcome is a channel that more closely mimics and works with natural processes.

Natural channel construction and relocation projects are gaining wider acceptance as our understanding of the river as a multi-purpose system continues to grow. Advancements in the fields of geomorphology, sediment transport, botany, fisheries, stream aquatics, and water quality have pointed to the importance of environmental factors and their interrelations. Although natural channel relocations or construction can offer many environmental benefits, their design is complex and should not be attempted based simply upon hydraulic designs for conveyance channels. The following list of design manual references for natural channel construction is offered to encourage thorough designs. The publishing dates of these references illustrate how recent this science and design methodology is:

- Biedenharn, Elliot, and Watson (1997)
- Copeland, et al.(2001)
- Federal Interagency Stream Restoration Working Group (1998)
- Inter-Fluve, Inc. (1998, 1999)

One shortcoming still found in some of the manuals listed above is the overreliance on "hard" protection in the upper banks of rivers. Materials such as riprap, concrete blocks, and other non-natural materials restrain the channel bank and prevent future lateral adjustments. Hard banks are a legacy from hydraulic conveyance channel design techniques of an earlier period. Stream corridors, natural channel geometry, and the informed use of vegetated bank protection techniques are more challenging to incorporate into the design, but they frequently offer the greatest environmental benefit and are most feasible over the long-term. A second common shortcoming to be avoided is to attempt natural channel construction without sufficient analysis and design of the system.

The complete design of a natural channel is beyond the scope of this manual, however the most important sediment transport consideration for a stable channel is "sediment in should equal sediment out" of the design reach. Yang's theory of minimum total stream power and unit stream power (presented in section 7.3.6.9) provides guidance on this rule. In many instances, this consideration will form the basis for the channel design.

7.4.3.11 Special Flow Releases From Dams

Downstream from dams, river channels are subject to a variety of impacts (Collier et al., 2000). In some cases, clear-water releases subject the downstream river channel to degradation, armoring, bank erosion, and increased sinuosity. In other cases, so much water is diverted from the downstream river channel that sediment from downstream tributaries aggrades the river channel, decreases sinuosity, and increases the rates of lateral migration.

One potential restoration tool for these downstream river channels is the scheduling of short-duration, high magnitude flows from the dam for a variety of purposes. For example, an experimental "beach-habitat-building flow" was released from Glen Canyon Dam during 7 days in March and April 1996 to rebuild sandbars along the Colorado River in the Grand Canyon (Webb et al, 1999). The beach-habitat-building flow increased riverflow from a normal fluctuating range between 5,000 and 20,000 ft^3/s to a steady 45,000 ft^3/s for 7 days. Low steady flows of 8,000 ft^3/s were released from the dam for 3 days prior to, and just after, the high steady flow, so that measurements could be easily conducted both before and after the experiment. Measurements were also conducted during the high steady flow. The experimental beach-habitat-building flow was highly successful in rebuilding sandbars throughout the river corridor in Grand Canyon, and about half the volume of sand deposited during this experiment still remained after 2 years.

For the Trinity River in California, short-duration, high steady flows have been proposed to temporarily mobilize the gravel-bed sediments and flush fine sediments from the gravels. Although upstream reservoirs trap nearly all of the upstream sediment supply, the reservoirs also reduce the annual flood peak, and fine sediment supplied from downstream tributaries has deposited on river gravels. The deposition of fine sediment on the river gravels reduces their suitability as fish spawning areas.

Short-duration, higher magnitude flows have been proposed for the Platte River in Nebraska to temporarily mobilize the riverbed to scour seedling vegetation and build sandbars for nesting birds (Murphy and Randle, 2003). Endangered birds that use the Platte River require wide, shallow, active channels, with wide open views that are unobstructed by vegetation. Annual peak flows have substantially reduced over the 20th century, in response to water resource development, and active channel widths and the associated open view have also decreased. Restoration plans call for mechanically clearing mature vegetation from selected areas and using short-duration, high releases from an upstream reservoir to annually scour seedlings.

Special flow releases from dams might also be used to increase the sediment transport capacity of the downstream river channel if sediment aggradation is a problem. Minimum releases to the downstream channel and limits on the rate of flow change may also have benefits for the aquatic ecosystem. For example, minimum flows will ensure some minimum level of aquatic habitat.

Limits on the rate that flow is decreased may help prevent the stranding of fish and seepage-based bank erosion. Limits on the rate that flow is increased may prevent injury to people that may be in the downstream river channel.

7.4.4 Biologic Function and Habitat

When using restoration treatments, consider their effect on the biologic function and habitats of the study reach. Such considerations include the channel cross-section shape, channel banks, channel planform characteristics, changes in channel grade, and flow and sediment designs.

7.4.4.1 Channel Cross-Section Shape

One of the most fruitful examples of habitat enhancement is the conversion of a conveyance channel to a natural channel. The geometric shape of a channel cross section heavily influences on biologic function. A conveyance channel is a highly efficient structure for transporting flow, yet it provides virtually no habitat that promotes biologic diversity. If the channel is constructed or managed with an emphasis on morphologic processes, more diverse and sustainable habitat for the biologic community can be generated.

A conveyance channel is normally trapezoidal, with a small width to depth ratio and hardened, steep sidewalls. Velocities are relatively high and consistent in the cross section and the profile, and most flows are contained within the channel. Steep walled conveyance channels are not easily accessible to the terrestrial biologic community, and the hard linings, required to protect against erosion, substantially reduce the development of riparian and fishery habitat. In northern Colorado and at other locations, deer attempting to reach water in concrete lined canals occasionally fall in and become trapped by the steep concrete walls. Atlantic salmon in Lake Michigan are hampered in their migration up concrete lined streams by the lack of slow velocity refuge areas. In contrast, the irregular bed of a natural channel offers resting pools and eddies where the salmon can shelter between bursts of greater swimming effort. In concrete lined channels, only the largest and strongest of the species can negotiate extended lengths of smooth channel, and even minor velocity variations at joints between the poured concrete are utilized for the passage.

In addition to an irregular bed, a natural channel provides refuge areas in a variety of flows by incorporating a low-flow channel, larger bankfull channel, and flood plain areas. The terrain and soil conditions are varied; moisture conditions and vegetation change with respect to proximity to the low-flow channel, and velocity and erosive conditions within the channel are lower and more variable in cross-section and profile. All aspects of a natural channel lead to a greater variety of niches for biologic communities. In particular, the streambanks support riparian vegetation that serves as a border between a terrestrial and aquatic habitat. This is classified as an "edge" habitat that sustains a particularly diverse range of wildlife.

7.4.4.2 Channel Banks

In areas with cohesive soils, steep and undercut banks in the outer bends of rivers offer multiple benefits, including habitat for fish, thermal cooling, and more vegetative material entering the system for macroinvertebrates. Fish lunkers, more commonly seen in the Midwest, are employed to mimic this beneficial habitat. Fish lunkers are cells constructed of heavy wooden planks and blocks, which are embedded into the toe of the streambanks at the channel bed level (Federal Interagency Stream Restoration Working Group, 1998). These structures provide shaded aquatic habitat and help prevent streambank erosion. In small stream systems, coarse or irregular stream beds and banks help to trap floating vegetation, thereby facilitating partial decomposition by macroinvertebrates which, in turn, fuels the stream ecosystem. Logjams and brush piles along the banks offer similar benefits. In addition, logjams and brush piles provide backwater and flow velocity diversity, which creates additional habitat niches. A movement in the mid-1900's encouraged the removal of all woody debris and jams from rivers for the sake of increased flow conveyance. However, natural logjams can still be found in many areas of the country, notably in the Pacific Northwest, where the construction of logjams for habitat generation and bank protection is also gaining support. In general, the number of fish a stream can support is a function of the available habitat produced by banks, logs, brush, boulders, and pools.

7.4.4.3 Channel Planform Characteristics

Wide, shallow channels provide habitat for a different biologic community than a narrow and deep channel configuration. Braided channels with higher sediment loads may be desirable for specific species. On other projects, backwaters and large pools could be the goals. In general, pools are desirable fish habitat and aid fish survival during periods of drought or extreme temperatures. A deeper pool offers more fish habitat and cooler water temperatures during hotter summer months. Therefore, the number and size of fish often increase with pool depth. Shallow flow depths have the potential to increase water temperatures and expose more fish to predators. Both results could be considered detrimental or beneficial, depending on the biologic community to be served. Interconnected lakes serve as desirable rearing habitat for sockeye salmon and attract migratory wildfowl. In the Rocky Mountain region, overbank areas that are periodically flooded provide riparian and wetland habitat, while steeper draining banks may be beneficial as upland browse for deer and elk. Along the Platte River in Nebraska, wide shallow channels without vegetated river islands are desirable habitat for migratory birds. In some cases, there are conflicting habitat needs among various species, especially when non-native species are present. Therefore, restoration planners may have to choose how to balance the competing habitat requirements.

7.4.4.4 Changes in Channel Grade

Profile breaks or grade controls can be barriers to fish passage and a negative consequence for desirable migratory salmonid fish in the Northeast, Northwest, and Alaska. However, a grade break can also be beneficial. An abrupt change in the channel grade can help to separate native and introduced fish species, such as the brown trout and rainbow trout in Colorado. The native brown trout remain only in the furthest upstream reaches of steep gradient streams, while the introduced rainbow have dominated and eliminated natives in the majority of downstream reaches. The configuration of the grade break needed to serve as a barrier is species dependent. An adult king salmon can negotiate grade breaks several feet in elevation, while smaller pink salmon cannot.

Grade breaks can also be beneficial in warm shallow systems by boosting depleted oxygen levels, providing mixing for water quality. In urban or park settings, grade breaks can increase audible and aesthetic values near pedestrian walkways. They can also provide areas of varying velocity that provide additional habitat diversity.

The construction of engineered riffles can be an effective means of providing fish and boat passage where a grade control structure or dam is needed. The engineered riffle is effective because it provides fish passage over the entire width of the channel. The slope of the riffle is generally less than 2 percent, but the required value is dependent on the fish species. For boat passage, special consideration is needed to provide navigable depths and the avoidance of strong hydraulic jumps.

7.4.4.5 Flow and Sediment Designs

Physical conditions that impact biologic function include not only geometric parameters, but also flow and sediment conditions. Options in flow regulation could also significantly impact species. For example, after the construction of storage reservoirs in the Platte River basin, peak riverflows along the Platte River in Nebraska are now about the same magnitude, whether they occur from the spring snowmelt in the Rocky Mountains or from summer thunderstorms in the Nebraska plains (Murphy and Randle, 2003). Therefore, sandbars that are deposited by the river during the reduced spring flows tend to be lower in elevation and are more easily inundated during the summer thunderstorms. Birds that attempt to nest on these sandbars in late spring may have their nests inundated by summer thunderstorms. Endangered birds along the Platte River, including the piping plover and least tern, may benefit from a special high-flow dam release that is higher than the typical peak flow from a summer thunderstorm. If such a special flow release results in the deposition of higher-elevation sandbars, the endangered birds may be able to build nests that are not inundated by summer thunderstorms.

Channels with a high sediment load tend to have a less diverse range of biologic communities, but they often contain members that have uniquely evolved to thrive in this environment, such as the humpback chub and pike minnow in the Colorado River or the pallid sturgeon in the Platte River. Higher sediment loads reduce sunlight penetration and inhibit the growth of aquatic macrophytes. Increased sediment can impact salmonid populations by smothering eggs laid in redds. Deposition of finer sediment over gravels can also decrease the production of aquatic insects. In addition, high turbidity can make it difficult for some fish species to find food.

Sediment loads can be used to promote bank stabilization through the construction or enhancement of bank configurations that produce eddies and encourage sediment deposition.

Deposition along banks produces riparian vegetation that serves to stabilize banks directly through root growth and indirectly by creating increased roughness along the banks that reduces flow velocities and shear forces.

Temperature can also play a role. Very cold temperatures reduce fish activity and feeding, while warm temperatures can reduce oxygen, impact biological oxygen demand (BOD) and chemical oxygen demand (COD) levels, and instigate fish kills.

7.4.5 Watershed Level Restoration

Many watersheds have had significant levels of logging, road building, and development that affect the volumes and delivery rates of water and sediment to the active channel. Restoration options involved in watershed scale recovery projects can be much more complicated and costly than projects within the river corridor because major changes in land use management may be needed. Where human-induced impacts on the watershed have created landslides, gully erosion, and mass wasting, it may be very difficult to restore hillslope features. Nonetheless, land treatment can have significant benefits to help restore watershed level natural processes. Treatments should be considered that minimize runoff and erosion from flood plains in agricultural areas, minimize bank erosion from grazing practices, and minimize water and sediment delivery as a result of unstable clear-cut or road failure areas along the hillside of the watershed.

As an example, several old logging roads in the Pacific Northwest are used solely as recreational access roads. Inadequate drainage across these roads often leads to hillslope failures and mass wasting. It is difficult to determine the exact quantity of additional sediment and water that is delivered to the river corridor from these hillslope failures. However, in watershed areas such as the Hoh River in Washington, the frequency of hillslope failures in logged areas has increased by nearly 200 percent (Lyon, 2003). The drainage systems that allow runoff to cross under the roads need to be improved through the installation of larger culverts and bridges to minimize the water and sediment delivery as a result of these failure areas. Also, the roads can be decommissioned, modified into foot trails, or hillslope stabilization techniques can be implemented.

7.4.6 Uncertainty and Adaptive Management

Although every restoration project has some degree of uncertainty that success will be achieved, the uncertainty for some projects may be significant. For these projects, an adaptive management program can be implemented to improve the chances for success. Policy agreements between resource managers and landowners should be developed to work with the restoration goals and objectives. Additionally, the adaptive management program should explicitly define how success of the project will be measured. Monitoring, research, and the willingness and ability to take corrective action are other key components to an adaptive management program. The following steps can lead to a successful adaptive management program:

1. Define the restoration goals and objectives and how success will be measured.

2. Formulate a set of hypotheses that explain how a restoration action will achieve a desired result. Even if some or all of the hypotheses are proven false, the information gained will ultimately help the restoration process.

3. Begin implementation of a restoration action, with monitoring designed to test the set of hypotheses. This is different than simply monitoring a list of resource parameters to test for possible trends. When monitoring is designed to test hypotheses, the causes and effects become better understood and at a much faster pace.

4. If the restoration action is not achieving the desired outcome, conduct research to discover why. Monitoring alone may not explain why a hypothesis is false, nor provide insight as to what new or modified management actions may be necessary to achieve the desired outcome.

5. Formulate a new set of hypotheses to explain how a new or modified restoration action will achieve the desired results.

6. Repeat steps 3 through 5 until the restoration success is achieved. An adaptive management program will be successful if resource managers are willing and able to implement corrective actions. Resource managers should not be expected to blindly accept recommendations for new or modified actions. However, objective presentations of study results should help resource managers to make informed decisions.

The actual implementation of an adaptive management program can be difficult if there are significantly different policy views among the many different resource management agencies and interested parties. However, the involvement of all parties is usually necessary for a successful adaptive management program, especially over the long term.

7.5 Summary

The restoration of river channels and flood plains is often complex, and every project has its own unique set of characteristics. River restoration typically requires a multi-disciplinary approach. The ideas presented in this chapter are intended to describe the sediment management aspects of river restoration. The development of a conceptual model of the physical processes and the linkages to biological processes is a good place to start. The conceptual model will help determine the data analysis needs and the feasible range of restoration options. Restoration projects will be most successful over the long term when they are designed to be consistent with natural processes and the concepts of dynamic equilibrium are incorporated. The degree of uncertainty about project performance can be reduced by the implementation of adaptive management, which incorporates monitoring, research, and the willingness to take new management actions.

7.6 References

Abbe, T.B. (2000). *Patterns, Mechanics and Geomorphic Effects of Wood Debris Accumulations in a Forest River System*, Ph.D. thesis, University of Washington, Department of Geological Sciences.

Andrews, E.D. (1980). "Effective and Bankfull Discharges of Streams in the Yampa River Basin, Colorado and Wyoming," *Journal of Hydrology*, vol. 46, pp. 37-330.

Bauer, T.R. and P.W. Makar (2003). *Regression relationships of width and discharge on the middle Rio Grande*. draft report, Sedimentation and River Hydraulics Group, Technical Service Center, Bureau of Reclamation, Denver, Colorado.

Beeson, C.E., and P.F Doyle. (December 1995). "Comparison of Bank Erosion at Vegetated and Non-vegetated Channel Bends," vol. 31, no. 6, *Water Resources Bulletin*, American Water Resources Association, pp. 983-990.

Best, J.L., and C.S. Bristow, eds. (1993). *Braided Rivers*, The Geological Society, London, UK.

Bettess, R., and W.R. White (1987). "Extremal Hypotheses Applied to River Regime," in *Gravel Bed Rivers,* John Wiley & Sons, New York.

Bevenger, G.S., and R.M. King, (May 1995). *A Pebble Count Procedure for Assessing Watershed Cumulative Effects.* Research Paper RM-RP-319, U.S. Department of Agriculture, Rocky Mountain Forest and Range Experiment Station, Fort Collins, Colorado.

Biedenharn, D.S., C.M. Elliot, and C.C. Watson (1997). *The WES Stream Investigation and Streambank Stabilization Handbook,* Waterways Experiment Station for EPA, U.S. Army Corps of Engineers, Vicksburg, Mississippi.

Biedenharn, D.S., R. Copeland, C.R. Thorne, P.J. Soar, R.D. Hey, and C.C. Watson (2000). *Effective discharge calculation: A practical guide*, U.S. Army Corps of Engineers, Engineer Research and Development Center, Coastal and Hydraulics Laboratory RECD/CHL Report TR-00-15, Vicksburg, Mississippi.

Birkeland, P. (1999). *Soils and Geomorphology*, Oxford University Press, New York.

Blench, T. (1957). *Regime Behavior of Canals and Rivers,* Butterworths Scientific Publications, London.

Bountry, J.A., T.J. Randle, L.A. Piety, and R.A. Link (May 2002). *Physical Processes, Human Impacts, and Restoration Issues of the Lower Dungeness River, Clallam County, Washington,* Technical Service Center, Denver, Colorado, and Pacific Northwest Regional Office, Boise, Idaho, Department of the Interior, Bureau of Reclamation.

Bountry, J.A., and T. Randle (March 2001). *Upstream Impacts after the 1976 Failure of Teton Dam.* presented at 7th Annual Federal Interagency Sedimentation Conference, Reno, Nevada.

Bray, D.I. (1982). "Regime Equations for Gravel Bed Rivers," in *Gravel-Bed Rivers*, John Wiley & Sons, New York, pp. 109-137.

Bridge, J.S. (1993). "The interaction between channel geometry, water flow, sediment transport and deposition in braided rivers" in *Braided Rivers*, J.L. Best and C.S. Bristow, eds., The Geological Society, London, UK., pp. 13-71.

Bunte, K., and S.R. Abt (2001). *Sampling surface and subsurface particle-size distributions in wadable gravel- and cobble-bed streams for analyses in sediment transport, hydraulics, and streambed monitoring*, General Technical Report RMRS-GTR-74. U.S. Department of Agriculture, Forest Service, Rocky Mountain Research Station. Fort Collins, Colorado

Bureau of Reclamation (November 1998). *Pilgrim Creek Maintenance Program, Jackson Lake Dam, Minidoka Project, Wyoming/Idaho, Technical Findings Report, Appendices*, U.S. Department of the Interior, Bureau of Reclamation, Boise, Idaho, pp. 3-1 to 3-10.

Bureau of Reclamation (September 2001). *Snake River at Fort Hall, Idaho Bank Erosion Study*, U.S. Department of the Interior, Bureau of Reclamation, Boise, Idaho.

Chang, H.H. (1988). *Fluvial Processes in River Engineering*, John Wiley & Sons, New York.

Cohn, T.A. (1995). "Recent advances in statistical methods for the estimation of sediment and nutrient transport in rivers," *Reviews of Geophysics*, Supplement, American Geophysical Union, pp. 717-723.

Colby, B.R. and C.H. Hambree (1955). "Computations of Total Sediment Discharge, Niobrara River near Cody, Nebraska," *U.S. Geological Survey Water Supply Paper No. 1357*, 1955.

Collier, M., R.H. Webb and J.C. Schmidt (1996). "Dams and Rivers, A Primer on the Downstream Effects of Dams," *U. S. Geological Survey Circular 726*, Reston, Virginia.

Collins, B.D., and D.R Montgomery (2002). "Forest Development, Wood Jams, and Restoration of Flood plain Rivers in the Puget Lowland, Washington," *Restoration Ecology*, vol. 10 no.2, pp.237-247.

Compton, R.R. (1962). *Manual of Field Ge*ology. John Wiley & Sons, New York, New York.

Copeland, R.R., D.N. McComas, C.R. Thorne, P.J. Soar, M.M. Jonas, and J.R. Fripp (2001). *Hydraulic Design of Stream Restoration Projects.* ERDC/CHL TR-01-28, U.S. Army Corps of Engineers, Vicksburg, Mississippi.

Danish Hydraulic Institute, Inc. (2003). *MIKE 11 Hydrodynamic Module Users Guide and Reference Manual - USA*, Eight Neshaminy Interplex, Suite 219, Trevose, Pennsylvania 19053

Derrick, D.L. (August 1998). "Four Years Later, Harland Creek Bendway Weir/Willow Post Bank Stabilization Demonstration Project," *Proceedings of the International Water Resources Engineering Conference*, vol. 1, pp.411-416.

Dodds, P.S., and D.H. Rothman (2000). "Scaling, universality, and geomorphology," *Annual Review of Earth and Planetary Sciences,* vol. 28.

Edwards, T.K., and G.D. Glysson (1988). "Field Methods for Measurement of Fluvial Sediment," *Techniques of Water Resources Investigations, Book 3, Chapter C2*, U.S. Geological Survey, Reston, Virginia, 89 p.

Federal Geographic Data Committee (1988). FGDC-STD-007.3-1998, "Geospatial Positioning Accuracy Standards," *Part 3: National Standard for Spatial Data Accuracy*, http://fgdc.er.usgs.gov/fgdc.html.

Federal Interagency Stream Restoration Working Group (1998). *Stream Corridor Restoration Handbook: Principles, Processes and Practices*, Natural Resource Conservation Service, Washington, DC.

Fonda, R.W. (1974). "Forest Succession in Relation to River Terrace Development in Olympic National Park, Washington," *Ecology* 55: pp. 927-942.

Gilroy, E.J., R.M. Hirsch, and T.A. Cohn (1990). "Mean Square Error of Regression-Based Constituent Transport Estimates," *Water Resources Research*, vol. 29, pp. 2069-2077.

Hey, R.D., and C.R. Thorne (1986). "Stable Channels with Mobile Gravel Beds," *Journal of Hydraulic Engineering*, American Society of Civil Engineers, vol. 72.

Ikeda, S, and G. Parker eds. (1989). *River Meandering,* Water Resources Monograph 12, American Geophysical Union, Washington D.C.

Inter-Fluve, Inc. (1998, 1999). *Design of Natural Stream Channels*. Short course handbook, Bozeman, Montana and Hood River, Oregon.

Julien, P.Y. (1995). *Erosion and Sedimentation*, Cambridge University Press, New York.

Julien, P.Y., and J. Wargadalam (1995). "Alluvial Channel Geometry: Theory and Application," *Journal of Hydraulic Engineering*, American Society of Civil Engineers, vol. 121.

Julien, P.Y. (2002). *River Mechanics*, Cambridge University Press, New York.

Kellerhals, R. (1967). "Stable Channels with Gravel Paved Beds," *Journal of the Waterways and Harbors Division*, American Society of Civil Engineers, vol. 93, pp. 63-84.

Knighton, D. (1998). *Fluvial Forms and Processes*, Arnold, London, England.

Lacey, G. (1929). "Stable Channels in Alluvium," *Proceedings of the Institute of Civil Engineers*, London, vol. 229, pp. 259-292.

Land Management Information Center, Minnesota Planning (October 1999). "Using the National Standard for Spatial Data Accuracy to measure and report geographic data quality," *Positional Accuracy Handbook*, <http://www.lmic.state.mn.us>.

Lane, E.W. (1955). "Design of stable channels," *Transactions of the American Society of Civil Engineers,* 120, p 1234-1260.

Leopold, L.B., and T. Maddock (1953). "The Hydraulic Geometry of Stream Channels and Some Physiographic Implications," *U.S. Geological Survey Professional Paper 252.*

Leopold, L.B. and M.G. Wolman (1957). "River channel patterns – braided, meandering, and straight," *U.S. Geological Survey Professional Paper 282A.*

Leopold, L.B., M.G. Wolman, and J.P. Miller. (1964). *Fluvial Processes in Geomorphology*, W.H. Freeman and Company, San Francisco, California.

Leopold, L.B. (1994). *A View of the River,* Harvard University Press, Cambridge, Massachusetts.

Lyon, E. (2003). Mass Wasting in the Upper Hoh River Watershed, Olympic National Park, Washington: Interpreted from 1939 and 2001 Aerial Photography, United States Department of the Interior, Bureau of Reclamation, Pacific Northwest Regional Office, Boise, Idaho.

MacDonald, L.H. (1991). Monitoring guidelines to evaluate effects of forestry activities on streams in the Pacific Northwest and Alaska. Environmental Protection Agency, Region 10, Water Division, Seattle, Washington.

Miller, C.R. (1951). *Analysis of Flow-Duration, Sediment-Rating Curve Method of Computing Sediment Yield*, Sedimentation and River Hydraulics Group, Technical Services Center, Bureau of Reclamation, Denver, Colorado.

Montgomery, D.R., and J.M. Buffington (1998). "Channel Processes, Classification, and Response," in *River Ecology and Management: Lessons from the Pacific Coastal Ecoregion*, Springer-Verlag, New York.

Murphy, P.J., and T.J. Randle (2003). *Platte River Sediment Transport and Riparian Vegetation Model, Technical Report*, U.S. Department of the Interior, Bureau of Reclamation, Denver, Colorado.

Murphy, P.J., and T.J. Randle (2003). *Platte River History and Restoration*, U.S. Department of the Interior, Bureau of Reclamation, Denver, Colorado.

Office of Management and Budget (2002). OMB Circular A-16, *Coordination of Surveying, Mapping, and Related Spatial Data Activities.* , Executive Office of the President, Office of Management and Budget, Washington, DC.

Osterkamp, W.R., E.J. Lane, and G.R. Foster (1983). "An analytical treatment of channel morphology relations," *U. S. Geological Survey Professional Paper 1288.*

Palmer, W.C. (1965). *Meteorological drought*, U.S. Weather Bureau, Research Paper no. 45.

Parker, G. (1978). "Self formed rivers with equilibrium banks and mobile bed, Part 1: the sand silt river and Part 2: the gravel river," *Journal of Fluid Mechanics*, vol. 89, pp. 109-148.

Parker, G. (1979). "Hydraulic Geometry of Active Gravel Rivers," *Journal of Hydraulic Engineering*, American Society of Civil Engineers, vol. 105, pp. 785-1201.

Pemberton, E.L., and J.M. Lara (January 1984). *Computing Degradation and Local Scour,* Technical Guideline for Bureau of Reclamation, Denver, Colorado.

Piety, L., J.A. Bountry, T.J. Randle, J.F. England, and E. Lyon (March 2003). *Geomorphic Reach Analysis of The Hoh River: Interim Report for River Miles 20.8 to 25.5 (Morgans Crossing Reach 7),* prepared for Jefferson County Department of Public Works, Port Townsend, Washington, by Technical Service Center, Denver, Colorado, and Pacific Northwest Regional Office, Boise, Idaho, Department of the Interior, Bureau of Reclamation.

Randle, T., J. Bountry, R. Klinger, A. Lockhart (May 2000). *Geomorphology and River Hydraulics of the Teton River Upstream of Teton Dam, Teton River, Idaho*, Bureau of Reclamation, Technical Service Center, Denver, Colorado.

Rosgen, D.L. (1996). *Applied River Morphology*, Wildland Hydrology, Pagosa Springs, Colorado.

Schumm, S.A, (1985). "Patterns of alluvial rivers," *Annual Review of Earth and Planetary Sciences* vol., 13. pp.5-27.

Schumm, S.A., J.E. Dumont, and J.M. Holbrook (2000). *Active Tectonics and Alluvial Rivers.* Cambridge University Press, New York, NY.

Simon, A., and C.R. Thorne (1996). "Channel adjustment of an unstable coarse-grained stream: Opposing trends of boundary and critical shear stress, and the applicability of extremal hypotheses," *Earth Surface Processes and Landforms*, vol 21, pp.155-180.

Simons, D.B. and M.L. Albertson (1963). "Uniform Water Conveyance Channels in Alluvial Materials," *Transactions of the American Society of Civil Engineers,* vol. 128, pp. 65-107.

Simons, D.B., and F. Sentürk (1992). *Sediment Transport Technology*, Water Resources Publications, Fort Collins, Colorado.

Soar, P.J., and C.R. Thorne (2001). *Channel Restoration Design for Meandering Rivers*, Coastal and Hydraulics Laboratory, U.S. Army Engineer Research and Development Center, ERDC/CHL CR-01-1.

Strand, R.I., and E.L. Pemberton (1982). *Reservoir Sedimentation,* Technical Guideline for Bureau of Reclamation, U.S. Department of the Interior, Bureau of Reclamation, Denver, Colorado.

Swanson, F.J., and G.W. Lienkaemper (1982). *Interactions among fluvial processes, forest vegetation, and aquatic ecosystems, South Fork Hoh River, Olympic National Park* U.S. Department of Agriculture, Forest Service, Pacific Northwest Forest and Range Experiment Station, Forestry Sciences Laboratory, Corvallis, Oregon.

Thorne, C.R. (2002). "Geomorphic analysis of large alluvial rivers," *Geomorphology* vol 44, pp 203-219.

Thorne, C.R. (1998). *Stream Reconnaissance Handbook,* John Wiley & Sons, New York.

U.S. Army Corps of Engineers - Hydrologic Engineering Center (August 1993). *HEC-6 Scour and deposition in rivers and reservoirs*, Users' Manual, Version 4.1.

U.S. Army Corps of Engineers (1999). *National Inventory of Dams, Water Control Infrastructure*, <http://crunch.tec.army.mil/nid/webpages/nid.cfm>.

U.S. Army Corps of Engineers - Hydrologic Engineering Center (November 2002). *HEC-RAS River Analysis System, Hydraulic Reference Manual, Version 3.1.*

U.S. Department of the Interior (1995). *Operation of Glen Canyon Dam, Final Environmental Impact Statement*, Bureau of Reclamation, Denver, Colorado.

Vanoni, V. ed. (1975). *Sedimentation Engineering*, American Society of Civil Engineers Manual no. 54, Reston, Virginia.

Webb, R.H., J.C. Schmidt, G.R. Marzolf, and R.A. Valdez (1999). *The Controlled Flood in Grand Canyon*, American Geophysical Union, Washington, DC., 367 p.

Williams, G.P. (1978) "Hydraulic Geometry of River Cross Sections – Theory of Minimum Variance." *U.S. Geological Survey Professional Paper 1029.*

Williams, G.P., and M.G. Wolman (1984). "Downstream Effects of Dams on Alluvial Rivers." *U.S. Geological Survey Professional Paper 1286.*

Yang, C.T. (1973). "Incipient Motion and Sediment Transport," *Journal of the Hydraulics Division*, ASCE, vol. 99, no. HY10, pp. 1679-1704.

Yang, C.T. (1976). "Minimum Unit Stream Power and Fluvial Hydraulics," *Journal of the Hydraulics Division*, ASCE, vol. 102, no. HY7, pp. 919-934.

Yang, C.T. (1979). "Unit Stream Power Equations for Total Load," *Journal of Hydrology*, vol. 40, pp. 123-138.

Yang, C.T., and C.C. Song (1979). "Theory of Minimum Rate of Energy Dissipation," *Journal of the Hydraulics Division*, ASCE, Vol. 105, No. HY7, Proceeding Paper 14677, pp. 769-784.

Yang, C.T., and C.C. Song (1984). "Theory of Minimum Energy and Energy Dissipation Rate," *Encyclopedia of Fluid Mechanics*, vol. 1, N.P. Cheremisinoff, ed, Gulf Publishing Company, Chapter 11.

Yang, C.T. (1986). "Dynamic Adjustments of Rivers," *Proceedings of the 3rd International Symposium on River Sedimentation*, Jackson, Mississippi, pp. 118-132.

Yang, C.T. (1996). *Sediment Transport: Theory and Practice*, McGraw-Hill Companies, Inc. (reprint by Krieger Publishing Company, 2003).

Yang, C.T., and F.J.M. Simões (2000). User's Manual for GSTARS 2.1 (Generalized Sediment Transport model for Alluvial River Simulation version 2.1). Bureau of Reclamation, Technical Service Center, Denver, Colorado.

Yang, C.T., and F.J.M. Simões (2002). *User's Manual for GSTARS3 (Generalized Sediment Transport model for Alluvial River Simulation version 3.0)*, Bureau of Reclamation, Technical Service Center, Denver, Colorado, in preparation.

Yang, C.T., J. Huang and B.P. Greimann (2004, 2005). *User's Manual for GSTAR-1D (Generalized Sediment Transport model for Alluvial River Simulation – One Dimensional),* Bureau of Reclamation, Technical Service Center, Denver, Colorado, in preparation.

Chapter 8
Dam Decommissioning and Sediment Management

Page

8.1 Introduction 8-1
8.2 Scope of Sediment Management Problems 8-2
8.3 Engineering Considerations of Dam Decommissioning 8-6
8.4 Sediment Management Alternatives 8-7
8.4.1 Integration of Dam Decommissioning and Sediment Management Alternatives 8-7
8.4.2 No Action Alternative 8-9
8.4.3 River Erosion Alternative 8-10
8.4.3.1 River Erosion Description 8-10
8.4.3.2 River Erosion Effects 8-12
8.4.3.3 Monitoring and Adaptive Management 8-13
8.4.4 Mechanical Removal Alternative 8-14
8.4.4.1 Sediment Removal Methods 8-15
8.4.4.2 Sediment Conveyance Methods 8-16
8.4.4.3 Long-Term Disposal 8-17
8.4.5 Stabilization Alternative 8-17
8.4.6 Comparison of Alternatives 8-19
8.5 Analysis Methods for River Erosion Alternative 8-20
8.5.1 Reservoir Erosion 8-21
8.5.1.1 Analytical Methods for Estimating Reservoir Erosion 8-23
8.5.1.2 Numerical Models 8-24
8.5.2 Downstream Impacts 8-26
8.5.2.1 Analytical Methods for Predicting Deposition Impacts 8-26
8.5.2.2 Numerical Modeling of Sediment Impacts 8-30
8.6 Summary 8-31
8.7 References 8-33

Chapter 8
Dam Decommissioning and Sediment Management

by

Timothy J. Randle and Blair Greimann

8.1 Introduction

This chapter will briefly discuss the engineering considerations associated with dam removal and then present the basic types of sediment management alternatives. Next, the chapter will discuss the potential impacts associated with dam decommissioning, data collection, analyses of the potential impacts, and case studies.

Over 76,000 dams (that are at least 6 feet in height) exist in the United States today, and they serve many different purposes. These purposes include water supply for irrigation, municipal, industrial, and fire protection needs; flood control; navigation; recreation; hydroelectricity; water power; river diversion; sediment and debris control; and waste disposal (Heinz Center, 2002 and American Society of Civil Engineers (ASCE), 1997). While the great majority of these dams still provide a vital function to society, some of these dams may need to be decommissioned for various reasons including:

- Economics
- Dam safety and security
- Legal and financial liability
- Ecosystem restoration (including fish passage improvement)
- Site restoration
- Recreation

Some dams no longer serve the purpose for which they were constructed. When a dam has significantly deteriorated, the costs of repair may exceed the expected benefits, and dam removal may be a less expensive alternative. For example, if a hydroelectric plant is old, the present operation and maintenance costs may exceed the project benefits. Also, the plant modernization costs may exceed the expected benefits, and decommissioning the hydroelectric plant may be a less expensive alternative. If the spillway of a dam needs to be enlarged, the costs may exceed the project benefits and dam removal may be a less expensive alternative. If fish cannot adequately pass upstream of the dam and reservoir, the cost of adequate fish passage facilities might exceed the project benefits and dam removal may be a less expensive alternative. Some dams and reservoirs may inundate important cultural or historic properties, and dam removal may restore those properties. Along some rivers, the demand for white-water recreation might be a compelling reason to remove a dam.

Three recent publications provide information on the overall considerations related to dam decommissioning and removal. The American Society of Civil Engineers (ASCE, 1997) publication describes the decisionmaking process, available alternatives, and the important considerations related to dam decommissioning and removal. The publication by the H. John Heinz III Center for Science, Economics, and the Environment (Heinz Center, 2002) summarizes the state of scientific knowledge related to dam removal and provides recommendations for additional research. The Aspen Institute (2002) "recommends that the option of dam removal be included in policy and decision making that affects U.S. dams and rivers."

This chapter of the *Erosion and Sedimentation Manual* focuses on the sediment management aspects of dam removal and avoids the discussion of the legal and institutional issues. This chapter also briefly describes the linkages between sediment management, dam removal engineering, and the effects on the aquatic ecosystem.

8.2 Scope of Sediment Management Problems

Rainfall runoff, snowmelt, and river channel erosion provide a continuous supply of sediment that is hydraulically transported and deposited in reservoirs and lakes (see Chapter 2, "Erosion and Reservoir Sedimentation" and Chapter 6 "Sustainable Development and Use of Reservoirs"). Because of the very low velocities in reservoirs, they tend to be very efficient sediment traps. Reservoir sediment disposal (through mechanical methods) can be very costly for large volumes of sediment. Therefore, the management of reservoir sediment is often an important and controlling issue related to dam removal (ASCE, 1997). The sediment erosion, transport, and deposition are likely to be among the most important physical effects of dam removal (Heinz Center, 2002).

The sediment related impacts associated with dam decommissioning could occur in the reservoir and in the river channel, both upstream and downstream from the reservoir. Depending on the local conditions and the decommissioning alternative, the degree of impact can range from very small to very large. For example, the removal of a small diversion dam that had trapped only a small amount of sediment would not have much impact on the downstream river channel. If only the powerplant of a dam were decommissioned, then sediment-related impacts would be very small. The top portion of a dam might be removed in such a way that very little of the existing reservoir sediment would be released into the downstream river channel. In this case, the impacts to the downstream river channel might be related only to the future passage of sediment from the upstream river channel through the reservoir. If dam removal resulted in a large quantity of sediment being released into the downstream river channel, then the impacts to both the upstream and downstream channels could be significant.

The extent of the sediment management problem can be estimated from the following five indicators:

1. The reservoir storage capacity (at the normal pool elevation) relative to the mean annual volume of riverflow.

2. The purposes for which the dam was constructed and how the reservoir has been operated (e.g., normally full, frequently drawn down, or normally empty).

3. The reservoir sediment volume relative to the mean annual capacity of the river to transport sediment of the same particle sizes within the reservoir.

4. The maximum width of the reservoir relative to the active channel width of the upstream river channel in an alluvial reach of river.

5. The concentration of contaminants present within the reservoir sediments relative to the background concentrations.

The first two of these indicators help to describe how much sediment could potentially be stored within the reservoir. The next three indicators (3, 4 and 5) help to scale the amount of reservoir sediment, and its quality, to the river system on which the reservoir is located.

The relative size of the reservoir (ratio of the normal reservoir capacity to mean annual flow volume) can be used as an index to estimate the reservoir sediment trap efficiency. The greater the relative size of the reservoir, the greater the sediment trap efficiency and the amount of reservoir sedimentation. The sediment trap efficiency primarily depends on the sediment particle fall velocity and the rate of waterflow through the reservoir (Strand and Pemberton, 1982). For a given reservoir storage capacity, the sediment trap efficiency would tend to be greater for a deeper reservoir, especially if riverflows pass over the crest of the dam. Brune (1953) developed an empirical relationship for estimating the long-term reservoir trap efficiency, based on the correlation between the relative reservoir size and the trap efficiency observed in Tennessee Valley Authority reservoirs in the southeastern United States. Using this relationship, reservoirs with the capacity to store more than 10 percent of the average annual inflow would be expected to trap between 75 and 100 percent of the inflowing sediment. Reservoirs with the capacity to store 1 percent of the average annual inflow would be expected to trap between 30 and 55 percent of the inflowing sediment. When the reservoir storage capacity is less than 0.1 percent of the average annual inflow, the sediment trap efficiency would be nearly zero.

The purpose for which a dam was constructed, along with legal constraints and hydrology, determines how the reservoir pool is operated. The operation of the reservoir pool will influence the sediment trap efficiency and the spatial distribution and unit weight of sediments that deposit within the reservoir. The reservoir trap efficiency of a given reservoir will be greatest if substantial portions of the inflows are stored during floods when the sediment concentrations are highest. If the reservoir is normally kept full (run of the river operation), floodflows would be passed through the reservoir and trap efficiency would be less. Coarse sediments would deposit as a delta at the far upstream end of the reservoir. When reservoirs are frequently drawn down, a portion of the reservoir sediments will be eroded and transported father downstream. Any clay-sized sediment that is exposed above the reservoir level will compact as they dry out (Strand and Pemberton, 1982).

The ratio of reservoir-sediment volume to the annual capacity of the river to transport sediment is a key index. This index can be used to estimate the level of impact that sediment release from a dam removal would have on the downstream river channel. When the reservoir sediment volume is small, relative to the annual sediment transport capacity, then the impact on the downstream channel likely will be small. Reservoirs have a finite capacity to trap and store sediments. Once that capacity is filled with sediment, the entire sediment load supplied by the upstream river channel is passed through the remaining reservoir. For example, the pool behind a diversion dam is typically filled with sediment within the first year or two of operation. Therefore, the relative volume of reservoir sediment may not be large, even if the dam is considered old. When a reservoir has a multiyear, sediment storage volume, the dam removal plan should consider staging dam removal over multiple years to avoid excessive aggradation of the downstream riverbed. The dam removal investigation should determine how much of the reservoir sediment would actually erode from the reservoir.

The width of the reservoir, relative to the width of the active river channel (in an alluvial reach) upstream from the reservoir can indicate how much sediment would be released from the reservoir both during and after dam removal. When a reservoir is many times wider than the river channel, then the river may not be capable of eroding the entire reservoir sediment volume, even long after dam removal (Morris and Fan, 1997 and Randle et al., 1996).

The presence of contaminants in the reservoir sediments, at concentrations significantly higher than background levels, would likely require mechanical removal or stabilization of the reservoir sediments prior to dam removal. Even if contaminants are not present in the reservoir sediments, the turbidity created by sediment erosion during dam removal may impact the aquatic environment of the downstream river channel. Increased turbidity could also be a concern for downstream water users.

As an example, these five indicators were applied to three dams in the Pacific Northwest that are being considered for removal to improve fish passage:

- Gold Hill Dam near Gold Hill, Oregon (Bureau of Reclamation, 2001a)

- Savage Rapids Dam near Grants Pass, Oregon (Bureau of Reclamation, 2001b)

- Glines Canyon Dam near Port Angeles, Washington (Randle et al., 1996)

These three dams range in size from small to large, and their potential effects on sediment management range from negligible to major (see Table 8.1).

The major issues associated with sediment management, related to dam removal, may include cost, water quality, flooding, operation and maintenance of existing infrastructure, cultural resources, the health of fish and wildlife and their habitats (including wetlands), recreation, and restoration of the reservoir area. Sediment management plans are important to prevent the following impacts:

- If a large volume of coarse sediment were eroded too quickly from a reservoir, then the sediment could aggrade the downstream river channel, cause channel widening and bank erosion, increase flood stage, plug water intake structures, and disrupt aquatic habitats.

- If large concentrations of fine sediment were eroded from the reservoir, then turbidity would increase in the downstream river channel and may significantly degrade water quality for the aquatic environment and for water users.

- If the reservoir sediment contains significant concentrations of contaminants, then these contaminants could be potentially released into the aquatic environment and into municipal water treatment plants and wells.

- If the reservoir sediment has to be mechanically removed, disposal sites can be difficult to locate and the sediment removal cost can be the most expensive portion of the dam removal project.

- If a delta is eroded from the upstream end of the reservoir, the erosion of sediment deposits can continue to progress along the upstream river channel. Sediment deposited along the backwater of the reservoir pool will begin to erode once the reservoir pool is drawn down.

Table 8.1. Sample application of reservoir sediment impact indicators to three dams in the Pacific Northwestern United States

Dam Properties	Gold Hill Dam near Gold Hill, Oregon	Savage Rapids Dam near Grants Pass, Oregon	Glines Canyon Dam near Port Angeles, Washington
River name and distance from mouth	Rogue River (river mile 121)	Rogue River (river mile 107.6)	Elwha River (river mile 13.5)
Active river channel width in alluvial reach	150 feet	150 feet	200 feet
Type of dam	Concrete gravity dam	Concrete gravity and multiple arch dam	Concrete arch dam
Hydraulic height	1 to 8 feet	30 to 41 feet	210 feet
Dam crest length	1,000 feet ("L" plan shape)	460 feet	150 feet
Reservoir Properties			
Reservoir length	1 mile	3,000 feet	2.3 miles
Reservoir width	150 to 350 feet	290 to 370 feet	1,000 to 2,000 feet
Reservoir capacity	100 acre-feet	290 acre-feet	40,500 acre-feet
Sediment Management Indicators			
Relative reservoir capacity	0.005 percent	0.01 percent	4.5 percent
Reservoir operations	Run-of-the-river	Reservoir pool raised 11 feet during the summer irrigation season	Run-of-the-river
Relative reservoir sediment volume	Negligible	1-to-2 year supply of sand and gravel	75-year supply of sand and gravel; 54-year supply of silt and clay
Relative reservoir width	2.3 (all sediment would be eroded from the reservoir)	2.5 (nearly all sediment would be eroded from the reservoir)	10 (about one-third of the sediment would be eroded from the reservoir)
Relative concentration of contaminants or metals	Less than background levels	Less than background levels	Only iron and manganese are above background levels
Sediment management problem	Negligible	Moderate	Major

The potential impacts from the erosion, transport, and deposition of reservoir sediment should be at least considered in all dam removal studies. If the impacts could be significant, then a sediment management plan should be developed. With an effective sediment management plan, potential impacts can be substantially reduced or avoided. In some cases, there may be benefits from the controlled release of reservoir sediments such as the introduction of gravel, woody debris, and nutrients for the restoration of downstream fish habitats.

8.3 Engineering Considerations of Dam Decommissioning

Dam decommissioning alternatives might include the discontinued use of a hydroelectric powerplant, partial removal of the dam, or complete removal of the dam and all associated structures (e.g., spillways, outlets, powerplants, switchyards, etc.). Partial removal of a dam could be planned in many different ways to achieve different purposes. For example, the portion of the dam that blocks the river channel and flood plains could be removed, while the abutments and other structures are left in place for historic preservation and to reduce removal costs. Any remaining structures would have to be left in a safe condition and may require periodic maintenance. In the case where a dam spans a valley width that is significantly wider than the river channel, a relatively narrow portion of the dam could be removed so that the remaining dam would help retain a significant portion of the reservoir sediments. A partial dam removal could also mean that the upper portion of the dam is removed, while the lower portion is left in place to retain reservoir sediments deposited below that elevation. This alternative might also help to reduce or eliminate any dam safety concerns by reducing the size of the reservoir, but fish passage facilities might still need to be provided.

The type of material used to construct a dam (concrete, masonry, rockfill, or earth) is important for determining how much of the dam to remove, the volume of material for disposal, and the removal process itself (ASCE, 1997). In addition, there are several other engineering considerations that influence the amount and rate of sediment erosion, transport, and deposition.

The rate of dam removal and reservoir drawdown has a strong influence on the rate that sediments are eroded and transported to the downstream river channel. The effects from releasing a large volume of reservoir sediment into the downstream channel can be reduced by slowing the rate of reservoir drawdown. This might be accomplished by progressively removing layers of the dam over a period of weeks, months, or years, depending on the size of the dam and the volume of the reservoir sediments. The rate of reservoir drawdown needs to be slow enough to avoid a flood wave of reservoir water spilling into the downstream river channel. Also, the rate needs to be slow enough to avoid inducing any potential landslides along the reservoir margins or a slide failure of any earthen dams.

The ability to drawdown the reservoir pool depends on how flows can be released through, over, or around the dam. If the dam has a low-level, high-capacity outlet works or diversion tunnel, the reservoir could be emptied at a prescribed rate and the dam could be removed under dry conditions. However, if the width of the outlet works is narrow relative to the reservoir sediment width, then a substantial portion of the sediments would remain in the reservoir until the dam is removed. A bypass channel could be constructed around the dam, but it would need the ability to at least partially drain the reservoir. For concrete dams, it may be acceptable to release flows over the dam or through notches cut into the dam (ASCE, 1997).

Dam removal and reservoir drawdown plans have to prepare for the possibility of floodflows occurring during dam removal. The occurrence of a flood may simply mean the temporary halt of dam removal and reservoir drawdown activities. However, an overtopping flood could cause a failure of the remaining structure and a downstream flood wave that would be many times larger than the reservoir inflow. If the remaining structure can withstand overtopping flows, then floods may help to erode and redistribute delta sediments throughout the reservoir. In a wide reservoir, a floodflow may help to leave the reservoir sediment in a more stable condition after dam removal.

8.4 Sediment Management Alternatives

The development of alternative sediment management plans for dam decommissioning requires concurrent consideration of engineering and environmental issues. Sediment management alternatives can be grouped into four general categories (ASCE, 1997):

> No action. Leave the existing reservoir sediments in place. If the reservoir-sediment storage capacity is not already full, then either allow future sedimentation to continue or reduce the sediment trap efficiency to enhance the life of the reservoir.
>
> River erosion. Allow the river to erode sediments from the reservoir through natural processes.
>
> Mechanical removal. Remove sediment from the reservoir by hydraulic or mechanical dredging or conventional excavation for long-term storage at an appropriate disposal site.
>
> Stabilization. Engineer a river channel through or around the reservoir sediments and provide erosion protection to stabilize the reservoir sediments over the long term.

A sediment management plan can also consist of a combination of these categories. For example, fine sediments could be mechanically removed from the downstream portion of the reservoir to reduce the impacts on water quality. At the same time, the river could be allowed to erode coarse sediments from the reservoir delta to resupply gravel for fish spawning in the downstream river channel.

8.4.1 Integration of Dam Decommissioning and Sediment Management Alternatives

The character of the sediment management alternative would depend on the dam decommissioning alternative. For example, the rate of river erosion is directly influenced by the rate of dam removal, and the amount of reservoir sediment eroded by riverflows would increase as more of the dam is removed. The cost of mechanically removing sediment from deep reservoirs (mean depth greater than 10 to 15 feet) would be less if the sediment can be removed as the reservoir is drawn down. The cost and scope of reservoir sediment stabilization would decrease as more of the dam is retained. The matrix of possible combinations of dam decommissioning alternatives and sediment management alternatives is shown in Table 8.2. There will be continual interplay between balancing the scope of the sediment management alternative, the requirements of dam decommissioning, acceptable environmental impacts, and cost. The steps to prepare a sediment management plan are shown in Table 8.3. Each sediment management alternative should include proper mitigation to make the alternative as feasible as possible.

Table 8.2. Relationship between dam decommissioning and sediment management alternatives (modified from ASCE, 1997)

Sediment management alternative	Dam decommissioning alternatives		
	Continued operation	Partial dam removal	Full dam removal
No action	• Reservoir sedimentation continues at existing rates, • Inflowing sediment loads are reduced through watershed conservation practices, or • Reservoir operations are modified to reduce sediment trap efficiency.	• Only applicable if most of the dam is left in place. • The reservoir sediment trap efficiency would be reduced. • Some sediment may be eroded from the reservoir.	• Not applicable.
River erosion	• Sluice gates are installed or modified to flush sediment from the reservoir. • Reservoir drawdown to help flush sediment.	• Partial erosion of sediment from the reservoir into the downstream river channel. • Potential erosion of the remaining sediment by sluicing and reservoir drawdown.	• Erosion of sediment from the reservoir into the downstream river channel. Erosion rates depend on the rate of dam removal and reservoir inflow. The amount of erosion depends on the ratio of reservoir width to river width.
Mechanical removal	• Sediment removed from shallow depths by dredging or by conventional excavation after reservoir drawdown.	• Sediment removed from shallow depths before reservoir drawdown. • Sediment removed from deeper depths during reservoir drawdown.	• Sediment removed from shallow depths before reservoir drawdown. • Sediment removed from deeper depths during reservoir drawdown.
Stabilization	• The sediments are already stable, due to the presence of the dam and reservoir.	• Retain the lower portion of the dam to prevent the release of coarse sediments or retain most of the dam's length across the valley to help stabilize sediments along the reservoir margins. • Construction of a river channel through or around the reservoir sediments.	• Construction of a river channel through or around the existing reservoir sediments. • Relocate a portion of the sediments to areas within the reservoir area that will not be subject to high-velocity riverflow .

Table 8.3. Steps to preparing alternative sediment management plans

1	Examine the possible range of dam decommissioning alternatives (continued operation, partial dam removal, and full dam removal).
2	Determine the reservoir sediment characteristics including volume, spatial distribution, particle-size distribution, unit weight, and chemical composition.
3	Investigate the existing and pre-dam geomorphology of the river channel upstream and downstream of the dam.
4	Inventory the existing infrastructure around the reservoir, along the downstream river channel, and along the upstream portion of the river channel influenced by the reservoir.
5	Determine the feasible range of sediment management alternatives and formulate specific alternatives.
6	Coordinate the details of each sediment management alternative with the other aspects of the dam decommissioning alternative.
7	Conduct an initial assessment of the risks, costs, and environmental impacts for each sediment management alternative.
8	Determine what mitigation measures may be necessary to make each alternative feasible and include these measures in the alternative.
9	Finalize the assessment of the costs, environmental impacts, and risks for each modified sediment management alternative.
10	Document the risks, costs, and environmental impacts of each alternative for consideration with the engineering and environmental components of the study. Provide technical support to the decisionmaking process.

8.4.2 No Action Alternative

Under this alternative, the dam, reservoir, and sediment would be left in place. For most diversion dams and other small structures, the sediment storage capacity of the reservoir pool is already full. In this case, floods, sluicing, and dredging can cause temporary changes in sediment storage, but the inflowing sediments are generally transported through the reservoir pool to the downstream river channel or into a canal. Under these conditions, a decision to leave the dam and reservoir in place will not change the existing impacts caused by the dam and its operation.

If the reservoir sediment storage capacity is not already full, future sedimentation could be allowed to continue or actions could be taken to reduce sedimentation rates and prolong the life of the reservoir. The life of the reservoir may be extended by reducing the upstream sediment loads, bypassing sediment through or around the reservoir, or removing the existing sediment (see Chapter 6 "Sustainable Development and Use of Reservoirs"). If the reservoir continues to trap sediment, the remaining reservoir capacity will eventually be filled with sediment, but this could take decades or centuries to occur, depending on the reservoir size and the upstream sediment loads.

Sediment deposited in the reservoir would have naturally been transported to the downstream river channel. Consequently, the clear-water releases from the reservoir tend to cause erosion of the downstream river channel (see Chapter 7 "River Processes and Restoration"). Continued long-term sedimentation of the reservoir would reduce the project benefits and perhaps even pose a threat to dam safety. Reservoir sedimentation can also cause deposition in the upstream river channel (especially for mild slope rivers) and increase river stage in the backwater reach upstream from the reservoir. Eventually, reservoir sedimentation will cause velocities through the reservoir to increase and subsequently decrease the sediment trap efficiency. Once the reservoir sediment storage capacity is full, the sediment load entering the reservoir would be transported through to the downstream river channel. Once coarse sediment (sand and gravel) passes through the reservoir, any erosion process of the downstream river channel would be reversed and sediment deposition would occur in the previously eroded river channel. Aggradation of the downstream riverbed may eventually increase water surface elevations to pre-dam levels (depending on the existing upstream sediment supply and downstream riverflows). New developments in the pre-dam flood plain may be flooded more frequently. Concentrations of fine sediment may also increase to pre-dam levels, which may affect downstream water users and the aquatic environment. In contrast, some reservoirs store and divert so much water that the downstream river channel aggrades. In this case, continued long-term sedimentation of the reservoir would tend to force more water into the downstream river channel and at least partially reverse the aggradation trend.

8.4.3 River Erosion Alternative

Sediment removal from the reservoir by river erosion can be applied to all dam decommissioning alternatives. River erosion is a frequently employed sediment management practice associated with dam removal of all sizes. In fact, this is the preferred alternative for the removal of the large Elwha and Glines Canyon Dams on the Elwha River in Washington (Olympic National Park, 1996). The reservoirs behind these two dams contain 18 million yd^3 of sediment (Gilbert and Link, 1995).

Allowing reservoir sediments to erode and discharge into the downstream river channel may be the least costly alternative if the downstream impacts can be accepted or mitigated. However, water quality considerations may make this alternative unacceptable if the reservoir sediments contain high concentrations of contaminants or metals. The advantage of the river erosion alternative is that the cost of physically handling the sediments is eliminated. However, these benefits must be weighed against the risks of unexpected riverbed aggradation or unanticipated increases in turbidity downstream.

8.4.3.1 River Erosion Description

In the case of continued dam operation, sluice gates with adequate discharge capacity can be used to initiate and maintain sediment transport through the reservoir. This is normally done in conjunction with reservoir drawdown to increase the flow velocities through the reservoir and increase the sediment transport (Morris and Fan, 1997). For partial dam removal, the amount of reservoir sediment eroded by riverflows would depend on how much of the dam is removed and how much of the reservoir pool is permanently and temporarily drawn down.

For small dams with relatively small reservoirs and sediment volumes (see Section 8.2), the rate of dam removal may not be critical. However, for dams that have relatively large reservoirs or sediment volumes, the rate of final reservoir drawdown (corresponding with dam removal) can be very important. Severe impacts to water quality and flooding can occur if the reservoir drawdown rate is too fast. However, the alternative would take too long to implement and perhaps cost too much if the reservoir drawdown rate were unnecessarily slow. The rate and timing of staged reservoir drawdown should meet the following general criteria:

- The reservoir discharge rate is slow enough that a downstream flood wave does not occur.

- The release of coarse sediment is slow enough so that severe riverbed aggradation does not cause flooding to people and property along the downstream river channel.

- The concentration of fine sediment released downstream is not too great, or its duration too long, so that it would not overwhelm downstream water users or cause unacceptable impacts to the aquatic environment.

These general criteria would need to be specifically defined for each local area. In order to reduce the downstream channel impacts, staged dam removal may need to be implemented over a period of months or years, depending on the size of the reservoir, height of the dam, and the volume of sediment. The structural and hydraulic stability of the partially removed dam must be analyzed at these various stages to ensure adequate safety and to prevent a large and sudden release of water or sediment. With the proper rate of reservoir drawdown, the magnitude of the downstream impacts can be reduced and spread out over time. In some cases, it may be more desirable to have the impacts occur over a shorter period of time, with higher magnitude, than over a longer period of time with lower magnitudes. For example, a shorter duration of high turbidity may affect only 1 or 2 year classes of fish, whereas a longer duration of impact with chronic levels of turbidity may affect multiple year classes of fish.

For reservoirs that are much wider than the upstream river channel, river erosion during dam removal may only result in a portion of the sediment being transported to the downstream river channel. This is because the river will tend to incise a relatively narrow channel through the reservoir sediments. This erosion channel would likely widen over time through channel migration, meandering, and flood plain development, but the entire erosion width may still be less than the initial reservoir sediment width. Also, riparian forests may naturally colonize the remaining sediment terraces and additionally prevent or slow their erosion. Vegetation could also be planted to speed up the natural process and prevent the establishment of non-native species.

Some reservoirs are many times wider than the river channel and have relatively thick delta deposits (more than 10 feet) at the upstream end of the reservoir. In this case, it may be desirable to induce lateral erosion of the delta sediments and redeposition across the receding reservoir. This would result in leaving the remaining delta sediments, as a series of low, stable terraces, rather than one high terrace that is potentially unstable. During a reservoir drawdown increment, the river would incise a relatively narrow channel through the exposed delta. As long as a reservoir pool continues to remain during dam removal, the eroded delta sediments would redeposit as a new delta across the upstream end of the lowered reservoir. As a new delta deposit forms across the receded lake, the erosion channel is forced to move laterally to meet deeper areas of the reservoir. Thus, the sediment erosion width is narrow at the upstream end, but it increases to the reservoir width where the channel enters the receded lake. This can be accomplished by holding the reservoir level at a constant elevation between drawdown increments. The duration of constant reservoir elevation between drawdown increments (a few days to a few weeks) corresponds to the length of time necessary for the river channel to re-deposit the eroding reservoir sediments across the width of the receded reservoir (Randle et al., 1996).

After enough of the dam and reservoir have been removed, the eroding delta sediments will have reached the dam and the reservoir pool will be completely filled in with sediment (Randle et al., 1996). At this critical point in time, further dam removal will result in the downstream release of coarse sediments. Also, the horizontal position of river erosion channel would be relatively fixed where the river channel passes the dam site and subsequent erosion widths through the reservoir sediment would be a function of riverflow and the bed material load.

8.4.3.2 River Erosion Effects

The amount and timing of reservoir sediment release and any resulting downstream impacts to water quality and flooding can be estimated through computer modeling, but thorough knowledge and experience with the model are required. The optimum rate of dam removal, for sediment management purposes, can be determined by modeling a range of dam removal rates.

Any sediment released downstream would deposit somewhere, either because of decreasing river channel slopes downstream or because the river enters a lake, estuary, or ocean. Depositional effects and sediment concentrations in the downstream river channel, lake, estuary, or ocean must be carefully studied to determine if the impacts from the sediment management alternative are acceptable or can be mitigated. Monitoring is essential during reservoir drawdown to verify these predictions and, if necessary, slow the rate of dam removal and reservoir drawdown.

The amount and rate of reservoir sediment that is eroded and released to the downstream river channel affect both short- and long-term impacts, the risk of unintended impacts, and cost. The period of short-term impacts might be considered the period of dam removal plus an additional 3 to 5 years. Over the short term, the release of fine lakebed sediment (silt and clay-sized material) would affect water quality, including suspended sediment concentration and turbidity. The release of coarse sediment (sand, gravel, and cobble-sized sediment) could increase flood stage, the rate of river channel migration, and deposition in a downstream lake or estuary. The release of gravel might improve existing fish spawning habitat. Over the long term, the amount and timing of sediment supplied to the downstream river channel would return to predam conditions. The predam conditions may be close to natural conditions if there are no other dams upstream. However, the presence of upstream dams may still leave the river system in an altered condition.

Floodflows may have different effects on sediment releases, depending on whether they occur during or after dam removal. Dam removal operations may have to be discontinued during floodflows. The temporary halt to dam removal during floods would tend to prevent large increases in the amount of sediment eroded from the reservoir. However, floods that occur immediately after dam removal could erode substantial amounts of reservoir sediment. After the first floodflow, significant channel widening in the former reservoir area would only occur during subsequently higher floodflows. Sediment releases downstream would rapidly decrease over time because higher and higher floodflows would be required to cause additional erosion. The time required to reestablish the natural river channel within the former reservoir area depends on the rate of final reservoir drawdown and future floodflows. If a period of drought occurs just after final reservoir drawdown and dam removal, the last phase of sediment erosion in the reservoir would be delayed. Conversely, if a major flood occurs just after reservoir drawdown and dam removal, large amounts of sediment could be transported downstream over a short period of time.

The short-term impact of full dam removal may be to temporarily aggrade the downstream river channel and increase suspended sediment concentration and turbidity. The long-term impact is to fully restore the upstream sediment supply to the downstream river channel. This may approach predam conditions, depending on the level of development in the upstream watershed.

8.4.3.3 Monitoring and Adaptive Management

For projects where the reservoir sediment volume is significant, monitoring and adaptive management are critical components to the river erosion alternative. The effects of the river erosion alternative should be predicted ahead of time. Monitoring is needed to confirm those predictions. If necessary, corrective actions should be taken before impacts could exceed these predictions. For example, the rate of dam removal could be temporarily slowed or halted to mitigate for unanticipated consequences.

Typically, the objectives of the sediment monitoring plan are to detect and avoid severe impacts related to flooding, erosion of infrastructure, and water quality. In addition, the monitoring program could assess project performance and provide scientific information applicable to other projects. A monitoring program could be designed to provide the following types of information:

- Real-time data on physical processes that would assist project management in decisions regarding the water treatment plant operations, bank erosion protection, flood protection, and the rate and timing of dam removal.

- Long-term data that would both identify and quantify physical processes associated with ecosystem restoration following dam removal.

Monitoring categories may include the following processes:

- Reservoir sediment erosion and redistribution
- Hillslope stability along the reservoir and downstream river channel

- Water quality (including suspended sediment concentration)
- Riverbed aggradation and flood stage along the downstream river channel
- Aquifer characteristics
- River channel planform and channel geometry
- Large woody debris
- Coastal processes including the delta bathymetry and turbidity plume

Not all of these processes may occur (or need to be monitored), and some processes may need detailed monitoring. The key is to determine if any of these processes could cause undesirable consequences and implement a monitoring program to provide early detection. In addition, a monitoring program could be used to assess project performance. The monitoring program could be divided into adaptive management and restoration monitoring categories. The adaptive management monitoring program could provide real-time information directly to project managers, verify or modify dam deconstruction scheduling, and trigger contingency actions required to protect downstream water quality, property, and infrastructure. The restoration monitoring program could provide a body of scientific knowledge applicable to understanding and interpreting natural river restoration processes. Such information could be used to guide management decisions over the long term and would be applicable to future dam removal projects in other locations.

The adaptive management responses could include the following actions:

- Modify monitoring techniques, locations, or frequencies
- Improve water treatment techniques
- Locally mitigate flooding and bank erosion
- Slow rate of dam removal
- Temporarily halt dam removal

The frequency and duration of monitoring activities depend on the local project conditions, including the relative volume of the reservoir sediment, rate of dam removal and time of year, hydrology, and budget. Measurement of initial conditions is necessary to establish a monitoring baseline for comparison. Monitoring should be conducted prior to dam removal, for a period long enough to test monitoring protocols and determine the range of variability in the data. As monitoring continues during dam removal, the results of certain parameters could be used to trigger the monitoring of additional parameters. For example, the monitoring of aggradation in the downstream river channel could be initiated after coarse sediment is transported past the dam site. Monitoring should continue after dam removal until either all of the reservoir sediments have eroded or stabilized in the reservoir and sediment has been flushed from the downstream river channel.

8.4.4 Mechanical Removal Alternative

Under this type of alternative, all or a portion of the reservoir sediment would be removed and transported to a long-term disposal site. This type of sediment management alternative can be used with any decommissioning scenario (continued operation, partial dam removal, or full dam removal). Sediment could be removed by conventional excavation, mechanical dredging, or hydraulic dredging. Transport to a disposal site could be through a slurry pipeline, by truck, or conveyor belt. Long-term disposal sites could include old gravel pits, landfills, or ocean disposal areas.

Mechanical removal would attempt to reduce the downstream concentration of sediment and turbidity by removing sediments from the reservoir before they could erode. This type of alternative is the most conservative and, potentially, the most costly. All costs are up-front construction costs, but the long-term risks would be relatively low (ASCE, 1997). Costs can be reduced by not removing all of the reservoir sediment. For example, only the sediments within the predam flood plain would need to be removed to prevent subsequent river erosion. The remaining portion could be allowed to stabilize within the reservoir. Coarse sediment (that may be present in a reservoir delta) could be allowed to erode downstream if it is considered to be a resource necessary to restore river gradient or spawning gravels for fish habitats. The coarse sediments, especially gravel, would likely be transported as bed load and would not increase turbidity as much as fine sediments (clay, silt, and fine sand). The three components of the mechanical removal alternative include: (1) sediment removal methods, (2) conveyance methods, and (3) long-term disposal methods.

8.4.4.1 Sediment Removal Methods

Several methods are available for removing the sediment. The main criteria for selecting a removal method are the size and quantity of sediments and whether the sediment would be removed under wet or dry conditions (ASCE, 1997). An overview of each method follows.

- Conventional excavation requires lowering of the reservoir or rerouting of the river so that sediment excavation and removal can be accomplished in dry conditions. After sediment has become dry enough to support conventional excavating equipment, the sediment can be excavated (by dozers and front-end loaders) and hauled (by truck) to an appropriate disposal site. The viability of this approach depends upon the facilities available, sediment volume, the amount of time required to dry the sediment, and the haul distance to the disposal site. If the sediment volume is small, and the sediments are not hazardous, this disposal process can be done economically. At a shallow 10-acre reservoir in northeastern Illinois, approximately 15,000 cubic yards of "special waste" sediment were removed and disposed of at a nearby landfill for total cost of $350,000 in 1989. The unit cost was about $25 per cubic yard.

- Mechanical dredging is performed using a clamshell or dragline, without dewatering the site, but it still requires that the excavated material be dewatered prior to truck transport to the disposal facility. Costs to dredge some 35,000 cubic yards of sediment from behind a low-head dam in northeastern Illinois were also estimated at $25 per cubic yard in 1987.

- Hydraulic dredging is often the preferred approach to removing large amounts of sediment, particularly if the sediments are fine-grained, because they are removed under water. The sediments are removed as a slurry of approximately 15 to 20 percent solids, by weight. Hydraulic dredging is normally conducted from a barge and can access most shallow areas of the reservoir. Dredging could begin in the shallow areas of the reservoir (5 to 30 feet) and continue to deeper areas as the reservoir is drawn down. If delta sediments are to be left to river erosion, dredges working from barges could pick up lakebed sediments immediately downstream from the eroding delta front. Submersible dredges could also be used to dredge deep areas of the reservoir before drawdown. Woody debris or tree stumps may prevent the removal of sediment from the lowest layer of the reservoir bottom. Design considerations would include volume and composition of material to be dredged, reservoir water depth, dredge capacity, and distance to and size of the disposal facility. For a 180-acre lake in central Illinois, 280,000 cubic yards were hydraulically dredged and disposed of at a facility constructed on the owner's adjacent property for a total cost of $900,000 in 1989, with a unit cost of approximately $3 per cubic yard.

8.4.4.2 Sediment Conveyance Methods

Some example methods of conveyance include transport through a sediment slurry pipeline, by truck, and by conveyor belt. A sediment slurry pipeline can be an efficient and cost-effective means of conveying sediment over long distances, especially under gravity-flow conditions. Conveyer belts may be efficient over short distances. Trucking is a conventional method that is often the most expensive because of the large quantities involved.

In the case of a sediment-slurry pipeline, the route and distance to the disposal site are an important design consideration. An alignment along the downstream river channel may allow for gravity flow and avoid pumping costs. However, construction in canyon reaches could be difficult and the pipeline would have to be protected from riverflows. The pipeline could be buried or secured above ground with lateral supports. These supports might consist of large concrete blocks or rock anchors. If gravity flow were not possible, then a pumping plant would be needed. Booster pumps also may be needed for slurry pipelines of long distance. The pipeline and any pumping stations could be removed after the sediments had been dredged from the reservoir.

A certain amount of water would be required to operate the slurry pipeline (80 to 85 percent water, by weight) and this amount would reduce downstream riverflows. If water is scarce, then the slurry pipeline operation may have to be temporarily curtailed or discontinued during low-flow periods to maintain minimum flows for the downstream water users and the aquatic environment.

Silt- and clay-sized sediments are expected to easily flow by gravity through the sediment slurry pipeline. However, sand-sized and larger sediment may abrade or clog the pipeline. Therefore, a settling basin or separator may be needed to prevent sand and coarser material from entering the slurry pipeline. The coarse sediment that is excluded could be discharged back into the reservoir or transported to the disposal site by conveyor belt or truck.

8.4.4.3 Long-Term Disposal

Disposal sites may include such places as old gravel pits, landfills, or ocean disposal areas. Distance from the reservoir is an important parameter in the selection of a disposal site, since conveyance costs increase with increasing distance to the disposal site. A land disposal site may have to be lined to prevent ground water contamination if the disposed sediments contained high concentrations of contaminants. In the case of a slurry pipeline, the sediment-water mixture is discharged into a settling basin at the disposal facility. The disposal facility should be sized to provide adequate settling times so that the return flow (effluent) meets regulatory criteria. Reservoir sediment volumes at the disposal site may be large (hundreds of thousands or millions of cubic yards) and require large land areas (tens or hundreds of acres). For example, disposal of the nearly 18 million cubic yards of sediment in two reservoirs on the Elwha River would require a 560-acre site if piled 20 feet high.

8.4.5 Stabilization Alternative

Under this type of alternative, sediment would be stabilized in the reservoir by constructing a river channel through or around the reservoir sediments. Stabilization of the reservoir sediments would prevent them from entering the downstream river channel. The cost for this alternative would typically be more expensive than river erosion, but less expensive than mechanical removal. This alternative may be desirable if the reservoir sediments are contaminated. One disadvantage of this alternative is that the reservoir topography would not be restored. If a river channel were constructed through the reservoir sediments (see Figure 8.1), then only some of the sediment would have to be moved and only short distances. Also, there would be a future risk that sediments could erode during floodflows and be transported into the downstream river channel. The challenge is to keep the reservoir sediments stable over the long term. A stable channel design should consider a range of river discharges and upstream sediment loads. The risk of erosion can be reduced if a flood plain is included in the design. If topographic conditions permit, the river channel and flood plains could be constructed around the reservoir sediments. Leaving the sediment in the reservoir may be an attractive alternative if restoring the reservoir topography is not an objective and the risk of erosion during floods is acceptable.

In the case of partial dam removal, the lower portion of the dam could be left in place to hold back the existing reservoir sediment. However, some fine sediment may be eroded downstream during drawdown of the upper reservoir. A portion of the dam could also be breached down to the predam riverbed, but the remaining length of the dam could be used to help retain sediment deposited on along the reservoir margins.

In the case of full dam removal, a stable channel to pass riverflows would have to be designed and constructed (either through or around the reservoir sediments). Mechanical or hydraulic dredging equipment can be used to excavate a new river channel through the reservoir sediments. The excavated sediments could be redeposited along the reservoir margins. The power of the river can also be used to excavate and transport sediment by controlling lake levels (similar to the river erosion alternative).

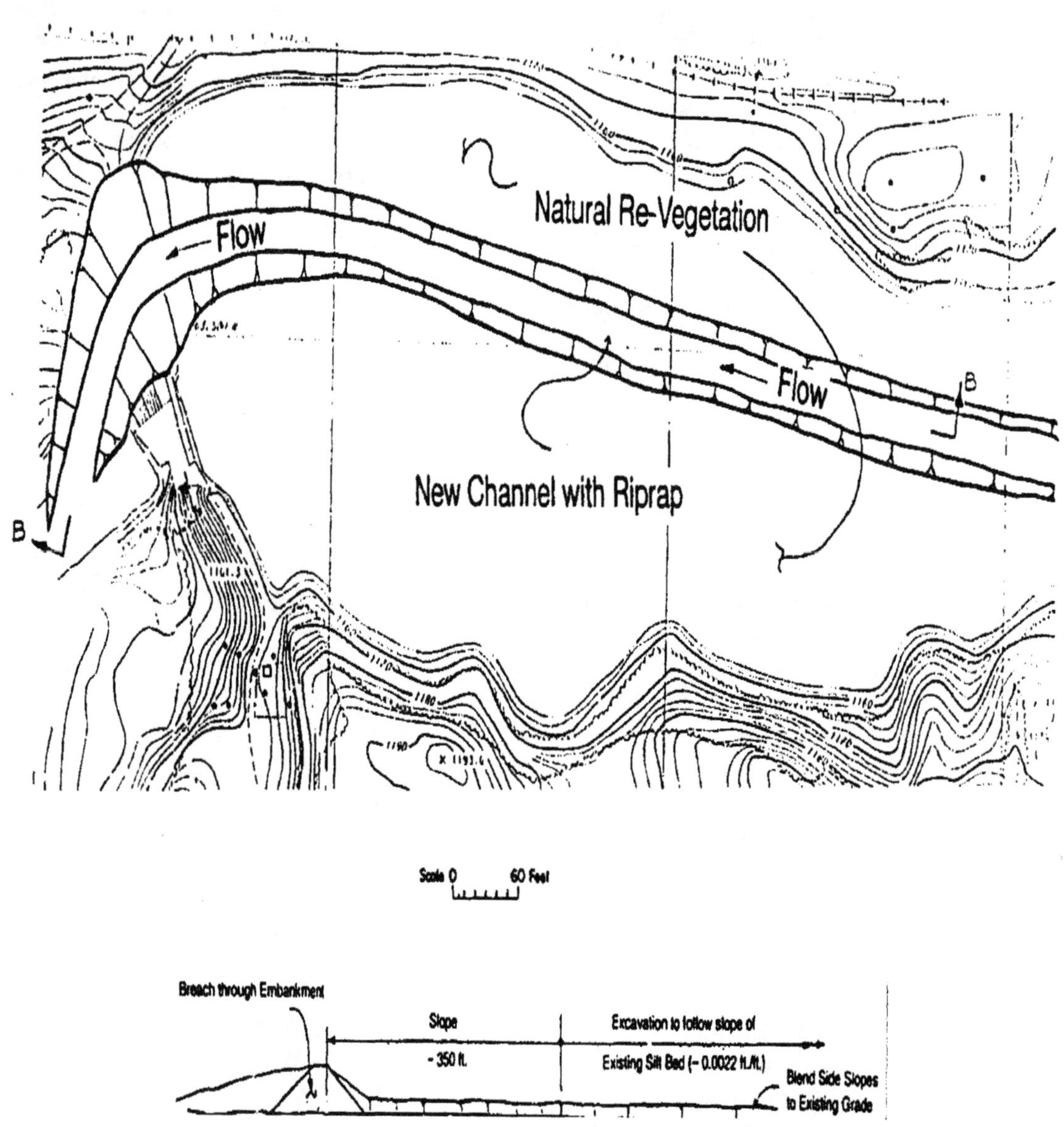

Figure 8.1. Example river channel constructed through the stabilized reservoir sediments (ASCE, 1997).

The size of the channel to be excavated is determined based on hydrologic, hydraulic, and sediment load characteristics of the river basin as well as an acceptable level of risk (e.g., the 100-year flood). Matching the alignment, slope, and cross section of a new river channel (excavated through the reservoir sediments) to that of the old predam river would help ensure a stable channel over the long term. A channel with relatively low velocity and slope would reduce the risk of bank erosion, but it may result in the deposition of the upstream sediment supply. A channel with relatively high velocity and slope would decrease the risk of sediment deposition, but it may result in erosion during floods. The width, depth, and slope for a stable channel can be computed for a given discharge, roughness, and upstream sediment supply. The procedure uses Manning's equation, the water conservation equation ($Q = VA$), a sediment transport equation, and the minimum unit stream power theory (VS = minimum) (see Chapter 7, "River Processes and Restoration").

Vegetation can be planted to help stabilize the remaining sediment from surface erosion. Bank protection structures may be required for the channel and the terrace banks at the edge of the flood plain. However, these bank protection structures would have to be maintained over the long term. If the bank protection failed during a flood, large quantities of sediment could be transported downstream. A diversion channel may be needed to route water around the work area while the channel and bank protection are constructed. This alternative can become quite costly if the channel to be excavated and protected extends a significant distance upstream of the existing dam.

The influences from tributary channels entering the reservoir area need to be considered in the stabilization alternative. Local storms may cause floods in these tributary channels, erode large amounts of the sediment, and damage the main channel protection. Channels may need to be excavated for these tributaries to prevent sediment erosion. To properly convey tributary inflow, the entire reservoir area must be mapped to identify these local inflow drainages, and erosion protection should be provided to contain the sediment on the flood plain.

A network of dikes could be constructed within the reservoir area to contain excavated sediment. A series of dikes could be constructed to contain the sediment so that, if one dike failed, only a portion of the stabilized sediment would be released downstream. If the dikes can be placed above the design flood stage, protection from riverflows would not be necessary. If the dikes are exposed to riverflows, stream bank protection is needed to prevent erosion. Stream bank protection structures could be constructed from natural materials such as rock, vegetation, or woody debris. For large volumes of sediment, the slope of the stabilized sediment or dikes is an important consideration. Although mild slopes are generally more stable than steep slopes, mild slopes would require a larger area of the reservoir to be occupied by the stabilized sediment.

8.4.6 Comparison of Alternatives

The best sediment management alternative will depend on the management objectives and design constraints, which depend on engineering, environmental, social, and economic considerations. Some of the basic advantages and disadvantages of the sediment management alternatives are listed in Table 8.4.

Table 8.4. Summary Comparison of Sediment Management Alternatives (ASCE, 1997)

Sediment management alternative	Advantages	Disadvantages
No action	• Low cost.	• Continued problems for fish and boat passage. • For storage reservoirs, continued reservoir sedimentation, loss of reservoir capacity, and reduced sediment supply to the downstream river channel.
River erosion	• Potentially low cost alternative. • Sediment supply restored to the downstream river channel.	• Generally, largest risk of unanticipated impacts. • Temporary degradation of downstream water quality. • Potential for river channel aggradation downstream from the reservoir.
Mechanical removal	• Generally low risk of reservoir sediment release. • Low impacts to downstream water quality. • Low potential for short-term aggradation of the downstream river channel.	• High cost. • Disposal site may be difficult to locate. • Contaminated sediments, if present, could impact ground water at the disposal site.
Stabilization	• Moderate cost. • Impacts avoided at other disposal sites. • Low to moderate impacts to downstream water quality. • Low potential for short-term aggradation of the downstream river channel.	• Long-term maintenance costs of the river channel through or around reservoir sediments. • Potential for failure of sediment stabilization measures. • Reservoir area not restored to natural conditions.

8.5 Analysis Methods for River Erosion Alternative

The river erosion alternative generally requires the most analysis from a sedimentation perspective. This section first describes methods appropriate to estimate the rate and volume of sediments eroded from the reservoir. It then describes the methods appropriate to estimate the downstream impacts.

In the river erosion alternative, the dam is removed either in several stages or all at once. As the reservoir is drawn down, the previously trapped sediment is now available for erosion. The rate at which the sediments are removed will be governed by the rate of reservoir drawdown, the flows, grain sizes, and the reservoir geometry. These sediments will be transported downstream eventually deposited in a downstream reach, reservoir, estuary, or ocean.

8.5.1 Reservoir Erosion

A general schematic of sediments within a reservoir is shown in Figure 8.2. The coarse sediments deposit in the upper reaches of the reservoir and form the delta, while the fine sediments are carried into the reservoir and deposited nearer the dam. In this figure, the reservoir still has a large portion of its original capacity. If, however, sediments have almost completely filled the reservoir, the delta deposits will have reached the dam and covered finer material below. Upon dam removal, the sediments will be eroded from the reservoir.

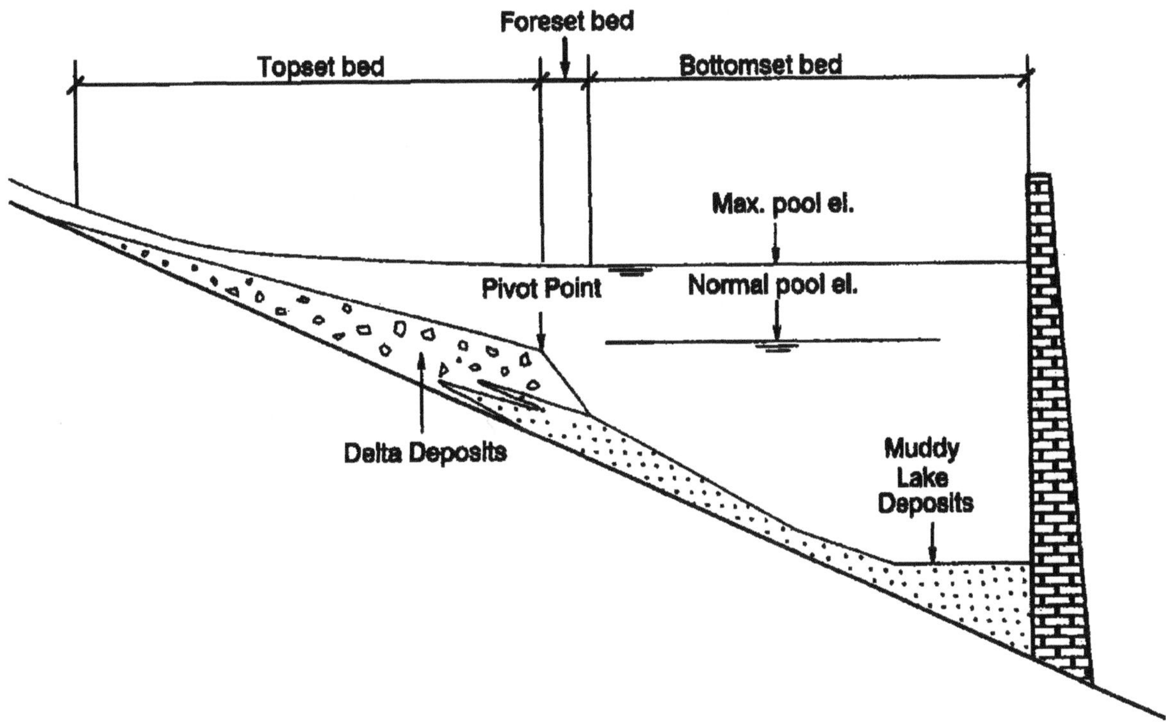

Figure 8.2. Generalized depositional zones in a reservoir (From Morris and Fan, 1998).

Doyle et al. (2003) used Figure 8.3 to describe the erosion of sediment from the reservoir. The geomorphic model was adapted from a headcut migration model of Shumm (1984). A summary of the model of Doyle et al. follows:

Stage A. This stage is the initial conditions before dam removal. Sediment has built up behind the dam.

Stage B. The dam is removed and/or the reservoir is drawn down.

Stage C. This stage is characterized by a rapid, primarily vertical erosion that begins at the downstream end and progresses upstream. Large amounts of sediment are released at this stage, and the downstream concentrations will be the highest of any stage. Depending upon the grain sizes present in the reservoir and the depth of the initial drawdown, this erosion may proceed as a headcut, or be primarily fluvial. The erosion is not expected to cut below the original bed elevation. The initial width of the channel formed by this erosion will be governed by the stability of the material in the reservoir.

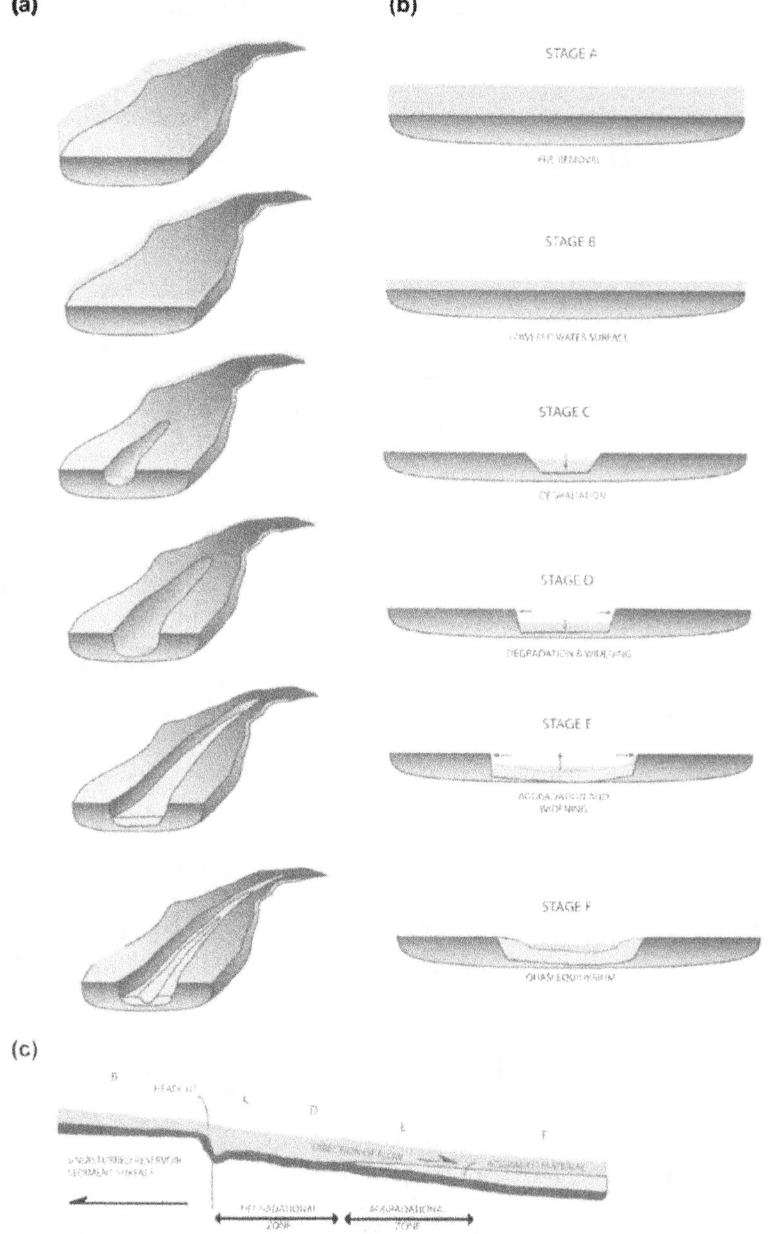

Figure 8.3. Schematic description of reservoir erosion process through delta deposits, from Doyle et al. (2003). (a) oblique view, (b) cross section view, (c) profile view.

Stage D. If the incision of Stage C produces banks that are too high or too steep to be stable, channel widening will occur by means of mass-wasting of banks.

Stage E. Sediment from the upstream reach starts to be supplied to the previously inundated reach. Some of this sediment is deposited in the reach, as the degradation and widening processes have reduced the energy slope within the reach. Some additional widening may occur during this stage, but at a reduced rate as compared to Stage D.

Stage F. This is the final stage and is the stage of dynamic equilibrium in which net sediment deposition and erosion in the reach are near zero.

It should be noted that most observations of dam removal have been for low head dams (i.e. less than 30 feet high) and correspondingly, the geomorphic model described above is most appropriate for smaller dams. The same processes will occur in larger dams, but the significance and magnitude of each stage may be much different. Also, the vertical and horizontal stratification of sediments may become more important. For example, if the dam shown in Figure 8.2 were removed in stages, the delta would progress as the dam is lowered and cover up finer sediments below. The delta would eventually reach the dam face, and sediment would begin to pass over the top of the dam. After the dam crest elevation is sufficiently lowered, the fine sediments would become exposed and quickly erode.

Another factor not considered in Figure 8.2 is the ability of the river to migrate laterally. Some rivers actively migrate laterally during storm events. Therefore, even though the initial channel formed through the reservoir may be small compared to the reservoir width, the river may eventually erode most of the reservoir sediments as it migrates across the valley floor. The lateral erosion process is expected to occur only during the larger storm events. These larger storm events usually carry large amounts of sediment under natural conditions, so the increase in sediment load due to the lateral erosion may not be significant.

8.5.1.1 Analytical Methods for Estimating Reservoir Erosion

Often, the best estimate for the final equilibrium profile of the river through the reservoir area is given by the pre-dam topography. However, predam surveys are not always available. If the necessary resources are available, drilling can identify the predam surface. For small dams, it is possible to estimate the stable bed from matching downstream and upstream slopes. An example of this is shown in Figure 8.4, where the equilibrium profile after dam removal is drawn through the stored reservoir sediments. A similar analysis was performed by Blodgett (1989), in which the downstream slope was projected through the reservoir region to determine the final channel bed thalweg profile after dam removal. For large dams, however, simply projecting the slope may introduce large errors. There may be hidden natural or manmade features that prevent a uniform slope. The resulting analyses and removal plans should take the possible uncertainty in the sediment volume and final profile into account.

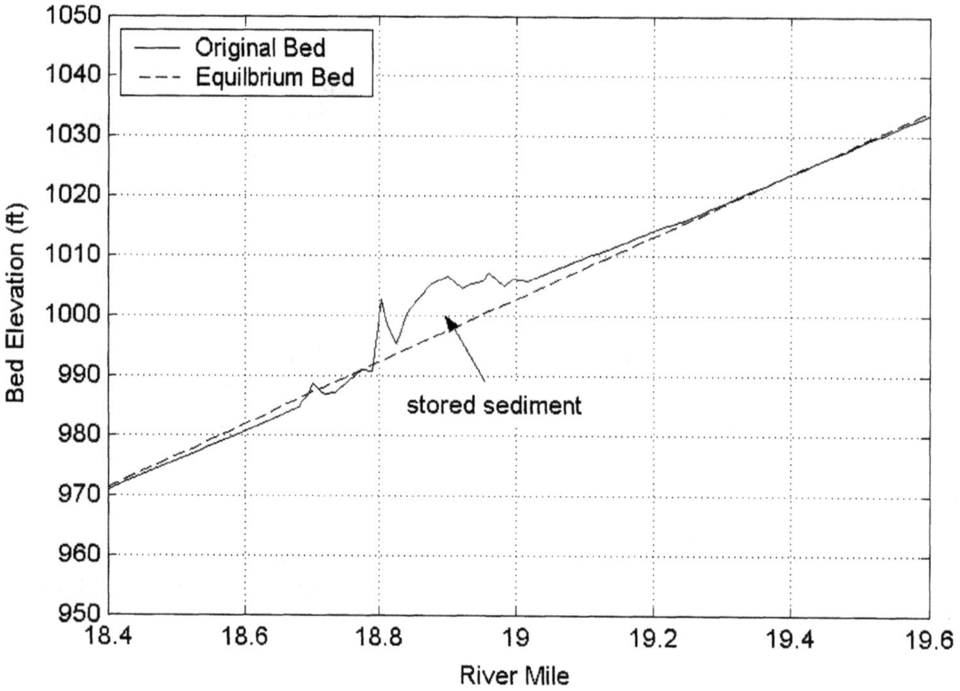

Figure 8.4. Example of estimating equilibrium profile after dam removal.

8.5.1.2 Numerical Models

There have been several applications of numerical models to the prediction of erosion following dam removal. These models can generally be divided into case-specific models and general application models.

Case-specific models are empirical in nature and generally need to be supported by field data. For example, a case-specific model was developed for Glines Canyon Dam. The model was based on physical reasoning and data from a drawdown experiment (Randle et al., 1996). The major components of the model were:

- Dam notching
- Assumption of stable slope, which can be calculated to be equal to the current delta slope
- Calculate new delta shape
- Calculate reservoir trap efficiency

- When delta meets the sill of the dam, start to move sediment out of the dam
- Continue until removal is complete

There are several general application sediment models that are able to simulate the transport of sediment in alluvial channels (see Chapter 5). Two basic categories of sediment models are one-dimensional (1-D) and two-dimensional (2-D) models. One-dimensional models generally solve steady-flow or unsteady equations of 1-D open channel flow. Unsteady flow effects are most likely not important to the erosion of the reservoir sediment and therefore steady flow models should be sufficient. The 1-D solution of the hydraulics is then coupled to tracking of the sediment movement and changes of the bed. One dimensional models have the general weakness that hydraulic properties are averaged over the cross section. In the river channel this is often a good assumption, but it tends to break down in wide reservoirs or meandering streams. The variation in velocity across the reservoir section may be large and cause the 1-D assumptions to not be valid. This can result in under prediction of erosion within the delta.

Two-dimensional hydraulic and sediment transport models have the ability to model the variation of hydraulic and sediment properties across the reservoir cross section. They could also model the failure of banks within the reservoir. At the present time, the author does not have a well-documented case of the application of a two-dimensional model to dam removal processes. Modeling the bank failure process in a two-dimensional model is a non-trivial exercise and would require advances to currently available models.

There are several other unique characteristics of erosion in reservoir deposits that may not be well represented with either one-dimensional or two-dimensional models. Some of the processes or features that are generally not well represented in sediment transport models are listed below:

- Headcut migration through cohesive material
- Bank erosion
- Large width changes
- Stratified bed sediment

Some more recently developed models have some ability to model these situations. Langendoen (2000) developed the CONCEPTS model to consider bank erosion by incorporating the fundamental physical processes responsible for bank retreat: fluvial erosion or entrainment of bank material particles by the flow and mass bank failure (for example, due to channel incision). It has not been applied to the case of dam removal, but has been applied to several rivers (Langendeon and Simon, 2000; Langendoen et al., 2002). The CONCEPTS model also accounts for stratified bed sediment.

MBH Software (2001) has made recent developments to the HEC-6T code to make it applicable to dam removal. In this model, the erosion width is determined by an empirical relationship between flow rate and channel width. Bank stability is modeled using a user input critical bank stability angle. If the bank becomes steeper than the input angle, the bank fails to that angle.

Stillwater Sciences has developed DREAM (Dam Removal Express Assessment Models), a model that is applicable to dam removal (Stillwater Sciences, 2002). The model assumes that the channel through the reservoir sediments has a simplified trapezoid shape. The user inputs the initial width and the model calculates the evolution of this channel based on transport capacity. The model ignores sediment that would travel as wash load (i.e., silts and clays).

The GSTARS-1D model (Yang et al., 2004, 2005) has been used to estimate the erosion of sediment from the reservoir. The channel formation is calculated based on the sediment transport capacity and bank slope stability criteria. In this way, it is similar to the DREAM model in the reservoir region.

More validation of these models with field data is required before their applicability and performance can be assessed.

8.5.2 Downstream Impacts

This section describes two basic methods for analyzing downstream impacts. The first is an analytical method that can be used to estimate deposition impacts of dam removal. The second method involves using more complicated numerical models to analyze a variety of impacts. The impacts include increase in suspended sediment concentrations and changes to the riverbed elevation.

8.5.2.1 Analytical Methods for Predicting Deposition Impacts

To predict the impacts associated with the movement of such accumulations, a model of the system needs to be constructed. The complexity of the model applied to the system should be consistent with the data and resources available. Most often, the prediction of the movement of these accumulations is accomplished by using a one- or two-dimensional hydraulic model coupled with a one- or two-dimensional sediment transport model (MBH Software, 2001; Stillwater Sciences, 2002; Reclamation, 2003). However, such models can be complex and require large amounts of input data. A simple method would be beneficial in providing initial estimates and for cases where complex models are not necessary.

Greimann et al. (2003) extended the analytical description of aggradation of Soni et al. (1980) to describe downstream aggradation following dam removal. A schematic of idealized representation of the movement of a sediment accumulation is shown in Figure 8.5. The sediment accumulation sits on top of the original bed material that is at a stable and uniform slope.

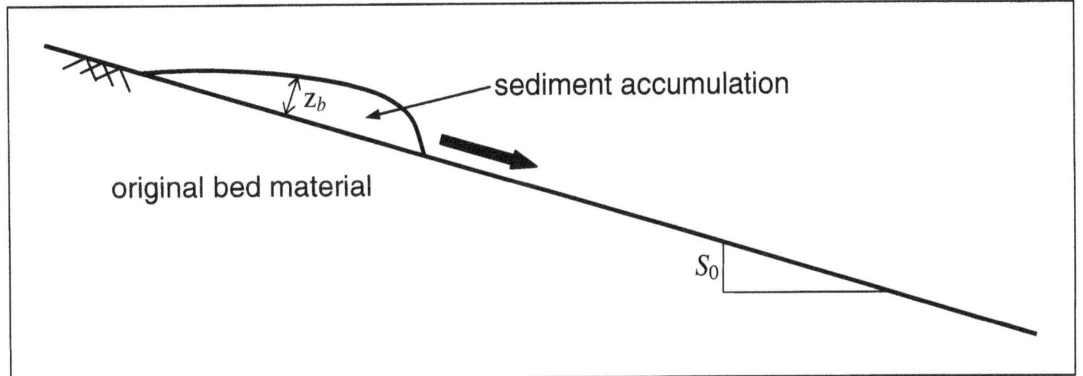

Figure 8.5. Schematic of idealized representation of the movement of a sediment accumulation.

The following equation was derived by Greimann et al. (2003):

$$\frac{\partial z_b}{\partial t} + u_d \frac{\partial z_b}{\partial x} = K_d \frac{\partial^2 z_b}{\partial x^2} \tag{8.1}$$

where

z_b = depth of the sediment originally trapped behind the dam, and
u_d, = velocity of sediment wave translation.

The variable u_d is defined as:

$$u_d = \frac{\left(G_d^* - G_0^*\right)}{h_d(1-\lambda)} \tag{8.2}$$

where

G_d^* = transport capacity in units of volume per unit width of the deposit material,
G_0^* = transport capacity of the original bed material,
h_d = maximum depth of the deposit, and
λ = sediment porosity.

The parameter, K_d, is the aggradation dispersion coefficient:

$$K_d = \frac{\left(b_d G_d^* + b_0 G_0^*\right)}{6 S_0 (1-\lambda)} \tag{8.3}$$

The transport rate of a particular sediment type is related to the flow velocity:

$$G_d = p_d G_d^* = p_d a_d U^{b_d}, \qquad G_0 = p_0 G_0^* = p_0 a_0 U^{b_0} \tag{8.4}$$

where

U = averaged flow velocity,

a_d , b_d = constants used to calculate the transport capacity of the deposit material, and
a_0, b_0 = constants used to calculate the transport capacity of the original bed material

The parameter b is generally bounded between 4 and 6 (Chien and Wan, 1999). Equation (8.1) can be solved analytically and can be applied to arbitrary initial deposits by dividing the stream into N segments,

$$z(x,t)=\sum_{i=1}^{N-1}\frac{(z_{1i}+z_{1i+1})}{4}\left[\begin{matrix}\operatorname{erf}\left(\frac{x-u_d t-x_i}{2\sqrt{K_d t}}\right)-\operatorname{erf}\left(\frac{x-u_d t-x_{i+1}}{2\sqrt{K_d t}}\right)-\\ \operatorname{erf}\left(\frac{x+u_d t+x_i}{2\sqrt{K_d t}}\right)+\operatorname{erf}\left(\frac{x+u_d t+x_{i+1}}{2\sqrt{K_d t}}\right)\end{matrix}\right] \tag{8.5}$$

where the function "erf" is the error function, and z_1 is the initial bed elevation.. There may have to be some trial and error in determining appropriate distances between stream segments. There should be enough segments so that the initial deposit and resulting bed profiles are adequately defined.

The error of this method is potentially great because of the simplifications made. A partial list follows:

- Assumes a prismatic channel
- Does not account for changes in channel geometry with distance along the channel
- Does not consider longitudinal slope breaks due to channel controls
- Assumes a steady flow rate
- Does not account for changes in roughness
- Is not applicable upstream of the sediment accumulation
- Assumes sediment accumulation is composed of single size fraction
- Assumes accumulation travels as bed load
- Ignores sediment sizes in the sediment accumulation that will travel as pure suspended load

Despite these shortcomings, this method holds promise as a simple assessment tool to determine impacts associated with aggradation. This method requires a minimal number of input parameters and can be completed in a fraction of the time required to complete a more complicated and time-consuming numerical model. The parameters that need to be estimated to use the model are listed in Table 8.5. All the parameters except for b_d are physical quantities that can be measured. The parameter b_d is the exponent in the sediment transport relation and based on results from several researches is generally bounded between 4 and 6 (Chien and Wan, 1999).

Table 8.5. Description of parameters necessary to use proposed model

Parameter	Range of values or method of obtaining value
S_0	Average natural stream slope. Measured from topographic maps.
G_d^* (L^2/T)	Transport capacity of sediment accumulation in units of volume per unit width
G_0^* (L^2/T)	Transport capacity of bed material in units of volume per unit width
λ	Sediment porosity, usually between 0.3 and 0.5
b_d	Exponent in sediment transport relation, usually between 4 and 6
h_d (L)	Maximum depth of sediment accumulation. Estimated from field surveys

A hypothetical case is considered to show how one would apply this methodology to the field. This hypothetical dam (No Name Dam) is approximately 4.5 m high and has approximately 120,000 m^3 of sediment deposited behind it. The average downstream river has an approximate bed slope of 0.005 and an average width of 30 m. The surface bed material downstream of the dam is mostly gravel with less than 20% sand. The trapped sediment in the reservoir consists of approximately 12,000 m^3 of silt, 42,000 m^3 of sand, and 66,000 $m^{3\ of}$ gravel and cobbles. The transport rates were taken from sediment transport calculations using the bed material. It was determined that the silt, clay, and fine sand would travel as suspended load, and the volume of those size fractions was subtracted from the total volume. After subtracting the fine sediments, the sediment wedge was determined to be approximately 3 m high near the dam and 0 m high approximately 1,200 m upstream of the dam.

Table 8.6 Problem parameters for simulations of hypothetical dam, No Name Dam

Parameter	Value for No Name Dam
S_0	0.005
G_d^* (m^2/s)	0.008
G_0^* (m^2/s)	0.0001
λ	0.4
b_d	5
h_d	3

The deposition along the downstream reach at various times is shown in Figure 8.6. For this problem, we have assumed the dam is suddenly and completely removed. The deposition decreases markedly in the downstream direction, and 2 km downstream deposition is less than 0.5 m.

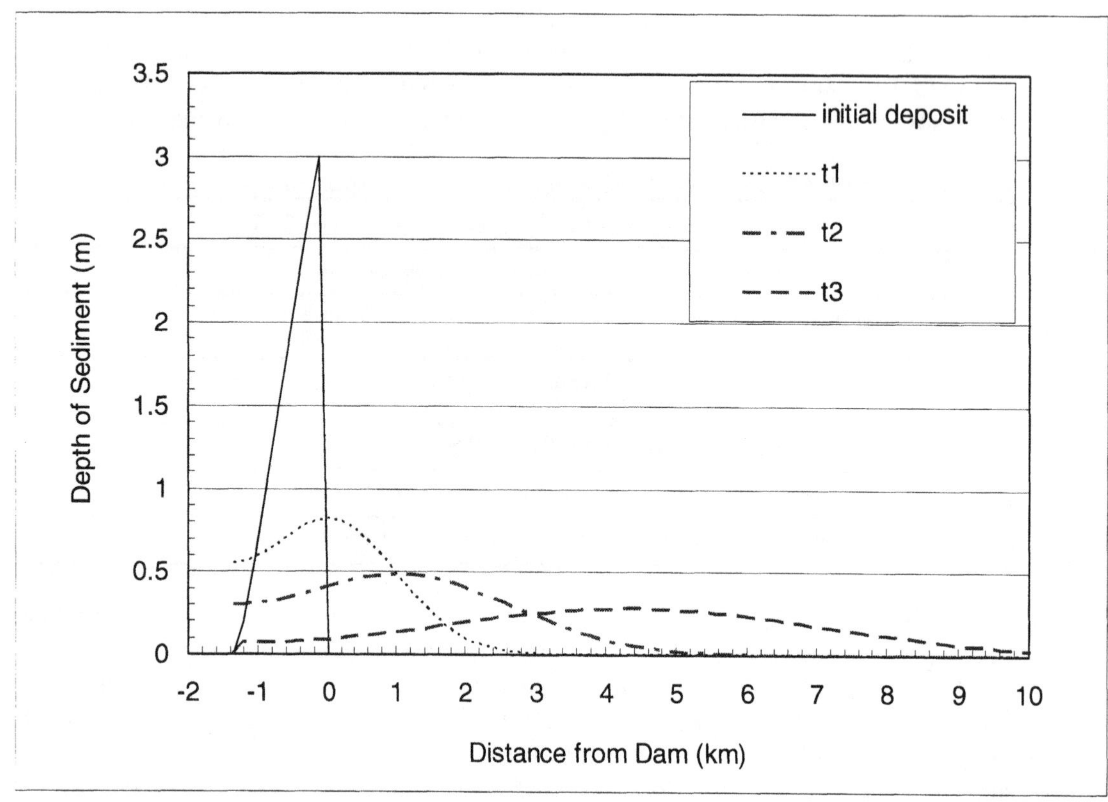

Figure 8.6. Predicted deposition downstream of No Name Dam using Equation (5). The deposition at different times is shown, $0 < t1 < t2 < t3$.

8.5.2.2 Numerical Modeling of Sediment Impacts

Numerical sediment models can be used to provide quantitative estimates of downstream impacts. However, dam decommissioning presents unique problems that some convention sediment transport models may not handle properly. First, the erosion in the reservoir must be estimated correctly. The previous section discusses issues involved in estimating the erosion in the reservoir. Two major considerations in modeling the downstream impacts include estimating the transport of sediment through pool-riffle systems and predicting the transport of fine sediment over a coarse bed.

Predicting the transport of sediment through pool-riffle river geometries is a general weakness of sediment transport models. Because the flow field in pool-riffle geometries has a three-dimensional structure, 2-D and 1-D models do not represent the transport of material well in these geometries. As a result, sediment transport models will generally overpredict the amount of material that is deposited in pool sections. Overprediction of pool filling is a serious problem when a goal of dam removal is to increase the amount of fish habitat in the downstream reaches.

Further work is necessary to develop better modeling techniques to better predict the transport of sediment through pool-riffle river geometries. Currently, the ad-hoc solution to predicting the transport through pool-riffle systems is to run a stabilization period before the dam removal is simulated. For example, the river is simulated with the dam in place until the entire river comes to equilibrium. During this stabilization period, the model would predict that many of the pools would fill with sediment. After the stabilization period, the dam removal is simulated and the additional deposition that occurs is assumed to be that caused by the dam removal. The incremental deposition is then superimposed upon the current bed to determine sediment elevations. Bountry and Randle (2001) used this method to predict downstream deposition. The DREAM model basically performs a similar computation by running what it calls a "zero process" before the dam is removed. This "zero process" will eliminate pool and riffles structures before the dam removal is simulated.

Another potential pitfall in simulating dam removal is predicting the transport of fine material over a coarse bed. Because most numerical sediment transport models are built upon an active layer concept, the transport capacity is proportional to the amount of sediment present in the bed. Therefore, the model will deposit fine sediment into the bed and mix the fine sediment with the coarse sediment. In reality, the fine sediment may just pass over the top of the coarse sediment, and very little may deposit. If the fine sediment does deposit, it may not mix with the coarser sediment of the original riverbed. To prevent this problem from becoming excessive, it may be necessary to have very small active layer thicknesses to limit the amount of mixing of fines with coarser bed sediment. Another option is to define the initial riverbed material as a very thin layer with a size gradation equal to the reservoir sediment gradation. The model is not allowed to erode material below the original bed elevations.

Hydraulic roughness also changes when a large amount of fine sediment is deposited on a coarse bed. When fines fill in the spaces between cobbles, the hydraulic roughness will decrease, and thereby, the transport capacity of the flow will increase. There are some empirical methods to account for this (Brownlie, 1982; Stillwater Sciences, 2002), but there has been little or no field verification of these models for large releases of fine sediment over coarse sediment.

8.6 Summary

While the great majority of dams still provide a vital function to society, some of these dams may need to be removed for various reasons such as economics, dam safety and security, legal and financial liability, ecosystem restoration (including fish passage improvement), site restoration, and recreation use.

The sediment effects related to dam removal may be significant if any of the following conditions apply:

- The reservoir storage, below the normal operating pool, is at least 1 percent of the average annual inflow.

- The reservoir sediment volume is equivalent to a multi-year sediment supply from the upstream river channel, or several years would be required to transport the reservoir sediment volume through the downstream river channel.

- The reservoir sediments are contaminated at concentrations significantly above background levels.

Portions of the dam can be left in place for historic preservation, to reduce dam removal costs, and to help stabilize reservoir sediments. The rate of reservoir sediment erosion and release to the downstream river channel is primarily controlled by the rate of dam removal and reservoir drawdown and by the upstream hydrology. Although headcuts may erode the reservoir sediments during periods of low flow, sufficient flow is necessary to provide transport capacity of reservoir sediments. The rate of reservoir drawdown needs to be slow enough to avoid a flood wave of reservoir water spilling into the downstream river channel. Also, the rate needs to be slow enough to avoid inducing any potential landslides along the reservoir margins or a slide failure of any earthen dams. The ability to draw down the reservoir pool depends on how flows can be released through, over, or around the dam. If the dam has a low-level, high-capacity outlet works or diversion tunnel, then the reservoir could be emptied at a prescribed rate and the dam could be removed under dry conditions. Otherwise, a diversion channel may have to be constructed around the dam or an outlet may have to be constructed through the dam.

The basic types of sediment management alternatives associated with dam removal include no action, river erosion, mechanical removal, and stabilization. River erosion is typically the least expensive and most commonly employed alternative. However, mechanical removal or stabilization may be required if the reservoir sediments are contaminated. If the reservoir is many times wider than the upstream river channel, then a significant portion of the reservoir sediments will remain stable in the reservoir over the long term, even without stabilization techniques.

The rate and extent of reservoir sediment erosion, and the possible redistribution and storage within the reservoir, need to be predicted before sediment transport can be predicted through the downstream river channel. The primary predictive tools include both numerical and physical modes. Physical models can provide accurate predictions if the model scales are properly selected and they can be used to calibrate numerical models. The numerical models tend to be more easily adaptable to simulate multiple management or hydrology scenarios. Most numerical sediment transport models are one dimensional and can simulate river conditions over many miles and over a time period of many decades. Two-dimensional models are also available, but their focus is normally limited to relatively short river lengths over periods of days or maybe weeks. A thorough understanding of the numerical model equations and limitations is necessary for proper application of the model to a dam removal problem. In addition, thorough understanding of the geomorphic, hydraulic, and sediment transport processes of the river is necessary for proper model application and interpretation of the results.

8.7 References

American Society of Civil Engineers (1997). "Guidelines for Retirement of Dams and Hydroelectric Facilities," New York, 222 p.

Aspen Institute (2002). *Dam Removal, A New Option for a New Century*, Aspen Institute Program on Energy, the Environment, and the Economy, Queenstown, Maryland, 66 p.

Blodgett, J.C. (1989). *Assessment of Hydraulic Changes Associated with Removal of Cascade Dam, Merced River, Yosemite Valley, California*, U.S. Geological Survey, Open File Report 88-733.

Bountry, J., and T. Randle. (2001). *Savage Rapids Dam Sediment Evaluation Study, Appendix B, Hydraulics and Sediment Transport Analysis and Modeling*, U.S. Department of the Interior, Bureau of Reclamation, Denver, Colorado.

Brownlie, W.R. (1982). *Prediction of Flow Depth and Sediment Discharge in Open Channels.* Doctoral dissertation, California Institute of Technology, Pasadena.

Brune, G.M. (1953). "Trap Efficiency of Reservoirs," *Transactions of the American Geophysical Union*, vol. 34, no. 3, June 1953, pp. 407-418.

Bureau of Reclamation (2000). *Matilija Dam Removal Appraisal Report*, Technical Service Center, Denver, Colorado, April.

Bureau of Reclamation (2001a). *City of Gold Hill Fish Passage Improvements at the Municipal Water Supply Diversion: Phase II*, Boise, Idaho, September 2001.

Bureau of Reclamation (2001b). Josephine County Water Management Improvement Study, Oregon, Savage Rapids Dam Sediment Evaluation Study, Denver, Colorado, February 2001.

Chien, N., and Z. Wan (1999). "Sediment Transport Capacity of the Flow," *Mechanics of Sediment Transport*, ASCE Press.

Doyle, M., Stanley, E.H., Harbor, J.M. (2003). "Channel Adjustments Following Two Dam Removals in Wisconsin," *Water Resources Research*, Vol. 39, No. 1, 2003.

Gilbert, J.D., and R. A. Link (1995). *Alluvium Distribution in Lake Mills, Glines Canyon Project and Lake Aldwell, Elwha Project, Washington*, Elwha Technical Series PN-95-4, Boise, Idaho, August 1995, 60 pp.

Greimann, B.P., T. Randle, and J. Huang, (2003). "Movement of Sediment Accumulations," *Submitted to ASCE Journal of Hydraulic Engineering.*

H. John Heinz III Center for Science, Economics, and the Environment (2002). *Dam Removal Science and Decision Making*, Washington D.C., 221 pages. Report, October.

Langendoen, E.J., (2000). *CONCEPTS – Conservational Channel Evolution and Pollutant Transport System*, USDA-ARS National Sedimentation Laboratory, Research Report No. 16, December.

Langendoen, E.J., and A. Simon (2000). *Stream Channel Evolution of Little Salt Creek and North Branch West Papillion Creek, eastern Nebraska.* Report, US Department of Agriculture, Agricultural Research Service, National Sedimentation Laboratory, Oxford, Mississippi.

Langendoen, E.J., R.E. Thomas, and R.L. Bingner (2002). "Numerical Simulation of the Morphology of the Upper Yalobusha River, Mississippi between 1968 and 1997." *Riverflow 2002*, D. Bousmar and Y. Zech, eds., Balkma, The Netherlands, 931-939.

MBH Software (2001). *Sedimentation in Stream Networks (HEC-6T): Users Manual,* http://www.mbh2o.com/index.html

Morris, G.L. and J. Fan (1997). *Reservoir Sedimentation Handbook, Design and Management of Dams, Reservoirs, and Watersheds for Sustainable Use*, McGraw-Hill, New York.

Olympic National Park (1996). *Elwha River Ecosystem Restoration Implementation, Draft Environmental Impact Statement*, Port Angeles, Washington, April 1996.

Olympic National Park (1996). *Elwha River Ecosystem Restoration Implementation, Final Environmental Impact Statement*, Port Angeles, Washington, November 1996.

Randle, T.J., C.A. Young, J.T. Melena,, and E.M. Ouellette (1996). *Sediment Analysis and Modeling of the River Erosion Alternative*, Elwha Technical Series PN-95-9, Denver, Colorado, October 1996, 136 pages.

Schumm, S.A., M.D. Harvey, and C.C. Watson (1984). *Incised Channels: Morphology, Dynamics, and Control*, Water Resource Public, Highlands Ranch, Colorado.

Strand, R.I. and E.L. Pemberton (1987). "Reservoir Sedimentation," *Design of Small Dams*, U.S. Bureau of Reclamation, Denver, Colorado.

Stillwater Sciences (2002). *Dam Removal Express Assessment Models (DREAM)*, Technical Report, October.

U.S. Army Corps of Engineers (1989). "Sedimentation Investigations of Rivers and Reservoirs," *Engineering Manual 1110-2-4000*, Washington, DC.

Yang, C.T., J.V. Huang, and B.P. Greimann (2004, 2005). *User's Manual for GSTAR-1D 1.0 (Generalized Sediment Transport for Alluvial Rivers – One Dimension, Version 1.0)*, Bureau of Reclamation, Technical Service Center, Denver, Colorado.

Chapter 9
Reservoir Survey and Data Analysis

Page

9.1 Introduction 9-1
9.2 Purpose of a Reservoir Survey 9-1
9.3 Sediment Hazards 9-3
9.4 Sediment Management 9-4
9.5 Frequency and Schedule of Surveys 9-5
9.6 Reservoir Survey Techniques 9-7
9.6.1 Shoreline Erosion 9-9
9.6.2 Data Density and Line Spacing 9-12
9.6.3 Cost of Conducting a Reservoir Survey 9-13
9.6.4 Selecting Appropriate Hydrographic Data Collection System and Software 9-13
9.7 Hydrographic Collection Equipment and Techniques 9-15
9.8 Global Positioning System 9-16
9.8.1 Absolute Positioning 9-17
9.8.2 Differential Positioning 9-18
9.8.3 Real-Time Kinematic GPS 9-21
9.8.4 GPS Errors 9-22
9.9 Horizontal and Vertical Control 9-23
9.9.1 Datums 9-24
9.10 Depth Measurements 9-24
9.10.1 Single Beam 9-24
9.10.2 Multibeam 9-29
9.10.3 Additional Sonar Methods 9-34
9.10.4 Single Beam Depth Records 9-35
9.11 Survey Accuracy and Quality 9-40
9.12 Survey Vessels 9-42
9.13 Survey Crew 9-43
9.14 Determination of Volume Deposits 9-43
9.14.1 Average-End-Area Method 9-45
9.14.2 Width Adjustment Method 9-45
9.14.3 Contour Method – Topographic Mapping 9-46
9.15 Final Results 9-46
9.15.1 Report 9-52
9.16 Reservoir Survey Terminology 9-56
9.17 Summary 9-62
9.18 References 9-63

Chapter 9
Reservoir Survey and Data Analysis

by
Ronald Ferrari and Kent Collins

9.1 Introduction

This chapter provides guidelines, techniques, and information that can be used by Reclamation and others for planning, collecting, analyzing, and reporting of reservoir and river survey studies; the ultimate goals are preservation of the information and uniformity of collection and analysis. This chapter mainly refers to the reservoir survey applications, but much of the equipment, techniques, and technology can be adapted for river and above water data collection.

This chapter mainly addresses the bathymetric or underwater field survey process, but the overall sedimentation analysis usually consists of data for the entire study area. In this guide, the term "bathymetric survey" specifically refers to the collection of water depths, while the term "hydrographic survey" refers to the entire survey, including the above and below water portions of the study area. The term "reservoir survey" implies a variety of field observations and measurements, data processing, analyses, and report preparation that can also be applied to river surveys.

The above water portion of the reservoir can be measured by several means. Conventional surveying techniques using stadia rods, transits, and total stations, and global positioning systems (GPS) can be used, along with photogrammetric mapping or aerial surveying, which provides a more automated means for above water collection. The Sedimentation Group usually coordinates with the Reclamation regional offices or other groups within the TSC to obtain the necessary above water data for an ongoing study. Some Reclamation offices have the capability of setting the necessary ground control and conducting the photo interpretation, some offices use a contracted professional surveying service for only the aerial flight or aerial field collection, and some offices contract the complete above water surveying services.

The survey technology has changed significantly over recent decades with the dramatic increase in the speed of data acquisition and computer system processing. GPS has significantly reduced the time and cost for data collection by changing the techniques of collection. Analysis procedures have also improved with the continued development of computers and data collection software. These trends of rapid technological advancements will likely continue well into the future. Many of the presently used collection and analysis techniques are addressed in this chapter, but it is the responsibility of the study manager to keep up with the latest technology and choose the proper methods for their study needs. This can be accomplished through publications, Internet Web sites, and attendance at conferences.

9.2 Purpose of a Reservoir Survey

Reservoirs come in all shapes and sizes and are designed for purposes such as retention for flood control, debris/sediment storage, irrigation, municipal water supply, power production, recreation, navigation, conservation, and water quality control. The reservoir size, shape, and operation affect the location and nature of the sediment depositions (figure 9.1). Reservoir sedimentation is an ongoing natural depositional process that can remain invisible for a significant portion of the life of a reservoir. However, lack of visual evidence does not reduce the potential impacts of reservoir sedimentation on functional operations of a reservoir (Lin, 1997). As sediment deposition depletes reservoir storage volume, periodic reallocation of available storage at various pool levels may be necessary to satisfy operational requirements of water users.

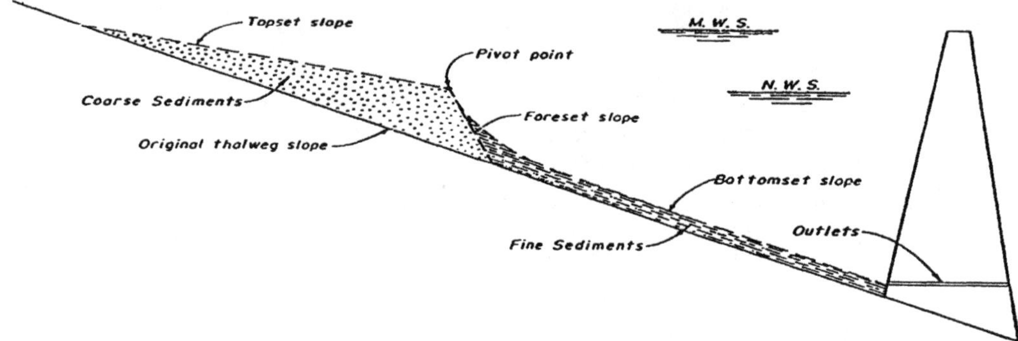

Figure 9.1. Profile of reservoir delta formation (Strand and Pemberton, 1982).

As rivers and streams enter a reservoir, the flow depth increases and the velocity decreases, causing a loss in the sediment transport capacity of the inflow. The loss of sediment transport capacity and the damming effect of the reservoir may cause deposition of sediment in the stream channels above the reservoir water surface and in the upper reservoir area. The sediment deposition process in reservoirs generally follows the same basic pattern, with coarser sediments settling first in the upper reservoir area as the river inflow velocities decrease, forming a delta. Deposition continues from upstream to downstream, with the sediment gradation becoming finer as the deposition progresses towards the dam until the inflowing sediment is deposited throughout the length of the reservoir. Some of the inflowing fine sediments (silts and clays) typically stay in suspension and may discharge through the dam outlets or spillways. As sediments deposit near the dam outlets, they eventually will be discharged downstream as releases are made from the dam.

In the United States, reservoir sedimentation seldom receives attention until the reservoir storage capacity has been significantly reduced or the reservoir operation or surrounding area is affected. The delta formation can cause local problems before sediment deposition significantly reduces reservoir capacity or causes operational problems at the dam. Some local problems that have been attributed to sediment deltas are increased elevation of the flood stage and ground water table, silting of pumping and intake structures, and blockage of navigation passages. Once at the dam, the released sediments have downstream impacts on river fisheries and municipal water systems.

The primary objective of a reservoir survey is to measure the current reservoir area and capacity. The main cause of storage capacity change is sediment deposition or erosion. Typical results from reservoir survey data collection and analysis include measured sediment deposition since dam closure and previous surveys, sediment yield from the contributing drainage, and future storage-depletion trends. Survey results can also include location of deposited sediment (lateral and longitudinal distribution), sediment density, reservoir trap efficiency, and evaluation of project operation.

The Sedimentation Group typically computes reservoir sediment accumulation by comparing the measured original capacity, prior to inundation, to the updated measured capacity. This method calculates a long-term sediment deposition value used for future sediment projections. Making comparisons to the original survey, rather than previous surveys, prevents errors that might exist in previous resurvey results from being included in the analysis. The calculations typically rely on accurate original reservoir topography available for many of Reclamation's reservoirs, but this must be evaluated on a case-by-case basis. Modifications to the analysis and study objectives must be made for cases where accurate original reservoir topography is not available. This was the case for the 1995 Theodore Roosevelt (Roosevelt) Reservoir survey (Lyons-Lest, 1996) and the 2002 Deadwood Reservoir survey (Ferrari, 2003).

The Roosevelt and Deadwood Reservoir resurveys measured finer detail than the original survey data. The 1995 Roosevelt survey was the eighth since dam closure in 1909, but the first to use aerial photography that provided more detail of the upper reservoir elevations than the original 1909 survey and the resurveys. Comparing the detailed 1995 survey with previous mapping information was not a means for computing sediment accumulation due to the precision differences between the surveys. The previous resurveys of Roosevelt Reservoir were valid for computing sediment inflow since they utilized a range line collection method that monitored the same range lines over the years. The changes at these locations were compared to the original topography for estimating the sediment deposition. The detailed 1995 Roosevelt Reservoir survey will be used as the basis for future comparisons. The same was true of the 2002 Deadwood Reservoir resurvey. The detailed aerial and multibeam data from the 2002 survey could not be compared to the less detailed original data for computing sediment accumulation.

Additional objectives of Reclamation's reservoir survey studies are to determine current reservoir topography, estimate the reservoir's economic life, and resolve storage capacity conflicts. The resulting study information is beneficial for describing existing conditions for a specific reservoir, monitoring upstream land management practices, evaluating current operation of a reservoir, and planning future reservoirs. The results from the study can provide insight for such operational objectives as sluicing sediment deposits to increase reservoir volume and possibly enhancing the downstream river environment, establishing bench marks for forecasting future reservoir depletion rates, revising intake or outlet design, assessing water quality control methods, and designing recreation facilities, structures, and operational schedules.

Reservoir sediment accumulation and distribution can be approximated theoretically. However, an accurate reservoir sedimentation survey is the best means for monitoring current reservoir sedimentation and for projecting future sediment inflow and deposition. Obtaining an accurate value requires measuring the complete reservoir area, or as much of the sediment delta as possible. As seen on figure 9.1, the majority of the delta may form in the very upper reaches of the reservoir but, eventually, the inflowing sediments can deposit throughout the reservoir. Full coverage requires both above and below water measurements that significantly increase the field collection time and cost. If the measurements are only under water, the survey should be scheduled when the reservoir is as full as possible. Although other types of factors must be considered, cost usually is the deciding factor in the data collection plan.

9.3 Sediment Hazards

As better understanding and management of sediments are achieved, a major issue will become how to deal with the associated hazards while maintaining the goal of preserving existing water resources. To sustain healthy management of the watershed, knowledge of sediment movement under different flow conditions is needed. This knowledge can assist in the evaluation of different sediment management options: ignoring, storing in place, or flushing downstream are a few examples. Regardless of options, the hazards must be addressed (Wohl, 1998). These hazards may only be local, or they may affect a large portion of the basin upstream and downstream of dams, resulting in excessive sediment deposits, contaminated sediments, or even decreased sediment concentrations. Excess sediment can fill a reservoir and change river channel patterns and features such as fish spawning sites. Contaminated sediments may include heavy metals from mine runoff or excessive levels of nutrients from urban and agricultural phosphorus. The excess phosphorus can create algal blooms that reduce dissolved oxygen and harm fish populations.

Sediment hazards have been indirectly created due to human activities such as timber harvest, crop cultivation, grazing, road construction, and urban development. They have also been directly created by dams and reservoirs, river channel altering from dam and dike diversions, channelization, and mining. A decrease of natural sediment supply downstream of a dam means the river is capable of transporting more sediment than the supply, resulting in channel erosion, riverbank collapse, bridge-pier scour, and channel downcutting.

There are many examples of problems associated with excess sediments. In some cases, excess sediments bypassed to maintain dam and reservoir functions resulted in immediate large fishkills, long-term effects on fish reproduction, and instantaneous and long-term effects on water quality. Possible contaminants attached to the sediments must be addressed if alternatives such as flushing and/or dredging disturb the sediments. There have been studies that measure the magnitude and duration of the effects these contaminants have on the surrounding environment for options such as leaving sediments in place or bypassing downstream. Past studies have collected reservoir sediment and determined the age of the material based on the amounts of DDT and lead measured and knowledge of when these contaminants were discontinued. The decrease of downstream sediment below dams has caused excessive erosion of the river channel and banks affecting the management of rivers. Loss of very fine sediment has affected the infiltration rate of downstream diversion ditches and the reproduction of fish that used this material to guard against predators.

9.4 Sediment Management

Reclamation's ability to manage current and future sediment hazards will be determined by knowledge of the problem and available options. A sediment management plan must address the social, environmental, and technical options with a goal of avoiding legal and political pressures in making important decisions. The sediment management plan must consider different alternatives such as ignoring sediment, allowing it to accumulate onsite for future generations to deal with, keeping it out of the reservoir with better upstream management practices, removing it from the reservoir, and flushing it downstream (beneficial in some cases). These management plans are difficult to develop with our present limited knowledge of the problems and hazards associated with the reservoir sediments. Aside from gaining a better understanding of the loss of reservoir capacity due to sediment accumulation, an understanding of possible contaminants within the sediment deposits is also needed. The knowledge of possible contaminants in both the deposits and within the mobilized sediments due to dredging, erosion, and flushing is needed. Increased knowledge of sediment transport in river channels and reservoirs is also needed, requiring the development of models that can describe how the sediments are transported in the different river channels and reservoirs. Along with other applications, these models could be used to determine the minimum flushing flows necessary to minimize reservoir sedimentation and downstream effects. Calibration and confirmation of these models can be obtained with accurate field data.

9.5 Frequency and Schedule of Surveys

The schedule and frequency of conducting reservoir surveys should depend on the estimated rate of reservoir sediment accumulation, along with the current operation and maintenance plan. However, the current need to address site-specific problems, along with available funding, usually determines if and when a survey is conducted. The frequency of resurveys may depend on the estimated rate of sediment accumulation in the reservoir. For example, some have used a 7.5-percent storage reduction between surveys or a 5- to 10-year interval. For Reclamation reservoir surveys, the decision on if and when a survey will be conducted is usually made by the responsible operations office. Influential factors in the decision include occurrence of a large flood, severe drawdown of the reservoir, planned construction of an upstream dam, loss of recreational area due to sediment encroachment, change in erosional characteristics of a basin due to land use or forest fires, raising of the dam, or changes to the reservoir operations. For Elephant Butte Reservoir in New Mexico, the frequency of surveys is set by a compact agreement between the states and Federal Government, using a projected 5-percent loss of capacity (Collins-Ferrari, 1999). The responsible office and available funding determine the method of collection. For example, the decision on whether or not an aerial survey for the above water portion of the data collection is conducted is usually based on cost and the amount of shoreline erosion. The Sedimentation Group works with the responsible field office to obtain the best study results within the allowable budget.

Reclamation has over 400 storage facilities, but only about 30 percent of these have had a resurvey conducted since initial filling. Of these resurveys, about 30 percent have had multiple surveys, with some reservoirs requiring 5 or more resurveys for monitoring high sediment inflow and for developing present area and capacity tables. The majority of these high sediment inflow sites are located in the southwestern United States and include Theodore Roosevelt in Arizona with 8 resurveys, Elephant Butte Reservoir in New Mexico with 11 resurveys, and Lake Mead in Arizona with 3 resurveys. There are Reclamation reservoirs in the state of Wyoming with high sediment yields that have had several resurveys, such as Buffalo Bill Reservoir with 3 resurveys and Guernsey Reservoir with 11 resurveys. All of these reservoirs are located in drainage basins with high sediment yields requiring multiple resurveys for effectively monitoring reservoir sedimentation rates and future impacts.

The schedule of the survey may be determined by methods of collection, weather, and reservoir operations. If aerial data are collected, it is recommended that collection take place when the reservoir is as low as possible and prior to the bathymetric survey. In most cases, this is in the fall, winter, or early spring and allows better coverage due to less vegetation. The bathymetric survey should be scheduled when the reservoir is as full as possible or with as much aerial coverage overlap as possible. This allows complete mapping of the reservoir and speeds up the underwater collection if the aerial collection covered the shallow water and underwater hazard areas. Due to cost, some Reclamation surveys are restricted to underwater collection and use existing above water maps to complete the analysis. For these types of surveys, all attempts are made to schedule the survey when the reservoir is as full as possible, requiring some survey delays during low runoff years. For some surveys, limited amounts of above water data are collected to complete the analyses. Usually, the data are collected in the upper tributaries where exposed sediment deltas had formed (Ferrari, 1996 and 2005).

The decision on when to schedule and at what frequency to conduct a reservoir survey must be made on a case-by-case basis. Current available equipment (i.e., GPS, digital depth sounders, and heave compensators) allows for year-round data collection and has significantly reduced actual field collection time. Advances in equipment technology and data collection techniques have also reduced the staff size and the amount of preliminary field work required by previous survey methods. Presently, collection systems are more compact and require less field staff for setup and operation, reducing the cost of downtime due to extreme weather conditions. However, each project contains unique conditions that must be considered when determining the timing, survey equipment, and frequency of the reservoir resurveys.

Many reservoirs are relatively small in size, requiring smaller survey vessels. Modern survey equipment can be more easily adapted for the smaller vessels. When available, an enclosed cabin on the survey vessel is a desirable option that protects the crew and equipment, and it allows surveys to be conducted safely throughout the year. Equipment can also be purchased that is weather proof, allowing open boat data collection in most weather conditions. Today's equipment has minimized the effect of rough water on data accuracy, but its effect on the collection crew must be considered. Although a nearly full reservoir during a nonrainy season is the best condition for conducting a reservoir survey, the equipment and survey vessel should be set up to cope with all conditions, since they can change at any time. The Sedimentation Group has collected data in all weather conditions (including snow and heavy rain) because the equipment and crew are usually housed in an enclosed cabin on a large, stable vessel.

Other means of determining frequency of reservoir sediment surveys include measured sediment rates from previous surveys and sediment records from inflow streams. In the United States, high operating costs have reduced the number of gauging stations that measure sediment inflow, requiring records from similar reservoirs, gauges, and drainages to be used. Observations of sediment deposition during a reservoir drawdown may also be used; however, as illustrated in figure 9.1, these observations may give a false impression of the severity of the problem if the exposed sediment delta is the majority of the deposition. A reconnaissance type survey can be conducted to periodically measure changes at a few previously established reservoir sediment range lines, but cost must be considered if the collection crew must travel an extended distance to the study site.

In general, larger reservoirs require less frequent resurveys. More frequent surveys are usually required if reservoirs are operating under conditions of greater risk, such as flood control or water supply storage, or if located in metropolitan areas. Small flood control reservoirs on the South Platte River in the metropolitan area of Denver, Colorado, have been resurveyed at about 5-year intervals. For similar type reservoirs, it is recommended that this short interval remain until enough years of data have been collected to determine if long-term sediment deposition trends exceed the original design projections. Fort Peck Lake on the Missouri River in Montana is situated in an isolated area, has a storage volume nearly 600 times larger than these Denver reservoirs, and has half the projected storage depletion rate. Initially, Fort Peck Lake was resurveyed at 5- and 10-year intervals until it became evident that the long-term depletion rate was substantially slower than originally expected. The resurveys are currently scheduled on 20-year intervals.

In Taiwan, capacity lost due to sedimentation is much more critical because of the limited capacity available within the total reservoir network and the relatively high percent of annual loss of capacity due to higher average sediment yields compared to the United States. For these types of situations, the reservoir resurvey interval should still be based on the individual reservoir, drainage basin characteristics, and need. In some cases, the interval may need to be as short as 1 year or following each major storm event that might generate a high sediment inflow and have a dramatic impact on a small capacity reservoir. In general, smaller reservoirs in Taiwan with higher sediment inflow would require shorter collection intervals than for reservoirs in the United States that are generally larger and have lower rates of sediment inflow (Yen, Pei-Hwa, 1999). Present collection systems and analysis software have made it possible to measure impacts from individual storms, but the need for and benefit of such information must be determined.

An additional factor in the survey schedule is the inflow of unconsolidated material that may create a soft reservoir bottom and erroneous echo sounder depths. The use of low frequency sounders, along with depth verification may provide quality assurance (QA) of the depth measurements. However, these additional verifications, add time to the collection and concerns about the accuracy. The lower frequency echo sounders can penetrate the soft layer and provide depths of the harder bottom, but these depths could be somewhat subjective to what the true bottom is. It would be best to avoid such conditions, but for some reservoirs, these soft bottom reservoir conditions always exist. For soft bottom reservoir surveys, echo sounder depths should be confirmed by manual measurement, despite the extra cost. However, manual measurements are somewhat subjective to individual judgment and are difficult in deeper reservoirs. The soft bottom fluff conditions appeared to be a factor during the December 2004 and May 2005 Lake Powell surveys (Clarke, 2005). In 2005, a multibeam survey was conducted from May 12-21, 2005, on the entire length of Lake Powell. During low and high frequency depth collection on the upper San Juan reach, the high frequency readings were, at times, several meters shallower then the low frequency readings, indicating the soft fluff bottom of the reservoir's inflowing sediments.

9.6 Reservoir Survey Techniques

Survey techniques have evolved around the development of equipment and analysis systems. Prior to computerized data collection and analysis systems, the **range line method** was commonly viewed as the only practical method for collection due to its relatively low field and analysis costs (Blanton, 1982). The range line method was used most often on medium to large reservoirs and on river modeling studies requiring underwater data collection for monitoring changes.

For reservoirs, the collection and analysis consists of determining sediment depths along predetermined range lines (usually established prior to inundation). Analysis requires detailed and accurate original reservoir topography. Various mathematical procedures have been developed to produce the revised reservoir contour areas at incremental elevations for the surveyed range lines. The range line method is still a valid means of conducting survey studies for certain reservoir conditions or if more modern collection and analysis systems are not available. For the 1986 Lake Powell survey (Ferrari, 1988), the range line collection and analysis method was used due to very deep (greater than 500 feet at the dam) vertical wall conditions and good original topographic maps. It now is possible to completely map Lake Powell in detail using a GPS, multibeam system, and aerial collection, but the range line method should still be considered, since it would be less costly for collection and analysis. Multibeam surveys on a large portion of Lake Powell in 2004 and 2005 covered many of the range lines surveyed in 1986. The multibeam surveys covered in days what took weeks to cover during the 1986 survey. A report, *Reconnaissance Technique for Reservoir Surveys*, presents the results from these surveys and a modified range line method to generate updated area–capacity tables (Ferrari, 2006).

For river collection, the field procedure is similar to reservoir collection where predetermined range lines are selected prior to the underwater collection. The range lines are usually established perpendicular to the riverflow and are used for monitoring and numerical modeling of the changes in the river over time. This method requires the survey vessel to run the underwater portion along the alignment of the selected range lines in a predetermined direction, but river conditions such as flow velocity and location of shallow water areas usually dictate the actual alignment. Advances in equipment and computer analysis systems now allow much more variance in river collection techniques, resulting in the collection of much more data in a safer manner. Many USACE river surveys are conducted in rivers utilized for navigation by large transport vessels. This large boat traffic hinders data collection by conventional range line method where the survey vessel must run from bank to bank perpendicular to the river alignment. Current collection systems allow data to be collected continuously in a diagonal direction or along the alignment of the river in a safer fashion and with enough density to generate detailed contours of the river channel. If needed for study purposes, there are computerized routines that can interpolate cross sections or range lines from these developed contours.

The **contour method** has become the preferred method for data collection and analysis with the development of electronic collection and analysis systems. It requires large amounts of data to be collected and stored, something that present systems can easily handle. The contour method results in more accurate reservoir topography and computed volumes than the range line method, but it usually takes more time for field data collection. This method revolves around computer and software packages that provide a means of organizing and interpreting large data sets. Contour development and analysis may be quicker than the range line method. The hydrographic survey data is usually collected in an x, y, z coordinate format conforming to a recognized coordinate system such as Universal Transverse Mercator (UTM), latitude/longitude, state plane, or other systems that represent the Earth's 3-dimensional features on a flat surface.

The most accurate contour map product is obtained when both the above and below water portions of the reservoir area are surveyed. The ideal contour map is developed by photogrammetry (aerial) when the reservoir is empty, exposing all areas to be measured, but this condition seldom occurs, making a combination of aerial and bathymetric survey necessary. To reduce the time and cost associated with underwater data collection, aerial data should be collected when the reservoir is as empty as possible, and the bathymetric survey should be conducted when the reservoir is as full as possible, providing maximum overlap of the two data sets. Surveying the underwater portion after the aerial survey with a large overlap reduces the time and cost, since the survey boat does not have to maneuver in shallow water portions already mapped by the aerial survey.

Due to cost of aerial data collection, some contour reservoir resurveys do not include an updated survey of the area above the existing reservoir water surface. For these surveys, the bathymetric survey should be scheduled when the reservoir is as full as possible. The above water area may be measured using the original or most recent contour map of the reservoir area. In this case, it is assumed that no change has occurred since the above water area was last mapped. Some Reclamation resurveys have used U.S. Geological Survey (USGS) quadrangle maps for the above water areas, since it was the best data available. It must be noted that an assumption of no change can cause computation errors for reservoirs with significant shoreline erosion or where the majority of the sediment settles in the shallow upper end not mapped by the bathymetric survey.

Recent improvements in conventional survey equipment (GPS) allow accurate measurement of point data and provide a cost-effective method for smaller reservoirs. A combination contour and range line method may be used where the range line method is used to measure the areas of exposed sediment deposition, as was done for the 1994 Boysen Reservoir Sedimentation Survey (Ferrari, 1996). This method does not accurately measure the surface area of the above water areas where significant reservoir changes have occurred due to bank erosion, but it is a viable alternative for measuring exposed sediment deltas in the upper reaches of the reservoir.

There are many contour development software packages on the market. The Sedimentation Group develops contours for the reservoir area from the compiled data using various methods; the most common is the triangular irregular network (TIN) package (Environmental Systems Research Institute, 2005 and HYPACK, 2005). A TIN is a set of adjacent, nonoverlapping triangles computed from irregularly spaced points with x, y, z values and was designed to deal with continuous data such as elevations. Triangles are formed among all collected data points, including boundary points, preserving each collected survey point. A digitized polygon enclosing the collected data can be developed such that interpolation is not allowed to occur outside the boundary. A linear interpolation option is then used to interpolate contours from the developed TIN.

9.6.1 Shoreline Erosion

The 2002 Tiber Reservoir underwater survey showed extensive shoreline erosion throughout the reservoir area. During collection, the GPS positions were found, at times, to be outside the digitized USGS quadrangle contour location, indicating that the boat was on dry ground. These USGS quadrangle contours were developed from aerial photography taken in the 1960s. At times, the position of the boat was found tens of feet outside their boundary. In addition, a major windstorm occurred during the 2002 survey, and the crew witnessed vertical sections of the shoreline collapsing into the reservoir area for days afterwards. Even with the shore erosion, the survey vessel was, at times, able to hug the vertical banks in deep water where previous collapses into the reservoir had occurred. It appears that, over time, the collapsed material washed further into the reservoir by wave action similar to shore ocean waves. This is possible because the shoreline material dissipated in the water and consisted of little to no rock or large cobble material. Figures 9.2 through 9.5 document these shoreline conditions at Tiber Reservoir (Ferrari and Nuanes, 2005).

The photographs show different stages of the shoreline erosion, along with the extent of occurrence. If the erosion were just below the reservoir high water mark, the total volume of the reservoir would not be greatly affected. What occurred in the upper reservoir elevations resulted in a gain in surface area and volume. This volume gain at the higher reservoir elevations offsets the loss of surface area and volume in the lower elevations of the reservoir due to the eroded shore material depositing at the lower elevations. The photographs show the large amount of the eroded material above the reservoir area, meaning that a portion of the loss of the original total reservoir volume is due to the shoreline erosion, along with the incoming river sediments. The only means to accurately measure the extent of the shoreline erosion would be an aerial and full bathymetric survey. Reconnaissance surveying techniques cannot be used in reservoirs with these types of conditions (Ferrari, 2006).

Figure 9.2. Eroded material depositing forming a shelf (photo by S. Nuanes).

Figure 9.3. Large areas of erosion above the reservoir maximum water surface (photo by S. Nuanes).

Figure 9.4. Recent eroded material that has not moved further into the reservoir (photo by S. Nuanes).

Figure 9.5. Eroded bank material depositing below the water line (photo by S. Nuanes).

9.6.2 Data Density and Line Spacing

The extent of data collection is determined by the project needs, reservoir conditions, cost of collection and analysis, and capability and limitations of the collection system. Typically, the GPS horizontal positions can be updated once per second, a single beam electronic depth sounder can provide continuous output of 20 or more depths per second, and a multibeam underwater collection system has the capability of several hundreds of thousands of points per minute. The advancement in the computer collection systems allows all of these data to be stored, but it is up to the study manager to determine what system and collection interval is necessary and practical. During collection, the most advanced available system should be used and the maximum amount of data should be stored. Filtering of the data that may be necessary for final computations should be conducted during data postprocessing.

For single beam collection systems, survey line spacing must be selected to provide the needed density for the study results. The study manager must understand the goals of the study and must determine the data density to meet the goals while staying within budget. The range line method assumes uniformity of the terrain between the survey lines, which is a valid assumption unless an abrupt change occurs. The challenge is knowing if and where abrupt changes occur and spacing the lines to best represent the bottom conditions. The survey crew needs to monitor the survey line during collection for possible changes and examine existing topographic maps that may warrant a modification of the line spacing during field collection. Typically, about 5 percent of the project study area is covered by the single beam collection method, which means care must be taken to collect adequate data to ensure accurate topography development.

The Sedimentation Group's single beam collection method typically begins with a 300-foot spacing and adjusts in the field to meet the study objectives. For smaller reservoirs and to show more bottom details, the data collections may be adjusted to 100- to 200-foot spacing. For some of the larger reservoirs, with flat bottom conditions with little or no detail, and when collection time and budget is limited, the spacing has been adjusted to 500, 600, and at times 2,000 feet. The upper delta of Canyon Ferry Reservoir in Montana was fairly flat with little to no channel detail in the deposited sediment. Those conditions permitted the collection crew to increase the profile spacing, allowing data collection during favorable weather conditions and reducing field collection time while maintaining the quality of the product (Ferrari, 1998). For the Salton Sea survey in California, the range line spacing was adjusted to 2,000 feet due to the limited budget for data collection and the relatively flat unchanging bottom conditions (Ferrari, 1997). The Canyon Ferry and Salton Sea surveys were conducted on large water masses with assumed uniformity of terrain between surveyed range lines justifying such large spacing. Parallel surveyed range lines and perpendicular survey lines confirmed the uniform bottom assumption for these large water surveys.

The use of a multibeam collection system provides the capability of full bottom coverage of the underwater reservoir areas, but it requires more time for collection and analysis than many budgets will allow. The multiple-transducer and multibeam collection systems can provide 100-percent coverage that removes the unknowns between the survey line spaces, but the costs and operation of such systems are more difficult to justify. It is up to the study leaders to determine the extent of collection to meet the study goals within the budget. For the 2001 Lake Mead study, the collection was limited to the original river channel areas where the majority of the sediment deposition was projected to occur. Only about 30 percent of Lake Mead was covered by the multibeam survey, but the 20 million data points mapped the majority of the submerged, deposited sediment elevations that could be collected by the survey vessel. This allowed the field collection to be accomplished in 3 weeks and within budget, while obtaining the needed detail to meet the study needs (Ferrari, 2006) (Twichell, Cross, and Belew, 2003).

9.6.3 Cost of Conducting a Reservoir Survey

Survey productivity has increased by a factor of 75 since the 1960s and a factor of 10 since the 1990s (USACE, 2004). The productivity increases are mainly related to electronics and computer development. Planning a survey is usually controlled by a budget that determines detail and method of collection, analyses methods, and who will be conducting the study. Before the use of electronic positioning systems, the collections were conducted using visual or manual distance tag lines. The manual method required significant setup time to establish range line locations and utilized large survey crews, often requiring two to three vessels and crews of five to eight people to conduct the survey. Depending on conditions, the crews were able to collect data from one to five range lines per day. Computer microwave system development reduced the crew size during collection, but it still required significant amount of time prior to the underwater collection to locate and establish control around the reservoir and river study areas. The field crew size during the underwater collection was usually around five, but as few as three could complete smaller jobs. Collection of sediment range line data increased from 5 to 10 range lines a day, but the major benefit was the possibility of detailed mapping of the reservoir bottom. The development of GPS hydrographic collection systems significantly reduced the time and cost of a survey by increasing field collection productivity and decreasing the number of staff days required to conduct the overalls study. The greatest cost savings occurred because the detailed control network required prior to the underwater survey was significantly reduced.

The Sedimentation Group conducts the majority of their surveys using sonic depth recording equipment interfaced with a real-time kinematic (RTK) GPS that gives continuous sounding positions throughout the underwater portion of the reservoir covered by the survey vessel. The RTK GPS system allows control to be established in hours, rather than days or weeks with conventional land surveys. The hydrographic crew size is usually two, compared to the previous three to five crew members. One result of using GPS and field computers is the automated collection and storage of massive amounts of data. For multibeam systems, the initial cost is significant, meaning workload and budgets should be sufficient to support the cost and necessary personnel for operation. The major benefit of this system is full bottom coverage with greater detail and less uncertainty in the results. For many studies, the mobilization and demobilization costs can exceed the actual survey cost. For small survey jobs, the Sedimentation Group attempts to schedule more than one survey per trip to reduce this cost. An experienced collection crew can significantly reduce the cost, since less time is needed for planning, preparation, and training, which allows the work to be conducted more efficiently and safely.

9.6.4 Selecting Appropriate Hydrographic Data Collection System and Software

The goal of any collection program is to obtain the highest quality data possible with the equipment available to the survey crew. Currently available hydrographic surveying equipment has the capability and flexibility to be used for bathymetric surveys in small and large reservoirs, and rivers, and for above water surveys. The equipment can be utilized for many different collection techniques, such as range line and contouring. The demand for data collection and required accuracy, along with the training and experience of the users, should dictate the type and quality of the equipment. Continual upgrades in location and sonic sounding instrumentation, along with computer collection system electronics, have greatly improved accuracy and speed of collection. These improvements have been so dramatic that resurvey data now has the potential of being more accurate than any of the previously collected data, even data collected prior to inundation, when most of the reservoir area was exposed. Technology has reduced the overall time and cost of data collection and analysis for hydrographic surveying projects and provided larger quantities of higher quality of data. Previous systems required more planning, extensive surveys to establish ground control, and larger field crews to complete the job.

When determining reservoir resurvey needs, selecting a reservoir survey data collection system, or establishing a reservoir survey program, several basic questions should be asked:

- What is the primary goal of the study?
- What are the needs or requirements of the final product?
- What is the desired or necessary level of detail?
- How often will the system be used?
- What experience, interests, and longevity do the operational personnel possess?

Other factors to consider are the cost of purchasing the system, the operational cost, and possible lease of the collection system, or components that would provide the latest technology. One potential problem with leasing may be the amount of time necessary to gain the expertise to use the system properly. Many of the commercially available software packages accept instrumentation from several manufacturers, so that after the user becomes proficient operating the software, adding and removing instrumentation becomes less of a concern. After addressing these issues and estimating associated expenses, it may be concluded that the more cost-effective solution is to contract out the study or studies.

The continual development of computer and electronic instrumentation has greatly changed the data collection and analysis methods for reservoir sedimentation surveys and is expected to bring additional changes in the future. This change requires collection programs to be flexible. There are no generic packages that meet all survey requirements. It is recommended that the system be built around the software that will be used for field data collection and analysis, requiring the equipment that is purchased to be adaptable to various software packages. There are some users that collect small amounts of data for one small project without hydrographic or survey software. Conducting the study without some form of software is very labor intensive during field collection and analysis and also increases the potential for collection and analysis errors.

Writing customized software for the system needs is a possibility since purchasing hydrographic software can be costly, ranging from $5,000 to $20,000. However, compared to the total cost of the hydrographic survey system and the time required to develop and maintain such programs, the cost of software procurement is less prohibitive. An additional concern with writing customized software is the availability and longevity of the programmer for making modifications to the software. While existing commercial software may not meet all of the needs initially, several software vendors provide customization of their software to meet the customer's needs.

Commercial software is often upgraded to incorporate the latest technology and upgrades are usually available to the users for an annual maintenance fee. Some hydrographic crews use the hydrographic software for the field collection and initial processing only, and then develop the final product using different software. Reclamation's Sedimentation Group collects and processes field data into x, y, z data sets with a portable computer and hydrographic software, then completes the contour mapping and final calculations with a desktop computer and commercial contouring software.

There are several versatile software packages capable of simultaneously receiving data from multiple devices during collection and processing the collected data for complete analyses. Some options include collection, postprocessing, and editing of single and multibeam data, geodesy transformations, tide corrections, TIN modeling, and volume computations. The packages vary but usually contain internal drivers that support equipment from numerous manufacturers for range-azimuth, range-range, mapping and survey grade GPS, and single and multibeam depth sounder instrumentation. In addition to collecting data from numerous instruments simultaneously, the computer and software can be set up to integrate data from various sensors such as gyros, acoustic systems, heave-pitch-roll indicators, magnetometers, and seabed identifiers. These computer programs are for the frequent user, due to the considerable time required to become proficient in the use of the software and survey instrumentation. These programs are very powerful, have a worldwide customer base, provide technical support by the phone and internet, and are usually adapted to the latest technology. Recent and continuous improvements of portable computers are making it easier to install and utilize software and numerous instruments on smaller survey vessels. The computer purchased that is purchased should be as advanced as possible with enough hard drive space for 5 to 50 megabytes (MB) of software and 1 to 2 MB of data storage per day for single beam data or as much as 10 MB per hour for multibeam data. Since computers may only have one serial port, additional hardware containing multiple ports will be needed to accommodate the number of measuring instruments in the system.

9.7 Hydrographic Collection Equipment and Techniques

Hydrographic survey equipment has transformed dramatically throughout its history, with the greatest changes occurring over the last decade. The latest major change in horizontal positioning is the GPS, which is more accurate and less costly to operate than past survey methods. No former positioning system has been so rapidly adapted to hydrographic collection systems. The most recent significant development in depth sounders is the multibeam system that allows massive amounts of data to be collected. The multibeam system provides the option of complete coverage of the underwater areas, thus removing the unknowns of previously unmapped underwater areas.

Equipment for hydrographic surveying varies, depending on reservoir size, field conditions, availability, familiarity to the collection crew, and cost. The positioning equipment for hydrographic surveys has varied over time, with the latest major change for measuring horizontal positioning being GPS. Although relatively new and still undergoing technological advancements, GPS is more accurate and less costly to operate than previously used conventional survey methods and has been rapidly integrated into hydrographic collection systems. GPS is a very versatile instrument for measuring horizontal positions, but it is not ideal for all reservoir and river situations. Past horizontal positioning equipment and techniques are still viable where site conditions may prohibit the use of GPS. Such systems include marked tag lines stretched along the range line, electronic distance meters that measure distances from a known point to the survey boat as it proceeds along the range line, range-azimuth positioning that involves the intersection of an angular and distance observation, and range-range positioning where survey vessel distances are measured from two or more shore stations. These previously used techniques and equipment are still viable positioning methods as described in more detail in other available publications (Blanton, 1982 and USACE, 2004):

- Constant boat speed, usually for reconnaissance type surveys
- Cutting in, or triangulation-intersection that uses two transits or theodolites
- Stadia rod and common transit for reconnaissance type surveys
- Tag line or calibrated cable for measuring range lines
- Electronic distance meter (EDM) for measuring distance from shore to boat on range lines
- Range-azimuth measuring of angular and distance observations from a reference station.
- Microwave range-range was the preferred method prior to GPS

9.8 Global Positioning System

The GPS has rapidly become the preferred positioning system for hydrographic surveying and does not require the time-consuming calibrations necessary for previously used systems. Although relatively new and continuously undergoing technological advancements and operational policy changes, GPS-based systems are the most accurate, least expensive to operate, and versatile positioning systems for obtaining positions of static monuments or moving platforms. Previous systems, such as range-range and range-azimuth, are limited to line-of-sight coverage from the shore-based earth stations, while GPS does not have that limitation. GPS can operate in all weather conditions requiring only a clear view of the sky.

Navigation Satellite Timing and Ranging (NAVSTAR) GPS is an all-weather, radio-based, satellite navigation system that enables users to accurately determine three-dimensional positions (x, y, z) worldwide. The NAVSTAR system's primary mission is to provide passive global positioning and navigation for land, air, and sea based strategic and tactical forces and is operated and maintained by the United States Department of Defense (DOD). The GPS receiver measures the distances from the satellites and determines the receiver's position from the intersections of the multiple range vectors. Distances are determined by accurately measuring the time a signal pulse takes to travel from the satellite to the receiver.

The NAVSTAR system consists of three segments:

1. The space segment is a network of 24 satellites maintained in precise orbits about 10,900 nautical miles above the earth, each completing an orbit every 12 hours. Satellites are spaced in orbit so that a minimum of 6 satellites is always in view anywhere in the world 24 hours a day. The satellites continuously broadcast the position and time data used by the receivers.

2. The ground control segment tracks the satellites and determines their precise orbits. The main control station is in the United States, in Colorado Springs, Colorado, with additional monitoring stations located throughout the world. The control stations determine and

periodically transmit ephemeris parameters (such as the satellite's position, correction, health status, and other system data) to each satellite, whereupon the data are retransmitted to the user segment receivers.

3. The user segment consists of the GPS receivers located worldwide for both military and civil activities for many different applications in many different conditions at air, sea, or land-based locations. The individual receivers process the NAVSTAR satellite broadcast signals and calculate their position.

The GPS receivers use the satellites as reference points for triangulating their position on earth from distance measurements to the satellites. To calculate the receiver's position on earth, satellite distance and position in space are needed (determined by ground control). The satellites transmit signals to the GPS receivers for distance measurements, along with data messages about their exact orbital location and operational status. The satellites transmit two "L" band frequencies for the distance measurement signals, called L1 and L2. Modulated on these frequencies are the coarse acquisition (C/A) and precise (P) codes. An additional message contains the satellite ephemeris and health status.

A minimum of four satellite observations is required to mathematically solve for the four unknown receiver parameters (latitude, longitude, altitude, and time). The time unknown is caused by the clock error between the expensive satellite atomic clocks and the imperfect clocks in the GPS receivers. For hydrographic surveying, the water surface elevation parameter may be measured by means other than GPS. This means that only three satellite observations are theoretically needed to track the survey vessel. However, to obtain the most highly accurate position, the survey vessel GPS receiver tracks all available satellites.

There are high-grade GPS collection systems with RTK surveying capability that accurately measure the position and altitude of the moving survey platform with obtainable centimeter accuracies for both horizontal and vertical measurements. RTK needs a minimum of five satellites for initialization, but after initializing, it can collect high precision data with a minimum of four satellites (Chisholm, 1998).

All GPS solutions depend on the accuracy of the known coordinate position of each observed satellite and the relative geometry of the satellites. The accuracy of a GPS-measured position can be characterized by its geometric dilution of precision (GDOP). GDOP describes the geometrical uncertainty and is a function of the relative geometry of the satellites and the user. Generally, the smaller the angle between the satellites and receiver, the greater the GDOP value, and the lower the measured precision. GDOP is broken into several components, such as position dilution of precision - x, y, z (PDOP) and horizontal dilution of precision - x, y (HDOP), where the components are based on the geometry of the satellites and should be monitored and recorded during the survey.

There are two basic operation methods to obtain GPS positions: absolute and differential.

9.8.1 Absolute Positioning

Absolute positioning normally involves only a single GPS receiver and is not accurate enough for use in most hydrographic positioning. A single GPS receiver's absolute position is not as accurate as it appears in theory because of range measurement precision and the geometric position of the satellites. Precision is affected by several inherent factors that cannot be eliminated but can be minimized. These factors are time (because of the clock differences), atmospheric delays (caused by the effect of the ionosphere on the radio signal), receiver noise (due to quality of GPS receivers), and multipath (errors caused by signal arrival by different paths, usually due to signal reflecting from obstacles). Due to these factors, estimated absolute real-time position accuracies of only ±10 to 16 meters are common. The absolute mode for positioning does not provide sufficient positional accuracy or precision for the majority of hydrographic survey studies.

Previously, the largest error source in GPS collection was caused by false signal projection, called selective availability (S/A). The DOD implemented S/A to discourage the use of the satellites as a guidance tool by hostile forces. Positions determined by a single receiver when S/A was active had errors of up to 100 meters 95 percent of the time horizontally and up to 180 meters 95 percent of the time vertically. S/A was eliminated in May 2000, but the absolute positioning of a single receiver caused by other error sources is still only around ±10 meters and usually will not satisfy the majority of hydrographic surveying requirements. The error sources can be eliminated or minimized by using Precision Positioning Units or differential GPS techniques.

The GPS satellites provide two levels of navigation services: Standard Positioning Service (SPS) and Precise Positioning Service (PPS). SPS receivers use available GPS information broadcast to anyone in the world, but security devices such as the anti-spoofing (A/S) factor guard against fake transmission data by encrypting the P code with a classified Y code denying the SPS user the higher P code accuracy. The position accuracies for SPS receivers are only around ±10 meters. None of these factors significantly affect the GPS users operating with differential positioning techniques.

Precise Positioning Service is an accurate worldwide positioning technique available to the military. With DOD authorization, nonmilitary government agencies can utilize PPS, which has the capability of deciphering the encrypted GPS signals. The encrypted or anti-spoofing P and Y codes guard against fake transmissions of satellite data and are available only to DOD authorized users. These PPS receivers can use the Y code and provide predictable autonomous horizontal positioning accuracies of ±4 meters. These positions are possible in real time, since they are not subjected to S/A. This is an absolute positioning mode with obtainable ±4 meters accuracy that, in some cases, meets the standards and requests of the current hydrographic survey systems. Prior to GPS, most hydrographic surveying systems operated with an accuracy of ±2 to 5 meters. Most studies now are calling for much greater accuracies and require the use of systems such as differential GPS (DGPS).

9.8.2 Differential Positioning

Differential positioning requires at least two receivers and can provide precisions necessary for real time hydrographic surveying. One method of collection to resolve or cancel the inherent errors of GPS (satellite position or S/A, clock differences, atmospheric delay, etc.) is called differential GPS. Differential surveying is the positioning of one point in reference to another. The basic principle is that errors calculated by multiple GPS receivers in a local area would all have common vectors. DGPS determines the position of one receiver in reference to another and is a method of increasing position accuracies by eliminating or minimizing the uncertainties. Differential positioning is not concerned with the absolute position of each unit, but with the relative difference between the positions of two units simultaneously observing the same satellites. The inherent errors in satellite positions and atmospheric delays are mostly canceled because the satellite transmission is essentially the same at both receivers.

The method includes setting one receiver over a known geographical benchmark programmed with known coordinates. This receiver, known as the master, base, or reference unit, remains over the known benchmark, monitors the movement of the satellites, and calculates its apparent geographical position by direct reception from the satellites. The inherent errors in the satellite position are determined relative to the master receiver's programmed position, and the error corrections or differences are applied to the mobile GPS receiver on the survey vessel.

The attainable accuracies using differential survey techniques usually depend on the grade or cost of the GPS receivers. The average grade-mapping receivers that determine differential positions from the L1 frequencies can obtain accuracies of ± 2 to 5 meters. The better grade-mapping receivers or low-end, survey grade receivers that determine differential positions from both the L1 and L2 satellite frequencies can obtain submeter accuracies. The high-end survey grade receivers can obtain subcentimeter accuracies. Hydrographic survey system needs to collect the most accurate survey information possible, but cost and need must be considered when assembling a system. For reservoir mapping, water capacity, and sediment volume situations, 1- to 2-meter position accuracy is generally considered an acceptable system at this time, but consideration must be given to collecting the most accurate data possible with the latest available technology.

There are different ways to apply DGPS collection methods to hydrographic surveying, including postprocessing and real-time DGPS. During postprocessing DGPS, the master and mobile GPS receivers record satellite tracking observations simultaneously as standalone units with no active data links between them. The master station is located at a known datum in the study area, or a community GPS base station is used. Differential correction software is used at a later time to combine and process the collected data. With this method, the mobile GPS receiver will be in the absolute mode and provide erratic positioning and tracking information during the time of collection. This method is not recommended if a precise range line survey method is used, but it is a valid means for collecting contour data if care is taken to ensure complete coverage.

To be valid for the study area, the master or reference station must be placed on a known survey monument located in an area having an unobstructed view of the sky. The GPS antenna should not be located near objects that would cause multipath or interference. Areas to avoid would be near other antennas, microwave towers, power lines, reflective surfaces, and other obstructions such as vertical walls, dams, and vegetation. In the United States, GPS community base station information is available from several sources including Federal, state, and county agencies along with local universities. There are also commercial services available for real-time differential data that may also provide data for postprocessing. For community GPS base stations, there are some limitations that need to be addressed using the provider's specifications.

Real-time DGPS is the current standard for hydrographic positioning. To collect data using real-time DGPS, a master receiver is stationed over a known datum where it computes, formats, and transmits correction information through a data link to the mobile GPS receiver on the survey vessel. The mobile GPS receiver requires the data link to receive the transmitted GPS corrections from the master receiver. Collecting in real time allows highly accurate positioning data to be merged with other collected data, such as depths, into a single file that significantly reduces the postprocessing.

In an attempt to standardize differential correction transmissions, the Radio Technical Commission for Maritime Services developed the RTCM-104 format for GPS that allows transmitted master station information from several sources. There are some community base stations maintained by United States Federal, state, and local government offices that transmit correction information that can be utilized by any manufacturer's mobile receiver. The U.S. Coast Guard operates high-frequency DGPS radio beacons positioned along coasts and major waterways, such as the Mississippi River and the east and west coasts. There are also commercial services that offer real-time correction information that is transmitted in RTCM-104 format from such devices as geostationary satellites and local radio station towers. One problem with these services is signal destruction when surveying in areas with canyon walls and vegetation, but radio repeaters can be used to resolve this issue. Currently, these sites may not provide the centimeter accuracy corrections needed for some studies. The Sedimentation Group uses DGPS with a master receiver set over a known geographical benchmark in the study area. It must be noted again that obtaining centimeter accuracies requires a real-time survey grade GPS with the master receiver located over a known datum near the survey vessel. The master GPS receiver transmits the calculated difference correction information through available communication data links such as high frequency (HF), very high frequency (VHF), and ultra-high frequency (UHF) radios. HF radios transmit over long distances but require a relatively large antenna. VHF and UHF radios are small, lightweight, and require smaller antennas, but the radio signals are somewhat limited to line-of-sight.

UHF and VHF systems are commercially available at a lower cost than other systems and are the recommended communication systems for DGPS. The disadvantages are the limited range due to the line-of-sight, licensing issues, and the effects of signal shadowing and multipath. All frequencies must be authorized for operation to avoid interference with other activities in the area. Allocations of the frequencies are handled through the National Telecommunications and Information Administration of the U.S. Department of Commerce, and it is the responsibility of the purchaser, user, and vendor to gain approval before any transmission occurs. Other possible means for transmitting the correction signal are microwave radios and cellular telephones.

The biggest weakness in all real-time collection systems is the communication link between the master and mobile GPS receivers. Surveying on open water removes many of the obstacles, but communication problems can occur with all systems when surveying in areas with obstructions such as mountains, cliffs, vegetation, and structures along the shoreline. When these situations occur, the flexibility of the hydrographic survey crew in being able to move the master receiver to new locations makes surveying more viable but, at times, more costly.

Plotting of position data during and after collection may indicate any problems with the GPS signals. Some collection software programs monitor parameters such as the differential correction signal and HDOP values where loss of correction signal or high HDOP values cause an error signal to be broadcast to the collection crew, indicating low-quality data is being measured. When these conditions occur, the collection can be delayed until the condition improves.

In theory, the master and mobile GPS receivers are simultaneously observing the same satellites, but, in practice, this is not always the case. The survey crew must be aware of this situation and take all practical measures to obtain simultaneous satellite observations. The best way to accomplish this is to place the master unit near the area to be surveyed. In the northern hemisphere, the satellite observations are obtained in the southern skies. For reservoirs with high bank and vegetation conditions, it may be best to place the master unit on the north bank of the reservoir, where the southern hemisphere is visible, allowing the master unit to obtain the majority of the satellite signals. The mobile GPS receiver on the boat should always be monitored, especially when surveying near obstructions that could affect the satellite and radio signals. Monitoring the mobile receiver includes checking PDOP or HDOP readings and tracking the survey vessel path on a helmsman display where a large jump in correction information may mean a loss of the differential signal, bad satellite geometry, or multipath.

9.8.3 Real-Time Kinematic GPS

RTK GPS in hydrographic surveying provides the highest precision of positioning. The major benefit of RTK versus DGPS is that precise heights can also be measured in real time. This is a major benefit for surveys in tidal and river conditions. The basic outputs from an RTK receiver are precise three-dimensional coordinates such as latitude/longitude/height with accuracies on the order of 2 centimeters horizontally and 3 centimeters vertically. Kinematic GPS employs at least two receivers that track the same satellites simultaneously, just like DGPS. The receivers track the L1 C/A code and full cycle L1 and L2 carrier phases observable even during periods of P-code encryption. The additional data logged from the second frequency facilitates faster resolution of the ambiguities that allow on-the-fly centimeter level measurements.

RTK GPS uses the carrier phase, rather than a code phase used by DGPS. Initializations of the RTK receivers are required at the start of survey and after continuous tracking of all available satellites signals are hindered. The hindrance or loss of a satellite signal in hydrographic surveying can occur while surveying under a bridge or near blockage by trees, dams, and surrounding topography such as vertical walls. RTK GPS receivers may take about 1 minute for initialization while satellite information is acquired for accurate positioning. Once initialization has been gained, the hydrographic survey can begin. With DGPS, the accuracy computations are instantaneous after acquiring the satellite signals and require a minimum of three satellites (if only horizontal positioning is measured). RTK needs a minimum of five satellites to start initialization, but only four satellites are needed when surveying. RTK may require 1 minute or more to reinitialize after reacquiring the satellite signal, but it can provide 2- to 3-centimeter accuracy if the base station is located close to the survey site. These real-time measurements allow accurate monitoring of the elevation changes of river, tide, and reservoir water surfaces, along with survey vessel elevation changes due to squat, weight changes, and water surface swells. The real-time centimeter accuracy also allows the survey crew to establish any necessary control around the study area, further reducing the field time for conducting the survey. By default, the RTK receives output precise 3-dimensional coordinates in latitude, longitude, and height in the GPS datum (WGS84), but there are options to output the coordinates and height for a selected local datum.

9.8.4 GPS Errors

GPS measurements are affected by similar sources of errors as previous collection systems (humidity and multipath, for example). GPS also must contend with the ionosphere and troposphere layers of the earth that delay the satellite signals that travel from 20,000 kilometers out in space. The tropospheric error, up to 3 meters, includes humidity that can delay the signal. Satellites low on the horizon send signals across the face of the earth through the troposphere, while satellites directly overhead have less troposphere to deal with. Masking the horizontal angle of the mobile receiver from 10 to 15 degrees above the horizon minimizes the tropospheric error. The ionospheric error includes sunspots and other electromagnetic phenomena that can cause GPS errors up to 30 meters during the day and 6 meters at night. The errors can be estimated, but the assumption that the error is the same at both the reference and mobile receivers eliminates it when using DGPS techniques.

Multipath occurs when the signal to the GPS receiver is reflected and is not a direct signal from the satellite. This reflection can occur below and above the GPS antenna and can be difficult to identify. The occurrence of multipath is less over water, but it is still present and can be constantly changing. Careful placement of the GPS receiver antenna can avoid areas of multipath such as rock outcrops, metal roofs, buildings, cars, ships, dams, and chain link fences. It is best to set the reference station away from these conditions, but another solution may be to increase the height of the GPS antenna to reduce the possibility of multipath. Masking out the satellite signals below 10 to 15 degrees above the horizon can also reduce multipath. Antenna ground planes are designed to reduce the effects of multipath and should be used for both the master and mobile GPS locations when practical. It must be noted that multipath occurrence from the satellites to a receiver, at a static location, can change over time, since the satellite location is constantly changing.

There are no prescribed calibration requirements for GPS as there were with previously used positioning systems, but there are items that can be monitored to check for operator errors. One item is an incorrect project or geodetic reference datum. Most GPS units output coordinates in latitude/longitude/height on the WGS84 ellipsoid, and the collection software converts the collected information into the project's coordinate system, such as state plane or UTM. The operator must confirm these conversion calculations prior to starting the data collection. It is recommended that, after programming, the GPS rover receiver be set over a known datum point to confirm the output coordinates, but, for many study locations, these known points are not easily available. It is also recommended that a map of the study area be available during collection to confirm locations during the survey by plotting the survey vessel at known locations on the map. Common errors that should be doubled checked are incorrect master station coordinate values, incorrect GPS antenna heights at either the master or rover receiver, or DGPS mode not operational due to a lost correction signal or radio not turned on at the base location. It must be noted that multipath effects are not eliminated by calibration, since they are dependent on the GPS antenna location for both the master and rover units.

9.9 Horizontal and Vertical Control

The basic horizontal control for many Reclamation projects varies from region to region and from project to project (Bureau of Reclamation, 1981). There were many project datums located and developed with conventional survey equipment on local horizontal and vertical coordinate systems. There are some projects that were tied to National Geodetic System (NGS) or USGS monuments but have not been referenced or adjusted with the current national network. Many of the Reclamation projects are tied to the national state plane coordinate system in NAD27 (North American Datum of 1927), and some projects cover several zones of the state plane coordinate system. Care must be taken to ensure the collected data and final results conform to the requested datum for the study.

It is recommended that all new surveys conform to the national network and that all study results clearly state on all maps and reports the horizontal and vertical datums used, along with the year the datum was established. For positions reported in state plane coordinates, the units of feet or meters should be stated with the state plane zone and year, such as NAD27 or NAD83. The UTM coordinate system is also an acceptable measuring projection that should be clearly labeled with proper zones and years.

Differential GPS collection systems are used for horizontal control for the majority of the hydrographic collection. The horizontal datum used for GPS is WGS84 that is essentially equivalent to NAD83. It is suggested that all new surveys be conducted and reported in NAD83 or WGS84, but NAD27 is also an acceptable horizontal datum for a project study. Even if the final results are to be reported in NAD27 or a local datum, it is recommended that all GPS data be collected in WGS84 or NAD83, then converted to NAD 27 during postprocessing. Collecting the data in WGS84 or NAD83 preserves the data in a raw format that can easily be used and imported into Geographic Information Systems (GIS) without worrying about datum conversion errors generated by the field crew and collection software.

The vertical controls for Reclamation projects cause more problems and confusion than the horizontal controls. This is mainly because most Reclamation projects have a vertical datum that was established during project design and construction and has been used since initial operation. There are some projects where Reclamation, USGS, Bureau of Land Management, state, water district, project, and NGS datums were all established over time, resulting in several different vertical elevations for the same point. The final results from the reservoir survey should clearly state the datum used and, if possible, reference it to permanent project features, such as top of dam, spillway crest, and outlet elevations, as a means of clarifying datum differences. Many recent studies have established new survey control with the hope of clarifying multiple vertical datums. However, some surveys failed because previous resurveys were not tied to any permanent reference object. Survey grade GPS, when used properly, will bring in the most accurate horizontal and vertical positions, but care must be taken to tie new control to previously established control, such as brass caps, sediment range line monuments, water surface gauge monuments, or top of the dam, spillway, and outlet works.

Ideally, the horizontal and vertical datum adjustments should be made on all projects in local datums in a uniform, orderly, and timely manner, but, realistically, they will only be performed on certain projects. In some cases, datums will not be adjusted at all because the time and expense would be far greater than the benefit of doing such adjustment. There are some Reclamation projects that have operated for nearly 100 years with a local datum, and all drawings and records for these projects were developed with these datums. Making any adjustments would be a great expense if it results in all past records having to be adjusted. At Elephant Butte Reservoir, which started initial operation in 1915, the project vertical elevations are 43.3 feet less than the National Geodetic Vertical Datum of 1929 (NGVD29).

9.9.1 Datums

Many Reclamation projects were established with horizontal and vertical control in a local project datum or referenced to the North American Datum of 1927, which was converted to the local state plane coordinate system. Since the majority of the hydrographic surveys are established and surveyed using DGPS in the control of the nationwide network, there is little need to adjust or establish supplemental horizontal control for the study. It is highly recommended that all GPS coordinates be collected in the WGS84 system and all conversions are to NAD83. There is a small difference between WGS84 and NAD83, but it is not significant for hydrographic applications. The WGS84 system can be converted on the fly to state plane or UTM with most hydrographic software. If the final product needs to be in NAD27, it is recommended that these conversions be conducted in the office environment during final processing. One reason for this is that the collected data should be in the cleanest format before conversion so it can be made available for other studies.

GPS satellite positions are based on the three-dimensional earth-centered WGS 84 ellipsoid. For study purposes, the WGS 84 ellipsoid information is usually converted to a user-defined ellipsoid/datum such as NAD27 or NAD83. The North American Datum of 1927 (NAD27) is a horizontal datum based on a comprehensive adjustment of a national network of traverse and triangulation stations. NAD27 was developed and best fitted for the continental United States with the fixed datum reference point located at Meades Ranch, Kansas. The original network adjustment used 25,000 stations referenced to U.S. survey feet. The North American Datum of 1983 (NAD83) used about 250,000 stations to readjust the national network. For practical purposes, NAD83 is equivalent to WGS84 and is currently the best available geodetic model of the worldwide shape of the earth's surface. The reference units are in meters.

9.10 Depth Measurements

9.10.1 Single Beam

Over the last 50 years, the majority of all hydrographic surveys have been conducted using some form of acoustic depth sounder. Manually operated sounding lines and poles may be considered outdated, but they are still viable means of depth measurement in reservoirs with thick vegetation and shallow depths. These manual measurements can also be used as confirmation of electronic depth soundings. Faulty or questionable readings from depth sounders may be caused by noise from vertical walls and structures or from silty bottoms containing "fluff" or light suspended material. Manual collection methods can be used to confirm "fluff" type conditions and possibly determine the type of material on reservoir and river bottoms. Brief summaries of manual collection techniques may be obtained from other publications (USACE, 2004 and ASTM, 2005).

The electronic method using **sonic (echo) soundings** has been the norm in hydrographic collection systems for measuring the bottoms of small and large reservoirs for several decades. The echo sounders have the capability of recording continuous profiles of the reservoir bottom, providing an analog bottom profile chart and digital records stored on the computer system. The computer system software matches these depths with other digital information such as horizontal positioning and heave components. Acoustic depth sounding equipment is preferred because it provides a continuous record and chart of the bottom profile. The basic components are the recorder, transmitting and receiving transducer, and power supply. With careful calibration and correct collection techniques, a high degree of bottom profile accuracy can be obtained and recorded.

The echo sounders are usually portable recorders that measure the time required for a sound wave to travel from its point of origin to the bottom and back to the origin. The time interval is converted to distance (depth) below the face of the sending plate or transducer. The transmission of sound is dependent on the properties of the water and reflecting surface and assumes a constant velocity throughout the depth measured. Since constant velocity is not the case, hydrographic system sounders usually are designed to permit adjustments (calibration) for the variations in the sound velocity in the water. Calibrations of the echo sounder are critical in assuring high-quality depth measurements by the hydrographic survey system. The largest and most critical correction results from the variability of the sound velocity in water due to temperature changes, but other factors such as water density, salinity, turbidity, and depth also affect sound velocity. In fresh water at 60 °F, echo sounders are generally calibrated for a sound velocity of 4,800 feet per second (1,463 meters per second), but water property variability can make it range from 4,600 to 5,000 feet per second. Most reservoirs and rivers exhibit large variations in temperature and water chemistry with depth, which means that the velocity of the sound wave will not be constant for the distance from the sounder's transducer to the bottom and back. The effect of the variation can be significant with a temperature change of 10 °F changing the velocity by about 70 feet per second or changing the depth measurement 0.8 feet per 50 feet of depth. For reservoirs such as Lake Mead and Lake Powell, the summer surface temperatures can be in the high 70s, while the bottom depths are still in the 40-degree range, causing a significant change in the sound velocity through the vertical temperature zones. A 10 parts per thousand salinity change can vary the velocity by around 40 feet per second or 0.4 feet in 50 feet.

For most single beam, shallow water, echo sounding work, an average velocity of sound is usually assumed. A bar check calibration determines the actual depth at the study area, and the sounder is adjusted to measure the correct depth. If the study is conducted in areas with known large variations in velocity by depth or location, the sounder should be set to measure the average or deeper depths that will be encountered during that time over the area being surveyed. For these types of conditions, more frequent calibrations are needed. The sound velocity can be determined by a bar check calibration or measured directly using a velocity probe. The velocity probe can measure the sound velocity at every foot of depth, and an average value can be computed from these measurements. Current hydrographic software allows the depth incremented velocity measurements to be recorded, stored, and used during postprocessing to adjust the sounder measurements to actual depths. The method of using a velocity probe for measuring depth-related sound velocities is more critical for the outer beam depth adjustments of multibeam systems when correcting raw field readings.

The velocity probe calibration method is being used more extensively with hydrographic systems, but depths/velocities should also be periodically checked using a bar check type system. A bar check consists of lowering an acoustic reflector, such as a flat metal plate or I-beam, to a known depth (below the transducer) and manually adjusting the sound velocity to produce an equivalent depth reading. Bar checks can be conducted in 1-foot and greater depths. For shallow conditions, a survey rod can be used in place of the reflector bar. The lines used to lower the bar should be of marked flexible steel wire or chain that does not stretch. The bar check suspension lines must be periodically checked to ensure accuracy of the line markings. The bar check should be conducted in relatively calm water with minimum wind conditions. Mild to strong wind will shift the sounding vessel so that the calibrating bar will be suspended at an angle from vertical, causing the narrow beam signal from the transducer to miss the bar or give false depth readings. In water deep enough for calibration, the survey vessel can be tied to an available buoy, pier, or other object during the calibration to stabilize the boat in windy conditions. The survey crew should conduct a bar check and record results on a depth chart or logsheet. Comparisons at predetermined intervals throughout the depth range of the survey should be recorded during both descent and ascent of the bar. Any adjustments to the speed of sound of the echo sounder should be noted. Figure 9.6 shows a typical chart produced during a bar check.

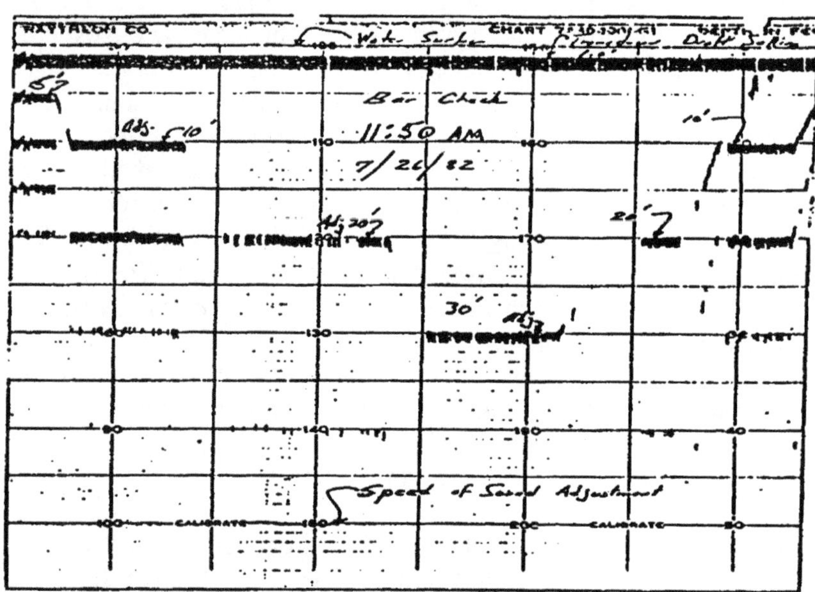

Figure 9.6. Bar check calibration sounding chart (Blanton, 1982).

The weight and structural design of the bar are dependent on the underwater currents and depths to be surveyed. The typical bar may weigh from 20 pounds to more than 100 pounds and can be designed to allow additional weight to be added. Some boats are designed with a center-mounted, spherical calibration ball where the ball is suspended by a single line, providing the advantage of less drift due to current. For small boat setups, a small plate suspended over the side by a single marked cable can be used.

The calmest water conditions for conducting bar checks usually occurs in the early morning or late evening. For areas where salinity and temperature of the water are unknown or are not well mixed, the echo sounder should, at a minimum, be calibrated before the start and end of each day's work. Additional calibrations should be conducted if the survey vessel is moved to a different location of the study area, such as from the main body to a significant tributary of the reservoir. For larger jobs, where stable water velocity conditions are known to exist, the number of bar checks can be reduced to one per day if previous calibrations confirm stable conditions. If previous calibrations indicate unstable conditions, more calibrations become necessary. Calibration of electronic echo sounding instruments is inherently an imprecise process. Bar check and velocity meter calibrations are performed while the vessel is stationary, so actual dynamic survey conditions and different study zones are not truly simulated.

There are different procedural methods of conducting a bar check calibration, but the goal is always to obtain the actual depths of the study area. All procedures require the lowering of a plate a fixed distance below the depth sounder transducer and adjusting the draft and sound velocity settings to match echo sounder depth measurements to know bar depths. A common bar check process is as follows:

- Turn echo sounder on 10 minutes prior to the calibration process to warm the machine up.
- Set initial settings (tide, draft, speed of sound) according to manufacturer's specifications.
- Lower the bar into the water to the 10-foot mark, ensuring the bar is directly underneath the transducer (may want to use a 5-foot mark if conducting shallow water work).
- Adjust the sounder's settings (draft) so that the depth tracing or digital reading matches the 10-foot depth reading on the sounder.
- Lower the bar to the cable increment mark closest to the greatest anticipated sounding depth.
- Adjust the speed of sound control setting so that the sounder's depth matches the bar's depth.
- Raise the bar to the 10-foot position and readjust the sounder's draft if necessary. If no adjustment is needed, then the echo sounder is calibrated.
- Repeat the above steps, starting with lowering the bar to 10 feet, as necessary, until the correct sounder's settings ($\pm$0.1 foot) are obtained for both deep and shallow water.
- Upon completion, intermediate readings between the maximum and minimum depth should be checked to compare the displayed value with the known bar depths.

When complete, the echo sounder should be calibrated at the shallow water (10-foot) and deep water depths. If the sound velocity is constant, the sounder should also be calibrated for the depths between the minimum and maximum depths. If the velocity of sound is not relatively constant throughout the working depth range, it will not be possible to adjust the instrument so that it reads equal to the bar check at each depth increment. One method of accounting for sound velocity variability is to not adjust the echo sounder, but to record the error at each depth and apply corrections during postprocessing. An alternative approach is to place the bar at the maximum study depth and set the depth sounder speed of sound for this observed bar reading. The bar is then raised at set increments as it is retrieved from the water, while sounder readings versus bar readings are recorded. The depth readings can be corrected during postprocessing.

The development of portable velocity meters has provided an additional acceptable means for calibration where the sound velocity in the water column is directly measured for different depth increments. This method provides a fast, reliable means for calibration of the depth sounders with speed of sound, but periodic verification of sounder readings using a standard bar check is still necessary. If repeated comparisons between the bar check and velocity probe provide consistent depth measurements, less checks are necessary. Major advantages of the velocity probe versus the bar check include its ability to perform rapid calibrations, calibrate in rough sea conditions, and perform more frequent calibrations. To ensure quality performance of the probe, it should be tested daily, simply by obtaining sound velocity readings in fresh water in a bucket. There are several companies that manufacture profilers consisting of a probe attached by a waterproof, depth-marked cable to a hand-held control unit. Some models use a pressure sensor for depth determination, which is necessary in deeper reservoirs to minimize the cable slant error. The velocity meter records the speed of sound at water depth increments (set at 1 foot or 1 meter) as it is lowered in the reservoir. These readings may be recorded and used by the hydrographic software for processing the actual depths, or the probe measurements can be used to provide an average sound velocity over the entire water depth that is then entered into the depth sounder.

The echo-sounding instrument consists of the recorder, the transmitting-receiving transducers mounted in the hull or side of the survey boat, and a power source from either a battery or generator. The depth of water is measured continuously at hundreds of soundings a minute and may be recorded on chart paper and/or computer at a prescribed interval. The chart becomes the official record of the depth measurements and should contain the file name, line identification, date, time of day, changes in chart speed, changes in vertical scale, water surface elevation, and any other information needed to interpret the charts. A typical range sounding chart is shown in figure 9.7. During data analysis, the charts are used to verify digitally recorded depths before final contour development. The trend now is for the hydrographic system to include paperless depth sounders. The key advantages of the paperless charts are their smaller size and fewer moving parts. There are several manufacturers of paperless sounders that provide a color sonogram display that looks similar to paper charts of the measured bottom. Some offer the option of storing the entire sonogram for future playback and printing if it becomes necessary during the analysis.

An echo sounder's transducer has many frequency options and should be selected to meet the majority of the study needs. The Sedimentation Group surveys are conducted using a 200-kilohertz, high-frequency echo sounder that also has a low-frequency, 24-kilohertz option. In general, the higher frequency transducers of 100 kilohertz and greater provide more precise and detailed bottom depth measurements due to the frequency characteristics and narrow beam width. The major disadvantage of higher frequency transducers is that they tend to reflect off of first signal change, which may provide false readings of the actual depth for such conditions as suspended sediment (fluff) and bottom vegetation. The lower frequency transducers of less than 40 kilohertz are less subject to attenuation and are capable of greater depth measurements since they can penetrate the suspended sediment or fluff type conditions. However, the lower-frequency transducers have a larger beam width, which may provide readings that are distorted due to smoothing of irregular bottom features and the side slopes. An experienced operator, along with some manual depth readings, is needed to assist in making a distinction between the fluff and actual bottom readings. For several years, the vendors have offered dual-frequency sounders that allow the operator to measure separate, simultaneous, high- and low-frequency depths. During the last few years, portable sounders offering variable and multiple frequency settings have been developed.

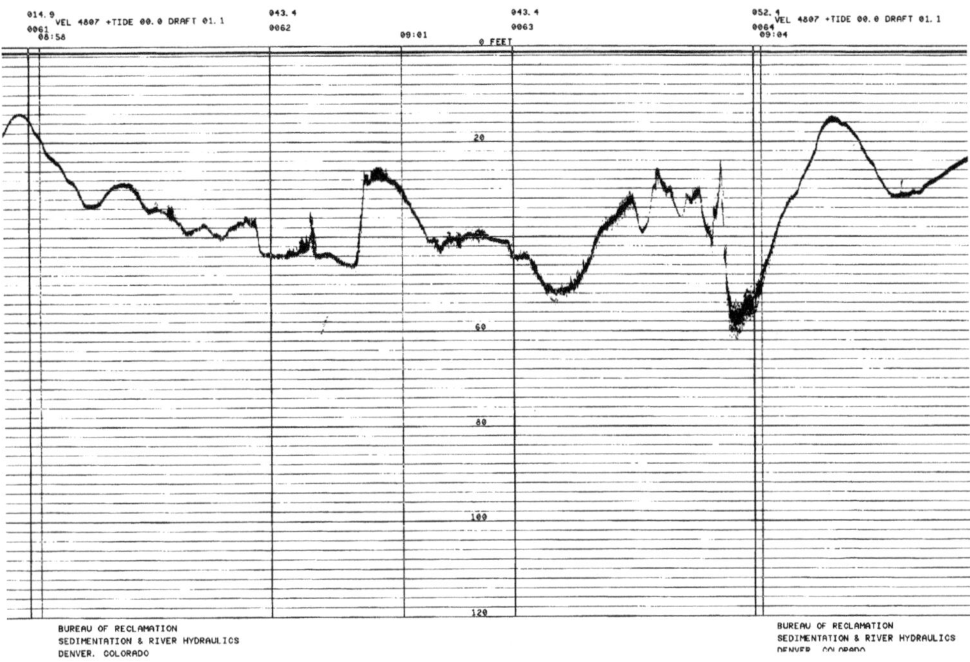

Figure 9.7. Range sounding chart produced from an Innerspace 448.

9.10.2 Multibeam

Multiple beam echo sounders have become more popular over the last few years. The multiple transducer method was mainly developed for situations where detection of navigational obstacles was the primary concern. Initial multiple transducer systems consisted of individual transducers mounted in the boat hull and along booms extending from each side of the boat (figure 9.8). Maneuvering the survey vessel was difficult with this system, and it was not suitable for rough water. The booms came in varying lengths, with some having a hydraulic design that could be automatically extended or withdrawn. The system used a vertical sounding beams system, just like the single beam system except there were multiple units. In theory, the system was simple to operate and the separation of the transducers and their width of coverage provided sweep coverage for full bottom mapping. The coverage depended on the beam length and number of transducers, and it is still used by the USACE for performing dredge measurement and payment surveys.

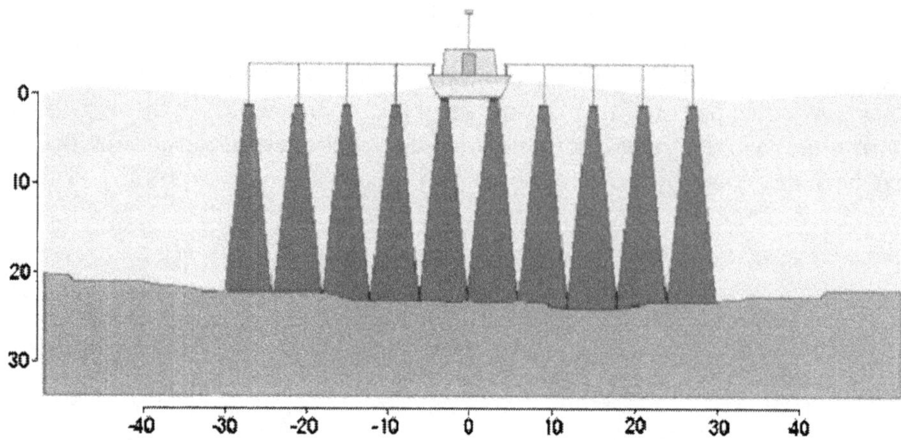

Figure 9.8. View of a multitransducer survey vessel (HYPACK, Inc.).

Recent improvements in collection and analysis techniques through multibeam survey technology have made it a standard for many survey groups, replacing several of the multiple beam systems that were in operation by the USACE. In 2001, the Sedimentation Group acquired a multibeam system, enhancing the capability their existing hydrographic collection system. The system was first used to survey the sediment deposition in Lake Mead from the dam to the upper shallow water areas of the reservoir. The end products of the Lake Mead survey were cross sections every 5 meters for the surveyed area, allowing detailed mapping of the sediment deposition. The detailed collection of the underwater portion of the Lake Mead sediment deposition was completed in less then 1 month with a two-person crew. Compared to the 6 months and 6-person crew required to collect 407 cross sections for the 1986 Lake Powell sedimentation survey using a single beam collection system, the potential time saved using a multibeam system becomes clear.

Multibeam technology was originally developed in the 1960s for deep water ocean mapping. In the 1990s, the technology was further developed and extended to shallow water applications. There are several vendors that manufacture these systems with the software needed for collection and processing of the data. The manufacturer's manuals need to be used for instructions on system operation and collection and processing of the acquired data. Multibeam systems are fan-beam acoustic sounding systems comprised of a number of narrow beam transducers mounted in close proximity and focused at equally spaced angles from one location under the survey boat (figure 9.9).

Each transducer acts as a separate acoustic-distance measuring unit, like a single beam, vertical mount system, except the multiple transducer beams are at a given angle with respect to the mounted single vertical transducer. Computations determine the depth of each beam from the slant-distance signal adjusted to incorporate the velocity profile data. Multibeam vessels can survey in rougher water and offer greater coverage. The coverage area is dependent on the water depths. For a fan of 120 degrees, the bottom sweep width is around 100 feet in 30 feet of water and around 350 feet in 100 feet of water. It must be noted that, for navigation type surveys, it is recommended that a 50-percent overlap of the survey sweeps is maintained for quality control

(QC). Most Reclamation surveys are not performed for navigation purposes, so the fan overlap can be reduced. The overlap should be enough to assure the outer beams of the two sweeps are collecting high-quality data. The fan angles for multibeam systems typically vary between 90 and 220 degrees. The primary justification for purchasing and using a multibeam system is that the majority of the water depths of the study area are greater then 30 feet. Some multibeam sonars can be tilted for mapping reservoir banks and features such as dam faces and outlet works; however, accuracy may be sacrificed, since the larger errors with a multibeam system occur in the outer beams (figure 9.10). Some multibeam systems and software provide the option of multibeam sidescan imagery that, in general, is not as good as the images from towed sidescans, but is a byproduct of the system that should be considered.

Figure 9.9. Multibeam collection system.

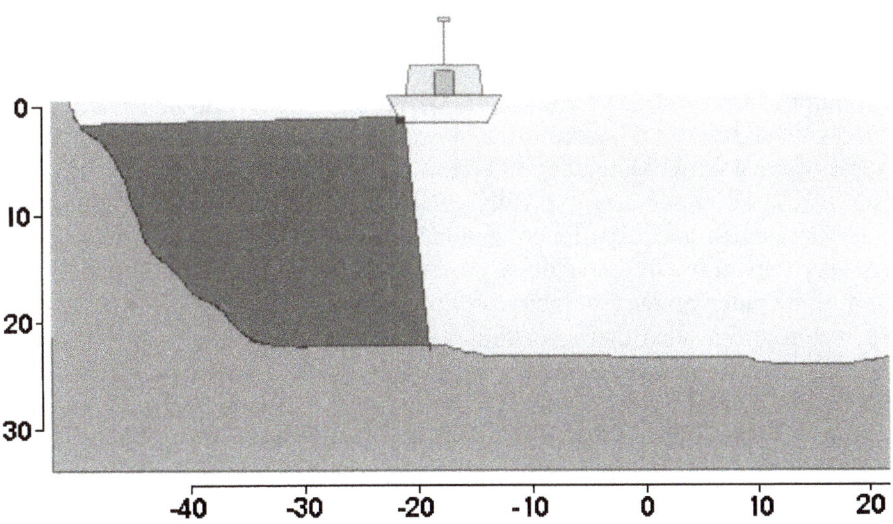

Figure 9.10. View of mounting options with multibeam survey boat (HYPACK, Inc.).

The horizontal and vertical accuracy of the collected data relies on many components necessary to complete a multibeam survey system. These components include the computer system with software, sensors for heave/pitch/roll and heading, positioning system, and velocity profiler for calibration. Multibeam systems have a very high data acquisition rate. With data collection rates in the thousands of depth points per second, manual editing in no longer feasible. With this in mind, an extensive calibration is necessary prior to starting the data collection. The calibration is necessary to determine the magnitude of such error sources as the vessel roll, pitch and yaw, mounting angles of the sonar, and incorrect x, y, z offset errors between positioning antenna and sonar.

In single beam soundings, the sounding or ping travels downward and upward along a vertical path with virtually no change in direction. The majority of the multibeam sonar beams are not vertical and encounter changes in sound velocity resulting in ping speed changes, along with slight changes in direction. When the sound velocity increases, the ray is bent upward; when the velocity decreases, the ray is bent downward (figure 9.11). Correction for these sounding refractions requires an actual velocity variation with depth table of values that are used in postprocessing of the depth data. This information does not come from the bar check procedure, but from a sound velocity probe that measures actual velocity variations with water depth. For the Lake Mead survey, a velocity probe was used that had a 100-meter cable length and readings were taken every meter of the depth zone. Other components of the system include a heave compensator for measuring the up and down motion of the boat, gyro for vessel heading, and motion reference unit (MRU) for measuring the pitch and roll data. All of these measurements are needed to correct the multibeam data and are monitored by the hydrographic collection software, and all data are stored and processed on the field collection computer. Proper system operation requires an experienced crew and good calibration practice. The cost of these systems is considerably more than for a single beam system, but, since it collects much more data, the final results may justify the greater expense.

Field calibration of multibeam systems is more critical and complicated than what is required for single beam systems. Periodic precise calibration is absolutely essential to ensure that the multibeam positions and elevations are accurate. The horizontal positioning accuracy is dependent upon the ability of the system to compensate for pointing errors caused by vessel roll, pitch, and yaw, where a small degree of roll can cause large errors in the outer beams. For high-accuracy surveys, restrictions are typically placed on the use of the outer beam data. Manufacturer suggestions and experience should be used to determine the use of these outer beams. It is very critical to collect velocity profile data for all beam measurements, but mainly for correction of the outer beams. Velocity profile readings should be taken a minimum of once per day, but it is recommended that a reading be collected several times per day and when the survey vessel relocates to a different portion of the study area. There are set QC calibrations and QA test procedures for multibeam systems to assure highly accurate data collection. These tests and procedures are generally available in the hydrographic software. The calibration of the system determines time latency, along with roll, pitch, and heading bias. Some calibrations are performed just once after the system is installed to measure sensor alignment and offsets, while other calibrations are performed on a more frequent basis, as recommended by the manufacturer, and are needed to ensure the validity of survey results.

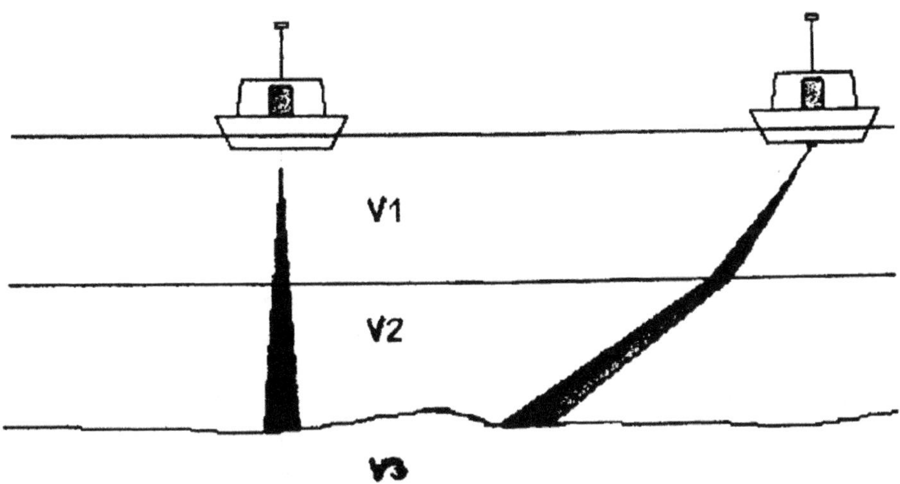

Figure 9.11. Sound beam bending due to change in sound velocity (sound velocity increase in zone V2) (HYPACK, Inc.).

One calibration method, called a "patch test," is performed after initial installation, after any sensor modification, and periodically to confirm previous system alignment. This comprehensive patch test includes a latency test for measuring position time delay and tests to determine the pitch, roll, and yaw or azimuth offsets. Velocity profile corrections are a must for a hydrographic survey using a multibeam system and need to be performed periodically during the day. It is recommended that the velocity profile correction be completed at least twice per day and more frequent in locations where physical changes in the water column are suspected or measured from observing previous collections. A traditional bar check can be used to verify the settings and corrections of the measured multibeam depths. The bar test can also be set up to check some of the outer beam depths, and it is an excellent method to confirm the draft settings of the system. The manufacturers and software vendors have manuals with detailed descriptions on performing the patch and performance tests necessary to ensure the quality of the survey data set. The performance test may compare overlapping survey data sets such as multibeam data overlapping a single beam data set in the same area.

Following is a brief description of a patch test that can be completed (including data collection and processing) in a few hours. The patch test should be conducted in calm conditions in a typical project area. The roll test is run in a flat bottom topography area where one line is run twice in opposite directions at typical survey speed. The latency test consists of running one line on a steep sloping bottom or well-defined feature twice in the same direction at two different speeds. The higher speed should be about double the slower speed. The pitch test consists of running one line on a steep sloping bottom or well-defined object twice in opposite directions at survey speed. The yaw test consists of running two adjacent parallel lines with a well-defined target or slope between them. The line spacing should be about two times the water depth where the swath overlap is between 30 and 70 percent. Each line is run once with reciprocal headings at survey speed.

Raw multibeam data sets are very large, usually with many data spikes, requiring a great deal of filtering and editing before they can be used for final map development. The hydrographic survey software packages contain many routines for manipulating and editing these large, raw data sets. Filtering and editing can be conducted automatically in the software editing processes but must be conducted with much caution to avoid accidental elimination of useful data. In the automated editing process, there are manual editing procedures where it is possible to view each cross section. This procedure can be very time consuming, but it should be performed on most of the data set as a means of conducting QC of the automated filtering process.

Even after the raw data sets are edited and filtered, additional depth data reduction may be necessary for the final data sets. The data sets can be so massive that generating the final map product may be impossible, or at least very time consuming. There have been filtering processes developed, tested, and incorporated into the software routines that allow final data filtering without sacrificing the quality of the study. These filtering routines must be used with caution, but the software saves the previous data set so the user can try different options before settling on a procedure for the filtering. The filtering can also be adjusted to identify important details for such areas as dam faces, possible sinkholes, and trashracks. One of the filtering methods is called gridding and has the option of saving one depth per set cell. The grid size depends on the study needs. It is common for the grid size to be set from 1 to 5 meters. There have been studies with irregular topography and underwater structures that were mapped with data from a grid size of 20 centimeters to obtain the necessary detail.

There are several other viable methods of mapping reservoir and river bottoms that will be mentioned, but manufacturer references and manuals by the USACE and others should be consulted for more detail (USACE, 2004). The Sedimentation Group acquired a multibeam depth sounder and incorporated it into their hydrographic survey system, mainly because it measures x, y, z data in real time, compared to the side scan option that produces only an image without the associated coordinates. The side scan image can be obtained from the multibeam system, but the quality is not as good as data collected with a side scan sonar system designed specifically to collect side scan data. The Sedimentation Group has leased equipment for site-specific studies such as low-frequency (24-kilohertz) transducers and a side scan sonar system with operator. The low frequency system was used in an attempt to locate soft bottom conditions on the Salton Sea. In an attempt to locate sinkholes at Horsetooth Dam in Colorado, an analog side scan sonar system was leased in 1998 and 1999. The collected images appeared to indicate a sinkhole in the left abutment area, which was confirmed once the reservoir level was dropped to expose it. It is the general conclusion that the combination multibeam and side scan system could have located and confirmed these conditions more easily.

9.10.3 Additional Sonar Methods

Side scan sonar is a high-resolution tool that provides a map on both sides of a survey vessel's path. The system does not provide absolute elevations of objects; however, it will provide relative elevations of the surrounding topography. The map images can be recorded as an analog image paper chart or a digital data image that allows mosaics to be produced and merged with other data sets such as multibeam data (Twichell, Cross, and Belew, 2004). The quality of sonar data is often a function of the height of the towfish above the bottom. Multibeam systems have the capability of providing a side scan image, but the quality is not as good as the towed side scan systems.

Airborne Light Detection and Ranging (LIDAR) hydrographic surveying method is a means of collecting above and below water data. The Sedimentation Group and other agencies have successfully used airborne LIDAR to conduct shallow water river surveys (Hilldale, 2005). The primary constraint of LIDAR is water clarity. LIDAR has been successful at collecting bottom data through as much as 40-meter depths of clear water. In less clear waters, LIDAR data collection has been successful at depths of two to three times the visible depth.

There are other systems and methods under development that should be monitored as they become more viable for survey applications. Hydrographic systems development can be monitored by attendance at conferences, through Internet research, and by belonging to technical groups associated with hydrographic and general surveying that provide periodic magazines and newsletters.

9.10.4 Single Beam Depth Records

The following information will deal with single beam systems. Knowledge and understanding of the single beam systems can be applied when working with the more advanced multibeam systems. When setting up and using any system, the manufacturer specifications and manuals should be consulted.

Interpreting depth records takes experience. The most reliable interpretation usually comes from the survey collection crew or someone who has survey collection experience. During interpretation, a plot of the digital record and bottom charts should be studied. In general, if any recorded traces on the graphic or digital record cannot be attributed with reasonable certainty to reflections from the reservoir bottom, the traces should not be part of the final recorded files. The echo sounder's trace on hard bottoms will reflect more strongly than on soft bottoms and will appear as a thin, dark trace on analog charts. In shallow conditions, multiple echoes may appear with the actual depth being the shallowest reading of the trace. Soft bottoms of unconsolidated materials may produce a broad trace, and sometimes the thickness of the fluff or soft layer can be determined by a split in the echo trace on the analog chart. These types of conditions usually need to be resolved in the field to confirm the true bottom versus the digital reading. This may be done by using the lead-line or sounding-cable method. Interpretations of the analog and digital depth data are sometimes made very difficult by the presence of heavy vegetation, floating objects, bottom projections and depressions representing sudden bottom changes, and steep bottom slopes (figure 9.12). Figures 9.13 through 9.15 are analog chart samples.

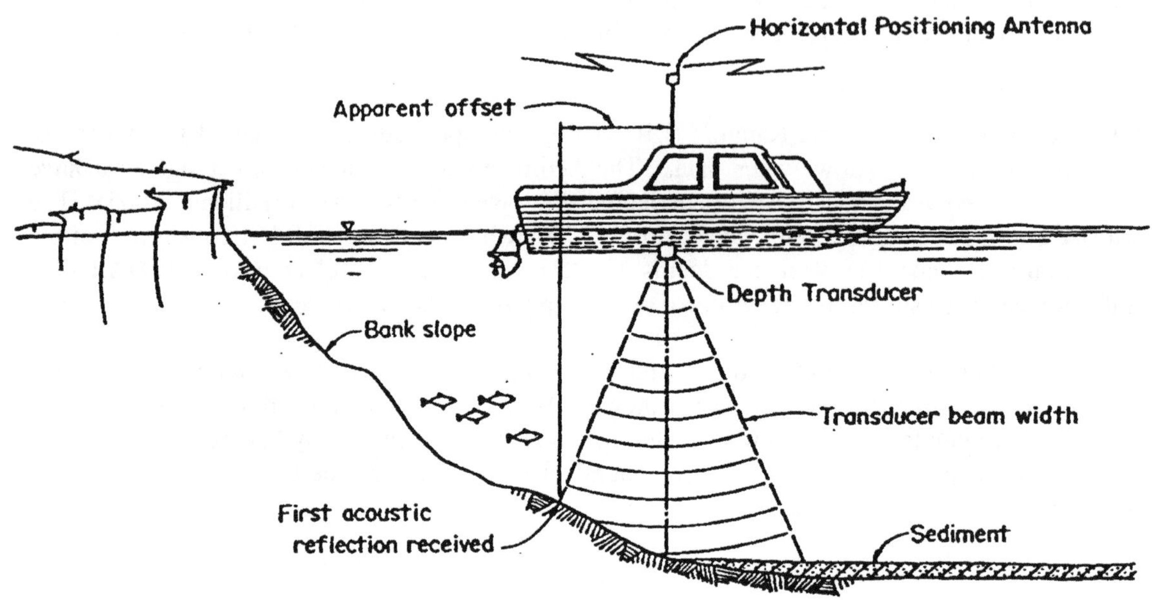

Figure 9.12. Position error due to side slope effect and sediment (Blanton, 1982).

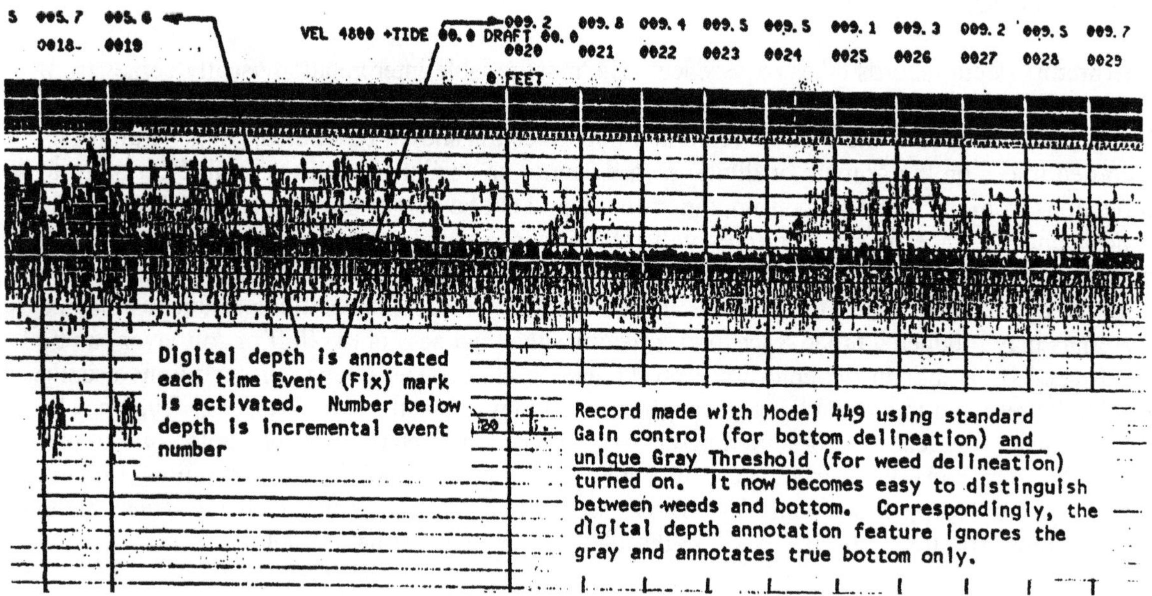

Figure 9.13. Chart from 27-kilohertz sounder, showing weeds and bottom (Innerspace, Inc.).

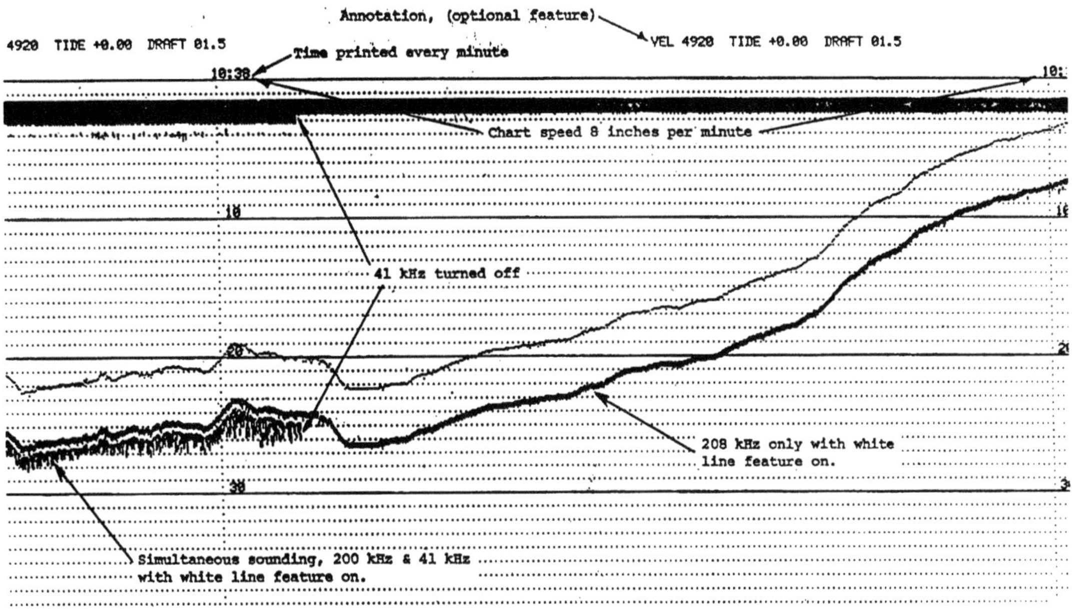

Figure 9.14 Simultaneous sounding chart, 41 kilohertz and 208 kilohertz (Innerspace, Inc.).

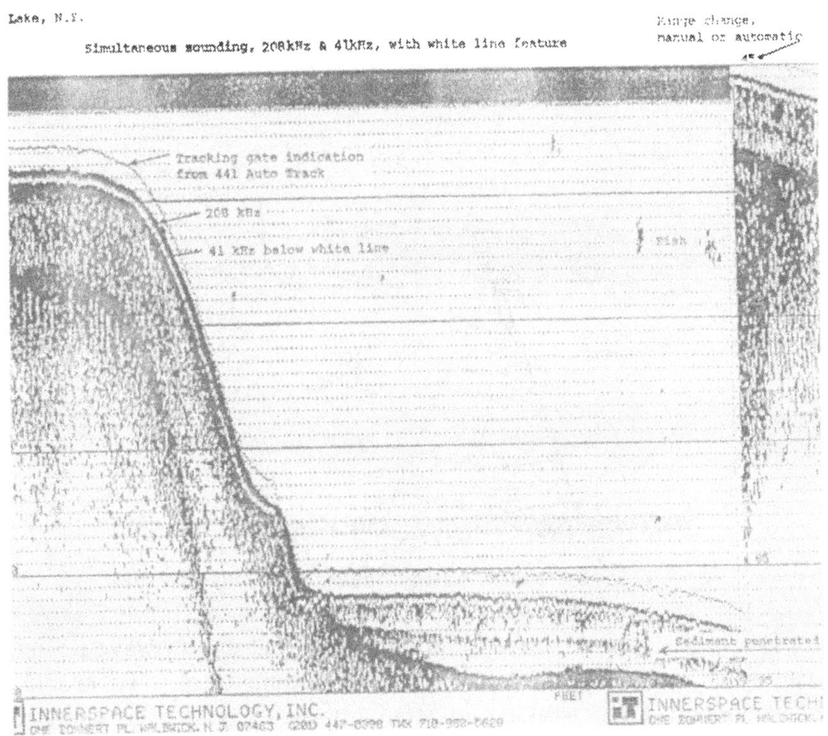

Figure 9.15. Chart from 41-kilohertz and 208-kilohertz sounder, showing fluff (Innerspace, Inc.).

Factors leading to possible errors in depth measurement need to be considered when conducting a reservoir survey and during the analysis of the survey depth data. Detailed descriptions of these factors are summarized by Hart and Downing (1977). Some of the most significant factors are as follows:

Acoustic velocity propagation. The velocity of sound through water varies with temperature and salinity. For many storage reservoirs, the most significant factor is reservoir water temperatures that can vary, in the summer season, by as much as 45 °F between the surface and deep water. Bar and velocity probe checks are designed to correct the measurement variation and provide necessary correction information that can be applied during field collection and data analysis. Multiple bar and velocity probe checks are essential where any significant variation exists.

Transducer location. The draft or vertical location of the transducer's bottom face with respect to the water surface can be set within the more advanced sounding instruments or within the hydrographic collection software. As shown on figure 9.16, the location of the transducer face with respect to a static water surface is different when the sounding boat is in motion. The effect of the boat motion on draft may be corrected in the calibration of the instrument. The effects of the boat speed on the transducer location can be measured by a squat calibration test and corrected during data processing (USCOE, 2004).

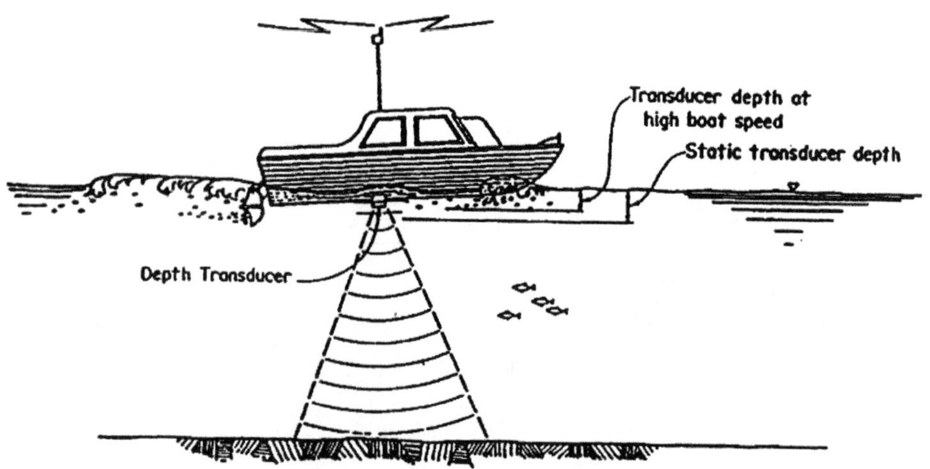

Figure 9.16. Motion effect on depth measurement (Blanton, 1982).

Wave action. The vertical and rotational motion of the boat due to wave action, as shown on figure 9.17, can result in severe fluctuations in the bottom trace. To ensure the safety of personnel and equipment when working in relatively small boats, the underwater collection should be temporarily halted when wind-generated waves affect the safety of the collection crew and compromise the data being collected. All wave-produced fluctuations in the bottom trace should be smoothed during data processing to produce acceptable data. However, more accurate data can be obtained if the survey is delayed in such conditions. Many errors due to wave action that causes survey vessel heave, roll, pitch, and yaw can be significantly reduced using accurate motion sensing instruments as part of the hydrographic collection system.

Figure 9.17. Wave effect on depth measurement (Blanton, 1982).

Bottom conditions. The reflective surface on the reservoir bottom may vary widely, as illustrated on figure 9.18. Vegetation attached to or suspended above the bottom and isolated boulders or manmade objects produce false bottom depth measurements that should be eliminated from the data. Very low-density sediment, suspended as a fluff above more compacted sediments, can result in dual depth readings (figure 9.15) unless the condition is anticipated and the sensitivity of the sound instrument is adjusted.

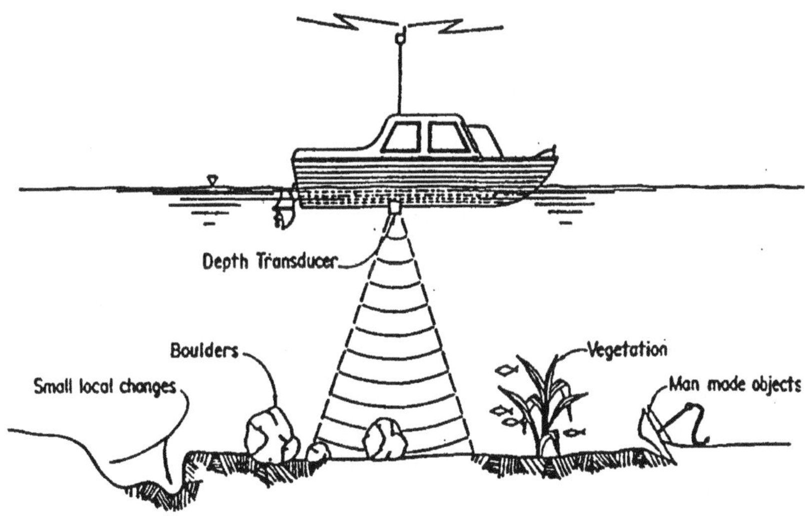

Figure 9.18. Bottom condition effect on depth measurement (Blanton, 1982).

Other errors in depth measurement may occur due to special circumstances not always encountered in all reservoir surveys:

- Reservoir level fluctuations due to inflow and outflow may change the vertical control during the survey period and introduce depth errors unless considered.

- Backwater effects in narrow canyon areas or in river portions above the main body of the reservoir may produce a water surface slope or change in the stage, which negates use of reservoir water surface for vertical control.

- Constant wind blowing from one side of a reservoir to another may cause reservoir level changes on each side of the reservoir and should be considered when operating in high wind conditions.

The use of a RTK GPS with centimeter vertical accuracies can be used to monitor and measure changes of the water surface during reservoir and river surveys, minimizing these errors. For a single beam survey in choppy water, mounting the positioning antenna above the transducer, along with using a heave compensator, reduces many of these errors, resulting in the collection of more reliable data.

9.11 Survey Accuracy and Quality

The objective of the hydrographic survey program is to measure the highest quality data possible. The degree of accuracy is determined by the need of the study: reconnaissance, partial, or complete survey. The partial and complete surveys require a higher degree of accuracy than a reconnaissance survey, since they usually are performed for construction or volume computation purposes. The Sedimentation Group typically uses their most accurate positioning and depth measuring equipment for surveys, so the accuracy difference between surveys is not a function of the equipment accuracy and limitations but, rather, a function of the difference in the field collection procedures. There are several publications that are good references when addressing the accuracy and quality of a hydrographic survey program. Good references include chapters 3 and 4 of USACE's "Hydrographic Surveying" manual (USACE, 2004) and the International Hydrographic Organization "Standards for Hydrographic Surveys" (IHO, 2005). Throughout this manual, procedures and cautions are listed to assist with the goal of obtaining the highest possible accuracy of the collected data, while achieving and maintaining QC and QA of the hydrographic program.

The QC process is where the quality of work and survey equipment are measured and controlled in order to minimize errors in the individual data points. There are a variety of QC procedures that are recommended for the survey system instrumentation and data collection techniques to minimize systematic and random errors in individual data points. Examples of QC procedures include bar checks, velocity casts, patch tests, instrument alignment tests, vessel velocity limitations, multibeam beam-width restrictions, and overlapping coverage. Some of the QC procedures are contained in this manual, but the equipment manufacturer's operating manuals should also be consulted. It must be noted that even performing all recommended QC procedures does not necessarily guarantee highly accurate collected data.

QA procedures are tests used to verify the accuracy of the collected hydrographic depth data. QA tests would typically compare two nearly independent sets of elevation data collected over the same area, but for many hydrographic surveys, QA tests are not practical and may even be impossible. QA tests are essential for multibeam surveys and typically compare single beam data with multibeam measurements, generally using the same positioning system. Since the data sets are obtained from the same instrument platform, the same position inaccuracies could be encountered.

The accuracy of a hydrographic survey is difficult to monitor relative to conventional land-base surveys, due to the lack of available control checks. Care must be taken in instrument calibration and collection procedures to ensure quality data collection, since adjusting the data during postprocessing is very difficult. Calibrations and verifications of the collection systems are time consuming but are necessary procedures to ensure the quality of the data. The accuracy of the measured bottom is dependent on the many inherent errors in the measuring process, and it is up to the collection crew to minimize these errors. Using experienced collection crews that utilize good collection techniques is one of the best means of ensuring proper survey methods, usually resulting in accurate survey data. The horizontal position of a hydrographic survey is usually established by an open-end survey method with no independent check, so the accuracy is totally dependent on the measuring process. The vertical measurement reference is usually the variable water surface and is independent of the horizontal measurement, except for the time of collection relationship, which allows the hydrographic survey software to merge each depth measurement (elevation) with its corresponding position (horizontal coordinates). The accuracy of the hydrographic survey is dependent on the accuracy of each instrument, calibration, correlation of all system components, collection method and techniques, corrections to the collected data, equipment selection and maintenance, and analysis techniques.

An important distinction exists between the accuracy and precision of the hydrographic survey measurements. The estimated accuracy of a hydrographic survey is usually based on results from the equipment calibration. Other techniques to determine the accuracy of the hydrographic survey include cross-checking lines and repeat surveys. The depth sounder may give repeated depth precisions of $\pm$0.1 feet in a stationary position, but the accuracy of the depth during the survey may be only $\pm$0.5 feet when all error components are included (i.e., multipath, water temperature and salinity changes, and one of the biggest factors, boat movement). Computer software packages have routines to compute position accuracy for the automated hydrographic survey instruments that give the survey crew the option of making adjustments to minimize collection errors. One such statistical software package allows the collection crew to compute and display differences between intersecting survey lines (HYPACK, 2005). The program provides a statistical report that shows the standard deviation distribution and average error. The output report contains detailed information for every intersecting point along with a three-dimensional view of the intersecting survey lines displaying the depth differences (figure 9.19).

An additional error with the automated hydrographic survey systems is the synchronization of the recorded data by the collection software (latency time). Latency is the time delay from the instant a measurement is taken by a survey instrument to the instant the instrument outputs the measurement data to the survey computer software. This time delay is usually measured in milliseconds (where 1 millisecond equals 1/1000 seconds). The position error will usually increase with increased velocity of the survey boat, and some systems have a time delay as long as 2 seconds. Current software has several methods of determining the time lag and correcting for it. Cross section plots illustrate the shift in the horizontal positions if a time lag exists. The adjustments can be made during the analysis process, but it is best to determine the lag time prior to data collection.

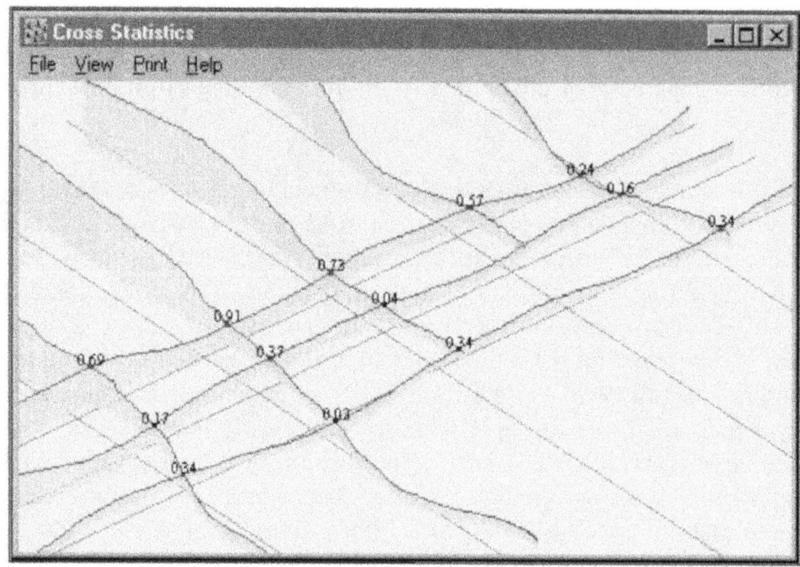

Figure 9.19. Cross-line check method (HYPACK, Inc.).

9.12 Survey Vessels

The type of survey boat is determined by the size of the reservoir and the equipment used. On small, shallow reservoirs, lightweight measuring equipment should be used on small, shallow-draft boats or rafts. Recent development of compact collection systems allows mounting on much smaller boats and easy transfer of equipment to different survey vessels. Reclamation has used vessels ranging from single-person, flat-bottom boats to large, multiperson, enclosed cabin vessels. For the survey of the sediment range lines in Lake Livingston in Texas, an airboat was used due to shallow lake conditions and many downed trees that made navigation impossible with conventional watercraft (Ferrari, 1993). For large reservoirs the travel time, safety, and housing for the equipment are important and require larger, faster, and more seaworthy survey vessels. The Sedimentation Group conducts many of its surveys in the 17 Western United States with their hydrographic survey equipment mounted in the cabin of a 24-foot, trihull aluminum vessel equipped with twin inboard motors. The hydrographic system contained on the survey vessel consists of the following equipment:

- RTK GPS receiver with an omnidirectional antenna
- Depth sounders (single and multibeam)
- Helmsman display for navigation
- Computers
- Monitors
- Gyro and motion reference sensors
- Hydrographic system software for collecting the underwater data
- An onboard generator (equipment can be powered by 12-volt batteries)

The cabin extends the life of the equipment and allows the collection of data in weather and reservoir conditions that are more difficult, or maybe even impossible, in smaller boats. The negative aspects of a larger boat are transporting the vessel to and from study sites and maneuvering in tight areas and shallow water reservoir conditions. Saltwater conditions warrant more care to protect the instruments and crew, further justifying an enclosed cabin. In the past, large survey vessels were required to house larger survey systems and reduce the effect of heave, pitch, and roll created by heavy seas, but state-of-the-art motion reference units have eliminated most of these effects, permitting the use of smaller survey vessels with more compact instrumentation.

9.13 Survey Crew

Personnel requirements vary by the size of the job, type of equipment, and method of collection. Qualified and experienced personnel are essential for an efficient and productive field operation, with a key being a hydrographic crew chief experienced in all phases of the field operations and knowledgeable about the computation and report needs of the study. The crewmembers must be capable of assisting in the operation and maintenance of the field instruments. Current systems allow data collection by as few as one person but, for safety purposes and assistance, a minimum of two field personnel is recommended for hydrographic survey vessel operations. For larger reservoirs, one or two additional crewmembers may be necessary for support of the survey, safety of the operation, operation of an auxiliary boat, transporting fuel and supplies to the survey boat, and setting up and maintaining any necessary shore-based equipment. Personnel should be trained in first-aid, cardiopulmonary resuscitation (CPR), and survey vessel operation.

There has been a push for a certification program for hydrographers. The American Congress on Surveying and Mapping (ACSM) offers a hydrographer certification program that is open to all persons to become certified. The applicant must demonstrate, to a certification board, the necessary knowledge to perform hydrographic surveys. The applicant must meet experience requirements by providing documentation of the understanding of hydrographic surveying and by passing an examination. To qualify for the examination the applicant must have 5 years of hydrographic surveying experience, with a minimum of 2 years being technically in charge and 2 years in the field. The applicants must submit an essay on the fundamentals of hydrography and their qualifications to perform hydrographic surveys.

9.14 Determination of Volume Deposits

Upon completion of the field surveys, the accumulated data must be assembled and analyzed for the purpose of calculating an updated elevation versus storage volume relationship for the reservoir. Comparing the new storage volume results with the original or previous baseline survey results can determine the volume of the accumulated sediment. It is recommended that all resulting long-term total sediment computations be compared to the original capacity.

For Reclamation reservoirs, the total volume of sediment deposited is generally determined as the difference between the original and updated reservoir capacity. This makes the sediment rate computation sensitive to the methods of measuring and calculating the reservoir volume. Ideally, identical computational methods should be used for each survey. For some reservoirs, advances in technology have made this difficult or impossible because the accuracy of the resurveys is better than the original collected data. In some cases, the accuracy difference is very pronounced, such as the Theodore Roosevelt Reservoir 1995 Sedimentation Survey (Lyons and Lest, 1996). The 1995 survey of Roosevelt Reservoir used GPS and aerial data collection to develop 5-foot contours. The original contour map was completed in the early 1900s using plane table techniques at 10-foot increments. Due to the method differences of the two surveys, comparing the results from the two contour surveys did not provide an accurate value of the total sediment deposition. For that study, the range line method would have provided a more accurate estimate of the sediment deposition since dam closure. The decision was made to use the 1995 contour map and resulting capacities as a baseline reference for computing future sediment deposition.

Many of Reclamation's reservoirs were originally measured by the contour method at 5-foot contour intervals prior to inundation, using aerial and plane table survey techniques. With this type of accuracy, the error differences due to the present collection methods used by Reclamation are somewhat minimized. With the development of RTK GPS and multibeam depth sounders, the resurveys could become more accurate than the original aerial developed reservoir contour maps. Due to complete bottom coverage, these new maps will give a better value of the actual capacity of the reservoir. However, when the sedimentation values are computed, it must be noted that a portion of the difference is due to the difference in collection methods.

As indicated before, in determining the *long-term* sedimentation rate, Reclamation figures the difference between the original recalculated capacity and the newly measured capacity of the reservoir. The original capacity is recalculated from the original surface areas using the same mathematical program used to calculate the present updated capacity tables. Reclamation computes the reservoir area-capacity using the computer program Area-Capacity Computation Program (ACAP) (Bureau of Reclamation, 1985). With all the improvements in collection and analysis methods and the accuracy of the survey instrumentation, it is possible to accurately measure even small sediment volume changes due to sediment inflow.

The range line survey and analysis methods are still used and are considered valid means for conducting reservoir survey studies. For some reservoirs, a combination of the contour and range line methods may be used. One example might be a large reservoir where the contour survey does not use photogrammetric collection to map the above water areas containing known sediment deposition. Range line surveys and analysis methods involve determining cross-sectional changes along range lines and applying the changes to the reservoir surface areas enclosed by the range lines. Water volumes or deposition volumes can be computed using the cross-sectional areas and the distance between the range lines, but these methods do not account for all volume changes occurring over irregularly shaped surface areas unless the range lines are closely spaced.

The range line method consists of laying out a system of representative ranges and determining the present sediment depths along those lines. The number and location of ranges depend on the shape and size of the reservoir being surveyed. Ranges subdivide the main body of the reservoir and its principal tributary arms so that sediment deposits in each subdivision or segment are represented by the average of conditions measured at the ranges. For the purpose of locating the sediment range lines during the resurveys, the end of each range may need to be marked above the normal shoreline with a permanent marker. With GPS survey capability, the range line end locations can be preserved by determining the range ends by digital geodetic locations and recording the information in a final report. The following methods have been used to compute the reservoir capacities when the cross-sectional areas of the ranges have been determined.

9.14.1 Average-End-Area Method

The average-end-area method for computing the capacity of a reservoir is an adaptation of the method commonly used in computing earthwork quantities. With contour maps, the average surface area enclosed by the contours is multiplied by the contour interval to compute the intermediate volume. With reservoir range line data, the cross-sectional areas of adjacent ranges are averaged and multiplied by the distance between ranges to compute the intermediate volume. This method does not account for the banks of the reservoirs that are indented with embayments and inlets, making it difficult to establish a range that is representative of any given reach. One solution is to increase the number of range lines surveyed, which increases the collection cost but may not proportionally increase the accuracy. The alignments of reservoirs are seldom straight, which makes it difficult to determine the distance to use between the range lines. Also, as the end area approaches zero (upper end of reservoir), the trapezoidal computation becomes a pyramid (or a triangle), and the error in using the average end area formula approaches 50 percent for this segment of the reservoir computation. Various methods have been developed to compensate for these problems, such as the width adjustment method.

9.14.2 Width Adjustment Method

In some earlier resurveys, new contour maps were drawn from range line survey data where all the contours between the resurveyed range lines for the new map were estimated by using the original contour map as a guide or control. The new contour locations were estimated based on changes that occurred at each range line. This method was abandoned for the constant factor method, which was further modified to the width adjustment method, as described by Pemberton and Blanton (1980).

In the width adjustment method, illustrated on figure 9.20, the new contour area, A_1, between any two ranges is computed by applying an adjustment factor to the original contour area, A_0, between the same two ranges. This adjustment factor is defined as the ratio of the new average width to the original average width for both upstream and downstream ranges at the specified contour. The revised segmented surface areas for each contour are then summed for the whole reservoir. The summarized segmented surface area versus elevation becomes the basic input for volume computations. The computation can be accomplished by hand or with commonly used electronic spreadsheet programs.

A simultaneous comparison of the plots of the original range profiles against the resurveyed range profiles displays the lateral distribution of the sediment at the measured points. Where these plots indicate that changes have occurred on the side slopes of the reservoir, engineering judgment is required to determine whether the change is due to survey inaccuracies or actual deposition or erosion.

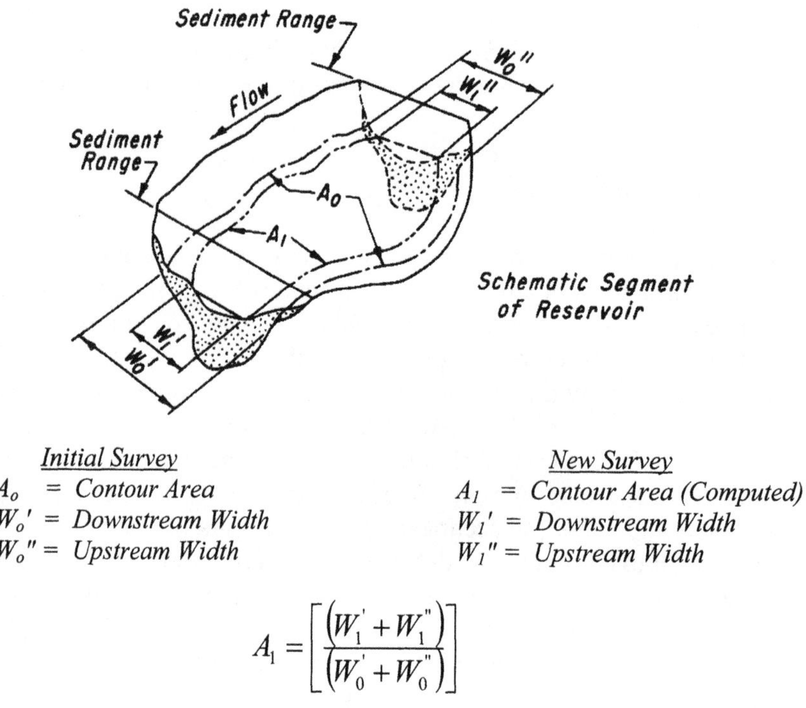

Initial Survey
A_o = *Contour Area*
W_o' = *Downstream Width*
W_o'' = *Upstream Width*

New Survey
A_1 = *Contour Area (Computed)*
W_1' = *Downstream Width*
W_1'' = *Upstream Width*

$$A_1 = \left[\frac{\left(W_1' + W_1''\right)}{\left(W_0' + W_0''\right)}\right]$$

Figure 9.20. Width adjustment method for revising contour areas (Blanton, 1982).

9.14.3 Contour Method – Topographic Mapping

The contour method, which creates a new reservoir topographic map, has become the preferred method for collecting and analyzing survey data. The development of electronic measuring and computerized collection and analysis systems has made it possible for collection and analysis of massive amounts of digital data (x, y, z coordinates), and the final product yields an accurate detailed contour map of the present reservoir conditions.

The contour method involves determining current water volumes and sediment deposition from newly developed reservoir topographic contours, allowing a three-dimensional view on a two-dimensional medium. The final results from the reservoir contour maps are generated surface contours at selected elevation intervals that are used to compute updated volumes. There are multiple computer contour packages and routines for personal or work station computer systems that can be used for this purpose.

9.15 Final Results

There are several computer programs that can generate elevation versus surface area and capacity for a reservoir. The computer program generally used by Reclamation for this purpose is ACAP, where basic elevation versus surface area data developed from the survey becomes the input for the computation of the revised area and capacity tables (Bureau of Reclamation, 1985). A procedure called segmented least squares-fit is used most often. In this procedure, surface areas at specified elevation increments between the basic data contours are derived by linear interpolation, creating a basic area curve/equation over that interval between the contours. The respective capacities and capacity equations are obtained by integration of these area equations. The resulting capacity curve is a series of equations applicable over the full range of the data set.

Table 9.1 contains an example set of these equations. The final result of the computer program is a set of area and capacity tables for use in allocating storage and for operation of the reservoir (see Tables 9.2 and 9.3). The tables can be produced at 0.01-, 0.1-, and 1.0-foot increments.

Table 9.1. ACAP Area and Capacity Equations

WILLOW CREEK RESERVOIR - SUN RIVER PROJECT, MONTANA
2002 AREA-CAPACITY TABLES

EQUATION NUMBER	ELEVATION BASE	CAPACITY BASE	COEFFICIENT A1(INTERCEPT)	COEFFICIENT A2(1ST TERM)	COEFFICIENT A3(2ND TERM)
1	4084.00	0	.0000	.0000	.4500
2	4086.00	1	1.8000	1.8000	.4500
3	4088.00	7	7.2000	3.6000	.7500
4	4090.00	17	17.4000	6.6000	5.0750
5	4092.00	50	50.9000	26.9000	5.8250
6	4094.00	128	128.0000	50.2000	4.9750
7	4096.00	248	248.3000	70.1000	6.3500
8	4098.00	413	413.9000	95.5000	7.7000
9	4100.00	635	635.7000	126.3000	6.7250
10	4102.00	915	915.2000	153.2001	8.1500
11	4104.00	1254	1254.1999	185.8000	11.4800
12	4106.00	1671	1671.7200	231.7200	10.2200
13	4108.00	2176	2176.0401	272.6000	11.0000
14	4110.00	2765	2765.2400	316.6001	14.3999
15	4112.00	3456	3456.0400	374.2001	16.1499
16	4114.00	4269	4269.0399	438.8002	19.2499
17	4116.00	5223	5223.6402	515.7999	20.8001
18	4118.00	6338	6338.4403	599.0003	20.2248
19	4120.00	7617	7617.3404	679.8999	18.4750
20	4122.00	9051	9051.0400	753.8003	18.1998
21	4124.00	10631	10631.4405	826.5999	19.6500
22	4126.00	12363	12363.2404	905.2000	18.4000
23	4128.00	14247	14247.2405	978.7995	18.0252
24	4130.00	16276	16276.9402	1050.9002	25.4100
25	4135.00	22166	22166.6912	1305.0001	11.1000
26	4140.00	28969	28969.1915	1416.0002	11.6000
27	4145.00	36339	36339.1913	1531.9999	14.0000
28	4150.00	44349	44349.1917	1672.0000	13.8000
29	4155.00	53054	53054.1914	1810.0000	17.5000
30	4160.00	62541	62541.6921	1984.9993	15.0000

Area and capacity data are usually plotted as illustrated on figure 9.21 to compare the revised data with the original data. These plot comparisons are valuable during analysis, since they can illustrate possible problems with the data set. Problems identified on an area and capacity plot can include, but are not limited to, datum or elevation shifts if the original versus new surface area curves do not match where little or no change is expected.

Table 9.2. ACAP Surface Area Computations

WILLOW CREEK RESERVOIR - SUN RIVER PROJECT, MONTANA

2002 AREA-CAPACITY TABLES

ACAP92) COMPUTED
7/30/2005
14:49: 8

THE AREA TABLE IS IN ACRES

THE ELEVATION INCREMENT IS IN ONE FOOT

ELEV. FEET	0	1	2	3	4	5	6	7	8	9
4080					0.	1.	2.	3.	4.	5.
4090	7.	17.	27.	39.	50.	60.	70.	83.	96.	111.
4100	126.	140.	153.	170.	186.	209.	232.	252.	273.	295.
4110	317.	345.	374.	406.	439.	477.	516.	557.	599.	639.
4120	680.	717.	754.	790.	827.	866.	905.	942.	979.	1015.
4130	1051.	1102.	1153.	1203.	1254.	1305.	1327.	1349.	1372.	1394.
4140	1416.	1439.	1462.	1486.	1509.	1532.	1560.	1588.	1616.	1644.
4150	1672.	1700.	1727.	1755.	1782.	1810.	1845.	1880.	1915.	1950.
4160	1985.	2015.	2045.	2075.	2105.	2135.	2165.	2195.	2225.	2255.
4170	2285.									

Table 9.3. ACAP Capacity Computations

WILLOW CREEK RESERVOIR - SUN RIVER PROJECT, MONTANA

(ACAP92) COMPUTED
7/30/2005
14:49: 8

2002 AREA-CAPACITY TABLES

THE CAPACITY TABLE IS IN ACRE FEET

THE ELEVATION INCREMENT IS ONE FOOT

ELEV. FEET	0	1	2	3	4	5	6	7	8	9
4080					0.	0.	2.	4.	7.	12.
4090	17.	29.	51.	84.	128.	183.	248.	325.	414.	517.
4100	636.	769.	915.	1077.	1254.	1451.	1672.	1914.	2176.	2460.
4110	2765.	3096.	3456.	3846.	4269.	4727.	5224.	5760.	6338.	6958.
4120	7617.	8316.	9051.	9823.	10631.	11478.	12363.	13287.	14247.	15244.
4130	16277.	17353.	18480.	19658.	20887.	22167.	23483.	24821.	26182.	27564.
4140	28969.	30397.	31848.	33322.	34819.	36339.	37885.	39459.	41061.	42691.
4150	44349.	46035.	47748.	49489.	51258.	53054.	54882.	56744.	58642.	60574.
4160	62542.	4542.	66572.	68632.	70722.	72842.	74992.	77172.	79382.	81622.
4170	83892.									

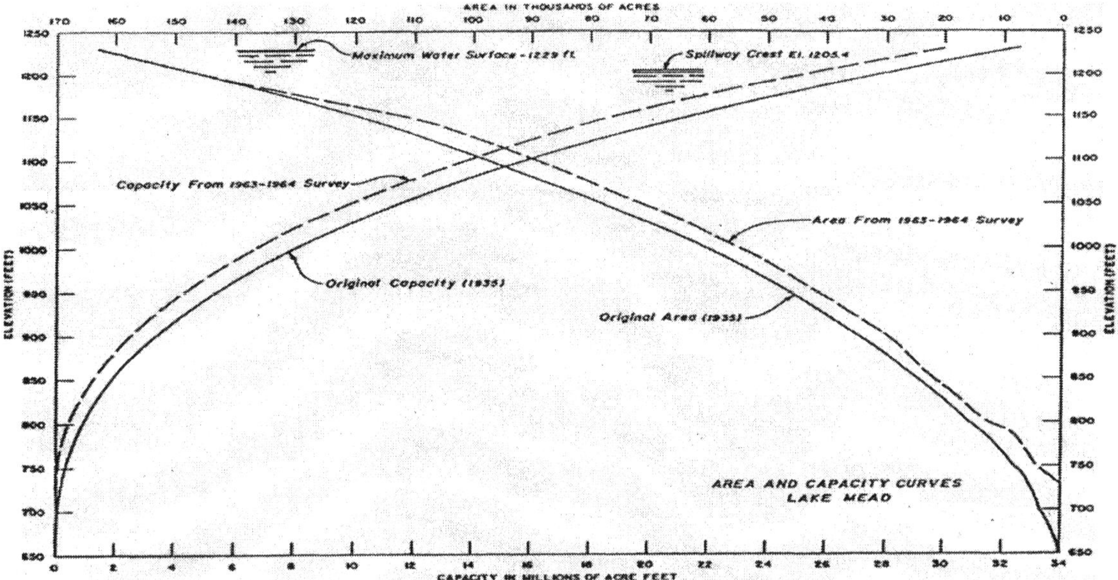

Figure 9.21 Area and capacity curves.

The tables for Willow Creek Reservoir were generated by means of the area-capacity program ACAP, using the least squares method of curve fitting developed by the Bureau of Reclamation's Technical Service Center. This program computes area at 1.0-, 0.1-, and 0.01-foot increments by linear interpolation between basic data contours. The respective capacities and capacity equations are then obtained by integration of the area equations. The initial capacity equation is tested over successive intervals to check whether it fits within an allowable error term. At the next interval beyond, a new capacity equation (integrated from the basic area equation over that interval) begins testing the fit until it too exceeds the error term. The capacity curve thus becomes a series of curves, each fitting a certain region of data. The final area equations are obtained by differentiation of the capacity equations. Capacity equations are of the form $y = a_1 + a_2x + a_3x^2$ where y is capacity and x is the elevation above an elevation base. The capacity equation coefficients for the Willow Creek Reservoir are shown below ($\varepsilon = 0.000001$).

Total volume of sediment deposited in the reservoir generally may be determined as the difference between the original reservoir capacity and the updated capacity computed from the resurvey. Both capacity computations should be made by the same method; that is, if the ACAP program is used to generate the updated capacity from the updated surface areas, it should also be used to regenerate the original capacity using the original measured areas. Even though all volume computation methods are basically similar, this eliminates any variation in calculations resulting from implicit differences between computational procedures. In reservoirs where significant compaction of sediment occurs between subsequent surveys, the difference in reservoir storage would not be truly representative of sediment deposition in the intervening time periods. For those reservoirs, the rate of sediment accumulation should be computed for the total storage period, based on the difference in capacity between the original and present capacity. The results should note the compaction. Ideally, bottom sampling should be part of these studies to provide density measurements for each survey.

Extreme caution must be taken in comparing survey results when it comes to vertical datums. With the use of GPS, which allows accurate horizontal and vertical control to be established, there are many new surveys conducted using the present vertical datum NAVD88 (North American Vertical Datum of 1988). Ideally, this procedure would always be followed. However, in doing so, all information such as survey control and sediment computations must take into account that original project or construction datums will more than likely require a shift to bring all values to a consistent reference datum. It must be noted that many of the Reclamation resurveys have updated the horizontal positions to presently used datums (such as state plane NAD83), but many of the vertical datums are tied to the original project datum. The main reason for retaining the original vertical datum is the cost required to inform all agencies of the elevation differences and the time and cost required to change all previous documentation to reflect the vertical datum shift. All new drawings and reports should state what vertical datum is being used and state the vertical shift needed to match the project datum to the present national vertical datum NAVD88.

Some reservoirs will be subject to significant bank erosion or bank collapsing caused by wave action or severe reservoir drawdown. In analyzing the survey data, the increase in contour surface areas at the higher reservoir elevations in areas where the bank line changes have taken place will offset some of the loss of surface area in the delta areas. In cases where the range line method is used for monitoring sediment, the increases in surface area of higher elevations will usually not be totally compensated for by a decrease in surface areas at lower elevations. The difference may be due either to compaction of sediment, movement of sediment downstream of the surveyed range line, or transport of fine sediment through the dam. If sediment inflow and outflow records are obtained for the period between surveys, it may be possible to roughly differentiate between the sediment accumulation due to inflow and that due to bank changes. If the bank changes are localized, not continuous through a segment of the reservoir, and the volume of material involved is relatively small, the bank changes may be voided in the analyses. The sediment yield rate developed for the drainage basin above a reservoir, where significant bank changes occur, will usually include those bank materials and should be explained in the report. The use of the contour method includes the capacity changes due to bank erosion and is the preferred method for a resurvey.

The trap efficiency of the reservoir may be determined when sediment inflow and outflow records are available for the period between surveys, but, in the United States, these data usually are not available for most reservoirs. When the records are available, the trap efficiency for the period between surveys should be computed and compared with predictive methods. Another useful and more common analysis tool is the sediment deposition profile plot extending the full length of the reservoir. Figure 9.22 shows the profile plot based on the original, 1948, 1963, and 2001 resurveys of Lake Mead. The deposition profile is useful in displaying delta growth and depth of sediment near the dam. The longitudinal profile may also be displayed in a dimensionless plot of percent distance from the dam versus percent reservoir depth, as is shown on figure 9.23 for Lake Powell. This plot permits the comparison of reservoir deposition profiles without scale interference.

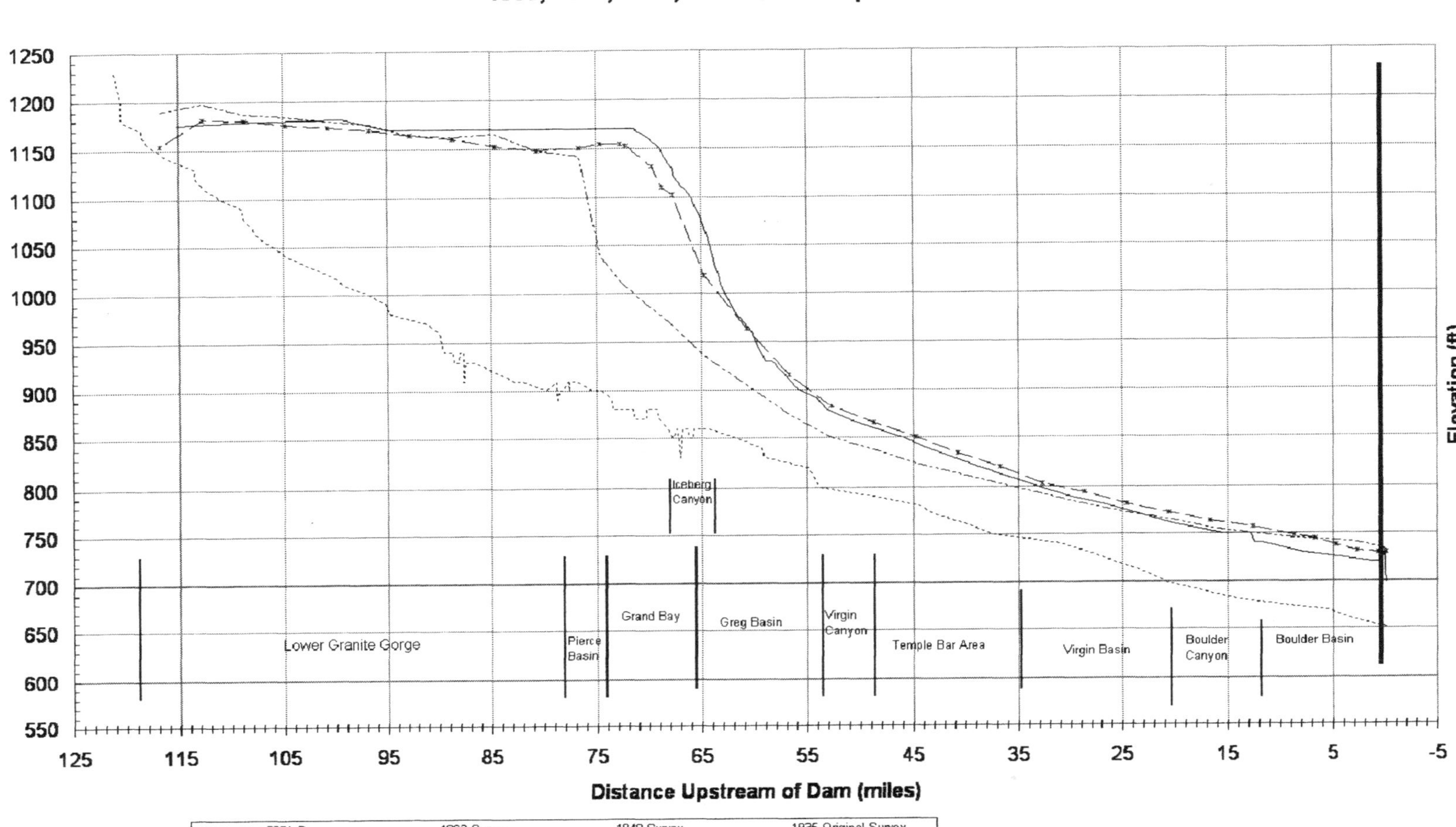

Figure 9.22 Colorado River profiles through Lake Mead.

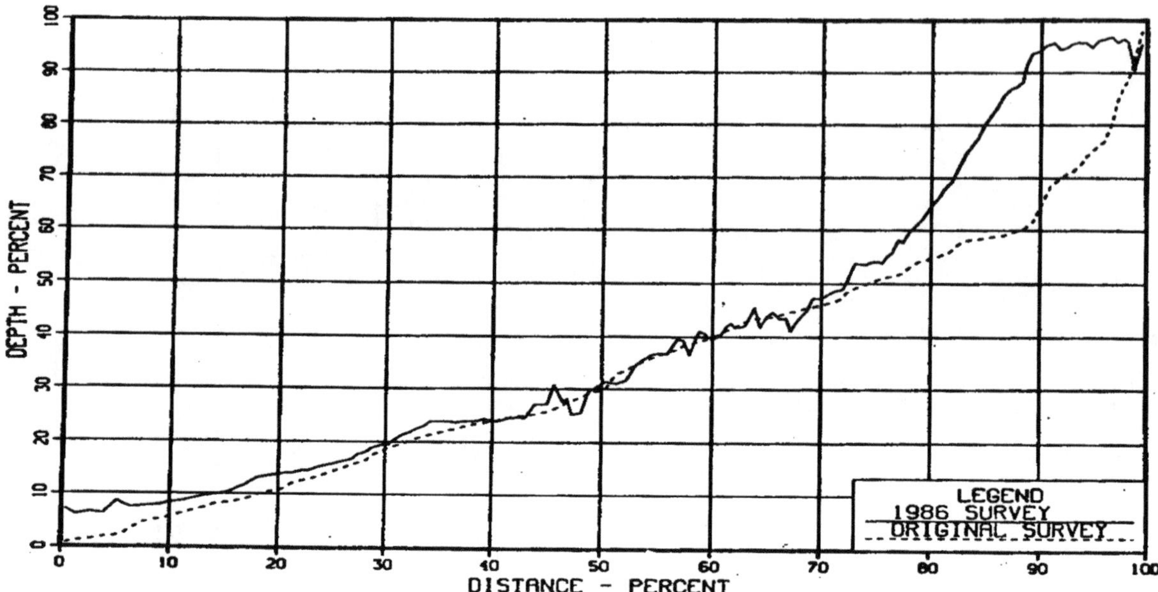

Figure 9.23 Dimensionless plot of Lake Powell sediment deposition profiles (Ferrari, 1988).

9.15.1 Report

An important feature of any reservoir survey is the report of results, so that others may benefit from the time and effort expended in the investigation. The information gathered during a detailed survey helps define the sedimentation characteristics of the contributing drainage basin as well as the reservoir. A well-prepared report serves the interests of the Federal, state, city, and county governments; water districts; and other engineers and scientists not involved in the investigation.

The report should be as inclusive as necessary to document the survey and provide the information useful for future surveys. The items included in individual reports may vary according to the problems and circumstances encountered. Some of the important items that may be included in a report are:

- General information on the dam, reservoir, and drainage basin.
- Information on all surveys, past and present, conducted on the reservoir.
- Description of the survey and sampling techniques for the present survey and the special equipment used.

- A reservoir map showing the location of all range lines. In many cases, the original range line survey map is sufficient. For contouring studies, a plot showing the areas surveyed would be beneficial, along with the newly developed contour map of the reservoir.

- A description of all major survey controls with a table listing horizontal and vertical information. Note: It is highly recommended that control datum used for the survey be reported. Include information such as WGS84, NGS datums used and year established, geoid model used (i.e., geoid99), and state plane coordinate used (i.e., NAD83 Colorado central).

- A plot or table showing the reservoir stage fluctuation or stage duration curve.

- Profiles of all reservoir and degradation ranges showing the latest survey superimposed on the original profile and, in some instances, plots from other surveys.

- Graphs of sediment distribution in a longitudinal profile, a percent depth versus percent sediment volume, and a percent depth versus percent distance from the dam plot

- Data in tabular and graphic form describing sediment densities and particle sizes

- Revised area and capacity tables and curves resulting from the new survey

- Any data on sediment inflow and outflow that may be available for estimating trap efficiency

- A completed reservoir sedimentation data summary sheet. The Subcommittee on Sedimentation, Interagency Advisory Committee on Water Data, provides some examples with instructions for filling out the form given in ASTM D4581 (ASTM, 2005).

Although a formal report is not essential for development of new area and capacity tables and curves, it does provide an excellent document for future surveys and for other reservoir sediment investigations. Figure 9.24 shows an example of a reservoir sediment data summary table.

RESERVOIR SEDIMENT
DATA SUMMARY

Bully Creek Reservoir
NAME OF RESERVOIR

1
DATA SHEET NO.

D	1. OWNER Bureau of Reclamation	2. STREAM Bully Creek	3. STATE Oregon
A	4. SEC 12 TWP. 18S RANGE 43 E	5. NEAREST P.O. Vale	6. COUNTY Maiheur
M	7. LAT 44 ° 00 ' 55 " LONG 117 ° 23 ' 45 "	8. TOP OF DAM ELEVATION 2529.0	9. SPILLWAY CREST EL 2494.0[1]

	10. STORAGE ALLOCATION	11. ELEVATION TOP OF POOL	12. ORIGINAL SURFACE AREA, AC-FT	13. ORIGINAL CAPACITY, AC-FT	14. GROSS STORAGE ACRE-FEET	
R E S	a. SURCHARGE	2523.0		7,300	38,950	15 DATE STORAGE BEGAN 2/63
E	b. FLOOD CONTROL					
R	c. POWER					
V	d. JOINT USE	2516.0	993	30,000	31,650	16 DATE NORMAL OPERATIONS BEGAN 2/63
O	e. CONSERVATION					
I	f. INACTIVE					
R	g. DEAD	2456.58	137	1,650	1,650	
	17. LENGTH OF RESERVOIR 3.4 MILES		AVG. WIDTH OF RESERVOIR 0.47 MILES			

B	18. TOTAL DRAINAGE AREA 547 SQUARE MILES	22. MEAN ANNUAL PRECIPITATION 9.1[2] INCHES
A	19. NET SEDIMENT CONTRIBUTING AREA 547 SQUARE MILES	23. MEAN ANNUAL RUNOFF 9.27[3] INCHES
S	20. LENGTH MILES / AVG. WIDTH MILES	24. MEAN ANNUAL RUNOFF 270,490[4] ACRE-FEET
I N	21. MAX. ELEVATION / MIN. ELEVATION	25. ANNUAL TEMP, MEAN 52 °F RANGE -27 °F to 106 °F

SURVEY DATA

26. DATE OF SURVEY	27. PER. YRS	28. PER. YRS	29. TYPE OF SURVEY	30. NO. OF RANGES OR INTERVALS	31. SURFACE AREA, AC.	32 CAPACITY ACRE - FEET	33 C/I RATIO AF/AF
2/63			Contour (D)	5 - ft	983[5]	31,590[5]	0.12
5/00	37.2	37.2	Contour (D)	5 - ft	911[6]	24,380[6]	0.09

26. DATE OF SURVEY	34. PERIOD ANNUAL PRECIPITATION	35. PERIOD WATER INFLOW, ACRE-FEET			36 WATER INFLOW TO DATE, AF	
		a. MEAN ANN.	b. MAX. ANN.	c. TOTAL	a. MEAN ANN.	b. TOTAL
5/00		182,855[7]	270,490	5,811,350	182,855	6,811,350

26. DATE OF SURVEY	37 PERIOD CAPACITY LOSS, ACRE-FEET			38. TOTAL SEDIMENT DEPOSITS TO DATE, AF		
	a. TOTAL	b. AVG. ANN.	c. $/MI^2$-YR.	a. TOTAL	b. AVG. ANN.	c. $/MI^2$-YR.
5/00	7210[8]	193.6	0.364	7210	193.6	0.364

26. DATE OF SURVEY	39 AVG. DRY WT. (#/FT³)	40. SED. DEP. TONS/MI^2-YR		41. STORAGE LOSS, PCT.		42 SEDIMENT INFLOW, PPM
		a. PERIOD	b. TOTAL TO DATE	a. AVG. ANNUAL	b. TOTAL TO DATE	a. PER. b. TOT.
5/00				0.613[9]	22.8[9]	

26. DATE OF SURVEY	43. DEPTH DESIGNATION RANGE BY RESERVOIR ELEVATION									
		2460-2425	2470-2460	2480-2470	2490-2480	2500-2490	2510-2500	2516-2510		
	PERCENT OF TOTAL SEDIMENT LOCATED WITHIN DEPTH DESIGNATION									
5/00		11.7	7.2	8.0		24.3	22.9	8.0		

26. DATE OF SURVEY	44. REACH DESIGNATION PERCENT OF TOTAL ORIGINAL LENGTH OF RESERVOIR												
	0-10	10-20	20-30	30-40	50-60	60-70	70-80	80-90	90-100	100-105	105-111	110-115	115-120	120-125
	PERCENT OF TOTAL SEDIMENT LOCATED WITHIN REACH DESIGNATION													

Figure 9.24. Reservoir sediment data summary for Bully Creek Reservoir (page 1 of 2).

45. RANGE IN RESERVOIR OPERATION [7]

YEAR	MAX. ELEV.	MIN. ELEV.	INFLOW, AF	YEAR	MAX. ELEV.	MIN. ELEV.	INFLOW, AF
				1963	2,491.2	2,452.2	32,720
1964	2,505.7	2,417.2	111,050	1965	2,515.3	2,476.0	253,410
1966	2,509.1	2,464.0	166,370	1967	2513.5	2,463.1	168,500
1968	2,510.2	2,478.3	168,040	1969	2,516.0	2,477.4	184,780
1970	2,515.5	2,483.4	228,580	1971	2,515.9	2,498.8	220,460
1972	2,513.4	2,481.4	163,060	1973	2,506.1	2,479.2	113,060
1974	2,516.0	2,488.1	205,950	1975	2,516.0	2,489.4	213,610
1976	2,512.5	2,498.1	188,960	1977	2,499.4	2,486.5	105,000
1978	2,518.9	2,461.0	183,050	1979	2,516.2	2,487.5	183,720
1980	2,514.4	2,504.2	221,990	1981	2,515.4	2,487.8	225,980
1982	2,516.8	2,486.5	256,000	1983	2,516.8	2,500.1	270,490
1984	2,516.3	2,494.1	243,930	1985	2,516.2	2,484.2	212,580
1986	2,516.0	2,480.8	195,110	1987	2,510.1	2,458.1	152,430
1988	2,497.1	2,458.5	52,960	1989	2,516.8	2,458.7	164,590
1990	2,500.1	2,458.7	105,630	1991	2,489.3	2,458.4	52,930
1992	2,496.6	2,458.8	61,230	1993	2,517.5	2,458.9	180,990
1994	2,507.0	2,458.8	154,430	1995	2,517.0	2,458.8	211,420
1996	2,516.4	2,493.3	237,570	1997	2,515.5	2,492.1	237,960
1998	2,517.2	2,491.1	251,390	1999	2,516.5	2,497.7	260,840
2000	2,516.3	2,495.4	170,330				

46. ELEVATION - AREA - CAPACITY - DATA FOR **2000 CAPACITY** [10]

ELEVATION	AREA	CAPACITY	ELEVATION	AREA	CAPACITY	ELEVATION	AREA	CAPACITY
2440.4	0	0	2445	15.0	38	2450	51.6	204
2455	89.3	556	2460	117.2	1,073	2465	141.0	1,718
2470	191.0	2,548	2475	244.1	3,636	2480	279.8	4,946
2485	319.9	6,445	2490	375.4	8,183	2495	466.8	10,289
2500	557.7	12,850	2505	635.2	15,832	2510	767.0	19,338
2515	889.9	23,480	2516	911.0	24,300	2520	995.7	28,194
2523	1,032.0	31,236	2525	1,056.9	33,325	2530	1,116.9	38,759

47. REMARKS AND REFERENCES

[1] Top of sluice gates, elevation 2,516.0.
[2] Bureau of Reclamation Project Data Book, 1981.
[3] Calculated using mean annual runoff value of 270,490 AF, item 24, 2/63 - 5/00.
[4] Computed annual inflows from 2/63 through 5/00.
[5] Original surface area and capacity at el. 2,516.0. For sediment computation purposes the original capacity was recomputed by the Reclamation ACAP program using the original surface areas.
[6] Surface area and capacity at el. 2,516.0 computed by ACAP program.
[7] Inflow values in acre-feet and maximum and minimum elevations in feet by water year from 2/63 through 5/00. Some months of missing records. Elevation data for 1963 through 1970 from USGS water records.
[8] Computed sediment volume at elevation 2516.0.
[9] Storage losses at elevation 2516.0.
[10] Capacities computed by Reclamation's ACAP computer program. At elevation 2,456.58 (dead capacity elevation) the calculated surface area was 98 acres with a capacity of 704 acre-feet.

48. AGENCY MAKING SURVEY Bureau of Reclamation
49. AGENCY SUPPLYING DATA Bureau of Reclamation DATE March 2001

Figure 9.24. Reservoir sediment data summary for Bully Creek Reservoir (page 2 of 2).

9.16 Reservoir Survey Terminology

Accuracy: Refers to how close a measurement is to the true or actual value.

Almanac: Data transmitted by a GPS satellite, which includes orbit information on all the satellites, clock correction, and atmospheric delay parameters. These data are used to facilitate rapid satellite acquisition. The orbit information is a subset of the ephemeris data with reduced accuracy.

Analog depth recorder: A graphical recording echo sounder showing profile view of channel section.

Anti-spoofing: For the NAVSTAR system, anti-spoofing is the process whereby the P code used for the precise positioning service is encrypted. The resulting encrypted code is called the Y code. The encryption data can only be decoded by GPS receivers with DOD-authorized special decryption circuitry that guards against fake transmissions of satellite data.

Automated hydrographic survey processing system: A computer system that combines positional and depth measurements into a single database.

Bar: A metallic channel, I-beam, pipe, plate, or ball that reflects sound waves produced by the fathometer.

Bar check: A method for calibrating a fathometer by setting a sound or acoustic reflector (bar) below a survey vessel to a known depth below a sounding transducer.

Baseline: The primary reference line for use in measuring azimuth angles and positioning distances.

Base station (master or reference station): A location that is known. For GPS, a receiver is located over and programmed with known coordinates and calculates error from satellites. With differential corrections, the rover (mobile) GPS position accuracy is improved.

Bathymetric survey (bathymetry): A survey of the underwater portion of the study area.

C/A code: The Coarse/Acquisition (or Clear/Acquisition; civilian or S code) modulated onto the GPS L1 signal. This code is a sequence of 1023 pseudo random binary biphase modulations on the GPS carrier at a chirping rate of 1.023 megahertz, thus having a code repetition period of 1 millisecond. This code was selected to provide good acquisition properties.

Carrier: A radio wave having at least one characteristic (such as frequency, amplitude, phase), which may be varied from a known reference value by modulation.

Carrier frequency: The frequency of the unmodulated fundamental output of a radio transmitter. The GPS L1 carrier frequency is 1575.42 megahertz.

Clock offset: Constant difference in the time reading between two clocks.

Cross section: Survey line normally run perpendicular to the flow direction in the original river channel.

Density: The mass of a substance per unit volume. It is usually expressed as ρ in kilograms per liter or kilograms per cubic meter. Use ρ_s for density of solid particles, ρ_w for water, ρ_d for dry sediment with voids, ρ_{sat} for saturated sediment, ρ_{wet} for wet sediment, and ρ_b for submerged sediment (buoyant weight).

DEM: Digital elevation model.

Depth: Vertical distance between a reference water surface elevation and grade below water.

Differential GPS (DGPS): Differential GPS is a technique for improving GPS solution accuracy. Determining the error at a known GPS location, then subtracting that error from the solution at an unknown GPS location to reduce error.

Differential (relative) positioning: Determination of relative coordinates of two or more receivers which are simultaneously tracking the same satellites.

Dilution of Precision (DOP): A mathematical quality description of the satellite geometry with Position Dilution of Precision (PDOP) the most common used with best-case value of 1. Standard precision terms used for GPS:

GDOP: Geometric (three position coordinates plus clock offset in the solution)
PDOP: Position (three coordinates)
HDOP: Horizontal (two horizontal coordinates)
VDOP: Vertical (height only)
TDOP: Time (clock offset only)
RDOP: Relative (normalized to 60 seconds)

Drainage basin: The area tributary to or draining to a lake, stream, or measuring site.

Draft (transducer draft): Vertical distance from the bottom of the transducer to the surface of the water.

Draft: Vertical distance from water surface to bottom point of survey vessel.

Dynamic positions: A position determined while in motion (kinematic positioning).

Electronic Distance Measurement (EDM): Measurement of distance using pulsing or phase comparison systems.

Electronic Positioning System (EPS): A system that receives two or more EDM signals to obtain a position.

Elevation: Height above mean sea level.

Elevation mask angle: That angle above the horizon below which it is not recommended to track satellites. Normally set to 15 degrees to avoid interference problems caused by buildings, trees, atmospheric, and multipath errors.

Ellipsoid height: The measure of vertical distance above the ellipsoid. Not the same as elevation above sea level. GPS receivers output position fix height in the WGS84 datum.

Ephemeris: The predictions of current satellite position that are transmitted to the user in the data message.

Erosion: The wearing away of the land surface by detachment and movement of soil and rock fragments through the action of moving water and other geological agents.

Fathometer: An electronic device for registering depths of water by measuring the time required for the transmission and reflection of sound waves between a sonic transducer and the lake or river bottom.

Fix: Instant the position of survey vessel is observed.

Fluff (suspended sediment): Lightweight particles in suspension.

Fundamental frequency: The fundamental frequency used in GPS is 10.23 megahertz. The carrier frequency's L1 and L2 are integer multiples of this fundamental frequency.
L1=1575.42 megahertz
L2=1227.60 megahertz

Gauging station: A selected cross section of a stream channel where one or more variables are measured continuously or periodically to index discharge and other parameters.

Geodetic surveys: Global survey to establish control networks for accurate land mapping.

GDOP: Geometric Dilution of Precision. The relationship between errors in user position, time, and satellite range.
$GDOP^2 = PDOP^2 + TDOP^2$

Global Positioning System (GPS): A satellite EDM system used in determining Cartesian coordinates (x, y, z) of a position by means of radio signals from NAVSTAR satellites.

GLONASS: Global Navigation Satellite System. Russian GPS.

HDOP: Horizontal Dilution of Precision.

Horizontal control: A series of connected lines whose azimuths and lengths have been determined by triangulation, trilateration, and traversing.

Hydrographic survey: A reservoir resurvey involving both above water and underwater surveys.

Ionosphere: A band of charged particles 80 to 120 miles above the earth's surface that changes the signal speed or causes a refraction as it passes through.

Kinematic surveying: A form of continuous differential carrier-phase surveying requiring only short periods of data observations. Operational constraints include starting from or determining a known baseline, and tracking a minimum of four satellites. One receiver is statically located at a control point, while others are moved between points to be measured (dynamic positioning).

Latitude: Angular distance, measured in degrees, north or south from the equator.

Longitude: Distance east or west on the earth's surface, measured as an arc of the equator in degrees between the meridian passing through Greenwich, England.

Multipath: Interference similar to "ghosts" on a television screen, which occurs when GPS signals arrive at an antenna having traversed different paths. The signal traversing the longer path will yield a larger pseudo range estimate and increase the error. Multiple paths may arise from reflections from structures near the antenna.

Multipath error: A positioning error resulting from interference between radio waves which have traveled between the transmitter and the receiver by two paths of different electrical lengths. Usually caused by path being reflected from surrounding objects.

Multibeam system: Channel sweep systems employing a single transducer.

Multiple transducer system: Channel sweep system using multiple, vertically mounted transducers.

NAVSTAR: Name given to United States GPS satellite system standing for Navigation Satellite Timing and Ranging.

NAD27: North American Datum of 1927. Older horizontal datum of North America using an approximate shape of Earth designed to fit only the shape of the United States.

NAD83: North American Datum of 1983. Official horizontal datum of North America that relies on the precise Geodetic Reference System of 1980 (GRS 80).

NAVD88: North American Vertical Datum of 1988. A fixed reference adopted as a standard geodetic datum for heights in North America.

NGVD29: National Geodetic Vertical Datum of 1929. NGVD is not the same as mean sea level (msl).

P code: Precise or protected code. A very long sequence of pseudorandom, binary biphase modulations on the GPS carrier at a chirp rate of 10.23 megahertz, which repeats about every 267 days. Each 1-week segment of the code is unique to one GPS satellite and is reset each week. This code is made available to U.S. Department of Defense authorized users only.

PDOP: Position dilution of precision measuring geometrical strength of the GPS satellite configuration.

Position: Latitude, longitude, and altitude of a point.

Postprocessed differential GPS: GPS where the base and rover receiver have no real-time data link between them. All GPS receiver data will be stored for differential correction processing at a later time.

Precision: Refers to how closely a set of measurements can be repeated. The amount by which a measurement deviates from its mean.

Pulse wave system: An electronic positioning system in which the signal from the transmitting station to the reflecting station travels in an electromagnetic wave pulse.

Range: Distance to a point measured by physical optical or electronic means.

Range line: An imaginary, straight line extending across a body of water between fixed shore markings.

Range line markers: Site poles or other identifiable objects used for positioning alignment on a range line.

Reservoir: An impounded body of water or controlled lake where water is collected and stored.

Real-time differential GPS: GPS where a base station computes, formats, and transmits corrections through a data line, such as VHF radio or cellular telephone, for each GPS observation. The rover or mobile GPS unit requires a data link to receive the transmitted correction where it is applied to the current satellite observation.

Reconnaissance survey: Minimal survey effort to determine approximate conditions of a study area.

RINEX: Receiver Independent Exchange format that is a standard to promote free exchange of GPS data from any manufacturer's GPS receiver. The format includes definitions for time, phase, and range.

RTK: Real Time Kinematic.

Satellite configuration: State of the satellite constellation at a specific time, relative to the user.

Satellite constellation: Arrangement of satellites in space.

Selective availability (S/A): Intentional degradation of the performance capabilities of the NAVSTAR system for civilian use by the U.S. military by creating a clock error in the satellites.

Sediment yield: The total sediment outflow from a drainage basin in a specific period of time. It includes bedload as well as suspended load.

Shore markings: Any object, natural or artificial, that can be used as a reference for maintaining boat alignment or establishing the boat's position as it moves along its course.

Site poles: Metal or wood poles used as a sighting rod.

Sounding: Subsurface depth measured by an acoustic device, echo sounder, or sounding pole.

Specific gravity: Ratio of the mass of any volume of a substance to the mass of an equal volume of water at 4°C.

Stadia: Telescopic instrument equipment with horizontal hairs used for measuring the vertical intercept on a graduated vertical rod held in front of the instrument at some distance.

State Plane Coordinate System: A reference coordinate system used by various states of the United States of America.

Thermocline: The middle layer of a water body, separating the upper warmer portion from the lower colder portion.

Total station: An electronic surveying instrument that digitally measures and displays horizontal distances and vertical angles to a distant object. Fully automated self-tracking or manual tracking total stations will digitally measure a moving target.

Transect: A sample area, cross section, or line chosen as the basis for studying one or more characteristics of a particular assemblage.

Transducer: A device for translating electrical energy to acoustical energy and acoustical energy back to electrical energy.

Triangulated Irregular Network (TIN): A linked network of x, y, z data points in a digital terrain model (DTM) from which volumes can be computed using the triangular prismoidal elements.

Universal Transverse Mercator (UTM) coordinate system: Worldwide metric military coordinate system.

WGS84: World Geodetic System 1984. Rotational ellipsoid reference model/datum whose surface is used to compute GPS coordinates. The WGS84 and the GRS80 use the same earth center, which makes NAD83 adjustment coordinates essentially the same.

Y-code (see anti-spoofing): Classified code similar to P-code with restricted access.

9.17 Summary

This chapter presents methodology used by the Bureau of Reclamation's (Reclamation) Sedimentation and River Hydraulics Group (Sedimentation Group) of the Technical Service Center (TSC) to measure reservoir topography for monitoring sediment deposition. The Sedimentation Group has monitored reservoir sediment[1] over the last century with closure of several dam structures in the early 1900s. The monitoring methodology has varied between reconnaissance to detailed field collection and analysis with the goal of accurately updating sedimentation and reservoir capacity information in a timely and cost-effective manner.

The Sedimentation Group continuously upgrades their technical procedures to reflect ever-changing technology. The majority of the techniques provided are from experience gained by Reclamation personnel in the planning and conducting of numerous reservoir and river surveys. Available publications with detailed descriptions such as hydrographic survey manuals by the U.S. Army Corp of Engineers (USACE, 2004), American Society of Testing and Materials (ASTM, 2005), International Hydrographic Organization (IHO, 2005), and manufacturer specifications should be used in conjunction with this guide.

The information presented is intended to assist in the planning and design of a data collection and analysis program. No deliberate endorsement of any particular procedure or manufacturer's equipment is either expressed or implied. The procedures and equipment presented have been utilized for numerous reservoir surveys by different U.S. Federal and state agencies, as well as various private companies. There are many worldwide vendors available with high-quality collection equipment and computer programs that should be reviewed prior to any selection.

All safety issues associated with hydrographic surveying are not discussed or addressed. It is the responsibility of the collection team and agency to establish appropriate safety and health guidelines, but the ultimate responsibility belongs to the individuals in the field who are using the equipment and executing the data collection procedures. Thorough evaluation of field conditions and available equipment, along with sound practical judgment, must be used in ensuring the safety and well-being of the personnel and equipment. Some safety decisions are made during the selection of equipment, such as the survey boat, that also determine the conditions in which crews can safely conduct surveys.

---

[1] The definition of numerous terms, such as "sediment," "sediment yield," "hydraulic height," "structural height," etc. may be found in manuals such as Reclamation's *Design of Small Dams, Guide for Preparation of Standing Operating Procedures for Dams and Reservoirs,* the American Society of Civil Engineers' (USASCE) *Nomenclature for Hydraulics, and ASTM D19 on water standards.*

9.18 References

American Society of Testing and Materials, 2005. Annual Book of Standards, Section 11, Volumes 11.01 and 11.02, ASTM, West Conshohocken, Pennsylvania.

Blanton, J.O. III, 1982. Procedures for Monitoring Reservoir Sedimentation: Technical Guideline for Bureau of Reclamation, Denver, Colorado.

Bureau of Reclamation, 1981. *Project Data*, Denver Office, Denver Colorado.

Bureau of Reclamation, 1985. Surface Water Branch, *ACAP85 User's Manual*, Technical Service Center, Denver Colorado.

Bureau of Reclamation, 1987(a). Guide for Preparation of Standing Operating Procedures for Bureau of Reclamation Dams and Reservoirs, U.S. Government Printing Office, Denver, Colorado.

Bureau of Reclamation, 1987(b). *Design of Small Dams*, U.S. Government Printing Office, Denver Colorado.

Chisholm, G., 1998. *Real-Time Kinematic (RTK) GPS in Hydrography,* POB (Point of Beginning) Magazine, USA. January 1998. http://www.bnp.com/pob/

Clarke Hughes, J.E., 2005. *Lake Powell Multibeam Mapping 2005,* University of New Brunswick, Canada.

Collins, K., and R. Ferrari, 2000. *Elephant Butte Reservoir 1999 Reservoir Survey,* U.S. Bureau of Reclamation, Denver, Colorado, August 2000.

Environmental Systems Research Institute, Inc., 2005. http://www.esri.com

Ferrari, R.L., 1988. *1986 Lake Powell Survey*, REC-ERC-88-6, U.S. Bureau of Reclamation, Denver, Colorado.

Ferrari, R.L., 1993. *Lake Livingston 1991 Sedimentation Survey*, U.S. Bureau of Reclamation, Denver, Colorado.

Ferrari, R.L., 1996. *Boysen Reservoir 1994 Sedimentation Survey*, U.S. Bureau of Reclamation, Denver, Colorado.

Ferrari, R.L., 1997. *Salton Sea 1995 Hydrographic GPS Survey*, U.S. Bureau of Reclamation, Sedimentation and River Hydraulics Group, Denver, Colorado.

Ferrari, R.L., 1998. *Canyon Ferry Lake 1997 Sedimentation Survey,* U.S. Bureau of Reclamation, Sedimentation and River Hydraulics Group, Denver, Colorado.

Ferrari, R.L., 2003. *Deadwood Reservoir 2002 Survey,* Bureau of Reclamation, Sedimentation and River Hydraulics Group, Denver, Colorado.

Ferrari, R.L., and S. Nuanes, 2005. *Tiber Reservoir – Lake Elwell 2002 Survey,* Bureau of Reclamation, Sedimentation and River Hydraulics Group, Denver, Colorado. www.usbr.gov/pmts/sediment

Ferrari, R.L., 2006. *Reconnaissance Technique for Reservoir Surveys,* U.S. Bureau of Reclamation, Denver, Colorado.

Hart, E.D., and G.C. Downing, 1977. Positioning Techniques and Equipment of U.S. Army Corps of Engineers Hydrographic Surveys: Technique Report No. H-77-10.

Hilldale, R.C., 2004. *Collection of River Bathymetry in Shallow Water Conditions Using Airborne Lidar,* Bureau of Reclamation, Sedimentation and River Hydraulics Group, Denver, Colorado.

HYPACK, Inc., 2005. *HYPACK for Windows*, Durham, Connecticut. http://www.coastalo.com

Innerspace Technology, Inc., 1999. *Operator's Manual Hydrographic Survey Software,* Waldwick, New Jersey.

International Hydrographic Organization (IHO), 2005. *Standards for Hydrographic Surveys,* Special Publication No. 44, Fourth Edition, Monaco.

Lin, S.S., 1997. *Strategies For Management of Reservoir Sedimentation.* Virginia Dam Safety Program, Richmond, Virginia.

Lyons, J., and L. Lest, 1996. *Theodore Roosevelt Reservoir 1995 Sedimentation Survey*, U.S. Bureau of Reclamation, Denver, Colorado.

Strand, R.I., and E.L. Pemberton, 1982. Reservoir Sedimentation, U.S. Bureau of Reclamation, Denver, Colorado, 48 p., October 1982.

Twichell, D.C., V.A. Cross, and S.D. Belew, 2003. *Mapping the floor of Lake Mead* (Nevada and Arizona): Preliminary discussion and GIS data release: USGS Open-File Report 03-320, 1 CD-ROM.

U.S. American Society of Civil Engineers (USASCE), 1962. *Nomenclature for Hydraulics,* USASCE headquarters, New York.

U.S. Army Corps of Engineers (USACE), 2004. *Engineering and Design Hydrographic Surveying*, Department of Army, Washington DC. www.usace.army.mil/inet/usace-docs/eng-manuals/em1110-2-1003/toc.htm

Wohl, E., R. McConnell, J. Skinner, and R. Stenzel, June 1998. Colorado Water Research Publication. *Inheriting our Past: River Sediment Sources and Sediment Hazards in Colorado,* Denver, Colorado.

Yen, Pei-Hwa, 1999. Brief Report on Reservoir Sedimentation Survey in Taiwan and Some Comments on Reservoir Sedimentation Monitoring Procedures, National Cheng Kung University, Tainan, Taiwan, Republic of China.

Appendix I
Notation

The following symbols are based on those used by the original authors:

A	parameter in Ackers and White's transport function; cross-sectional area; or soil loss in (ton/acre)/year in the universal soil loss equation
A_1, A_2, B_1, B_2	constants
A_c	a function used in Toffaleti's method
A_g	a volume of material to be degraded per unit channel width
A_h	reservoir area at a given elevation h
A_Ld	average step length
A_m	amplitude of sand waves
A_o, A_{ijk}, B_o, B_{pqr}	constants used in Karim and Kennedy's equation
a	distance from the bed where bed-load is transported; rill width–depth ratio; or relative sediment area
a and a_s	thicknesses of bed layer and suspended layer, respectively
a'	distance from the bed to the sediment sampler inlet
B	roughness function or channel bottom width
b_f	bed form shape factor
C	Chezy's roughness coefficient or sediment concentration; parameter in Ackers and White's transport function; or cropping management factor in the universal soil loss equation
C and C_i	total sediment concentration and concentration for size i, respectively
C_b, C_s, and C_t	sediment concentrations (in ppm by weight) for bed-load, suspended load, and total bed-material load, respectively
C_D and C_L	drag and lift coefficients, respectively
C_f	friction coefficient; or fine sediment concentration

C_t	total sediment concentration, with wash load excluded (in ppm by weight)
C_{tg}	Total gravel concentration (in ppm by weight)
C_{ts}	total sand concentration (in ppm by weight)
C_{ui}, C_{mi}, and C_{Li}	sediment concentrations for size fraction i in the upper, middle, and lower zones defined by Toffaleti, respectively
C_v	sediment concentration by volume
C_{vy}	time-averaged sediment concentration by volume at a distance y above the bed
C'_s	measured sediment concentration in the sampled zone
$\overline{C}$ and $\overline{C_a}$	time-averaged sediment concentrations at a given cross-section and at a distance a above the bed, respectively
D	average flow depth or pipe diameter
D_c	critical depth required at incipient motion
D_g	depth of degradation
D_s and D_m	depths of sampled and unsampled zones, respectively
D_{gr}	dimensionless particle size
D_v	mean depth at a vertical where suspended sediment samples were taken
D' and D''	hydraulic depths for grain roughness and form roughness, respectively
d	sediment particle diameter
d_{35}, d_{50}, d_{65}, and d_{90}	sediment diameters where 35, 50, 65, and 90 percent of the materials are finer, respectively
d_{gr}	dimensionless grain diameter
E and E'	parameters used in the Einstein and modified Einstein procedures, respectively
e	dimensionless coefficient
e_b and e_s	transport efficiency coefficients for bed-load and suspended load, respectively
F	dimensionless function of total reservoir sediment deposition, capacity, depth, and area

F_D, F_L, and F_R	drag, lift, and resisting forces acting on a sediment particle, respectively
F_{gr}	Ackers and White's mobility number
F_r	Froude number
f	Darcy–Weisbach friction coefficient
f and f_o	resistance coefficients of sediment-laden flow and clear water, respectively
f' and f''	Darcy–Weisbach friction coefficients for grain and form roughness, respectively
f'	Engelund and Hansen's transport function
G_{gr}	Ackers and White's sediment transport function
g	gravitational acceleration
H	original depth of reservoir
h_f	friction loss
I, J	dimensionless parameters
I_1, I_2	parameters in Einstein's and Chang's transport functions
i_{BW}	percentage of bed-load by weight of size d
i_{bw}	number of particles available on the bed
J	number of rills
J_1, J_2	parameters used in the modified Einstein procedure
K	soil erodibility factor in the universal soil loss equation
K, K', K'', K_1, K_2, K_3	parameters or constants
K_r, K_s	coefficients in the Meyer-Peter and Müller formula
K_t	Chang, Simons, and Richardson's bed-load discharge coefficient
k	von Kármán–Prandtl universal constant (= 0.4); or other constant
k_s	equivalent sediment diameter for roughness computation; average height; or roughness element
k_1, k_2, k_3	correction factors in Colby's approach; or parameters

L	slope-length factor in the universal soil loss equation; or length
$M, N, M_1, N_1, M_2, N_2, M_3, N_3$	dimensionless parameters
M_O and M_R	overturning and resisting moments, respectively
m	exponent in the universal soil loss equation; or parameter in Ackers and White's transport function
N_d and N_e	rates of number of sediment particles deposited and eroded, respectively
n	Manning's roughness coefficient; or transition exponent in Ackers and White's mobility number
P	erosion-control particle factor in the universal soil loss equation; total power available per unit channel width; or wetted perimeter
P_B	parameter used in the Einstein procedure
P_E	parameters used in the Einstein's transport function
$\overline{P}_i$ and p	time-averaged and fluctuating part of pressure, respectively
P_1, P_s, P_b, and P_2	power expenditure per unit channel width to overcome resistance, transport suspended load, bed-load, and other causes, respectively
p	relative depth of reservoir measured from the bottom; porosity; or probability
p_c, p_m, and p_s	percentages of clay, silt, and sand, of the incoming sediment to a reservoir, respectively
p_i	percentage of material available in size i
Q	water discharge
QS	stream power
Q_s	suspended load
Q_{ti}	total sediment discharge for size fraction i
Q'	water discharge in the sampled zone
q, q_b, and q_s	water discharge, bed-load, and sediment discharge per unit channel width, respectively
q_B	bed-load discharge per unit channel width

q_{Bi}, q_{sli}, q_{smi}, *and* q_{sLi}	sediment load per unit channel width in the bed-load, upper, middle, and lower zones defined by Toffaleti, respectively
q_{bv} *and* q_{bw}	bed-load by volume and by weight per unit channel width, respectively
q_c	critical discharge per unit channel width required at incipient motion
q'_s	sediment discharge per unit channel width in the sampled zone
q_{sv} *and* q_{sw}	suspended sediment load per unit channel width by volume and by weight, respectively
q_t	total bed-material load per unit channel width
q^2	$u_i u_i$
R	rainfall factor in the universal soil loss equation; or hydraulic radius
R_e	Reynolds number
R_s	parameter containing integrals I_1 and I_2
R' *and* R''	hydraulic radii due to grain roughness and form roughness, respectively
r	sediment particle radius
S	water surface or energy slope; or slope-steepness factor in the universal soil loss equation
S_d	total reservoir sediment deposition
S_O	energy slope of clear water
S_p	shape factor
S' *and* S''	friction slopes due to grain roughness and form roughness, respectively
T	time; or temperature
t	time
$\tan \alpha$	ratio of tangential to normal shear force
$\overline{U}_l$ *and* u_i	time-averaged and fluctuating part of the velocity in the i direction, respectively
U_*	shear velocity
u *and* v	local velocities in the x and y directions, respectively
u_b *and* u_s	velocities of bed-load and suspended load

u_x and u_y	fluctuating parts of the velocity in the *x* and *y* directions, respectively
$\overline{u}$	time-averaged local velocity
V and V_{cr}	average flow velocity and critical velocity at incipient motion, respectively
V_b	bottom velocity
VS and $V_{cr}S$	Yang's unit stream power and critical unit stream power required at incipient motion, respectively
V_y	time-averaged flow velocity at a distance *y* above the bed
W	unit weight of sediment deposit (in lb/ft3); channel top width; or rill shape factor
Wc, Wm, and Ws	initial weights of clay, silt, and sand, respectively, based on reservoir operation
W_O and W_T	initial and average reservoir sediment densities after *T* years of operation, respectively
W_s, W'	submerged weight of sediment
W_i^*	Parker's dimensionless bed-load
X	sediment concentration flux by weight in Ackers and White's transport function; or Einstein's characteristic grain size of sediment mixture
X_i, X_j, X_k, X_p, X_q, X_r	dimensionless variables used in Karim and Kennedy's equation
x	Einstein's correction factor, which is a function of ks/δ
Y	parameter used in Shen and Hung's equation; or Einstein's lifting correction factor
Y_a and Y_d	thickness of armoring layer and depth of degradation, respectively
y	potential energy per unit weight of water
Z	rill or channel side slope; or ω/kU_*(a parameter in Rouse's equation)
Z, Z_1	parameters used in the Einstein procedure

α	coefficient in Ackers and White's mobility number (=10); or longitudinal angle of inclination of a channel
β	angle of inclination of shear stress due to secondary motion; or coefficient
β_1	correction factor for non-uniform bed layer
γ, γ_m, γ_s, and γ_f	specific weights of water, sediment-laden flow, sediment, and fluid, respectively
γ_1, γ_2	discrepancy ratios
$\Delta = k_s/x$	Einstein's apparent roughness of bed surface
$\Delta = (\rho_s - \rho)/\rho$	relative density
δ	boundary layer thickness
ε_m and ε_s	momentum diffusion coefficients for fluid and sediment, respectively
ζ_s	specific gravity of sediment (= 2.65)
η	parameter for the fluctuation of velocity
η_1, η_2, η_3, η_v	exponents used in Toffaleti's method
θ	dimensionless shear stress used in Engelund and Hansen's transport function, and in Karim and Kennedy's equation; angle of slope in the universal soil loss equation; angle of inclination of channel bank; Engelund and Hansen's roughness function; or Shield's parameter
θ' and θ''	Engelund and Hansen's roughness functions for grain roughness and form roughness, respectively
θ_{cr} and θ_c	critical Shield's parameters for initiation of suspension and incipient motion, respectively
λ	slope length (in ft) in the universal soil loss equation; or porosity of bed material
μ	dynamic viscosity
μ, μ_m, and μ_r	dynamic viscosities of water, sediment-laden flow, and relative dynamic viscosity, respectively
v and v_m	kinematic viscosities of water and sediment-laden flow, respectively
ξ	relative depth = y/D; or Einstein's hiding correction factor

ρ, ρ_f, ρ_m, and ρ_s	densities of water, fluid, sediment-laden flow, and sediment, respectively
σ	standard deviation
τ and τ_c	shear stress and critical shear stress at incipient motion, respectively
τ_o	shear stress at the bed
τ_{xy}	turbulent shear stress
τ' and τ''	shear stresses due to grain roughness and form roughness, respectively
τ_{*ri}	Parker's reference shear stress
τV	Bagnold's stream power
φ	Engelund and Hansen's transport functions; angle of repose; or velocity potential
φ_i	Parker's dimensionless shear stress for size d_i
φ_*	parameters used in Einstein's transport function
ψ, ψ', ψ_*	Einstein's transport functions
ω and ω_m	sediment fall velocities in clear water and sediment-laden flow, respectively

Appendix II
Conversion Factors

To convert	To	Multiply by
Length (L)		
inches (in.)	centimeters (cm)	2.54
feet (ft)	meters (m)	0.304 8
miles (miles)	kilometers (km)	1.609
meters (m)	inches (in.)	39.37
meters (m)	feet (ft)	3.281
kilometers (km)	miles (miles)	0.621 4
Area (L^2)		
square inches (in^2)	square centimeters (cm^2)	6.452
square feet (ft^2)	square meters (m^2)	0.092 90
square miles (sq miles)	square kilometers (km^2)	2.590
acres (acre)	square meters (m^2)	4047
square centimeters (cm^2)	square inches (in^2)	0.155 0
square meters (m^2)	square feet (ft^2)	10.76
hectares (ha)	acres (acre)	2.471
square kilometers (km^2)	square miles (sq miles)	0.3861
Volume (L^3)		
cubic inches (in^3)	cubic centimeters (cm^3)	16.39
cubic feet (ft^3)	cubic meters (m^3)	0.028 32
cubic yards (yd^3)	cubic meters (m^3)	0.764 6
gallons (gal)	liters (l)	3.785
cubic centimeters (cm^3)	cubic inches (in^3)	0.061 02
cubic meters (m^3)	cubic feet (ft^3)	35.31
liters (l)	cubic feet (ft^3)	0.035 31
liters (l)	gallons (gal)	0.264 2

To convert	To	Multiply by
Velocity (*L/T*)		
feet per second (ft/s)	meters per second (m/s)	0.304 8
meters per second (m/s)	feet per second (ft/s)	3.281
Discharge (L^3/T)		
cubic feet per second (ft3/s)	cubic meters per second (m^3/s)	0.028 32
cubic feet per second (ft3/s)	liters per second (l/s)	28.32
cubic meters per second (m^3/s)	cubic feet per second (ft^3/s)	35.31
liters per second (l/s)	cubic feet per second (ft^3/s)	0.035 31
Mass (*M*)		
pounds (lb)	kilograms (kg)	0.453 6
kilograms (kg)	pounds (lb)	2.205
Density (M/L^3)		
pounds per cubic foot (lb/ft^3)	kilograms per cubic meter (kg/m^3)	16.02
kilograms per cubic meter (kg/m^3)	pounds per cubic foot (lb/ft^3)	0.02 43
kilograms per cubic meter (kg/m^3)	grams per cubic centimeter (g/cm^3)	0.001 00
Force (ML/T^2) †		
pounds (lb)	kilograms (kg)	0.453 6
pounds (lb)	newtons (N)	4.448
kilograms (kg)	pounds (lb)	2.205
kilograms (kg)	newtons (N)	9.807
newtons (N) ‡	kilograms (kg)	0.102 0
newtons (N)	pounds (lb)	0.224 8
dynes (dyn)	newtons (N)	0.000 01

To convert	To	Multiply by
Pressure (M/LT^2) †		
pounds per square inch (lb/in^2)	kilograms per square meter (kg/m^2)	703.1
pounds per square inch (lb/in^2)	newtons per square meter (N/m^2)	6895
pounds per square foot (lb/ft^2)	kilograms per square meter (kg/m^2)	4.882
pounds per square foot (lb/ft^2)	newtons per square meter (N/m^2)	47.88
kilograms per square meter (kg/m^2)	pounds per square inch (lb/in^2)	0.001 422
kilograms per square meter (kg/m^2)	pounds per square foot (lb/ft^2)	0.204 8
kilograms per square meter (kg/m^2)	newtons per square meter (N/m^2)	9.807
Specific weights (M/L^2T^2) †		
pounds per cubic foot (lb/ft^3)	kilograms per cubic meter (kg/m^3)	16.02
pounds per cubic foot (lb/ft^3)	newtons per cubic meter (N/m^3)	157.1
kilograms per cubic meter (kg/m^3)	pounds per cubic foot (lb/ft^3)	0.062 43
kilograms per cubic meter (kg/m^3)	newtons per cubic meter (N/m^3)	9.807
Kinematic viscosity (L^2/T)		
square feet per second (ft^2/s)	square centimeters per second (cm^2/s)	929.0
square feet per second (ft^2/s)	square meters per second (m^2/s)	0.092 90
square meters per second (m^2/s)	square feet per second (ft^2/s)	10.76
square meters per second (m^2/s)	square centimeters per second (cm^2/s)	1000

† The factors relating pounds of force, kilograms of force, and newtons are based on the standard value of the gravitational acceleration, $g = 32.174$ ft/s^2 = 9.806 65 m/s^2.

‡ 1 N = 1 kg-m/s^2.

Appendix III
Physical Properties of Water

IMPERIAL (ENGLISH) UNITS

Temperature (°F)	Specific weight γ (lb/ft^3)	Density ρ (slugs/ft^3)	Viscosity $\mu \times 10^5$ (lb-s/ft^2)	Kinematic viscosity $\nu \times 10^5$ (ft^2/s)
32	62.42	1.940	3.746	1.931
40	62.43	1.941	3.229	1.664
50	62.41	1.940	2.735	1.410
60	62.37	1.938	2.359	1.217
70	62.30	1.936	2.050	1.059
80	62.22	1.934	1.799	0.930
90	62.11	1.931	1.595	0.826
100	62.00	1.927	1.424	0.739
110	61.86	1.923	1.284	0.667
120	61.71	1.918	1.168	0.609
130	61.55	1.913	1.069	0.558
140	61.38	1.908	0.981	0.514
150	61.20	1.902	0.905	0.476
160	61.00	1.896	0.838	0.442
170	60.80	1.890	0.780	0.413
180	60.58	1.883	0.726	0.385
190	60.36	1.876	0.678	0.362
200	60.12	1.868	0.637	0.341
212	59.83	1.860	0.593	0.319

METRIC UNITS

Temperature (°C)	Specific weight γ (kN/m^3)	Density ρ (kg/m^3)	Viscosity $\mu \times 10^3$ (N-s/m^2)	Kinematic viscosity $\nu \times 10^6$ (m^2/s)
0	9.805	999.8	1.781	1.785
5	9.807	1000.0	1.518	1.519
10	9.804	999.7	1.307	1.306
15	9.798	999.1	1.139	1.139
20	9.789	998.2	1.002	1.003
25	9.777	997.0	0.890	0.893
30	9.764	995.7	0.789	0.800
40	9.730	992.2	0.653	0.658
50	9.689	988.0	0.547	0.553
60	9.642	983.2	0.466	0.474
70	9.589	977.8	0.404	0.413
80	9.530	971.8	0.354	0.364
90	9.466	965.3	0.315	0.326
100	9.399	958.4	0.282	0.294

Author Index

Abbe, T.B., 7-6, 7-50
Abt, S.R., 7-5, 7-15, 7-33
Ackers, P., 3-23, 3-25 to 3-28, 3-36, 3-64, 3-67, 3-68, 3-70, 3-83, 3-87, 3-88, 3-99, 3-101, 3-103, 3-104, 6-26
Akiyama, J., 5-49, 5-50
Albertson, M.L., 7-15
Alonso, C.V., 3-63, 3-67, 3-68
American Society of Civil Engineers, 1-1, 3-63, 3-67, 3-68, 5-35, 5-38 to 5-40, 8-1, 8-2, 8-6 to 8-8, 8-15, 8-18, 8-20, 9-62
American Society for Testing and Materials, 9-62, 9-63
Amos, C.L., 4-14, 4-32
Andrews, E.D., 7-30
Annandale, G.W., 2-78 to 2-80, 2-83, 2-84
Arcement, G.J., 5-20
Ariathurai, R., 4-12, 4-20, 4-21, 4-32, 6-26
Armanini, A., 5-13, 5-14, 5-25
Arulanandan, K., 4-12, 4-16, 4-17
ASCE (see American Society of Civil Engineers)
Ashida, K., 5-56
Aspen Institute, 8-1
ASTM (see American Society for Testing and Materials)
Atkinson, E., 5-45, 5-46, 5-47, 5-48
Atterberg, A., 4-11, 4-22, 4-23
Bagnold, R.A., 3-23 to 3-25, 3-27, 3-28, 3-33, 3-36, 3-64, 3-67, 3-68, 3-88, 3-101, 3-104
Bagnold, A., 5-14
Bakry, M.F., 5-21
Barbarossa, N.L., 3-49, 3-50, 3-51
Barnes, H.H., 5-19
Basson, G., 2-1
Bathurst, J.C., 2-29, 2-31, 5-22
Batina, J.T., 5-33
Bauer, T.R., 7-1, 7-11
Beasley, D.B., 2-29
Beasley, R.P., 4-22
Beeson, C.E., 7-48
Belew, S.D., 9-13, 9-34
Bell, R., 5-14
Bennett, J., 5-25, 5-64, 5-69, 5-70
Bennett, R.H., 4-12
Best, 7-34
Best, J.L., 7-34
Bestawy, A., 5-23
Bettess, R., 7-15
Bevenger, G.S., 7-33
Biedenharn, D.S., 7-30, 7-50, 7-51
Bingner, R.L., 2-30, 2-31
Birkeland, P., 7-11, 7-14
Bishop, A.A., 3-64, 3-67
Black, K.S., 4-13, 4-15
Blanton, J.O., 2-17, 9-7, 9-16, 9-26, 9-36, 9-38, 9-39, 9-45, 9-46
Blench, T., 3-12, 3-13, 7-15
Blodgett, J.C., 8-23
Boates, J.S., 4-14
Borah, D., 5-25
Borland, W.M., 2-65
Bountry, J.A., 7-1, 7-4, 7-22, 7-46, 8-31
Bouraoui, F.B., 2-31
Bourget, E., 5-40
Bray, D.I., 5-22, 7-15
Brebbia, C.A., 5-33
Brekhovskikh, V.F., 4-13, 4-18
Bridge, J.S., 7-13, 7-24, 7-47
Brimberg, J., 5-50
Bristow, C.S., 7-34
Britter, R.E., 5-56
Brooks, N.H., 3-4
Brown, L.C, 2-10
Brownlie, W.R., 3-63, 8-31
Brune, G.M., 2-58, 2-59, 6-12, 8-3
Buffington, J.M., 7-15
Bunte, K., 7-5, 7-15, 7-33
Burban, P.Y., 4-3
Burch, G.J., 2-24, 2-25 to 2-28, 2-34, 2-46, 3-73, 3-74
Bureau of Reclamation, 1-1, 2-61, 2-62, 2-65 to 2-67, 2-69, 2-71 to 2-74, 2-76, 3-5, 3-6, 3-78, 3-83, 4-35, 5-44, 5-62, 6-12, 6-25, 6-28, 7-5, 7-14, 7-32, 8-4, 8-26, 9-1, 9-3 to 9-5, 9-9, 9-15, 9-23, 9-24, 9-31, 9-42 to 9-44, 9-46, 9-49, 9-50, 9-53, 9-62
Burt, T.N., 4-24
Campbel, 5-22
Canuto, C.M., 5-33
Cassie, H.J., 3-71, 3-72
Chang, H.H., 1-1, 2-78, 3-82 to 3-84, 5-38, 5-62, 5-73, 7-15
Chapuis, R.P., 4-16
Chien, N., 3-33, 8-28, 8-29
Chow, V.T., 5-19, 5-83
Chung, T.J., 5-31, 5-32

Churchill, M.A., 2-58, 2-59, 6-12
Clarke Hughes, J.E., 9-7
Cohn, T.A., 7-30, 7-37
Colby, B.R., 3-17, 3-42 to 3-44, 3-64, 3-103, 7-36,
Cole, P., 4-2, 4-32
Collier, M., 7-20, 7-52
Collins, B.D., 7-9, 7-48
Collins, K., 9-1, 9-5
Compton, R.R., 7-14
Copeland, R., 7-18, 7-51
Cormault, P., 4-21
Cornelisse, J.M., 4-20
Coussot, P., 5-83
Cowan, W.L., 5-19, 5-20
Crabe, A.D., 3-64
Cross, V.A., 9-13, 9-34, 9-41, 9-42, 9-57
Cunge, J., 5-28
Daborn, G.R., 4-13, 4-14
Danish Hydraulic Institute, 7-32
Daraio, J.A., 2-1, 2-32, 2-33
Darby, S.E., 5-62
Davar, K.S., 5-22
Deletic, A., 5-23
Delo, E.A., 4-18
DeLong, L., 5-11
Dennett, K.E., 4-12, 4-16, 4-17
Derrick, D.L., 7-50
Di Silvio, G., 5-13, 5-14
Dillaha, T.A., 2-31
DiToro, D.M., 4-9
Dixit, 4-30
Dodds, P.S., 7-14
Dorough, W.C., 6-10, 6-11
Dou, G., 3-36
Downer, C.W., 2-29
Downing, G.C., 9-38
Doyle, M., 8-21, 8-22
Doyle, P.F., 7-48
DuBoys, M.P., 3-78
Edwards, T.K., 7-36
Egashira, S., 5-56
Egiazaroff, I., 5-24
Einstein, H.A., 3-16, 3-17, 3-33, 3-41, 3-43 to 3-45, 3-49 to 3-51, 3-53, 3-64, 3-67, 3-75 to 3-77, 3-83, 3-87, 3-88, 3-99, 3-102, 3-103, 5-23
Elliot, C.M., 7-51
Engelund, F., 3-23, 3-25, 3-36, 3-54 to 3-56, 3-60, 3-64, 3-67, 3-68, 3-70, 3-82, 3-83, 3-87, 3-88, 3-99, 3-101 to 3-104, 6-26
Environmental Systems Research Institute, 9-9
Fan, J., 2-57, 5-49, 5-50, 5-52, 5-56, 5-57, 8-4, 8-10, 8-21
Fan, S.S., 5-62
Fander, M., 2-30
Federal Interagency Stream Restoration Working Group, 5-62, 7-1, 7-10, 7-13, 7-50, 7-51, 7-54
Ferrari, R.L., 9-1, 9-3, 9-5 to 9-7, 9-9, 9-10, 9-12, 9-13, 9-42, 9-52
Ferziger, J., 5-26
Fischer, H., 5-13
Fix, G., 5-33
Folly, A., 2-31
Fonda, R.W., 7-8
Ford, D., 5-50
Fortier, S., 3-7, 3-8
Fortin, M.J., 5-40
Franzini, J.B., 2-57
Fredlund, D.G., 5-36
Fukuda, M.K., 4-18
Gadian, A.M., 2-32
Gailani, J., 4-4, 4-32
Garcia, M.H., 5-55
Gatien, T., 4-16
German Association for Water and Land Improvement, 3-64, 3-67
Ghosh, S.N., 3-12
Gilbert, K.G., 3-18, 3-21, 3-23, 3-36, 3-70, 3-71
Gilbert, J.D., 8-10
Gilroy, E.J., 7-37
Givoli, G., 5-61
Glysson, G.D., 7-36
Govers, G., 2-22 to 2-24, 2-34, 3-4, 3-5, 3-12
Graf W., 1-1, 5-23
Gray, W., 5-5
Greimann, B., 8-1, 8-26, 8-27
Griffiths, G.A., 5-22
Gschwend, P.M., 4-15, 4-19, 4-28
Gularte, R.C., 4-30
Gunzburger, M., 5-33
Gust, G., 4-15, 4-19
Guy, H.P., 3-18, 3-48
Gwinn, W.R., 5-21
Hagerty, D., 5-35
Hairsine, P.B., 2-34

Hall, K.R., 4-10, 4-17, 4-23
Hambree, C.H., 7-36
Hamrick, J., 4-32
Han, Q., 3-63, 5-71, 6-26
Hansen, E., 3-23, 3-25, 3-36, 3-54 to 3-56, 3-64, 3-67, 3-68, 3-70, 3-82, 3-83, 3-87, 3-88, 3-99, 3-101 to 3-104, 6-26
Harleman, D., 5-51, 5-54
Harrison, L.L., 6-23, 6-24
Hart, E.D., 9-38
Hayter, E.J., 4-1, 4-5, 4-32, 4-34, 4-35
He, G., 2-83
He, M., 3-63, 5-71
Hebbert, B., 5-50
Heilig, A., 2-31 to 2-33
H. John Heinz III Center for Science, Economics, and the Environment, 8-1, 8-2
Hembree, C.H., 3-17, 3-64
Henderson, F., 5-19, 5-60, 5-65, 5-66, 5-83
Hey, R.D., 7-15
Hicklin, P.W., 4-14
Hill, R.D., Jr., 3-63
Hill, J., 5-38
Hilldale, R.C., 4-1, 4-10, 4-16, 4-17, 9-35
Hjulstrom, F., 3-7
Holly, F.M., 4-34
Holly, M.F., 5-24, 5-62
Hong, S., 2-32
Horton, R.E., 2-21
Houwing, E.J., 4-19, 4-20
HR Wallingford, 3-27, 5-21
Hsu, S., 5-24
Huang, C., 3-64, 3-83, 3-85 to 3-87, 3-99, 3-101, 3-103, 4-1
Hubbell, 3-67
Hung, C.S., 3-14, 3-64, 3-67, 3-103
Hwang, K.N., 4-21, 4-23, 4-24
HYPACK, Inc., 9-9, 9-30, 9-31, 9-33, 9-41, 9-42
IHO (see Innerspace, Inc.)
Ikeda, S., 5-71, 5-72, 7-34
Innerspace, Inc., 9-29, 9-36, 9-37, 9-40, 9-62
Inter-Fluve, Inc., 7-51
Ippen, A.T., 5-51
Jain, S.K., 2-31
Jain, S.C., 5-50
Jirka, G., 5-50
Johannesson, 5-62
Johansen, C., 4-29
Johanson, R.C., 2-38
Johnson, B.E., 2-29, 2-32
Johnson, M., 5-50
Julien, P.Y., 1-1, 2-29, 3-85, 5-34, 7-15, 7-38
Juraschek, M., 5-45
Jürgens, C., 2-30
Kalinske, A.A., 3-37, 3-39, 3-83, 3-87, 3-88, 3-99
Kamphuis, J.W., 4-10, 4-17, 4-23
Karahan, E., 3-4
Karim, M., 3-15, 3-103, 5-25
Keller, J.B., 5-61
Kellerhals, R., 7-15
Kelly, W.E., 4-30
Kennedy, J., 3-15, 3-103, 5-25
Kennedy, R.G., 3-12
Keulegan, G.H., 5-51
Khan, H.R., 3-20, 4-12
Kikkawa, M., 5-71
Kilinc, M., 2-32
Kinsel, W.G., 2-15, 2-29
Kirnak, H., 2-31
Klaassen, G., 5-19, 5-25
Klumpp, C., 4-42
Knighton, D., 7-12
Kong, 3-21
Kothyari, U.C., 2-31
Kouwen, N., 5-21
Kramer, L.A., 2-22
Krishnappan, B.G., 4-18
Krone, R.B., 4-2, 4-5, 4-6, 4-21, 4-32, 6-26
Kuijper, C., 4-30
Kundzewicz, Z.W., 6-2
Lacey, G., 3-12, 3-13, 5-34, 7-15
Lane, E.W., 3-5, 3-6, 6-4, 7-16, 7-17
Langendoen, E.J., 8-25
Lara, J.M., 2-61, 2-62, 2-65, 2-67, 2-73, 7-33, 7-39
Latteux, 4-7, 4-21, 4-23, 4-32
Laursen, E.M., 3-41, 3-42, 3-68, 3-75 to 3-78, 3-83, 3-87, 3-99, 3-103, 6-26
Leavesley, G.H., 2-38
Lee, 4-30
Leopold, L.B., 3-13, 3-18, 7-9, 7-12, 7-15, 7-29, 7-33, 7-34, 7-40
Lest, L., 9-3, 9-44
Letter, J.V., 4-8, 4-32, 4-41
Li, J., 5-43
Li, M.Z., 4-32

Li, R.M., 5-21
Lick, J., 4-4
Lick, W., 4-4, 4-18, 4-32
Lienkaemper, G.W., 7-8
Lin, S.S., 9-2
Linden, P.F., 5-56
Link, R.A., 8-10
Linsley, R.K., 2-47, 2-57
Liu, H.K., 3-4
Lumley, J., 5-4
Lyon, E., 7-56
Lyons, J., 9-3, 9-44
Maa, J.P.Y., 4-18
MacDonald, L.H., 7-25
Maddock, T., 3-13, 3-18, 7-15
Makar, P.W., 7-1, 7-11
Malcomb, R.L., 4-12
Manzenrider, H., 4-13
Marzolf, R., 5-58
Masch, F.D., 4-16
Matejke, 3-67
MBH Software, 8-25, 8-26
McAnally, W.H., 4-1, 5-62
McNeil, J., 4-16, 4-17
Meadows, A., 4-13
Meadows, P.S., 4-13
Mehta, A.J., 4-1, 4-5 to 4-7, 4-18, 4-21, 4-23, 4-24, 4-29, 4-30
Meyer, L.D., 2-22, 2-75, 3-21, 3-22, 3-39, 3-64, 3-67, 3-68, 3-70, 3-82, 3-83, 3-87, 3-88, 3-99, 3-101 to 3-103
Meyer-Peter, E., 3-21, 3-22, 3-39, 3-64, 3-67, 3-68, 3-70, 3-82, 3-83, 3-87, 3-88, 3-99, 3-101 to 3-103, 6-26
Migniot, C., 5-46
Miles, G.V., 4-2, 4-32
Millar, R.G., 4-10
Miller, C.R., 2-63, 2-65, 7-37
Milli, H., 3-64
Mitasova, H., 2-30
Mitchell, J.K., 2-31
Molinas, A., 2-78, 3-21, 3-29, 3-63, 3-78, 3-83, 4-35, 5-62, 5-66
Montes, S., 5-9
Montgomery, D.R., 7-9, 7-15, 7-48
Moody, L.F., 4-20, 5-54
Moore, I.D., 2-24, 2-25 to 2-28, 2-34, 2-46, 3-73, 3-74
Morgan, R.P.C., 2-29
Morris, G.L., 2-12, 2-57, 5-49, 5-50, 5-56, 5-57, 8-4, 8-10, 8-21
Morris, M.J., 4-15, 4-19
Moss, A.J., 2-27
Mostaghimi, S., 2-32
Müller, R., 2-75, 3-21, 3-22, 3-39, 3-64, 3-67, 3-68, 3-70, 3-82, 3-83, 3-87, 3-88, 3-99, 3-101, 3-102, 3-103, 6-26
Murphy, P.J., 7-1, 7-32, 7-43, 7-52, 7-55
Murthy, B.N., 2-59
Nachtergaele, J., 4-28
Nakagawa, H., 4-2, 5-83
Naot, D., 5-21
Nash, J.E., 2-30
National Research Council, 5-62
Nearing, M.A., 2-29, 2-31, 2-33
Nelson, J.M., 5-62
Nezu, I., 4-2, 5-83
Nicholson, J., 4-3, 4-7, 4-21, 4-34
Nikuradse, J., 5-17
Nisbet, B.S., 4-32
Noh, W.F., 5-33
Nordin, C., 5-25, 5-64, 5-69, 5-70
Nuanes, S., 9-9 to 9-11
Odd, N.V.M., 4-32
Office of Management and Budget, 7-14
Ogden, F.L., 2-29, 2-31, 2-32, 2-33
Olmstead, F.H., 2-41
Olympic National Park, 8-10
Onishi, Y., 4-32
Orlob, G.T., 4-9, 4-22, 4-32, 4-35
Osman, A.M., 5-62
Osterkamp, W.R., 7-14
Overbeek, J.T.G., 4-12
Owen, M.W., 4-32
Pacheco-Ceballos, P., 3-32, 3-34, 3-104
Pacific Southwest Interagency Committee, 2-19
Pakala, C.V., 4-34
Palmer, W.C., 7-10, 7-11
Palmieri, A., 6-29
Parchure, T.M., 4-29, 4-30
Parker, G., 3-59, 3-60, 3-82, 3-83, 3-103, 5-52 to 5-55, 5-62, 6-26, 7-34
Parsons, A.J., 2-31, 2-32
Parsons, J.G., 2-24, 4-30
Partheniades, E., 4-5, 4-6, 4-18, 4-20, 4-23, 4-30
Patera, A., 5-32
Paterson, D.M., 4-10, 4-13

Pemberton, E.L., 2-1, 2-18, 2-19, 2-57, 2-60, 2-62, 2-64, 2-67, 2-75, 7-33, 7-36, 7-39, 8-3, 9-2, 9-45
Peric, M., 5-26
Piety, L.A., 7-4
Pinder, G., 5-5
Pitlo, R.H., 5-21
Poesen, J., 4-28
Preissman, A., 5-28, 5-29
Quick, M.C., 4-10
Randle, T.J., 2-1, 2-20, 2-42, 7-1, 7-4, 7-32, 7-43, 7-52, 7-55, 8-4, 8-12, 8-24, 8-31
Rastogi, A., 5-4
Raudkivi, A.J., 4-11
Rauws, G., 2-22, 2-23, 2-24, 2-34
Ravens, T.M., 4-15, 4-19, 4-28
Ravisanger, V., 4-12, 4-25
Reclamation (see Bureau of Reclamation)
Ree, W.O., 5-21
Renard, K.G., 2-8, 2-9, 2-10, 2-11, 2-12, 2-13, 2-14, 2-15, 2-16, 2-29, 2-33, 2-43
Richardson, E.V., 2-32, 3-64, 5-53, 5-56, 6-13
Roberts, J., 4-23
Rodi, W., 5-4, 5-8, 5-83
Rooseboom, A., 5-44
Rose, C.W., 2-34
Rosgen, D.L., 7-15
Rothman, D.H., 7-14
Rottner, J., 3-40, 3-64, 3-67, 3-83, 3-87, 3-88, 3-99
Rouse, H., 3-3, 5-50, 5-51
Rubey, W., 3-45
Samuels, P.G., 5-60
Sanders, H.I., 2-61
Savage, S., 5-50
Savat, J., 2-22
Scarlatos, P.D., 4-32
Schneider, V.R., 5-20
Schnitzer, M., 4-12
Schoklitsch, A., 2-75, 3-36, 3-64, 3-83, 3-99
Schröder, A., 2-31
Schulits, S., 3-63
Schumm, S.A., 3-20, 7-10, 7-12
Scobey, F.C., 3-7, 3-8
Senarath, S.U.S., 2-33
Sentürk, F., 1-1, 7-38, 7-39
Sharma, K.D., 2-30
Shen, H.W., 3-14, 3-64, 3-67, 3-103
Sheng, Y.P., 4-18, 4-34
Sheppard, J.R., 2-75
Sherard, J.L., 4-12
Shields, A., 3-2, 3-3, 3-4, 3-5, 3-7, 3-15
Shimizu, Y., 5-8, 5-21
Shrestha, P.L., 4-9, 4-22, 4-32, 4-35
Schumm, S.A., 8-21
Simöes, F.J.M., 2-33, 2-36, 2-37, 2-38, 3-27, 3-32, 3-45, 3-46, 3-47, 3-63, 3-82, 3-83, 4-31, 4-35, 5-5, 5-23, 5-62, 5-75, 6-10, 6-26, 6-27, 7-32
Simon, A., 7-18, 8-25
Simons, D.B., 1-1, 3-64, 7-15, 7-38, 7-39
Singh, S., 2-30
Singh, V.J., 2-38
Singh, V.P., 5-43
Smerdon, E.T., 4-22
Smith, D.D., 2-2 to 2-7, 2-29
Smith, J.D., 5-62
Smith, P.C., 4-14
Smith, R.E., 2-31
Smith, S.J., 2-17
Soar, P.J., 7-1, 7-15, 7-16, 7-30
Song, C.C., 2-19, 2-27, 2-33, 2-78, 3-14, 3-58, 3-62, 3-82, 5-38, 6-9, 6-26, 7-39
Song, T., 5-23
Soni, 8-26
Southard, J.B., 4-19
Spasojevic, M., 4-34
Stefan, H., 5-49, 5-50
Stein, R.A., 3-20
Stevens, H.H., 3-36, 3-83
Stillwater Sciences, 8-26, 8-31
Strand, R.I., 2-1, 2-18, 2-19, 2-57, 2-60, 2-64, 2-67, 2-75, 7-36, 8-3, 9-2
Straškraba, M., 5-58, 5-59
Sutcliffe, J.V., 2-30
Sutherland, A., 5-14
Swamee, P., 2-37, 5-77, 5-78
Swanson, F.J., 7-8
Tait, J., 4-13
Takahashi, T., 5-83
Takeuchi, K., 6-2
Talapatra, S.L., 3-12
Tan, Y., 6-13, 6-14, 6-15
Tannehill, J., 5-28
Teisson, C., 4-7, 4-21, 4-23, 4-28, 4-32
Tennekes, H., 5-4

Tetra Tech, Inc., 4-34
Thomann, R.V., 4-9
Thomas, W.A., 5-62
Thompson, S.M., 5-22
Thorn, M.F.C., 4-3, 4-4, 4-24, 4-30
Thorne, C.R., 5-62, 7-1, 7-9, 7-15, 7-16, 7-18, 7-30
Thurman, E.M., 4-12
Tiwari, A.K., 2-30, 2-31
Toffaleti, F.B., 3-17, 3-44, 3-45, 3-64, 3-67, 3-75 to 3-78, 3-83, 3-99, 3-102, 3-103, 6-26
Tomasi, L., 6-21
Toro, E.F., 5-83
Tsujimoto, T., 5-21
Turner, J.S., 5-47, 5-56
UNCED (see United Nations Conference on Environment and Development)
USACE (see U.S. Army Corps of Engineers)
U.S. Army Corps of Engineers, 3-77, 3-78, 3-79, 3-83, 4-31, 4-34, 5-39, 5-62, 6-11, 6-12, 7-14, 7-20, 7-32, 9-1, 9-14, 9-25, 9-30, 9-34, 9-40
U.S. Committee on Water Resources, 2-49
U.S. Department of Agriculture, 2-64, 3-78
U.S. Department of the Interior, 7-42
U.S. Environmental Protection Agency 1-3, 2-7, 6-28
U.S. Interagency Committee on Water Resources, 3-45, 3-46
United Nations Conference on Environment and Development (UNCED), 6-2, 6-3
Van Leussen, W., 4-2
van Rijn, L.C., 4-4, 4-19, 4-20, 4-23, 4-24, 4-25, 4-28, 5-14, 5-73
Vanoni, V.A., 1-1, 2-61, 3-3, 3-4, 3-7, 3-8, 3-20, 3-64
Velikanov, M.A., 3-34, 3-35, 3-36, 3-104
Veltrop, J.A., 6-24
Vermeyen, T., 4-15, 4-16, 4-17, 4-23
Verwey, E.J.W., 4-12
Vetter, M., 3-63
Vreugdenhil, C., 5-9
Wainwright, J., 2-31
Wan, S., 3-64, 3-73, 3-76, 3-77, 3-88
Wan, Z., 6-16, 6-17, 8-28, 8-29
Wargadalam, J., 5-34, 7-15
Warsi, Z., 5-3
Watanabe, M., 5-50
Watson, C.C., 7-51
WCED (see World Commission on Environment and Development)
Webb, R.H., 7-52
Wesseling, P., 5-26
Westrich, B., 4-17, 4-28, 5-45
White, C.M., 3-4
White, F., 5-3
White, W., 3-23, 3-25 to 3-28, 3-36, 3-63, 3-64, 3-67, 3-68, 3-70, 3-83, 3-87, 3-88, 3-99, 3-101, 3-103, 3-104, 6-26, 7-15
Wicks, J.M., 2-29, 2-31
Williams, J.R., 2-16, 2-17, 2-30
Williams, G.P., 7-15, 7-20
Winterwerp, J.C., 4-10, 4-11, 4-30
Wischmeier, W.H., 2-2, 2-3, 2-4, 2-5, 2-6, 2-7, 2-29
Wohl, E., 9-4
Wolman, M.G., 7-12, 7-20, 7-33, 7-34
Woolhiser, D.A., 2-29
World Commission on Environment and Development (WCED), 6-1
Wu, C.M., 6-10
Wu, T.H., 2-31, 2-34
Yalin, M.S., 1-1, 3-4, 3-68
Yang, C.T., 1-1, 2-1, 2-8, 2-19, 2-21, 2-25, 2-27 to 2-29, 2-33, 2-34, 2-36 to 2-38, 2-46, 2-55, 2-77 to 2-79, 2-81, 2-83, 2-85, 3-1 to 3-4, 3-8, 3-10 to 3-12, 3-14 to 3-23, 3-25, 3-27 to 3-32, 3-34, 3-36, 3-41, 3-44 to 3-48, 3-51, 3-56, 3-58 to 3-88, 3-99 to 3-104, 4-31, 4-35, 4-36, 4-42, 5-1, 5-38, 5-62, 5-66, 5-75, 5-80, 6-1, 6-4, 6-5, 6-9 to 6-11, 6-26 to 6-28, 7-15, 7-17, 7-32, 7-33, 7-39, 7-52, 8-26
Yeh, K.C., 6-5, 6-9
Yen, Pei-Hwa, 9-7
Young, R.A., 2-29, 4-19
Zhang, R., 2-31, 3-36, 3-104
Zhou, Z., 6-19, 6-20, 6-22
Ziegler, C.K., 2-32, 4-32
Zreik, D.A., 4-10, 4-18, 4-23

Subject Index

active layer, 4-34, 4-36 to 4-41, 5-12, 5-69, 5-70, 8-31
adsorption, 4-11, 4-35
aggregation, 4-1, 4-2, 4-36, 6-28
armor layer, 5-25, 5-82, 7-8
Atterberg limits, 4-11, 4-22, 4-23
backward difference, 5-27
backwater computations, 5-64, 5-65, 5-66,
bank protection, 6-17, 7-1, 7-9, 7-19, 7-24, 7-25, 7-28, 7-32, 7-43, 7-44, 7-46, 7-47, 7-49 to 7-51, 7-54, 8-19
bank stabilization, 7-33, 7-49, 7-50, 7-55
bathymetric survey, 9-1, 9-5, 9-8, 9-9, 9-10, 9-13
bed merge, 4-41
bedload, 2-17, 2-18, 2-36, 2-75, 3-16, 3-17, 3-21, 3-23, 3-24, 3-32 to 3-45, 3-64, 3-67, 3-70, 3-82, 3-83, 3-85, 3-87, 3-88, 3-99, 3-102, 3-103, 5-11, 5-12, 5-14, 5-23, 5-61, 5-73, 5-79, 6-15, 7-6, 7-36, 7-37, 9-61
bed-material load, 2-18, 2-20, 3-17, 3-28, 3-32, 3-63, 3-67, 3-83, 3-87, 3-88, 3-101, 3-102, 5-13, 5-24, 6-5, 6-8
bed-material sorting, 5-11
bend-type intakes, 6-14
biologic function, 7-41, 7-53, 7-55
biological factors, 4-10, 4-14, 4-19
bridges, 2-75, 4-36, 6-28, 7-1, 7-19, 7-23, 7-24, 7-47, 7-56
bypass, 1-1, 4-42, 8-7
Cation Exchange Capacity, 4-12
channel adjustments, 5-73, 7-17
channel banks, 7-11, 7-29, 7-48, 7-49, 7-53
channel forming discharge, 7-29, 7-30
channel geometry, 2-36, 3-12, 3-78, 3-82, 4-35, 5-13, 5-38, 5-62, 5-64, 5-70, 5-76, 6-27, 7-6, 7-13, 7-15, 7-16, 7-30, 7-31, 7-51, 8-14, 8-28
channel grade, 7-53, 7-54
channel morphology, 7-9, 7-11
channel planform, 7-6, 7-18, 7-20, 7-25, 7-53, 8-14
Chézy roughness coefficient, 5-19, 5-45
clay, 2-4, 2-12, 2-18, 2-27, 2-28, 2-42, 2-60 to 2-63, 2-67, 3-8, 3-13, 3-46, 3-73, 3-74, 4-1, 4-11, 4-12, 4-18, 4-23, 4-30, 4-37, 4-42, 5-70, 5-80, 7-22, 8-3, 8-5, 8-12, 8-15, 8-16, 8-29
climate, 2-22, 2-23, 2-57, 2-58, 2-71, 2-72, 5-40, 7-9 to 7-11, 7-18, 7-28
cohesive sediment transport, 1-3, 1-5, 3-1, 4-7, 4-9, 4-31, 4-32, 4-36, 4-42, 4-43, 4-46, 6-26, 6-28
cohesive, 1-3, 1-5, 2-36, 3-1, 3-12, 4-1 to 4-23, 4-30 to 4-37, 4-42 to 4-46, 5-23, 5-35, 5-37, 5-44, 6-26 to 6-28, 7-54, 8-25
combined intakes, 6-14
computer model, 1-1 to 1-3, 2-1, 2-29, 2-34, 2-38, 2-77, 3-17, 3-27, 3-45, 3-63, 3-64, 3-77, 3-78, 3-82, 3-83, 3-88, 4-8, 5-1, 5-3, 5-11, 5-59, 5-64, 5-82, 6-10, 6-13, 6-25, 6-27, 6-30, 7-30, 8-12
conceptual model, 4-34, 4-36, 4-37, 7-2, 7-3, 7-31, 7-45, 7-57
conservation of mass, 4-17
conservation of mass, 5-9, 5-33, 5-51, 5-75
consolidation, 2-36, 2-61, 2-62, 4-7, 4-8, 4-11, 4-28, 4-31, 4-32, 4-36, 4-40, 4-41, 5-46, 6-28
continuity equation, 3-37, 3-47, 3-51, 5-4 to 5-6, 5-9, 5-15, 5-51
conveyance, 5-10, 5-16, 5-18, 5-61, 5-65, 5-67, 7-25, 7-48, 7-49, 7-51, 7-53, 7-54, 8-15 to 8-17
Corophium volutator, 4-14
cover-management factor, 2-14
critical dimensionless unit stream power, 3-29
critical shear stress, 3-3 to 3-5, 3-41, 4-5 to 4-7, 4-11, 4-13 to 4-15, 4-17, 4-18, 4-20 to 4-25, 4-28, 4-31, 4-37, 4-43, 4-46, 5-72
cropping-management factor, 2-2, 2-4
dam decommissioning, 1-4, 1-5, 8-1, 8-2, 8-7 to 8-10, 8-30
dam removal, 7-21, 8-1 to 8-15, 8-17, 8-21, 8-23 to 8-26, 8-30 to 8-32

dam, 1-4, 1-5, 2-1, 2-57, 2-58 to 2-60, 2-68, 2-70, 2-72, 2-73, 3-78, 5-40, 5-42, 5-47, 5-56 to 5-58, 5-61, 5-75, 5-77, 5-80, 5-83, 6-3, 6-5, 6-13, 6-17, 6-19 to 6-21, 6-22, 6-24, 6-29, 7-1, 7-16, 7-19 to 7-21, 7-32, 7-38, 7-52, 7-55, 8-1 to 8-15, 8-17, 8-19 to 8-21, 8-23 to 8-27, 8-29 to 8-32, 9-2 to 9-5, 9-8, 9-23, 9-30, 9-31, 9-34, 9-44, 9-50, 9-52, 9-53, 9-62
Darcy-Weisbach coefficient, 5-19
Darcy-Weisbach friction factor, 3-16, 3-48, 4-20
de Saint Venant equation, 5-9, 5-28, 5-29, 5-42
dead storage, 6-1
debris basin, 6-10, 6-12, 6-22
degradation, 2-20, 3-1, 3-12, 4-35, 5-25, 5-35, 6-4, 6-23, 7-10, 7-16, 7-22, 7-32, 7-33, 7-38, 7-39, 7-43, 7-52, 8-20, 8-23, 9-53, 9-61
delta, 2-37, 2-59, 2-73, 2-75, 2-77, 4-32, 5-8, 5-81, 6-12, 6-17, 8-3, 8-5, 8-7, 8-11, 8-12, 8-14 to 8-16, 8-21 to 8-25, 9-2, 9-3, 9-6, 9-12, 9-50
density current, 5-41, 5-44, 5-47, 5-56, 5-58, 6-16, 6-19, 6-20, 6-22
density, 1-2, 2-1, 2-11, 2-22, 2-49, 2-57, 2-60 to 2-63, 3-2 to 3-4, 3-9, 3-31, 3-33, 3-34, 3-38, 3-41, 4-1, 4-2, 4-7, 4-8, 4-11, 4-13, 4-14, 4-21 to 4-24, 4-28, 4-29, 4-32, 4-35, 4-36, 4-37, 4-39, 4-41, 5-3, 5-7, 5-8, 5-18, 5-19, 5-21, 5-30, 5-41, 5-44 to 5-52, 5-56 to 5-58, 6-16, 6-19, 6-20, 6-22, 7-5, 7-40, 9-2, 9-8, 9-12, 9-25, 9-39, 9-49, 9-57
deposition, 1-1, 1-2, 1-4, 2-8, 2-17, 2-32, 2-33, 2-36, 2-38, 2-57, 2-60, 2-61, 2-64, 2-65, 2-67, 2-70, 2-73, 2-76, 2-83, 2-85, 3-7, 3-63, 3-77, 3-78, 3-82, 3-83, 4-5 to 4-7, 4-10, 4-11, 4-14 to 4-18, 4-21, 4-28, 4-31 to 4-40, 4-43, 4-44, 4-46, 5-11, 5-15, 5-38, 5-39, 5-44, 5-53, 5-64, 5-67, 5-69, 5-70, 5-71, 5-74, 5-77, 6-9, 6-12, 6-16, 6-17, 6-23, 6-28, 7-1, 7-3, 7-8, 7-11, 7-12, 7-19, 7-21 to 7-24, 7-31, 7-32, 7-46, 7-47, 7-49, 7-52, 7-55, 8-2, 8-6, 8-10, 8-12, 8-19, 8-23, 8-26, 8-29 to 8-31, 9-2, 9-3, 9-6, 9-9, 9-13, 9-30, 9-44 to 9-46, 9-49, 9-50, 9-62
desorption, 4-35
DGPS, 9-18 to 9-24, 9-57
deterministic approach, 1-3, 3-12, 3-17, 3-104
diffusion coefficient, 5-8, 5-13
diffusion wave, 5-42
dimensionless particle diameter, 3-85, 3-101
dimensionless unit stream power, 2-41, 2-55, 2-56, 3-20 to 3-23, 3-29 to 3-32, 3-58, 3-63, 3-69, 3-85, 3-87, 3-101, 3-102, 3-104
dimensionless unit stream power equation, 3-29 to 3-31, 3-69
discrepancy ratio, 3-64, 3-67 to 3-71, 3-75, 3-85 to 3-88, 3-99, 3-100
disturbance, 2-15, 4-2, 4-14, 7-2, 7-3, 7-6, 7-16, 7-19, 7-42, 7-48
diversions, 7-1, 7-19, 7-21, 7-28, 7-32, 7-38, 9-4
drag force, 3-1, 3-4, 3-8
dredging, 6-19, 6-20, 6-21, 6-22, 6-23, 6-24, 8-8, 8-9, 8-15, 8-16, 8-18, 9-4
dynamic adjustment, 2-36, 6-1, 6-3, 6-4, 7-17
dynamic equilibrium, 1-1, 2-36, 2-77, 2-83, 3-12, 3-13, 5-38, 6-9, 6-10, 7-1, 7-15, 7-16, 7-32, 7-44, 7-57, 8-23
dynamic waves, 5-42, 5-43
echo sounder, 9-7, 9-25 to 9-29, 9-35, 9-56, 9-61
economic model, 1-4, 6-28, 6-29, 6-30
energy equation, 5-65, 5-66
entrainment coefficient, 5-51, 5-53, 5-55
erodibility, 2-2, 2-11, 2-12, 4-10, 4-11, 4-12, 4-14, 4-15, 4-17, 4-19, 4-22, 4-23, 7-13
erosion, 1-1 to 1-5, 2-1, 2-2, 2-6 to 2-8, 2-10 to 2-15, 2-17, 2-19 to 2-21, 2-24, 2-27, 2-29 to 2-46, 2-55 to 2-57, 2-85, 3-7, 3-73, 4-8, 4-10 to 4-25, 4-28 to 4-41, 4-43, 4-44, 4-46, 5-15, 5-25, 5-34, 5-35, 5-37 to 5-39, 5-44, 5-53, 5-63, 5-64, 5-70, 5-73 to 5-75, 6-1, 6-5, 6-10, 6-11, 6-12, 6-19, 6-24 to 6-26, 6-28, 6-29, 7-1 to 7-3, 7-6, 7-8, 7-11, 7-14, 7-15, 7-18 to 7-27, 7-31, 7-32, 7-39, 7-42 to 7-50, 7-52 to 7-54, 7-56, 8-2, 8-4 to 8-14, 8-17, 8-19 to 8-26, 8-30, 8-32, 9-2, 9-4, 9-5, 9-9, 9-10, 9-45, 9-50
erosion index map, 2-41
erosion-control practice factor, 2-2, 2-6
Eulerian coordinates, 5-33

fall velocity (ies) 2-21, 2-27, 2-29, 2-41, 2-46, 2-49, 2-55, 2-57, 3-9, 3-14, 3-24, 3-29 to 3-33, 3-41, 3-45, 3-46, 3-48, 3-63, 3-85, 3-104, 4-2, 4-33, 4-44, 5-14, 5-45, 5-53, 5-54, 5-61, 5-71, 6-5, 7-33, 7-40, 8-3
finite difference methods, 5-28
finite difference, 1-3, 5-2, 5-26, 5-28 to 5-30, 5-32, 5-82
finite element methods, 5-30 to 5-33
finite element, 1-3, 4-32, 5-2, 5-26, 5-30 to 5-33
finite volume methods, 1-3, 5-32
finite volume, 1-3, 5-2, 5-26, 5-32
flood frequency, 7-28
flow duration, 2-38, 2-47, 6-5, 7-28
flow resistance, 3-34, 5-19, 5-21, 5-37, 5-54, 7-49
flushing, 5-45, 5-46, 6-16, 6-17, 6-19 to 6-21, 6-24, 7-31, 9-3, 9-4
fluvial process, 5-3, 5-34, 7-19
forestry practices, 7-25
forward difference, 5-27
Froude number, 3-20, 3-26, 3-54, 3-60, 3-64, 3-75, 3-85, 3-87, 3-99, 3-101, 5-48 to 5-51, 7-32
Geographic Information System (GIS), 7-14
geology, 2-19, 2-20, 2-45, 7-9 to 7-14
geomorphic mapping, 7-13, 7-14
geomorphic processes, 1-4, 7-22
geomorphology, 7-14, 7-41, 7-51, 8-9
GPS, 7-4, 7-5, 9-1, 9-6, 9-8, 9-9, 9-12, 9-13, 9-15 to 9-24, 9-40, 9-42, 9-44, 9-45, 9-50, 9-56 to 9-63
gradation coefficient, 3-86
grade control structures, 7-50
gravel mining, 7-1, 7-20, 7-26
gravitational power, 1-3, 3-34, 3-36, 3-104
grazing, 2-20, 2-45, 5-34, 7-1, 7-20, 7-37, 7-49, 7-56, 9-4
groundwater, 7-27, 7-29, 7-32
GSTARS, 1-2 to 1-4, 2-33, 2-36 to 2-38, 2-39, 2-85, 3-27, 3-78, 3-82, 3-83, 4-31, 4-35, 5-62, 5-63, 5-75, 6-10, 6-25 to 6-28, 7-32, 8-26
GSTARS 2.0, 3-82, 3-83, 6-10, 6-25 to 6-27
GSTARS 2.1, 1-2 to 1-4, 1-3, 2-33, 2-36 to 2-39, 3-27, 3-82, 3-83, 4-31, 4-35, 5-62, 5-63, 5-75, 6-10, 6-26 to 6-28, 7-32
GSTAR-1D, 1-8, 4-4, 4-5, 4-31, 4-35, 4-36, 4-42 to 4-44, 6-28, 7-32, 8-26
GSTAR-1D model, 1-3, 8-26
GSTARS3, 1-2 to 1-4, 2-33, 2-36 to 2-39, 2-85, 3-27, 3-45, 3-82, 3-83, 4-31, 4-35, 5-62 to 5-82, 6-10, 6-27, 6-28, 7-32
GSTAR-W, 1-2, 1-4, 2-33, 2-34, 2-38, 2-39, 6-28
habitat, 2-35, 2-42, 6-25, 7-1, 7-2, 7-6, 7-21, 7-25, 7-28, 7-31, 7-41, 7-42, 7-48 to 7-55, 8-12, 8-30
hydraulic analysis, 7-30, 7-31
hydraulic dredging, 15
hydraulic geometry, 3-13, 3-14, 7-15, 7-30
hydrologic analysis, 7-27
inactive layer, 4-36 to 4-41, 5-69, 5-70
incipient motion, 1-3, 2-21, 2-25, 2-27, 2-46, 3-1 to 3-5, 3-8 to 3-10, 3-15, 3-18, 3-29 to 3-31, 3-83, 3-86, 3-88, 3-102, 3-103, 7-33
intake structure, 1-3, 6-10, 6-13, 6-14, 8-4, 9-2
joint operation, 6-18, 6-20
kinematic waves, 5-42
Lagrangian coordinate, 5-33
lake circulation, 5-58
Lake Powell, 2-60, 9-7, 9-25, 9-30, 9-50, 9-52, 9-63
lateral erosion, 6-19, 6-20, 7-8, 7-45, 7-49, 8-11, 8-23
law of average stream fall, 6-9
law of least rate of energy dissipation, 6-9, 6-10
levees, 1-1, 7-1, 7-9, 7-19, 7-22, 7-32, 7-46, 7-47
LIDAR, 7-4, 7-5, 9-35
lift force, 3-1, 3-4, 3-9, 5-71
liquid limit, 4-22
mathematical model, 3-102, 5-1, 5-2, 5-11, 5-26, 5-34, 5-42
mechanical dredging, 7, 15
minimum energy dissipation rate theory, 2-22, 2-41, 5-38, 6-30
minimum energy dissipation rate, 2-19, 2-27, 2-33, 2-36, 2-78, 2-85, 5-38, 6-10, 6-30

minimum energy dissipation, 2-19, 2-27, 2-33, 2-36, 2-78, 2-85, 3-82, 5-38, 6-10, 6-30
minimum rate of energy dissipation theory, 2-19, 2-33, 7-39
minimum rate of energy dissipation, 3-58, 7-39
minimum stream power, 1-2, 2-36, 2-78, 2-79, 2-83, 2-85, 3-82, 5-38, 5-82, 6-10, 7-15
minimum total stream power, 1-3, 5-64, 5-74, 6-26, 6-27, 6-30
minimum unit stream power theory, 1-2, 2-81, 3-61, 8-19
minimum unit stream power, 1-2, 2-1, 2-77 to 2-79, 2-81, 2-85, 3-14, 3-58 to 3-62, 3-73, 6-9, 6-30, 7-39, 8-19
mobility number, 3-25
momentum equations, 5-5, 5-7, 5-64, 5-82, 6-26
Moody diagram, 4-20, 5-54
Navier-Stokes equations, 4-16, 5-3, 5-4, 5-7
noncohesive sediment transport, 1-3, 1-5, 3-1, 3-104
noncohesive, 1-3, 1-5, 2-36, 3-1, 3-104
nonequilibrium sediment transport, 1-3, 3-1, 3-63, 3-104, 5-71, 6-27
numerical modeling, 1-3, 5-1, 5-24, 7-2, 7-4, 9-8
numerical models, 4-31, 4-33, 4-35, 4-46, 5-1, 5-15, 7-3, 7-6, 7-32, 8-24, 8-26, 8-32
one-dimensional models, 2-32, 5-15, 5-18, 5-37, 5-41, 5-60 to 5-62, 7-32
physical model, 5-1, 7-6, 7-31
plastic limit, 4-11, 4-22
plasticity index, 4-22
plunge point, 5-48 to 5-50
pool and riffle, 7-18, 8-31
pools, 5-58, 7-9, 7-12, 7-20, 7-33, 7-50, 7-53, 7-54, 8-31
power balance, 1-3, 3-32, 3-34, 3-104
Preissman scheme, 5-28, 5-29
probabilistic approach, 1-3, 3-16, 3-17
rainfall factor, 2-2
range line, 9-3, 9-6 to 9-9, 9-12 to 9-14, 9-16, 9-19, 9-23, 9-42, 9-44, 9-45, 9-50, 9-53, 9-60
recreation, 1-4, 6-3, 6-22, 7-1, 7-20, 8-1, 8-4, 8-31, 9-1, 9-3
regime equation, 3-12, 3-13, 5-35, 7-15, 7-40
regime theory, 3-12, 5-34, 5-40
regime, 1-1, 1-3, 2-32, 3-9, 3-10, 3-12, 3-13, 3-25, 3-58, 3-103, 3-104, 5-18, 5-25, 5-34, 5-40, 5-66, 7-15, 7-20, 7-21, 7-30, 7-40
regression, 1-3, 2-17, 2-18, 2-21, 2-48, 3-12, 3-14, 3-15, 3-16, 3-40, 3-88, 3-103, 3-104, 4-22, 4-23, 6-8, 7-15
reservoir circulation, 5-40, 5-58
reservoir drawdown, 2-62, 7-41, 8-6 to 8-8, 8-10 to 8-13, 8-20, 8-32, 9-6, 9-50
reservoir hydraulics, 5-41
reservoir sediment, 1-1, 1-4, 1-5, 2-1, 2-18, 2-57 to 2-60, 2-66, 2-73, 2-83, 2-85, 5-13, 5-40, 5-44, 5-47, 5-62, 5-75, 5-80, 6-3, 6-13, 6-16 to 6-18, 6-20 to 6-23, 6-25, 6-27, 6-29, 7-20, 7-38, 8-2 to 8-15, 8-17 to 8-20, 8-23, 8-25, 8-26, 8-31, 8-32, 9-2 to 9-6, 9-14, 9-53, 9-62
reservoir sedimentation modeling, 5-62
reservoir sedimentation, 1-4, 1-5, 2-1, 2-57 to 2-59, 2-85, 5-13, 5-40, 5-44, 5-47, 5-62, 5-75, 5-80, 6-16 to 6-18, 6-20 to 6-23, 6-25, 6-27, 6-29, 7-38, 8-3, 8-10, 8-20, 9-2 to 9-5, 9-14, 9-53
resistance force, 3-1, 3-8
resistance to flow, 3-32, 3-33, 3-48, 3-49, 3-85, 3-104
Revised Universal Soil Loss Equation, 2-8, 2-29, 2-42
Reynolds number, 2-22, 3-4, 3-9, 3-10, 3-11, 3-12, 5-51, 5-54, 5-55
Richardson number, 5-53, 5-56
riffles, 7-12, 7-33, 7-55
river corridor, 1-4, 7-1, 7-2, 7-4, 7-18, 7-19, 7-23, 7-26, 7-48, 7-51, 7-52, 7-56
river erosion, 1-4, 7-23, 8-7, 8-10 to 8-13, 8-15 to 8-18, 8-20, 8-32
river restoration, 1-3, 5-1, 7-1, 7-2, 7-19, 7-28, 7-41, 7-57, 8-14
scour, 1-1, 1-2, 2-36, 2-83, 2-85, 3-63, 3-77, 3-78, 3-80, 3-82, 3-83, 5-11, 5-25, 5-37, 5-61, 5-64, 5-67, 5-69, 5-74, 5-76, 6-12, 7-9, 7-11, 7-12, 7-16, 7-32, 7-38, 7-50, 7-52, 9-4

sediment concentration, 2-17, 2-18, 2-21 to 2-23, 2-27, 2-28, 2-32, 2-36, 2-48, 2-54, 2-56, 2-59, 2-83, 3-14, 3-20, 3-21, 3-28 to 3-31, 3-33 to 3-36, 3-43, 3-48, 3-58, 3-60, 3-61, 3-63, 3-74, 3-75, 3-85 to 3-88, 3-101, 3-102, 4-2 to 4-4, 4-6, 4-9 to 4-11, 4-14, 4-19, 4-32, 4-35, 4-44, 5-12 to 5-14, 5-44, 5-53, 5-61, 5-71, 6-13, 6-16, 6-19, 6-21, 6-23, 7-15, 7-36, 7-37, 8-3, 8-12 to 8-14, 8-26, 9-4
sediment continuity, 5-12, 5-64, 5-74, 6-27
sediment management, 1-4, 1-5, 6-16 to 6-18, 6-20, 6-29, 7-2, 7-57, 8-1, 8-2, 8-4, 8-6 to 8-10, 8-12, 8-14, 8-19, 8-32, 9-4
sediment particle size, 2-49, 2-56, 3-26, 3-45, 3-54, 3-86, 3-87, 5-1, 5-37, 7-34
sediment routing, 2-36, 2-38, 3-1, 3-63, 3-77, 3-78, 3-88, 5-15, 5-62, 5-64, 5-68, 5-74, 5-80, 5-82, 6-26
sediment transport, 1-1 to 1-4, 2-21, 2-29, 2-33, 2-34, 2-36, 2-38, 2-39, 2-41, 2-85, 5-1, 5-6, 5-9, 5-11, 5-13 to 5-16, 5-23, 5-25, 5-35, 5-37, 5-41, 5-44, 5-45, 5-48, 5-54, 5-61 to 5-64, 5-70 to 5-73, 5-78, 5-82, 6-5, 6-8, 6-9, 6-12, 6-22, 6-26, 6-27, 6-30, 7-4 to 7-6, 7-8, 7-9, 7-15, 7-16, 7-19 to 7-22, 7-24, 7-26, 7-28 to 7-33, 7-36, 7-38 to 7-41, 7-43, 7-47 to 7-49, 7-51, 7-52, 8-3, 8-10, 8-19, 8-25 to 8-26, 8-29 to 8-32, 9-2, 9-4
sediment transport formula, 2-36, 3-1, 3-17, 3-36, 3-45, 3-61, 3-63, 3-67, 3-68, 3-82, 3-83, 3-85, 3-86, 3-102, 3-104, 5-23, 5-82, 6-26
sediment yield, 1-2, 2-1, 2-16, 2-20, 2-29 to 2-33, 2-42 to 2-45, 2-49 to 2-56, 2-85, 6-11, 6-12, 6-13, 6-22, 6-29, 9-2, 9-5, 9-7, 9-50, 9-62
settling velocity, 2-25, 2-26, 2-28, 3-3, 3-4, 3-9, 3-10, 3-15, 3-26, 3-29, 3-30, 3-38, 3-41, 3-49, 3-51, 3-57, 3-85, 3-87, 3-101, 4-1 to 4-5, 4-8, 4-11, 4-18, 4-45, 5-45
shear stress, 2-22, 2-78, 3-2, 3-3, 3-4, 3-5, 3-6, 3-17, 3-18, 3-19, 3-21, 3-22, 3-23, 3-25, 3-30, 3-41, 3-54, 3-85, 3-88, 3-104, 4-2, 4-3, 4-5, 4-6, 4-15, 4-16, 4-17, 4-18, 4-19, 4-20, 4-21, 4-23, 4-24, 4-25, 4-28, 4-29, 4-30, 4-37, 4-43, 5-8, 5-16 to 5-18, 5-23, 5-24, 5-45, 5-71, 5-72, 5-78, 5-79, 7-30 to 7-32, 7-49, 7-50
shear velocity, 2-21 to 2-23, 2-78, 2-79, 2-81, 5-17, 5-52, 5-54, 5-72, 5-79
sheet erosion, 2-27, 2-38, 2-42, 2-46, 2-55, 2-56
side channel, 7-9, 7-13, 7-24, 7-28, 7-44, 7-48
silt, 2-2, 2-4, 2-8, 2-18, 2-27, 2-42, 2-61, 2-63, 2-67, 3-8, 3-13, 3-46, 3-99, 3-101, 3-103, 4-1, 4-11, 4-42, 5-45, 5-70, 5-80, 6-17, 7-9, 7-22, 8-5, 8-12, 8-15, 8-29
similarity principle, 3-25
single beam, 9-12, 9-15, 9-25, 9-29, 9-30, 9-32, 9-33, 9-35, 9-40, 9-41
siphoning, 6-20
slope-length factor, 2-2, 2-3
slope-steepness factor, 2-2, 2-3
Sodium Adsorption Ratio (SAR), 4-12
soil conservation, 1-1, 6-10, 6-11, 6-17, 6-18, 6-22, 6-24
soil-erodibility factor, 2-2
soils, 2-4, 2-12, 2-14, 2-19, 2-27, 2-33, 2-42, 2-49, 2-55, 4-10 to 4-15, 4-17 to 4-20, 4-22 to 4-24, 4-28, 4-29, 4-31, 7-9, 7-11, 7-13, 7-14, 7-27, 7-54
spectral element method, 5-32
stable channel design, 3-1, 3-5, 3-7, 7-33, 7-39, 8-17
standard-step method, 5-65, 5-66
stratification, 4-28, 5-35, 5-40, 8-23
stream classification, 7-15
stream corridor, 5-76, 7-1, 7-49
stream functions, 5-64
stream power, 1-2, 1-3, 2-1, 2-19, 2-21 to 2-29, 2-33, 2-36, 2-37, 2-41, 2-42, 2-49, 2-55, 2-56, 2-77, 2-79, 2-81, 2-83, 2-85, 3-14, 3-17 to 3-23, 3-25, 3-27 to 3-32, 3-34 to 3-36, 3-48, 3-58 to 3-64, 3-69, 3-71 to 3-74, 3-82, 3-85, 3-87, 3-88, 3-101 to 3-104, 4-36, 5-38, 5-73, 5-74, 5-76, 6-10, 7-13, 7-15, 7-29, 7-33
stream power minimization, 2-37, 3-82, 4-36, 5-76
stream tubes, 1-3, 5-62, 5-64 to 5-66, 5-71, 5-74, 6-26, 6-27
streamlines, 5-5, 5-60, 5-64, 5-65, 5-71
suspended-load transport, 5-23
sustainability, 6-1, 6-2, 6-24, 6-28, 6-29

sustainable development, 1-3, 1-5, 6-1, 6-2, 6-3, 6-4, 6-25, 6-29
sustainable use, 2-57, 6-1, 6-2, 6-3, 6-29, 6-30
terraces, 7-4 to 7-6, 7-8, 7-9, 7-11 to 7-14, 7-23, 7-49, 8-11
theory of minimum energy dissipation rate, 2-27, 2-36, 2-78, 5-38, 5-64, 6-9
theory of minimum stream power, 5-38
theory of minimum total stream power, 7-52
three-dimensional models, 1-3, 5-3, 5-15, 5-16, 5-30, 5-60, 5-61, 5-62, 5-82, 7-38
tiered intakes, 6-14
TMDL (see total maximum daily load)
topographic data, 2-75, 7-5
topography, 1-4, 2-19, 2-32, 2-39, 2-41, 2-55, 4-20, 5-60, 5-83, 6-12, 6-16, 6-18, 7-4, 7-5, 7-9, 7-11, 7-14, 7-17, 7-18, 7-23, 7-30, 8-17, 8-23, 9-2, 9-3, 9-7, 9-8, 9-12, 9-21, 9-33, 9-34, 9-62
total stream power minimization, 5-73
total stream power, 1-3, 5-64, 5-73, 7-33
trap efficiency, 1-2, 2-41, 2-57 to 2-60, 2-70, 6-12, 7-20, 7-38, 8-3, 8-7, 8-8, 8-10, 8-24, 9-2, 9-50
trapping efficiency, 5-23, 5-41, 5-44
two-dimensional models, 2-32, 5-13, 8-25
uncertainty, 2-32, 2-35, 3-88, 4-46, 7-3, 7-56, 7-57, 8-23, 9-13, 9-17
unit stream power, 1-2, 1-3, 2-1, 2-19, 2-21 to 2-29, 2-33, 2-41, 2-49, 2-55, 2-56, 2-77 to 2-79, 2-81, 2-85, 3-14, 3-17 to 3-23, 3-28 to 3-32, 3-34 to 3-36, 3-48, 3-58 to 3-64, 3-69, 3-71 to 3-74, 3-85, 3-87, 3-88, 3-101 to 3-104, 5-44, 6-4, 6-10, 7-17, 7-30 to 7-32, 7-40, 7-52
unit stream power equation, 2-21, 2-25, 2-27 to 2-29, 3-29 to 3-31, 3-61, 3-64, 3-69, 3-71, 3-74, 6-4, 7-17
unit stream power theory, 1-2, 2-21, 2-29, 2-33, 2-81, 2-85, 3-14, 3-61, 3-88, 3-101, 5-44
universal soil loss equation, 1-2
urbanization, 6-11, 7-1, 7-20
vegetation, 2-8, 2-20, 2-41, 2-45, 2-55, 5-1, 5-19 to 5-21, 5-23, 5-34, 5-37, 5-40, 6-11, 6-24, 7-1, 7-2, 7-4, 7-5, 7-9, 7-11, 7-13, 7-14, 7-16, 7-21, 7-23, 7-25, 7-26, 7-29, 7-32, 7-41, 7-42, 7-46, 7-48, 7-49, 7-52 to 7-54, 7-56, 8-19, 9-5, 9-19, 9-20, 9-21, 9-24, 9-28, 9-35
velocity, 2-3, 2-19, 2-21 to 2-23, 2-29, 2-41, 2-42, 2-46, 2-49, 2-55, 2-57, 2-58, 2-77 to 2-79, 2-81, 3-2 to 3-4, 3-7, 3-9 to 3-12, 3-14 to 3-19, 3-24 to 3-52, 3-56, 3-57, 3-61, 3-63, 3-85, 3-87, 3-101, 3-104, 4-2 to 4-4, 4-13, 4-17, 4-18 to 4-20, 4-30 to 4-34, 5-3, 5-5, 5-6, 5-8, 5-12, 5-14, 5-17, 5-18, 5-25, 5-37, 5-42 to 5-44, 5-48, 5-51, 5-52, 5-54, 5-56, 5-60, 5-65, 5-66, 5-70, 5-72, 6-4, 6-12, 6-15 to 6-17, 7-6, 7-8, 7-22 to 7-32, 7-34, 7-38 to 7-40, 7-47 to 7-49, 7-53 to 7-55, 8-8, 8-19, 8-25, 8-27, 9-2, 9-8, 9-25, 9-27, 9-28, 9-30, 9-32, 9-33, 9-39 to 9-41
venting, 5-56, 5-57, 6-19, 6-20, 6-22
volume, 1-3, 2-16, 2-17, 2-20, 2-31, 2-33, 2-38, 2-41, 2-44 to 2-61, 2-64, 2-65, 2-68, 2-70, 2-73, 2-75, 2-77, 2-83, 3-31, 3-35, 4-1, 4-16, 4-31, 4-33 to 4-42, 5-3, 5-26, 5-31, 5-42, 5-45, 5-67, 6-3, 7-21, 7-26, 7-38, 7-52, 8-2 to 8-6, 8-9, 8-11, 8-13 to 8-16, 8-20, 8-23, 8-27, 8-29, 8-32, 9-2, 9-3, 9-6, 9-9, 9-15, 9-19, 9-40, 9-43 to 9-45, 9-49, 9-50, 9-53, 9-57, 9-61
von Kármán's constant, 5-17
warping, 6-18, 6-20
wash load, 1-3, 2-18, 3-1, 3-29, 3-31, 3-32, 3-63, 3-64, 3-71, 3-86, 3-102 to 3-104, 5-44, 8-26
watershed, 1-2, 2-1, 2-17, 2-29 to 2-39, 2-41, 2-57, 2-58, 2-85, 5-40, 5-58, 6-13, 6-14, 6-17, 6-18, 6-22, 6-24, 6-28, 7-3, 7-6, 7-9, 7-12 to 7-14, 7-19, 7-20, 7-25, 7-27, 7-28, 7-32, 7-33, 7-37, 7-44, 7-56, 8-8, 8-13, 9-4
weighting parameters, 5-71
woody debris, 7-1, 7-6, 7-8, 7-9, 7-11, 7-19, 7-22 to 7-25, 7-31, 7-46 to 7-48, 7-54, 8-6, 8-14, 8-19
Yellow River, 2-83, 3-31, 3-32, 3-71 to 3-73, 6-16

☆ U.S. GOVERNMENT PRINTING OFFICE: 2007—618-472

www.ingramcontent.com/pod-product-compliance
Lightning Source LLC
Chambersburg PA
CBHW061321190326
41458CB00011B/3857

* 9 7 8 1 7 8 0 3 9 3 5 9 9 *